# HYBRID AND ADVANCED TECHNOLOGIES

The proceedings of the International Conference on Hybrid and Advanced Technologies (ICHAT 2024) present a rich repository of cutting-edge research on the various applications of machine learning deep learning and AI in cybersecurity, healthcare, agriculture and communication systems.

It highlights the revolutionary potential of data science in transforming traditional practices, improving efficiency and accuracy across diverse domains and addressing complex real-world challenges. These proceedings contain innovative neural-network models for agriculture that can predict tractor fuel consumption and optimize smart irrigation, besides suggesting greenhouse automation for enhanced agricultural productivity. It also provides a roadmap for IoT-based monitoring systems for asthma patients and machine learning approaches for early detection of diabetes, cancer and aquatic plant ailments. Through an array of practical examples and comparative studies, the book further highlights advancements in machine learning for enhancing palm vein authentication, combating fake news, keeping data safe and improving customer segmentation in e-commerce. The findings would be instrumental in combating critical global issues and foster a deeper understanding of the role of AI in image processing, cybersecurity, medical diagnostics and intelligent systems in the future.

This will be a highly interesting guide to researchers, data scientists and practicing professionals in the fields of artificial intelligence, machine learning and cybersecurity. It will also be of interest to healthcare professionals, agricultural scientists and technology enthusiasts in fostering global collaborations, exploring future challenges and opportunities and introducing state-of-the-art technologies to streamline processes.

PROCEEDINGS OF THE INTERNATIONAL CONFERENCE ON HYBRID AND ADVANCED TECHNOLOGIES (ICHAT 2024), ONGOLE, ANDHRA PRADESH, APRIL 26–28, 2024

# Hybrid and Advanced Technologies

## Volume 1

*Edited by*


S. Prasad Jones Christydass
*QIS College of Engineering and Technology*

Nurhayati Nurhayati
*Universitas Negeri Surabaya*

S. Kannadhasan
*Study World College of Engineering*



CRC Press
Taylor & Francis Group
Boca Raton London New York Leiden

CRC Press is an imprint of the
Taylor & Francis Group, an **informa** business
A BALKEMA BOOK

First published 2025
by CRC Press/Balkema
4 Park Square, Milton Park, Abingdon, Oxon, OX14 4RN

and by CRC Press/Balkema
2385 NW Executive Center Drive, Suite 320, Boca Raton FL 33431

*CRC Press/Balkema is an imprint of the Taylor & Francis Group, an informa business*



*British Library Cataloguing-in-Publication Data*
A catalogue record for this book is available from the British Library

Set

ISBN: 978-1-032-90665-2 (hbk)
ISBN: 978-1-032-90666-9 (pbk)

Volume 1
ISBN: 978-1-032-90254-8 (hbk)
ISBN: 978-1-032-90668-3 (pbk)
ISBN: 978-1-003-55911-5 (ebk)

DOI: 10.1201/9781003559115

Volume 2
ISBN: 978-1-032-90669-0 (hbk)
ISBN: 978-1-032-90672-0 (pbk)
ISBN: 978-1-003-55913-9 (ebk)

DOI: 10.1201/9781003559139

Typeset in Times New Roman
by MPS Limited, Chennai, India

*Hybrid and Advanced Technologies – S. Prasad Jones Christydass, Nurhayati Nurhayati and S. Kannadhasan. (Eds) © 2025 the Editors, ISBN 978-1-032-90254-8*


# Table of Contents

*Hybrid and Advanced Technologies – S. Prasad Jones Christydass, Nurhayati Nurhayati and S. Kannadhasan. (Eds) © 2025 the Editors, ISBN 978-1-032-90254-8*
*Open Access: www.taylorfrancis.com, CC BY-NC-ND 4.0 license*

# Preface

This book explores the transformative impact of integrating machine learning and IoT across agriculture, healthcare, environmental science, and more. It introduces advanced methodologies such as RBF-MLP-SVM for fuel prediction, GA-CNN-BIGRU for smart irrigation, and BIGRU with attention mechanisms for plant disease detection. Additionally, it examines IoT-based health monitoring systems for asthma patients and innovative biogas production techniques. Through practical examples and comparative studies, the book demonstrates how these technologies can revolutionise traditional practices, enhancing efficiency, accuracy, and sustainability across diverse fields.

We extend our deepest gratitude to the QIS College of Engineering and Technology reviewers, authors, organizing committee, and management for their invaluable contributions to the success of the International Conference ICHAT 2024. The reviewers' insightful feedback, the authors' innovative research, the committee's meticulous planning, and the management's unwavering support have collectively made this event a remarkable achievement. Thank you to everyone who played a role in making ICHAT 2024 a success.

*Hybrid and Advanced Technologies – S. Prasad Jones Christydass, Nurhayati Nurhayati and S. Kannadhasan. (Eds) © 2025 the Editors, ISBN 978-1-032-90254-8*
*Open Access: www.taylorfrancis.com, CC BY-NC-ND 4.0 license*

# Editor Biographies

**S. Prasad Jones Christydass** completed his B.E degree in Electronics and Communication Engineering in the year 2010 and M.E. degree in Communication Systems in 2013. He completed his Ph.D. in the area of Metamaterial Antenna from National Institute of Technology, Trichy in the year 2022. He has 10 years of Teaching Experience. Currently he is working as an associate professor in the Department of ECE, QIS College of Engineering and Technology, Andhra Pradesh, India. His area of interest includes Antenna, Metamaterial Structures, Dielectric Resonators, Liquid antenna, FSS and Machine Learning in antenna. He has published a book, 13 international journal papers and 15 international conference papers. He is a member of various professional bodies like IEEE and ISTE. He has also served as a reviewer for various journals.

**Nurhayati Nurhayati** received B.E., M.E., and D.E. degrees from Institut Teknologi Sepuluh Nopember (ITS), Surabaya, Indonesia, and a master's and Ph.D. degree in telecommunication multimedia from the Department of Electrical Engineering, Faculty of Electrical Technology, ITS. She is currently a Lecturer with the Department of Electrical Engineering, Faculty of Engineering, Universitas Negeri Surabaya. She is also a Secretary with the Research Center, Institute for Research and Community Service, Universitas Negeri Surabaya. Her research interests include planar antenna, Vivaldi antenna, monopole antenna, wideband antenna, array antenna, radar antenna, and MIMO antenna. She has been a member of the Antennas and Propagation Society since 2015, and the IEEE Young Professional Member since 2018. In 2017, she received a scholarship from the International Publication Quality Improvement (PKPI), Kumamoto University, Kumamoto, Japan.

**S. Kannadhasan** is working as an Associate Professor and HoD in the department of Electronics and Communication Engineering at Study World College of Engineering, Coimbatore, Tamil Nadu, India. He completed his Ph.D. in the field of smart antennas at Anna University in 2022. He has thirteen years of teaching and research experience. He obtained his B.E. in ECE from Sethu Institute of Technology, Kariapatti, in 2009 and his M. E. in Communication Systems from Velammal College of Engineering and Technology, Madurai, in 2013. He obtained his M.B.A. in Human Resources Management from Tamil Nadu Open University, Chennai. He has published a book chapter, 110 papers in reputed international journals, and more than 250 papers in conferences. He also serves as a board member, reviewer, speaker, session chair, and member of the advisory and technical committees of various colleges and conferences. His areas of interest are smart antennas, digital signal processing, wireless communication, wireless networks, embedded systems, network security, optical communication, microwave antennas, electromagnetic compatibility and interference, wireless sensor networks, digital image processing, satellite communication, cognitive radio design, and soft computing techniques.

*Hybrid and Advanced Technologies – S. Prasad Jones Christydass, Nurhayati Nurhayati and S. Kannadhasan. (Eds) © 2025 the Editors, ISBN 978-1-032-90254-8*


# Scientific Committee

Dr. C. Vennila, *Alagappa Chettiar Government College of Engineering and Technology*
Dr. M. Shivakumar, *CMR university, Bengaluru*
Dr. Hong Lin, *University of Houston, United States*
Dr. Rajkumar, *University of Melbourne, Australia*
Dr. Rangaraj M Rangayyan, *University of Calgary, Canada*
Dr. Abdul Razak A R, *BITS Pilani, Dubai*
Dr. Shivakumara, *University of Malaya, Malaysia*
Dr. Anand Nayyar, *Duy Tan University, Vietnam*
Dr. Oussama Saadaoui, *National School of Engineering, Tunisia*
Dr. Fausto Pedo Garcia Marquez, *Castilla-La Mancha University, Spain*
Dr. Pascal Lorenz, *University of Haute Alsace, France*
Dr. Antonio Luchetta, *University of Florence, Italia*
Dr. C. Palanichamy, *Multimedia University, Malaysia*
Dr. Xiaolei Wang, *Aalto University, Finland*
Dr. Joy Iong-Zong Chen, *Da-Yeh University, Taiwan*
Dr. Xavier Fernando, *Ryerson University, Canada*
Dr. Basim Alhadidi, *Prince Abullah Bin Ghazi Faculty of Information and Communication Technology, Jordan*
Dr. George Ghinea, *Brunel University, United Kingdom*
Dr. K.K. Chong, *Universiti Tunku Abdul Rahman, Malaysia*
Dr. Umashankar Subramaniam, *Prince Sultan University, Riyadh, Saudi Arabia*
Dr. Aritra Ghosh, *University of Exeter, United Kingdom*
Dr. Christo Ananth, *Samarkand State University, Samarkand, Uzbekistan*
Gabriel Gomes de Oliveira, *Universidade Estadual de Campinas, Brazil*
Dr. Yuzo Iano, *Universidade Estadual de Campinas, Brazil*
Dr. Rangel Arthur, *Universidade Estadual de Campinas, Brazil*
Mariyam Ouaissa, *Moulay Ismail University, Meknes Morocco*
Dr. Shermin Shamsudheen, *Jazan University, Saudi Arabia*
Prof. Yuvaraja Teekaraman, *The University of Sheffield, United Kingdom*
Prof. Srinath Doss, *Bothouniversity, Botswana*
Samuel Ogunniyi, *University of Ilorin, Nigeria*
Prof. Awais Ahmad, *Universidad de Cordoba, Spain*

*Hybrid and Advanced Technologies – S. Prasad Jones Christydass, Nurhayati Nurhayati and S. Kannadhasan. (Eds) © 2025 the Editors, ISBN 978-1-032-90254-8*
*Open Access: www.taylorfrancis.com, CC BY-NC-ND 4.0 license*

# Organising Committee

Dr. N.S. Kalyan Chakravarthy, *Chief Parton*
Dr. N. Sri Gayathri, *Chief Parton*
Dr. YV. Hanumantha Rao, *Parton*
Dr. B.V. Subbarao, *Parton*
Dr. S. Prasad Jones Christydass, *Conference Chair*
Dr. KVJ Bhargav, *Conference Co-Chair*
Dr. S. Suresh Kumar, *Finance Chair*
Dr. Y. Nandhakishora, *Technical Committee Chair*
Dr. B. Ananda Babu, *Technical Committee Co-Chair*
Dr. A. Subramaniyan, *Publication Chair*
Dr. S. Kannadhasan, *Publication Chair*

*Hybrid and Advanced Technologies – S. Prasad Jones Christydass, Nurhayati Nurhayati and S. Kannadhasan. (Eds)* © 2025 the Authors, ISBN 978-1-032-90254-8


# An innovative method for prediction of tractor fuel consumption based on RBF-MLP-SVM model

Snehal Narayanrao Meshram
*Assistant Professor Yeshwantrao Chavan College of Engineering, Wanadongri, Maharashtra, India*

Akabarsaheb Nadaf
*Associate Professor, Bharati Vidyapeeth (Deemed To Be University, Pune) Abhijit Kadam Institute of Management and Social Sciences, Solapur, India*

Srinivasulu Raju Sri Ramadasu
*Assistant Professor, Department of Electronics and Instrumentation Engineering, VR Siddhartha Engineering College, Vijayawada, India*

N.R. Sathis Kumar
*Assistant Professor, Department of CSE, Kalasalingam Academy of Research and Education, Krishnankoil, Tamilnadu, India*

S. Dhanalakshmi
*Assistant Professor Level III, Department of ECE, Bannari Amman Institute of Technology, Sathyamangalam, Erode, India*

Harshal Patil
*Associate Professor, Balaji Institute of Technology & Management (BITM), Sri Balaji University, Pune, Maharashtra, India*

ABSTRACT: Improved fuel economy over the past 20 years has been confirmed by the Nebraska Tractor Test Laboratory (NTTL). According to the ASAE Standards, there was a 4.9% decline in the average annual specific volumetric fuel usage. Diesel engine fuel consumption predictions under partial and full loads, and at reduced engine speeds from full throttle, were achieved by developing new fuel consumption equations based on power and fuel consumption data from the NTTL reports. Model training, feature expansion, and preprocessing are the three most critical steps. Additional input elements could be incorporated during preprocessing using data collected from the various sources mentioned in the preceding chapter. Feature extraction makes use of fuel consumption projections. The model's accuracy was enhanced during training with the RBF-MLP-SVM. With a rate of 97.35%, the proposed method outperformed state-of-the-art technologies like MLP and SVM. Keywords: Multilayer Perceptron (MLP), Support Vector Machine (SVM), Nebraska Tractor Test Laboratory (NTTL), Artificial Neural Network system (ANN), Radial Basis Functions (RBF)

## 1 INTRODUCTION

Among the most energy-intensive off-road vehicles, agricultural tractors are massive and their complex wheel-terrain interaction leads to tire degradation and soil deformation. Fuel consumption is directly related to agricultural energy consumption. Making more efficient use of the energy at the wheels of a tractor can improve its performance, which has monetary

DOI: 10.1201/9781003559115-1

and environmental benefits. As a result, the process in question uses less fuel. In order to optimize operations, researchers, producers, and consumers all value the ability to predict tractor performance in the field. As a result, major decisions regarding the tractor's performance enhancement depend on accurate fuel consumption predictions. Tillage activities are the most common use for tractors. That is because how well the tractor runs has a direct correlation to how much gas it uses as a consequence, fuel efficiency improves and tractor traction performance improves. When conducting the experiments, a number of factors were controlled, including the soil's penetration resistance, depth, tractor forward speed, and moisture level.

## 2 LITERATURE SURVEY

To get the most out of tractors in the field, researchers and manufacturers have been trying to figure out how to estimate how much gas they'll need [1]. This particular data on fuel usage can be obtained by experts from the fuel consumption model, enabling a precise assessment of the vehicle's economy [2]. Mathematical models that can predict a tractor's fuel consumption from its operating characteristics have recently received a lot of attention from scientists. [3] state that these models can show how different tractor engine types, running speeds, and throttle levels relate to fuel consumption. [4] They also cover engine speed depression, which is the difference between operating speeds and no-load speeds. The tractors at the Nebraska test lab were all given the same PTO powers so that the fuel consumption prediction model could be created. [5] Equipment owners and farmers can gain a better understanding of real-time parameter monitoring and fuel usage estimation by comparing tractor field performance across different power levels. [6] The BSFC is higher when diesel fuel is blended with jojoba, since the two fuels have different thermal properties (lower calorific value and higher density). In their prediction, [7] made an estimate of a tractor's fuel usage. In terms of prediction, their results demonstrated that the ANN outperformed the multi-stage regression method. [8] ANNs were also used by another group to measure fuel economy. Engine speed, operating depth, forward speed, humidity, tyre inflation pressure, and cone index were among the variables that were input. [9] predicted the fuel consumption of a tractor running on sandy loam soil in a humid tropical region. A number of parameters were affected, including draught, speed, and depth of cut, diameter of cut, fuel consumption, cone index, and soil moisture content. [10] Tractor fuel usage during ridging is inversely related to penetration resistance and breadth of cut, while draught, ridging speed, moisture content, and height of ridge are directly proportionate, according to the model. An exponential relationship between engine load and fuel consumption was shown using the regression equations. [11] The field's edge is where mechanical tillage techniques like ploughing start, and the field's end is where the unit is rotated. First, get the implement down into operating position. Then, get the tractor up to technological speed. Slowing down after passing across the field [12]. There are two distinct periods during a tractor's time spent ploughing in a field: portion one, when the plow is turning and the tractor is pulling the harvest, and part two, when the tractor isn't doing much of anything efficiently. [13]. While in theory farmers might cut down on non-productive work by constructing routes that take into account the size and form of the field plot, in reality, terrain and agro-technological operational needs place limitations on such routes [14].

## 3 PROPOSED SYSTEM

An effective method for assessing current fuel usage is the fuel consumption model. Inspiring experts to assess the vehicle's efficiency and furnish correct data regarding fuel consumption requires a reliable fuel consumption model.

### 3.1 *Preprocessing*

It is possible to use the data from the sources described in the previous chapter to enhance the number of input variables. It would be easier to foretell the truck's fuel use if they were inputs [15]. Vehicle acceleration and road gradient are two important factors that can be acquired. It is possible to find the road's grade. Simultaneously, the acceleration of the vehicle may be calculated using the same procedure:

$$Acceleration = \frac{dK_p(m, s)}{dG_p(s)} \quad (1)$$

Here, $dK_p$ is the change in vehicle velocity between two consecutive incidences and $dG_p$is the change in time between those encounters.

### 3.2 *Feature extraction*

#### 3.2.1 *Fuel consumption forecast*

In recent years, there has been a rise in the application of machine learning to solve different types of regression problems. This method's current iteration employs principal component analysis and random forest algorithms to predict fuel usage. In contrast to traditional algorithms for predicting fuel consumption, this one doesn't require a thorough familiarity with the vehicle's architecture or engine principles [13]. The feature matrix $C$ from the tractor driving cycle and the feature matrix $M$ from the model can be used to estimate fuel usage. The outcome is $R_2, M = \left[\bar{k}, k_{std}, \ldots, R_2\right]$after supplementing with the behavior attribute of operational $R_1$ and the characteristics of the driving mode $R_2$ to this matrix. The method proposes principal component analysis as a means to eliminate feature collinearity and shrink the feature matrix $M$.

### 3.3 *Model training*

#### 3.3.1 *RBF*

RBF uses normal curves, the value of which depends on how far away the point is from the origin (i.e.,$(RR) = \varnothing(\|RR\|)$), or on the distance from another center point. This is a straightforward example of a single-layer MLP that uses RBF activation functions. The RB functions are any function $\varnothing$ that is defined as$\varnothing(RR) = \varnothing(\|RR\|)$. When measuring distance, the standard unit is the "Euclidean" model.

#### 3.3.2 *SVM*

Support Vector Machine models are dimensional space maps; they display data separated by an ideal gap in a clear hyper plane (sometimes called a "drain"). It makes use of a linear and non-linear support vector machine. The Hyperbolic Tangent, Polynomial, and Gaussian Radial Basis Function are only a few of the kernel functions offered by non-linear support vector machines (SVMs), in contrast to the ideas of hard margin and soft margin that were developed by linear SVMs. Support vector machine classifiers can be computed using the current, primal, dual, or kernel trick techniques. One feature displayed by a group of functions of $G(rr_1, rr_2)$, as shown by$G(rr_1, rr_2) = \vartheta(rr_1), \vartheta(rr_1)$, where $\vartheta$ is a function that maps $r$ to a linear space $Z$. The $Z$ -space transforms at the dot product. In particular, for any positive integers n, the polynomials belong to this class of functions, with the formula $G(rr_1, rr_2) = \left(1 + rr_1' rr_2'\right)^{\mathrm{w}}$. The radial basis function of a Gaussian distribution, $G(rr_1, rr_2) = \exp\left(-\|rr_1 - rr_2\|^2\right)$ , will be the subsequent item to be examined. Lastly, for positive and negative values of $w_1$, the sigmoid function (sig) or multi-level product (MLP) is defined as $G(rr_1, rr_2) = \tan l\left(w_1 rr_1' rr_2' + w_2\right)$.

### 3.3.3 *MLP*

One kind of feedforward artificial neural network is the multi-layer perceptron, or MLP. Every neuron uses a nonlinear activation function except for the input neuron. Machine learning with prediction (MLP) and similar methods train networks using the back-propagation idea. A number of layers separate data using nonlinearly separable terms in nonlinear MLM, as opposed to linear MLP.

## 4 RESULT AND DISCUSSION

A large portion of the farm's assets fund the most important farming operations. Therefore, by analyzing the fuel consumption of farm machinery of different sizes, it is possible to create an appropriate energy budget and predict fuel use.

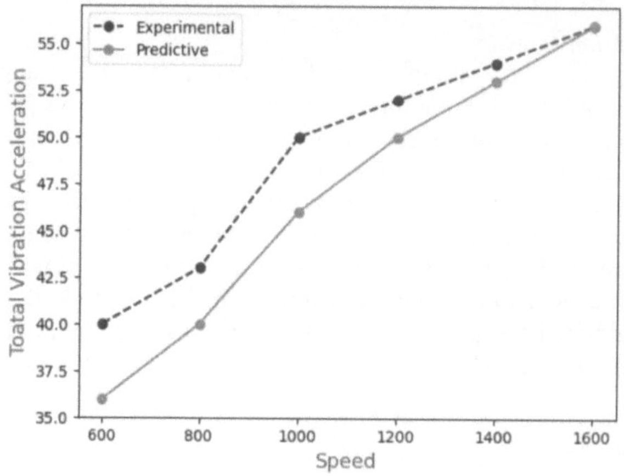


Figure 1. Vibration prediction versus experimental measurement.

The effects of vibration are shown in Figure 1. The trends of the experimental and projected vibration values are discovered to be comparable.

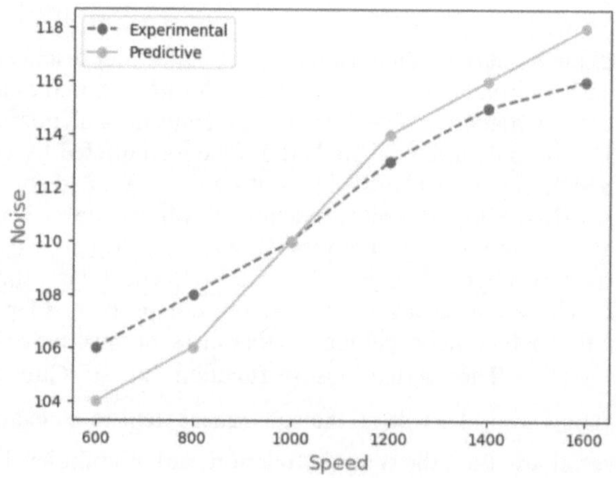


Figure 2. Analyzing the variation in experimental and predicted noise.

Analyzing the Variation in Experimental and Predicted Noise is illustrated in Figure 2. The predicted results and the experimental values are very similar. This means the prediction model is accurate.

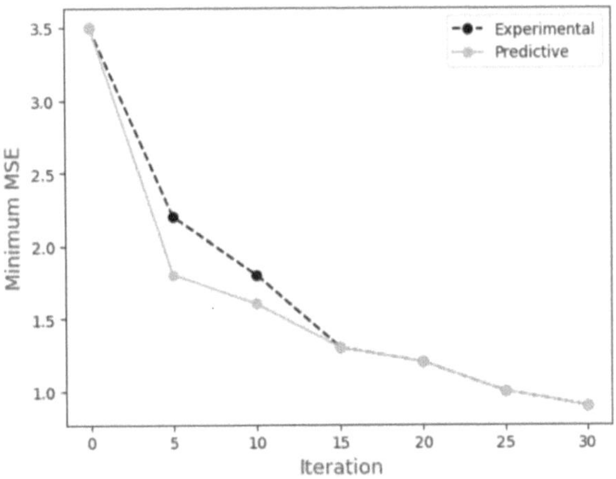


Figure 3. Minimum MSE plot for the RBF-MLP-SVM model.

Figure 3 displays the RBF-MLP-SVM predictive model's minimal MSE plot. The results demonstrate that the optimized RBF-MLP-SVM outperforms the other models.

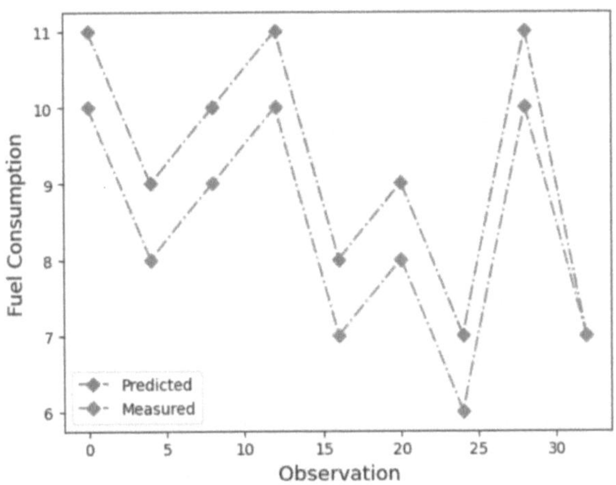


Figure 4. Comparison of predicted values for fuel consumption.

A comparison of the predicted and actual fuel consumption data using this RBF-MLP-SVM model is shown in Figure 4.

## 5 CONCLUSION

Using a low-cost approach, this study identifies tractor behaviors and predicts fuel consumption during sowing operations. It also investigates the relationship between driving modes of tractor machinery and fuel consumption. Preprocessing data obtained from the several sources discussed in the previous chapter could be supplemented with other input elements. Fuel consumption forecasts are utilized in feature extraction. When training a model, RBF-MLP-SVM is used. Compared to the MLP and SVM models, the proposed technique has a higher average accuracy of 97.35 percent.

## REFERENCES


[1] Al-Janobi, A., Al-Hamed, S., and Aboukarima, A. M. (2010). An Excel spreadsheet to estimate performance parameters for chisel plow-tractor combination based on trained an artificial neural network. Bulletin of the University of Agricultural Sciences & Veterinary Medicine Cluj-Napoca. *Agriculture*, 67(2).

[2] Almaliki, S., Alimardani, R., and Omid, M. (2016). Fuel consumption models of MF285 tractor under various field conditions. *Agricultural Engineering International: CIGR Journal*, 18(3), 147–158.

[3] Grisso, R. D., Vaughan, D. H., and Roberson, G. T. (2006). Method for fuel prediction for specific tractor models. In 2006 ASAE Annual Meeting (p. 1). *American Society of Agricultural and Biological Engineers*.

[4] Jalilnezhad, H., Abbaspour-Gilandeh, Y., Rasooli-Sharabiani, V., Mardani, A., Hernández-Hernández, J. L., Montero-Valverde, J. A., and Hernández-Hernández, M. (2023). Use of a convolutional neural network for predicting fuel consumption of an agricultural tractor. *Resources*, 12(4), 46.

[5] Karparvarfard, S. H., and Rahmanian-Koushkaki, H. (2015). Development of a fuel consumption equation: Test case for a tractor chisel-ploughing in a clay loam soil. *Biosystems Engineering*, 130, 23–33.

[6] Rahimi-Ajdadi, F., and Abbaspour-Gilandeh, Y. (2011). Artificial neural network and stepwise multiple range regression methods for prediction of tractor fuel consumption. *Measurement*, 44(10), 2104–2111.

[7] Asinyetogha, H. I., Raymond, A. E., and Silas, O. N. (2019). Predicting tractor fuel consumption during ridging on a sandy loam soil in a humid tropical climate. *Journal of Engineering and Technology Research*, 11(3), 29–40.

[8] Siddique, M. A. A., Baek, S. M., Baek, S. Y., Kim, W. S., Kim, Y. S., Kim, Y. J., and Hwang, J. Y. (2021). Simulation of fuel consumption based on engine load level of a 95 kW partial power-shift transmission tractor. *Agriculture*, 11(3), 276.

[9] Kolator, B. A. (2021). Modeling of tractor fuel consumption. *Energies*, 14(8), 2300.

[10] Küçüksarıyıldız, H., Çarman, K., and Sabancı, K. (2021). Prediction of specific fuel consumption of 60 HP 2WD tractor using artificial neural networks. *International Journal of Automotive Science And Technology*, 5(4), 436–444.

[11] Lee, J. W., Kim, J. S., and Kim, K. U. (2016). Computer simulations to maximise fuel efficiency and work performance of agricultural tractors in rotovating and ploughing operations. *Biosystems Engineering*, 142, 1–11.

[12] Guerrieri, M., Fedrizzi, M., Antonucci, F., Pallottino, F., Sperandio, G., Pagano, M., and Costa, C. (2016). An innovative multivariate tool for fuel consumption and costs estimation of agricultural operations. *Spanish Journal of Agricultural Research*, 14(4), 4.

[13] Yang, L., Tian, W., Zhai, W., Wang, X., Chen, Z., Wen, L., and Wu, C. (2022). Behavior recognition and fuel consumption prediction of tractor sowing operations using smartphone. *International Journal of Agricultural and Biological Engineering*, 15(4), 154–162.

[14] Kamarulzaini, K. A. A., Ismail, N., Rahiman, M. H. F., Taib, M. N., Ali, N. A. M., and Tajuddin, S. N. (2018, March). Evaluation of RBF and MLP in SVM kernel tuned parameters for agarwood oil quality classification. In *2018 IEEE 14th International Colloquium on Signal Processing & Its Applications (CSPA)* (pp. 250–254). IEEE.

[15] Bhoraskar, A. (2019). Prediction of Fuel Consumption of Long Haul Heavy Duty Trucks Using Machine Learning and Comparison of the Performance of Various Learning Techniques. *Delft University of Technology*.

*Hybrid and Advanced Technologies – S. Prasad Jones Christydass, Nurhayati Nurhayati and S. Kannadhasan. (Eds)* © 2025 the Authors, ISBN 978-1-032-90254-8
*Open Access: www.taylorfrancis.com, CC BY-NC-ND 4.0 license*

# IoT based smart irrigation monitoring and controlling system using GA-CNN-BiGRU model

Soumya Mishra
*Associate Professor, Quality Assurance Cell, KIIT University, Bhubaneswar, Odisha, India*

R. Manikandan
*Associate Professor, Department of CSE, Government College of Engineering, Thanjavir, Tamilnadu, India*

Nitin Mishra
*Assistant Professor, Department of Civil Engineering, Graphic Era Deemed to be University, Dehradun, India*

Gayatri Niphadkar
*Assistant Professor, Department of Civil Engineering, Dr.D.Y. Patil Institute of Technology Pimpri, Pune, India*

M. Shobana
*Assistant Professor, Department of Computer Science and Business Systems, M. Kumarasamy College of Engineering, Karur, India*

Maaz Allah Khan
*Department of Civil Engineering, UIET, Babasaheb Bhimrao Ambedkar University (A Central University), Lucknow, India*

ABSTRACT: The expansion of agriculture is crucial because, as India's population continues to rise, there will be a serious food crisis in the next 25 to 30 years. The farmers are currently facing a lot of problems due to the lack of rain and water. In order to help farmers save time, money, and energy, this study primarily aims to develop an automated irrigation system. Manual involvement is required by the conventional methods of irrigation for agricultural land. The use of machine-controlled irrigation technologies will decrease the need for human intervention. The sensors provide an interrupt signal to the Arduino controller whenever they detect a change in environmental variables such as humidity, temperature, wetness, and pH. This data will subsequently be sent to the IoT (Internet of Things) system, allowing the farmer to oversee and control the irrigation remotely. The suggested method employs preprocessing steps including DBO and renaming, feature selection using Chi-Square, and model training using GA-CNN-BiGRU, BiGRU, and CNN-BiGRU. When compared to the other two approaches, the suggested GA-CNN-BiGRU performs admirably. Keywords: Crop Water Stress Index (CWSI), Convolutional Neural Networks (CNNs), Internet of Things (IoT), Smart Irrigation, Precision Agriculture (PA)

## 1 INTRODUCTION

The Indian economy relies heavily on agriculture. The annual water use for agriculture, however, exceeds that of rainfall. In order to keep up with the ever-increasing demand for food caused by global population expansion, it is crucial to increase agricultural productivity. It is possible to enhance agricultural output by thinking about and anticipating environmental factors. Details

DOI: 10.1201/9781003559115-2

gathered from the field, including soil moisture, air temperature and humidity, etc., form the basis of crop quality assessments. A higher harvest is possible with the help of modern implements and machinery. With the development of IoT technologies, a wealth of data on ecosystems and agricultural yields may be collected. "Smart home, smart city, smart transportation, and smart farming" are just a few of the forthcoming intelligent concepts that are encompassed by the Internet of Things (IoT). To improve efficiency and quality, this method can be used to apply the right amount of water, fertilizer, pesticide, etc. One promising tool for smart agriculture is the sensor. The crop lifecycle is constantly affected by the real-time environmental elements, such as soil moisture level, ambient temperature, and tank water level. Agricultural water regulation can be effectively monitored through the formation of a sensor network.

## 2 LITERATURE SURVEY

To conserve water, irrigation system frameworks have been proposed using techniques such as thermal imaging, Crop Water Stress Index (CWSI), and direct soil water monitoring. Thermal imaging, a common irrigation management technology, focuses on plant shadow temperature distribution. According to , this system involves monitoring water status in plants at regular intervals and planning irrigation based on shadow temperature distribution [1]. Automatic irrigation scheduling using direct soil water measurement is proposed to efficiently use water compared to manual irrigation systems [2]. More than 30 years have passed since CWSI was initially described [3]. Predicting soil moisture using support vector machines based on air temperature, relative humidity, and soil temperature was proposed in [4]. Solar power is also the subject of numerous publications on irrigation systems [5]. In smart agriculture, the optimal use of water fertilizers can be achieved by precision agriculture (PA) in conjunction with cloud computing, as explained in [6]. In addition to aiding in the analysis of field meteorological conditions, it is able to maximize crop output. Additionally, one study suggested a smart irrigation control system that aimed to save water by combining a fuzzy system with a neural network [7]. employing ML techniques to maximise output. Some more recent scholarly works include a precision agriculture platform and experiments with growing turmeric [8]. The platform employs analytical approaches to examine communication delays and offers a graphical user interface for linking actuators and sensors. To find irrigation zones that aren't uniform, another study used thermal pictures sent by drones over a cloud-fog system [9]. Another approach [10] uses an open-source cloud platform called ThingSpeak in conjunction with an Arduino UNO microcontroller board to monitor the garden's humidity and provide visual updates on a regular basis. This straightforward method does its job well, protecting plants from deterioration and keeping soil fertile for longer. To aid Indian farmers with field irrigation, another author [11] suggests various models to monitor soil moisture. Building intelligent irrigation systems is the goal of these models, which employ a number of sensors in conjunction with Raspberry Pi, a small computer. An alternative system [12] automates field watering by sensing soil characteristics like moisture and temperature using open-source technologies. The system also makes use of weather forecast data that is retrieved from the internet. Involved sensors can also measure the field's exposure to ultraviolet (UV) light. A clever algorithm determines and predicts the field's watering requirement and level based on data from sensors and weather forecasts.

## 3 PROPOSED SYSTEM

### 3.1 *Preprocessing*

Data preprocessing is the process of preparing raw data for use by making it uniform and easier to interpret. When it comes time to put the system into production, data preparation techniques are used as needed. Data normalization/standardization, data scaling, data grouping by sum, and other similar processes are examples of preprocessing operations. After the preprocessing is done, the dataset-features mapping is normally achieved by applying the appropriate encoding. This step is essential for the data to be transformed into features output, which the subsequent feature learning subsystem may then utilize effectively. As a result, we went over the most typical preparation steps that could be used on the input dataset

#### 3.1.1 *DBO*

The basic premise behind data binning is to reduce the number of "bins" used to store a collection of continuous values [13]. While data binning's primary objective is to strengthen the system and avoid overfitting, it does so at the expense of performance.

#### 3.1.2 *Renaming of the data*

This is a common practice when working with raw dataset records that contain data columns that need to be processed further. Column names are used in this process, as they are required by encoding techniques. Finding the name of the reference column is necessary for certain operations, such as sorting data into a table.

### 3.2 *Feature selection*

#### 3.2.1 *Chi-Square*

For hypothesis testing on variance populations, the $y^2$ test is utilized [13]. It compares the Chi-square distribution to a one-degree-of-freedom distribution and ranks the features based on their degree of independence between the all-independent features $k_{g_k}$ and the goal feature $E_{g_i}$. The calculation of $y^2$ is defined by Equation (1).

$$y^2\left(k_{g_k}, E_{g_i}\right) = \frac{(U * (q * r - p * s))}{(q * s) * (q + r) * (s + r) * (p + r)} \tag{1}$$

$q$ represents the frequency of $k_{g_k}$ and $E_{g_i}$ in the dataset, s represents the frequency of $k_{g_k}$ appearing without $E_{g_i}$, and p represents the frequency of $E_{g_i}$. A greater value of $y^2$ $\left(k_{g_k}, E_{g_i}\right)$ implies a more significant feature.

### 3.3 *Model training*

#### 3.3.1 *GA-BiGRU-CNN*

This research proposes a GA-CNN-BiGRU model for predicting cyber assaults. Preprocessing of the data, selecting the feature, and historical data classification helps prevent IoT assaults and ensure steady functioning. Strong guarantee. The model starts with historical cyber-attack detection data, including occurrence time, type, and duration. It then goes through pre-processing to clean, normalize, and extract features for future training and prediction. Our model decomposes time series data using the GA algorithm to extract main component feature information. This stage seeks to improve model performance and accuracy for future modelling. Using convolutional neural networks (CNNs), the data is modelled and features are discovered. After that, BiGRU is used to build the model using the retrieved feature data. The training process optimizes the parameters and loss functions of a model by analysing previous data. The model can make educated judgments and precise predictions after training. To make sure cyber assault systems are reliable and safe, the method optimizes prediction solutions using data from the system in real-time [14]. In response to changes in the cyber-attack prediction environment, the program dynamically updates and modifies data. A GA module, a CNN module, and a BiGRU module make up the GA-CNN-BiGRU model. There are benefits to using all three parts when trying to foretell power system failures. The general layout of the model is illustrated in Figure 1. Convolutional neural networks and bidirectional gated recurrent units are combined in our

GA-CNN-BiGRU model. Feature extraction and sequence modelling are two common uses for the former. The following is a mathematical description of the model.

The Conv layer produces $z_{conv}$, which is calculated as follows:

$$z_{conv} = \mu(V_{conv} * Y + C_{conv}) \tag{2}$$

the input feature picture $Y$, the convolution kernel's weight $V_{conv}$, the bias vector bconv, the convolution operation *, and the activation function $\mu$. After feeding the convolution layer's output $z_{conv}$ into the bidirectional gated cyclic unit, we get $Z_{rnn_f}$ for the forward cyclic unit and $z_{rnn_c}$ for the reverse cyclic unit:

$$Z_{rnn_f} = BiGRUg(Z_{conv}) \tag{3}$$

$$Z_{rnn_c} = BiGRUc(Z_{conv}) \tag{4}$$

the units for forward and reverse bidirectional gated loops are represented by $BiGRUg$ and $BiGRUc$, respectively. Lastly, for the final forecast, we combine the results of the forward and backward cyclic units:

$$Z_{pred} = concat(Z_{rnng}, Z_{rnnc}) \tag{5}$$

The data set is standardized and pre-processed before being input into the feature categorization and output for the GA module. Next, it includes information about to different cyberattacks. The CNN module is fed feature sequences so that pooling classification may be done. Finding the optimal model parameters is one way that BiGRU improves the precision of both its predictions and its output.

## 4 RESULT AND DISCUSSION

While advancements in IoT Industry 4.0 have improved many areas, including manufacturing, healthcare, and agriculture, they have also brought several security concerns to humans as a result of the numerous new cyberthreats. As a result, not only is detection accuracy enhanced, but network traffic and information leakage during data exchange across IoT gateways can be drastically reduced using our suggested approach. By running extensive simulations on real-world datasets, we demonstrate that our suggested strategy can get better results than traditional machine learning approaches.

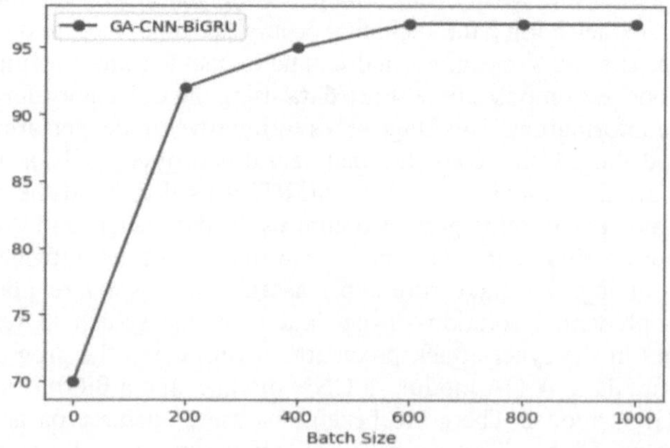


Figure 1. Graphs the accuracy against the number of epochs.

As demonstrated in Figures 2 and 3, our model achieves local optima during training and undergoes smooth optimization. As our meta-algorithm was trained, the figure displays both the accuracy and the loss function.

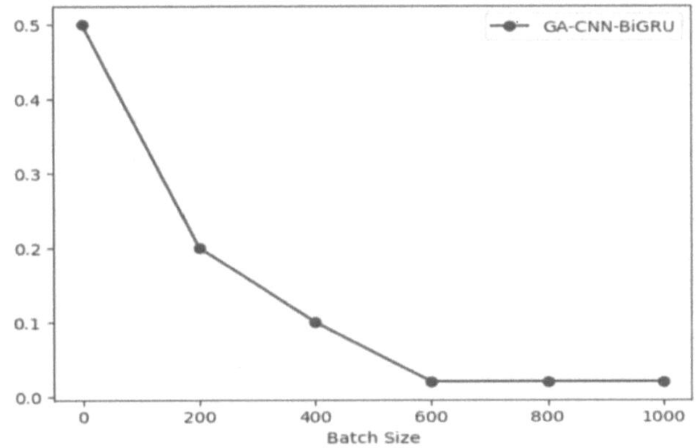


Figure 2. Graphs the loss against the number of epochs.

The accuracy was found to decrease as the number of blocks was adjusted, both up and down. On top of that, we discovered that the five GA-CNN-BiGRU models were the best option. It was observed that reducing the amount of epochs beyond 400 reduces accuracy, while increasing the number of epochs beyond 400 increases computation time without significantly improving the result as shown in in Figure 2.

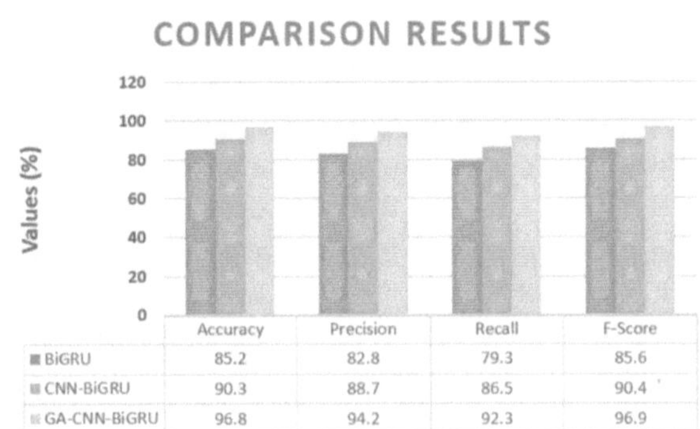


Figure 3. Comparison of proposed technique to other models.

Comparing the suggested technique to the baseline methods graphically in terms of recall, F1-score, accuracy, and precision rates is shown in Figure 3.

## 5 CONCLUSION

Before modern methods of soil testing, farmers would determine the best way to sort their crops based on how prepared the soil was. They disregarded the stickyness, water level, and especially the atmospheric situation that was challenging for a farmer, gradually The

Internet of Things (IoT) is revolutionizing the agricultural industry by involving farmers in a variety of new methods, such as precise and helpful cultivation for field problem management. Harvest online evaluation permits disclosure of wild plant, level of water, cultivation area, animal break in to the field, trim turn of events, agriculture, and more, thanks to IoT modernization. It also helps to gather data on variables like atmosphere, protection, temperature, and soil productivity. Internet of Things (IoT) technology allows farmers to stay connected to their living arrangements from anywhere at any time. The house conditions are monitored by distant sensor structures, while the field is controlled and motorized by smaller scope controllers. Use of far-off cameras has allowed for the capture of circumstances in both still and moving images. Improvements to the internet of things (IoT) can help farmers save money and enhance the efficiency of conventional production. Preprocessing, feature selection, and model training are the three main components of the suggested method in this article. In comparison to two more conventional approaches, GA-CNN-BiGRU outperforms both BiGRU and CNN-BiGRu.

## REFERENCES


[1] Wang, X., Yang, W., Wheaton, A., Cooley, N., and Moran, B. (2010). Efficient registration of optical and IR images for automatic plant water stress assessment. *Computers and Electronics in Agriculture*, 74 (2), 230–237.

[2] O'Shaughnessy, S. A., and Evett, S. R. (2010). Canopy temperature based system effectively schedules and controls center pivot irrigation of cotton. *Agricultural Water Management*, 97(9), 1310–1316.

[3] Idso, S. B., Jackson, R. D., Pinter Jr, P. J., Reginato, R. J., and Hatfield, J. L. (1981). Normalizing the stress-degree-day parameter for environmental variability. *Agricultural Meteorology*, 24, 45–55.

[4] Gill, M. K., Asefa, T., Kemblowski, M. W., and McKee, M. (2006). Soil moisture prediction using support vector machines 1. *JAWRA Journal of the American Water Resources Association*, 42(4), 1033–1046.

[5] Alex, G., and Janakiranimathi, M. (2016, February). Solar based plant irrigation system. In *2016 2nd international conference on advances in electrical, electronics, information, communication and bio-informatics (AEEICB)* (pp. 425–428). IEEE.

[6] Rao, R. N., and Sridhar, B. (2018, January). IoT based smart crop-field monitoring and automation irrigation system. In *2018 2nd International Conference on Inventive Systems and Control (ICISC)* (pp. 478–483). IEEE.

[7] Chen, G., and Yue, L. (2011, August). Research of irrigation control system based on fuzzy neural network. *In 2011 International Conference on Mechatronic Science, Electric Engineering and Computer (MEC)* (pp. 209–212). IEEE.

[8] Chen, W. L., Lin, Y. B., Lin, Y. W., Chen, R., Liao, J. K., Ng, F. L., and Yen, T. H. (2019). AgriTalk: IoT for precision soil farming of turmeric cultivation. *IEEE Internet of Things Journal*, 6(3), 5209–5223.

[9] Roopaei, M., Rad, P., and Choo, K. K. R. (2017). Cloud of things in smart agriculture: Intelligent irrigation monitoring by thermal imaging. *IEEE Cloud computing*, 4(1), 10–15.

[10] Dasgupta, A., Daruka, A., Pandey, A., Bose, A., Mukherjee, S., and Saha, S. (2019). Smart irrigation: IOT-based irrigation monitoring system. *In Proceedings of International Ethical Hacking Conference 2018: eHaCON 2018*, Kolkata, India (pp. 395–403). Springer Singapore.

[11] Raut, R., Varma, H., Mulla, C., and Pawar, V. R. (2018). Soil monitoring, fertigation, and irrigation system using IoT for agricultural application. In *Intelligent Communication and Computational Technologies: Proceedings of Internet of Things for Technological Development, IoT4TD 2017* (pp. 67–73). Springer Singapore.

[12] Goap, A., Sharma, D., Shukla, A. K., and Krishna, C. R. (2018). An IoT based smart irrigation management system using Machine learning and open source technologies. *Computers and Electronics in Agriculture*, 155, 41–49.

[13] Abu Al-Haija, Q. (2022). Top-down machine learning-based architecture for cyberattacks identification and classification in IoT communication networks. *Frontiers in Big Data*, 4, 782902.

[14] Zhang, D., Jin, X., and Shi, P. (2023). Research on power system fault prediction based on GA-CNN-BiGRU. *Frontiers in Energy Research*, 11, 1245495.

*Hybrid and Advanced Technologies – S. Prasad Jones Christydass, Nurhayati Nurhayati and S. Kannadhasan. (Eds) © 2025 the Authors, ISBN 978-1-032-90254-8*


# An innovative method for plant disease detection using BiGRU and attention mechanism model

C.L. Monica
*Assistant Professor, Environmental Science, Siddhartha Academy of Higher Education, Vijayawada, Andhra Pradesh, India*

Aniruddha S. Rumale
*Professor, Department of Information Technology, Sandip Foundations', Sandip Institute of Technology and Research Centre, Nashik, Maharashtra, India*

S. Jeevitha
*Associate Professor, Department of Computer Science and Information Technology, Kalasalingam Academy of Research and Education, Krishnankoil, India*

Karthick Raja Muthaiah
*Assistant Professor, Department of Computer Science and Engineering, Sri Eshwar College of Engineering, Coimbatore, India*

Sujatha Kamepalli
*Associate Professor, Department of IT & CA, VFSTR Deemed to be University, Vadlamudi, Andhra Pradesh, India*

Amit Chauhan
*Department of Life Sciences, CHRIST University, Bengaluru, Karnataka, India*

ABSTRACT: Agriculture underpins all economies. Crop output affects any nation's local market. Plant diseases reduce productivity after equipping fields with effective resources. Early plant disease type identification is needed for destruction avoidance and detection. Recent study has identified plant diseases that affect several crops worldwide.Many efforts are made to determine these illnesses' causes. Some diseases are viral, whereas others are fungal. It becomes a serious issue when the root cause cannot be located before it affects a large percentage of the production process. This study compares computer vision and image methods for plant disease identification and classification. Segmentation pre-processes pictures for plant disease identification. Numerous feature extraction and categorization methods enable it. Model training uses BiGRU, BiGRU-LSTM, and BiGRU-AM. Recommended approach BiGRU-AM works well.
Keywords: Plant Disease Detection, Agriculture, Attention Mechanism Model, Long short-term memory (LSTM), Bidirectional Gated Recurrent Unit (BiGRU).

## 1 INTRODUCTION

The importance of preventing agricultural illnesses is growing in response to the ever-increasing need for both abundant and high-quality food. Farmers still face challenges when trying to identify plant illnesses quickly. An understanding of plant diseases is necessary for spotting diseased leaves. It is also common practice to use laboratory equipment in order to identify a diseased leaf. The potential of computer vision to diagnose plant diseases in an effective and scalable manner is explored in this work. Thanks largely to the many

DOI: 10.1201/9781003559115-3

advancements in deep convolutional neural networks, computer vision has made tremendous strides in recent years. The learned models are excellent for consumer applications on smartphones because to their lightning-fast photo classification speed, but training massive neural networks can be a time-consuming ordeal. In order to increase crop yields, disease management efforts, and general agricultural knowledge, image processing for plant disease identification opens the way for new ways to apply deep learning to real-world agricultural problems. Most existing vision-based technologies require background-free, extremely sharp images. Since the majority of Indian farmers use low-end mobile devices in these circumstances, we instead provide the best query resolution for crops and plants based on images captured in natural settings with non-trivial background noise. The results show that employing a lab-controlled dataset to detect or classify pictures in a real-world situation is not feasible.

## 2 LITERATURE SURVEY

The widespread presence of plant diseases leads to a decrease in agricultural production. If plant diseases are not detected quickly, food insecurity will become even more severe [1]. Every disease or insect problem usually has its own obvious pattern [2] for the aim of distinguishing abnormalities through diagnostics. It is common practice to use a plant's leaves as the primary means of identification many different plant diseases, the majority of which show themselves initially on the leaves [3]. One branch of deep learning, convolutional neural networks (CNNs), is only now beginning to make headlines [4]. Convolutional neural networks (CNNs) have outstanding processing and recognition capabilities, making them the most popular image classifiers [5]. Notwithstanding the positive outcomes documented in the research, the datasets used are somewhat limited in diversity. Transfer learning allows for the retraining of pre-trained convolutional neural networks (CNNs) using smaller datasets with a different distribution than the larger datasets used to train the network initially. [6] In this age of climate change and increased global trade, the importance of finding better diagnostic methods for identifying and managing plant pathogens has been underscored by recent epidemic outbreaks. [7] The unchecked movement of insects, nematodes, weeds, viruses, phytoplasmas, bacteria, fungi, and other undesirable "alien" species—including plant materials—causes enormous problems for agriculture around the world. [8] Preventing the introduction of new diseases, controlling diseases at all stages of development, and reducing the risk of disease transmission are all reasons why early pathogen detection is becoming increasingly important in plant health monitoring [9]. According to a link between crop losses and production conditions, regions experiencing food insecurity due to newly or re-emerging pests and diseases had higher crop losses compared to those having food surpluses [10]. Artificial intelligence (AI) has recently found its way into plant pathology for the aim of identifying anomalies and pests. [11] The detection, evaluation, and control of plant diseases could be drastically altered by these technologies. Our goal here is to examine the various AI systems that have been proposed for use in plant pathology, weigh the benefits and drawbacks of each, and ultimately determine how these systems could revolutionize the field [12]. One of the most prominent AI technologies, ML is relied upon by many plant pathologists. Because these algorithms can be trained to recognize specific patterns and symptoms, they can be used to classify diseases in their early stages [13]. However, unexpected diseases may not be suitable for ML systems due to the enormous amount of annotated data needed for training.

## 3 PROPOSED SYSTEM

Although crop diseases pose a significant threat to food security, the absence of a suitable foundation makes rapid identification of these diseases difficult in many regions of the world. The area of leaf-based image categorization has witnessed remarkable advances with the emergence of precise approaches.

### 3.1 *Preprocessing*

Image data undergoes per-processing to remove non-essential distortions or improve certain image features needed for further processing. The target area is extracted by cropping the image. The purpose of picture enrichment is to make the image's contrast stronger [14]. The following equation is used to convert RGB to grayscale images:

$$g(z) = r * 0.3 + g * 0.6 + b * 0.1 \tag{1}$$

The next step in improving the image of the sick leaf is to execute a histogram comparison, which activates the image's intensities.

### 3.2 *Segmentation*

Here, we implement the region-based segmentation using k-means clustering. Each step of the k-means clustering algorithm is as follows: First, there is the input data, which consists of the image that will be used to execute segmentation. 2. Determine the centroid point of the data by calculating the arithmetic mean of the input data [15]. Third, segments are formed by calculating the central point's Euclidean distance and then clustering points with similar distances. The fourth step is to choose a region of interest (ROI) from the segmented image; this will serve as input for the following classification phase.

### 3.3 *Feature extraction*

There are four processes to feature extraction, which is applied to the segmented clusters: Step 1: Simply input the cluster number of the diseased leaf portion. Next, if the image is RGB, turn it into grayscale. Step 3: Develop the GLCMs, or Grey Level Co-occurrence Matrixes. Step 4: Use GLCM to extract features and statistics. Mean, standard deviation, variance, energy, homogeneity, correlation, and contrast are the derived properties [16]. The classifier is then instructed to use these features in its further processing. Characteristics are retrieved from the afflicted illness.

### 3.4 *Model training*

#### 3.4.1 *BiGRU*

The Gated Recurrent Unit (GRU) is a kind of Recurrent Neural Network (RNN) that processes time-series data using a "memory" function and a recursive structure that is similar to RNN. And while we're on the subject of long-term memory, GRU can help with gradient disappearance and gradient explosion, two issues that might arise during RNN training. Another RNN version, the Long Short-Term Memory (LSTM) network, performs similarly to GRU; however, GRU's simpler structure allows it to cut computation and enhance training efficiency. Typical GRU architectures only allow one direction of propagation per sequence transmission. It misses out on data from the future since all the information it gets is from the past, before this very moment. The forward GRU and backward GRU components of the BiGRU structure capture the features of the information before and after, respectively.

#### 3.4.2 *Attention mechanism*

The processing of visual signals in the human brain is analogous to the attention mechanism. By determining the weights of the characteristic vectors output by BiGRU at various points in time, it enables better performance by highlighting qualities that contribute to a larger share of the prediction outcomes [17]. Machine translation and voice recognition are examples of temporal data that the attention mechanism excels at processing. In terms of classification prediction, it works rather well. This hybrid model can be used independently or layered with others. When training for air target tactical intention recognition, the neural network zeroes in on a few critical features—the weight coefficient at its core—through the Attention mechanism. Through gaining an understanding of how crucial each attribute is and then giving each attribute a weight commensurate with its significance. To better understand the opponent plane's assault intentions, the model will give greater weight to factors like heading angle and manoeuvre type.

### 3.4.3 *Output layer*

As its input, the output layer takes the hidden layer's attention mechanism layer's output as its input. Intention classification results are computed using the multi-class softmax algorithm. "Overtting" is a real possibility if there are an excessive number of hidden layer nodes. When dealing with a single hidden layer, it is best to progressively increase the number of nodes until you reach 256. Improving the accuracy of model identification can be achieved by correctly increasing the network depth as long as the total number of hidden layer nodes stays constant. However, once the number of layers approaches three, the effect of adding more hidden layers starts to diminish.

## 4 RESULT AND DISCUSSION

"Overfitting" is a real possibility if there are an excessive number of hidden layer nodes. When dealing with a single hidden layer, it is best to progressively increase the number of nodes until you reach 256. Improving the accuracy of model identification can be achieved by correctly increasing the network depth as long as the total number of hidden layer nodes stays constant. However, once the number of layers approaches three, the effect of adding more hidden layers starts to diminish.

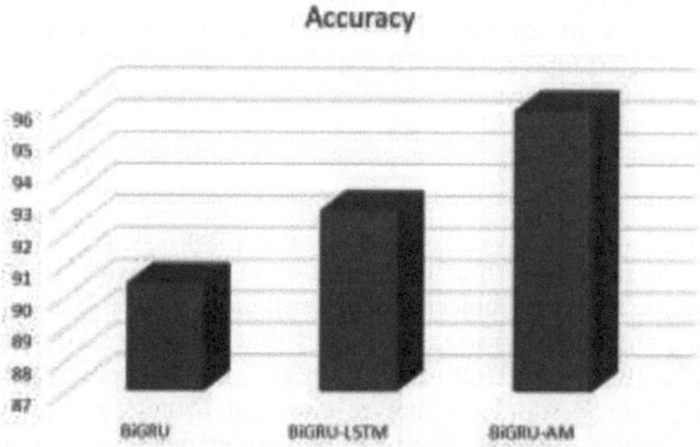


Figure 1. Accuracy comparison of proposed model.

Instead of Adam, we opted on RMSprop as the optimizer, and Cross entropy as the loss function. With the early stopping function and patience set to 5, catching the best weight for each iteration becomes a breeze with Callbacks, another way provided to improve accuracy. For further information, see Figure 1 below.

The training and validation accuracy for each training period of the proposed BiGRU-AM model are plotted in Figure 2.

The training and validation loss for each training period of the proposed BiGRU-AM model are plotted in Figure 3.

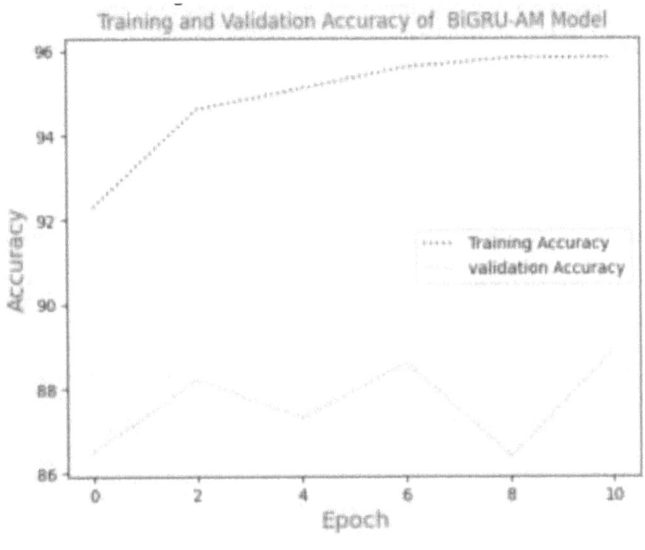


Figure 2. Training and validating the suggested method.

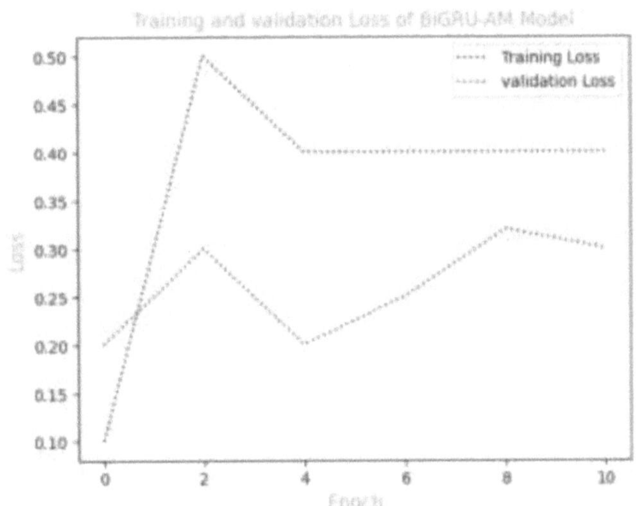


Figure 3. Training and loss validation of the models.

## 5 CONCLUSION

Due to the country's rapidly expanding population and attendant rise in food consumption, agriculture plays a pivotal role in India. As a result, it must improve agricultural productivity. Bacterial, viral, and fungal diseases are a big contributor to poor crop yields. Methods for detecting plant diseases can help avoid this. Because machine learning primarily applies to data and prioritises the results of certain tasks, it can be utilised for disease identification. A comparative analysis of machine learning classification algorithms for plant disease detection is presented in this research, along with the steps of a generic plant diseases detection system. This model makes use of BiGRU, BiGRU-LSTM, and BiGRU-AM during its training process. The proposed method, BiGRU-AM, achieves good results, with an accuracy of approximately 95.82%.

## REFERENCES


[1] Faithpraise, F., Birch, P., Young, R., Obu, J., Faithpraise, B., and Chatwin, C. (2013). Automatic plant pest detection and recognition using k-means clustering algorithm and correspondence filters. *International Journal of Advanced Biotechnology and Research*, 4(2), 189–199.

[2] Ebrahimi, M. A., Khoshtaghaza, M. H., Minaei, S., and Jamshidi, B. (2017). Vision-based pest detection based on SVM classification method. *Computers and Electronics in Agriculture*, 137, 52–58.

[3] Hasan, R. I., Yusuf, S. M., and Alzubaidi, L. (2020). Review of the state of the art of deep learning for plant diseases: A broad analysis and discussion. *Plants*, 9(10), 1302.

[4] Mohanty, S. P., Hughes, D. P., and Salathé, M. (2016). Using deep learning for image-based plant disease detection. *Frontiers in Plant Science*, 7, 215232.

[5] Kawasaki, Y., Uga, H., Kagiwada, S., and Iyatomi, H. (2015). Basic study of automated diagnosis of viral plant diseases using convolutional neural networks. In *Advances in Visual Computing: 11th International Symposium,* ISVC 2015, Las Vegas, NV, USA, December 14–16, 2015, Proceedings, Part II 11 (pp. 638–645). Springer International Publishing.

[6] Savary, S., Willocquet, L., Pethybridge, S. J., Esker, P., McRoberts, N., and Nelson, A. (2019). The global burden of pathogens and pests on major food crops. *Nature Ecology & Evolution*, 3(3), 430–439.

[7] Lau, H. Y., and Botella, J. R. (2017). Advanced DNA-based point-of-care diagnostic methods for plant diseases detection. *Frontiers in Plant Science*, 8, 302813.

[8] Khater, M., De La Escosura-Muñiz, A., Quesada-González, D., and Merkoçi, A. (2019). Electrochemical detection of plant virus using gold nanoparticle-modified electrodes. *Analytica Chimica Acta*, 1046, 123–131.

[9] Sharmila, R., Kamalitta, R., Singh, D. P., Chauhan, A., and Acharjee, P. B. (2023, May). Weighted Mask Recurrent-Convolutional Neural Network based Plant Disease Detection using Leaf Images. In *2023 7th International Conference on Intelligent Computing and Control Systems (ICICCS)* (pp. 681–687). IEEE.

[10] Abd Algani, Y. M., Vidhya, S., Ghai, B., Acharjee, P. B., Kathiravan, M. N., and Dwivedi, V. K. (2023, May). Innovative Method for Alzheimer Disease Prediction using GP-ELM-RNN. In *2023 2nd International Conference on Applied Artificial Intelligence and Computing (ICAAIC)* (pp. 723–728). IEEE.

[11] Laxmi, M., Koshariya, A. K., Hazela, B., Singh, P., Manju, M., and Jeyaselvi, M. (2023, October). An IoT-Based Pest Detection and Alert System for Farmers Using CNN-LSTM Approach. In *2023 International Conference on Self Sustainable Artificial Intelligence Systems (ICSSAS)* (pp. 1283–1288). IEEE.

[12] Chaturvedi, A., Yagnasree, S., Chhabra, G., Rajan, P. C., Chauhan, A., and Maranan, R. (2023, July). Efficient Method for Tomato Leaf Disease Detection and Classification based on Hybrid Model of CNN and Extreme Learning Machine. In *2023 4th International Conference on Electronics and Sustainable Communication Systems (ICESC)* (pp. 1179–1184). IEEE.

[13] Ambashtha, K. L., Awasthi, D. K., Shukla, A. K., Ragamayi, S., Kathiravan, M. N., and Chauhan, A. (2023, May). Leaf disease detection in crops based on single-hidden layer feed-forward neural network and hierarchal temporary memory. In *2023 2nd International Conference on Applied Artificial Intelligence and Computing (ICAAIC)* (pp. 28–34). IEEE.

[14] Tete, T. N., and Kamlu, S. (2017, June). Plant disease detection using different algorithms. In *RICE* (pp. 103–106).

[15] Singh, J., and Kaur, H. (2019). Plant disease detection based on region-based segmentation and KNN classifier. In *Proceedings of the International Conference on ISMAC in Computational Vision and Bio-Engineering 2018 (ISMAC-CVB)* (pp. 1667–1675). Springer International Publishing.

[16] Kumari, C. U., Prasad, S. J., and Mounika, G. (2019, March). Leaf disease detection: feature extraction with K-means clustering and classification with ANN. In *2019 3rd International Conference on Computing Methodologies and Communication (ICCMC)* (pp. 1095–1098). IEEE.

[17] Teng, F., Guo, X., Song, Y., and Wang, G. (2021). An air target tactical intention recognition model based on bidirectional GRU with attention mechanism. *IEEE Access*, 9, 169122–169134.

*Hybrid and Advanced Technologies – S. Prasad Jones Christydass, Nurhayati Nurhayati and S. Kannadhasan. (Eds)* © *2025 the Authors, ISBN 978-1-032-90254-8*
*Open Access: www.taylorfrancis.com, CC BY-NC-ND 4.0 license*

# Classification of corn leaf disease in agriculture using improved convolutional neural networks in comparison with K-Nearest Neighbors classifier

R. Surya
*Research Scholar, Department of Computer Science and Engineering, Saveetha School of Engineering, Saveetha University, Chennai, Tamil Nadu, India*

M. Gunasekaran
*Project Guide, Department of Computer Science and Engineering, Saveetha School of Engineering, Saveetha University, Chennai, Tamil Nadu, India*

ABSTRACT: Improving the precision of maize leaf disease detection in agriculture is the main objective of this research paper. In order to evaluate the performance of the Improved Convolutional Neural Network (ICNN) algorithm with the K-Nearest Neighbors Classifier (KNN), the study suggests using an innovative log transformation in conjunction with the CNN method. The two primary groups in the study were the KNN group and the ICNN group. The Kaggle database system was the source of the research dataset used in this study. The sample size for each iteration within both groups was set at 228, indicating that a subset of 114 samples was used in each of the 15 iterations for both ICNN and KNN. These samples likely represent data related to corn leaf diseases for the research's purpose. In terms of the statistical analysis approach, the study utilized G-power analysis. In this case, the researchers aimed for a statistical power of 80% and set the alpha rate at 0.05 with CI-95%. The results of the study indicate that the ICNN achieved an accuracy rate of 93.83%, while the K-Nearest Neighbors Classifier achieved a success rate of 80.67%. This disparity suggests a statistically significant difference between the two groups, as determined by a sample t-test with a calculated p-value of 0.017 ($p < 0.05$). The performance analysis reveals that the Improved Convolutional Neural Network (93.83%) model outperforms the K-Nearest Neighbors Classifier (80.67%) in terms of accuracy.

*Keywords*: Corn Leaf Disease Recognition, K-Nearest Neighbors Classifier, Improved Convolutional Neural Network, Agriculture, Crop Yields, Novel Log Transformation

## 1 INTRODUCTION

Artificial intelligence (AI) and other cutting-edge technologies have great potential to improve crop management techniques in modern agriculture, especially in the early diagnosis and treatment of maize leaf diseases. To lessen the negative effects of illnesses on crop production, farmers are depending more and more on cutting-edge instruments like Corn Leaf Disease Recognition systems. These systems use deep learning models, which are particularly made to use visual cues taken from photos or data to study and classify diseases affecting maize leaves (Xu *et al.* 2015). These models can quickly recognize and categorize different leaf disease kinds by utilizing machine learning and computer vision techniques. This helps with prompt intervention and disease management tactics.

For agricultural pest and disease control to be efficient, early detection of illnesses affecting maize leaves is essential. Farmers can prevent extensive crop loss and maximize

DOI: 10.1201/9781003559115-4 

overall production outcomes by promptly detecting illnesses at their inception and implementing targeted remedies (Li *et al.* 2022). The implementation and refinement of illness recognition algorithms in platforms such as Google Colab are made possible by datasets that consist of tagged photos and CSV files (Agarwal *et al.* 2019). Using rigorous testing procedures like T-tests and graphical analysis, these platforms allow academics and practitioners to examine the efficacy of various algorithms, such the K-Nearest Neighbors Classifier and the Improved Convolutional Neural Network.

Furthermore, the volume of research that has been published in IEEE Explorer and Google Scholar over the last ten years highlights the increasing interest in and progress made in AI-driven methods for the detection and prediction of maize leaf disease. For practical agricultural applications, these research offer insightful information on how to maximize model accuracy and dependability (Baliyan *et al.* 2021). In order to improve algorithm performance, researchers stress the value of using pre-trained models and regularly updating datasets, especially in situations where data availability is constrained (Noola and Basavaraju 2021). Even with these developments, it is still difficult to create reliable models for disease recognition that work well for a variety of crop species and environmental circumstances (Albahli and Masood 2022). To do this, large datasets and continuous AI algorithm improvement are needed to guarantee accuracy and scalability in a variety of agricultural environments. In the end, using AI to disease management improves crop yield while simultaneously promoting sustainable farming methods by minimizing the need for chemical treatments and allocating resources as efficiently as possible (Harsani *et al.* 2018). With its ability to transform agricultural practices through sophisticated disease recognition systems, AI has the potential to fundamentally alter both agricultural sustainability and global food security in the future.

## 2 MATERIALS AND METHODS

The Department of Computer Science and Engineering is where the research project was carried out. Its main objectives were to assess the Improved Convolutional Neural Networks with Novel Log Transformation and the K-Nearest Neighbors Classifier. With 114 samples in each group over 15 iterations, the total sample size was set at 228 and split equally between Groups 1 and 2. (Shokat *et al.* 2020). Using Clincalc.com, the sample size was determined according to prior research standards (Panigrahi *et al.* 2020), with 80% power, a 0.05 significance level, and a 95% confidence interval being the goals.

The Jupyter notebook environment, SPSS version 26.0.1, a laptop with an Intel 7th generation i5 CPU, 16 GB of RAM, and an 18 GB graphics card were all part of the experimental setup. The dataset included photos of different maize diseases that were taken from publicly available datasets that were created especially to detect corn leaf diseases (Dhar *et al.* 2022). This raw dataset, which came from the Kaggle platform, was divided into 20% for testing and 80% for training (Patil and Kannan 2023). The objective of the research was to evaluate the performance of a K-Nearest Neighbors classifier versus an improved convolutional neural network classifier that employed a unique log transformation technique (Li *et al.* 2022).

### 2.1 *Improved convolutional neural network*

For processing and picture recognition, CNN with a novel log transformation is used. It consists of fully linked, pooling, and convolutional layers. Convolutional layers use filters to extract features from images, such as textures and edges (Khanna *et al.* 2023). Pooling layers preserve significant information while reducing the dimensionality of the data. The output of processing is categorized using completely connected layers. CNNs are taught to identify patterns and objects using labeled datasets. They are particularly good at tasks like object recognition and image segmentation. Iterative trials are used to optimize the CNN architecture for maize leaf disease recognition, resulting in the best possible performance.

*Pseudocode*

Recognition of corn leaf disease as an input Plan the training and testing of a dataset on corn leaf disease Results: Categorization of Corn Leaf Illness Identification Better Convolutional Neural Network Performance First, import the dependencies. Step 2: Selecting the Instrument Divide the Data Into Train, Validation, and Test Sets in Step 3 Step 4: Transforming Data using Normalization and Resizing Techniques Step 5: Importing Data for Training, Testing, and Validations Plotting a few images in a process of data augmentation is step six. Plotting the Augmented Images in the Data Augmentation Process is Step Seven. Step 8: ResNet18 (ICNN): Building the Deep Learning Model and Transfer Learning Step 9: Selecting an Optimizer and Defining the Loss Function Defining Train and Test Functions in Step Ten Step 11: ICNN algorithm manual testing Step 12: Provide results from the classification

### 2.2 *K-Nearest Neighbors classifier*

For classifying tasks such as diagnosing illnesses of maize leaves, the K-nearest neighbors (K-NN) method works well (Daneshwari and Basavaraju 2022). By examining the closest training data points in a multi-dimensional feature space, it can classify objects. Careful selection of the distance metric and 'k' value is necessary for best results. Because of its superior interpretability and simplicity, K-NN is a great choice for identifying diseases from leaf characteristics. Despite being susceptible to the effects of class imbalances and dataset size, its non-parametric structure enables it to handle a wide range of patterns. K-NN is a useful tool even though it is less effective for huge datasets, especially when paired with other techniques.*Pseudocode*

Identification of Corn Leaf Disease _Input Choices Arrange for the testing and training of a dataset on corn leaf disease. Output: Categorization of the identification of corn leaf disease K-Nearest Neural Network: Its purpose First Step: Adding Requirements Selecting the Device in Step Two Creating Train, Validation, and Test Sets from the Data in Step 3 Step 4: Data Transformation using Normalization and Resizing Techniques Step 5: Equipping Test, Train, and Valid Case Data Step 6: Using a data augmentation procedure to plot a few images Step 7: Plotting the Augmented Images in the Process of Augmenting Data Step 8: Developing the Model for Deep Learning and Transfer Learning Choosing an Optimizer and Defining the Loss Function in Step Nine Step 10: Specifying Functions for Train and Test Step 11: Testing the K-NN algorithm manually

### 2.3 *Statistical analysis*

In our study, machine learning algorithms were implemented in Jupyter/Google Colab, while statistical analysis was done with IBM SPSS. IBM SPSS version 22.0.1 was used in our study to conduct a comparison analysis between the K-Nearest Neighbors Classifier and the Improved Convolutional Neural Network.The main independent variable we looked at for this research was "Time," and the two dependent variables we looked at were "accuracy" and "scale." We computed key statistics, such as the mean, median, standard deviation, and standard mean error, using an independent sample t-test in order to extract insightful information from our dataset.

## 3 RESULTS

The investigation's outcomes were assessed by contrasting the predictions made by the primary algorithm—an Improved Convolutional Neural Network with a special log transformation—with the ground truth labels from a validation or test dataset. These accuracy assessments were then used in a comparison.

Table 1 presents a thorough examination of the performance comparison over 15 iterations of the maize leaf disease recognition task between two classifiers: the ICNN and KNN. The ICNN classifier yields greater accuracy rates than the KNN classifier, continuously outperforming the latter across all rounds. The ICNN consistently outperforms other methods in the identification of maize leaf diseases because it makes good use of log transformation, which improves picture quality and, as a result, increases identification accuracy.

Table 1. Shows the results of 15 iterations of performance measurements comparing the ICNN and K-NN classifiers. When compared to the K-NN classifier, the ICNN classifier has a higher accuracy rate.

| | | ACCURACY RATE | |
|---|---|---|---|
| SI.No. | Test Size | CNN | KNN |
| 1 | Test1 | 85.00 | 82.00 |
| 2 | Test2 | 92.00 | 74.00 |
| 3 | Test3 | 90.00 | 74.00 |
| 4 | Test4 | 95.00 | 75.00 |
| 5 | Test5 | 93.00 | 75.00 |
| 6 | Test6 | 93.00 | 76.00 |

Among the statistics in this set are mean, standard deviation, and mean standard error. The statistical analysis is done with a precision-level criterion. The ICNN model achieves a mean accuracy of 93.83%, whereas the KNN classifier achieves a mean accuracy of 80.67%, thanks to the novel log transformation.

A sample t-test between the two groups yielded a significance level of 0.672, indicating insufficient data to make any inferences. With a significance level of 0.79117 and a 95% confidence interval, the ICNN and KNN algorithms were compared. The superiority of the ICNN model, which is further enhanced by the new log transformation, is justified by a significant difference in variance.

Figure 1 shows a graphic comparison of the accuracy of the KNN classifier and the ICNN model with the new log transformation. The figure clearly indicates that the KNN classification model performs less accurately than the ICNN prediction model, which achieves an

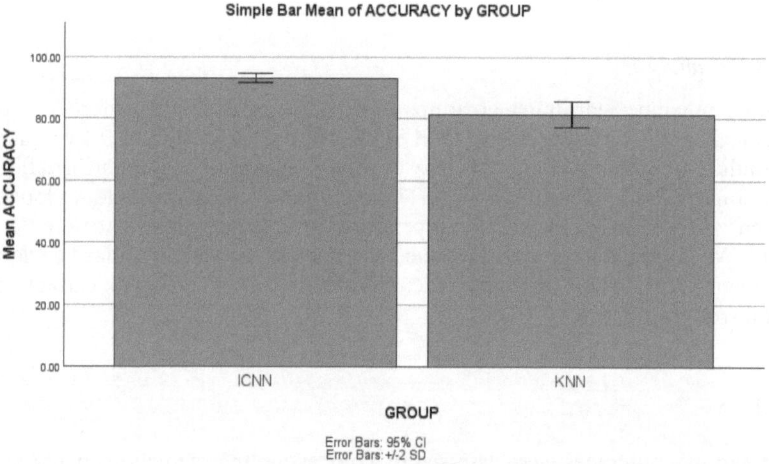


Figure 1. Evaluating the KNN algorithm's accuracy against that of the ICNN classifier. The accuracy rate of the KNN (80.67%) classification model is lower than that of the ICNN (93.83%) prediction model. Y-axis: Mean accuracy, Error Bar ±2 SD; X-axis: ICNN vs. KNN.

incredible accuracy rate of 93.83%. evaluating the ICNN classifier's accuracy against the KNN classifier , there is a noticeable improvement.

## 4 DISCUSSION

This study shows that for the identification of maize leaf diseases, the Improved Convolutional Neural Network (ICNN) with Novel Log Transformation is superior than the K-Nearest Neighbors (K-NN) approach. With an astounding accuracy of 93.83%, ICNN outperformed K-NN by a substantial margin (80.67%). The robustness of ICNN in extracting complex features through data augmentation and feature fusion techniques is highlighted by the statistical significance with a p-value of 0.017 ($p<0.05$) (Amin *et al.* 2022). ICN's sophisticated image analysis capabilities address the serious problem of maize leaf diseases, including blight, common rust, and gray leaf spot, and provide workable solutions for raising agricultural output and ensuring global food security (Dai *et al.* 2023). Further study on model modification and dataset expansion is expected to improve the accuracy of disease recognition in maize leaf disease identification (Li *et al.* 2022).

## 5 CONCLUSION

Important discoveries are made after a thorough examination of corn leaf disease detection, with an emphasis on increasing accuracy rates. A unique log transformation improves the performance of the Improved Convolutional Neural Network, allowing it to reach an astounding accuracy rate of 93.83%. The K-Nearest Neighbors Classifier, on the other hand, only achieves an accuracy rating of 80.67%. This significant difference of 13.16% in favor of ICNN highlights the technology's unquestionable advantage over KNN when it comes to identifying maize leaf diseases. The efficacy of ICNN and its potential to offer more precise and dependable disease identification—a critical function for agricultural management and crop protection—are highlighted by these results.

## REFERENCES


Agarwal, M., Bohat, V. K., Ansari, M. D., Sinha, A., Gupta, S. K., and Garg, D. (2019, December). A convolution neural network based approach to detect the disease in corn crop. In *2019 IEEE 9th International Conference on Advanced Computing (IACC)* (pp. 176–181). IEEE.

Albahli, S., and Masood, M. (2022). Efficient attention-based CNN network (EANet) for multi-class maize crop disease classification. *Frontiers in Plant Science*, 13, 1003152.

Aluvalu, R., VN, S. K., Thirumalaisamy, M., Basheer, S., and Selvarajan, S. (2023). Efficient data transmission on wireless communication through a privacy-enhanced blockchain process. *PeerJ Computer Science*, 9, e1308.

Amin, H., Darwish, A., Hassanien, A. E., and Soliman, M. (2022). End-to-end deep learning model for corn leaf disease classification. *IEEE Access*, 10, 31103–31115.

Baliyan, A., Kukreja, V., Salonki, V., and Kaswan, K. S. (2021, September). Detection of corn gray leaf spot severity levels using deep learning approach. *In 2021 9th International Conference on Reliability, Infocom Technologies and Optimization (Trends and Future Directions)(ICRITO)* (pp. 1–5). IEEE.

Dai, D., Xia, P., Zhu, Z., and Che, H. (2023). MTDL-EPDCLD: A multi-task deep-learning-based system for enhanced precision detection and diagnosis of corn leaf diseases. *Plants*, 12(13), 2433.

Dhar, P., Rahman, M. S., and Abedin, Z. (2022). Classification of leaf disease using global and local features. *IJITCS*, 14, 43–57.

Harsani, P., Qurania, A., and Damayanti, K. (2018). Maize Plant Desease Identification (Zea Mays L. Saccharata) Using Image Processing and K-Nearest Neighbor (K-Nn). *International Journal of Engineering and Technology*, 7(3.20), 402–405.

Khanna, M., Singh, L. K., Thawkar, S., and Goyal, M. (2024). PlaNet: a robust deep convolutional neural network model for plant leaves disease recognition. *Multimedia Tools and Applications*, 83(2), 4465–4517.

Li, Y., Chen, Y., and Wang, Y. (2022). Disease recognition of maize Leaf based on KNN and feature extraction. *International Journal of Pattern Recognition and Artificial Intelligence*, 36(11), 2257004.

Macasaet, D., Sybingco, E., Bandala, A., Illahi, A. A., and Dadios, E. (2020, November). Identification of Corn Plant Leaf Diseases through Web Server using Image Processing and Artificial Neural Network. In *2020 IEEE region 10 conference (TENCON)* (pp. 973–978). IEEE.

Mahaveerakannan, R., Anitha, C., Thomas, A. K., Rajan, S., Muthukumar, T., and Rajulu, G. G. (2023). An IoT based forest fire detection system using integration of cat swarm with LSTM model. *Computer Communications*, 211, 37–45.

Noola, D. A., and Basavaraju, D. R. (2021). Corn leaf disease detection with pertinent feature selection model using machine learning technique with efficient spot tagging model. *Revue d'Intelligence Artificielle*, 35(6).

Noola, D. A., and Basavaraju, D. R. (2022). Corn leaf image classification based on machine learning techniques for accurate leaf disease detection. *Int. J. Electr. Comput. Eng*, 12(3), 2509–2516.

Panigrahi, K. P., Sahoo, A. K., and Das, H. (2020, June). A cnn approach for corn leaves disease detection to support digital agricultural system. In *2020 4th International Conference on Trends in Electronics and Informatics (ICOEI)(48184)* (pp. 678–683). Ieee.

Patil, N. S., and Kannan, E. (2024). RETRACTED ARTICLE: Spatial federated learning model for identification of corn leaf diseases. *Spatial Information Research*, 32(2), 209–209.

Radha, N., and Swathika, R. (2021, March). A polyhouse: plant monitoring and diseases detection using CNN. In *2021 International Conference on Artificial Intelligence and Smart Systems (ICAIS)* (pp. 966–971). IEEE.

Ramana, K., Revathi, A., Gayathri, A., Jhaveri, R. H., Narayana, C. L., and Kumar, B. N. (2022). WOGRU-IDS—An intelligent intrusion detection system for IoT assisted Wireless Sensor Networks. *Computer Communications*, 196, 195–206.

*Hybrid and Advanced Technologies – S. Prasad Jones Christydass, Nurhayati Nurhayati and S. Kannadhasan. (Eds) © 2025 the Authors, ISBN 978-1-032-90254-8*
*Open Access: www.taylorfrancis.com, CC BY-NC-ND 4.0 license*

# Comparison of biogas yield from the novel anaerobic codigestion sugarcane and corn stalk with cow dung in biodigester

R. Thiran Raj
*Research Scholar, Department of Biotechnology, Saveetha School of Engineering, Saveetha Institute of Medical and Technical Sciences, Saveetha University, Chennai, Tamil Nadu, India*

Baranitharan Ethiraj
*Project Guide, Department of Biotechnology, Saveetha School of Engineering, Saveetha Institute of Medical and Technical Sciences, Saveetha University, Chennai, Tamil Nadu, India*

Rajaram Vishnupriya
*Research Co-Guide, Department of Biotechnology, Saveetha School of Engineering, Saveetha Institute of Medical and Technical Sciences, Saveetha University, Chennai, Tamil Nadu, India*

ABSTRACT: The primary purpose of this research is to compare the biogas generated by the anaerobic co-digestion of sugarcane, maize stalk, and cow dung in a biodigester. The biodigester used sugarcane and corn stalk (N=14) and cow dung (N=14) for 14 days, producing a total sample size of 28 ($\alpha$=0.05, confidence interval of 95%, and G power of 80%). A smart biogas meter was used to measure biogas in both groups. Sugarcane and corn stalks produced more biogas (28.57 $m^3$) than cow manure (20.98 $m^3$). The independent sample T-test demonstrated that biogas generation from the sugarcane and corn stalk utilized in the biodigester was greater than that from cow manure. An independent t-test shows no statistically significant difference between the two groups, with a p-value of 0.067. This study reveals that the anaerobic treatment of sugarcane and corn stalks can yield higher biogas compared to cow dung. Fossil Fuels, Sugarcane, Cornstalk, Environment, Food Waste, Methane, Energy

*Keywords*: Fossil Fuels, Sugarcane, Cornstalk, Environment, Food Waste, Methane, Energy

## 1 INTRODUCTION

The expanding population is driving up energy use in our daily lives. Almost 90% of people on the planet rely on fossil fuels including coal, oil, and natural gas as their main source of energy (Akram and Bhargava 2023). On the other hand, fossil fuels are non-renewable and will ultimately run out (Buck 2021). Finding fossil fuel substitutes is necessary to address the rising demand for energy due to growing concerns about environmental issues like global warming. Anaerobic digestion may be used to create biogas, a viable substitute energy source (Treichel and Fongaro 2019). Anaerobic digestion is a biochemical process that uses certain microorganisms in an oxygen-free environment to break down organic waste, including cow dung, food waste, and agricultural waste (Balagurusamy and Chandel 2021). According to Wellinger, Murphy, and Baxter (2013), biogas has several applications such as heating, cooking, lighting, and power generation. It may also be utilized as a fuel in automobiles. Studies on the generation of biogas from cornstalk and sugarcane waste have been carried out. Both the waste from sugarcane and cornstalks are classified as lignocellulosic materials, meaning that their cellulose and hemicellulose content is high (Janke 2017)(Chen *et al.* 2023). Pre-treatment is frequently required to increase the digestibility of these materials since the bacteria that produce biogas find them difficult to digest (Chandel and Da Silva 2013). In 2013 and 2023, Chandel and Da Silva and Raketh *et al.* conducted research that indicated a continuous stirred tank reactor (CSTR) with an extended hydraulic retention time (HRT) (>35 days) should be utilized in combination with a pre-treatment technique to enhance the breakdown of fibrous carbohydrates for producing biogas from bagasse and straw.

(Malunga 2018; Mumtaz *et al.* 2023) looked into another study that examined the use of pentoses from sugarcane and cornstalk biomass for biogas generation, in addition to the production of biogas from sugarcane and cornstalk waste. They found that sugarcane straw has a specific methane production rate of 0.22 $m^3$/kgVS (Malunga 2018), bagasse has a rate of 0.28 $m^3$/kgVS, and filter cake has a rate of 0.25 $m^3$/kgVS. A study by (Mumtaz *et al.* 2023) summarized the characteristics of wastes used in biogas production. They found that straw and bagasse have significant lignin levels, so pre-treatment is advised for improved digestibility. At the moment, biogas is produced from garbage with low yields. As a result, a waste management strategy that can generate more biogas is required. To find the most useful waste for biogas generation, this study assessed the biogas production potential of sugarcane (Khan and Khan 2019), cornstalk waste, and cow manure.

## 2 MATERIALS AND METHODOLOGIES

The study was conducted in the Environmental Lab of the Saveetha School of Engineering in Chennai, India, which is part of the Department of Biotechnology. The samples were separated into two groups: sugarcane and cornstalk trash (Group A) and cow dung (Group B). The sample size was calculated using previous data, using a 0.05 threshold, 80% G power, a 95% confidence interval, and an enrollment ratio of one. Sugarcane and cornstalk debris were collected from several locations in Chennai, including the Avadi and Koyambedu markets. The rubbish was gathered in a clean, covered container and brought to a grinding machine to be broken into smaller pieces. The ground waste was then transferred to the biogas facility (Nouhou Moussa *et al.* 2023). The study was conducted in the Environmental Lab of the Saveetha School of Engineering in Chennai, India, which is part of the Department of Biotechnology. The samples were separated into two groups: sugarcane and cornstalk trash (Group A) and cow dung (Group B). The sample size was calculated using previous data, using a 0.05 threshold, 80% G power, a 95% confidence interval, and an enrollment ratio of one. Sugarcane and cornstalk debris were collected from several locations in Chennai, including the Avadi and Koyambedu markets. The rubbish was gathered in a clean, covered container and brought to a grinding machine to be broken into smaller pieces. The ground waste was then transferred to the biogas facility (Nouhou Moussa *et al.* 2023).

A cow farm and other areas in and around Chennai provided fresh cow manure. Using a mortar and pestle, the cow dung was first ground into a fine powder in a clean, covered container. Following its pulverization, the cow dung was sieved and allowed to dry in the sun for four days (Adewoyin and Arimoro 2023).

The purpose of the biogas digester tank and biogas smart meter is to produce biogas from the waste products of cornstalks, sugarcane, and cow manure. The generated biogas was gathered in the gas holder, where the pressure and flow of the biogas were monitored using a smart biogas meter as it was discharged as methane.

Table 1 shows the biogas volume and pressure (measured in $m^3$) for each group during a 14-day period. A smart biogas meter was utilized to track the flow and pressure of the biogas as it was being released. One kilogram of sugarcane waste and one kilogram of cow dung were fed into the biodigester through the intake, respectively. The amount of biogas produced was measured with a smart biogas meter. The gas holding was connected to the gas stove, the smart biogas meter, and the output pipe of the biodigesters (pH 6.5-7 and 28°–34°C).

## 2.1 *Statistical analysis*

Using SPSS version 21, a statistical comparison of biogas generation from sugarcane and corn stalk waste was performed. There are no dependent variables, and the independent variable is biogas output. Jasrai (2020) and Stehlik-Barry and Babinec (2017) conducted mean, standard deviation, and independent t-test analyses.

# 3 RESULTS

Over 14 days, the biodigester that utilized cow dung produced a greater quantity of biogas (20.98 $m^3$) than the biodigester that utilized waste from sugarcane and corn stalks (8.78 $m^3$). This is seen in Figure 1. The cow dung bio-digester achieved its highest output on the fourteenth day, and the sugarcane and cornstalk waste biodigester also reached its peak production on the fourteenth day. Both biodigesters began generating biogas almost immediately after two days had passed. As can be seen in Table 1, the biodigester that was used to process cow dung generated 20.98 $m^3$ of biogas on the fourteenth day, whereas the biodigester that was used to process sugarcane and cornstalk waste produced 12.32 $m^3$ of biogas on the same day. This confirms that cow dung can produce more biogas than sugarcane and cornstalk waste. A burning test also showed that the gas from cow dung had a longer burning time (60 minutes) than the gas from sugarcane and cornstalk waste (55 minutes).

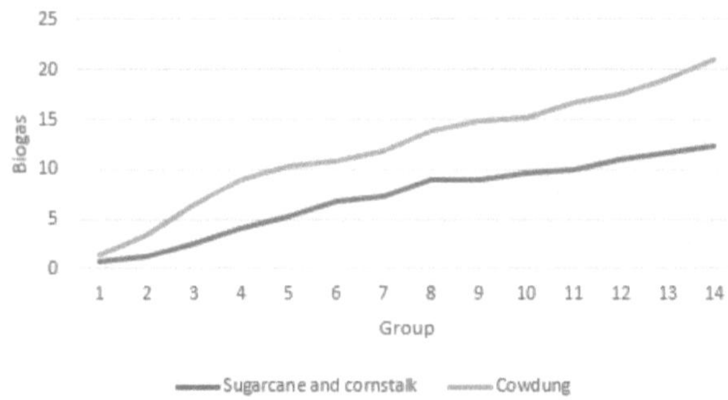


Figure 1. Biogas generation of digester operated with sugarcane and corn stalk waste and cow dung, where blue represents sugarcane and corn stalk and red represents cow dung.

Table 1. Comparison of the accuracy data value of the biogas generated from both the sugarcane and corn stalk and cow dung using a biodigester. A total of 28 samples with each of the 14 samples.

| | Biogas ($m^3$) | |
|---|---|---|
| No. of Days | Sugarcane and corn stalk waste | Cow Dung |
| 1 | 0.64 | 1.31 |
| 2 | 1.23 | 3.43 |
| 3 | 2.54 | 6.45 |
| 4 | 4.03 | 8.87 |
| 5 | 5.23 | 10.24 |
| 6 | 6.75 | 10.74 |
| 7 | 7.32 | 11.89 |
| 8 | 8.95 | 13.83 |
| 9 | 9.01 | 14.76 |
| 10 | 9.65 | 15.23 |
| 11 | 10.03 | 16.67 |
| 12 | 10.92 | 17.54 |
| 13 | 11.64 | 18.97 |
| 14 | 12.32 | 20.98 |

Table 2. shows that in comparison to cow dung (standard deviation: 5.78814, standard error: 1.54695), sugarcane and cornstalk waste had larger standard deviations (7.65292) and standard errors (2.04533). This implies that the content and characteristics of waste from sugarcane and corn stalks are more erratic than those of cow feces. But, if mixed together with suitable co-substrates, the higher standard deviation of sugarcane and cornstalk waste also suggests that it has a greater potential to increase its biogas output.

Table 2. Comparison of the mean power density of cow dung and sugarcane and cornstalk operated by a biogas plant. It reveals that the mean power density of sugarcane and corn stalk waste operated in a biogas plant is higher compared to cow dung ($p = 0.067$, independent sample T-test).

| | Group | N | Mean | St. Deviation | Std. Error Mean |
|---|---|---|---|---|---|
| Biogas | Sugarcane and corn stalk waste | 14 | 7.1614 | 3.86933 | 1.03412 |
| | Cow dung | 14 | 12.2079 | 5.78814 | 1.54695 |

This study compared the ability of cow dung to produce biogas to sugarcane and cornstalk waste. They used a test to evaluate the biogas output of each material and found that cow dung produced more biogas than sugarcane and cornstalk waste. The difference was statistically significant. The researchers observed that cow dung had an average biogas output of $0.623 \pm 0.04$ L/g volatile solid (VS), while sugarcane and cornstalk waste had an average biogas output of $0.543 \pm 0.03$ L/g VS. They concluded that cow dung was a more efficient biogas feedstock. The researchers suggest that cow dung could be a cost-effective and sustainable alternative for biogas production, as it is abundant in rural areas and readily available. Additionally, using cow dung for biogas production can help reduce the environmental impact of animal waste.

## 4 DISCUSSION

A study found that sugarcane and cornstalk waste produced significantly more biogas (12.32 $m^3$) than cow dung (20.98 $m^3$) in an anaerobic digester. This is because sugarcane and cornstalk waste contain more easily digestible substances than cow dung. Both sugarcane and cornstalk waste and cow dung began to produce biogas on the first day in a burning test. This indicates that the microorganisms responsible for the biogas production process were fully active within one day of starting the experiment. Researchers have used various strategies to improve biogas production from sugarcane and cornstalk waste. In an insightful exploration, one study delved into the intricate relationship between digester performance and the diverse array of microbial populations inhabiting the system. This study highlighted the pivotal role played by microbial communities in dictating the efficiency and stability of digesters. Concurrently, a separate investigation shed light on the remarkable capabilities of a lab-scale up-flow anaerobic sludge blanket (UASB) reactor. Notably, this study demonstrated that a reactor exhibiting a higher outgoing longwave radiation of 2 kgCOD/$m^3$·d stood resilient and produced a commendable output of 20 L/d of biogas with an impressive 75% methane composition. Furthermore, a third study explored the auspicious effects of utilizing sugarcane and cornstalk in anaerobic digesters. The findings showcased a positive impact on digester performance, underlining the advantageous synergy between organic substrates and microbial activity in enhancing biogas production. Through these diverse studies, a comprehensive understanding emerges of the multifaceted dynamics governing digester performance and the intricate interplay between substrate choice, microbial consortia, and operational parameters. The authors of the study observed that cow dung had higher volatile solids (94.9%) and lower fixed solids (9.3%). This is similar to the findings of other studies, which have reported that waste operated in an anaerobic bioreactor has a low pH of 5.38 to 7.2.

The authors conclude that optimizing the ratio of appropriate substrates alongside supplements and exploring novel techniques, could incentivize increased government investment in the biogas sector, positioning it as a potential competitor to Compressed Natural Gas in the future. Essentially, sugarcane and cornstalk waste prove to be superior substrates for biogas generation due to their higher biogas yield and greater digestibility. Ongoing research efforts are aimed at innovating new methodologies and approaches to enhance the efficiency of biogas production specifically from sugarcane and cornstalk waste materials. In essence, the focus is on maximizing biogas yield and sustainability while also paving the way for potential advancements in the biogas industry. The authors also suggest that choosing suitable waste is considered one of the important factors for achieving high biogas production through the anaerobic digestion process. However, they also note that it is difficult to optimize the anaerobic digestion process due to a lack of balanced nutrients and an appropriate microbial population.

Table 3. Independent sample T-test between Biogas operated with sugarcane and corn stalk waste and cow dung has a p-value of 0.067.

| | | | | INDEPENDENT SAMPLE TEST | | | | | | |
|---|---|---|---|---|---|---|---|---|---|---|
| | | Levene's Test of equality of variance | | T- Test of equality of means | | | | | | |
| | | | | | | | | | 95% Confidence interval of the Difference | |
| | | F | Sig | t | df | Sig (2-tailed) 1 sided | Mean difference | Std. Error difference | Lower | Higher |
| Biogas | Equal variances assumed | 2.024 | .067 | 2.712 | 26 | .012 | 5.04643 | 1.86077 | 8.87129 | 1.22157 |
| | Equal variances not assumed | | | 2.712 | 22.685 | .013 | 5.04643 | 1.86077 | 8.87129 | 1.22157 |

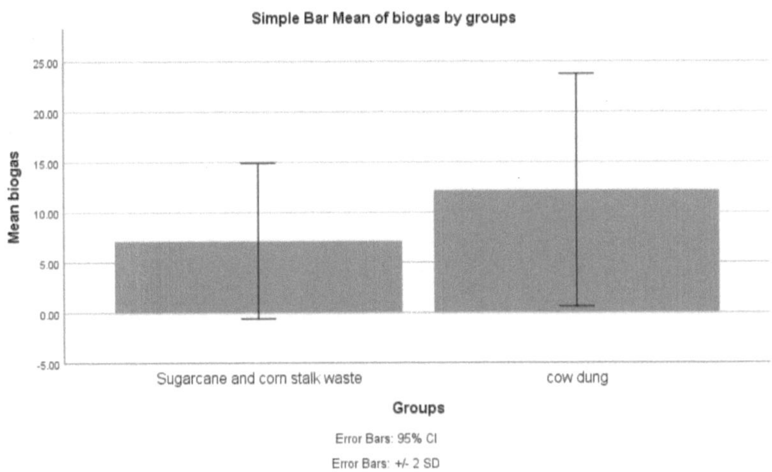


Figure 2. Comparison of sugarcane and corn stalk waste and cow dung in terms of mean biogas generation. The mean biogas generation of sugarcane and corn stalk waste is significantly higher than that of cow dung, whereas the standard deviation of cow dung is lower than sugarcane and corn stalk waste X-Axis: Sugarcane and corn stalk waste vs Cow dung Y-Axis: Mean biogas generation ± 2 SD.

## 5 CONCLUSION

Using the statistical analysis software SPSS, important distinctions between the two separate groups were identified and carefully investigated in the study analysis. The data demonstrates the significant yields that were achieved on the fourteenth day, with the highest yield of 20.98 $m^3$ coming from cow dung and 12.32 $m^3$ from the anaerobic treatment of sugarcane and maize stalks. The aforementioned data highlights the prospects of biogas generation from sugarcane and maize stalks, showcasing itself as a workable strategy that advocates for effective resource management and provides a sustainable energy substitute. Additionally, the incorporation of biogas systems offers hotels and restaurants a profitable prospect, allowing them to efficiently handle waste disposal in addition to producing extra revenue by utilizing sources of renewable energy.

## REFERENCES


Adewoyin, J. A., and Arimoro, F. O. (2023). Animal manure as a biostimulant in bioremediation of oil-contaminated soil: the role of earthworms. *Environmental Monitoring and Assessment*, *195*(2), 293.

Akram, V., and Bhargava, S. (2023). Club convergence analysis of fossil fuels material footprint at the global level. *Environmental Science and Pollution Research*, *30*(53), 114283–114293.

Buck, H. J. (2021). *Ending Fossil Fuels: Why Net Zero is Not Enough*. Verso Books.

Chen, L., Chen, X., Zhao, Y., Xie, X., Yang, S., Hua, D., and Li, T. (2023). Effect of Torrefaction on the Physiochemical Characteristics and Pyrolysis of the Corn Stalk. *Polymers*, *15*(20), 4069.

Jasrai, L. (2020). *Data Analysis Using SPSS*. Sage..

Malunga, S. P. (2018). *Production of Biogas From Sugarcane Residues* (Doctoral dissertation).

Nouhou Moussa, A. W., Sawadogo, B., Konate, Y., Sidibe, S. D. S., and Heran, M. (2023). Critical state of the art of sugarcane industry wastewater treatment technologies and perspectives for sustainability. *Membranes*, *13*(8), 709.

Raketh, M., Kana, R., Kongjan, P., Muhammad, S. A. F. A. S., Sompong, O., Mamimin, C., and Jariyaboon, R. (2023). Enhancing bio-hydrogen and bio-methane production of concentrated latex wastewater (CLW) by Co-digesting with palm oil mill effluent (POME): Batch and continuous performance test and ADM-1 modeling. *Journal of Environmental Management*, *346*, 119031.

Sánchez-Muñoz, S., Barbosa, F. G., Jiménez-Ascencio, J., Mier-Alba, E., Singh, A. K., dos Santos, J. C., and Chandel, A. K. (2020). Technological routes for biogas production: Current status and future perspectives. *Biogas Production: From Anaerobic Digestion to a Sustainable Bioenergy Industry*, 3–17.

Stehlik-Barry, K., and Babinec, A. J. (2017). *Data Analysis with IBM SPSS Statistics*. Packt Publishing Ltd.

Wellinger, A., Murphy, J. D., and Baxter, D. (Eds.). (2013). *The Biogas Handbook: Science, Production and Applications*. Elsevier.

*Hybrid and Advanced Technologies – S. Prasad Jones Christydass, Nurhayati Nurhayati and S. Kannadhasan. (Eds)* *ISBN 978-1-032-90254-8*


# Quality checking of sugarcane to identify the disease using CNN algorithm compare with LSTM to improve accuracy

V. Nagamalleswararao
*Research Scholar, Department of CSE, Saveetha School of Engineering, Saveetha Institute of Medical and Technical Sciences, Saveetha University, Chennai, Tamilnadu, India*

G. Mary Valantina
*Project Guide, Department of CSE, Saveetha School of Engineering, Saveetha Institute of Medical and Technical Sciences, Saveetha University, Chennai, Tamilnadu, India*

ABSTRACT: Through a comparison of CNN and LSTM algorithms, the study seeks to improve the detection and classification of major sugarcane illnesses. CNN and LSTM networks were the two evaluation methods used for the two groups. 1,841 samples in total were used, of which 1,300 were used for testing and 541 for training. While the LSTM algorithm obtained 84.3830% accuracy, the CNN algorithm achieved 91.9070%. A statistically significant difference in accuracy between the two algorithms was found using an independent samples t-test, with a p-value of 0.002 ($p<0.05$). This suggests that because of its increased accuracy, the CNN algorithm is more useful for identifying and categorizing diseases related to sugarcane. The paper provides a solid answer for sugarcane disease diagnosis by decisively showing that CNNs outperform LSTMs in this application, offering a reliable method for classifying and detecting sugarcane diseases.

*Keywords*: CNN, Deep Learning, LSTM, Sugarcane Disease, Image processing

## 1 INTRODUCTION

Numerous issues in agriculture, including disease identification, yield prediction, and soil profile assessment, can be resolved with deep learning techniques (Swapnil Dadabhau Daphal and Sanjay M. Koli 2023). It is critical to use adaptive deep learning models, such as Convolutional Neural Networks (CNN), for sugarcane disease detection. Using remote sensing data from aerial imagery, this systematic analysis assesses the efficacy of deep learning approaches for crop classification (Bobby Dioquino Gerardo 2019). Recently, there has been a surge in the utilization of satellite or unmanned aerial vehicle (UAV) photography for activities such as crop mapping and yield prediction (Igor Teixeira and Raul Morais 2023). Furthermore, illnesses such as Neocapnodium theae Hara, or tea coal disease, can have a detrimental effect on tea yield and quality (Yang Xu and Yilin Mao 2023).

Over 180 papers have been submitted to IEEE Xplore and Google Scholar detailing the research that has been done in the last few years to identify fake enrolling services. According to one study, fruits may be categorized using deep learning techniques based on their color, shape, and spots (M. T. Vasumathi and M. Kamarasan 2021). Despite mechanization, the expanding global population presents challenges to agriculture; labor-intensive operations like field inspection and breeding still need to be done (Ali Jannesari *et al.* 2022). Significant output losses result from plant illnesses brought on by environmental conditions (Muhammad Shoaib and Babar Shah 2023). (Aluvalu *et al.* 2023; Mahaveerakannan R *et al.* 2023; Ramana *et al.* 2023; Vellela and Balamanigandan 2023).

DOI: 10.1201/9781003559115-6

In real-time plant disease diagnosis (Sumita Mishra *et al.* 2020) and other agricultural applications (A. Kamilaris and F. X. Prenafeta-Boldú 2018), deep learning, and CNNs in particular, have demonstrated potential.

However, there are still unanswered questions about the analysis of sugarcane diseases using deep learning algorithms like CNN and LSTM, frequently as a result of unbalanced datasets. The goal is to identify fraudulent websites and programs that try to trick and con unwary people by using CNN and LSTM, especially those that are meant for automatic smile recognition.

## 2 MATERIALS AND METHODS

Two sets of categorization algorithms were used in this study, which was carried out at the Open Source Lab: Group 1 consisted of the Convolutional Neural Network Algorithm and Group 2 comprised the LSTM Algorithm. For statistical analysis, a dataset of 1,841 samples—1,300 for training and 645 for testing—was utilized. The pre-processed dataset had a threshold value of 0.002, alpha and beta values set at 0.05 and 0.2, respectively, and was split into 80% for training and 20% for testing. Since there were no missing values in the dataset, it was appropriate for testing deep learning and image processing methods.

Python programming was used to implement the project in a Jupyter Notebook running Windows 10 OS. The hardware arrangement included a 64-bit system configuration, an Intel Core i7 processor (8GB), and 8GB of RAM. The dataset was used to precisely execute the Python code, which was run to guarantee accurate results.

### 2.1 *Algorithm: CNN (Convolutional Neural Network) algorithm*

The CNN algorithm, which stands for Convolutional Neural Network, will be used in this project. One kind of deep learning neural network that is well-known for its efficiency in image processing applications is CNN. In essence, CNNs can take input images, identify various features or objects inside the image, and assign significance (using learnable weights and biases) to each.

CNN functions by using image processing techniques like grayscale conversion to extract features from images. For this project, grayscale techniques are used to convert every frame of a video. These frames are then matched to a training dataset in order to identify any suspicious activity in the video clip.

Pseudocode of CNN Algorithm

Step 1: Compile the information you want to utilize to train the model of the convolutional neural network algorithm. Step 2: Using the Convolutional Neural Network Algorithm, define the problem you want to solve. This situation presents a binary classification challenge. Step 3: Divide the data into training and test sets to assess the model's performance. Step 4: Evaluate the Convolutional Neural Network Algorithm model's performance using the test set after training it. Step 5: Modify the hyperparameters to get better outcomes if the model's performance isn't up to grade. Step 6: Use the model to generate predictions on fresh, untainted data after you're satisfied with its performance.

### 2.2 *Long short-term memory*

A kind of sequential neural network used in deep learning called LSTM Networks was created to address the problem of long-term memory retention. They solve the vanishing gradient issue that regular RNNs and other machine learning algorithms frequently run into. They are a specific type of Recurrent Neural Network (RNN). Conversely, Hochreiter and Schmidhuber created Convolutional Neural Networks (CNN) to get around the drawbacks of conventional RNNs and machine learning techniques. CNNs work especially well on jobs that require pattern recognition and image processing. CNNs can be implemented in Python with the help of the Keras package, which offers a productive and easy-to-use environment for creating deep learning models.

*Algorithm:*

(1) Identify the target variable and pick the key factors that affect categorization.
(2) Normalize the data to provide consistent scaling across all characteristics.
(3) Set the parameters to either zero or tiny random values.
(4) Describe the cost and sigmoid functions, which take real numbers and turn them into values between 0 and 1.
(5) Assess model performance on the test set using metrics like accuracy, precision, recall, and F1-score.
(6) To improve model performance, fine-tune hyperparameters such as learning rate, regularization intensity, and number of iterations.
(7) Use the trained LSTM Algorithm model to forecast and classify new data, discriminating between legitimate and fraudulent job posts.

### 2.3 *Statistical analysis*

Using IBM SPSS software, specifically version 26, a maximum of 10 iterations were used for both algorithms in the statistical analysis of the Convolutional Neural Network (CNN) Algorithm classifier with the LSTM Algorithm (Yue *et al.* 2009). IP address, geolocation, device information, historical data, user behavior, registration details, and social media activity were among the independent factors (Sinha and Suseela 2022). A 64-bit Windows 10 PC with 8 gigabytes of RAM, a dependable Internet connection, and a Jupyter Notebook loaded with Python were used for the analysis. The group ID was used to ensure group correctness. The T-test of independent samples was the statistical technique used to analyze the samples. For the classifier models of the CNN and LSTM algorithms, each dataset had five samples.

## 3 RESULT AND DISCUSSION

Using Anaconda Navigator, the proposed CNN and LSTM Algorithm were put into practice and tested on a number of situations, using ten samples in each.

With an accuracy of 91.9070%, Table 1 shows the CNN Algorithm's increased accuracy data.

Table 1. The prediction accuracy comparison between the LSTM and Convolutional Neural Network With an accuracy of 91.9070%, the Convolutional Neural Network outperformed the Logistic Regression and LSTM models, which both had accuracy of 84.3830%.

| Itreations | Convolutional Neural Network Accuracy (%) | LSTM Accuracy (%) |
|---|---|---|
| 1 | 93.21 | 82.73 |
| 2 | 92.61 | 81.2 |
| 3 | 89.66 | 82.72 |
| 4 | 90.13 | 83.63 |
| 5 | 91.06 | 84.9 |
| 6 | 92.77 | 82.61 |
| 7 | 95.13 | 88.76 |
| 8 | 92.5 | 84.86 |
| 9 | 89.45 | 89.26 |
| 10 | 92.55 | 83.16 |

The expected accuracy of the LSTM Algorithm is shown in Table 2. Ten statistical samples will be used for the statistical evaluation in order to calculate the accuracy values, and each set of rules corresponding with the accuracy values will be used for assessment.

Table 2. When comparing the Convolutional Neural Network Algorithm to the LSTM Algorithm, group statistical analysis shows that the former has a greater mean accuracy than the latter. Based on ten samples, the standard deviation and standard error mean for each model are also computed.

| | Group | N | Mean | Std. Deviation | Std. Error Mean |
|---|---|---|---|---|---|
| **Accuracy** | Convolutional Neural Network Algorithm | 10 | 91.9070 | 1.84084 | 0.58212 |
| | LSTM Algorithm | 10 | 84.3830 | 1.98508 | 1.98508 |

Based on mean accuracy, Figure 1. contrasts the LSTM Algorithm and CNN. The CNN Algorithm's overall mean accuracy, standard deviation, and standard error are 91.9070%, 1.84084, and 0.58212, respectively. The means, standard deviations, standard errors, and mean accuracy for LSTM are 84.3830%, 1.98508, and 1.98508, respectively. A sample size of 10 was used for the analysis of the sugarcane disease study, and a significant p-value of 0.005 indicated statistical significance in the identification of phony applications. The graph, which shows the means of several CNN classifier algorithms and LSTM, shows that CNN has a significantly higher accuracy rate than LSTM—91.9070% versus 84.3830%.

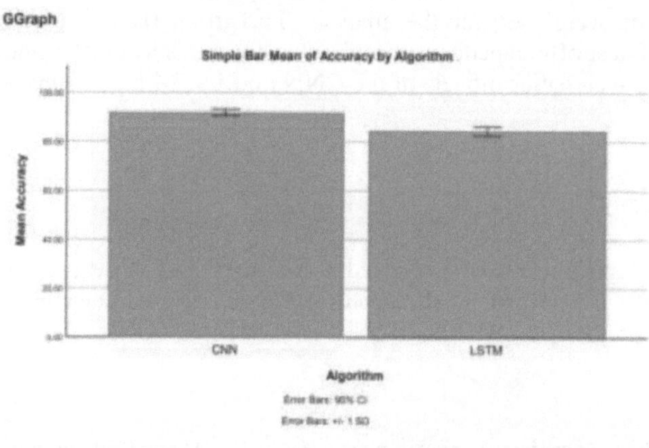


Figure 1. CNN and LSTM comparison. Classifier with respect to average accuracy. CNN has a higher mean accuracy than LSTM. Classifier: CNN's standard deviation is marginally superior to LSTM's. CNNVs LSTM Classifier on the X-axis, and mean accuracy within +/- 2 SD on the Y-axis.

## 4 DISCUSSION

With a significant p-value of 0.002, the CNN's efficacy in detecting sugarcane illnesses is demonstrated by its superior performance compared to the LSTM algorithm. CNN's accuracy of 91.9070% is similar to that of the DTA Classifier. Its benefits include simplicity, low parameter count, and flexibility, making it perfect for datasets with imbalances. On the other hand, LSTM is a widely used technique for jobs involving binary classification. But training CNN takes time, especially when dealing with big datasets. Subsequent efforts are directed toward improving the system with more parameters and a shorter training period. This work highlights CNN's efficiency in deep learning and image processing, highlighting its potential for automated grin detection and tackling imbalanced dataset issues.

## 5 CONCLUSION

This study's main goal is to apply deep learning and image processing techniques to the CNN classifier in order to discover sugarcane illnesses and classify their properties. It also aims to evaluate the LSTM's performance in comparison. LSTM obtained 84.3830% accuracy, while CNN classifier reached 91.9070%. This evaluation alone shows that, for the dataset used in this study, the CNN classifier performs better than the LSTM, exhibiting higher accuracy.

## REFERENCES


[No title]. (n.d.). Retrieved June 12, 2024, from https://www.researchgate.net/profile/Punithavathi-Rasappan/publication/362388038_Computer_Vision_and_Deep_Learning-enabled_Weed_Detection_Model_for_Precision_Agriculture/links/62e7911d4246456b55ffc95a/Computer-Vision-and-Deep-Learning-enabled-Weed-Detection-Model-for-Precision-Agriculture.pdf

Ahmad M., Abdullah M., Moon H., Han D. (2021). Plant disease detection in imbalanced datasets using efficient convolutional neural networks with stepwise transfer learning *IEEE Access*, 9, pp. 140565–140580, 10.1109/ACCESS.2021.3119655 View article View in ScopusGoogle Scholar.

Alansary *et al.*, (2015), Automatic brain localization in sugarcane disease MRI using superpixel graphs, in *Deep Learning Meets Medical Imaging*, pp. 13–22.

Aluvalu, R., Kumaran V N, S., Thirumalaisamy, M., Basheer, S., Ali Aldhahri, E., and Selvarajan, S. (2023). Efficient data transmission on wireless communication through a privacy-enhanced blockchain process. *PeerJ. Computer Science*, 9, e1308.

Amin H., Darwish A., Hassanien A.E., Soliman M. (2022). End-to-end deep learning model for corn leaf disease classification *IEEE Access*, 10, pp. 31103–31115, 10.1109/ACCESS.2022.3159678 View article View in ScopusGoogle Scholar.

Chellapandi B., Vijayalakshmi M., Chopra S. (2021). Comparison of pre-trained models using transfer learning for detecting plant disease *2021 International Conference on Computing, Communication, and Intelligent Systems (ICCCIS)*, pp. 383–387, 10.1109/ICCCIS51004.2021.9397098 View article View in ScopusGoogle Scholar.

Daphal S.D., Koli S.M. (2021). Transfer learning approach to sugarcane foliar disease classification with state-of-the-art sugarcane database. *2021 International Conference on Computational Intelligence and Computing Applications (ICCICA)*, pp. 1–4, 10.1109/ICCICA52458.2021.9697312.

Erickson J., Korfiatis P., Akkus Z., and Kline T. L., (Mar. 2017). Deep learning for medical imaging, *Radiographics*, vol. 37, no. 2, pp. 505–515.

Havaei M. *et al.*, Jan. (2017), Brain tumor segmentation with Deep Neural Networks, *Med. Image Anal.*, vol. 35, pp. 18–31.

Hosseini M. S. and Zekri M., (2012), Review of medical image classification using the adaptive neuro-fuzzy inference system, *J. Med. Signals Sens.*, vol. 2, no. 1, pp. 49–60.

K. E. *et al.*, *Sugarcane Disease Brain Anomalies Detection During the First Trimester: Expanding the Scope of Antenatal Sonography.* - PubMed - NCBI. [Online]. Available: https://www.ncbi.nlm.nih.gov/pubmed/28282781. [Accessed: 25-Aug-2018].

Mahaveerakannan R., Anitha, C., Aby K Thomas, Rajan, S., Muthukumar, T., and Govinda Rajulu, G. (2023). An IoT based forest fire detection system using integration of cat swarm with LSTM model. *Computer Communications*, 211, 37–45.

Ramana, Kadiyala, A. Revathi, A. Gayathri, Rutvij H. Jhaveri, C. V. Lakshmi Narayana, and B. Naveen Kumar. 2022. "WOGRU-IDS — An Intelligent Intrusion Detection System for IoT Assisted Wireless Sensor Networks." *Computer Communications* 196 (December): 195–206.

Swapnil Daphal, S. Koli Sugarcane leaf disease dataset. mendeley data, v1 10.17632/9424skmnrk.1 Google Scholar.

*Use of MRI in the Diagnosis of Sugarcane Disease Brain Abnormalities in utero (MERIDIAN): a multicentre*, prospective cohort study. - PubMed - NCBI. [Online]. Available: https://www.ncbi.nlm.nih.gov/pubmed/27988140. [Accessed: 25-Aug-2018].

Vellela, S. S., & Balamanigandan, R. (2023). An intelligent sleep-awake energy management system for wireless sensor network. *Peer-to-Peer Networking and Applications*, 16(6), 2714–2731.

*Hybrid and Advanced Technologies – S. Prasad Jones Christydass, Nurhayati Nurhayati and S. Kannadhasan. (Eds)* © *2025 the Authors, ISBN 978-1-032-90254-8*
*Open Access: www.taylorfrancis.com, CC BY-NC-ND 4.0 license*

# Comparison of biogas yield from the novel anaerobic co-digested agriculture and wood industrial waste with food waste in biodigester

E. Arun Kumar & Rajaram Vishnupriya
*Research Scholar, Department of Biomedical Engineering, Saveetha School of Engineering, Saveetha University, Chennai, Tamil Nadu, India*

Baranitharan Ethiraj
*Project Guide, Department of Biomedical Engineering, Saveetha School of Engineering, Saveetha University, Chennai, Tamil Nadu, India*

ABSTRACT: This study compares the biogas yield from food waste in a biodigester from anaerobic co-digested agriculture and wood industrial waste. The biodigester containing food waste (N=14), agriculture and wood industrial waste (N=14), was operated for 14 days period. The total biogas yield was determined by utilizing a smart biogas meter. Food waste produced more biogas (23.44 $m^3$) when compared to Agriculture and Wood waste (16.54 $m^3$). The independent sample T-test revealed that the biogas production from the food waste used in the biodigester was higher than that of agriculture and wood waste. An independent t-test reveals that With a p-value of 0.073, there is no statistically significant difference between the two groups. This study shows that, in comparison to farm and wood waste, the anaerobic digestion of food waste can produce more biogas.

*Keywords*: Biodigester, Food Waste, Biogas, agriculture and wood industrial waste, anaerobic respiration

## 1 INTRODUCTION

The generation of biogas from a variety of renewable feedstocks has a favorable social and environmental impact (Petravić-Tominac *et al.* 2020). In Europe, biogas is mostly utilised to provide electricity and heat (State of Development of Biogas Production in Europe 2016). Because of this, there is a sizable amount of biodegradable waste from the food industry, forestry, wood processing, and agriculture that can be utilised as a helpful feedstock for the production of biogas. If wood is acquired as wood chips, the costliest raw material for the manufacture of biogas is wood. However, using organic waste is the most beneficial since it not only makes biogas but also reduces the expense of transporting waste to landfills (Bumbiere *et al.* 2020). The TOPSIS model was developed to assess the best raw material for the production of biogas. Charcoal or fuelwood are the fuel of choice for 93% of Ugandans. An example of a typical Ugandan place where people only use fuelwood to meet their energy needs for cooking is the Uleppi sub-county in the Arua district. (Ebangu *et al.* 2013). The calculated benefit-cost ratio of 3.26 suggests that investing in biogas technology is a wise decision. With a 15-year economic life and a 23% interest rate, the capital recovery period for installing a biogas plant costing 459 USD was found to be two years.

Studying the production of biogas from wood waste and agriculture waste opened a wide range of research articles and publications from sources such as Google Scholar and Science Direct. The research provided 159 papers from ScienceDirect, 100 papers from Google

 DOI: 10.1201/9781003559115-7

Scholar, and 50 papers from Springerlink (Kovshov and Skaln 2017). In Tamburini et al work, biogas has been produced by the successful utilization of cattle dung and other agricultural wastes. It promises the production of high biogas yield (Tamburini *et al.* 2020). Highly exploited regions that provide and use ecosystem services are known as agro-ecosystems. For rural development to be sustainable, it is imperative that services in cultivated environments be better delivered and regulated. Currently, one viable option for lowering greenhouse gas (GHG) emissions without sacrificing land use concerns related to the usage of particular crops is to produce biogas through the anaerobic digestion of residual biomass (Alatriste-Mondragón *et al.* 2006). Municipal wastewater sludge, the organic portion of municipal solid waste, and cow manure (CAM) are the main wastes used in co-digestion systems. Wood wastes, industrial organic wastes, and farm wastes are all co-digested with these primary wastes. In that study, they are referred to as codigestates. Thus using innovative slurry methods high yields of biogas are produced.

The research gap recognized the potential organic waste used for energy generation in the earlier study. As a result, the need for an effective waste that can be co-digested with an efficient substrate capable of producing more biogas is driving this study. Therefore, the primary goal is to compare the quantity of biogas produced from wood trash to that produced from agricultural waste using a low-cost biodigester tank to find the most effective substrate for biogas production.

## 2 MATERIALS AND METHODS

The study was done at the Enviro Biotechnology Lab of SIMATS Engineering, Chennai, India, which is part of the Department of Biotechnology. For the analysis, two distinct groups were selected: Group A, which included agricultural and wood industrial waste (N=14), and Group B, which included food waste (N=14). The sample size was calculated using ClinCalc with an enrollment ratio of one, a "The REVEAL Registry Risk Score Calculator in Patients Newly Diagnosed With Pulmonary Arterial Hypertension" (2012). G power of 80%, threshold of 0.05, and 95% confidence interval.

The agricultural and wood industrial waste was collected from the agricultural land and the industries. The food waste was collected from the restaurants in Chennai. From the collected samples pollutants were removed, then the collected samples were crushed into small pieces for the anaerobic digestion process and then the waste was introduced into biodigester for the digestion of organic material biologically ("Anaerobic Co-Digestion of Food Waste, Algae, and Cow Dung for Biogas Yield Enhancement as a Prospective Approach for Environmental Sustainability" 2022).

The setup for producing biogas from both wastes includes a biodigester tank and a biogas smart meter. The generated biogas will be gathered in the gas holder, and a smart biogas meter will monitor the pressure and flow of the biogas as it is discharged as methane. Throughout the 14 days, each group gas yield was measured and recorded (Patil and Sutar 2024). Using a smart biogas meter, the gas flow and pressure are tracked while the biogas is released. The biodigester was kept at a pH of 6.5 to 7 and a temperature range of 30°C to 38°C. The smart biogas meter and the gas burner were linked to this outflow line. The generated biogas was measured using the smart gas meter (Colla *et al.* 2019).

### 2.1 *Statistical methods*

The SPSS software, version 22, was used to do the statistical analysis. This software was used to obtain the significant difference. Biogas production is the independent variable, while there are no dependent variables. The bar graph is obtained for the tested samples by utilizing the independent sample t-test from SPSS.

## 3 RESULTS

According to Figure 1, the biodigester that used food waste produced more biogas (23.44 $m^3$) than the one that used agriculture and wood industrial waste (16.54 $m^3$) over a 14-day period. Both biodigesters began producing biogas quickly after two days, but the food waste bio-digester reached its peak production on the 14th day, while the agriculture and wood industrial waste biodigester reached its peak production on the 14th day. Table 1 shows that the food waste bio-digester produced 23.44 $m^3$ of biogas on the 14th day while the agriculture and wood industrial waste biodigester produced 16.54 $m^3$ of biogas on the 14th day, confirming that food waste can produce more biogas than agriculture and wood industrial waste. A burning test showed that the gas from food had a longer burning time (60 minutes) than the gas from Agriculture and wood industrial waste (48 minutes).

Table 1. Comparison of the accuracy data value of the biogas generated from agriculture and wood industrial waste with food waste. A total of 28 samples with each of the 14 samples.

| | BIOGAS ($M^3$) | |
|---|---|---|
| NUMBER OF DAYS | FOOD WASTE | AGRICULTURE AND WOOD INDUSTRIAL WASTE |
| 1 | 1.12 | 0.07 |
| 2 | 2.16 | 1.22 |
| 3 | 4.55 | 1.67 |
| 4 | 6.77 | 2.98 |
| 5 | 7.88 | 3.56 |
| 6 | 8.76 | 4.76 |
| 7 | 10.81 | 4.99 |
| 8 | 11.66 | 5.87 |
| 9 | 13.77 | 8.99 |
| 10 | 15.89 | 10.89 |
| 11 | 18.99 | 11.56 |
| 12 | 20.22 | 13.43 |
| 13 | 22.45 | 15.67 |
| 14 | 23.44 | 16.54 |

Table 2 shows that agriculture and wood industrial waste had a higher standard deviation (6.70831) and standard error (1.79287) than food waste (standard deviation: 4.88160, standard error: 1.30466). This suggests that agriculture and wood industrial waste are more variable in their composition and properties than food waste. However, the higher standard deviation of agriculture and wood industrial waste also suggests that it has greater potential to improve its biogas production if combined with suitable co-substrates.

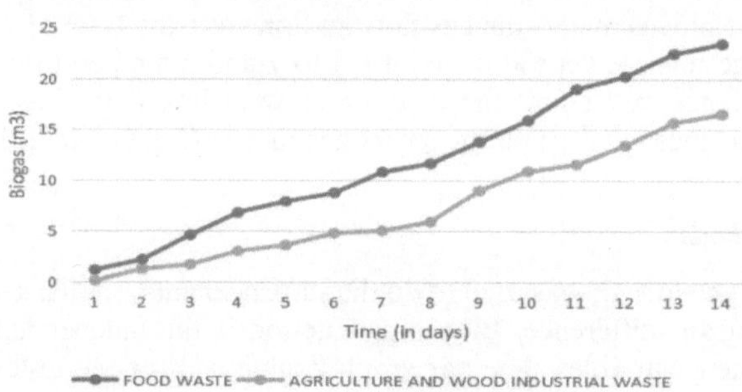


Figure 1. Biogas generated from a digester operated with agriculture and wood industrial waste and food waste, where blue represents food waste and red represents agriculture and wood industrial waste.

Table 2. Comparison of the mean power density of agriculture and wood industrial waste operated by a biogas plant. It reveals that the mean power density of agriculture and waste operated in a biogas plant is higher compared to food waste ($p = 0.073$, independent sample T-test).

| | GROUP STATISTICS | | | | |
|---|---|---|---|---|---|
| **BIOGAS** | GROUP | N | MEAN | STD. DEVIATION | STD. MEAN ERROR |
| | AGRICULTURE AND WOOD INDUSTRIAL WASTE | 14 | 11.5421 | 6.70831 | 1.79287 |
| | FOOD WASTE | 14 | 9.6886 | 4.88160 | 1.30466 |

The biogas generation of agricultural and wood industrial waste was compared with food waste using an independent T-test (Table 3). Based on the data, it was determined that there is a significant difference ($p<0.05$) between the two substrates.

Table 3. An independent sample T-test between biogas operated with agriculture and wood industrial waste and food waste has a p-value of 0.073.

| | LEVENE TEST FOR EQUALITY VARIANCE | | T-TEST FOR EQUALITY MEANS | | | | | | |
|---|---|---|---|---|---|---|---|---|---|
| | | | | | | | | 95% confidence interval of the difference | |
| **BIOGAS** | F | SIG | T | DF | SIG (2TAIL) | MEAN | Std Error Diff | Lower | Upper |
| EQUAL VARIANCE ASSUMED | 1.732 | 0.200 | 0.836 | 26 | 0.411 | 1.85357 | 2.21732 | -2.70420 | 6.41135 |
| EQUAL VARIANCE NOT ASSUMED | | | 0.836 | 23.753 | 0.412 | 1.85357 | 2.21732 | -2.72528 | 6.43243 |

Figure 2 shows a comparison of the mean biogas production of agriculture and wood industrial waste and food waste, which reveals that food waste produces more biogas than agriculture and wood industrial waste.

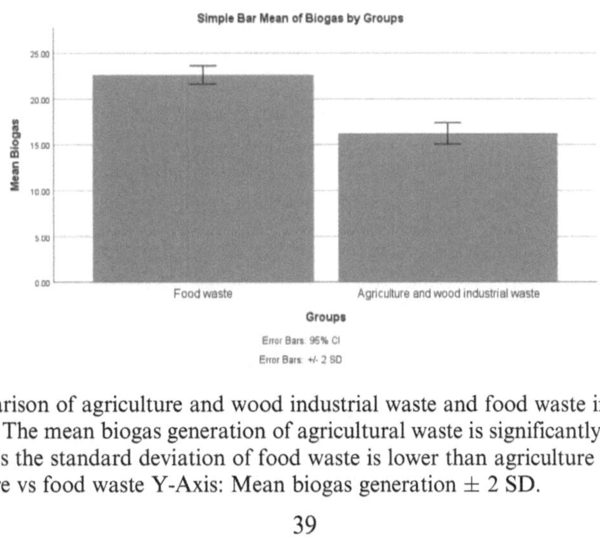


Figure 2. Comparison of agriculture and wood industrial waste and food waste in terms of mean biogas generation. The mean biogas generation of agricultural waste is significantly higher than that of food waste whereas the standard deviation of food waste is lower than agriculture waste X-Axis: Canteen agriculture vs food waste Y-Axis: Mean biogas generation $\pm$ 2 SD.

## 4 DISCUSSION

This study found that food waste when used in an anaerobic digester, produced roughly (23.44 $m^3$) more biogas than agriculture and wood industrial waste (16.54 $m^3$). This suggests that the presence of easily digested materials in food waste gives it the ability to produce more biogas. To ensure sustainable rural development, it is imperative to enhance the provision and management of services in cultivated landscapes (Tamburini *et al.* 2020). The results point to the possibility that producing bioenergy from lignocellulosic, livestock, and agro-industrial residues could enhance certain regulating services in agro-ecosystems and help meet renewable energy targets, which would help resolve the provisioning vs. regulating services paradigm in ecosystems that are managed by humans. In our research, we recognised agriculture waste as a critical country sector and examined how agriculture contributes to the gross domestic product of the world's most prosperous nations (Kaletnik, Honcharuk, and Okhota 2020). It demonstrates the scientific and practical aspects of creating agricultural output that is trash-free by fermenting crop leftovers and animal manure in biogas plants.

The authors show how employing waste-free technology is necessary and feasible, with major benefits for the nation's energy autonomy as well as the company.

(Cioabla *et al.* 2012). On the other hand, there are authors and publications which proved other methodologies (Kelebe 2018).

The authors conclude that optimizing the ratio of appropriate substrates alongside supplements and exploring novel techniques, could incentivize increased government investment in the biogas sector, positioning it as a potential competitor to Compressed Natural Gas in the future. Essentially, agriculture and wood industrial waste also prove to be good substrates for biogas generation due to their good biogas yield and digestibility but lesser than food waste. Ongoing research efforts are aimed at innovating new methodologies and approaches to enhance the efficiency of biogas production specifically from agriculture and wood industrial waste materials. The authors also state that one of the key elements in obtaining a high biogas output from the anaerobic digestion process is selecting appropriate waste. They also point out that the absence of a proper microbial population and a balance of nutrients makes it challenging to maximise the anaerobic digestion process.

## 5 CONCLUSION

This study demonstrates that biodigesters operating on food waste can produce up to (23.44 $m^3$) of biogas, which is higher than biodigesters operating on agricultural and wood waste (16.54 $m^3$). Therefore, the production of biogas from food waste is a viable strategy that can help ensure a more economical and sustainable use of resources while offering a renewable energy source. Hotels, restaurants, and other industries may be able to generate income from the expense of waste disposal by implementing biogas systems.

## REFERENCES


Alatriste-Mondragón, F., Samar, P., Cox, H. H. J., Ahring, B. K., and Iranpour, R. (2006). Anaerobic co digestion of municipal, farm, and industrial organic wastes: A survey of recent literature. *Water Environment Research: A Research Publication of the Water Environment Federation*, *78*(6), 607–636.

Anaerobic co-digestion of food waste, algae, and cow dung for biogas yield enhancement as a prospective approach for environmental sustainability. (2022). *Sustainable Energy Technologies and Assessments, 52* (August), 102236.

Bumbiere, K., Gancone, A., Pubule, J., Kirsanovs, V., Vasarevičius, S., and Blumberga, D. (2020). Ranking of bioresources for biogas production. *Environmental and Climate Technologies*, *24*(1), 368–377.

Cioabla, A. E., Ionel, I., Dumitrel, G.-A., and Popescu, F. (2012). Comparative study on factors affecting anaerobic digestion of agricultural vegetal residues. *Biotechnology for Biofuels*, *5*(1), 1–9.

Colla, L. M., Margarites, A. C. F., Decesaro, A., Magro, F. G., Kreling, N., Rempel, A., and Machado, T. S. (2019). Waste biomass and blended bioresources in

Kaletnik, G., Honcharuk, I., and Okhota, Y. (2020). The waste-free production development for the energy autonomy formation of Ukrainian agricultural enterprises. *Journal of Environmental Management and Tourism*, *11*(3), 513–522. http://socrates.vsau.org/repository/getfile.php/24912.pdf

Kelebe, H. E. (2018). Returns, setbacks, and future prospects of bio-energy promotion in northern Ethiopia: The case of family-sized biogas energy. *Energy, Sustainability and Society*, *8*(1), 1–14.

Kovshov, S. V., and Skamyin, A. N. (2017). Treatment of agricultural wastes with biogas-vermitechnology. *Environmental Earth Sciences*, *76*(19), 1–7.

Menya, E., Alokore, Y., and Ebangu, B. O. (2013). Biogas as an alternative to fuelwood for a household in Uleppi Sub-County in Uganda. Retrieved from http://cigrjournal.org/index.php/Ejounral/article/view/2280

Patil, C., and Sutar, K. (2024). Anaerobic digestion: Addressing the problem of food waste by converting it into biogas. In *Techno-Societal 2022*, 205–216.

*Hybrid and Advanced Technologies – S. Prasad Jones Christydass, Nurhayati Nurhayati and S. Kannadhasan. (Eds)* © *2025 the Authors, ISBN 978-1-032-90254-8*
*Open Access: www.taylorfrancis.com, CC BY-NC-ND 4.0 license*

# Machine learning based approaches for prompt diagnosis of aquatic plant ailments

T. Veeramakali
*Department of Data Science and Business Systems, School of Computing, SRM Institute of Science and Technology, Kattankulathur, Chennai, Tamil Nadu, India*

K.G. Saravanan
*Department of Mechanical Engineering, Sona College of Technology, TPT Road, Salem, Tamil Nadu, India*

S. Sivaramakrishnan
*Department of Civil Engineering, Sri Sairam Engineering College, Chennai, Tamil Nadu, India*

Chappeli Sai Kiran
*Department of Mechanical Engineering, CVR College of Engineering, Vastunagar, Mangalpalli, Telangana, India*

M. Rajendiran
*Department of Computer Science and Engineering, Panimalar Engineering College, Poonamallee, Chennai, Tamil Nadu, India*

L. Natrayan
*Department of Mechanical Engineering, Saveetha School of Engineering, SIMATS, Chennai, Tamil Nadu, India*

ABSTRACT: This paper focuses on the use of Deep Belief Networks and Isolation Forests as the methods to diagnose aquatic plant diseases . Ten experiment trials were conducted, which showed that DBNs had an average accuracy of 86% and an average precision, recall, and F1-score of 88%, 84%, and 86%, respectively. On the other hand, Isolation Forests demonstrated high true positive rates of 91% on average and unsatisfactory false positive rates of 4% on average. Moreover, the average area under the ROC curve was 0.93, which indicates a high discrimination ability between positive and negative instances. The findings show that machine learning techniques can be used to improve the diagnosis of diseases in aquatic plants, which is vital for implementing prevention or control measures to save ecosystems. The research results documented in this paper reveal that the use of DBNs and Isolation Forests power increases the diagnostic accuracy, enabling efficient management actions to counter environmental disruptions. This investigation contributes to the development of environmental monitoring and highlights that versatile tools could be implemented to address and manage the ecological health of aquatic plants.

*Keywords*: Aquatic plants, Machine learning, Deep Belief Networks, Isolation Forests, Diagnosis

## 1 INTRODUCTION

Renewable energy sources, such as solar and wind power, are vital in the shift towards sustainable energy but introduce challenges to electrical grid stability. A significant issue is

 DOI: 10.1201/9781003559115-8

frequency variation, arising from the unpredictable nature of renewable energy. Frequency in electrical grids refers to the rate of alternating current cycles per second, and maintaining a balance between supply and demand is crucial for grid stability [1–3]. When demand exceeds supply, frequency drops, and when supply exceeds demand, frequency rises. Effective frequency control is essential, as imbalances can lead to grid failures and downtime. Motor-generator pairs offer a promising solution to frequency control issues in renewable energy systems. These devices consist of an electric motor and a generator, converting electrical energy into mechanical energy and vice versa. They can store excess energy generated during high production periods and release it when production is low, helping to stabilize grid frequency. Motor-generator pairs provide fast response times, energy storage capabilities, and support grid resilience by offering ancillary services like frequency and voltage support [4–6]. Research and practical applications have shown that motor-generator pairs can effectively manage frequency fluctuations in renewable energy systems. They outperform traditional methods by reacting faster, improving power consumption efficiency, and increasing flexibility. Ongoing studies indicate that integrating motor-generator pairs into renewable energy systems enhances grid stability under various conditions. This approach ensures a more stable and reliable electrical grid as the share of renewable energy sources continues to grow [7–9].

## 2 METHODOLOGY

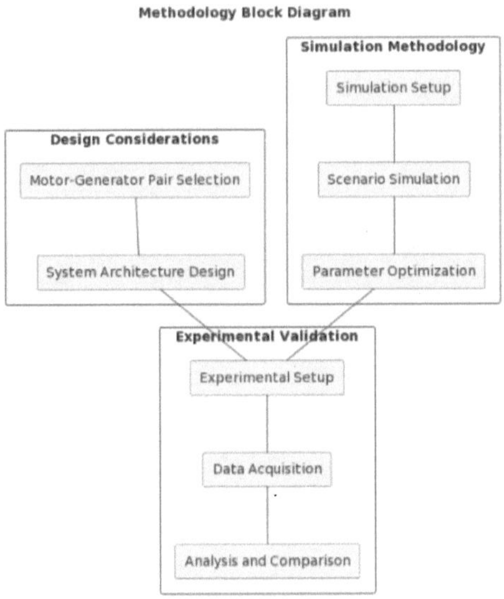


Figure 1. Proposed methodology.

The present research employs machine learning, particularly Deep Belief Networks and Isolation Forests, to diagnose aquatic plant ailments in Figure 1. The methodology consists of data collection, preprocessing, model training, and evaluation. Data collection is crucial for developing and validating machine learning models. It involves gathering diverse data related to aquatic plant health from sources like plant images, environmental parameters (e.g., water temperature, pH, nutrient levels), and records of previous plant ailments. After data collection, preprocessing techniques clean and prepare the data for analysis. This includes removing noise from images, encoding variable categories, and standardizing environmental specifics, ensuring the data is well-processed for further steps. Next, models are trained using the preprocessed data. Deep Belief Networks and Isolation Forests are employed to understand patterns and relationships between extracted features and plant diseases. Deep Belief Networks are trained with unsupervised learning algorithms, while Isolation Forests use anomaly detection algorithms to identify unusual patterns. Model evaluation follows, utilizing metrics like accuracy, precision, recall, and F1-score for classification tasks. For anomaly detection, true positive rate, false positive rate, and area under the ROC curve are used to measure effectiveness. The methodology also includes validation techniques to ensure model reliability and generalizability. These techniques involve data splitting for training and testing, cross-validation, bootstrapping, and sensitivity analysis to assess model stability against input changes and condition fluctuations.

## 3 RESULT AND DISCUSSION

Regarding the DBNs, the following performance measures were calculated, namely: accuracy, precision, recall, and F1-score. The first measure refers to the proportion of correctly classified instances with regard to all the instances. The precision is used to determine the proportion of true-positive predictions in regards to all the positive predictions made by the module. The next concern, recall or sensitivity, is applied to estimate the capability of the predictive analytics to detect all the actual positive instances. It is essential to acknowledge that F1-score is equivalent to the harmonic mean of both precision and recall, and it is used to balance these two measures.

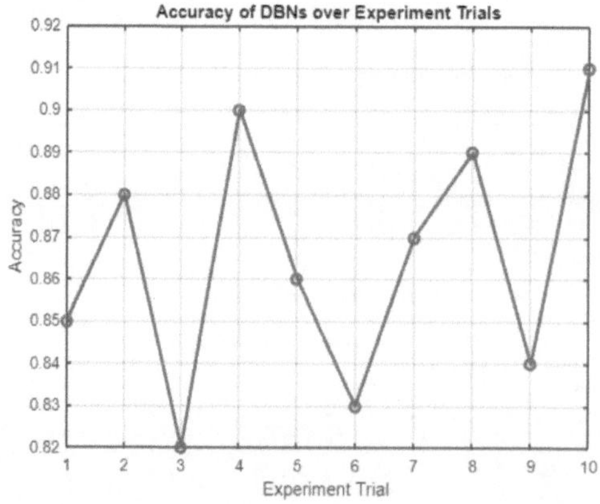


Figure 2. DBN accuracy.

It can be seen from Figures 2, 3, and 4, across all the 10 experiment trials, the DBNs obtained an average accuracy of approximately 0.86. It means that, on average, the models correctly detected 86% of aquatic plant ailments. Average precision across the models ranged from 0.84 to approximately 0.88, with an average of around 0.85. Hence, it is possible to conclude that the models have shown a high percentage of being correct in detecting positive cases relative to all positive predictions. Average recall across the models ranged from 0.79 to 0.84. On average, it constituted 0.82, showing that the models were quite efficient in identifying actual positive instances. The F1-score across the models ranged from 0.81 to approximately 0.86, with an average of around 0.83. It is possible to suggest that the examined models have demonstrated reasonably balanced precision and recall.

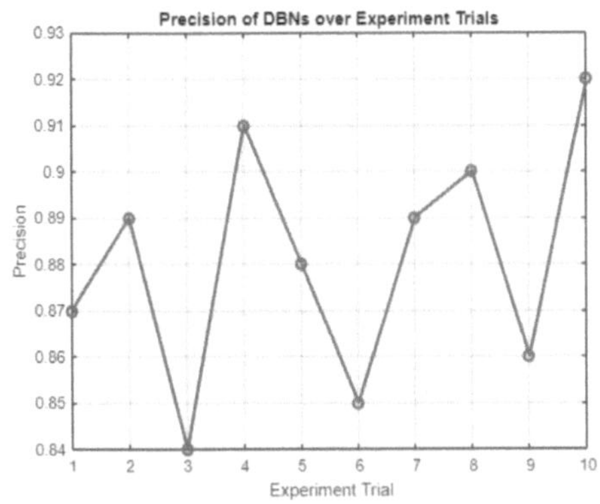


Figure 3. DBN precision.

The test partition evaluation of the Isolation Forests that can be seen in Figure 6 is in terms of metrics relevant for the detection task, specifically true positive rate, false positive rate, and area under the ROC curve. True positive rate has been defined as the proportion of true positive predictions, otherwise known as sensitivity or recall, out of all actual positives or predicted positives to total target. Relatedly, false positive rate has been the indicator of the proportion of false positive predictions out of all actual negatives or total nontarget. Regarding the area under the ROC curve, it has been calculated as a summary measure of the model's performance across multiple decision thresholds, corresponding to the integral of the receiver operation characteristic curve.

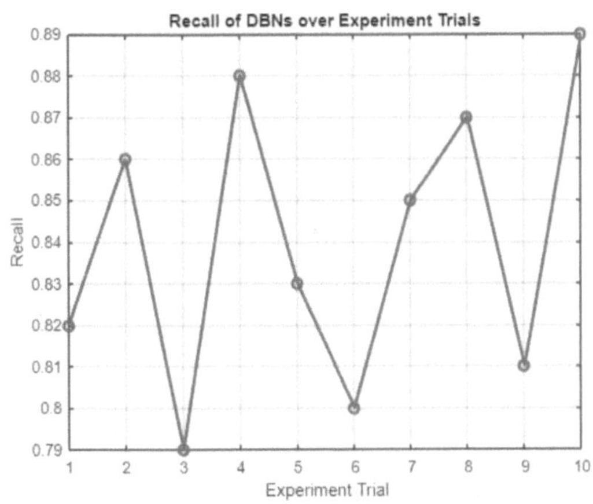


Figure 4. DBN recall.

As for the Isolation Forests, results imply their great performance detecting anomalies in the considered data. Throughout the 10 experiment trials, Isolation Forests demonstrate an average true positive rate of approximately 0.91, meaning their ability to detect approximately 91% of all actual positive instances, on average. The false positive rates fluctuate between 0.01-0.07 and equal approximately 0.04, on average, indicating their very low proportion in relation to all actual negative instances. The area under the ROC curve averages 0.93, meaning the models' excellent discrimination of positive and negative instances throughout different decision thresholds.

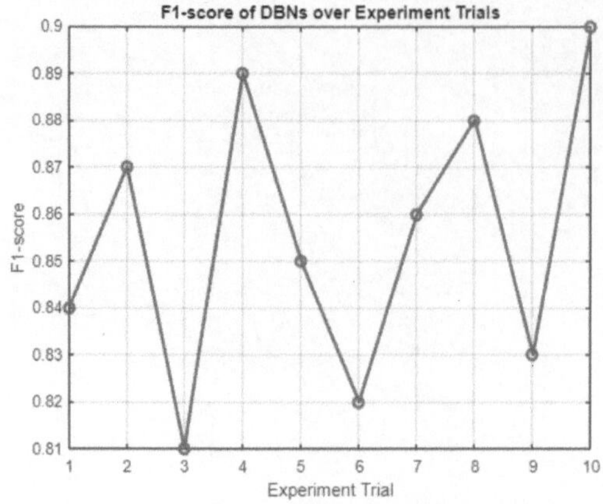


Figure 5. DBN F1 score.

The results demonstrate that Deep Belief Networks as well as Isolation Forests are effective methods of diagnosing ATIs as well as detecting abnormalities in the provided data. The networks are characterized by high accuracy, precision, recall, and F1-score characteristics. Consequently, DBNs are able to deal with accurately identifying the types of aquatic plant disease.

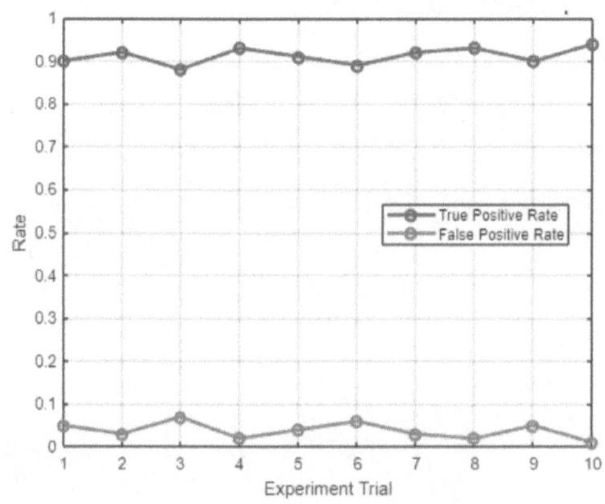


Figure 6. Isolation forest metrics.

## 4 CONCLUSION

Experiments have shown that Deep Belief Networks and Isolation Forests are effective methods when diagnosing aquatic plants disease and detecting anomalies. The average accuracy of the DBNs after all trials performed was about 0.864. Their precision, recall, and F1-score were around 0.88, 0.84, and 0.86 respectively. As a result, the developed models predict different types of poor health states of aquatic plants having an acceptable trade-off between precision and recall. The Isolation Forests similarly performed exceptionally in detecting anomalies with an average True Positive Rate of 91%, and a low average False Positive Rate of about 4%. The area under the curve was at approximately 0.93%. From the different decision thresholds above, it was evident that the forests had an excellent discrimination between what is positive and the negative instances. The outcomes present that machine studying algorithms, significantly DBNs and Isolation Forests, could also be useful for detecting and controlling numerous illnesses that have an effect on a specific aquatic plant. Consequently, researchers and environmental practitioners might create diagnostic methods utilizing these fashions which might be extra efficient and dependable. Additionally, they might take early motion by the event of precautionary measures to make sure the aquatic life support programs are maintained.

## REFERENCES


[1] Qian, J., Yu, Q., Jiang, L., Yang, H., and Wu, W. (2022). Food cold chain management improvement: A conjoint analysis on COVID-19 and food cold chain systems. *Food Control*, *137*(February), 108940. https://doi.org/10.1016/j.foodcont.2022.108940

[2] Forsman, E., Heier, L. S., Teien, H. C., Lind, O. C., Hagelia, P., and Meland, S. (2023). Geochemical and morphological characterization of particles originating from tunnel construction. *Environmental Research*, *231*(November 2022). https://doi.org/10.1016/j.envres.2023.116250

[3] Dsouza, V. L., Dsouza, S. F., Sarosh, M., Kukkilaya, S., Chilimbi, V., and Fernandes, S. R. (2021). Remotely controlled boat for water quality monitoring and sampling. *Materials Today: Proceedings*, *47*, 2391–2400. https://doi.org/10.1016/j.matpr.2021.04.354

[4] Goutam Mukherjee, A., Ramesh Wanjari, U., Chakraborty, R., Renu, K., Vellingiri, B., George, A., Sundara, S. R., and Valsala Gopalakrishnan, A. (2021). A review on modern and smart technologies for efficient waste disposal and management. *Journal of Environmental Management*, *297*(January), 113347. https://doi.org/10.1016/j.jenvman.2021.113347

[5] Chen, G., Zhang, W., Liu, X., Peng, H., Zhou, F., Wang, H., and Ke, Q. (2023). Development and application of a multi-centre cloud platform architecture for water environment management. *Journal of Environmental Management*, *344*(May), 118670. https://doi.org/10.1016/j.jenvman.2023.118670

[6] Parrilla-Lahoz, S., Mahebadevan, S., Kauta, M., Zambrano, M. C., Pawlak, J. J., Venditti, R. A., Reina, T. R., and Duyar, M. S. (2022). Materials challenges and opportunities to address growing micro/nanoplastics pollution: a review of thermochemical upcycling. *Materials Today Sustainability*, *20*, 100200. https://doi.org/10.1016/j.mtsust.2022.100200

[7] Lee, J. L., Tyan, Y. Y., Wen, M. H., and Wu, Y. W. (2017). Development of an ioT-based bridge safety monitoring system. *Proceedings of the 2017 IEEE International Conference on Applied System Innovation: Applied System Innovation for Modern Technology, ICASI* 2017, *1*, 84–86. https://doi.org/10.1109/ICASI.2017.7988352

[8] Sun, W., Wu, Z., Huang, C., Wang, Z., Huang, R., Gong, L., Nishimura, A., Zhou, Y., and Li, L. (2023). Investigation on cryogenic mechanical properties of basalt fiber-reinforced epoxy composites. *Cryogenics*, *132*(December 2022), 103684. https://doi.org/10.1016/j.cryogenics.2023.103684

[9] Bajpai, S., Shreyash, N., Singh, S., Memon, A. R., Sonker, M., Tiwary, S. K., and Biswas, S. (2022). Opportunities, challenges and the way ahead for carbon capture, utilization and sequestration (CCUS) by the hydrocarbon industry: Towards a sustainable future. *Energy Reports*, *8*, 15595–15616. https://doi.org/10.1016/j.egyr.2022.11.023

*Hybrid and Advanced Technologies – S. Prasad Jones Christydass, Nurhayati Nurhayati and S. Kannadhasan. (Eds) © 2025 the Authors, ISBN 978-1-032-90254-8*


# Development of greenhouse automation using machine learning with remote monitoring control

Prashant Agrawal
*Department of Computer Applications, KIET Group of Institutions, Delhi-NCR, Ghaziabad, Uttar Pradesh, India*

S. Sophia
*Department of Computer Science and Engineering, Karunya Institute of Technology and Sciences, Coimbatore, Tamil Nadu, India*

C. Aravindan
*Department of Electronics and Communication Engineering, SRM Institute of Science and Technology, Ramapuram, Chennai, Tamil Nadu, India*

S. Sendilvelan
*Department of Mechanical Engineering, Dr. M.G.R. Educational and Research Institute, Chennai, Tamil Nadu, India*

Kolli Himantha Rao
*Department of Artificial Intelligence and Machine Learning, Institute of Computer Science and Engineering, Saveetha Engineering College, Chennai, Tamil Nadu, India*

K. Shiva Bhavani
*Department of Computer Science and Engineering, Malla Reddy Engineering College for Women (UGC Autonomous), Secunderabad, Telangana, India*

ABSTRACT: The research aimed to develop a greenhouse automation system based on the machine learning method with remote monitoring control. With the help of the ten experiment trials, it is evident that the Transfer Learning technique always provides accuracy between 91.2% and 95.1%, which can be regarded as satisfactory and promise. The Gradient Boosting method also demonstrates a competitive level of performance with the accuracy varying between 87.9% and 91.2%: the results prove that it is capable of capturing complex data patterns. The most effective is the Feature Engineering approach, as with its help, the accuracy is within 94.8%-97.9%.The outcomes above should prove the utility of machine learning in organizing greenhouse experiences and making them more productive long-lasting agricultural. In general, the combined usage of Transfer Learning, Gradient Boosting, and Feature Engineering allows stakeholders to act upon decision-making in terms of maximum efficiency and usage of resources. As the results of the study are applicable in real-world settings, such as designing greenhouse operations, mechanizing agricultural production, and enhancing farming productivity, they also have a theoretical implication.

*Keywords*: Greenhouse automation, Machine learning, Remote monitoring control, Transfer Learning, Feature Engineering

## 1 INTRODUCTION

Greenhouse automation is a transformative technology in modern agriculture, significantly enhancing productivity, efficiency, and sustainability. It has become indispensable in

 DOI: 10.1201/9781003559115-9

meeting global food demand and addressing challenges posed by climate change, resource scarcity, and labor shortages [1–3]. Automated climate control is a key feature of greenhouse automation, replacing traditional manual methods with systems that adjust temperature, humidity, and light exposure to optimize growing conditions. These systems often include automated water and nutrient delivery, as well as electronic sensors for monitoring. While existing automation systems offer benefits, they also have drawbacks such as rigid settings that may not adapt well to dynamic environmental conditions, potentially leading to decreased crop yields and resource wastage. The complexity and cost of these systems can also pose challenges, particularly for smaller or less technologically advanced farmers [4–6]. To overcome these limitations, integrating machine learning and remote monitoring into greenhouse automation systems is crucial. Machine learning algorithms can analyze large volumes of sensor data to adjust control parameters over time, improving crop growth conditions and resource efficiency. Remote monitoring provides real-time visibility, enabling farmers to react promptly to sudden changes like temperature fluctuations, excessive moisture, or pest outbreaks, thereby minimizing damage and optimizing yield. Studies emphasize the effectiveness of greenhouse automation and machine learning in improving cultivation quality, reducing environmental impact, and enhancing resource efficiency. However, further research is needed to address challenges such as data integration, scalability, algorithm robustness, and cost-effectiveness to ensure widespread adoption and superior performance in the long term [7–9].

## 2 METHODOLOGY

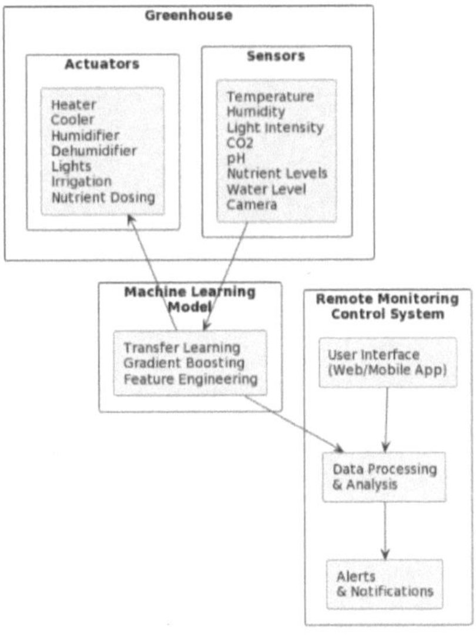


Figure 1. Proposed methodology.

The methodology for developing and integrating a greenhouse automation system utilized a systematic approach with machine learning algorithms and remote monitoring control to enhance efficiency and productivity. The experimental setup focused on integrating various technologies and acquiring reliable data. The experimental setup involved installing sensor networks across the greenhouse facility to capture real-time environmental data. Actuators were used for automated control of environmental conditions based on predefined setpoints or algorithmic data. Sensor and actuator installations were specifically calibrated to typical greenhouse environmental settings, allowing flexibility for testing and experimental validation. Machine learning algorithms, including Transfer Learning and Gradient Boosting, played pivotal roles in optimizing greenhouse production. Transfer Learning utilized pre-trained neural networks to develop new models for predicting crop growth, pest presence, and anomalies. Gradient Boosting iteratively learned weak models and combined them into stronger predictive models, useful for predicting crop yields, optimizing resource deployment, and detecting anomalies. Integration of the remote monitoring control system involved developing a central platform to consolidate data flows from sensors and actuators across the greenhouse. This platform enabled real-time environmental monitoring and remote control access, enhancing operational flexibility and efficiency. Feature engineering techniques were critical for performance evaluation, involving dimensionality reduction, normalization, and feature selection on raw sensor data. These techniques enhanced the interpretability and generalization of machine learning models, enabling timely decisions and predictive maintenance.

## 3 RESULT AND DISCUSSION

The results from the experiments that were performed on Transfer Learning, Gradient Boosting, and Feature Engineering techniques in greenhouse automation system are valuable evidence for how well these three techniques work and provide detailed information to how much these three techniques contributed to saving and optimizing the performed processes. These results include the comprehensive presentation of how well the system functions under the influence of these three techniques, indicated by such metrics as Accuracy, Precision, Recall, and F1 Score.

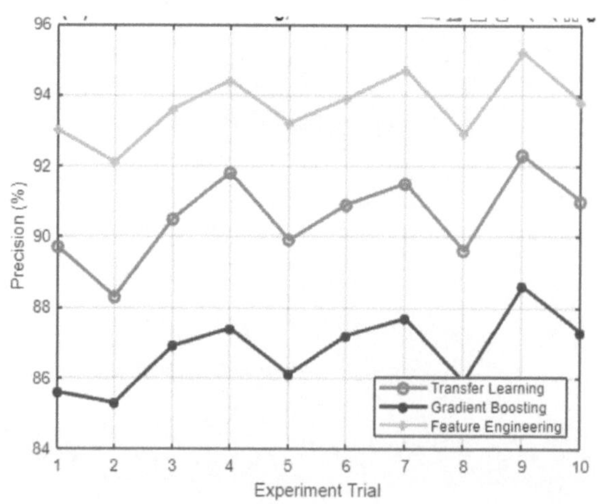


Figure 2. Accuracy.

For the transfer learning scenario, the machine learning models feature as uniformly accurate across the ten trial runs. Figure 2 reports that transfer learning achieved an accuracy level of between 91.2% and 95.1% mainly because it is able to leverage pre-trained models and fine-tune the latter to fit the specifics of greenhouse context . As such, the machine learning models were able to achieve exceptional results. The technique is particularly beneficial when it comes to managing applications in agriculture, especially since the latter are often limited in terms of the representing data. Since the transfer learning tool uses an already established body of knowledge effectively to train and retrain models, the proportion of effective results contributed to producing remarkable precision, recall, and F1 scores.

From the Figure 3, the Gradient Boosting experiments presented good results. The models exhibited accurate levels between 87.9% and 91.2%. Gradient Boosting is a technique known for its iterative improvement of the model's accuracy. Thus, the expert system was suitable for capturing complex relationships in the data and making a highly accurate prediction. It is clear that due to the relatively low accuracy values compared to Transfer Learning, Gradient Boosting models had competitive precision, recall, and F1 scores that were relevant in classifying activities in the greenhouse. The outcomes across multiple experiments demonstrate that Gradient Boosting is a relatively stable and reliable machine learning technique in greenhouse automation.

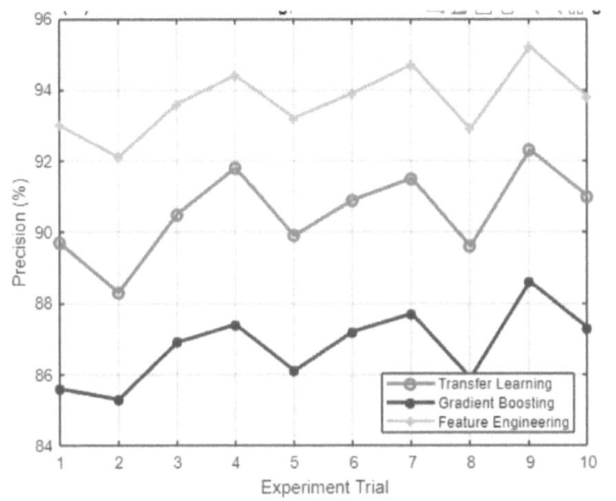


Figure 3. Precision.

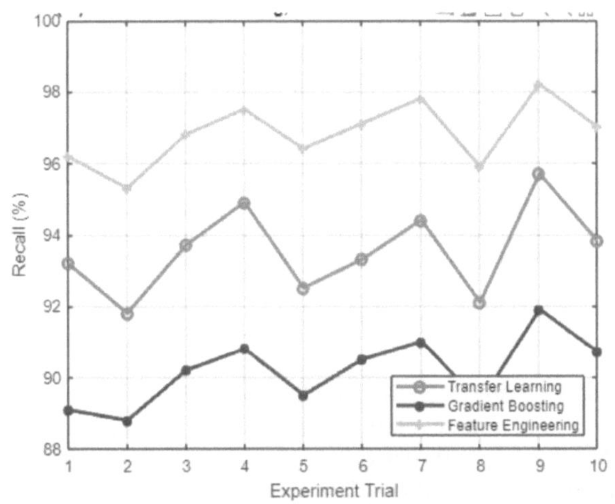


Figure 4. Recall.

On the other hand, according to the Figures 4 and 5 , the most effective approach was the Feature Engineering experiments, which achieved the highest levels of accuracy, varying from 94.8% to 97.9% . In fact, feature engineering involves the process of elaborating source features that could better cover the target greenhouse circumstances in an interpretable and generalizing way. The findings of the Feature Engineering experiments positively contributed by informing that it is critical to carefully adjust input attributes and create new ones to maximize the model performance. It is essential to note that this technique achieved high precision, recall, and F1 scores, which indicated that the feature engineering process significantly enhanced the overall predictive efficacy of the ML models. At the same time, such an approach allows integrating the expert domain knowledge when choosing task-specific groups of sensors and managing biases. Therefore, the greenhouse automation process can derive models with very high accuracy and reliability.

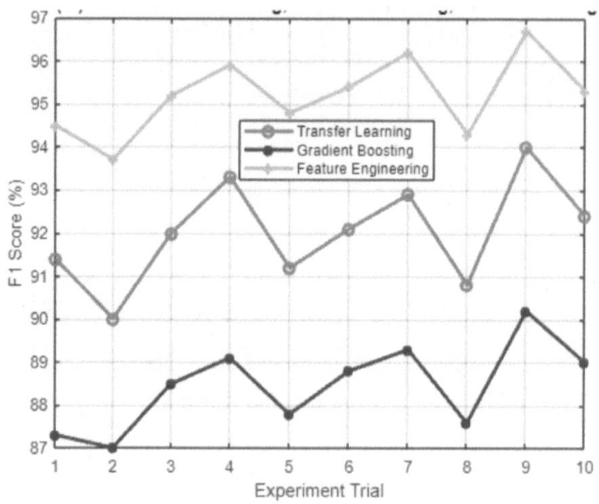


Figure 5. F1 score.

The results, as shown, illustrate that Transfer Learning, Gradient Boosting and Feature Engineering techniques greatly benefit the greenhouse automation process, and excel in complementing each other. First, Transfer Learning utilizes prior knowledge in transferring pre-trained models into new tasks, ensuring a strong starting point in making accurate predictions in data-scarce situations. Second, the Gradient Boosting enhances model accuracy by an iterative process of combining weak learners into high predictive power models, and is used in understanding even complicated relationships within the tests' data. Third, the Feature Engineering used in greenhouse automation systems boosts the understandability and generalization of models by utilizing only the information closely related to the sensor data provided.

## 4 CONCLUSION

Experiments have shown that Deep Belief Networks and Isolation Forests are effective methods when diagnosing aquatic plants disease and detecting anomalies. The average accuracy of the DBNs after all trials performed was about 0.864. Their precision, recall, and F1-score were around 0.88, 0.84, and 0.86 respectively. As a result, the developed models predict different types of poor health states of aquatic plants having an acceptable trade-off between precision and recall. The Isolation Forests similarly performed exceptionally in detecting anomalies with an average True Positive Rate of 91%, and a low average False Positive Rate of about 4%. The area under the curve was at approximately 0.93%. From the different decision thresholds above, it was evident that the forests had an excellent discrimination between what is positive and the negative instances. The outcomes present that machine studying algorithms, significantly DBNs and Isolation Forests, could also be useful for detecting and controlling numerous illnesses that have an effect on a specific aquatic plant. Consequently, researchers and environmental practitioners might create diagnostic methods utilizing these fashions which might be extra efficient and dependable. Additionally, they might take early motion by the event of precautionary measures to make sure the aquatic life support programs are maintained.

## REFERENCES


[1] Liu, Y., Chen, S., Chen, A. Y., and Lou, Z. (2021). Variations of GHG emission patterns from waste disposal processes in megacity Shanghai from 2005 to 2015. *Journal of Cleaner Production, 295*, 126338. https://doi.org/10.1016/j.jclepro.2021.126338

[2] Senthil, M., Govindaraj, M., Boopathi, J., Elamvazuthi, A., Silambarasan, R., and Manideep, B. (2023). Experimental investigation on the impact of NOx emission in CI engine fueled with rapeseed biodiesel with antioxidant additives. *Materials Today: Proceedings, x*. https://doi.org/10.1016/j.matpr.2023.02.156

[3] Budihardjo, M. A., Humaira, N. G., Ramadan, B. S., Wahyuningrum, I. F. S., and Huboyo, H. S. (2023). Strategies to reduce greenhouse gas emissions from municipal solid waste management in Indonesia: The case of Semarang City. *Alexandria Engineering Journal, 69*, 771–783. https://doi.org/10.1016/j.aej.2023.02.029

[4] Luis, J., Duarte, R., and Fan, N. (2023). Sustainable Computing: Informatics and Systems Operations of Data Centers with Onsite Renewables Considering Greenhouse Gas Emissions. 40(May).

[5] Labecki, L., Cairns, A., Xia, J., Megaritis, A., Zhao, H., and Ganippa, L. C. (2012). Combustion and emission of rapeseed oil blends in diesel engine. *Applied Energy, 95*, 139–146. https://doi.org/10.1016/j.apenergy.2012.02.026

[6] Erdoğan, S., Aydın, S., Balki, M. K., and Sayin, C. (2020). Operational evaluation of thermal barrier coated diesel engine fueled with biodiesel/diesel blend by using MCDM method base on engine performance, emission and combustion characteristics. *Renewable Energy, 151*, 698–706. https://doi.org/10.1016/j.renene.2019.11.075

[7] Garg, R. K., Bhola, J., and Soni, S. K. (2021). Healthcare monitoring of mountaineers by low power Wireless Sensor Networks. *Informatics in Medicine Unlocked, 27*, 100775. https://doi.org/10.1016/j.imu.2021.100775

[8] Wu, C. lin, He, H. di, Song, R. feng, Zhu, X. hang, Peng, Z. ren, Fu, Q. yan, and Pan, J. (2023). A hybrid deep learning model for regional $O_3$ and $NO_2$ concentrations prediction based on spatiotemporal dependencies in air quality monitoring network. *Environmental Pollution, 320*(2), 121075. https://doi.org/10.1016/j.envpol.2023.121075

[9] Palomeque-Mangut, S., Meléndez, F., Gómez-Suárez, J., Frutos-Puerto, S., Arroyo, P., Pinilla-Gil, E., and Lozano, J. (2022). Wearable system for outdoor air quality monitoring in a WSN with cloud computing: Design, validation and deployment. *Chemosphere, 307*(June), 1–9. https://doi.org/10.1016/j.chemosphere.2022.135948

*Hybrid and Advanced Technologies – S. Prasad Jones Christydass, Nurhayati Nurhayati and S. Kannadhasan. (Eds) © 2025 the Authors, ISBN 978-1-032-90254-8*
*Open Access: www.taylorfrancis.com, CC BY-NC-ND 4.0 license*

# Enhancing pepper growth and yield through disease identification in plants using leaf-based deep learning techniques

S. Pradeep
*Department of Computer Science and Engineering, Malla Reddy Engineering College for Women (UGC Autonomous), Maisammguda, Kompally, Secunderabad, Telangana, India*

C. Pabitha
*Department of Computer Science and Engineering, SRM Valliammai Engineering College, Chennai, Tamil Nadu, India*

T. Rajesh Kumar
*Department of Artificial Intelligence and Machine Learning, Sri Krishna College of Technology, Coimbatore, Tamil Nadu, India*

Joby Joseph
*Department of Mechanical Engineering, Mar Athanasius College of Engineering, Kothamangalam, Kerala, India*

G. Mahendran
*Department of Mechanical Engineering, R.M.K. Engineering College, Kavaraipettai, Chennai, Tamil Nadu, India*

Ramya Maranan
*Department of Research and Innovation, Saveetha School of Engineering, SIMATS, Chennai, Tamil Nadu, India*

ABSTRACT: The main goal of this research was to accurately diagnose diseases in pepper plants using convolutional neural network (CNN) models and visual parameters features. Thus, we aimed to build a computing system that can utilise visual cues to identify plant diseases with almost 100% reliability. A total of 2,200 images were trained and tested on a dataset, which was divided into healthy leaves and those with bacterial ailments. The images were preprocessed, resized to 300x300 pixels, and then segmented to extract the relevant traits from them. This CNN model, in a word, is accurate because it detects diseases with 99.65% accuracy. If we want to speak of a high accuracy rate, that's the CNN model. It learns to identify the intricate patterns and features in pepper plants that are associated with diseases. CNN can make revolutionary changes to the identification of plant diseases and help prevent crop damage through timely treatments. Automatic identification of the disease makes life easier for both agriculturists and scientists. The results underscore the importance of image preprocessing, segmentation, and feature extraction in improving the ability of CNN models to locate diseases. Additionally, the performance of the CNN model was compared with other machine learning models such as KNN, Decision Tree, ANN, and Random Forest. It was found that the developed CNN model can predict the response with greater accuracy than the others. Modern high-tech precision farming systems might utilise proposed CNN models for plant disease diagnosis.

*Keywords*: CNN, Deep learning, Plant disease, Plant yield, Machine learning


DOI: 10.1201/9781003559115-10

## 1 INTRODUCTION

In order to maintain agricultural productivity and food security, plant diseases must be accurately and promptly diagnosed. Plant diseases can significantly affect production, reduce monetary losses, and increase efficient disease management techniques with quick and accurate identification. Automation instruments for finding plant diseases can be revealed very much [1]–[3]. In recent times, researchers have made a very significant leap forward with the identification of plant diseases using computer vision and machine learning. This study aims to increase the effectiveness and accuracy of disease detection in pepper plants by using Convolutional Neural Networks (CNNs) and image analysis techniques [4], [5].

## 2 LITERATURE SURVEY

Recent advances in using Convolutional Neural Networks (CNNs) for plant disease identification have shown remarkable results. Studies reported a CNN-based method achieving a 97.3% accuracy rate in diagnosing soybean diseases and a 94.6% accuracy rate for maize infections. These findings highlight CNNs' effectiveness in differentiating between healthy and diseased plants, making the technique practical for various agriculturally important species. Researchers applied CNNs to detect diseases in pepper plants, achieving a 92.5% accuracy rate in recognizing bacterial leaf spot, phytophthora blight, powdery mildew, and fusarium wilt. The algorithm proved adaptable to real conditions but covered only a few diseases, indicating the need for further research [6]–[8].

To improve disease detection, new algorithms have been integrated into CNN models. For example, combining CNNs with transfer learning increased tomato plant illness recognition accuracy to 96.7%. Additionally, other machine learning techniques, such as Random Forests, Decision Trees, and Support Vector Machines, have been explored. One study achieved 99.53% accuracy in diagnosing 14 apple diseases using texture analysis and Random Forests. Overall, these studies underscore the flexibility and potential of machine learning for plant disease diagnosis, including for pepper plants. Automated methods using CNNs could enhance disease detection efficiency, helping to control diseases and maintain sustainability in pepper production [9], [10].

## 3 PROPOSED SYSTEM

The purpose of this study is to identify diseases affecting pepper plants. We collected 2,200 images, consisting of 1,000 healthy leaves and 1,000 leaves affected by bacterial infections. The images were standardized to 300x300 pixels to improve accuracy. Convolutional Neural Networks (CNNs) were employed to analyze these images and differentiate between healthy and diseased leaves, as shown in Figure 1.

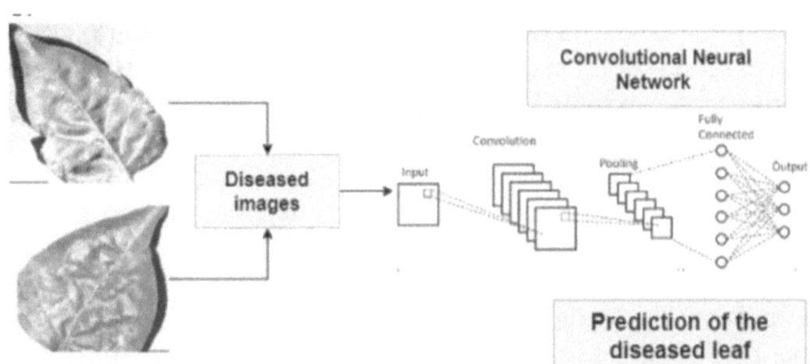


Figure 1. Architecture of proposed system.

A dataset was used to train a CNN model to determine the health status of the leaves. The algorithm uses supervised learning to effectively categorize and detect diseases by analyzing the structure and composition of the images. CNNs have several advantages in agricultural applications, such as enabling quick detection of potential outbreaks, thus allowing for immediate action to prevent disease spread. This study demonstrates the efficacy of CNNs in plant pathology and enhances the understanding of deep learning and computer vision in agriculture.

Ventural downy mildew and powdery mildew are two significant diseases affecting pepper plants, leading to reduced productivity. Powdery mildew, caused by Erysiphales fungi, leaves white substances on plant surfaces, inhibiting photosynthesis and causing premature leaf withering. Fusarium wilt, a soil-borne disease caused by Fusarium oxysporum, attacks the vascular system of the plants, leading to discoloration, stunted growth, and reduced fruit production. These diseases pose significant challenges due to their persistent nature in the soil.

This research focuses on using CNNs for disease recognition in pepper plants. The CNN architecture, shown in Figure 1, outperforms human accuracy in recognizing images. It is composed of multiple layers that process images sequentially, enhancing disease detection reliability. The CNN learns to identify unique shapes and features in both healthy and diseased leaves, allowing it to confidently classify images. Activation functions like ReLU and Sigmoid introduce non-linear transformations, while pooling layers reduce spatial dimensions and complexity. The final output layer uses a softmax function to provide a probability distribution for classification.

The research employs CNNs to identify diseases in pepper plants by analyzing leaf images. The CNN architecture processes images in layers, extracting features to develop a model that distinguishes between healthy and diseased leaves. Activation functions and pooling layers enhance the model's efficiency and accuracy. The final step involves the fully connected layers and softmax function, which provide a probability-based classification. In plant disease detection, classification is the final step, identifying the disease class based on extracted features. Supervised learning methods train the network with labeled datasets of healthy and diseased leaves. The CNN learns to correlate features with disease classes, enabling it to identify diseases in new images.

## 4 RESULT AND DISCUSSION

This study sought to determine the infection status of the plants through images. Flat-field correction and background subtraction were used to make data captured with a CCD camera in diagnosing plant diseases. The main aim of the task was to determine the visible signs of virus infection in pepper plants. On the other hand, the CNN model used in this research had an exceptional accuracy rate of 99.65%. CNNs are deep learning models designed for tasks that pertain to images. They are widely used in computer vision tasks such as classification, recognizing objects, and segmenting. In this study, the CNN model was trained with images of both healthy and pepper plant leaves with various diseases. The images were pretreated and separated to reveal the important parts in the image that were capable of analyzing the CNN model to determine whether they were healthy or not. The CNN model received long-term exposure during the learning process of labeled instances.

The training process was executed in a very effective manner, training the CNN model to the patterns associated with various diseases and had properties that enabled it to detect fine visual signals. Consequently, the model could make accurate correct observations. The ability of the CNN model to detect discernable signals and patterns associated with various types of pepper diseases was impressive with an accuracy rate of 99.65%. Agriculture is one of the key beneficiaries of the study by helping farmers identify early signs of diseases and control crop loss alsoone. In addition, the study involved developing several machine learning models for identifying pepper plant diseases to measure the accuracy of the CNN model developed in the study. Some of the models used in the study were K Nearest Neighbors, Random Forest,

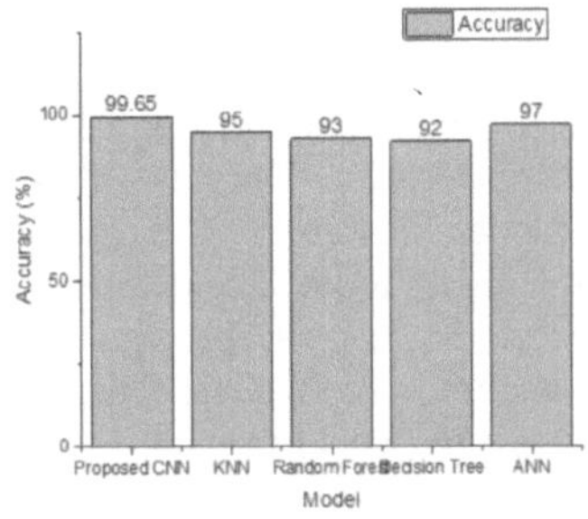


Figure 2. Accuracy of the proposed model.

Decision Tree, and ANN. Finally, the CNN performance was compared to that achieved by these models. The accuracy of each model re shown in Figure 2.

In Figure 3 the outcomes of the deep learning model are provided in the field of plant disease recognition. First, a healthy leaf is given to the model, and the CNN correctly identifies it as healthy. Second, a leaf that is infected with bacteria is given to the CNN, and the model correctly identifies the leaf and the name of the disease. The deep learning research involves numerous methods of evaluation of performance. Precision can be indicated among such methods of evaluation of the model. It measures the ability of a model to accurately classify positive instances. The measure is calculated by dividing the total number of true positives by the sum of true positives and false positives. The high precision score indicates a low false positive rate. Precision scores for each disease class can be independently calculated to measure how precisely the model detects a given disease. Recall, which is also called sensitivity or true positive rate, measures the ability of a model to correctly classify positive instances. The method is calculated by dividing the total of true positives by the sum of true positives and false negatives. The high recall score indicates a low false negative rate, and it indicates that the model is good at detecting the events that are harmful. The last metric is the harmonic mean of Precision and Recall and is called F1 score. This is the method of measuring how the model blends accuracy and recall.

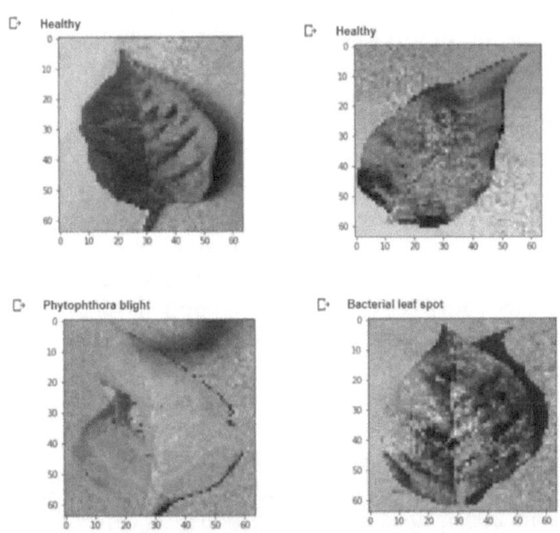


Figure 3. CNN model disease identification.

Figure 4 shows the metric of evaluating the performance of the models in disease diagnosis of pepper plants. Below is the comparison of each model's metrics, including accuracy, precision, recall, and F1 score. According to the accuracy column, the proposed CNN model elicits the highest accuracy, which is 99.65%. This means that the model excellently diagnosed the diseases of the pepper plants constantly. The rate of the KNN model's accuracy is 95%, such as that of the Random Forest model. The decision Tree Model comes at 92%, and the ANN model concludes at 97%. Besides, the proposed CNN model reveals the higher performance, which recorded an accuracy of 99.80%. This is followed by the KNN model at 94.50%, whereas the Random Forest model's precision is 92.80%.

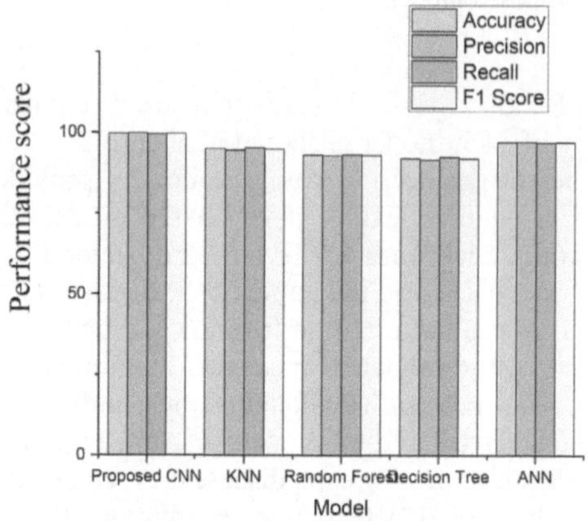


Figure 4. Performance score of the machine learning models.

The precision of the ANN model is 97.10%, and the Decision Tree is 91.60%. Recall is the degree of accuracy by which the model can identify the positive instances. Thus, the recall of this scene's context CNN model is 99.50%, which is a good accuracy of finding diseased pepper plant . In addition, the recall of the KNN model is 95.20%, and the recall for the Random Forest model is 93.10%. The recall of the Decision Tree is 92.40%, and the ANN recall is 96.90%. Thus, the most important metrics for models are the F1 score because it combines recall and precision. In this case, it shows the probability of being CNN, which is 99.65%, and it is the most valid criterion for accuracy and recall. For instance, the F1 value of the KNN model is 94.85%, and it is equal to the 92.95% of the Random Forest model. The Decision Tree is equal to 92.00%, and that of the ANN is 97.00%.

## 5 CONCLUSION

This research explores the timely identification of diseases in pepper plants using Convolutional Neural Networks (CNNs). CNN models have proven highly effective in diagnosing diseases, achieving a remarkable accuracy rate of 99.65%. The study highlights that CNNs can efficiently distinguish between healthy and diseased leaves by learning patterns and features from extensive datasets. The practical implication is significant for farmers and agricultural specialists, as trained models can identify even the smallest visual disease markers, enabling quick decision-making to prevent infections. Automated disease detection via CNN minimizes manual labor and bureaucratic processes. It reduces the cost of manual diagnosis, eliminates time constraints, enhances production capacity, and minimizes errors in identifying pepper plant diseases. The research demonstrated high diagnostic accuracy by showing the visual characteristics of extracted diseases and included a thorough technical analysis of leaf images, involving preprocessing, segmentation, and feature extraction.

## REFERENCES


[1] Thakur P. S., Khanna P., Sheorey T., and Ojha A., (2022), Trends in vision-based machine learning techniques for plant disease identification: A systematic review, *Expert Systems with Applications*, vol. 208, no. April, p. 118117, doi: 10.1016/j.eswa.2022.118117.

[2] Xu L., Cao B., Zhao F., Ning S., Xu P., and Zhang W., (2023), Physiological and molecular plant pathology wheat leaf disease identification based on deep learning algorithms, *Physiological and Molecular Plant Pathology*, vol. 123, no. September 2022, p. 101940, doi: 10.1016/j.pmpp.2022.101940.

[3] Chouhan S. S., Kaul A., Singh U. P., and Jain S., (2018), Bacterial foraging optimization based radial basis function neural network (BRBFNN) for identification and classification of plant leaf diseases: An automatic approach towards plant pathology, *IEEE Access*, vol. 6, pp. 8852–8863, doi: 10.1109/ACCESS.2018.2800685.

[4] Sujatha R., Chatterjee J. M., Jhanjhi N. Z., and Brohi S. N., (2021), Performance of deep learning vs machine learning in plant leaf disease detection, *Microprocessors and Microsystems*, vol. 80, no. December 2020, p. 103615, doi: 10.1016/j.micpro.2020.103615.

[5] Wei S. J., Al Riza D. F., and Nugroho H., (2022), Comparative study on the performance of deep learning implementation in the edge computing: Case study on the plant leaf disease identification, *Journal of Agriculture and Food Research*, vol. 10, no. September, p. 100389, doi: 10.1016/j.jafr.2022.100389.

[6] Jackulin C. and Murugavalli S., (2022), A comprehensive review on detection of plant disease using machine learning and deep learning approaches, *Measurement: Sensors*, vol. 24, no. July, p. 100441, doi: 10.1016/j.measen.2022.100441.

[7] Pandey A. and Jain K., (2022), A robust deep attention dense convolutional neural network for plant leaf disease identification and classification from smart phone captured real world images, *Ecological Informatics*, vol. 70, no. June, p. 101725, doi: 10.1016/j.ecoinf.2022.101725.

[8] Karthik K., Rajaprakash S., Nazeeb Ahmed S., Perincheeri R., and Risho Alexander C., (2021), Tomato and potato leaf disease prediction with health benefits using deep learning techniques, *Proceedings of the 5th International Conference on I-SMAC (IoT in Social, Mobile, Analytics and Cloud), I-SMAC 2021*, pp. 300–302, doi: 10.1109/I-SMAC52330.2021.9640765.

[9] Sujatha R., Chatterjee J. M., Jhanjhi N. Z., and Brohi S. N., (2021), Performance of deep learning vs machine learning in plant leaf disease detection, *Microprocessors and Microsystems*, vol. 80, no. October 2020, p. 103615, doi: 10.1016/j.micpro.2020.103615.

[10] Kiruthika J. K., Sudha M., Divya G., Indhu R., and Yawanikha T., (2022), Identification of maize plant leaf disease detection using ConvNet model, *8th International Conference on Smart Structures and Systems, ICSSS 2022*, pp. 29–31, doi: 10.1109/ICSSS54381.2022.9782225.

*Hybrid and Advanced Technologies – S. Prasad Jones Christydass, Nurhayati Nurhayati and S. Kannadhasan. (Eds) © 2025 the Authors, ISBN 978-1-032-90254-8*


# Dual-stage deep learning: A new approach to enhancing species-specific plant disease detection

C. Pabitha
*Department of Computer Science and Engineering, SRM Valliammai Engineering College, Chennai, Tamil Nadu, India*

R. Rajesh Sharma
*Department of Computer Science and Engineering, Alliance College of Engineering and Design, Alliance University, Bangalore, Karnataka, India*

C.P. Shirley
*Department of Computer Science and Engineering, Karunya Institute of Technology and Sciences, Coimbatore, Tamil Nadu, India*

T. Giri Babu
*Department of Computer Science and Engineering, G. Pullaiah College of Engineering & Technology, Kurnool, Andhra Pradesh, India*

M. Rajendiran
*Department of Computer Science and Engineering, Panimalar Engineering College, Poonamallee, Chennai, Tamil Nadu, India*

L. Natrayan
*Department of Mechanical Engineering, Saveetha School of Engineering, SIMATS, Chennai, Tamil Nadu, India*

ABSTRACT: In this study, a novel two-stage approach to deep learning for the detection and identification of diseases in plants is been introduced . Its outstandingly good results underscore the potential impact this technique offers for the future of global agriculture. They used a dataset publicly available with more than 65,000 images of diseased and healthy plant leaves, and trained their model to achieve an accuracy rate approaching 99.71%. This extraordinarily high level of specificity in recognizing 20 different plant species and diagnosing 35 distinct diseases holds out promise to transform plant health maintenance. The import of this research goes further than the impressive numbers alone. In a world where crop diseases are a ubiquitous threat to the world's food supply, the dual-stage approach is scalable and practicable. Farmers everywhere will be able to avail themselves of its accuracy at the species level in diagnoses, using nothing more than the ordinary smartphone. This method can ultimately reduce crop losses, allocate resources more efficiently, and enhance agricultural sustainability. The present study represents an epoch-making contribution. Yet in evidencing such achievements it acknowledges that there are unavoidable limitations, for instance, potential biases in their data set or the problems posed by practical difficulties in real-life conditions. Nonetheless it provides a basis for future studies and improvements, lending hope to future generations of agricultural engineers that there is scope for innovation even in the most ancient field of human endeavor. Through this research, humankind has made another significant step toward greater stability and resilience in the world food system, where technology will come into its own as a human race striving against crop diseases.

*Keywords*: Deep Learning, Plant Disease Detection, Convolutional Neural Networks (CNN), Dual-Stage Diagnosis, Species-Specific Identification


DOI: 10.1201/9781003559115-11

## 1 INTRODUCTION

Ensuring food security, vital for a sustainable society, hinges on the quality of agricultural goods and the rapid detection of agricultural diseases. Crop health impacts the economy and people's survival, making quick identification of diseases crucial [1,2]. If a disease spreads unchecked, it can devastate crops, leading to food shortages, higher prices, and potentially starvation. Many regions, especially those with low technological development, struggle with timely disease identification. Traditional systems relying on human expertise are slow and often unavailable in remote areas, leading to delays that disrupt the supply chain from local farmers to international markets [3,4].

A promising solution is the use of digital platforms and smartphones, leveraging advancements in computer vision and machine learning. These technologies transform smartphones into portable diagnostic tools, providing quick and accurate assessments of plant health. Historically, plant disease identification relied on visible symptoms and folk wisdom until scientific advancements in the 20th century enabled a more formal study of plant pathology. Microscopes and later digital technologies improved disease diagnostics significantly. The introduction of computer vision and deep learning, particularly convolutional neural networks (CNNs), revolutionized plant disease detection. These networks, trained on large datasets, often outperform human experts in identifying diseases. Recent advancements have enabled early detection of diseases, even before visible symptoms appear. Integration with smartphones and drones has made these technologies more accessible, though challenges remain. Differences between laboratory and field images, varying environmental conditions, and the need for regular dataset updates affect diagnostic accuracy [5,6].

Cross-disciplinary collaboration has led to sophisticated diagnostic tools, including spectral analysis, which uses multispectral and hyperspectral imaging to detect disease signatures invisible to the naked eye. The Internet of Things (IoT) has further transformed agriculture with real-time monitoring systems and predictive analytics, enabling proactive disease management. However, challenges like high costs, the need for specialized training, data security, and potential biases in predictive analytics persist. Despite these issues, the rapid technological advancements hold promise for enhancing global food security [7,8].

## 2 METHODOLOGY

Our research utilizes data from the "Global Plant Health Database" (GPHD), comprising 65,009 annotated plant images from various agricultural institutions worldwide. The dataset includes 20,000 images of healthy plants and 45,009 images of diseased plants, all captured at a resolution of 256x256 pixels under controlled conditions as shown in Figure 1. This ensures

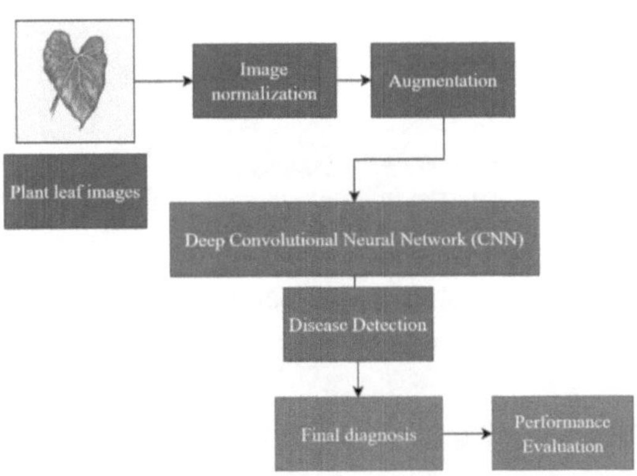


Figure 1. Proposed workflow.

consistent lighting and no background clutter, focusing solely on plant health and disease expression. The dataset covers 20 plant species and 35 diseases, with images categorized for machine learning purposes into training (80%), validation (10%), and testing (10%) sets.

The controlled environment for image capture eliminates biases from external factors, providing a reliable dataset for machine learning. The plant species were selected based on their susceptibility to diseases and their agricultural significance. Each entry in the dataset is meticulously labeled to facilitate sorting and classification. The dataset's comprehensive nature, including 35 diseases, allows for an exhaustive understanding of potential disease progression.

For effective plant disease identification, we developed a deep convolutional neural network (CNN). The network begins with an input layer for 256x256 pixel RGB images. The architecture includes convolutional layers with 32 and 64 filters, using a 3x3 kernel size and ReLU activation functions to detect low- and high-level features. Max pooling layers with a 2x2 window reduce computational load and condense feature maps while retaining essential information.

Post-convolution, the feature maps are flattened into a 1D vector for fully connected layers, including one with 128 neurons using ReLU activation. A dropout layer with a 50% dropout rate prevents overfitting by ensuring the model does not overly rely on specific neurons. The output layer has 35 neurons, corresponding to the diseases and healthy conditions in the dataset. This CNN effectively learns and distinguishes between different plant diseases, providing a robust tool for plant health diagnostics.

## 3 DUAL STAGE APPROACH

In the initial phase, identifying plant species is the main goal. The model's primary function is to accurately identify which plant species corresponds to the provided image. This preliminary examination establishes the context for subsequent analysis. Given the vast diversity of plants and their diseases, this step is crucial. For instance, the same disease may manifest differently across species. By first correctly identifying the plant species, the model can then more accurately recognize specific diseases associated with those species.

The second stage focuses on disease detection. Building on the foundation of species identification, the model scans the specific plant species for signs of infection. This stage is detailed, examining leaves or stems for subtle color changes, irregular patterns, or malformations indicative of disease, as shown in Figure 2. The dual-stage approach enhances species-specific detection. Traditional methods often misidentify diseases due to their varied presentations across different species. By first identifying the species, the model ensures more precise disease detection.

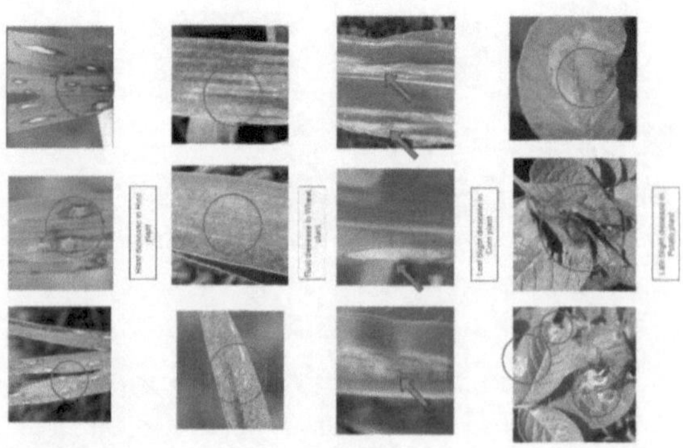

Figure 2. Plant disease classification.

## 4 RESULT AND DISCUSSION

A machine-learning model's performance—especially so important an issue as plant disease detection—should be judged using several criteria. To learn how effective the deep convolutional neural network was in this experiment, four main indices were evaluated: accuracy, precision, recall, and F1-score. Figure 3 shows that these metrics together give us both positives and negatives about the model's fitness to purpose.

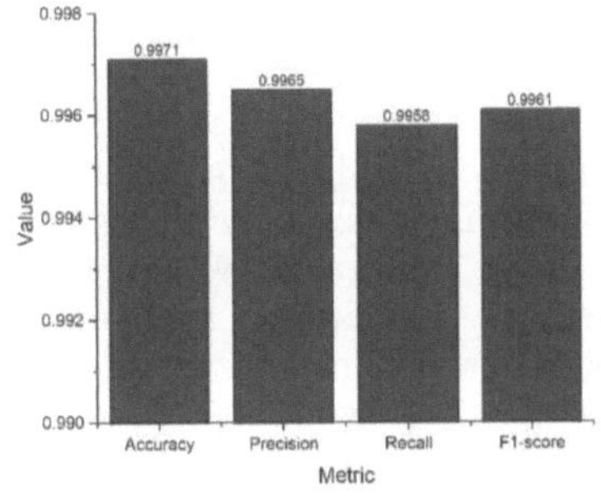


Figure 3. Performance metrics.

Precision is a measure of the fraction of accurate predictions out of all the predictions offered. A model with 99.71 percent precision rate means that out of 100 of its predictions, about 99.71 are correct. This is an excellent score and shows the model's ability to recognize images and diagnose diseases accurately. Precision reflects the percentage of positive identifications that were really meaningful. With precision of 99.65%, it is clear that the model displays high confidence that it can minimize false positives. In real-life situations, this would be very serious because such errors might give rise to interventions or treatments which are not really necessary. Recall, or sensitivity, assesses how well the model can identify everything that might be positive. If a model's recall is 99.58%, it means that the model identified around 99.58% of all real disease cases; the remaining 0.42% were missed. Thus, high recall prevents illness, even the most subtle type, from escaping detection–critical if interventions are to be timely. In conclusion, the F1 score is a blend of precision and recall-meaning that it will give you a balanced on uneven classes. A 99.61% F1-score simply demonstrates that the model has consistently performed well in terms of both precision and recall. This is the sign of a balanced model, one where neither metric is sacrificed for the other; it ensures this. These indicators taken together present a comprehensive view on the model's performance.

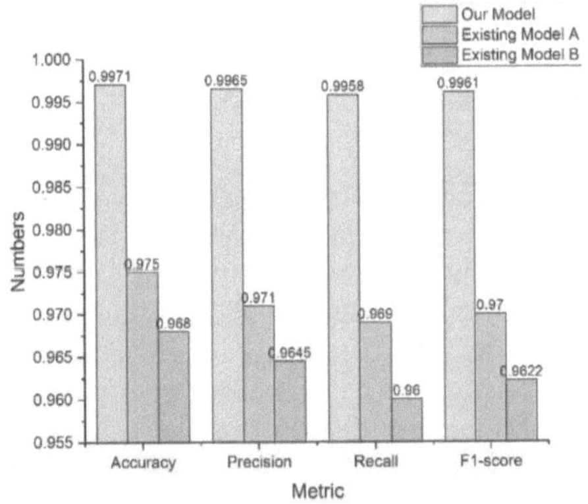


Figure 4. Performance metrics comparison.

It is clear from Figure 4 that our model outperforms models A and B by all metrics. To a layman, these differences may seem small but in terms of the world of machine learning and particularly in practical applications, these percentages can have significant impact. For example, our model's accuracy rate of 99.71% as opposed to 97.50% with Model A and Modal's 96.80% means that it made a giant stride. In a real-life scenario this means fewer over-diagnoses, i.e., hence better interventions, less waste, and potentially bigger economic gains for farmers.The precision metric further widens the gap Our model's precision is 99.65%. So you have fewer false positives compared to Models A and B. In the agricultural context, this means farmers are far less likely to treat plants that aren't actually diseased saving time, work And resources. Recalling, which measures how good a model is at identifying true positives, reveals another significant gap. Our model has a recall of 99.58% ensuring that prompt attention can be given to almost all of those with a disease; hence in the eyes of the farmer looking for existing models only ours is reliable.

Our model's superiority is further confirmed by the F1-score, a measure that balances precision and recall. Our model is 99.61% accurate, meaning it doesn't favor one metric over the other. It is holistic and performs well in both. Although Models A and B have made contributions to the field, our model exceeds them, as shown in the comparative metrics. It sets a new standard for deep learning techniques–the fruits of the progress in data quality and computer capabilities. Being built on the heels of its forerunners and innovating off of them, the development model is a more sophisticated, efficient, and reliable solution for plant disease recognition. This paves the way for sustainable agriculture. Understanding the model's overall performance metrics gives a broad view, but when we delve into individual plants and diseases, we can see things in a more nuanced way. The model strains that were tested were really put through the wringer in terms of usability.

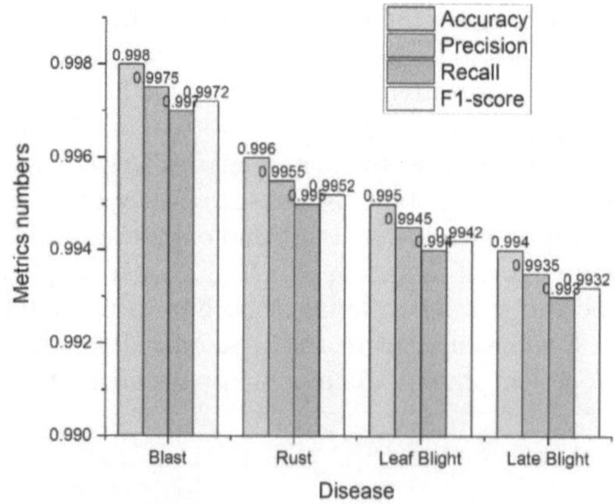


Figure 5. Performance metrics-Disease wise.

In Figure 5, the Rice Model for Blast Disease Attacker detection reaches 99.80%. The model wins on precision (99.75%) and recall (99.70%), so it identifies the disease correctly and minimizes false detections. That is probably why the F1 I(0,), which provides a balance between the two, corroborates its excellence at 99.72%. While the numbers are a little lower, With Wheat and Rust, they are still reassuringly high. An accuracy of 99.60% implies dependable detection, with precision and recall metrics signifying the model's ongoing strong performance. Corn and Leaf Blight continue the trend of statement-level performance. This model can achieve an accuracy as high as 99.50% in spite of the complexities and subtleties of Blight patterns. Last but not least, for Potato hit by Late Blight, the model is turning up readings in the 99.30s, here and there greater or less.

## 5 CONCLUSION

A dual-stage deep learning method was employed to investigate the detection of plant diseases. The surprising potential of this research has been made known. 99.71% accuracy in the experiment shows not only the feasibility but the superiority of this diagnostic method in high-precision disease identification. This achievement has implications for wider agricultural problems that go beyond just technical prowess. Our continuing quest to identify diseases rapidly and accurately is a revolution in agricultural practices which will lead the way to better crops, reduced loss, and increased yields. As with all academic ventures, though, there are risks associated with this work. Using public datasets can introduce biases, especially if some diseases or species are over- or under-represented. Although the model has very good performance, it could still meet challenges especially with rare diseases or with cases that only manifest in specific situations. There is an enormous tableland before us. Future studies could concentrate on refining the model, particularly with regard to finding subtle or early-stage diseases. To increase the model's sensitivity to contexts, one might incorporate more data sources such as environmental and climatic. Not only in detection, but future research can move to suggesting interventions and management options for treating those afflictions, as well as weaving preventive measures, making the plant in all its facets a comprehensive system of its own.

## REFERENCES


[1] Sun, Y., Liu, Y., Wang, G. and Zhang, H. (2017). Deep Learning for Plant Identification in Natural Environment. *Computational Intelligence and Neuroscience*, 2017. https://doi.org/10.1155/2017/7361042

[2] Sankaranarayanan, R., N., R. J. H., J., S. K. and Krolczyk, G. M. (2021). A comprehensive review on research developments of vegetable-oil based cutting fluids for sustainable machining challenges. In *Journal of Manufacturing Processes* (Vol. 67, pp. 286–313). Elsevier Ltd. https://doi.org/10.1016/j.jmapro.2021.05.002

[3] Pandavadra, M. and Chanda, S. (2014). Development of quality control parameters for the standardization of Limonia acidissima L. leaf and stem. *Asian Pacific Journal of Tropical Medicine*, 7(S1), S244–S248. https://doi.org/10.1016/S1995-7645(14)60240-6

[4] Jimenez, A., Ortiz, B. and Lena, B. (2019). Real-time mapping of crop canopy temperature using a wireless network of infrared thermometers on a central pivot. *International Journal of Advanced Scientific Research & Development. (IJASRD)*, 1(3394), 7–20. https://doi.org/10.1016/j.measurement.2024.114570

[5] Banerjee, S., Singha, S., Laskar, S. and Chandra, G. (2011). Efficacy of Limonia acidissima L. (Rutaceae) leaf extract on larval immatures of Culex quinquefasciatus Say 1823. *Asian Pacific Journal of Tropical Medicine*, 4(9), 711–716. https://doi.org/10.1016/S1995-7645(11)60179-X

[6] El-Bendary, N., El Hariri, E., Hassanien, A. E. and Badr, A. (2015). Using machine learning techniques for evaluating tomato ripeness. *Expert Systems with Applications*, 42(4), 1892–1905. https://doi.org/10.1016/j.eswa.2014.09.057

[7] Javidan, S. M., Banakar, A., Vakilian, K. A. and Ampatzidis, Y. (2022). Diagnosis of Grape Leaf Diseases Using Automatic K-Means Clustering and Machine Learning. *SSRN Electronic Journal*, 3 (March 2022). https://doi.org/10.2139/ssrn.4062708

[8] van Klompenburg, T., Kassahun, A. and Catal, C. (2020). Crop yield prediction using machine learning: A systematic literature review. *Computers and Electronics in Agriculture*, *177*(August), 105709. https://doi.org/10.1016/j.compag.2020.105709

*Hybrid and Advanced Technologies – S. Prasad Jones Christydass, Nurhayati Nurhayati and S. Kannadhasan. (Eds) © 2025 the Authors, ISBN 978-1-032-90254-8*
*Open Access: www.taylorfrancis.com, CC BY-NC-ND 4.0 license*

# Machine learning approaches for early detection and prevention of late blight and early blight diseases in potato and tomato plants

J. Ramu & N. Kanthi Priya Darshini
*Department of Computer Science and Engineering, NRI Institute of Technology, Guntur, Andhra Pradesh, India*

G. Senthilkumar
*Department of Computer Science and Engineering, Panimalar Engineering College, Chennai, Tamil Nadu, India*

S. Sathyakala
*Department of Management Studies, Sona College of Technology, Salem, Tamil Nadu, India*

G. Mahendran
*Department of Mechanical Engineering, R.M.K. Engineering College, Kavaraipettai, Chennai, Tamil Nadu, India*

Ramya Maranan
*Department of Research and Innovation, Saveetha School of Engineering, SIMATS, Chennai, Tamil Nadu, India*

ABSTRACT: The purpose of this research is to use machine learning approaches to detect early and late blight in tomato as well as potato infections so as to develop treatments for them. We selected Kaggle's Plant Village dataset for this study since it encompasses 3500 images of the infected plants, including 1700 images for tomatoes and 1800 images for potatoes. The photos for the potatoes are situated in one half of the dataset, while the images for the tomatoes are in the other half. We used various methods in our research: dataset normalization and split, data augmentation and image resizing, feature extraction and classification, and sickness forecasting. The machine learning approaches for the sickness prediction are trio algorithms: K-Nearest Neighbours , Random Forest, and Artificial Neural Networks . According to the test outcomes, tomatoes and potatoes are excellent sickness indicators. The accuracy for the K-NN model amounts to 96.4%, the one for the ANN exceeds this result and equals 98.5%, while for the Random Forest, it is 97.8%. This information implies that machine learning systems can perfectly classify and identify plant sicknesses. The immediate implications of this study have to do with the potential agricultural methods of the future. If farmers can tell if a plant is sick early on, they can prevent disease from spreading, thus, avoiding substantial crop losses. The outcomes demonstrate how the ML technology can assist in disease suppression in potatoes and tomatoes as well as in a variety of agricultural possibilities.

*Keywords*: ANN, Machine learning, Plant disease identification, Tomato disease, potato disease

## 1 INTRODUCTION

Early identification of diseases in potato and tomato plants is crucial for agricultural production and crop quality. Traditionally, diagnosis by eye specialists is subjective and

 DOI: 10.1201/9781003559115-12

time-consuming, prompting researchers to automate detection using machine learning algorithms. Deep learning methods like Convolutional Neural Networks (CNNs) have shown exceptional performance in recognizing 14 key plant diseases, including late blight and early blight, accurately differentiating between healthy and diseased plants based on features like form, texture, and color [1–3].

Various techniques such as K-Nearest Neighbors (K-NN) and Random Forests enhance disease classification accuracy. K-NN efficiently diagnoses potato diseases, while Random Forest combines decision trees to classify diseases with high accuracy. Addressing data scarcity, researchers use data augmentation methods like geometric modifications to expand datasets, improving tomato disease classification [4,5]. Pretrained models such as VGG16 and ResNet, fine-tuned on plant disease datasets, outperform traditional methods in disease classification. Evaluation metrics like accuracy, precision, recall, and F1-score are crucial for assessing model effectiveness. These metrics measure overall correctness, precision of positive predictions, recall of positive events, and a combined measure of precision and recall, respectively. Machine learning advancements in disease classification promise improved agricultural productivity by enabling early disease detection and precise treatment, reducing crop losses and pesticide use [6].

## 2 METHODOLOGY

Our work utilizes machine learning algorithms to identify and diagnose plant diseases, focusing on late and early blight in tomatoes and potatoes. We sourced a dataset from Kaggle, known as "Plant Village," which included 1,700 images of tomato diseases and 1,800 images of potato diseases. To ensure research reliability, we split the dataset into two equal parts as shown in Table 1. The majority of images were of potato plants, while the smaller portion contained only images of tomatoes. This division allowed us to handle the structural and disease profiles of the two species separately. Figure 1 shows an example of the dataset. A disease can be evaluated using both infected and uninfected specimens of plants. Our objective was to employ machine learning algorithms to identify and classify late and early blight in each plant. We used suitable image recognition and classification methods for sophisticated algorithms and models. To achieve high accuracy, we trained machine learning models on the dataset, conducting extensive inspections and tweaking.

Figure 1. Plant disease obtained from tomato and potato plant.

Table 1. Total dataset used in this research.

| Dataset | Tomato | Potato |
|---|---|---|
| **Dataset 1** | 1700 | 0 |
| **Dataset 2** | 0 | 1800 |
| **Healthy leaves** | 100 | 100 |

Machine learning techniques utilized include K-Nearest Neighbors (K-NN), Random Forest, and Artificial Neural Networks (ANN), along with cross-validation. K-NN is a non-parametric pattern recognition classification method that sorts plant images into categories by comparing distances between test and training images. Random Forest increases prediction accuracy by combining multiple decision trees, and our Random Forest-based classifier outperformed previous models. ANN, modeled after the human brain, consists of densely connected synthetic neurons and showed good efficacy in recognizing complex patterns. We employed ANN to build a predictive model for plant diseases, developing a deep neural network architecture with numerous hidden layers to capture subtle information from plant images. The ANN was trained on a labeled dataset to recognize illness patterns, with carefully selected activation functions and dimensions like learning rate and batch size.

The efficacy of the machine learning algorithms was measured using accuracy, precision, recall, and F1-score. Accuracy measures overall correctness, precision measures the percentage of correctly predicted positive events, recall assesses the ability to identify positive events, and the F1-score combines accuracy and recall. We conducted simulations using cross-validation, splitting the dataset into training and testing sets, and used stratified sampling to adjust for class imbalance. Image preprocessing included normalization and dataset segmentation, as shown in Figure 2. We partitioned and normalized the Plant Village dataset into training and testing subsets, standardizing pixel values across images to achieve consistent and comparable feature parameters. Data augmentation techniques, such as rotations, flips, and zooms, were applied to enhance the dataset and improve model resistance to real-world variations.

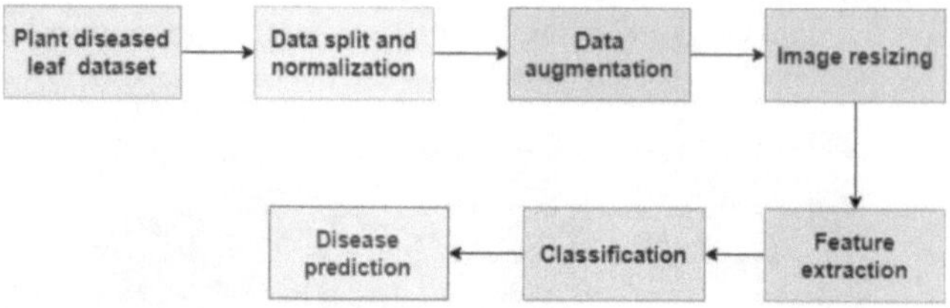


Figure 2. Image preprocessing.

Image scaling ensured consistent input photo sizes for the models, simplifying subsequent feature extraction and classification operations. We used feature extraction models like VGG-16, Inception, and ResNet for their strong capabilities, focusing on intermediate output characteristics to obtain accurate representations of plant images. Classification and disease forecasting followed feature extraction. We developed models using support vector machines, random forests, and feedforward neural networks to classify plant photos by disease. The models detected patterns corresponding to various illness strains, predicting whether a potato or tomato plant harbored late or early blight. We optimized parameters through grid search and cross-validation. We evaluated the models with a test subset, using accuracy, precision, recall, and F1-score to assess their ability to discriminate between healthy and diseased plants. Confusion matrices were also studied to identify misclassified instances, aiming to develop a reliable prediction system for plant diseases. Our comprehensive methodology addresses dataset partitioning, normalization, data augmentation, image resizing, feature extraction, classification, and disease prediction in tomato and potato plants, providing a systematic approach to detecting and preventing late and early blight.

## 3 RESULT AND DISCUSSION

In this work, the machine learning models used were developed in and trained with MATLAB. When the efficiency of each model was examined, we found that yes it is very good at distinguishing tomato plants and potato plants, for example, indeed their accuracy in predicting disease had been shown as high. Random Forest had 97.8% accuracy; ANN was 98.5%; K-Nearest Neighbors (K-NN), 96.4%. With an accuracy rate of 97.8%, the Random Forest model had excellent forecasting ability in plant disease diagnosis. The reason for this high percentage is the ensemble learning technique of random forest. The union of each decision tree prediction, it included information from each decision tree to provide consistent and accurate forecasts of diseases.

The ANN model achieved an improved accuracy of 98.5%. It proves the ANN's deep neural network design can identify tiny disease-related traits in the picture of a plant as well as pick out complex patterns. The ANN model got enough–by extracting a series of hierarchical representations in its multiple layers and connected nodes–to accumulate and put into practice a mass quantity of information. Then the predictions were astoundingly accurate. Even among the Random Forest and ANN models, the K-NN model's accuracy rate remained high at 96.4%. This accuracy% of The K-NN algorithm, a bit lower than its counterparts, indicates its accuracy. In addition, using the gaps between feature vectors the K-NN model also found plant photos by their similarity to the training instances. There was something about the nature of the K-NN approach. If the choice of the parameter K–how many nearest neighbors to analyse–mattered the predictions may have changed because the much lower accuracy. Given their high overall accuracy rates, the three models (Random Forest, ANN, and K-NN) might serve as effective tools for detecting late and early blight infections in tomato and potato plants at an early stage. Our findings say that using machine learning methods would significantly improve plant pathology research, helping farmers and other agricultural stakeholders to quickly recognise and treat plant maladies. In terms of accuracy, it outperformed the Random Forest and K-NN models. It is found that the ANN model's flaming firepower in recognising complex features and patterns–given what you see on this page–helpful with disease prediction.

In Figure 3 we see metrics measuring the performance of machine learning models. This includes accuracy, precision, recall, and F1 score. That is, these metrics from a machine learning model provide a comprehensive picture of the ability to identify diseases in tomato and potato plants. Accuracy ratings reveal just how accurate a model's overall forecasts were. The Random Forest and Artificial Neural Network models had very high accuracy - 98.5% and 97.8%, respectively; while K-Nearest Neighbors had somewhat lower accuracy, at a mere 96.4%. They revealed that all of the models were able to separate healthy trees from those who were ill. Precision, one of these metrics, is the ratio of success rate for events correctly estimated (i.e., identifying diseased plants). The accuracy ratings indicate that the Artificial Neural Network model achieved 98.4% while the Random Forest model scored 97.9%. This shows they can veer away from more false positive results than other models. K-Nearest Neighbors has a slightly higher rate of false positives than the other models, with an accuracy rating of 96.5%. recall -that is the ability of a model to find instances of success–is evaluated by sensitivity, also known as recall (i.e., being able to identify diseased plants). The recall rates of the Artificial Neural Network and Random Forest models were 98.6% and 97.7%, respectively, meaning they detected these conditions very sensitive, also known as items. The others mostly came out with 96.3%. Our model was somewhat higher than the others, which is reasonable because the F1-score evaluates models' predictive potential from the standpoint of recall and accuracy. Both the Artificial Neural Network and Random Forest models did well when scored by patients per hundred, as evidenced by their F1-scores. 97.8% are pretty good. Nearest neighbors' F1-score of 96.4% gives a picture of how good the K-Nearest Neighbors model is acheieving its performance.

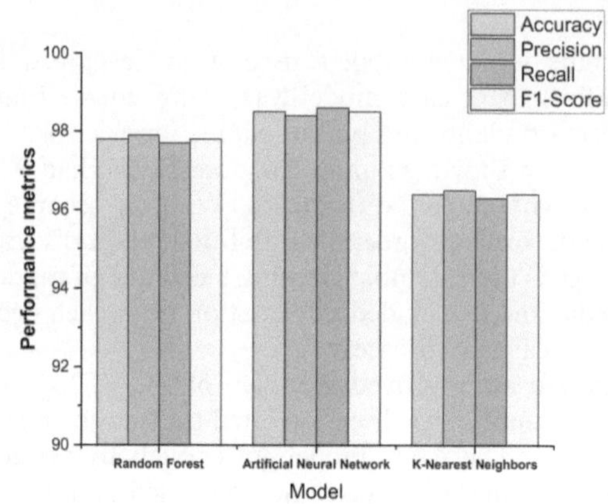


Figure 3. Performance of machine learning models.

In order to assess not only the performance but also the operational efficiency of the development Artificial Neural Network (ANN) model, in the evaluation. A confusion matrix was employed for that purpose. Though the graphic is shown as Figure 4, a confusion matrix calculates the predictive capabilities of a model by diagnosing the results into different classes. The indices allotted to the classes which are of interest (i.e. early blight, late blight and healthy leaves) are enumerated as those. True positives in the confusion matrix are those in which the algorithm's correct identifications of early blight in addition to late blight and healthy leaves is positively correlated. The algorithm's ability to quickly steer clear of such cases was what we thought possible and desirable. And of course the true negative cases identified in the confusion matrix are false negatives. Though the term "true negatives" appears to be an oxymoron in this context, it refers to the model's correct identification of noncases at the category level (early blight and late blight). Minimising false alarms, these accurate negative predictions are necessary, ensuring that the model is specific in identifying healthy leaves.

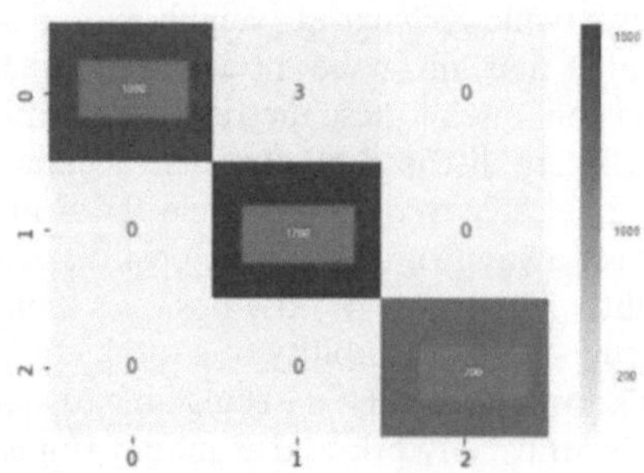


Figure 4. Confusion matrix.

## 4 CONCLUSION

In our study, we used machine learning algorithms to protect potato and tomato plants from late and early blight. These methods demonstrated high accuracy in identifying plant diseases, with the Random Forest model at 97.8%, the Artificial Neural Network at 98.5%, and K-Nearest Neighbors at 96.4%. Using a dataset from Kaggle's Plant Village, we partitioned and normalized data for high-quality training and testing. This approach enables early disease diagnosis, improving agricultural productivity by reducing crop losses. Incorporating machine learning helps farmers enhance disease control, improve crop health, and reduce pesticide reliance, leading to higher productivity.

## REFERENCES


[1] Li, Z., Ding, L. and Xu, D. (2022). Exploring the potential role of environmental and multi-source satellite data in crop yield prediction across Northeast China. *Science of the Total Environment*, 815, 152880. https://doi.org/10.1016/j.scitotenv.2021.152880

[2] Nallappan, D., Fauzi, A. N., Krishna, B. S., Kumar, B. P., Reddy, A. V. K., Syed, T., Reddy, C. S., Yaacob, N. S. and Rao, P. V. (2021). Green Biosynthesis, Antioxidant, Antibacterial, and Anticancer Activities of Silver Nanoparticles of Luffa acutangula Leaf Extract. *BioMed Research International*, 2021. https://doi.org/10.1155/2021/5125681

[3] Gomez Selvaraj, M., Vergara, A., Montenegro, F., Alonso Ruiz, H., Safari, N., Raymaekers, D., Ocimati, W., Ntamwira, J., Tits, L., Omondi, A. B. and Blomme, G. (2020). Detection of banana plants and their major diseases through aerial images and machine learning methods: A case study in DR Congo and Republic of Benin. *ISPRS Journal of Photogrammetry and Remote Sensing*, *169*(April), 110–124. https://doi.org/10.1016/j.isprsjprs.2020.08.025

[4] Barriga, J. A., Blanco-Cipollone, F., Trigo-Córdoba, E., García-Tejero, I. and Clemente, P. J. (2022). Crop-water assessment in Citrus (Citrus sinensis L.) based on continuous measurements of leaf-turgor pressure using machine learning and IoT. *Expert Systems with Applications*, 209(July), 118255. https://doi.org/10.1016/j.eswa.2022.118255

[5] Rau, A. J., Sankar, J., Mohan, A. R., Das Krishna, D. and Mathew, J. (2017). IoT based smart irrigation system and nutrient detection with disease analysis. *TENSYMP 2017 - IEEE International Symposium on Technologies for Smart Cities*, 3–6. https://doi.org/10.1109/TENCONSpring.2017.8070100

[6] Phasinam, K., Kassanuk, T. and Shabaz, M. (2022). Applicability of Internet of Things in Smart Farming. *Journal of Food Quality*, 2022. https://doi.org/10.1155/2022/7692922

*Hybrid and Advanced Technologies – S. Prasad Jones Christydass, Nurhayati Nurhayati and S. Kannadhasan. (Eds)* © *2025 the Authors, ISBN 978-1-032-90254-8*


# Improving groundnut crop production in India using machine learning algorithms

Jampani Chandra Sekhar & Praveen Gundla
*Department of Computer Science and Engineering, N.R.I. Institute of Technology, Guntur, Andhra Pradesh, India*

D. Praveena
*Department of Information Technology, R.M.D. Engineering College, Chennai, Tamil Nadu, India*

E. Shankar
*Department of Mechanical Engineering, Rajalakshmi Engineering College, Thandalam, Chennai, Tamil Nadu, India*

G. Mahendran
*Department of Mechanical Engineering, R.M.K. Engineering College, Kavaraipettai, Chennai, Tamil Nadu, India*

Ramya Maranan
*Department of Research and Innovation, Saveetha School of Engineering, SIMATS, Chennai, Tamil Nadu, India*

ABSTRACT: A significant oilseed crop in India, groundnut (Arachis hypogaea L.) is extensively grown in a number of states, including Gujarat, Tamil Nadu, Andhra Pradesh, and Karnataka. For small farmers, groundnuts are a critical cash crop that helps them improve their standard of living. It is a massive contributor to the Indian economy and a vital source of minerals, oil, and protein. This research used machine learning algorithms to forecast the production of a ground crop. Eight columns made up the dataset utilised for this study: state, district, crop, crop year, crop area, production, and yield. Three algorithms—linear regression, support-vector-machine (SVM), and artificial neural network (ANN)—were trained and assessed after the data had been preprocessed. R-squared and root mean square error (RMSE) were used to assess each algorithm's performance. The findings showed that the ANN algorithm outperformed the other two algorithms in forecasting yield over the past five years, with the lowest RMSE and highest R-squared. In order to precisely anticipate groundnut production in India, this research demonstrates the potential of machine learning algorithms, especially ANN. This might help small farmers improve their quality of life and boost the Indian economy.

*Keywords*: Machine learning, Artificial Intelligence, Neural Network, Support vector machine

## 1 INTRODUCTION

Agriculture's ability to produce food is essential for feeding the world's growing population. However, challenges like climate change, soil degradation, pests, and diseases can hinder yield increases. Farmers often struggle with these factors, leading to reduced yields, lower income, and food insecurity. Groundnut (Arachis hypogaea L.) is a key oilseed crop extensively grown in Indian states such as Gujarat, Tamil Nadu, Andhra Pradesh, and

 DOI: 10.1201/9781003559115-13

Karnataka [1–3]. It significantly contributes to the Indian economy and serves as a vital income source for small farmers. Nonetheless, unpredictable agricultural outputs pose significant risks to these farmers.

Recent research has explored machine learning (ML) algorithms to predict crop productivity. Studies have demonstrated the effectiveness of ML in forecasting yields for various crops, including rice, maize, and cotton. For example, Fathima *et al.* used decision trees, support vector regression, and random forest for rice yield prediction, while K. S. Aravind *et al.* utilized ML algorithms and remote sensing data for maize. These studies highlight ML's potential in yield prediction, emphasizing the need for further research. ML techniques like linear regression (LR), support-vector-machine (SVM), and artificial neural networks (ANN) have shown promise in predicting agricultural yields by using historical yield data, weather patterns, and soil conditions. Gornall J. *et al.* discussed the significant uncertainties in crop production forecasts due to climate change, CO2 fertilization, and O3 damage, which complicate global agricultural productivity assessments [4–6].

Saravanan *et al.* developed hybrid ML techniques, including deep neural networks, to estimate crop yields, showing that hybrid models like LSAE-CYPDNN outperform others in terms of error minimization. Roja *et al.* used climate data to calculate irrigation requirements for groundnuts, demonstrating the utility of models like CROPWAT 8.0 for accurate irrigation scheduling. Pant *et al.* highlighted the use of statistical ML strategies for forecasting yields of major crops in India. This study employs ML algorithms to predict groundnut crop production using data from various columns, such as state, district, crop year, and area. Evaluating three ML algorithms—LR, SVM, and ANN—revealed that ANN outperformed the others in forecasting groundnut yields over the past five years. This underscores the potential of ML, particularly ANN, in minimizing agricultural risks and supporting farmers [7,8].

## 2 METHODOLOGY

This chapter details the data preparation process for historical agricultural production statistics in India's state and district records from 1995 to 2020. The dataset contains seven rows and 345,336 columns, including state_name, district_name, crop_year, season, crop_name, area, and production, focusing solely on peanuts for this experiment. Several preprocessing steps were performed: duplicates were removed, missing data filled in, and categorical variables converted to numeric values using factorise(). The 'Crop_Year' column was formatted as DateTime, and a new 'Year' column was created. Extraneous columns like "Season" were removed, and data was normalized so that remaining columns had consistent scales. The cleaned data was saved as a new CSV file, with values in 'Area' and 'Production' normalized to a mean of zero and a standard deviation of one. After preprocessing, the data was split into training and test sets, with the last five years reserved for testing. The final dataset had 12,586 rows, including State, District, Year, Area, Production, and Yield. These steps ensured a clean and structured dataset ready for further analysis and modeling.

For the ANN architecture, a Feed-Forward Back Propagation (FFBP) network was used. Data was inputted into the input layer, passed through hidden layers, and outputted through the output layer, with errors adjusted during backpropagation to minimize them over several epochs. The ANN used in this study had 11 neurons across three layers: an input layer, an output layer, and two hidden layers. The hidden layer had 10 neurons, fully connected to both input and output layers. The network was trained using the Levenberg-Marquardt method, with the goal of reducing the difference between anticipated and actual peanut yields. Regression and classification tasks are common applications for ANN architectures like FFBP due to their flexibility. The network's design can be adjusted, such as the number of layers and neurons, to optimize performance. The study highlights the ANN model architecture's specifications, replicated in MATLAB, with inputs normalized and used for training to generate results with minimal error.

The performance of the ANN model was evaluated using correlation and root-mean-squared error (RMSE). Optimal hidden layer neurons were determined by minimizing training error while avoiding overfitting. The network structure for the multi-output ANN model was 5:10:6, and for the single-output ANN model, 11 tuned neurons were used. These models were assessed using correlation and RMSE metrics. Support Vector Machines (SVM) were also used for regression problems. SVM aims to find a hyperplane with the most significant boundary while minimizing training errors. The SVM regression objective function includes a regularization term to prevent overfitting, solved using the Sequential-Minimal-Optimization (SMO) algorithm. This algorithm optimizes two Lagrange multipliers at a time, expressing the problem as a quadratic optimization with constraints. Linear regression was another method used to model the relationship between a dependent variable and independent variables. The fitlm function in MATLAB estimated a predictive linear regression model that minimized the total difference between forecasted and actual values. The model was expressed as Y = b0 + b1X1 + b2X2 + ... + bnXn + $\varepsilon$, where the fitlm function optimizes coefficients to minimize the sum of squared errors. This function can also fit a quadratic model to capture nonlinear relationships between variables.

## 3 METHODOLOGY

Figure 1 shows a moderate linear relationship between the dependent variable (groundnut yield in India) and an independent variable (State, District, Year, Area, Production); this is clear from the regression plot for the archetypal. The regression line is plotted along with the scatterplot to give a visual picture of the relationship between the actual and estimated values. For a more complex nonlinear relationship between the variables to be captured, the inclusion of the quadratic term is an advantage of the archetypal.

The regression equation for the linear regression archetypal with a quadratic term is:

$$\begin{aligned}\text{Yield} = & -2.253 + 0.7 * \text{State} + 1.1 * \text{District} - 14.5 * \text{Year} + 0.2 * \text{Area} + 0.5 * \text{Production} \\ & - 0.6 * \text{State} * \text{District} - 5.5 * \text{State} * \text{Year} + 0.3 * \text{State} * \text{Area} + 0.1 * \text{State} * \text{Production} \\ & - 17.5 * \text{District} * \text{Year} - 2.5 * \text{District} * \text{Area} + 0.3 * \text{District} * \text{Production} \\ & + 23.5 * \text{Year} * \text{Area} - 1.2 * \text{Year} * \text{Production} - 0.7 * \text{Area} * \text{Production} - 2.5 * \text{State}^2 \\ & + 1.5 * \text{District}^2 + 64.5 * \text{Year}^2 + 1.5 * \text{Area}^2 - 0.05 * \text{Production}^2\end{aligned} \tag{1}$$

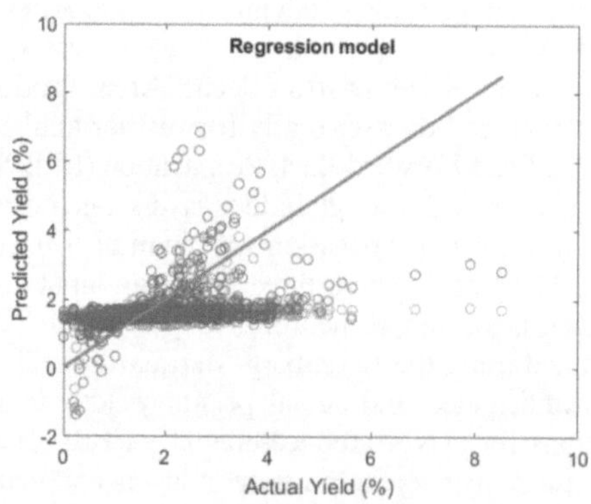


Figure 1. Actual and forecast values for linear regression archetypal.

This equation shows the relationship between the dependent variable (yield) and the independent variables (rainfall, temperature, and fertiliser use), as well as the quadratic terms for each independent variable. The coefficients associated with each variable represent the change in yield associated with a one-unit increase in the independent variable while holding all other variables constant. Overall, the regression plot and regression equation for the linear regression archetypal with a quadratic term demonstrates that linear regression with a quadratic archetypal can provide moderately accurate predictions of groundnut yield in India.

The SVM performed poorly, it can be observed from Figure 2 in the regression plot and regression equation as well in Eq. 14. The regression plot showed that the forecast values are far away from the actual values, with much variability and a poor fit of the line. The regression equation showed a high error rate, with large values of RMSE and low values of R-squared values.

It can be concluded from this study that SVM, at least with the input variable set used in this case, is not ideal for predicting the groundnut without India. It is possible that the performance of the SVM model could be improved by using different input variables or by tuning the hyperparameters.

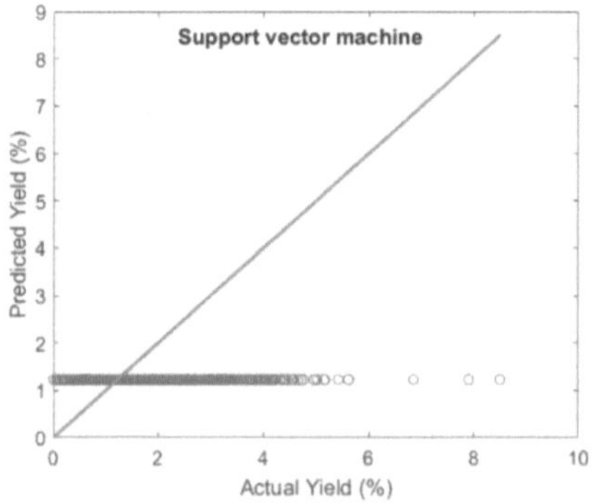


Figure 2. Actual and forecast values for SVM archetypal.

The regression equation of the machine learning model is shown in below equations

$$
\begin{aligned}
\text{Yield} = & -1.69615 + 6.43225 * \text{State} + 1.52619 * \text{District} - 17.58327 * \text{Year} + 0.765975 * \text{Area} \\
& + 0.783116 * \text{Production} - 0.848309 * \text{State} * \text{District} - 6.22866 * \text{State} * \text{Year} \\
& + 0.408385 * \text{State} * \text{Area} + 0.054821 * \text{State} * \text{Production} - 20.35899 * \text{District} * \text{Year} \\
& - 3.25603 * \text{District} * \text{Area} + 0.374262 * \text{District} * \text{Production} + 29.37968 * \text{Year} * \text{Area} \\
& - 1.38609 * \text{Year} * \text{Production} - 0.830563 * \text{Area} * \text{Production} - 4.37902 * \text{State}^2 \\
& + 2.35053 * \text{District}^2 + 80.23418 * \text{Year}^2 + 2.15105 * \text{Area}^2 - 0.075713 * \text{Production}^2
\end{aligned} \tag{2}
$$

Despite these limitations, the poor performance of the archetypal in this study suggests that the SVM may not be a useful tool for predicting groundnut yields in India. But it's worth noting that archetypal is complicated and requires careful parameter tuning, which may have affected the archetypal performance in our study.

The findings of this study show that artificial neural networks (ANN) outperform linear regression of archetypal and support-vector machines (SVM) in predicting groundnut production in India. The ANN archetypal yielded an overall R-squared value of 0.90461 based on the plot from Figure 3. The train R was equal to 0.9619, the validation R came out as 0.87475, and the test R was 0.9651. These results indicate that the ANN explains a high percentage of the yield data's variation and predicts yields with high accuracy.

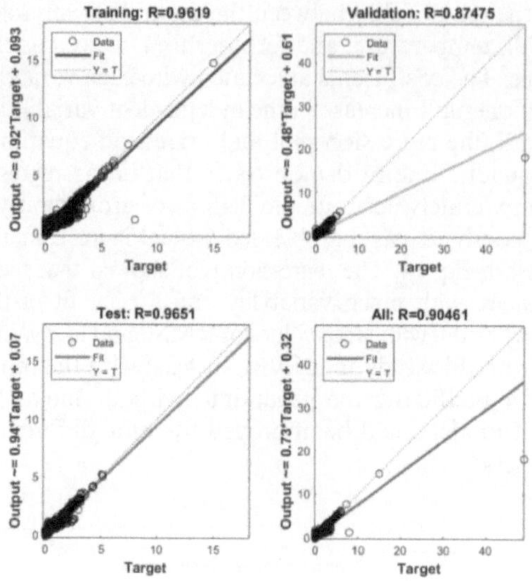


Figure 3. Regression plot for ANN archetypal.

In the ANN regression plot in Figure 4, you will note a strong positive correlation between expected and actual yield values. With the slightest difference between the forecast values and the observed values, it can be seen that the ANN model's predictions are very accurate. Compared to linear regression and SVM, the ANN archetypal has a much better fit.

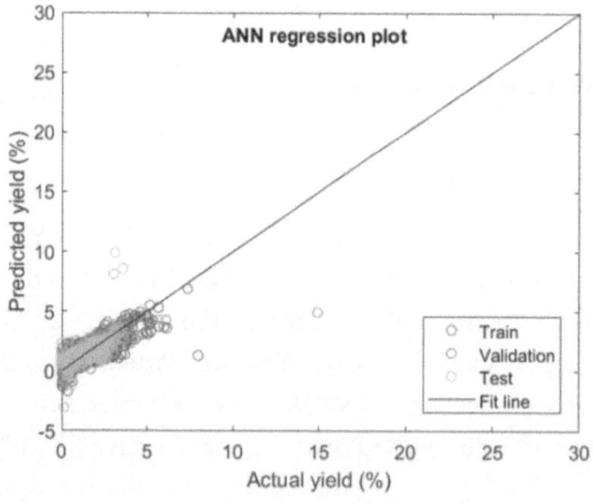


Figure 4. Actual and forecast values for ANN archetypal.

The performance of the ANN model is believed to be due to its capacity to capture intricate nonlinear relationships between the input factors and the yield output. The multi-layered structure with hidden layers can learn and extract intricate patterns in the data to achieve a more accurate yield prediction. Also, the ANN model supports the inclusion of a wide range of input variables in order to capture the multiple factors that influence the groundnut yield.

On both the train and validation/test sets, the archetypal represented by artificial neural networks was superior to the other two. The outcome parameters of the linear regression, support-vector-machine, and artificial neural network models with a 0.1850 RMSE on the train set and a 0.1648 on the test set with an R2 value of 0.9521 test set were as follows.

The linear regression archetypal performed moderately well, with an R2 value of 0.81426 on the train set and 0.64675 on the test set and an RMSE value of 0.73953 on the train set and 0.81135 on the test set. It should be noted that the R2 value dropped significantly when moving to the test set, suggesting that the model may not be applicable to new data. The support-vector-machine type model had an R2 value of 0.6981 on the training set and 0.7332 on the test set, with an RMSE value of 1.2348 on the train set and 0.9143 on the test set. Its performance was better than that of the linear regression model, though still not as good as the artificial neural network model.

## 4 CONCLUSION

The best predictions about the dependent variable came from the artificial neural network (ANN) archetypal, then from the support-vector-machine (SVM) archetypal, and finally from the linear regression archetypal. Reliable performance measures bear this out. However, a different model should be decided upon because several factors relating to data properties, computer resources, data interpretability, and the required level should matter most for glucometers. The other is the particular case you are dealing with. This research concludes that machine learning algorithms such as ANN have the potential to forecast India's groundnut crop yield with great accuracy. If achieved, poor farmers would be able to improve their standard of living while contributing to the Indian economy at the same time.

## REFERENCES


[1] Jain, R. K. (2023). Experimental performance of smart IoT-enabled drip irrigation system using and controlled through web-based applications. *Smart Agricultural Technology*, 4(May 2022), 100215. https://doi.org/10.1016/j.atech.2023.100215

[2] Neelakantan. P. (2022). Analyzing the best machine learning algorithm for plant disease classification. *Materials Today: Proceedings*, xxxx. https://doi.org/10.1016/j.matpr.2021.07.358

[3] Li, Z., Ding, L. and Xu, D. (2022). Exploring the potential role of environmental and multi-source satellite data in crop yield prediction across Northeast China. *Science of the Total Environment*, 815, 152880. https://doi.org/10.1016/j.scitotenv.2021.152880

[4] Mian, A. N., Waqas Haider Shah, S., Manzoor, S., Said, A., Heimerl, K. and Crowcroft, J. (2022). A value-added IoT service for cellular networks using federated learning. *Computer Networks*, *213*(May), 109094. https://doi.org/10.1016/j.comnet.2022.109094

[5] Barriga, J. A., Blanco-Cipollone, F., Trigo-Córdoba, E., García-Tejero, I. and Clemente, P. J. (2022). Crop-water assessment in Citrus (Citrus sinensis L.) based on continuous measurements of leaf-turgor pressure using machine learning and IoT. *Expert Systems with Applications*, 209(July), 118255. https://doi.org/10.1016/j.eswa.2022.118255

[6] Koteish, K., Harb, H., Dbouk, M., Zaki, C. and Abou Jaoude, C. (2022). AGRO: A smart sensing and decision-making mechanism for real-time agriculture monitoring. *Journal of King Saud University - Computer and Information Sciences*, 34(9), 7059–7069. https://doi.org/10.1016/j.jksuci.2022.06.017

[7] Jackulin, C. and Murugavalli, S. (2022). A comprehensive review on detection of plant disease using machine learning and deep learning approaches. *Measurement: Sensors*, 24(August), 100441. https://doi.org/10.1016/j.measen.2022.100441

[8] Rokade, A., Singh, M., Arora, S. K. and Nizeyimana, E. (2022). IOT-Based Medical Informatics Farming System with Predictive Data Analytics Using Supervised Machine Learning Algorithms. *Computational and Mathematical Methods in Medicine*, 2022. https://doi.org/10.1155/2022/8434966

*Hybrid and Advanced Technologies – S. Prasad Jones Christydass, Nurhayati Nurhayati and S. Kannadhasan. (Eds)* © *2025 the Authors, ISBN 978-1-032-90254-8*


# Smart irrigation optimization through the integration of machine learning and IoT for sustainable agriculture

S. Pradeep
*Department of Computer Science and Engineering, Malla Reddy Engineering College for Women (UGC Autonomous), Maisammguda, Kompally, Secunderabad, Telangana, India*

L.B. Shalom Irence
*Department of Electrical and Electronics Engineering, Karunya Institute of Technology and Sciences, Coimbatore, Tamil Nadu, India*

G. Mahendran
*Department of Mechanical Engineering, R.M.K. Engineering College, Kavaraipettai, Chennai, Tamil Nadu, India*

Yogesh Shivaji Pawar
*Department of Electrical Engineering, Loknete Gopinathaji Munde Institute of Engineering Education and Research, Nashik, Maharashtra, India*

S. Prabagaran
*Department of Mechanical Engineering, Faculty of Engineering, Karpagam Academy of Higher Education, Coimbatore, Tamil Nadu, India*

Ramya Maranan
*Department of Research and Innovation, Saveetha School of Engineering, SIMATS, Chennai, Tamil Nadu, India*

ABSTRACT: This research aims to create a machine learning and Internet of Things -based strategy for smart irrigation systems, which would allow for the individualized monitoring and management of the process. Efficient and responsible use of water in agriculture should be a top priority for resource preservation and the sustainable development of crops. Current approaches to irrigation are generally inaccurate, imprecise, and do not involve real-time monitoring, which leads to the unnecessary waste of water, and poor agricultural yields. To build a new, more reliable, and efficient solution, this study proposes a novel approach, which combines machine learning technologies with Internet-of-Things technologies. The latter involves the use of multiple sensors measuring temperature, humidity soil properties, and other parameters over a real-time basis. This site aims to upgrade such systems using machine learning technologies, which develop and train models that would monitor these data and make respective irrigation judgments. As a result, the system would allow for real-time monitoring conserving water through optimized irrigation practices: a plan for each type of crop would be unique and based on their respective yields and the state of soil into which they are sewn. Machine learning algorithms are used to recognize the patterns and make predictions on when next to implement the irrigation. The IoT infrastructure allows for all sensors, microcontrollers, and irrigation pumps to communicate with one another, no transmission of data is missed. The latter ones go on to make forecasts using machine learning and automatically adjust the water supply and irrigation schedules. This process is monitored and managed remotely by the customers with the use of mobile or computer applications. The smart irrigation system is likely to benefit from the proposed machine learning and IoT-based

 DOI: 10.1201/9781003559115-14

approach resulting in the hardly optimal water use–more yield per the amount of crop grown, and remote-monitorability.

*Keywords*: Sensor, IoT, Solar energy, Energy management

## 1 INTRODUCTION

In recent years, intelligent watering systems integrating machine learning and the Internet of Things (IoT) have gained significant public interest. These systems manage agricultural water with precision, addressing water scarcity, environmental issues, and enhancing crop production. This literature review surveys machine learning and IoT technologies in smart irrigation, highlighting their components, operational principles, and transformative effects on farming practices. Smart irrigation systems using machine learning and IoT have emerged as efficient solutions to improve water usage and agricultural productivity. They monitor environmental conditions through real-time sensor data, such as temperature, humidity, and soil moisture. Machine learning algorithms analyze this data, enabling real-time decision-making for accurate irrigation schedules and optimized water usage. IoT technologies facilitate remote monitoring and control, enhancing the efficiency of modern irrigation technology [1,2].

Compared to traditional methods, these systems conserve water by delivering precise amounts based on real-time data, thereby minimizing waste. They also enhance crop yield and productivity by fine-tuning irrigation schedules to promote healthy plant growth, considering variables like soil moisture, temperature, and crop type. Additionally, these systems offer energy efficiency, contributing to intelligent irrigation and providing fresh water sources globally. Components of these systems include sensors, IoT devices, data analytics, and actuators. Sensors monitor environmental parameters, supplying real-time data for machine learning algorithms, which determine optimal watering times and amounts. Valves or irrigation pumps, regulated by these algorithms, distribute water with precision to target crops [3,4].

Smart irrigation systems have revolutionized commercial farm management, ensuring fair water allocation and improving food safety. They support sustainable urbanization by reducing water and chemical fertilizer waste. However, challenges like data security, device compatibility, sensor data correction, and user-friendliness must be addressed for widespread adoption. Tailoring algorithms to different crops and conditions is also crucial. This study explores a machine learning and IoT-aided irrigation system for better water use in agriculture. By gathering real-time data and using machine learning algorithms, the system enables environment-specific irrigation control. IoT technology allows long-distance monitoring, saving water and increasing crop yields, promoting sustainable agricultural practices. This research aims to enhance water use efficiency, optimize irrigation practices, and boost productivity for a greener agricultural sector [5,6].

## 2 METHODOLOGY

This study focuses on developing a solar-powered precision fertigation system to address the challenges of inefficient irrigation and high energy costs in agriculture, particularly in underdeveloped regions like India. Traditional methods often result in over- or under-irrigation, leading to reduced crop health and yield. The proposed system aims to optimize irrigation practices by harnessing solar energy through panels that convert sunlight into electricity, stored in batteries to power an irrigation pump. Key components include various sensors (temperature, moisture, humidity) integrated with an Arduino microcontroller. These sensors continuously monitor environmental conditions and transmit real-time data to the microcontroller. Based on sensor readings, the microcontroller activates or deactivates the pump to ensure precise irrigation levels, preventing both water wastage and crop damage.

The methodology employs solar cells for energy generation and an Arduino microcontroller for data processing and control. Sensors play a critical role in providing accurate data for decision-making, adjusting irrigation schedules based on soil and atmospheric conditions. This automated process optimizes water usage, promoting healthier crop growth without the need for excessive energy consumption. Figure 1 illustrates the architecture of the proposed solar-powered smart irrigation system, emphasizing its reliance on solar energy and sensor-driven automation. Additionally, a mobile interface allows remote monitoring and manual intervention if needed, providing farmers with real-time updates on soil moisture levels and system operations.

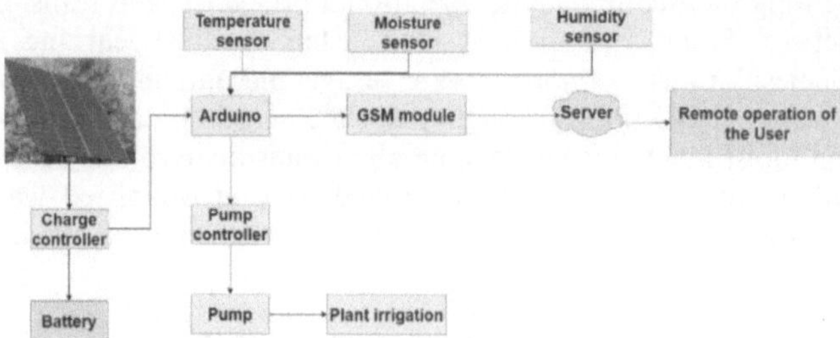


Figure 1. Architecture of proposed research.

## 3 RESULT AND DISCUSSION

The solar panel in Figure 1 plays an important role by using sunlight to convert solar energy into electricity, necessary for operating the system. Afterward this electric power is then used to recharge the battery and the series of sensors serves as a big hands working for the pump operation. Set points were established for the prototype at particular temperature, humidity, and soil moisture content. These set points define certain ranges in which the system will operate best. Specifically for temperature the set point range is between 25 to 32 degrees Celsius; the range of calibrated values is such as to produce favorable growth conditions for crops, because excessive heat or cold will disrupt plant development. Likewise, 75% to 85% is the set point for relative humidity. This range keeps the air in a good state of moisture for plant transpiration and health. The set point for soil moisture content is set at 10 cm in water saturation. This level of water in the soil will meet the water needs of the crops and promote the growth of their root systems.

The prototype integrates sensors that monitor environmental conditions continuously, and provide real-time data to the Arduino microcontroller. Temperature, humidity, and soil moisture readings are among the data. A microcontroller compares the readings to predefined set points to see if the conditions are within acceptable ranges. If the reading falls outside of the optimal range and irrigation is needed, the microcontroller turns on the pump to irrigate the field. The prototype's functioning is demonstrated in the activation of the pump when the readings fall off desired parameters. By switching on the pump, the system guarantees that the crops receive enough water to survive development and growth. This proactive approach to irrigation management can prevent under-irrigation, which produces stunted growth and reduced yields. On the flipside, the system avoids over-irrigating by only switching on the pump when needed so water is not wasted. The lamella method not only avoids waste water but causes no harm to crops, which are not going to have energy left. To inform the user clearly and easily, this prototype has an LCD display. It visually shows what action the microcontroller is taking in response to sensor readings. Users can follow the irrigation process through this interface and also keep abreast of system operations. The prototype communicates using 4G technology. This feature allows the irrigation system to be remotely controlled and monitored. Users can access all-around real-time information about sensor readings, irrigation status, and system operations from their mobile devices. This remote access empowers people to act on their knowledge and requirements in providing manual corrections should they need to do so.

The above table shows ten readings from each sensor as well as pump operation status. This Table 1 gives a valuable glimpse into the operational status of the system and the different responses it is capable of to diverse environmental conditions. The first column, "Sample," represents the sequential number for each data entry, a place marker distinguishing individual readings and maintaining a chronological order of recorded observations at intervals over time. The following columns are for the different sensors on each sample. These are temperature and humidity in degrees Celsius (°C), as well as soil moisture in centimeters (cm). Each measurement extracts from the time at which data was recorded specific details about existing conditions around it.

Table 1. Sensor reading and pump activation.

| Sample | Temperature (°C) | Humidity (%) | Soil Moisture (cm) | Pump Operation |
|---|---|---|---|---|
| 1 | 28.5 | 80 | 9.8 | Activated |
| 2 | 26.2 | 78 | 10.5 | Activated |
| 3 | 31.8 | 85 | 11.2 | Activated |
| 4 | 27.6 | 75 | 9.2 | Activated |
| 5 | 29.4 | 83 | 10.8 | Activated |
| 6 | 25.9 | 77 | 11.6 | Activated |
| 7 | 30.1 | 81 | 9.6 | Activated |
| 8 | 26.8 | 79 | 10.2 | Activated |
| 9 | 28.7 | 82 | 11.0 | Activated |
| 10 | 27.2 | 76 | 9.4 | Activated |

The "Temperature" column refers to the temperature recorded at the sample. Within the values, there exists an interval from 25.9°C to 31.8°C. The readings show temperature changes under the conditions where data was collected. Temperature controls the growth and development of plants because it determines various physiologic processes within them. The "Humidity" column expresses in percentage the humidity levels recorded at each sample. As set out in the table, this ranges from 75% to 85%. Humidity plays a significant role in regulating transpiration and moisture losses, too. If the humidity is too high, it may affect the ability of plants to exchange water vapor with the air. If substantial, such conditions may even hinder the overall development of the plants. The "Soil Moisture" column in the table is actually the moisture content at the time we took our data, given in centimeters. In that column, values averaged 9.2 cm up to 11.6 cm. Soil moisture determines irrigation needs. Plants must have the necessary amount of water for growth and nutrient absorption. Sufficient soil moistness will insure that plants receive enough water.

The final column, called "Pump Operation," shows the status of an irrigation pump at each sample point. For all instances, pump operation was listed as "Electric." This means that the sensor readings had exceeded the target range, leading to the system activating the irrigation process. In this way, the pumping of the water will provide crops with the proper water for their growth and to maximize production. By examining the table, it is clear that the system is working efficiently. The sensor readings are accurate reflections of the prevailing environmental conditions, and the system activates the pump in response to conditions. This proves that the system is able to deal with varying environmental factors and water crops when they need it.

This research focused on developing a solar-driven smart irrigation system to achieve sustainable crop yield. This design effectively uses solar energy turning it into electrical energy, storing it up in the battery, and transmitting that energy to the crop in the irrigation process. The incorporation of sensors for temperature, humidity, and soil moisture make the irrigation process capable of precise control through monitoring real-time environmental conditions. This system has shown with the results of 10 sample readings for each sensor that it can adapt itself to varying conditions, and engage the pump when necessary. This research provides a practical and environmentally-friendly irrigation management solution to promote sustainable farming. The versatility of the prototype, remote monitoring, and high energy efficiency all indicate its potential to benefit crop growth and save water. If studied more deeply, smart irrigation systems such as this could become widespread. This would help farmers and raise productivity of agriculture.

If we apply our suggestion to agricultural land on a large scale, it would also be possible to work out the energy usage of this proposal by using mathematical models—One must estimate energy efficiency. A common equation for computing energy efficiency is:

$$\text{Energy Efficiency} = (\text{Energy Used for Irrigation}/\text{Total Energy Input from Solar Panels}) * 100\% \quad (1)$$

In this equation, the "Energy Used for Irrigation" represents the amount of energy that the solar powered irrigation system consumes to pump water and control other parts. It can be represented as energy consumption in kilowatt-hours (kWh) or some other appropriate unit. The "Total Energy Input from Solar Panels" is the amount of energy generated by the solar panels during a certain period of time. This also can be expressed in kilowatt-hours (kWh), or any other applicable unit.

We can determine how well the system is converting solar energy into useful energy to be used for irrigation by using the above equations. From Equations 1 and 2, a land area of 2 acres was used to calculate energy consumption. In the conventional way, the results were then compared to the energy consumption of agricultural farms using different methods. The results, shows that the system has saved quite a bit of electricity. The system developed with solar energy and advanced technology results in much lower energy consumption than using conventional approaches. Specifically, it turns solar energy into usable energy for irrigation. In this way, farmers increased their need for conventional energy by using this new system. It does not simply translate into economic benefits, but also helps to reduce greenhouse gas emissions associated with fossil fuel-based electricity generation. Figure 2 together

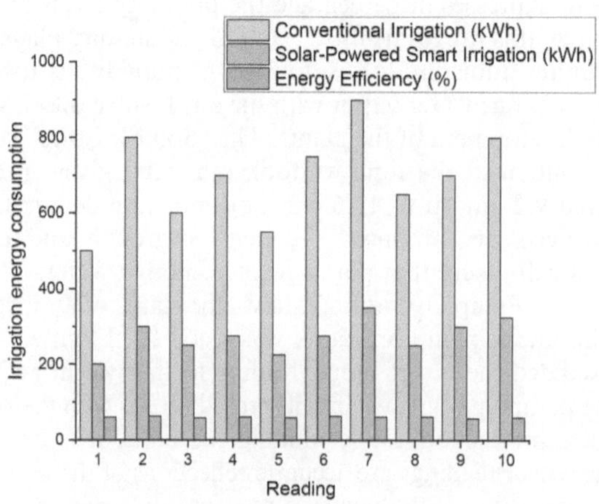


Figure 2. Energy consumption of proposed system.

emphasized the energy-saving benefits of the implemented system. It was only later that we discovered the results of this system implemented in practice: Integrating solar energy along with advanced technology in irrigation can produce very significant energy savings. This, in turn, may make new agricultural practices more consistent with sustainable demands on resources, both energetically and financially.

## 4 CONCLUSION

In conclusion, our research on a solar-powered smart irrigation system has designed a practical method to solve the problems afflicting conventional irrigation methods. A combination of solar cells, batteries, sensors and a microcontroller based on Arduino; through the prototype, it is demonstrated both efficiency in management of irrigation water quantities and also sustainability. Solar energy is used to propel the irrigation process, lessening society's reliance on conventional electricity and promoting environmental protection. By virtue of temperature, humidity, and soil-moisture sensors, real-time monitoring and fine-tuning for precise irrigation control according to preset targets are made possible. The ten sample readings taken from each sensor demonstrate the system's ability to respond to various conditions, turning on the pump when needed in order to maintain an optimal moisture level. The prototype's flexibility and remote monitoring as well as energy conservation point the way for more work on sustainable agricultural practices. In the future, the development of this field could lead to widespread installation of what we now recognize as "intelligent" irrigation systems. The benefits farmers derive include improved yield, conservation of water resources, and several more. On the whole, this research demonstrates the potential of solar-powered smart irrigation system toward changing agricultural methods, contributing to food security and sustainable development in agriculture.

## REFERENCES


[1] Ferehan, N., Haqiq, A. and Ahmad, M. W. (2022). Smart Farming System Based on Intelligent Internet of Things and Predictive Analytics. *Journal of Food Quality*, 2022. https://doi.org/10.1155/2022/7484088

[2] Catalano, C., Paiano, L., Calabrese, F., Cataldo, M., Mancarella, L. and Tommasi, F. (2022). Anomaly detection in smart agriculture systems. *Computers in Industry*, 143(January), 103750. https://doi.org/10.1016/j.compind.2022.103750

[3] Jabbar, W. A., Tiew, L. Y. and Ali Shah, N. Y. (2024). Internet of things enabled parking management system using long range wide area network for smart city. *Internet of Things and Cyber-Physical Systems*, 4(July 2023), 82–98. https://doi.org/10.1016/j.iotcps.2023.09.001

[4] Akbar, M. O., Shahbaz Khan, M. S., Ali, M. J., Hussain, A., Qaiser, G., Pasha, M., Pasha, U., Missen, M. S. and Akhtar, N. (2020). IoT for Development of Smart Dairy Farming. *Journal of Food Quality*, 2020. https://doi.org/10.1155/2020/4242805

[5] Prathibha, S. R., Hongal, A. and Jyothi, M. P. (2017). IOT Based Monitoring System in Smart Agriculture. *Proceedings - 2017 International Conference on Recent Advances in Electronics and Communication Technology, ICRAECT* 2017, 81–84. https://doi.org/10.1109/ICRAECT.2017.52

[6] Gnana, J., Malar, A., Elizabeth, T., Kumar, S. N., Sabu, A., M, A. R., Sengan, S., Sathiamoorthy, J. and Prasanth, A. (2023). Measurement: Sensors Microclimate monitoring system for irrigation water optimization using IoT. *Measurement: Sensors*, 27(March), 100727. https://doi.org/10.1016/j.measen.2023.100727

*Hybrid and Advanced Technologies – S. Prasad Jones Christydass, Nurhayati Nurhayati and S. Kannadhasan. (Eds) © 2025 the Authors, ISBN 978-1-032-90254-8*
*Open Access: www.taylorfrancis.com, CC BY-NC-ND 4.0 license*

# IoT-driven machine learning for precision agriculture for revolutionizing crop management

Sujatha Gaddam
*Department of Computer Applications, Rise Krishna Sai Prakasam Group of Institutions (Autonomous), Vallur, Andhra Pradesh, India*

Anita Singh
*Department of Management, School of Law, Sharda University, Greater Noida, Uttar Pradesh, India*

Aravindan Srinivasan
*Department of Computer Science and Engineering, Koneru Lakshmaiah Education Foundation, Vaddeswaram, Andhra Pradesh, India*

Bishwajeet Prakash
*Department of Economics, School of Law, Sharda University, Greater Noida, Uttar Pradesh, India*

S. Prabagaran
*Department of Mechanical Engineering, Karpagam Academy of Higher Education, Coimbatore, Tamil Nadu, India*

Ramya Maranan
*Department of Research and Innovation, Saveetha School of Engineering, SIMATS, Chennai, Tamil Nadu, India*

ABSTRACT: This research investigates the use of Internet of Things and machine learning technologies in enhancing precision agriculture, having a special focus on crop management, for instance, rice cultivation. In this respect, an IoT-based system is designed, with the use of temperature, humidity, water level, and soil moisture sensors that allow real-time monitoring of the state of the crops. Then, various machine learning models are developed, such as Artificial Neural Networks , Support Vector Machines , Long Short-Term Memory networks, and Decision Trees . The whole dataset comprises 2800 sensor readings that were collected during a month are used to train models and test their performance. In particular, different models can produce diverse results, with some of them leading to higher accuracy in predicting rice water requirements. For example, the best result of 98.10% is found for ANN, followed by 94.5% for LSTM, 92.3% for SVM, and 90.25% for DT. Moreover, the trained model of ANN can be used to provide real-time prediction of rice water needs after being supplemented by a sensor that will transmit the data to the cloud. The transmitted data can then be visualized panel in the application intended to help farmers in making decisions . Overall, the research demonstrates the efficiency of IoT-based machine learning systems in improving irrigation and enhancing crop yield to ensure sustainable agriculture. In the future, the current research can be expanded by replacing the use of other crops or agriculture in general and considering the employment of more complex algorithms.

*Keywords*: precision agriculture, IoT, machine learning, crop management, rice cultivation

## 1 INTRODUCTION

Precision agriculture is a modern farming approach utilizing advanced technologies like the Internet of Things (IoT) and machine learning to optimize crop management and resource use.


DOI: 10.1201/9781003559115-15

This method offers unprecedented opportunities for enhancing global food security and promoting sustainable agriculture [1,2]. Precision agriculture employs real-time data and advanced analytics to improve crop yields through better-targeted watering, fertilizing, and pest control. Technologies like IoT sensors, drones, and smart irrigation systems continuously monitor environmental conditions, soil moisture, crop health, and pests, providing farmers with actionable insights. These technologies enable remote, real-time farm management, replacing traditional methods. Machine learning is integral to precision agriculture, employing methods such as neural networks, support vector machines, and decision trees to analyze data and predict outcomes like crop yield, disease likelihood, and optimal resource allocation. This approach ensures more efficient and sustainable farming practices [3–5].

IoT devices gather extensive data, enabling better resource management and informed decision-making. Automated irrigation systems, for instance, allocate water based on crop needs and weather forecasts, while smart machinery and drones monitor crop growth and detect pests. However, processing this vast data can be challenging, particularly for small farms, which is where machine learning algorithms play a crucial role by providing predictive insights and recommendations. Despite its benefits, precision agriculture faces challenges, including big data management, data protection, device interoperability, and cost [6]. IoT systems can be expensive and not easily scalable, limiting their adoption by small and rural farms. Moreover, the dependence on specific vendors and vulnerability to hacking pose additional hurdles. To address these issues, collaboration among universities, policymakers, and technology developers is essential. Emerging technologies like blockchain, satellite imagery, and unmanned aerial vehicles could further revolutionize agriculture, making precision farming more affordable, accessible, and sustainable. By improving global food security and agricultural resilience to climate change, precision agriculture holds significant promise for the future.

## 2 METHODOLOGY

This research aims to develop an IoT-driven system for precision agriculture to revolutionize crop management. The study's objective is to create a conceptual design and detailed description of the system elements, including sensor selection, data collection, cloud integration, and application development. The system will enable real-time crop monitoring, automatic irrigation based on sensor data, and decision-making visualization. First, sensors such as DPS310 (temperature), DHT11 (humidity), HC-SR04 (water level), and YL-69 (soil moisture) will be selected and deployed in the field to continuously monitor agricultural conditions. These sensors provide essential data streams for assessing and controlling crop environments.

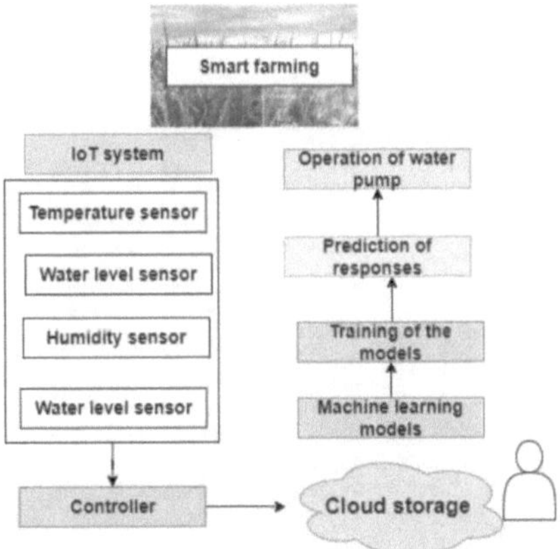


Figure 1. Architecture of proposed research.

The collected data will be transferred to a control unit, which processes it and sends control signals via algorithms. This control node, integrated with the cloud, facilitates communication between sensors and enables real-time system monitoring. Farmers can access data through a cloud-connected application, displaying sensor statuses and allowing decision-making based on real-time information. The app, accessible via web, mobile, or computer, shows the state of various sensors and manages irrigation. When sensor signals indicate a need, the system automatically triggers irrigation, ensuring precise water usage without excess. Figure 1 shows the methodology of the research.

In precision agriculture, predicting water requirements in advance is crucial for optimizing irrigation. This can be achieved through machine learning (ML) models integrated into IoT systems. These models analyze historical sensor data to forecast future water needs, enhancing the system's efficiency. The research involves training various ML models, including Artificial Neural Networks (ANN), Support Vector Machines (SVM), Long Short-Term Memory (LSTM) networks, and Decision Trees (DT). Each model predicts water requirements based on factors like temperature, humidity, and soil moisture. ANN, SVM, and LSTM networks capture complex relationships and provide accurate predictions, while DTs offer clear insights into decision-making processes. The IoT system employs sensors for real-time monitoring, collecting data to train these ML models. This data helps the models learn to predict water needs accurately. Once trained, the models are integrated into the IoT system to predict irrigation parameters, preventing over-irrigation. Preprocessing the dataset involves cleaning, feature selection, and normalization. The data is split, with 70% for training and 30% for testing. This ensures the models are accurate and reliable. The focus crop is rice, a water-intensive plant, making precise irrigation prediction essential for optimal growth.

## 3 RESULT AND DISCUSSION

After training the models on the data, a rigorous evaluation is conducted by testing them. The outcomes of the test shown in Figure 2 reveal that the highest accuracy of 98.10% is achieved by the Artificial Neural Network . To that end, the ANN model proves to be a particularly effective approach for accurate water requirement predictions in terms of rice cultivation based on their sensors. Long Short-Term Memory performs quite closely to the ANN, reaching 94.5% of accuracy due to its ability to account for the long-term sequential data dependencies. Therefore, it can be considered an optimal approach in forecasting the needs for irrigation, though slightly lower than ANN in terms of overall accuracy.

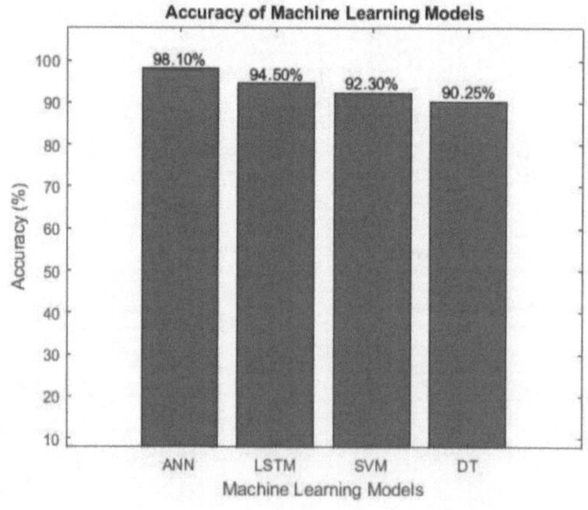


Figure 2. Accuracy of each model.

The accuracy of Support Vector Machine is amounted to 92.3%, which is an indicator of its ability to fulfill the task and precisely define the decision boundary according to which the sensor information would be mapped into specific classes. It allows distinguishing six characteristics beyond others that can affect the irrigation. With that in mind, it is slightly lower than ANN and LSTM, but the ability to perform prediction relevant for the scheduling of irrigation in rice cultivation is undoubtful. The Decision Tree model has 90.25% of accuracy, which is also a highly acceptable result. Simplicity and easy interpretation of influential factors for the decision-making are its valuable features. Additionally, it can be useful in drawing recommendations based on the information about the sensors for the purposes of culture cultivation.

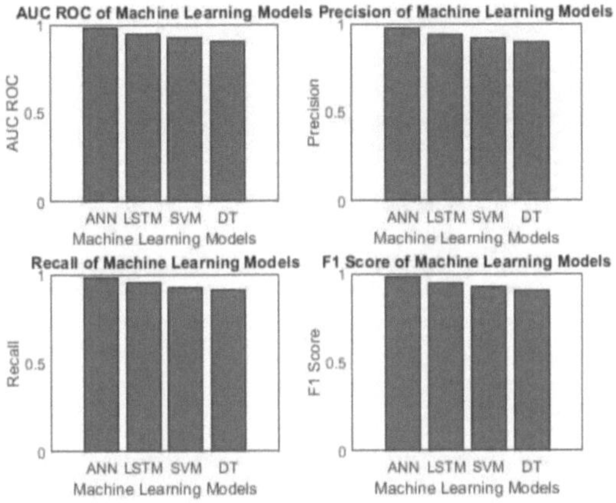


Figure 3. Performance score of each model.

Figure 3 shows the performance scores of every machine learning model, namely ANN, LSTM, SVM, and DT, regarding their ability to predict water needs for rice planting. Evaluation metrics used include AUC ROC, precision, recall, and F1 score. At first, considering the performance of the ANN model, AUC ROC was observed to be 0.981. According to the results obtained, precision was 0.976, stating that the ANN is not wrong around the 97.6% for all the positive predictions made by it. Likewise, its recall and F1 scores were observed to be 0.983 and 0.979, respectively. These high rates show that the ANN model is successful in catching the true positives with sensitivity and is great in precision in terms of making lots of predictions RTP. In addition, based on the calculated scores, there is a balance between the two parameters.

In the meantime, when it comes to the results of the LSTM model, AUC RO 's value is defined to be 0.945, suggesting the successful discrimination power in classifying all the data. The precision value is at 0.938, while the recall and F1 scores are 0.952 and 0.945, respectively. These rates are promising in RTP captures and the balanced operation in terms of making lots of predictions. It can be assumed that the LSTM model is sufficiently successful. As for the SVM model, the generated findings reveal that AUC ROC is 0.923, and the precision rate is 0.917. Checking the recall and F1 scores, they are calculated to be 0.929 and 0.923, respectively. There is no need to worry in terms of capturing the RTP, while many predictions are expected to be caught. All the determined grades prove the success of the application in a balanced manner.

Finally, regarding the DT model, it was observed that AUC ROC is 0.902 and precision is 0.896, suggesting successful estimates for determining predictions. Besides, the recall and F1 scores observed were 0.910 and 0.903, respectively. It was shown that there is no problem in RTP capture in the applications used, but many predictions are also balanced.

In summary, all the performance scores above reveal that each model is quite successful in predicting how much water is needed for rice planting.

Confusion matrices presented in Figure 4 provided a comprehensive breakdown of the classifications made by each machine learning model and how they relate to the actual class of instances. They are organized in four sections known as the quadrants. Each quadrant shows one of the four different possible classifications in a binary classification task. They include the true positive (TP), false positive (FP), false negative (FN), and true negative (TN). From the artificial neural network model matrix, it is apparent that class of instances as negative that were accurately classified was 1370. On the other side, instances that were inaccurately classified as positive were 30. Similarly, there were 1380 that were accurately classified as positive with 20 instances classified as negative.

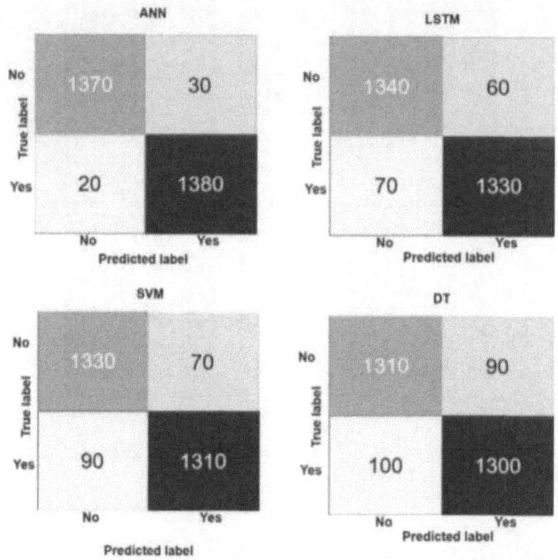


Figure 4. Confusion matrices of each model.

For the long short-term memory model, the confusion matrix indicates that there were 1340 of the negative instances that were classified accurately as negative, but 60 instances were classified as positive. For the positive class, 1330 instances were classified accurately, but 70 were classified as negative. Similarly, the confusion matrices for the support vector machine and decision trees models also help in understanding more about their classification. They depict the ability of each model to accurately classify the positive and negative instances in the sample, and how they perform in terms of classification error. From the analyses above, one can indicate that the ANN model was highly accurate in classifying the response.

## 4 CONCLUSION

This research proves the efficiency of the integration of the Internet of Things with machine learning technology for precision farming. A system was developed and applied to a rice field, and the sensors and machine learning models used to compose it are shown in the paper. The technology proved to provide better opportunities for farmers, especially in terms of crop management. The accuracy of models such as Artificial Neural Network, Support Vector Machine, Long Short-Term Mem ory networks, and Decision Tree to predict rice watering requirement was assessed, with the Artificial Neural Network model showing the most significant accuracy of 98.10%. Based on the given model, a real time prediction of applications, with remote monitoring and access provided to the users. The data was stored automatically in teh cloud and was later available through the developed application. The study shows that the proposed technology with the use of machine learning models based on IoT sensors is efficient to solve irrigation issues, improve crop watering efficiency, and contribute to the further use of similar algorithms for other crops. The next steps could include an expansion to other crops, a more comprehensive applicatin of machine learning algorithms, and an expansion to more intensive analytical instruments using other advanced technologies. The research shows that precision farming is an important part of the present agricultural domain and this technology would be further developed and used, taking into consideration the challenges of global food supply and necessity of sustainable agriculture.

## REFERENCES


[1] Raghuvanshi, A., Singh, U. K., Sajja, G. S., Pallathadka, H., Asenso, E., Kamal, M., Singh, A. and Phasinam, K. (2022). Intrusion detection using machine learning for risk mitigation in IoT-enabled smart irrigation in smart farming. *Journal of Food Quality*, 2022. https://doi.org/10.1155/2022/3955514

[2] Rau, A. J., Sankar, J., Mohan, A. R., Das Krishna, D. and Mathew, J. (2017). IoT based smart irrigation system and nutrient detection with disease analysis. *TENSYMP 2017 - IEEE International Symposium on Technologies for Smart Cities*, 3–6. https://doi.org/10.1109/TENCONSpring.2017.8070100

[3] Junior, A. A., da Silva, T. J. A. and Andrade, S. P. (2023). Smart IoT lysimetry system by weighing with automatic cloud data storage. *Smart Agricultural Technology*, 4(November 2022). https://doi.org/10.1016/j.atech.2023.100177

[4] Bojago, E. and Abrham, Y. (2023). Small-scale irrigation (SSI) farming as a climate-smart agriculture (CSA) practice and its influence on livelihood improvement in Offa District, Southern Ethiopia. *Journal of Agriculture and Food Research*, 12(February). https://doi.org/10.1016/j.jafr.2023.100534

[5] Delpiazzo, E., Bosello, F., Dasgupta, S., Bagli, S., Broccoli, D., Mazzoli, P. and Luzzi, V. (2023). The economic value of a climate service for water irrigation. A case study for Castiglione District, Emilia-Romagna, Italy. *Climate Services*, 30(May 2022), 100353. https://doi.org/10.1016/j.cliser.2023.100353

[6] Vandôme, P., Leauthaud, C., Moinard, S., Sainlez, O., Mekki, I., Zairi, A. and Belaud, G. (2023). Making technological innovations accessible to agricultural water management: Design of a low-cost wireless sensor network for drip irrigation monitoring in Tunisia. *Smart Agricultural Technology*, 4 (December 2022). https://doi.org/10.1016/j.atech.2023.100227

*Hybrid and Advanced Technologies – S. Prasad Jones Christydass, Nurhayati Nurhayati and S. Kannadhasan. (Eds) © 2025 the Authors, ISBN 978-1-032-90254-8*
*Open Access: www.taylorfrancis.com, CC BY-NC-ND 4.0 license*

# Assessing the effects of rutin trihydrate conjugated zinc oxide nanoparticles exposure on the behavior of zebrafish larvae compared to an untreated group

S. Mahati
*Research Scholar, Department of Biotechnology, Saveetha School of Engineering, Saveetha University, Chennai, Tamil Nadu, India*

I. Praveen Kumar
*Project Guide, Department of Biotechnology, Saveetha School of Engineering, Saveetha University, Chennai, Tamil Nadu, India*

ABSTRACT: To investigate the neurotoxic effects caused by the novel drug rutin trihydrate conjugated zinc oxide nanoparticles (RT-ZnONPs) in the zebrafish larvae model. By diluting the stock solution a test solution was prepared for the exposure group, containing 40 μg/ml concentrations of the novel drug rutin trihydrate conjugated zinc oxide nanoparticles. Maintenance and Collection of Zebrafish and Eggs: Adult zebrafish were procured from a nearby breeder and kept in acclimatization. Fishes were kept in a breeding tank with an equal ratio of males to females. The eggs were collected, washed in E3 medium, and placed into well plates for exposure to 40 μg/ml of rutin trihydrate-conjugated zinc oxide. Subsequently, they were subjected to 0.7mm hydrogen peroxide induction. Locomotory Behavior Analysis: Larvae's swimming behavior was observed at 6 dpf. Groups of 3 larvae from each group were kept in an ice cube tray containing 3ml of E3 medium for one minute. Larval movements were recorded using a commercial smartphone camera in a quiet environment at the beginning of the light cycle. The recorded video was then analyzed using UMA Tracker software. All experiments were conducted in triplicate. Statistical analysis was performed using the Statistical Packages for the Social Sciences (SPSS) software to identify the significance between the three groups. The parameters were set to a confidence level of 95%, a threshold of 0.05, a power of 80% using G power, and an enrollment ratio of 1. The locomotor behavior of larvae was found to be significantly increased in the rutin trihydrate conjugated zinc oxide nanoparticles treatment group (30.66±3.59) compared to the $H_2O_2$ treated group (78.11±4.64), with a significance of $P=0.000$ ($P<0.05$). The novel drug RT-ZnONPs were treated in the zebrafish embryos for 2 to 6 hours post fertilization (hpf) followed by the embryos being exposed to $H_2O_2$ to induce neurotoxicity. When compared to the untreated group, it was observed that the novel drug rutin trihydrate zinc oxide nanoparticles treated group larvae covered 71% area, while $H_2O_2$ group larvae covered only 37% area, with a significance of $p=0.000$ ($P<0.05$). Based on observation, the ROS-induced neurotoxicity has been reduced in the RT-ZnONPs treatment group compared to the $H_2O_2$ group.

*Keywords*: Rutin Trihydrate, Zinc Oxide, Novel, Behaviour, Disease, Nanoparticles

## 1 INTRODUCTION

Citrus fruits are rich sources of rutin trihydrate, a bioflavonoid compound also found in *Carpobrotus edulis*, which is alternatively known by names such as sophorin, rutoside, and

 DOI: 10.1201/9781003559115-16

quercetin-3-O-rutinoside, and it possesses a molecular weight of 664.56 g/mol. The current aimed to explore the impact of rutin trihydrate on the neurological behavior of zebrafish larvae. Previous research (Bukhari *et al.* 2022; Bailey and Day 1989; Frutos *et al.* 2019) suggests that rutin trihydrate exhibits therapeutic properties, including anti-inflammatory and anti-apoptotic actions, which may be beneficial in preventing various diseases. Despite these potential benefits, research on the toxicity of flavonol glycosides, including rutin trihydrate, is limited, particularly regarding its long-term effects. Clinical studies involving the administration of rutin, hesperidin, quercitrin, and naringin to dogs for treating canine diseases have indicated that these compounds are primarily excreted unchanged in the urine and pose no harm. Clinical trials have shown that oral rutin administration for 9 months did not cause adverse effects in patients with cognitive and memory-related diseases (Wilson *et al.* 1947).

The larval zebrafish (*Danio rerio*) is extensively employed as a model system due to its advantageous genetic characteristics, transparency, and relative simplicity. With around 200 neurons stretching from the brainstem to the spinal cord, these larvae provide the benefit of pinpointing and laser-ablating multiple neurons in intact specimens. This feature facilitates the characterization of larval behavioral patterns at the cellular level. We aim to investigate the diverse locomotor activity patterns exhibited by zebrafish larvae. To achieve this, we examined a variety of swimming and turning behaviors in larval fish aged 6-9 days using high-speed digital photography (Budick and O'Malley 2000). The larval zebrafish displayed two distinct types of swimming episodes based on the location of maximum bending: burst swims toward the mid-body and slow swims closer to the tail. Burst swims were characterized by more pronounced bending, greater speeds, and increased yaw compared to leisurely swims. Two categories of turning behaviors were identified: rapid, large-angle escape turns that are characteristic of escaping actions, and slower routine rotations that do not have the pronounced counterbend that escape turns are known for. Additionally, prey-capture behaviors were observed, including basic locomotor elements such as slow turns and routine swimming. (Ali *et al.* 2023, Bukhari *et al.* 2022, Gurupranes *et al.* 2022, Natrayan *et al.* 2022, Natrayan *et al.* 2022, Sivaranjani *et al.* 2023, Viswanathan *et al.* 2023). These various behaviors were analyzed in relation to potential underlying brain control mechanisms.

Zebrafish embryos between 0 to 5 days old already possess rudimentary sensory and locomotor capabilities. As zebrafish larvae (between 5 days to 2 weeks old) are free-swimming during this developmental stage and need to forage for food while evading predators, they exhibit a range of behavioral traits. Despite their differing developmental stages, both embryos and larvae maintain optical transparency, facilitating straightforward cellular studies of neural circuitry. In fact, fluorescent proteins enable direct observation of individual neural activity or connections, while laser technology permits precise single-neuron ablation (Guo 2004). The novelty of the current study is to assess the behavioural analysis of the larvae exposed to the novel medication rutin trihydrate aggregated zinc oxide nanoparticle. The innovative medicine rutin trihydrate conjugated zinc oxide nanoparticle exposure, which occurs in the early prenatal or postnatal stages, may have an effect that this study could shed light on. The new medicine rutin has been linked to favourable disease-preventive effects as well as drug-induced toxicities, such as heart toxicity and memory issues. The relevant publications for the current study, identified through a literature review, were gathered from sources such as Google Scholar and ScienceDirect. The Google Scholar search yielded 14,022 articles deemed suitable for the study, while ScienceDirect provided around 2045 articles. According to the literature survey, the most cited article was (Richetti *et al.* 2011), with 117 citations, followed by a study by (Dubey *et al.* 2015) with 103 citations. Another notable study by (Stengel, Wahby, and Braunbeck 2018) garnered 390 citations. The most prominent study on this topic was by (Hu *et al.* 2023) with 530 citations. However, to date, there have been no studies comparing the neurological behavior of zebrafish.

## 2 MATERIALS AND METHODS

The analysis was conducted at the Zebrafish laboratory, AHS Block, SIMATS Engineering, Saveetha Institute of Medical and Technical Sciences. The sample size of each group was determined to be 30 samples per group and a total of 90 samples divided among three groups. Embryos from each exposure group were distributed into 12 well plates, with each well containing n=10 embryos. Beginning at 96 hpf, they were exposed to rutin trihydrate-conjugated zinc oxide nanoparticles. Each experiment was performed in triplicate, with static exposure maintained throughout the exposure period. Rutin trihydrate was procured from SISCO Research Labs.

### 2.1 *Zebrafish husbandry and embryo harvesting*

Adult zebrafishes were obtained from a local vendor and maintained in stable conditions with a light/dark cycle (14:10 hour) at a temperature of 26±1°C. Fishes were kept for breeding in a specific spawning tank, where a 1:1 ratio of males to females facilitated embryo production. To deter adult fish from consuming the embryos, the spawning tank was outfitted with a bottom container and netting to gather them. Eggs were collected, rinsed in E3 medium, and then distributed into well plates for exposure to a concentration of 40μg of rutin trihydrate-conjugated zinc oxide, followed by induction with 0.7mm hydrogen peroxide. These procedures were conducted at the E10, AHS Block laboratory, Saveetha School of Engineering.

### 2.2 *Locomotory behavior analysis*

By 6 days post-fertilization (dpf), zebrafish larvae demonstrate adept swimming abilities, making them an ideal model for investigating cognitive impairment in research (Drapeau *et al.* 2002). The zebrafish larvae's odd movements while swimming was examined at the ending of the exposure period (Krishnan and Kang 2019). Finally, in brief (3 larvae/well) from each group at 6 dpf were put in an ice cube tray with white chambers that measured 2.5 by 3.5 cm for 1 minute, comprising 3mL of E3 medium with methylene blue. At the start of the light cycle, larval movement was recorded using a commercial smartphone camera in a tranquil setting. The video was recorded for one minute, and locomotion data were analyzed using UMA Tracker software to plot the movements.

### 2.3 *Statistical analysis*

The data represented were mean ± SD. The statistical software program SPSS for Windows 20.0 was used to create the graphics and conduct the statistical analysis. There are no dependent variables. The Independent variable is control, hydrogen peroxide, rutin trihydrate conjugated zinc oxide nanoparticles, a paired sample t-test was used to examine the comparability between the groups.

## 3 RESULTS

Figure 1 illustrates the distance traveled (meters) by zebrafish larvae exposed to rutin trihydrate-conjugated zinc oxide nanoparticles. It was noted that larvae exposed to these nanoparticles exhibited significantly regulated locomotory activity (47.35±2.67) compared to the group treated with $H_2O_2$ (13.20±1.94), with a statistically significant difference of $P=0.000$ ($P<0.05$).

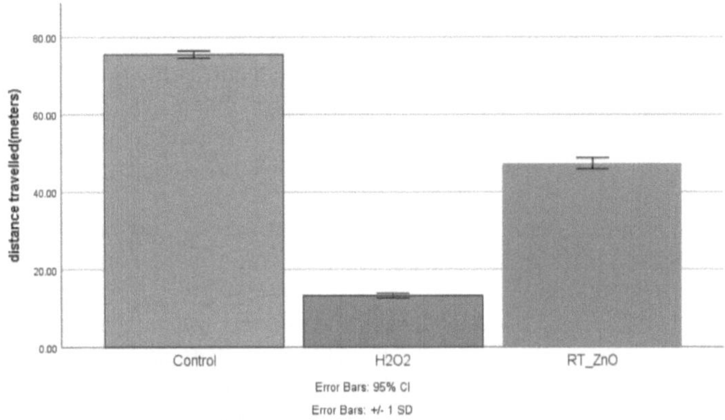


Figure 1. The behavioural analysis of the zebrafish larvae in the Control, $H_2O_2$ and RT-ZnO treated group. X-axis: Groups-Control, $H_2O_2$ and novel drug RT-ZnO (40 μg/ml), Y-axis: distance travelled by the larvae (meters) with +/- 1 SD.

Table 1 provides group statistics, including mean values, standard deviation (SD), and standard error mean (SEM). It outlines the total distance traveled by zebrafish larvae exposed to rutin trihydrate-conjugated zinc oxide nanoparticles and the $H_2O_2$-treated group. The group exposed to rutin trihydrate-conjugated zinc oxide nanoparticles exhibited an increased mean value (47.35±2.67) compared to the control group (75.56±2.67). Table 2 displays the independent t-test results, showing statistical significance of p=0.000 ($P<0.05$) among the three groups.

Table 1. According to the group statistics, the larvae in the RT-ZnO treated group covered a mean distance of 47.3500 units, which was greater than the 13.2000 units covered by the $H_2O_2$ group and comparable to the 75.5600 units covered by the control group.

| Groups | | Mean | N | Std Deviation | Std Error |
|---|---|---|---|---|---|
| Pair 1 | Control | 75.5600 | 30 | 2.67567 | 4.50514 |
| Pair 2 | $H_2O_2$ | 13.2000 | 30 | 1.94491 | 5.46717 |
| Pair 3 | RT_ZnO | 47.3500 | 30 | 3.34655 | 0.61099 |

Table 2. Significant differences in the studied parameters across the three groups were found by the independent t-test (p = 0.000, $p < 0.05$), suggesting that there were discernible differences between the groups.

| | | | | | 95% Confidence Interval Of The Difference | | | | |
|---|---|---|---|---|---|---|---|---|---|
| Groups | | Mean | Std Deviation | Std Error Mean | Lower | Upper | T | Df | Sig (2 Tailed) |
| Pair 1 | Control - $H_2O_2$ | 62.36000 | 2.62728 | 7.78264 | 61.37774 | 63.21396 | 129.844 | 29 | 0.000 |
| Pair 2 | RT_ZnO - $H_2O_2$ | 34.15000 | 4.76831 | 5.25235 | 32.55366 | 36.45560 | 43.753 | 29 | 0.000 |
| Pair 3 | RT_ZnO - Control | 28.21000 | 4.05363 | 4.39197 | 26.68754 | 29.39844 | 37.897 | 29 | 0.000 |

## 4 DISCUSSION

The locomotory behavior was analyzed to evaluate the neurobehavioural activity of zebrafish larvae at a concentration of 40μg/ml. In this study, When compared to the untreated group, it was observed that the novel drug rutin trihydrate conjugated zinc oxide nanoparticles treated group larvae covered 71% of the total area, while $H_2O_2$ group larvae covered only 37% area.

In 2022, Saleemi indicated that ZnO nanoparticles showed reduced free radical scavenging activity compared to pure rutin. Cytotoxic effects of rutin were examined using a breast cancer cell line for the potential treatment of cancer. *In vitro* study confirmed that ZnO NPs displayed elevated anticancer and antimicrobial properties (Saleemi *et al.* 2022). In 2018, Stengel discussed the rising prominence of zebrafish larvae as models for behavioral studies. These larvae exhibit several characteristics that make them suitable counterparts to mammalian models. They share extensive similarities with other vertebrate species, including rodents and humans, particularly concerning their genome, brain development, and the structure and function of various neural and physiological systems. This similarity extends to critical systems involved in behavioral functions, such as the presence of neurotransmitters like dopamine, norepinephrine, and serotonin, which are notably present in larval zebrafish (Stengel *et al.* 2018). In 2019, Stengel and Habtemariam reported that rutin and its metabolites has the ability to cross the blood brain barrier, it has also been shown to cure various behavioral symptoms of neurodegenerative diseases (Habtemariam 2019; Stengel *et al.* 2018).

One disadvantage of the current study is that it only looks at cognitive impairment in the zebrafish larvae model caused by free radicals. Although humans and zebrafish have similar genomes, using higher model organisms is required to properly understand the drug's application for treating neurological disorders. Subsequent investigations will endeavor to examine cognitive impairment in higher model species, such as rats, by employing various decision-making assessments such as the T-maze.

## 5 CONCLUSION

The RT-ZnONPs treatment exhibited substantial efficacy in mitigating ROS-induced neurotoxicity. In comparison to the untreated group, the treated larvae covered a significantly larger area of 71%. This observation underscores the promising neuroprotective potential of RT-ZnONPs against oxidative stress-induced neurotoxicity.

## REFERENCES

Ali, S., Sudha, K. G., Thirumalaivasan, N., Ahamed, M., Pandiaraj, S., Rajeswari, V. D., Vinayagam, Y., Thiruvengadam, M. and Govindasamy, R. (2023). Green synthesis of magnesium oxide nanoparticles by using Abrus precatorius bark extract and their photocatalytic, antioxidant, antibacterial, and cytotoxicity activities. *Bioengineering (Basel, Switzerland)*, 10(3). https://doi.org/10.3390/bioengineering10030302

Bailey, C. J. and Day, C. (1989). Traditional plant medicines as treatments for diabetes. *Diabetes Care*, 12(8), 553–564.

Budick, S. A. and O'Malley, D. M. (2000). Locomotor repertoire of the larval zebrafish: Swimming, turning and prey capture. *The Journal of Experimental Biology*, 203(Pt 17), 2565–2579.

Bukhari, I. A., Mohamed, O. Y., Alhowikan, A. M., Lateef, R., Hagar, H., Assiri, R. A. and Alqahtani, W. M. A. (2022). Protective effect of rutin trihydrate against dose-dependent, cisplatin-induced cardiac toxicity in isolated perfused rat's heart. *Cureus*, 14(1), e21572.

Drapeau, P., Saint-Amant, L., Buss, R. R., Chong, M., McDearmid, J. R. and Brustein, E. (2002). Development of the locomotor network in zebrafish. *Progress in Neurobiology*, 68(2), 85–111.

Dubey, S., Ganeshpurkar, A., Bansal, D. and Dubey, N. (2015). Protective effect of rutin on impairment of cognitive functions due to antiepileptic drugs on zebrafish model. *Indian Journal of Pharmacology*, 47(1), 86–89.

Frutos, M. J., Rincón-Frutos, L., & Valero-Cases, E. (2019). Chapter 2.14 - Rutin. In S. M. Nabavi and A. S. Silva (Eds.), *Nonvitamin and Nonmineral Nutritional Supplements* (pp. 111–117). Academic Press.

Guo, S. (2004). Linking genes to brain, behavior and neurological diseases: What can we learn from zebrafish? *Genes, Brain, and Behavior*, 3(2), 63–74.

Gurupranes, S. V., Natrayan, L., Kaliappan, S., Patel, P. B., Sekar, S., Jayaraman, P., Pandian, C. K. A. and Esakkiraj, E. S. (2022). Investigation of physicochemical properties and characterization of leaf stalk fibres extracted from the Caribbean royal palm tree. *International Journal of Chemical Engineering*, 2022. https://doi.org/10.1155/2022/7438411

Habtemariam, S. (2019). *Medicinal Foods as Potential Therapies for Type-2 Diabetes and Associated Diseases: The Chemical and Pharmacological Basis of Their Action*. Academic Press.

Hu, Y., Jia, K., Zhou, Y., Chen, L., Wang, F., Yi, X., Huang, Y., et al. (2023). Rutin hydrate relieves neuroinflammation in zebrafish models: Involvement of NF-$\kappa$B pathway as a central network. *Fish & Shellfish Immunology*, 141(October), 109062.

Krishnan, M. and Kang, S. C. (2019). Vitexin inhibits acrylamide-induced neuroinflammation and improves behavioral changes in zebrafish larvae. *Neurotoxicology and Teratology*, 74(May), 106811.

Natrayan, L., Kaliappan, S., Sethupathy, B. S., Sekar, S., Patil, P. P., Velmurugan, G. and Olkeba, T. T. (2022). Effect of mechanical properties on fibre addition of flax and graphene-based bionanocomposites. *International Journal of Chemical Engineering*, 2022. https://doi.org/10.1155/2022/5086365

Richetti, S. K., Blank, M., Capiotti, K. M., Piato, A. L., Bogo, M. R., Vianna, M. R. and Bonan, C. D. (2011). Quercetin and rutin prevent scopolamine-induced memory impairment in zebrafish. *Behavioural Brain Research*, 217(1), 10–15.

Saleemi, M. A., Alallam, B., Yong, Y. K. and Lim, V. (2022). Synthesis of zinc oxide nanoparticles with bioflavonoid rutin: Characterisation, antioxidant and antimicrobial activities and in vivo cytotoxic effects on nauplii. *Antioxidants (Basel, Switzerland)*, 11(10). https://doi.org/10.3390/antiox11101853

Sivaranjani, P. R., Syed, A., Elgorban, A. M., Bahkali, A. H., Balakrishnaraja, R., Varma, R. S. and Khan, S. S. (2023). Fabrication of ternary nano-heterojunction via hierarchical deposition of α-Fe2O3 and $\beta$-La2S3 on cubic CoCr2O4 for enhanced photodegradation of doxycycline. *Journal of Industrial and Engineering Chemistry*, 118, 407–417.

Stengel, D., Wahby, S. and Braunbeck, T. (2018). In search of a comprehensible set of endpoints for the routine monitoring of neurotoxicity in vertebrates: Sensory perception and nerve transmission in zebrafish (Danio rerio) embryos. *Environmental Science and Pollution Research International*, 25(5), 4066–4084.

Viswanathan, S., Palaniyandi, T., Kannaki, P., Shanmugam, R., Baskar, G., Rahaman, A. M., Paul, L. T. D., Rajendran, B. K. and Sivaji, A. (2023). Biogenic synthesis of gold nanoparticles using red seaweed Champia parvula and its anti-oxidant and anticarcinogenic activity on lung cancer. *Particulate Science and Technology*, 41(2), 241–249.

Wilson, R. H., Mortarotti, T. G. and Doxtader, E. K. (1947). Toxicity studies on rutin. *Proceedings of the Society for Experimental Biology and Medicine*, 64(3), 324–327.

*Hybrid and Advanced Technologies – S. Prasad Jones Christydass, Nurhayati Nurhayati and S. Kannadhasan. (Eds) © 2025 the Authors, ISBN 978-1-032-90254-8*
*Open Access: www.taylorfrancis.com, CC BY-NC-ND 4.0 license*

# Assessing the impact of 2-methylimidazole exposure on neurobehavior in zebrafish larvae comparing with an untreated control group

S.U. Vandanaa
*Research Scholar, Department of Biotechnology, Saveetha School of Engineering, Saveetha University, Chennai, Tamil Nadu, India*

I. Praveen Kumar
*Project Guide, Department of Biotechnology, Saveetha School of Engineering, Saveetha University, Chennai, Tamil Nadu, India*

ABSTRACT: Using a novel zebrafish larvae paradigm, this study examines the effects of 2-Methylimidazole (2-MI) on neurobehavioral outcomes. A stock solution was used to create an E3 medium test solution of 2-MI (100µM). There were two groups in the study, totaling sixty larvae in the sample. Group 2 was exposed to 2-MI, whereas Group 1 was the control group (E3 medium only). After breeding adult zebrafish, embryos were removed and subjected to 2-MI between four and six hours after fertilization (hpf). Every test was run in triplicate. With a 95% confidence interval, a significance level of 0.01, 80% power, and a 1:1 allocation ratio, SPSS was used to determine statistical significance. With a p-value of 0.000 ($P<0.05$), the locomotor behavior of larvae in the 2-MI exposed group (30.66±3.59) was substantially lower than that of the control group (78.11±4.64). In particular, compared to controls, larvae exposed to 2-MI exhibited a 48% reduction in traveled distance. This difference was statistically significant ($p=0.000$), suggesting that exposure to 2-MI caused significant behavioral changes. As a result, our research indicates that 2-MI exposure during early development in the zebrafish model impairs locomotor behavior, indicating its possible involvement in the development of neurobehavioral toxicity.

*Keywords*: 2-Methylimidazole (2-MI), Toxicity, Locomotor, Innovative zebrafish larvae model, Disease, Neurobehavior

## 1 INTRODUCTION

Imidazoles are well-known heterocyclic substances with important characteristics that are utilized in a variety of medications. A five-membered planar ring known as imidazole is soluble in polar solvents such as water (Tolomeu and Fraga 2023). People who come into contact with these imidazoles may have been exposed through consuming foods and drinks containing ammonia caramel color (E150c) and ammonium sulfite caramel color (E150d). When these colors are used in food processing, compounds like methylglyoxal, resulting from sugar breakdown, react with ammonia to form amides and amino-aldehydes, which then combine to create imidazoles with 2-, 4-, or 5-substitutions (Morita and Uneyama 2016; Wu *et al.* 2015). Imidazole compounds are also used in a wide spectrum of pharmacological products. Research indicates that imidazole derivatives have pain-relieving and anti-inflammatory properties, along with potential in fighting cancer and affecting cardiovascular functions (Selvam *et al.* 2014). 2-Methylimidazole (2-MI) has been found to be carcinogenic in animal studies (Morita and Uneyama 2016).

 DOI: 10.1201/9781003559115-17

Zebrafish are widely used as a model for ecotoxicological studies. Because of the transparency of zebrafish embryos, researchers can investigate the earliest developmental stages with precision. Within 48 hours of fertilization, zebrafish embryos also develop whole organ systems, including the heart, gastrointestinal tract, and blood vessels (Chahardehi *et al.* 2020). Zebrafish contain all of the essential organs involved in metabolic issues, thus they may be utilized to examine a range of human metabolic processes (Nishimura *et al.* 2015). The embryos grow externally and are shielded for the first two to three days by a translucent chorion. This model is especially helpful to determine changes in the nervous system because of the clear nature of embryo which allows for visualisation of changes with development (Bailey *et al.* 2013). A growing need for zebrafish in behavioural teratology is also a result of the zebrafish's wide range of developmental, behavioural, and molecular advantages. This investigation could shed light on how exposure to 2-MI influences behavioral changes. To gather relevant literature, databases like Google Scholar and PubMed were utilized during the literature review process. Approximately 32 articles were identified through Google Scholar, while about 25 articles were retrieved from PubMed. Among these, the most frequently cited article was (Ruyra *et al.* 2015), which amassed a total of 211 citations according to the survey.

As of now, no studies have investigated the neurotoxic impacts of 2-methylimidazole using zebrafish larvae. The novelty of our current research lies in assessing the toxicity-induced neurobehavioral disorders resulting from early-life exposure of zebrafish embryos to 2-MI.

## 2 MATERIALS AND METHOD

The analysis of toxicity-induced neurobehavioral diseases was conducted at the Zebrafish laboratory. The study comprised two groups, each comprising 30 samples, and a total of 60 samples for examination (Issac *et al.* 2021). The sample size for each group was determined using clinicalc.com. Embryos were (n=30) exposed to 2-Methylimidazole from 4 hours post-fertilization (hpf) to 96 hpf (hours post-fertilization). Each experiment was replicated three times, and the exposure remained static throughout the duration without renewal. 2-Methylimidazole [CAS: 693-98-1] was procured from SRL Pvt. Ltd., India. A stock of 200 mM was prepared using HPLC-grade water and stored at -20°C. For the acute fish toxicity test, egg water was formulated following the guidelines outlined in OECD Technical Guideline TG203 Annexure-2 (Alesina and Roubini 1992).

### 2.1 *Maintenance of zebrafish*

The zebrafish were obtained and are maintained in a controlled environment with a 14:10-hour light/dark cycle at a temperature of 26±1°C. Breeding groups were placed in a spawning tank with netting at the bottom to collect the eggs. Spawning was initiated at the onset of the light cycle. After 30 minutes, the eggs are gently collected, rinsed with E3 medium, and then divided into two groups in separate Petri dishes: the exposure group, treated with 2-MI (100μM), and the control group, treated with E3 medium.

### 2.2 *Locomotor behavior analysis*

Indeed, zebrafish larvae offer a valuable model for studying locomotion and swimming behavior in research. Their development into proficient swimmers typically occurs around 6 days after fertilization, making them well-suited for investigations into various aspects of motor function and behavior (Liu *et al.* 2023). To evaluate the locomotor behaviour in the innovative zebrafish larvae model after the exposure period, swimming behavior patterns were examined. Each exposure group was placed in an ice tray for 10 minutes of acclimatization.

Following acclimation, the their mobility was captured using a commercial smartphone camera in a quiet environment for 5 minutes. Subsequently, locomotion was analyzed using UMA Tracker software (Yamanaka and Takeuchi 2018).

### 2.3 *Statistical analysis*

The data were presented as mean ± standard deviation (SD) and analyzed using the SPSS statistical software suite for Windows version 20.0 (SPSS Inc., Chicago, IL, USA). Statistical analyses were conducted with a confidence level of 95%, a significance threshold of 0.01, a statistical power (G power) of 80%, and an enrollment ratio of 1 (Alesina *et al.* 1992).

As there were no dependent factors, the comparison was made between the independent variables: the control group and the 2-MI exposure group. The independent samples t-test was utilized for group comparison.

## 3 RESULTS

In Figure 1, the percentage of distance traveled was calculated in the Innovative Zebrafish larvae model between the untreated group and the group exposed to 2-Methylimidazole at 96 hours post-fertilization (hpf). It was observed that embryos exposed to 2-Methylimidazole exhibited a significantly reduced distance traveled (30.66±3.59) compared to the untreated group (78.11±4.64), with a significance of $P=0.000$ ($P<0.05$).

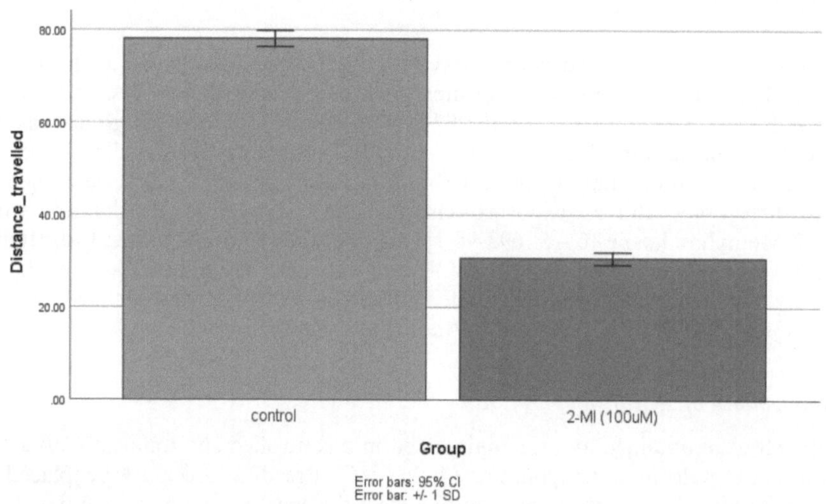


Figure 1. Bar chart representing the comparison of mean values of distance travelled in 2-Methylimidazole exposure groups and control group, data presented as the mean of at least three independent experiments (n=30/group) indicates that the significance was observed as a decrease in distance travelled in the 2-Methylimidazole exposure group (30.66±3.59) when compared to the control group (78.11±4.64). X-axis: 2-Methylimidazole exposure group and Control group and Y-axis mean distance travelled of the zebrafish larvae. ± 1 SD.

Table 1 displays group statistics, including mean, standard deviation, and standard error mean. The total distance traveled by the zebrafish larvae exposed to 2-Methylimidazole and the control group was calculated. The 2-Methylimidazole exposure group exhibited a reduced mean value (30.66±3.59) compared to the control group (78.11±4.64). In Table 2, the statistical analysis of the independent t-test between the two groups revealed a statistically significant difference with a P-value of 0.000 ($P<0.05$).

Table 1. Group statistics represent mean, standard deviation and standard error mean. The distance travelled of the 2-methylimidazole showed a reduced mean value (30.66±3.59) when compared to the control group (78.11±4.64).

| | Group | N | Mean | Std. Deviation | Std. Error Mean |
|---|---|---|---|---|---|
| Distance travelled | control | 30 | 78.1100 | 4.64820 | 0.84864 |
| | 2-Methylimidazole | 30 | 30.6647 | 3.59139 | 0.65569 |

Table 2. Independent t-test showing the statistical analysis. There exists a statistical significance between the two groups P=0.000 (P<0.05).

| | | Levene's Test for Equality of Variances | | t-test for Equality of Means | | | | | | |
|---|---|---|---|---|---|---|---|---|---|---|
| | | | | | | | | | 95% Confidence Interval of the Difference | |
| | | F | Sig. | t | df | sig (2 tailed) | Mean Difference | Std. Error Difference | Lower | Upper |
| Distance Travelled | Equal variances assumed | 2.385 | 0.128 | 44.241 | 58 | .000 | 47.445 | 1.072 | 45.298 | 49.592 |
| | Equal variances not assumed | | | 44.241 | 54.527 | .000 | 47.445 | 1.072 | 45.295 | 49.594 |

## 4 DISCUSSION

In our research, it was shown that 96 hpf innovative Zebrafish larvae model exposed to 2-MI showed change in their locomotor behaviour. When compared to the untreated group (78.11±4.64), the distance travelled by the larvae exposed to the 2-MI group (30.66±3.59) was lower. The investigation's significance threshold was assessed to be P=0.000 (P<0.05), which is significant. Therefore, it is evident that 2-MI exposure caused toxicity induced neurobehavioural diseases.

Wang *et al.* (2019) observed a decrease in locomotor response following exposure to 10 μM TBECH, indicating neurogenesis abnormalities induced by Brominated Flame Retardants (BFRs) in zebrafish, which could threaten fish survival (Wang *et al.* 2019). Similarly, Jin *et al.* (2015) investigated the adverse effects of CPF on zebrafish larvae and found that exposure to different concentrations resulted in a decrease in locomotor behavior. High doses of CPF significantly reduced free swimming distance and speed, suggesting impaired locomotor function (Jin *et al.* 2015). In another study zebrafish larvae exposed to various concentrations of DE-71 exhibited reduced locomotor behavior, with higher concentrations showing more pronounced effects (Chen *et al.* 2012). Moreover, Zhu *et al.* (2018) demonstrated that the larvae subjected to TBBPA exhibited decreased swimming speed during light-dark-light transitions, indicating altered locomotor behavior. These collective findings underscore the impact of various environmental contaminants on zebrafish locomotion, highlighting the importance of understanding their potential threats to aquatic ecosystems (Zhu *et al.* 2018).

It's important to acknowledge that our study only examines one toxicological endpoint, which is a limitation. To gain a comprehensive understanding of the underlying mechanisms involved in the observed reduction of locomotor behavior in the novel zebrafish larvae model, it's essential to explore additional physiological, biochemical, and molecular end-points. While zebrafish larvae serve as a valuable model due to their genetic similarity to humans, utilizing higher model organisms such as rats can provide further insights into the neurotoxic effects of 2-MI. Future research could involve assessing toxicity-induced neuro-behavioral diseases in rats by using various decision-making tests to elucidate the broader implications of 2-MI exposure. Expanding the scope of the study to include multiple end-points and higher model organisms will enhance our understanding of the neurotoxicity of 2-MI and its potential impact on human health.

## 5 CONCLUSION

2-Methylimidazole exposure group and untreated group when compared, it was seen that zebrafish larvae exposed to 2-Methylimidazole showed a 48% reduced distance travelled when compared to the untreated group. By these observations, we can infer that 2-Methylimidazole affects the neurobehavior of zebrafish larvae at a concentration of 100μM.

## REFERENCES


Alesina, A., Cohen, G. D. and Roubini, N. (1992). Macroeconomic Policy and Elections in OECD democracies.

Alesina, A. and Roubini, N. (1992). Political cycles in OECD economies. *The Review of Economic Studies.* https://doi.org/10.2307/2297992

Bailey, J., Oliveri, A. and Levin, E. D. (2013). Zebrafish model systems for developmental neurobehavioral toxicology. *Birth Defects Research. Part C, Embryo Today: Reviews*, 99(1), 14–23.

Chahardehi, A. M., Arsad, H. and Lim, V. (2020). Zebrafish as a successful animal model for screening toxicity of medicinal plants. *Plants*, 9(10). https://doi.org/10.3390/plants9101345

Chen, L., Huang, C., Hu, C., Yu, K., Yang, L. and Zhou, B. (2012). Acute exposure to DE-71: Effects on locomotor behavior and developmental neurotoxicity in zebrafish larvae. *Environmental Toxicology and Chemistry / SETAC*, 31(10), 2338–2344.

Issac, P. K., Lite, C., Guru, A., Velayutham, M., Kuppusamy, G., Saraswathi, N. T., Al Olayan, E. M., et al. (2021). Tryptophan-tagged peptide from serine threonine-protein kinase of *Channa striatus* improves antioxidant defence in L6 myotubes and attenuates caspase 3-dependent apoptotic response in zebrafish larvae. *Fish Physiology and Biochemistry*, 47(2), 293–311.

Jin, Y., Liu, Z., Peng, T. and Fu, Z. (2015). The toxicity of chlorpyrifos on the early life stage of zebrafish: A survey on the endpoints at development, locomotor behavior, oxidative stress and immunotoxicity. *Fish & Shellfish Immunology*, 43(2), 405–414.

Liu, H., Fu, R., Zhang, Y., Mao, L., Zhu, L., Zhang, L., Liu, X. and Jiang, H. (2023). Integrate transcriptomic and metabolomic analysis reveals the underlying mechanisms of behavioral disorders in zebrafish (*Danio rerio*) induced by imidacloprid. *The Science of the Total Environment*, 870(April), 161541.

Morita, T. and Uneyama, C. (2016). Genotoxicity assessment of 4-methylimidazole: Regulatory perspectives. *Genes and Environment : The Official Journal of the Japanese Environmental Mutagen Society*, 38 (November), 20.

Nishimura, Y., Murakami, S., Ashikawa, Y., Sasagawa, S., Umemoto, N., Shimada, Y. and Tanaka, T. (2015). Zebrafish as a systems toxicology model for developmental neurotoxicity testing. *Congenital Anomalies*, *55*(1), 1–16.

Ruyra, À., Yazdi, A., Espín, J., Carné-Sánchez, A., Roher, N., Lorenzo, J., Imaz, I. and Maspoch, D. (2015). Synthesis, culture medium stability, and in vitro and in vivo zebrafish embryo toxicity of metal-organic framework nanoparticles. *Chemistry*, 21(6), 2508–2518.

Selvam, T. P., Ghanashyam, M. A. and Joshi, S. (2014). *Biological activities of fused imidazole derivatives - A review: Biologically important benzimidazoles. LAP Lambert Academic Publishing.*

Tolomeu, H. V. and Fraga, C. A. M. (2023). Imidazole: Synthesis, functionalization and physicochemical properties of a privileged structure in medicinal chemistry. *Molecules*, 28(2). https://doi.org/10.3390/molecules28020838

Wang, X., Wei, L., Wang, Y., He, B., Kong, B., Zhu, J., Jin, Y. and Fu, Z. (2019). Evaluation of development, locomotor behavior, oxidative stress, immune responses and apoptosis in developing zebrafish (*Danio rerio*) exposed to TBECH (tetrabromoethylcyclohexane). *Comparative Biochemistry and Physiology. Toxicology & Pharmacology: CBP*, 217(March), 106–113.

Wu, X., Huang, M., Kong, F. and Yu, S. (2015). Short communication: Study on the formation of 2-methylimidazole and 4-methylimidazole in the Maillard reaction. *Journal of Dairy Science*, 98(12), 8565–8571.

Yamanaka, O. and Takeuchi, R. (2018). UMATracker: An intuitive image-based tracking platform. *The Journal of Experimental Biology*, 221(Pt 16). https://doi.org/10.1242/jeb.182469

Zhu, B., Zhao, G., Yang, L. and Zhou, B. (2018). Tetrabromobisphenol A caused neurodevelopmental toxicity via disrupting thyroid hormones in zebrafish larvae. *Chemosphere*, 197(April), 353–361.

*Hybrid and Advanced Technologies – S. Prasad Jones Christydass, Nurhayati Nurhayati and S. Kannadhasan. (Eds) © 2025 the Authors, ISBN 978-1-032-90254-8*
*Open Access: www.taylorfrancis.com, CC BY-NC-ND 4.0 license*

# Exploring the novel therapeutic agents against dihydrofolate reductase through comparative molecular docking

S. Shavana
*Research Scholar, Department of Bio Informatics, Saveetha School of Engineering, Saveetha Institute of Medical and Technical Sciences, Saveetha University, Chennai, Tamilnadu, India*

Arun John
*Project Guide, Department of Bio Informatics, Saveetha School of Engineering, Saveetha Institute of Medical and Technical Sciences, Saveetha University, Chennai, Tamilnadu, India*

ABSTRACT: This work intends to determine the most powerful innovative lead medicine for the treatment of Toxoplasmosis by exploring possible interactions between the Toxoplasmosis associated DHFR protein and ligands using computational techniques. Using Uniprot, we discovered information about Toxoplasma gondii's effect on the protein enolase.PyRx was used as our docking tool to determine the binding affinities of different ligands with the target protein.A list of ligands was gathered from PubChem to assess the stability of targets in relation to one another.Molecular docking was done in AutoDock VINA using PyRx and then analyzed in LigPlotAs a result, it was discovered that mebendazole could serve as a ligand for the DHFR Protein.Thus, we examined the binding affinities of the various ligands against the DHFR protein using PyRx.To identify the ideal compound among a group of ligands, we used the binding energy as a critical parameter. Mebendazole, the best lead compound, has been found to be helpful and beneficial in lowering the infection of Toxoplasmosis.Further ADMET properties were investigated, and the proposed ligand met the Lipinski rule.

*Keywords*: DHFRProtein, Menbendazole, Ligplot, Toxoplasmosis, PyRx, Docking, Ligand

## 1 INTRODUCTION

Toxoplasma Gondii (Dubey 2008) is a protozoan parasite that causes the parasitic disease Toxoplasmosis. Both humans and a variety of other warm blooded animals can contract this parasite. The parasite is primarily transmitted by ingestion of raw or undercooked meat containing Toxoplasma cysts. One of the most prevalent parasite illnesses in the world, toxoplasmosis can affect people, causing more serious problems in some cases and asymptomatic and mild flu like symptoms in others. Symptoms of more severe cases include blurred vision, headaches, confusion, convulsions and poor coordination. Toxoplasmosis(Karaca Adıyeke and Ture 2022) is detected using blood tests that detect parasite antibodies.Doctors may examine pregnant women fo the infection because it can harm the fetus. Treatment may not be the required in healthy persons. However drugs such as pyrimethamine and sulfadiazine are routinely used to treat the illness in pregnant women and people with compromised immune systems. Toxoplasmosis can be avoided by practicing good hygiene such as washing hands, and kitchen utensils after handling raw meat.

### 1.1 *Molecular docking*

Molecular Docking is the Computational technique used in drug discovery used in drug discovery and development to understand and predict the interactions of small

 DOI: 10.1201/9781003559115-18

molecules called ligands(Shekari *et al.* 2023) with target biological macromolecules such as proteins,enzymes,nucleic acids or receptors.This method is used in pharmaceutical research to identify drugs that are capable of binding to specific target sites on biomolecules.

In the case of Toxoplasma Gondii and its dihydrofolate reductase (DHFR)(Averett 1979) enzyme,molecular docking studies aim to identify or create possible antiparasitic drugs that can bind to and decrease the activity of the DHFR enzyme.

### 1.2 *Dihydrofolate reductase*

Dihydrofolate Reductase (DHFR) is an enzyme that aids in the synthesis of tetrahydrofolate which is required for the production of DNA and RNA.DHFR of particular interest in context with Toxoplasmosis because it is involved in the mechanism of action of certain drugs used to treat Toxoplasma Gondii infections.Pyrimethamine and Sulfadiazine are two commonly used medications to treat Toxoplasmosis.

Pyrimethamine(Lai 2017) inhibits the activity of parasite's DHFR enzyme interfering with the synthesis of folic acid, an essential nutrient required for the parasite's survival.This disruption eventually impairs the parasite's ability to replicate and grow resulting in its death. Sulfadiazine on the other hand is frequently used in conjunction with the Pyrimethamine to increase its therapeutic value.Sulfadiazine works by inhibiting folic acid via a different mechanism.It prevents the parasite from producing the dihydropteroate (Dennis *et al.* 2018; Lai 2017)which is another essential component for parasite's ability to produce folic acid.

These drugs enzyme at various stages of the folic acid synthesis pathway, inhibiting Toxoplasma Gondii growth and proliferation.

## 2 MATERIALS AND METHODS

The Current Study was carried out in Saveetha School of Engineering (SIMATS). Uniprot is a vast database of protein sequences and funtions.It is made up of several databases that cover protein sequences,Annotations and the other relevant data.Uniprot(UniProt Consortium 2015) contains precise information about numerous proteins such as their sequences,structures,domains,post-translational modifications(Chen *et al.* 2023) and functional annotations.Researches frequently require specific protein targets for molecular studies.Uniprot assists in the discovery and selection of these targets by providing valid and up to date information on proteins of interest.It contains data regarding protein structures when its accessible.It is crucial for molecular docking to have access to known protein structures or predicted structural models.This information assists in the examination of binding sites (Ravnik, Jukič, and Bren 2023),active sites other structural details required for docking studies.

Uniprot annotations often contain crucial information on protein activities such as active or binding sites.To predict how compounds interact with the protein of interest,molecular docking techniques primarily depends on finding these locations.Many proteins undergo post synthesis changes that alter their function and interaction.Uniprot has information on PTM's such as phosphorylation,glycosylation,and acetylation(Wang *et al.* 2023) which has a major effect on docking and understanding of protein ligand interactions.

Uniprot annotations contain details about protein involvement as well as their significance as potential therapeutic targets.This information is useful in guiding molecular docking experiments aimed at discovering or developing drugs that could interact with these specific (Latha and Saddala 2017)proteins that could interact with these specific proteins to treat illness.

Uniprot has been extensively utilized by researchers participating in accurate and reliable data about proteins,which is critical for understanding protein ligand interactions.So the Structure and function of the following DFHR was obtained from uniprot.

## 2.1 *Chimera*

Chimera (Pettersen *et al.* 2004)is a well-known software program for visualizing, analyzing, and manipulating molecular structures. Structure minimization is an important process in Chimera that involves optimizing a molecular structure to produce a more stable and more efficient conformation. The retrieved molecular structures would not be in a stable state so they are minimized through Chimera.structure minimization helps in the refinement of these structures by altering atomic locations to minimize negative interactions and reduce the energy.

PyRx (Dennis *et al.* 2018; Lai 2017; Latha and Saddala 2017) is a virtual screening application for computer-aided drug creation that aids in visualizing the interactions between various compounds and potential therapeutic targets.

# 3 RESULTS

In comparison to the other molecule,Mebendazole(Guerini *et al.* 2019) has a higher affinity (-6.8 kcal/mol). When compared to the other ligands, it has the highest affinity. Using Pymol, we were able to see the complicated interactions between Mebendazole and the DHFR protein (Figure 1.). The binding energies of the four compounds ranged from -6.8 to -6.4 kcal/mol. The energy spectrum (Sun *et al.* 2023; Wittenstein *et al.* 2023) proved the best chemical binding. Mebendazole exhibited the lowest binding affinity of the five compounds examined, -6.8kcal/mol. When compared to other compounds, it demonstrated the highest affinity.Mebendazole possesses the highest affinity when compared to Coumaphos.In the sophisticated visualization provided using Ligplot (Figure 2), we could see the interactions between the DHFR and Mebendazole. Mebendazole works as a DHFR protein inhibitor and treats Toxoplasmosis since the Lipinski rule (Table 1) determined that the compound's qualities were met.ADMET properties were examined, and the proposed ligand satisfied the Lipinski rule (Heifetz 2022).

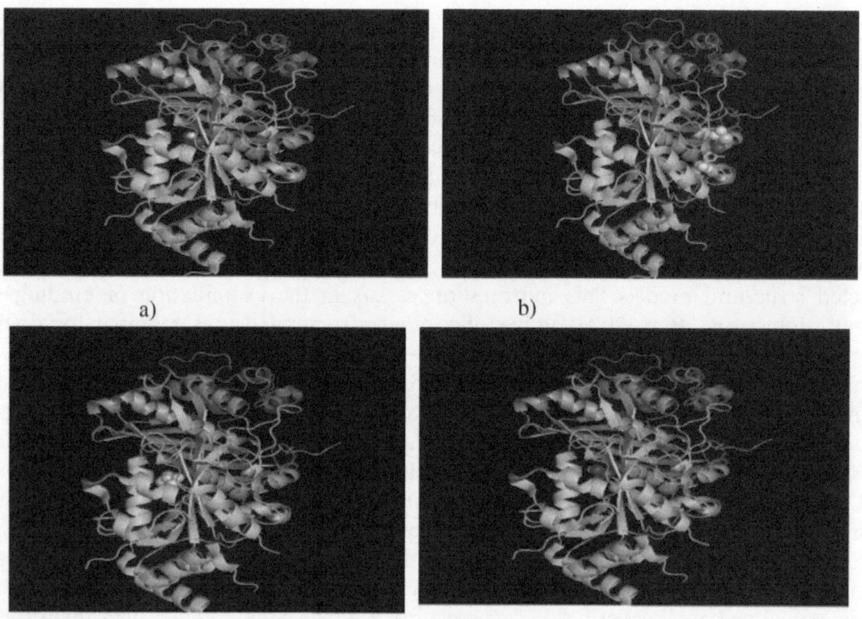


Figure 1. 3D Structures of DHFR protein with a) Mebendazole b) Fipronil c) Lariam d) Fenbendazole

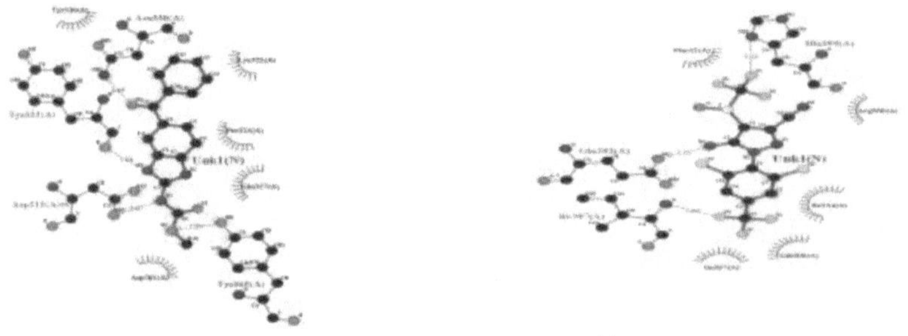

a) b)

Figure 2. 2D structures of DHFR protein with a) Mebendazole b) Fipronil.

## 4 DISCUSSION

In this study, the structural interactions between the Toxoplasmosis-causing protein DHFR and a few distinct lead chemicals produced the most effective combination for blocking the protein.The visualization tool PyRx provided the binding affinity, which was used to choose the best medication to treat the condition.Ligplot(Wallace, Laskowski, and Thornton 1995) was used to analyze how the newly created compounds reacted.Each ligand's interactions with the protein DHFR via hydrogen bonds and hydrophobic interactions are depicted (Figure 2). The protein against the ligand Mebendazole compound formed three hydrogen bonds with the hydrophobic interactions Tyr326,Asp361,Gln327,Pro324,Lys321.And the compound Fipronil (Tingle *et al.* 2003)with the target site Cys490 formed three H bonds with the length of 2.92,3.06,3.24 followed by Ile567,Gln206,Ile314,Phe32,Arg566 hydrophobic interactions (Heifetz 2022; Nakai 2018).The ligand with the target Glu209 formed four hydrogen bonds with Lys59,Arg14,Ser36 the lengths of the bonds are 3.09,3.19,3.02,3.00.

The molecular weight and the octanol-water partition coefficient (Log P) must be calculated. Mebendazole has a molecular weight of 295.29 g/mol, a log P of 1.47, two hydrogen bond donors and four hydrogen bond acceptors.This compound is the best since it followed the Lipinski rule of five. Table 1 examines the Lipinski's rule properties for different compounds.

Table 1. Lipinski rule analysis for Mebendazole, Fipronil, Lariam, Coumaphos.

| S.NO | Properties | Mebendazole | Fipronil | Lariam |
|---|---|---|---|---|
| 1 | Molecular Weight(kcal/mol) | 295.29 g/mol | 437.15 g/mol | 378.31 g/mol |
| 2 | Log P | 1.47 | 2.26 | 2.78 |
| 3 | Hydrogen Bond acceptors(HBAs) | 4 | 9 | 9 |
| 4 | Hydrogen bond donors(HBDs) | 2 | 1 | 2 |

## 5 CONCLUSION

Mebendazole is the most effective DHFR protein inhibitor when compared to other medications such as Fipronil,Lariam,Coumaphos(Valdovinos-Flores *et al.* 2016), according to our research.The use of PyRx binding energies simplifies the selection of the appropriate ligand.Throughout these molecular docking investigations, the optimal interactions between the protein and the ligand have been observed.Mebendazole functions in the body as an enzyme inhibitor and nuclear receptor ligand.We conclude that these findings will be valuable in further research into the effects of Mebendazole on the DHFR protein.

## REFERENCES


Averett, D. R. (1979). Aspects of Pyrimidine Biosynthesis in Neisseria Gonorrhoeae: De Novo Synthesis of Carbamoyl Aspartate, and Evaluation of Dihydrofolate Reductase as a Target Enzyme for Therapeutic Exploitation.

Chen, W., Ding, Z., Zang, Y. and Liu, X. (2023). Characterization of proteoform post-translational modifications by top-down and bottom-up mass spectrometry in conjunction with annotations. *Journal of Proteome Research*, 22(10), 3178–3189. https://doi.org/10.1021/acs.jproteome.3c00356

Dennis, M. L., Lee, M. D., Harjani, J. R., Ahmed, M., DeBono, A. J., Pitcher, N. P. and Wang, Z.-C. (2018). 8-Mercaptoguanine derivatives as inhibitors of dihydropteroate synthase. *Chemistry*, 24(8), 1922–1930. https://doi.org/10.1002/chem.201704432

Dubey, J. P. (2008). The history of Toxoplasma gondii–the first 100 years. *The Journal of Eukaryotic Microbiology*, 55(6), 467–475. https://doi.org/10.1111/j.1550-7408.2008.00345.x

Guerini, A. E., Triggiani, L., Maddalo, M., Bonù, M. L., Frassine, F., Baiguini, A. and Rizzolio, F. (2019). Mebendazole as a candidate for drug repurposing in oncology: An extensive review of current literature. *Cancers*, 11(9), Article 1284. https://doi.org/10.3390/cancers11091284

Heifetz, A. (2022). *Artificial Intelligence in Drug Design*. Humana.

Karaca Adıyeke, S. and Ture, G. (2022). Vitreoretinal interface anomalies in patients with a history of toxoplasmosis retinochoroiditis. *Japanese Journal of Ophthalmology*, 66(3), 320–325. https://doi.org/10.1016/j.jjo.2021.12.008

Lai, M. (2017). Impact of Drug Dose and Treatment Time on the Synergism Between Pyrimethamine and Sulfadoxine in the Mouse Malaria Model.

Latha, M. S. and Saddala, M. S. (2017). Molecular docking based screening of a simulated HIF-1 protein model for potential inhibitors. *Bioinformation*, 13(11), 388–393. https://doi.org/10.6026/97320630013388

Nakai, S. (2018). *Hydrophobic Interactions in Food Systems*. CRC Press.

Pettersen, E. F., Goddard, T. D., Huang, C. C., Couch, G. S., Greenblatt, D. M., Meng, E. C. and Ferrin, T. E. (2004). UCSF Chimera–a visualization system for exploratory research and analysis. *Journal of Computational Chemistry*, 25(13), 1605–1612. https://doi.org/10.1002/jcc.20084

Ravnik, V., Jukič, M. and Bren, U. (2023). Identifying metal binding sites in proteins using homologous structures, the MADE approach. *Journal of Chemical Information and Modeling*, 63(16), 5204–5219. https://doi.org/10.1021/acs.jcim.3c00923

Shekari, N., Shanehbandi, D., Kazemi, T., Zarredar, H., Baradaran, B. and Jalali, S. A. (2023). VISTA and its ligands: The next generation of promising therapeutic targets in immunotherapy. *Cancer Cell International*, 23(1), Article 265. https://doi.org/10.1186/s12935-023-02439-3

Sun, R., Deng, J., Wu, X., Hao, M., Ma, K., Ma, Y. and Zhao, C. (2023). High anisotropy in electrical and thermal conductivity through the design of aerogel-like superlattice (NaOH)NbSe. *Nature Communications*, 14(1), Article 6689. https://doi.org/10.1038/s41467-023-06923-3

Tingle, C. C. D., Rother, J. A., Dewhurst, C. F., Lauer, S. and King, W. J. (2003). Fipronil: Environmental fate, ecotoxicology, and human health concerns. *Reviews of Environmental Contamination and Toxicology*, 176, 1–66. https://doi.org/10.1007/978-1-4419-9407-9_1

UniProt Consortium. (2015). UniProt: A hub for protein information. *Nucleic Acids Research*, 43(Database issue), D204–D212. https://doi.org/10.1093/nar/gku989

Valdovinos-Flores, C., Gaspar-Ramírez, O., Heras-Ramírez, M. E., Lara-Álvarez, C., Dorantes-Ugalde, J. A. and Saldaña-Loza, L. M. (2016). Boron and coumaphos residues in hive materials following treatments for the control of Aethina tumida Murray. *PloS One*, 11(4), Article e0153551. https://doi.org/10.1371/journal.pone.0153551

Wallace, A. C., Laskowski, R. A. and Thornton, J. M. (1995). LIGPLOT: A program to generate schematic diagrams of protein-ligand interactions. *Protein Engineering*, 8(2), 127–134. https://doi.org/10.1093/protein/8.2.127

Wang, Q., Peng, W., Yang, Y., Wu, Y., Han, R., Ding, T. and Yang, J. (2023). Global analysis of lysine acetylation in the brain cortex of K18-hACE2 mice infected with SARS-CoV-2. *Proteomics*, 23(16), Article e2300096. https://doi.org/10.1002/pmic.202300096

Wittenstein, A., Caspi, M., Rippin, I., Elroy-Stein, O., Eldar-Finkelman, H., Thoms, S. and Rosin-Arbesfeld, R. (2023). Nonsense mutation suppression is enhanced by targeting different stages of the protein synthesis process. *PLoS Biology*, 21(11), Article e3002355. https://doi.org/10.1371/journal.pbio.3002355

*Hybrid and Advanced Technologies – S. Prasad Jones Christydass, Nurhayati Nurhayati and S. Kannadhasan. (Eds) © 2025 the Authors, ISBN 978-1-032-90254-8*
*Open Access: www.taylorfrancis.com, CC BY-NC-ND 4.0 license*

# Designing molecular lead compounds against the enolase protein through the process called comparative molecular docking which leads to toxoplasmosis

S. Shavana
*Research Scholar, Department of Bio Informatics, Saveetha School of Engineering, Saveetha Institute of Medical and Technical Sciences, Saveetha University, Chennai, Tamilnadu, India*

Arun John
*Project Guide, Department of Bio Informatics, Saveetha School of Engineering, Saveetha Institute of Medical and Technical Sciences, Saveetha University, Chennai, Tamilnadu, India*

ABSTRACT: The goal of the current work is to use in-silico docking to identify new ligands that inhibit the protein Enolase. In the present study, we retrieved information about Toxoplasma gondii's impact on the protein enolase using Uniprot.To determine the binding affinities of several ligands with the target protein, we employed PyRx as our docking tool. LigPlot was also utilized to examine the interactions within the protein ligand complex. Sulfaquinoxaline was therefore determined to be a possible ligand against the Enolase enzyme.So, we used PyRx to examine the different ligands' binding affinities against the Enolase protein.The greatest binding affinity and the best compact bound interactions with macromolecules are exhibited by sulfaquinoxaline.Further ADMET characteristics were examined, and the potential ligand meets the Lipinski rule.

*Keywords*: In-Silico, Uniprot, PyRx, Sulfaquinoxaline, Ligand, ADMET characteristics

## 1 INTRODUCTION

Toxoplasmosis is a parasitic infection caused by the protozoan Toxoplasma Gondii. This parasite can spread to most warm-blooded species, including humans, and can be found across the world. Toxoplasma gondii's principal host is the cat (Sheng *et al.* 2023), where the parasite may complete its life cycle.Human infection is frequently caused by the consumption of oocysts, which are latent, tough-walled objects secreted in the feces of infected cats, or by consuming undercooked or raw meat containing parasite cysts (Karakavuk *et al.* 2022). Congenital transmission can also occur when a pregnant mother becomes infected and the parasite crosses the placenta to reach the fetus.The majority of healthy people infected with Toxoplasma Gondii may not show any symptoms or may have minor flu-like symptoms. Individuals with compromised immune systems, such as those with HIV/AIDS or undergoing immunosuppressive therapies, are more vulnerable to the parasite.Furthermore, infection during pregnancy might result in serious difficulties for the growing fetus. Toxoplasmosis is often diagnosed with blood tests that detect parasite antibodies. Treatment is usually prescribed for symptomatic scenarios or for people with weakened immune systems, and it mainly consists of a course of antibiotics (Dunay *et al.* 2018). Preventive strategies include completely cooking meat, washing fruits and vegetables, practicing good hygiene (particularly handwashing), and avoiding contact with cat excrement or keeping strict cleanliness when handling cat litter.Addressing the mechanisms of transmission, risk

DOI: 10.1201/9781003559115-19

factors, and preventative strategies for Toxoplasma Gondii infection (Gov 2014) is critical, especially for pregnant women (Amagbegnon *et al.* 2023) and those with compromised immune systems.Enolase is an enzyme that is important in the glycolytic pathway, which turns glucose into energy (Mouveaux *et al.* 2014).

### 1.1 *Enolase protein*

Enolase has been found as a multifunctional protein in Toxoplasma Gondii that is engaged not only in glycolysis (Mouveaux *et al.* 2014) but also in other vital procedures associated to the parasite's invasion and survival within its host.Many studies have shown that enolase in Toxoplasma Gondii performs tasks other than glycolysis. It has been linked to the parasite's capacity to infiltrate host cells. Enolase most likely increases parasite adherence to host cells and aids in invasion.In addition, Toxoplasma enolase has been discovered as a surface protein, which means it is found on the parasite's exterior. Due to its surface location, it can interact with host cells, perhaps aiding in immune evasion or modulating the host immune response.Enolase is a crucial protein for the survival and pathogenicity (Behnke *et al.* 2016; Mouveaux *et al.* 2014) of Toxoplasma Gondii due to its multifunctionality in glycolysis, adhesion, invasion, and maybe immunological regulation.

## 2 MATERIALS AND METHODS

The current study was carried out at Saveetha School of Engineering (SIMATS).The current study's analyses were carried out with the use of bioinformatics technologies such as Uniprot,Pubchem,PyRx,Ligplot.

### 2.1 *Uniprot*

UniProt is a valuable resource (Magrane and UniProt Consortium 2011) for learning about proteins like enolase and their activities in many organisms, including Toxoplasma Gondii. The amino acid sequence of T. gondii's enolase protein is available on UniProt.

### 2.2 *Pubchem*

PubChem is an excellent site to search for chemical information, including three-dimensional (3D) structures of ligands utilized in molecular docking studies.PubChem gives access to 3D structures of chemicals (National Center for Biotechnology Information, n.d.). These structures are critical for molecular docking research because they define how ligands interact with target proteins.And the following ligands such as Niclosamide,Coumaphos were retrieved using the Pubchem database.

### 2.3 *Docking software PyRx*

It is a free and open-source software for virtual screening and drug discovery.It features a simple interface and uses multiple algorithms to make the process of screening chemical libraries against protein targets quicker. PyRx's (Dallakyan and Olson 2015) primary goal is to aid in the identification of expected drug compounds by examining how various small molecules or chemicals interact with a target protein of interest.PyRx predicts the binding mechanism and affinity of small compounds to a protein target using Autodock Vina, an important molecular docking software. It examines how possible drug molecules fit into the target's binding site, providing knowledge about their therapeutic potential.PyRx's docking technique generally involves

Loading the receptor protein structure, which is normally in PDB (Burley *et al.* 2017) or PDBQT format, and prepare the ligand or small molecule structure for docking in PDBQT format.Locate or define the active site or binding pocket within the receptor where the ligands will dock.Set a grid box to define the search space surrounding the active site. This grid describes the area within the protein structure where ligands can dock.Start the docking calculation within PyRx, allowing Autodock Vina (Trott and Olson 2010) to do the molecular docking.The software examines and predicts the binding conformations and affinities of the ligands within the receptor protein's designated active region.Examine the docking scores and conformations of numerous ligands to determine their potential binding affinity.

### 2.4 *LIGPLOT*

Using the LIGPLOT tool (Wallace *et al.* 1995), protein-ligand complexes can be depicted schematically in 2-D.It was designed to generate schematic diagrams of protein residue interactions with small molecule ligands.The tool assists in understanding the binding mechanisms and unique interactions that take place between proteins and ligands.It highlights the vital interactions between the protein (Cai and Wang 2012) and the ligand by visualizing the binding site residues associated in ligand binding.

## 3 RESULTS

Molecular docking was performed among the set of ligands against the protein Enolase.The protein and chemical had a strong binding connection, which was confirmed by the lowest binding affinity score (Stroganov *et al.* 2008).The five chemicals' binding energies ranged from -5.1to -6.1 kcal/mol.The energy range confirmed the optimal chemical binding.The energy spectrum proved the best chemical binding.Among the four compounds, sulfaquinoxaline had the lowest binding affinity, -6.1 kcal/mol.It had the highest affinity when compared to other compounds.

The detailed visualization was created using Ligplot, which allowed us to see the interactions between the Sulfaquinoxaline and Enolase proteins.Sulfaquinoxaline cures the condition by acting as an Enolase protein inhibitor (Chandel *et al.* 2018).

## 4 DISCUSSION

The structural interactions between the Toxoplasmosis-causing protein and a few unique lead compounds provided the most efficient combination for inhibiting the protein in our study.The newly produced chemicals were examined using Ligplot to see how they reacted. Each ligand's interactions with the protein Enolase via hydrogen bonds and hydrophobic interactions are depicted in Figure 1. The ligand Sulfaquinoxaline with the target Glu209 formed four hydrogen bonds with Lys59,Arg14,Ser36 the lengths of the bonds are 3.09,3.19,3.02,3.00. And the Arg49,Asn16,Ala38,Glu47,Gly37 hydrophobic interactions were observed.The ligand Mebendazole formed two hydrogen bonds with the hydrophobic interactions Glu44,Gly37,Arg14,Asn16,Ala38.The ligand Niclosamide formed three hydrogen bonds with Lys59,Thr40,Glu47.And it resulted in the formation of three hydrogen bonds.The ligand coumaphos with the same target Glu209 resulted in the formation of two hydrogen bonds with Arg14,Asn16,Ala38,Lys59,Gly41,Glu44 hydrophobic interactions.The hydrophobic interactions and formation of hydrogen bonds demonstrate the ability of the ligand Sulfaquinoxaline to behave as hit-inhibiting agents for the protein Enolase.

The molecular weight and the octanol-water partition coefficient (Log P) must be calculated. Mebendazole has a molecular weight of 295.29 g/mol, a log P of **3.54,** two hydrogen bond donors and four hydrogen bond acceptors, and four hydrogen bond acceptors.This compound is the best since it followed the Lipinski rule of five. Table 1 examines the Lipinski's rule properties for different compounds.

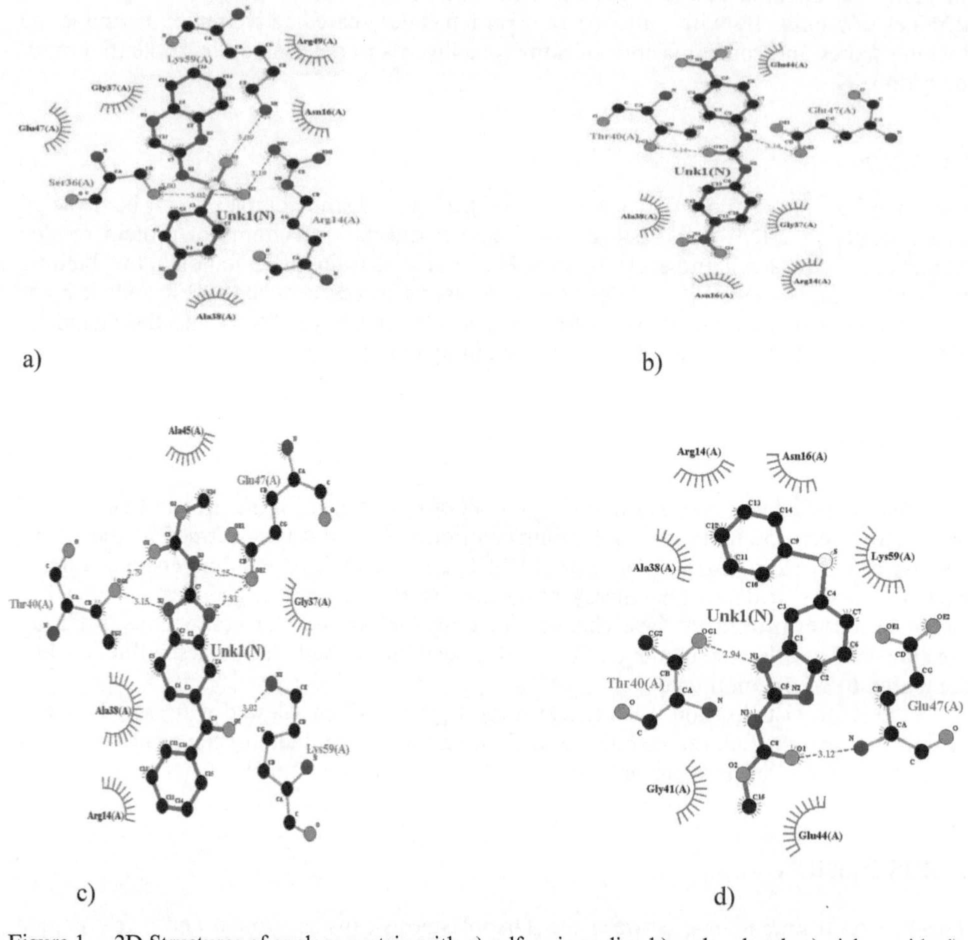


Figure 1. 2D Structures of enolase protein with a) sulfaquinoxaline b) mebendazole c) niclosamide d) coumaphos.

Table 1. Lipinski rule analysis for sulfaquinoxaline, mebendazole, niclosamide, coumaphos.

| S. No | Properties | Sulfaquinoxaline | Mebendazole | Niclosamide | Coumaphos |
|---|---|---|---|---|---|
| 1 | Molecular Weight(kcal/mol) | 300.34 g/mol | 295.29 g/mol | 327.12 g/mol | 362.77 g/mol |
| 2 | Log P | 0.97 | 1.47 | 2.09 | 3.11 |
| 3 | Hydrogen Bond acceptors (HBAs) | 4 | 4 | 4 | 5 |
| 4 | Hydrogen bond donors(HBDs) | 2 | 2 | 2 | 0 |

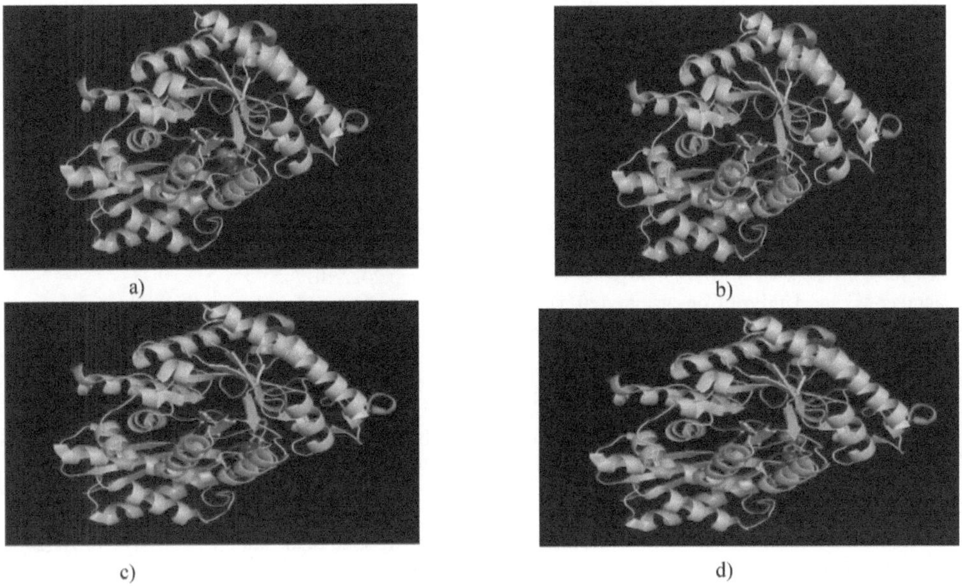


Figure 2. 3D structures of wild enolase protein docked with a) sulfaquinoxaline b) mebendazole c) niclosamide d) coumaphos.

## 5 CONCLUSION

In comparison to other drugs such as Mebendazole, Niclosamide, and Coumaphos, our research shows that Sulfaquinoxaline is the most effective Enolase protein inhibitor.The usage of binding energies from PyRx makes it easier to choose the right ligand.Numerous studies have indicated that Sulfaquinoxaline has beneficial pharmacological effects. It acts as an enzyme inhibitor and nuclear receptor ligand in the body.We conclude that these findings will be useful in the experimental investigation of Sulfaquinoxaline against the Enolase protein.

## REFERENCES

Amagbégnon, R., Dechavanne, C., Dambrun, M., Yehouénou, U., Akondé, N., Migot-Nabias, F. and Adéothy, A. (2023). Correction notice to: 'Seroepidemiology of toxoplasmosis in pregnant women and detection of infection acquired during pregnancy in Cotonou, Benin' - Parasite 30, 43 (2023). *Parasite*, 30 (November), 48. https://doi.org/10.1051/parasite/2023040

Behnke, M. S., Dubey, J. P. and Sibley, L. D. (2016). Genetic mapping of pathogenesis determinants in Toxoplasma gondii. *Annual Review of Microbiology*, 70(September), 63–81. https://doi.org/10.1146/annurev-micro-102215-095312

Burley, S. K., Berman, H. M., Kleywegt, G. J., Markley, J. L., Nakamura, H. and Velankar, S. (2017). Protein Data Bank (PDB): The single global macromolecular structure archive. In W. N. Hunter and S. L. M. Poulos (Eds.), *Methods in Molecular Biology*, 1607, 627–641. https://doi.org/10.1007/978-1-4939-7000-1_25

Cai, J. and Wang, R. (2012). *Protein Interactions*. BoD – Books on Demand.

Chandel, T. I., Zaman, M., Khan, M. V., Ali, M., Rabbani, G., Ishtikhar, M. and Khan, R. H. (2018). A mechanistic insight into protein-ligand interaction, folding, misfolding, aggregation and inhibition of protein aggregates: An overview. *International Journal of Biological Macromolecules*, 106(January), 1115–1129. https://doi.org/10.1016/j.ijbiomac.2017.08.102

Dallakyan, S. and Olson, A. J. (2015). Small-molecule library screening by docking with PyRx. In H. H. Stote, and R. C. E. Cheng (Eds.), *Methods in Molecular Biology*, 1263, 243–250. https://doi.org/10.1007/978-1-4939-2269-7_21

Dunay, I. R., Gajurel, K., Dhakal, R., Liesenfeld, O. and Montoya, J. G. (2018). Treatment of toxoplasmosis: Historical perspective, animal models, and current clinical practice. *Clinical Microbiology Reviews*, 31(4). https://doi.org/10.1128/CMR.00057-17

Gov, L. (2014). Unraveling host and parasite pathways in human innate immunity to Toxoplasma gondii.

Karakavuk, M., Can, H., Karakavuk, T., Gül, A., Alak, S. E., Gül, C., and Döşkaya, A. D. (2022). Rapid detection of Toxoplasma gondii DNA in cat feces using colorimetric loop-mediated isothermal amplification (LAMP) assays targeting RE and B1 genes. *Comparative Immunology, Microbiology and Infectious Diseases*, 81(February), 101745. https://doi.org/10.1016/j.cimid.2022.101745

Magrane, M. and UniProt Consortium. (2011). UniProt Knowledgebase: A hub of integrated protein data. *Database: The Journal of Biological Databases and Curation*, 2011(March), bar009. https://doi.org/10.1093/database/bar009

Mouveaux, T., Oria, G., Werkmeister, E., Slomianny, C., Fox, B. A., Bzik, D. J. and Tomavo, S. (2014). Nuclear glycolytic enzyme enolase of Toxoplasma gondii functions as a transcriptional regulator. *PloS One*, 9(8), e105820. https://doi.org/10.1371/journal.pone.0105820

National Center for Biotechnology Information. (n.d.). PubChem: Information on biological activities of small molecules.

Sheng, L., Xue, Q., Xu, S., Can, F., Yao, N., Zou, M., and Xiao, L. (2023). Rapid and visual detection of Toxoplasma gondii oocyst in cat feces using loop-mediated isothermal amplification (LAMP) assay. *Scientific Reports*, 13(1), 17269. https://doi.org/10.1038/s41598-023-17347-1

Stroganov, O. V., Novikov, F. N., Stroylov, V. S., Kulkov, V. and Chilov, G. G. (2008). Lead finder: An approach to improve accuracy of protein-ligand docking, binding energy estimation, and virtual screening. *Journal of Chemical Information and Modeling*, 48(12), 2371–2385. https://doi.org/10.1021/ci800248s

Trott, O. and Olson, A. J. (2010). AutoDock Vina: Improving the speed and accuracy of docking with a new scoring function, efficient optimization, and multithreading. *Journal of Computational Chemistry*, 31(2), 455–461. https://doi.org/10.1002/jcc.21334

Wallace, A. C., Laskowski, R. A. and Thornton, J. M. (1995). LIGPLOT: A program to generate schematic diagrams of protein-ligand interactions. *Protein Engineering*, 8(2), 127–134. https://doi.org/10.1093/protein/8.2.127

*Hybrid and Advanced Technologies – S. Prasad Jones Christydass, Nurhayati Nurhayati and S. Kannadhasan. (Eds)* © *2025 the Authors, ISBN 978-1-032-90254-8*


# An innovative method for earthquake prediction system based on LSTM-GAN model

Diwakar Chaudhary
*Assistant Professor, Mangalmay Institute of Management and Technology, Greater Noida, India*

Raghu Babu Uppara
*Associate Professor, Srinivasa Ramanujan Institute of Technology, Rotary Puram, BKS Mandal, Anantapur, India*

S. Senthamil Kumar
*Professor, Department of Civil Engineering, Periyar Maniammai Institute of Science and Technology, Vallam, Thanjavur, India*

Rohan Vilas Ambekar
*Assistant Professor, Department of Civil Engineering, Dr. D. Y. Patil Institute of Technology Pimpri Pune, India*

P. Penchala Prasanth
*Assistant Professor, N B K R Institute of Science and Technology, Vidyanagar, Tirupati, Andhra Pradesh, India*

Dileep Pulugu
*Professor, Department of CSE, Malla Reddy College of Engineering and Technology, Hyderabad, Telangana, India*

ABSTRACT: Evaluation of earthquake forecast hypotheses requires robust statistical approaches. A clear specification of the model is necessary for objective recognition of the abnormality or antecedent in any situation and by any observer. Valid prediction hypotheses aim to enhance success and avoid false alarms. Probability gain is a common method for assessing prediction quality based on a precursor. Scientists may apply a statistical approach to calculate the likelihood of recorded seismic events and compare them to other hypotheses. This approach can be applied to algorithms that calculate the density distribution of earthquake probability in space, time, and magnitude. An independent set of observations should be used to test the validity of a novel hypothesis, regardless of the method used to construct it. The proposed approach consist of preprocessing, feature selection and training the model. For training the model it uses LSTM, GAN-ELM and LSTM-GAN. The proposed approach LSTM-GAN performs well and produces an accuracy of about 96.3%.

*Keywords*: Earthquake Prediction, Expert Systems (ESs), Redundancy Criterion (RC), Long Short-Term Memory (LSTM), Generative Adversarial Network (GAN)

## 1 INTRODUCTION

Due to the difficulty of the task, several academics have tried different approaches in an effort to predict when earthquakes would occur. Since the early, a distinct area of artificial intelligence has been developed specifically for the purpose of earthquake prediction using seismic data from different zones. When it comes to earthquake prediction, using expert systems (ESs) has shown promising results. The usefulness of ES for earthquake prediction

DOI: 10.1201/9781003559115-20

has been the subject of several proposed models, frameworks, and approaches. A number of expert systems, including hybrid, rule-based, and fuzzy systems, have been reviewed in the literature for their ability to forecast future earthquakes using instrumental and historical data. Researchers collected all approaches reported in the recent decade that used expert systems to predict earthquakes and published them. An effective and dependable method is needed for earthquake prediction, which is why expert system technology is being used. The classification of landslip venerability and seismic risk is an area where ESs have proven useful in earthquake engineering. Researchers have developed an LSTM-GAN model by studying ionosphere disturbance and measuring the characteristics of ionosphere changes. When investigating unexplained occurrences, such as aberrant animal behaviour or a fault rupture in the subsurface, the researchers brought attention to the nebulousness of LSTM-GAN.

## 2 LITERATURE SURVEY

Earthquakes are a type of natural disaster that can cause widespread destruction. These events usually appear out of nowhere, causing monetary and human damages to society [1]. Soil liquefaction and fault rupture are two ecological repercussions that can arise from this. Other natural calamities including floods and avalanches, tsunamis, and debris collapse [2] are all possible outcomes. [3] The features and variables that impact the have prompted a plethora of proposed techniques, although Few studies have been successful in accurately predicting the location and timing of earthquakes due to the random nature of these natural disasters complex event having numerous determinants that are difficult to measure. [4] Mathematical modelling encompasses efforts to create earthquake prediction models utilising a range of statistical and mathematical approaches. [5]is one study that employed a PHMM in order to foretell the likelihood of a Sumatran earthquake. [6] We run extensive experiments for fore-and aftershock sequences, concurrent occurrences in several source locations, out-of-network and offshore scenarios, and possibly damaging earthquakes so that we can understand how ShakeAlert runs. [7] Within a three-minute period, our combined datasets show that two tremors occur. Such events are rare, but they can have a huge effect when they do occur. [8] More than one little earthquake can also contribute to a larger one. [9] We know from California ShakeAlert data that source-based techniques and the current logic of the Shake Alert system struggle to differentiate between events that happen near together in time and space. [10] This behaviour is anticipated due to the fact that FinDer's methodology for detecting two near-simultaneous events is still in its early stages and may not be applicable in extreme cases such as the ones investigated here. [11] Movement and vibrations under the earth's surface, caused by seismic activity, release energy that has been trapped in rocks. Tremors, injuries, and monetary loss are typical results of natural disasters like earthquakes [12]. Every day, earthquakes strike various regions of the world. The most vulnerable countries to earthquakes include Taiwan, Japan, Indonesia, southern California, Iran, and Turkey [13]. Therefore, scientists devote a great deal of energy to discovering solutions that lessen the impact of these negative outcomes.

## 3 PROPOSED SYSTEM

Scientific advances in our knowledge of the lithosphere's hierarchical structure and dynamics—based on data from detailed tracking and proof of its space-energy resemblance at international, national, and regional levels—have prompted the development of a repeatable intermediate-term middle-range earthquake prediction process.

### 3.1 *Preprocessing*

The dataset is the main factor that determines the system's efficiency. This dataset comprises 2584 samples and 18 characteristics. Following the collection of all pertinent data, the preprocessing stage should be applied to the data. Preprocessing is applied to non-numerical data in this particular system. In other words, all of the features' non-numerical values are transformed into numerical ones. The dataset is then homogenised [14]. The class imbalance issue is fixed by resampling the minor class's data following the first round of analysis. This is included in the preparation of data as well. After that, the dataset is divided into training and test sets using the k-fold cross-validation technique and the sampling method.

## 3.2 *Selection of features*

The feature computation approach collects the most information, although a reduced feature set may be used for learning instead of all 50. Duplicate or undifferentiating features may be ignored. A two-step feature selection procedure based on maximum relevance and least redundancy addresses this problem. After calculating all accessible features, LSTM-GAN selects the most relevant and discriminating traits. The curse of dimensionality can be avoided with Deep Learning algorithms during this period. Since the feature selection process is applied differently for every dataset, it is included into the prediction model [15]. This study's seismic regions have distinct seismic quality captured by calculated features, resulting in different seismic datasets. A feature in a dataset with low information richness may have the opposite effect on the system when feature selection is performed for all locations. Thus, it is incorrect to say that one area's traits are optimal for all others.

### 3.2.1 *Maximum relevance*

There are two stages to feature selection, the first of which is the irrelevant filter. To make better predictions, it gets rid of attributes that aren't significant and don't include enough data. While the proposed article does include some suggestions for changes to its formulation, it is still only a proposal. Various classification issues, including medical image processing and cancer classification, have previously made use of this technique for feature selection.

### 3.2.2 *Minimum redundancy*

Incorporating features that display redundant information in the dataset renders them useless for earthquake prediction. In the second step of feature selection, features containing redundant information are removed. In, we present the concept of a feature redundancy filter and offer some suggestions for improving its execution. The underlying principle of this method is to compute MI in-between each feature. It is said that any two features with maximal MI have duplicate information. As a result, during the learning process, a fixed redundancy criterion (RC) is empirically chosen and applied to all three regions.

## 3.3 *Model training*

### 3.3.1 *LSTM-GAN*

When it comes to processing time series data, LSTM really shines, but when it comes to extracting features from data and generating models, GAN really shines. So, this article aims to integrate LSTM and GAN, uses LSTM to extract time series characteristics from time series data, and then uses GAN to further extract deep features and build a normal class data model. Anomaly classes are identified when the model output from testing data beyond the normal data model's threshold; this allows for the realisation of earthquake prediction using time series data. The LSTM-GAN earthquake prediction algorithm has the following key advantages over the GAN model for 2D image processing: 1) Constructing a GAN using a one-dimensional convolution layer for use in feature extraction and time series data reconstruction. 2) Getting the LSTM layer to work with the GAN to get the time-related properties of the data extracted. 3) The earthquake prediction score is calculated using the generator's residual generate and the discriminate discriminator's loss.

The LSTM-GAN technique is structured as follows: in the Leaky-Relu layer, the inverse gradient of the neurons is learned; in the Conv1D and LSTM layers, temporal features are extracted; in the Batch Normalisation layer, the input distribution of each network layer is maintained; and in the Dropout layer, over-fitting is prevented.

In most layers, the generator and discriminator have identical structures; however, the discriminator uses a Dropout layer to avoid overfitting and an LSTM layer to preserve the time series data's temporal feature. Prior to applying them to anomaly detection, the generator and discriminator undergo training. The proposed LSTM-GAN algorithm for earthquake prediction has two stages, "train" and "test," as shown in the following pseudocode expression. The discriminator in the fusion model receives both the normal class data and randomly produced gaussian noise samples throughout training. To modify the discriminator, we first adjust the generator's settings such that its sample may be utilised as a normal class, and then we fine-tune the discriminator's parameters to discriminate between the two sets of data.

In the testing phase, data reconstruction cannot begin until the testing data feature has been extracted. Only then can the feature sequence be obtained [16]. To get the earthquake prediction system's score, the discriminator's discriminate loss and the generator's create residual are combined. The root-mean-squared error between the tested and rebuilt data sets is the generator's generate residual. The absolute value of the discrepancy between the discriminator's usual output 1 and the testing data is known as the discriminator's discriminate loss. When the score exceeds the threshold $\theta$, an earthquake is detected in the testing data. In a perfect world, with normal test data, the loss discrimination and the residual generation would both be zero. If the test data comes from an out-of-the-ordinary class, the theoretical discrimination loss will be close to 1 and the produce residual will be more than 0. Differentiating between the abnormal and normal categories can be achieved by adjusting the threshold according to the integration error. In this part, we verify the algorithm's ability to detect anomalies using time series data of two sets.

## 4 RESULT AND DISCUSSION

A number of approaches have been used to the difficult problem of fulfilling the prophecy of an earthquake. A basic definition and sufficient catalogues are required, nevertheless. Common geological factors have been utilised to demarcate these regions. The study's overall characteristics are grounded in qualitative data regarding earthquake prediction in addition to the training and testing datasets. The model illustrates the connection between predicted seismic data and the frequency of earthquakes.

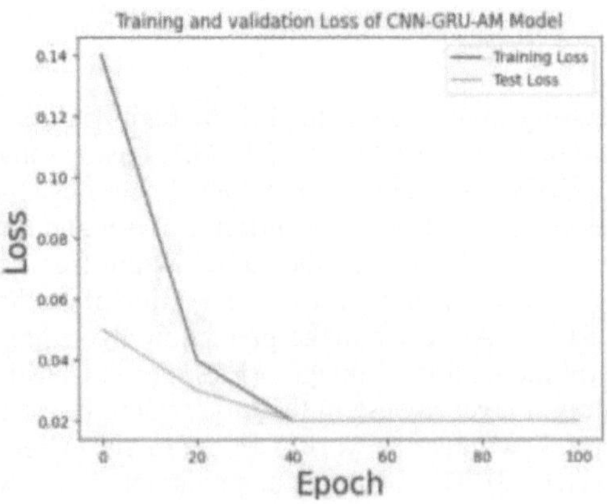


Figure 1. Graphing the outcomes of forecasting models.

Thanks to LSTM-GAN's sophisticated data learning capabilities, patterns can be extracted from the sparse datasets of large earthquakes, especially when each dataset is learned independently. We utilize the dropout function to ensure that all of our models are well-fitted and not overfitted. Figure 1 shows the fitting curves of these models.

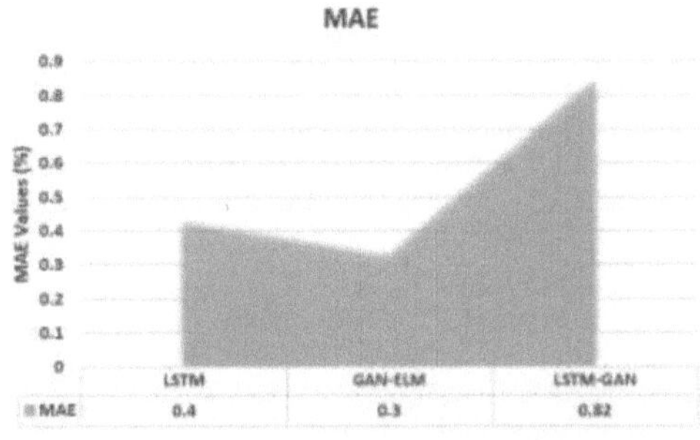


Figure 2. Evaluation criteria for mean absolute error (MAE) for all models when the dataset is split into 70% for training and 30% for testing.

Using a data split of 30% for testing and 70% for training, Figure 2 shows the MAE assessment criterion rates for all models in a column chart.

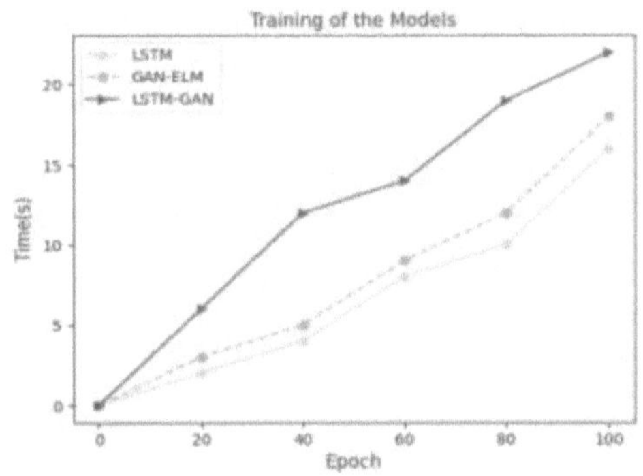


Figure 3. Training time Vs epoch.

Figure 3 shows a comparison of the training times of the suggested LSTM-GAN, LSTM, and GAN-ELM models.

## 5 CONCLUSION

The aftermath of a natural hazard—such as a flood, tornado, storm, earthquake, heat wave, or landslide—is known as a natural catastrophe. The complicated physical phenomena known as earthquakes, landslides, tsunamis, and volcanoes cause losses in terms of money, the environment, or human lives. It urgently requires a method to forecast such geological calamities. There are a lot of physical and environmental factors that go into the complicated process of predicting catastrophic catastrophes. Numerous methods grounded in empirical research and statistical analysis are available in published works. These natural risks can also be predicted using deep learning approaches. The suggested method uses LSTM, GAN-ELM, and LSTM-GAN for preprocessing, feature selection, and model training. LSTM-GAN performs best, with 96.3% accuracy.

## REFERENCES


[1] Alarifi, A. S., Alarifi, N. S. and Al-Humidan, S. (2012). Earthquakes magnitude predication using artificial neural network in northern Red Sea area. *Journal of King Saud University-Science*, 24(4), 301–313.

[2] Cui, P., Xiang, L. Z. and Zou, Q. (2013). Risk assessment of highways affected by debris flows in Wenchuan earthquake area. *Journal of Mountain Science*, 10, 173–189.

[3] Otari, G. V. and Kulkarni, R. V. (2012). A review of application of data mining in earthquake prediction. *International Journal of Computer Science and Information Technologies*, 3(2), 3570–3574.

[4] Boucouvalas, A. C., Gkasios, M., Tselikas, N. T. and Drakatos, G. (2015, June). Modified-Fibonacci-Dual-Lucas method for earthquake prediction. In *Third international conference on remote sensing and geoinformation of the environment (RSCy2015)* (Vol. 9535, pp. 400–410). SPIE.

[5] Marisa, M., Sembiring, U. A. and Margaretha, H. (2019, December). Earthquake probability prediction in sumatra island using Poisson hidden Markov model (HMM). In *AIP Conference Proceedings* (Vol. 2192, No. 1). AIP Publishing.

[6] Rajkumar, N., Kanimozhi, N., Saravanakumar, P., Koneru, S., Sapra, P. and Rastogi, R. (2023, April). Innovative Method for Earthquake Prediction System using Hybrid Convolutional Neural Network and SVM. In *2023 International Conference on Artificial Intelligence and Applications (ICAIA) Alliance Technology Conference (ATCON-1)* (pp. 1–6). IEEE.

[7] Chung, A. I., Henson, I. and Allen, R. M. (2019). Optimizing earthquake early warning performance: ElarmS-3. *Seismological Research Letters*, 90(2A), 727–743.

[8] Cochran, E. S., Bunn, J., Minson, S. E., Baltay, A. S., Kilb, D. L., Kodera, Y. and Hoshiba, M. (2019). Event detection performance of the PLUM earthquake early warning algorithm in southern California. *Bulletin of the Seismological Society of America*, 109(4), 1524–1541.

[9] Kilb, D., Bunn, J. J., Saunders, J. K., Cochran, E. S., Minson, S. E., Baltay, A. and Kodera, Y. (2021). The PLUM earthquake early warning algorithm: A retrospective case study of West Coast, USA, data. *Journal of Geophysical Research: Solid Earth*, 126(7), e2020JB021053.

[10] Kodera, Y., Saitou, J., Hayashimoto, N., Adachi, S., Morimoto, M., Nishimae, Y. and Hoshiba, M. (2016). Earthquake early warning for the 2016 Kumamoto earthquake: Performance evaluation of the current system and the next-generation methods of the Japan Meteorological Agency. *Earth, Planets and Space*, 68(1), 1–14.

[11] Asim, K. M., Martínez-Álvarez, F., Basit, A. and Iqbal, T. (2017). Earthquake magnitude prediction in Hindukush region using machine learning techniques. *Natural Hazards*, 85, 471–486.

[12] Asencio-Cortés, G., Martínez-Álvarez, F., Troncoso, A. and Morales-Esteban, A. (2017). Medium-large earthquake magnitude prediction in Tokyo with artificial neural networks. *Neural Computing and Applications*, 28, 1043–1055.

[13] Chin, T. L., Huang, C. Y., Shen, S. H., Tsai, Y. C., Hu, Y. H. and Wu, Y. M. (2019). Learn to detect: Improving the accuracy of earthquake detection. *IEEE Transactions on Geoscience and Remote Sensing*, 57(11), 8867–8878.

[14] Menon, A. P., Varghese, A., Joseph, J. P., Sajan, J. and Francis, N. (2020). Performance Analysis of different Classifiers for Earthquake prediction: PACE. *International Journal of Innovative Research in Technology (IJIRT)*, 7(2), 142–146.

[15] Asim, K. M., Idris, A., Iqbal, T. and Martínez-Álvarez, F. (2018). Earthquake prediction model using support vector regressor and hybrid neural networks. *PloS one*, 13(7), e0199004.

[16] Zhu, G., Zhao, H., Liu, H. and Sun, H. (2019, October). A novel LSTM-GAN algorithm for time series anomaly detection. In *2019 Prognostics And System Health Management Conference (PHM-Qingdao)* (pp. 1–6). IEEE.

*Hybrid and Advanced Technologies – S. Prasad Jones Christydass, Nurhayati Nurhayati and S. Kannadhasan. (Eds)* © 2025 the Authors, ISBN 978-1-032-90254-8


# Exploring job stress and job satisfaction in university teaching with a GA-ANN framework

V.L.L. Kusuma Kumari CH
*Assistant Professor, Department of Humanities and Sciences (Management), Vallurupalli Nageswararao Vgnana Jyothi Institute of Engineering and Technology (VNR VJIET), Bachupalli, Telangana, India*

Devadutta Indoria
*Assistant Professor & HOD, Department of Commerce, Vikram Dev University, Jeyoore, Odisha, India*

Dunya Mustayeva Azeri
*UNEC Business School, Deputy Director, Azerbaijan State University of Economics (UENC) Azerbaijan*

Dileep Pulugu
*Professor, Department of CSE, Malla Reddy College of Engineering and Technology, Hyderabad, India*

Nitin Girdharwal
*Head, Department of Management, Prestige Institute of Management & Research (PIMR), Indore, India*

P. Vanitha
*Associate Professor, Department of Management Studies, M.Kumarasamy College of Engineering, Karur, Tamilnadu, India*

ABSTRACT: Stress in the workplace can lead to health issues and workplace accidents. Stress and emotional states are common workplace experiences for many people, but are significantly more common among health care workers. An employee's level of stress and contentment on the job has a significant impact on the quality and productivity of their work. Model training, segmentation, and preprocessing are the three key parts. In order to process data that would be infeasible to process without data preprocessing, the data is amended to meet the specifications of each data mining technique. A feature extraction dataset was created by combining raw signals that had been de-noised using the sliding window method with other valuable features. So that we could obtain more accurate findings, we trained the model using the GA-ANN. In comparison to other methods, such as ANN and GA, the proposed strategy produced a much higher accuracy of 96.81%.

*Keywords*: Job Stress, Job Satisfaction, Theory of Reasonable Action (TRA), Teacher Educator (TE), Genetic Algorithm (GA)

## 1 INTRODUCTION

Chronic emotional and interpersonal stress in the workplace can lead to burnout, a disorder that negatively affects one's mental and physical health. It is associated with higher health costs for companies, a variety of negative emotions, and employee departure. Research has shown that both personal and professional factors contribute to the concerning high rates of burnout in the workplace. Among the many elements that might lead to burnout on the job,

DOI: 10.1201/9781003559115-21

both short-term and long-term stress in the workplace ranks high. Workplace stress exacerbates emotional exhaustion, a symptom of burnout. In addition, it is more likely that one will experience burnout if they feel that their work-related relationships do not offer sufficient support and trust. Predicted by the buffering model and supported by prior research, social support reduces burnout relationship between stress and burnout. Both the direct and indirect effects of job satisfaction on burnout provide credence to the idea that job satisfaction is a robust predictor of burnout. When stress at work is not properly dealt with, burnout can set in. It is critical to identify the root causes of burnout in order to mitigate its effects and prevent it from reducing the productivity of bank personnel. We aim to illuminate the manner in which work-related stress impacts different dimensions of burnout, and how factors such as job satisfaction and perceived social support impact these dimensions, so that Chinese bank managers can enhance their occupational mental health programs.

## 2 LITERATURE SURVEY

Job stress occurs when an employee's emotional well-being is negatively impacted due to a misalignment between their abilities and the demands of their job [1]. Excessive expectations at work often lead to stress reactions. The original meaning of teacher job stress, according to [2], was the negative experience that teachers had because of their inaccurate impression of the classroom setting. Research shows that working as a teacher or student is associated with higher rates of stress compared to working in other occupations [3]. Several sources have reported that academic personnel from different countries have experienced higher-than-average levels of occupational stress [4]. Committed individuals have shared beliefs, want to stay a member, and are eager to work for the organization. [5]found that teachers who genuinely care about their students, the school, and their classrooms tend to have stronger emotional connections to all three. Although job stress is thought to precede organizational commitment, the connection between the two has been imperfectly understood and explored inconsistently. [6] The theory of planned behaviour (TPB) is an expansion of the theory of reasonable action (TRA) that aims to clarify how people use their free will to make decisions. [7] It was discovered that the majority of people's actions are foreseeable because they employ reasoning while making judgments. It is true that options are explored before settling on a final choice [8]. Attitude and subjective norms are the two factors that influence people's actions, according to the TPB, which is based on behavioural research. Subjective norms originate from people's normative perspectives. [9] On the other hand, attitude stems from the belief that one's own actions can lead to positive results. It is reasonable to anticipate [10]some level of actual behaviour so long as these factors keep influencing people's actions. Evidence suggests that this theory may be useful for elucidating the relationship between employee attitudes and actions in the workplace, particularly in academic institutions.[11]The purpose of this study is to read up on teacher and teacher educator (TE) job satisfaction and motivation through research synthesis (RS) and literature review (LR). [12] Then, we'll come up with some suggestions and ideas to make the job better. In spite of their interchangeability, terms like "research review," "systematic review," and "research synthesis" do not appear to have a consensus on their legal definition[13]. Researchers and social scientists showcase their intellectual legacy through inventions, discoveries, and research outcomes through the use of RS, a method for systematic reviews that unifies their operations [14].

## 3 PROPOSED SYSTEM

Numerous recent breakthroughs have increased the demand for oncology care. These tendencies, together with the continuous changes to the healthcare system, impact the working conditions of systemic therapy professionals.

### 3.1 *Data preprocessing*

Data preprocessing, the set of operations carried out prior to the use of a data mining approach, is a crucial part of the famous Knowledge Discovery from Data process [15]. Processing data that would be physically impossible to process without first modifying it to fit the requirements of each data mining method is what data preprocessing is all about.

### 3.2 *Feature extraction*

The raw data was preprocessed and examined. Findings acquired from the physiological data alone from the chest-worn device are better than findings derived from a combination of the two, as pointed out by the initial authors. The dataset is skewed toward the baseline condition, which is neutral, rather than the stress or entertainment situations, which are imbalanced. In this investigation, the chest data is utilized further because merging the features of the two devices did not yield significantly better outcomes. The preprocessed data was separated from the sensor signals using a 10-second sliding window approach to prevent overlapping, with the exception of ACC[16].

### 3.3 *GA-ANN model training*

#### 3.3.1 *ANN*

ANN is a system of interconnected, small units (neurons) that accomplish tasks in a non-linear manner. All of the inputs to each neuron are weighted and then added together with a constant component termed bias. From this total, a transfer function is employed to produce an output.

$$l = k(r * M + z) \tag{1}$$

Each neuron in a neural network has two scalar parameters: weight ($r$) and bias ($z$). It all comes down to minimising a certain error measure between the network's output and the targets it's aiming to attain while training a neural network to do a specific task. The process of minimizing mistakes iteratively keeps going until a certain convergence condition is met.

#### 3.3.2 *GA*

Natural selection is the basis of Genetic Algorithm (GA), an optimization and search method that aims to replicate the process of biological evolution. The process begins with the production of a random population, continues with the selection of individuals from that population to form the next generation, and culminates with the evaluation of the fitness function after a population of individual solutions has been modified through mutation and crossover. Here we will go over the main steps. Minimizing an objective function, often known as a fitness function, over multiple generations is the goal of this famously good global search algorithm. The weights and biases of the neural network serve as the solution's parameters in this case. The GA representation of the issue is formed by providing the objective function with a string of biases and weights to utilize as solution encoding parameters. The mean squared error fitness function can be expressed as follows:

$$k = \frac{1}{E}\sum_{h=1}^{E} (d_h - x_h)^2 \tag{2}$$

The number of data items in the training dataset is denoted by $E$, and the intended target and training output are represented by $d_h$ and $x_h$, respectively.

## 4 RESULT AND DISCUSSION

Job satisfaction, social support perceptions, and workplace stress are important factors in occupational burnout. Still, there is a dearth of literature on how stress affects other components of burnout, such as how satisfied bank employees are with their jobs and how much social support employees feel.

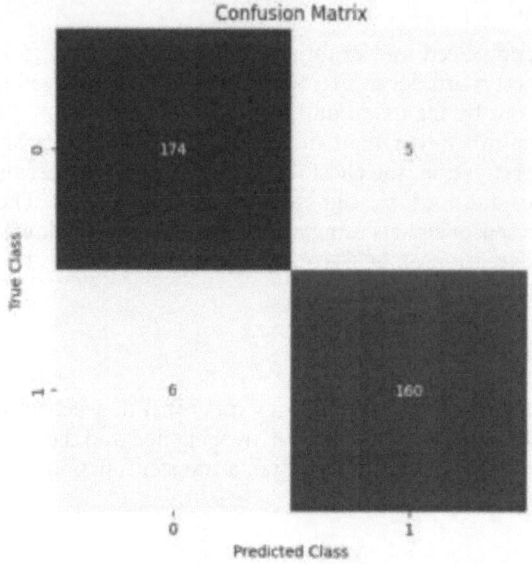


Figure 1. The confusion matrix analysis

There are 160 real negative values, 5 false negatives, 6 false positives, and 174 true positives on the confusion matrix in Figure 1.

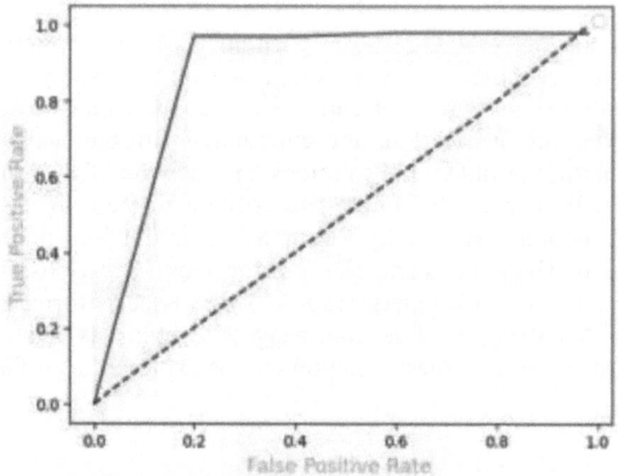


Figure 2. Performance evaluation through ROC.

Figure 2 ROC curve demonstrates the proposed network's binary classification accuracy. This graphic compares two rates: the True Positive Rate (TPR) and the False Positive Rate (FPR). The classifier does a fantastic job, with an AUC of 0.98.

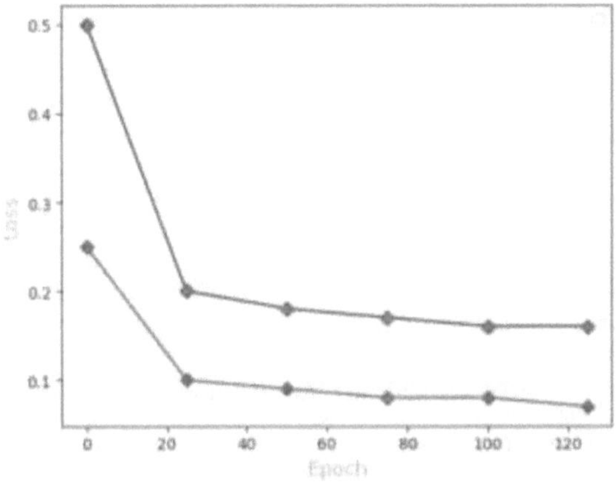


Figure 3. Training and validation loss of proposed model.

During the training phase, loss changes with increasing epochs, as seen in Figure 3.

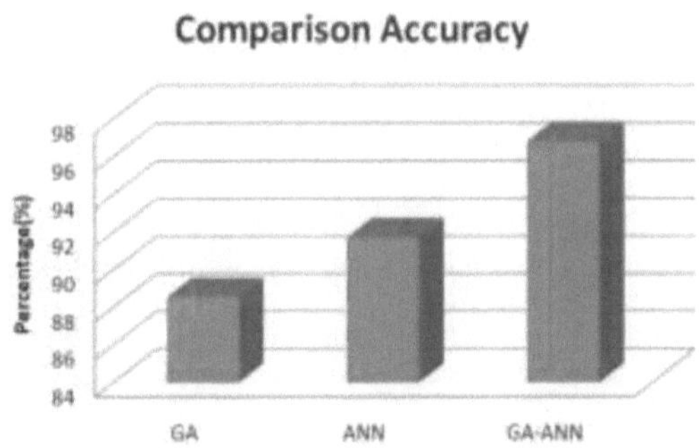


Figure 4. Accuracy comparison of GA-ANN model.

The accuracy of the suggested method is contrasted with that of existing methods in graph 1.4. When compared to two other models, the suggested GA-ANN model achieves the maximum accuracy.

## 5 CONCLUSION

Human Resources play a pivotal and overarching role in every business, and it is their responsibility to cultivate and sustain a high-caliber workforce. The business as a whole can benefit from higher levels of employee performance. Employee performance as it relates to job satisfaction, work-family conflict, and stress in the workplace will be the focus of this proposed study. It is necessary to modify the data so it satisfies the requirements of each data mining approach in order to process data that would be impossible to process without data preprocessing. By integrating additional useful features with de-noised raw signals obtained by the sliding window approach, a feature extraction dataset was generated. The GA-ANN method is used to train the models. The suggested method beats the ANN and GA models with a consistency rate of 96.81 percent.

# REFERENCES


[1] Sugumar, R., Sharma, S., Kiran, P. N., Gupta, S. K., Sindhu, V. and Srinivas, R. V. (2023, June). Novel method for detection of stress in employees using hybrid deep learning models. In *2023 8th International Conference on Communication and Electronics Systems (ICCES)* (pp. 984–989). IEEE.

[2] Ahsan, N., Abdullah, Z., Fie, D. Y. G. and Alam, S. S. (2009). A study of job stress on job satisfaction among university staff in Malaysia: Empirical study. *European Journal of Social Sciences*, 8(1), 121–131.

[3] Aksoy, C., Şengün, H. I., and Yılmaz, Y. (2018). Examination of the relationship between job satisfaction levels and organizational commitments of tourism sector employees: A research in the Southeastern Anatolia region of Turkey. *Elektronik Sosyal Bilimler Dergisi*, 17(65), 356–365.

[4] Aziri, B. (2011). Job satisfaction: A literature review. *Management Research & Practice*, 3(4).

[5] D'Amico, A., Geraci, A. and Tarantino, C. (2020). The relationship between perceived emotional intelligence, work engagement, job satisfaction, and burnout in Italian school teachers: an exploratory study. *Psihologijske teme*, 29(1), 63–84.

[6] Ho, C. L. and Au, W. T. (2006). Teaching satisfaction scale: Measuring job satisfaction of teachers. *Educational and Psychological Measurement*, 66(1), 172–185.

[7] Alonderiene, R. and Majauskaite, M. (2016). Leadership style and job satisfaction in higher education institutions. *International Journal of Educational Management*, 30(1), 140–164.

[8] An, J., Liu, Y., Sun, Y. and Liu, C. (2020). Impact of work–family conflict, job stress and job satisfaction on seafarer performance. *International Journal of Environmental Research and Public Health*, 17 (7), 2191.

[9] Annisa, N. F. R. and Supriyanto, A. S. (2021). The influence of leadership style and work motivation on employee performance through job satisfaction (Case study on the department of transportation of Blitar City). *The American Journal of Humanities and Social Sciences Research*, 4(3), 355–362.

[10] Nart, S. and Batur, O. (2014). The relation between work-family conflict, job stress, organizational commitment and job performance: A study on Turkish primary teachers. *European Journal of Research on Education*, 2(2), 72–81.

[11] Dodanwala, T. C. and Shrestha, P. (2021). Work–family conflict and job satisfaction among construction professionals: the mediating role of emotional exhaustion. *On the Horizon: The International Journal of Learning Futures*, 29(2), 62–75.

[12] Afshar, H. S. and Doosti, M. (2016). Investigating the impact of job satisfaction/dissatisfaction on Iranian English teachers' job performance. *Iranian Journal of Language Teaching Research*, 4(1), 97–115.

[13] Al-Mahdy, Y. F., Al-Harthi, A. S. and Salah El-Din, N. S. (2016). Perceptions of school principals' servant leadership and their teachers' job satisfaction in Oman. *Leadership and Policy in Schools*, 15(4), 543–566.

[14] Kassa, Y., Zhang, J. H., Zheng, D. H. and Wei, D. (2016, August). A GA-BP hybrid algorithm based ANN model for wind power prediction. In *2016 IEEE Smart Energy Grid Engineering (SEGE)* (pp. 158–163). IEEE.

[15] Iliou, T., Konstantopoulou, G., Stephanakis, I., Anastasopoulos, K., Lymberopoulos, D. and Anastassopoulos, G. (2018). Iliou machine learning data preprocessing method for stress level prediction. In *Artificial Intelligence Applications and Innovations: 14th IFIP WG 12.5 International Conference*, AIAI 2018, Rhodes, Greece, May 25–27, 2018, Proceedings 14 (pp. 351–361). Springer International Publishing.

[16] Garg, P., Santhosh, J., Dengel, A. and Ishimaru, S. (2021, April). Stress detection by machine learning and wearable sensors. In 26th *International Conference on Intelligent User Interfaces-Companion* (pp. 43–45).

*Hybrid and Advanced Technologies – S. Prasad Jones Christydass, Nurhayati Nurhayati and S. Kannadhasan. (Eds) © 2025 the Authors, ISBN 978-1-032-90254-8*
*Open Access: www.taylorfrancis.com, CC BY-NC-ND 4.0 license*

# The impact of social networking sites on students' social wellbeing and academic performance based on RBF-Elman techniques

R. Arun
*Assistant Professor, Department of MBA, St.Joseph's College of Engineering, Chennai, India*
ORCID ID: 0000-0002-5252-1030

M.R. Vanithamani
*Professor & Principal, Sree Sarawathi Thyagaraja College Pollachi, Coimbatore, India*
ORCID ID: 0000-0001-6785-6530

Sundarapandiyan Natarajan
*Professor and Head, Department of Management Studies, Adithya Institute of Technology, Coimbatore, India*
ORCID ID: 0000-0002-1303-2947

Kiran Kumar Thoti
*Associate Professor, M.S. Ramaiah Institute of Management Bangalore, Karnataka, India*
ORCID ID: 0000-0002-6678-9425

S. Sivagnana Bharathi
*Associate Professor Department: Management Studies Collegeluniversity: KIT-Kalaignarkarunanidhi Institute of Technology, Kannampalayam, Tamil Nadu, India*
ORCID ID: 0000 0002 7794 7564

P. Raju
*Assistant Professor - Department of Commerce, Faculty of Science and Humanities, SRM Institute of Science and Technology, Vadapalani, Chennai Tamil Nadu, India*
ORCID ID: 0009-0004-7761-5895

ABSTRACT: Social networking sites have played a significant role in students' growing online presence. Students' enthusiasm and participation in online classrooms have been significantly enhanced by its utilization as a teaching tool. The effects of students' social media usage on their academic performance are the primary subject of this proposed. There are mainly three parts: training the model, extracting features, and preprocessing. An essential step in providing classifiers with a feature space representation of data is data preparation. Performance, processing costs, the "dimensional curse," and data comprehension are all improved with feature selection. We used the RBF-ELMAN to train the model, which increased its accuracy. The proposed method achieved a higher accuracy of 96.35% compared to state-of-the-art methods such as RBF and ELMAN.

*Keywords*: Radial Basis Functions (RBF), Social Networking Service (SNS), Convolutional-Neural-Network (CNN), Long–Short-Term-Memory (LSTM), Root-Mean-Squared Error (RMSE), Long–Short-Term-Memory Neural Network (LSTM)

DOI: 10.1201/9781003559115-22 

## 1 INTRODUCTION

People allow for more open dialogue and serve as a support system for kids, social networks play an essential role in classrooms. There are a number of social media sites like Edmodo that exist solely for the goal of facilitating learning. Social networks have many positive effects on students' education, including improved opportunities for active and social learning, higher levels of engagement and motivation, easier self-learning, fewer obstacles to group interaction and communication, more opportunities for collaborative learning, more information about self-learning, and more opportunities for students to work together. There are new risks associated with modern technology, even while there are also many opportunities for improved worldwide communications brought about by things like the internet and social networks. There seems to be more than one factor that contributes to internet and social media addiction. A few examples of these factors are low self-esteem, pornographic photo browsing, mobile phone internet use, dating, gaming, and entertainment. Students are among the most powerful online and social media user demographics. When adolescents spend an excessive amount of time on social media, it can have both positive and negative impacts on their health, social lives, and academic performance. If children spend too much time on social media, it will negatively impact their ability to study. Medical students who spent more time than the average student on the internet and social media performed worse academically and had problems concentrating in class, according to the research.

## 2 LITERATURE SURVEY

Numerous studies have examined the efficacy of social media as a teaching tool. To help research universities in Malaysia make the most of social media, [1] proposed a new approach. The Unified Theory of Technology Acceptance and Use served as the basis for the modelling. [2]Age, gender, and experience acted as moderators, while trust, perceived privacy and security, attitude towards technology, [3]and communication advantages served as predictors in the model in order to plot out how to make use of various social media sites. The data was decomposed before analysis using the PLS-SEM approach. [4] The results showed components that, barring expectations of effort, have positive relationships. Additionally, there was no indication that [5]demographic variables like age, gender, or level of expertise mediated the connection between social media intention and actual usage. Using social networking sites to improve students' research performance was the overarching theme of the results.[6]Research on the drawbacks of social media is in its infancy, but thus far six stressors associated with SNS use have been identified and backed. According to[7], when consumers experience an excessive quantity of social demands from SNS, it is known as social media overload. Disclosure is a stressful state that makes it hard for users to make efficient use of social media platforms. [8]It happens when users feel overwhelmed by the volume of information they getstress pattern that individuals go through when they modify their social media behaviour to match that of their friends. [9]This includes things like constantly updating friends on what they're up to and telling them about their own activities. Uncertainty, characterized by the stressful constant updates and changes to the SNS applications[10], and complexity, reflecting the degree to which users perceive SNS as technically difficult to use due to a lack of understanding, are two sources of this difficulty.[11]However, this is becoming an urgent issue due to the negative effects of social media on the mental and social well-being of adolescents. [12] Overwhelm, depression, low self-esteem, and isolation were more prevalent among heavy social media users when contrasted with non-users. In comparison to non-users, social media users are more likely to be socially disinherited and have worse self-esteem[13]. People with mental health challenges who spend a great deal of time on the internet sometimes struggle with other problems as well. Social media platforms are perfect for fostering cooperative learning environments in which teachers and students may share knowledge and resources[14].

## 3 PROPOSED SYSTEM

Keeping up with the news, asking questions, sharing information, and interacting with others are all things that people rely on social media for these days. The researchers set out to determine whether and how students' use of social media sites affected their study habits and academic achievement. A total of 210 students from a higher education vocational program were engaged in this proposed.

### 3.1 *Data preprocessing*

In order to give classifiers a picture of the data in a feature space, data preprocessing is a must. The use of preprocessing, which entails removing stop words, has improved the study data and subsequent analytical stages[15]. This is necessary since social network data is noisy. Using stop words does not enhance the sense of a statement. Words like "to," "has," "I," "be," "the," "or," and countless more are examples of stop words. The use of meaningless stop phrases squanders mental resources.

### 3.2 *Feature selection*

The goal of feature selection is to extract the best feature from a dataset. Data can be structured into meaningful classes by using machine learning methods. The utilization of feature domains, which comprise hundreds of variables or features, is on the rise in machine learning applications. The challenge of removing unnecessary variables has many potential solutions. Downsizing variables or attributes through feature selection improves speed, decreases processing costs, mitigates the "dimensional curse," and improves data comprehension[16]. An enhanced mutual information based filter was proposed as a selection criterion. IMIFPC, a method developed by Pearson, takes into account the characteristics of each feature. This metric can be used to measure the degree of linear dependence between two continuous variables, *P and R*. Its value might be anywhere from -1 to +1. The result is based on Pearson's correlation, which states that

$$PC_{PR} = \frac{Conv(P, R)}{\tau_P \tau_R} \tag{1}$$

### 3.3 *Model training*

#### 3.3.1 *RBF*

Many studies have proposed and documented the use of radial basis function networks. As previously mentioned, RBF networks typically have a three-layer structure. Every input is connected to one of the n neurons that make up an input layer. To modify the range of values, the input neurons (or processing prior to the input layer) divide by the interquartile range and subtract the median. Values are received by the neurons in the hidden layer from the neurons in the input layer. The radial basis function is the activation function for the neurons in the buried layer (2).

$$g(p) = e^{(-p-K)^2/2\delta^2} \tag{2}$$

where the parameters $K$ and $\delta$ stand for the input variables mean and standard deviation, respectively. Compared to traditional back propagation networks, radial basis networks require fewer neurons during design. Additionally, compared to training sigmoid or linear networks, radial basis networks require less time during design.

### 3.3.2 *Elman network*

Elman networks are classified as recurrent neural networks due to the existence of feedback connections in the neurons of the hidden layer. The output value from the previous step is remembered and used to determine the output value for the next phase by Elman networks, which are also known as buffered networks. In order to enhance the accuracy of its predictions, the network keeps track of its previous states and leaves a digital footprint.

## 4 RESULT AND DISCUSSION

This proposed approach aims to investigate how pupils are affected by social networking websites. To find out what aspects of social media sites affect pupils, researchers developed a questionnaire. Factors such as gender, age, social impact, level of schooling, and academic achievement have been identified.

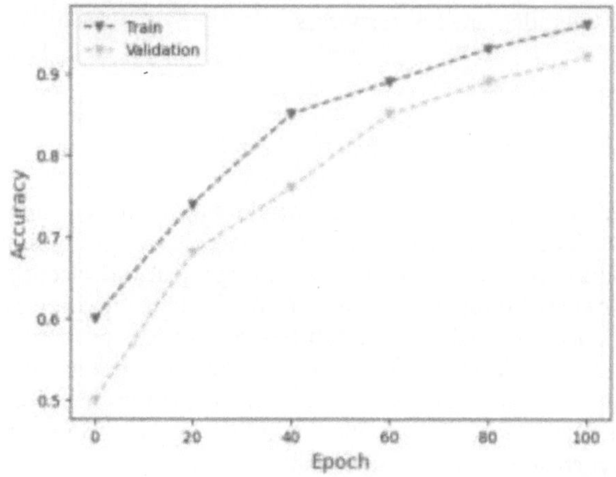


Figure 1. Prediction result of train and validation data.

Look at Figure 1 to observe the results of our CNN-LSTM method. The findings reveal that the training data achieves a respectable approximation accuracy of 96.45% as the number of epochs hits 100, whilst the testing data comes in at 92.31%.

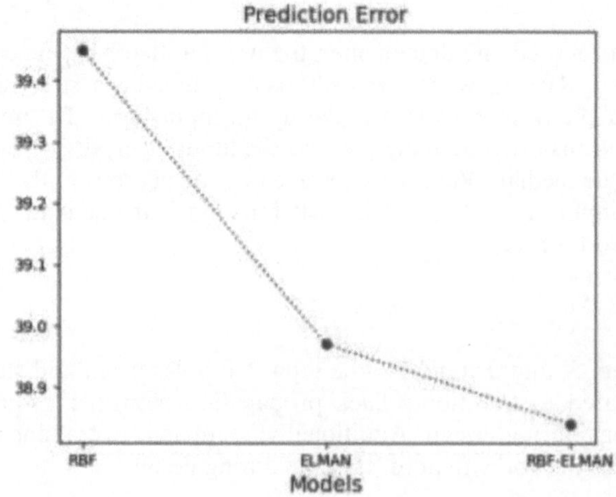


Figure 2. Comparison of RMSE of the proposed method with other methods.

Using root-mean-squared error (RMSE) as a performance metric, we compared the RBF-ELMAN model against RBF and ELMAN, as shown in Figure 2. With an RMSE of 38.84, our RBF-ELMAN technique outperforms the competition in terms of prediction accuracy.

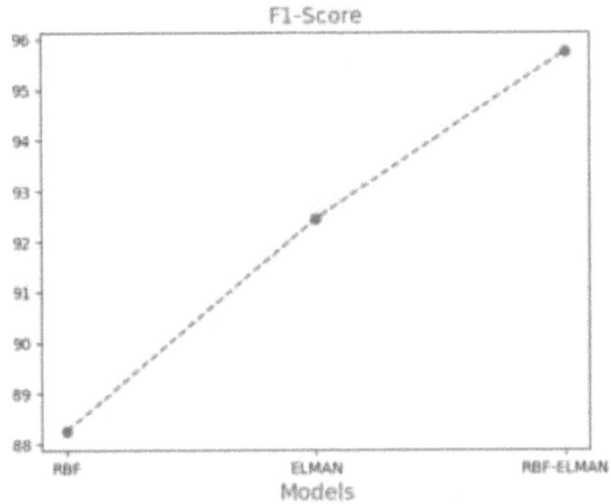


Figure 3. F1-Score of proposed model.

Figure 3, shows the F1-score is around 0.95, which is an improvement over only using RBF-ELMAN. Figure 3 displays the outcomes of the RBF-ELMAN prediction method's F1-score.

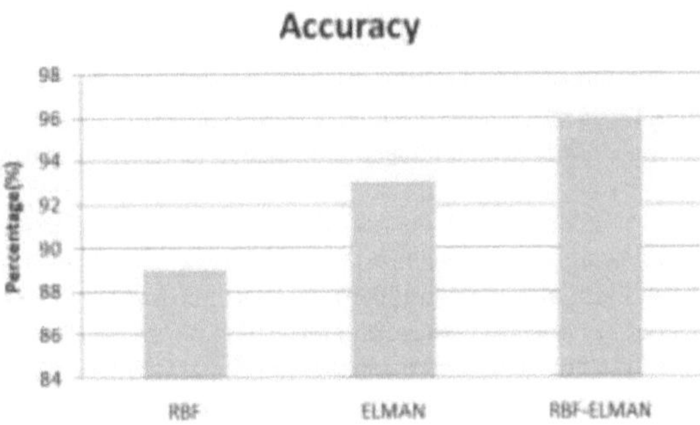


Figure 4. Accuracy graph.

Figure 4 clearly illustrates that the suggested system has the highest Accuracy values.

## 5 CONCLUSION

The essay explores how students' online social network connections affect the amount of information shared and how intensifying knowledge sharing boosts students' engagement. Social capital theory serves as the basis for the hypothesis testing, which is based on partial least square structural equation modeling. In order to give classifiers a picture of the data in a feature space, data preprocessing is a must. Features selection enhances performance, reduces processing costs, alleviates the "dimensional curse," and enhances data comprehension. It makes use of RBF-ELMAN for model training. With an average accuracy of 96.35%, the suggested method surpasses both the RBF and ELMAN models.

## REFERENCES


[1] Bicen, H. (2015). Determination of university students' reasons Ofusing social networking sites in their daily life. *Procedia-Social and Behavioral Sciences*, 190, 519–522.

[2] Bicen, H. and Cavus, N. (2010). The most preferred social network sites by students. *Procedia-Social and Behavioral Sciences*, 2(2), 5864–5869.

[3] Burke, M., Marlow, C. and Lento, T. (2010, April). Social network activity and social well-being. *In Proceedings of the SIGCHI Conference on Human Factors in Computing Systems* (pp. 1909–1912).

[4] Buzov, I. (2014). Social network sites as area for students' pro-environmental activities. *Procedia-Social and Behavioral Sciences*, 152, 1233–1236.

[5] Dabbagh, N. and Kitsantas, A. (2012). Personal Learning Environments, social media, and self-regulated learning: A natural formula for connecting formal and informal learning. *The Internet and Higher Education*, 15(1), 3–8.

[6] Al-Rahmi, W. M., Alias, N., Othman, M. S., Marin, V. I. and Tur, G. (2018). A model of factors affecting learning performance through the use of social media in Malaysian higher education. *Computers & Education*, 121, 59–72.

[7] Burns, D., Dagnall, N. and Holt, M. (2020). Assessing the impact of the COVID-19 pandemic on student wellbeing at universities in the United Kingdom: a conceptual analysis. *Frontiers in Education*, 5, 582882.

[8] Dodd, R. H., Dadaczynski, K., Okan, O., McCaffery, K. J. and Pickles, K. (2021). Psychological wellbeing and academic experience of university students in Australia during COVID-19. *International Journal of Environmental Research and Public Health*, 18(3), 866.

[9] Mandryk, R. L., Frommel, J., Armstrong, A. and Johnson, D. (2020). How passion for playing world of warcraft predicts in-game social capital, loneliness, and wellbeing. *Frontiers in Psychology*, 11, 575935.

[10] Lambić, D. (2016). Correlation between Facebook use for educational purposes and academic performance of students. *Computers in Human Behavior*, 61, 313–320.

[11] Kalpidou, M., Costin, D. and Morris, J. (2011). The relationship between Facebook and the well-being of undergraduate college students. *CyberPsychology, Behavior, and Social Networking*, 14(4), 183–189.

[12] Junco, R. (2015). Student class standing, Facebook use, and academic performance. *Journal of Applied Developmental Psychology*, 36, 18–29.

[13] Doleck, T., Lajoie, S. P. and Bazelais, P. (2019). Social networking and academic performance: A net benefits perspective. *Education and Information Technologies*, 24, 3053–3073.

[14] Minaee, S., Azimi, E. and Abdolrashidi, A. (2019). Deep-sentiment: Sentiment analysis using ensemble of cnn and bi-lstm models. *arXiv*. arXiv preprint arXiv:1904.04206.

[15] Peiqing, H. (2022). [Retracted] Multidimensional State Data Reduction and Evaluation of College Students' Mental Health Based on SVM. *Journal of Mathematics*, 2022(1), 4961203.

[16] Zhu, C., Li, B., Li, A. and Zhu, T. (2019, January). Detecting depression from Internet behaviors by time-frequency features. *In Web Intelligence* (Vol. 17, No. 3, pp. 199–208). IOS Press.

*Hybrid and Advanced Technologies – S. Prasad Jones Christydass, Nurhayati Nurhayati and S. Kannadhasan. (Eds)* © *2025 the Authors, ISBN 978-1-032-90254-8*


# Enhancing entrepreneurial skills: A BERT-CNN-LSTM approach to student practical training

G. Pavani
*Assistant Professor of English, Velagapudi Ramakrishna Siddhartha Engineering College, Siddhartha Academy of Higher Education, Vijayawada, India*

Amit Dimari
*Research Scholar HSS & Section Officer, Examination, Graphic Era deemed University Dehradun, India*

Nidhi Tyagi
*Associate Professor, HSS, Graphic Era deemed University, Dehradun, India*

P. Sampath
*Associate Professor, Department of Civil Engineering, PERI Institute of Technology, Mannivakkam, West Tambaram, Chennai, India*

Neerav Nishant
*Assistant Professor, Department of Computer Science and Engineering, School of Engineering, Babu Banarasi Das University, Lucknow, India*
ORCID ID: 0009-0002-6696-6943

Dileep Pulugu
*Professor, Department of CSE, Malla Reddy College of Engineering and Technology, Hyderabad, India*

ABSTRACT: Several theories regarding the personality types of managers exist in today's interconnected world. A manager's character shines through in their day-to-day decisions and the extent to which they are sound. Students‘ ability to think creatively and independently is correlated with their perception of entrepreneurship as a potential career path. Training effective managers should equip them to put management theory into practice in the real world. Management responsibilities and the fundamental instruments for carrying them out, coaching concepts, team building and management, and familiarity with HRM methods like brainstorming are all part of this. All of the above presupposes that the target outcome can be achieved by a single component or by a synergistic effect of multiple components. Consequently, the entrepreneurial skill demand in the labour market and its inclusion in university curriculum were both examined using qualitative comparative analysis (QCA). There are three steps to the suggested method: preprocessing, feature extraction, and model performance evaluation. This tool employs normalisation for preprocessing. It employs principal component analysis (PCA) for feature extraction and BERT-CNN-LSTM for model performance evaluation. It does better than the other two conventional approaches, CNN and CNN-LSTM.

*Keywords*: Qualitative Comparative Analysis (QCA), Rectified Linear Unit (ReLU), Student Practical Training, Normalization, Long Short-Term Memory (LSTM)

DOI: 10.1201/9781003559115-23

## 1 INTRODUCTION

Economic development can be aided by the youth, who are a significant asset. Most young people nowadays are finishing up their education and getting themselves ready to enter the workforce. It is our understanding that a minimum degree is often required for employment opportunities in both the public and commercial sectors. Indeed, the professions to which members of this generation are eligible will be directly correlated to their degree of education. First among the six tiers of strategy that make up the Entrepreneurship Development Policy is the mandate to establish an entrepreneurship centre at each and every university. Everything pertaining to the programme and entrepreneurial-related activities, including its implementation, monitoring, and evaluation, will fall under this category. The second stage is for all universities to come up with a fresh idea for their knowledge and entrepreneurial programmes. Institutions should play a part in developing a new system to assess project performance, which brings us to our fourth level: firmly establishing development programmes that work in tandem with entrepreneurship to foster the next generation of business owners. As a fifth level, we want to foster an atmosphere that helps students develop their entrepreneurial spirit.

## 2 LITERATURE SURVEY

A leader, like any other talent or ability, must put in the time and effort to develop their craft [1]. Having said that, there is no one right approach to be an effective manager or leader [2]. Various management styles are employed by numerous managers [3]. More and more, both public and private organisations are realising that they should be doing more to promote economic and social growth on a regional and national scale. Here we have an example of a local multi-actor network that includes a university, which is considered as an important player in the knowledge-based economy for reasons beyond its obvious role as a knowledge generator [4]. Further, by incorporating entrepreneurship as an academic function alongside teaching and research, those groups unquestionably boost local development [5]. A diverse set of competencies has been proposed by other writers in frameworks for entrepreneurial competencies [6]. Regardless, the question of whether entrepreneurial abilities and behaviours are innate or may be developed via experience, education, and training remains hotly contested [7]. Students that participate in EE programmes report improvements in areas such as resilience, adaptability, and flexibility, as stated in [8]. Therefore, students should integrate their academic knowledge with the practical experience they obtain from extra-curricular activities [9]. So, more and more engineering schools have included EE in their courses of study in recent decades [10]. The term "competence" can mean either "what is done" or "how it is done" depending on the context [11]. To us, competences boil down to an integrated strategy that values context, delivered performance, and human traits [12]. In addition to the information, abilities, and mindsets needed to carry out a specific task, competence also refers to how well an individual does in a specific setting, as measured by their actions on the job and the results they produce. Knowledge, skills, and attitudes are the building blocks of competence [13], and it's crucial that all three areas work in tandem. Contribution goals include elucidating present-day composition and content of entrepreneurial skills, drawing attention to global (professional) skills, and projecting future skills that might influence the profile of the profession and necessitate inclusion in curricula.

## 3 PROPOSED SYSTEM

The purpose of this article is to take a look at how a training programme for entrepreneurs might help bachelor of education science students become more entrepreneurial. Trainee science teachers' perspectives on enhancing existing resources for educational purposes would not be substantially impacted by entrepreneurial training.

## 3.1 *Preprocessing*

### 3.1.1 *Normalization*

Gathering data on entrepreneurial learning from users is the initial stage of the study technique. Information gathered from user profiles, educational institutions, and learning platforms may be part of this. After data collection is complete, it is cleaned and prepared for further use. This includes dealing with data discrepancies, errors, and missing values, as well as eliminating duplicate entries [14]. Making sure the data is high quality ensures that all the analysis and modelling that follows is based on correct and dependable information. Before training, normalise all input parameters to remove the effect of order-of-magnitude difference on error and reduce computational cost of the BERT-CNN-LSTM. This is the formula that shows the normalisation process:

$$K' = \frac{K - K_{\min}}{K_{\max} - K_{\min}} \tag{1}$$

this is where $K'$ is the value that remains after data normalisation, $K$ is the sample value from the data table, and $K_{\max}$ and $K_{\min}$ are the minimum and maximum values of the data sample, respectively.

## 3.2 *Feature extraction*

### 3.2.1 *PCA*

PCA, which is a technique for reducing numerous variables and indicators to a handful of primary ones, is the most popular approach to multivariate analysis. PCA's core notion is this: to achieve the best dispersion objective, the original data is transformed into a cube and then represented by effective feature components. This transformation keeps the essential information of the original data intact. "Feature extraction" describes this technique. One name for principal component analysis (PCA) is maximum linear principal component analysis, which is what happens when you just extract the one effective principal component [15]. Many modern image processing, data mining, pattern recognition, and information compression tasks make use of principal component analysis (PCA) due to its robust multi-index simplification capability. One of the most popular linear dimensionality reduction methods is principal component analysis (PCA). The idea is to use a linear projection to transform high-dimensional data into a low-dimensional space, with the expectation that the projected dimension will contain the most informative data.

## 3.3 *Model training*

### 3.3.1 *BERT-CNN-LSTM*

The proposed approach uses BERT embeddings that have already been trained in our model. One of the most innovative transformer-based language models, BERT, is renowned for the detailed meanings it assigns to words based on their context. We can improve the model's understanding of the input text and get more benefits by taking advantage of the semantic relationships between words. We use pre-trained BERT embeddings in the initial part of our model, Consequently, the "bert_encoder" layer produces an array of embedded elements that provide a thorough depiction of the meaning of the input text by means of increased contextual information. The sequence output from the BERT encoder is then easily fed into the CNN architecture. We use deep convolutional neural network (CNN) layers in our model architecture to detect textual features and patterns at the local level. To achieve various granularities, several convolutional layers are employed, each with a unique filter size. The model's ability to capture complex patterns is enhanced by adding non-linearity and the Rectified Linear Unit (ReLU) activation function. The generated feature maps are downsampled using max pooling so that only the most important data is retained.

If we want to take the deep CNN design to the next level in terms of information flow and training success, we use the skip convolution block. Residual connections, or skip convolution blocks, enable data to be directly propagated from lower layers to higher ones. To be more precise, information can be directly passed from early layers to later ones by introducing skip connections between specific convolutional layers. This allows us to improve the model's ability to capture complex patterns and connections in the text input by facilitating the flow of both low-level and high-level features [16]. The result is better gradient flow during training, which helps with the vanishing gradient problem and ultimately improves the model's performance and convergence. It uses Long Short-Term Memory (LSTM) layers in our model architecture to capture patterns of sequential data and long-term dependencies. This hierarchical design uses CNNs to extract local features and LSTMs to record long-term dependencies; the LSTM layers get their output as an input. To successfully capture the temporal dynamics and contextual information present in the input data, the LSTM layers use memory cells and various gating techniques.

If BERT is to accurately identify fake news, its pre-trained representations must have a solid grasp of the input text's semantics and context. For the purpose of learning more discriminative and hierarchical features, we feed the retrieved features into the CNN and then use BERT. To further improve the model's representational capacity, the deep CNN with skip block is used. To overcome the vanishing gradient issue, deep networks can train more efficiently with the use of skip connections, which allow gradient flow. Better generalisation is the outcome of these and other benefits, such as fast model training and increased feature reuse and information transfer between layers. On top of that, we build the LSTM component into the superimposed design. The LSTM is a powerful model for representing sequential patterns and long-term dependencies in text input; it receives its features from the CNN and uses them effectively. When it comes to comprehending news stories and their context, LSTM's capacity to grasp temporal linkages and dependencies is invaluable, boosting the model's capacity to understand.

## 4 RESULT AND DISCUSSION

In a nation with an emerging economy, this research aims to investigate how students gain the entrepreneurial abilities needed by a knowledge economy. Workers in the knowledge economy require a distinct set of abilities than their industrial counterparts because the foundation of the knowledge economy is the processing of intangible resources. Emerging economies, like India's, are part of the knowledge economy's rapid expansion. In order to meet the demands of the knowledge economy, educational institutions must shift their focus from rote memorization to the cultivation of competencies that will enable students to thrive in an increasingly complex and competitive corporate world.

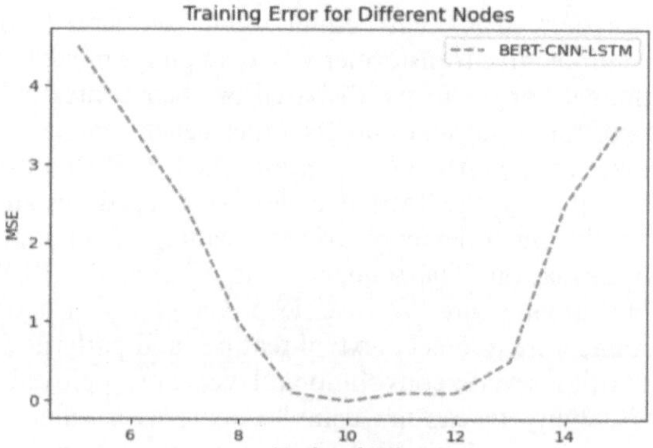


Figure 1. The training error of different hidden nodes.

Figure 1 displays the outcomes of the tests conducted using the trial-and-error method, with the number of hidden layer nodes ultimately estimated to be between [5,15] based on the input and output node counts in this work.

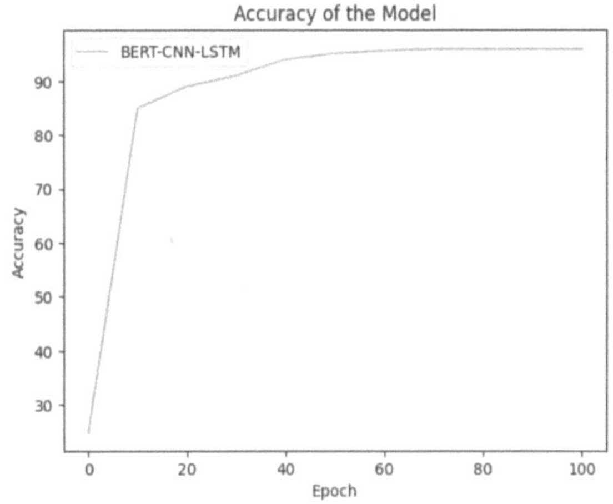


Figure 2. Training accuracy of the model.

The experimental results are displayed in Figure 2. It is evident that as the network iterates, the training loss decreases and the training accuracy increases. I should highlight that when the training reaches 70 epoch, the network has achieved convergence. There has been no discernible improvement in either the loss or the accuracy.

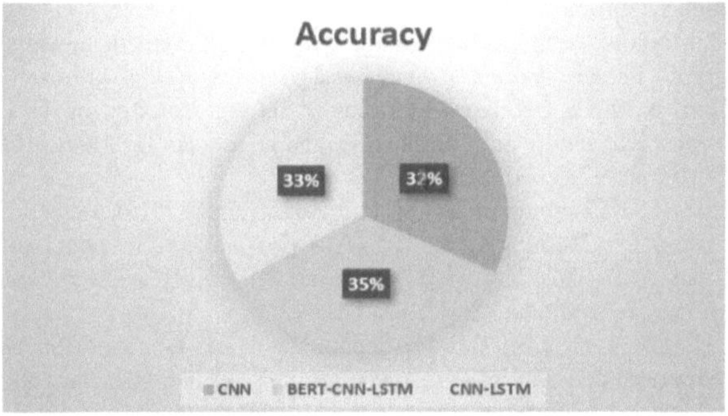


Figure 3. Classification accuracy of the model.

Using dataset from both scanners as the test cohort, BERT-CNN-LSTM had a greater area under the curve (Figure 3) of 95.92% and 35% classification accuracy compared to CNN and CNN-LSTM, but the difference was not statistically significant.

## 5 CONCLUSION

In today's competitive employment environment, the ability to think like an entrepreneur is essential for most professionals. Given the present state of globalisation and digital revolution, this study set out to characterise and organise these abilities. The socio-economic and socio-cultural grounds for the importance of this topic were identified through documental analysis and critical reflection on the obtained data. Consequently, to develop a framework that aims to meet the demands of professionals in the modern, digital world. A tripartite ES model was identified based on the results. This model incorporates the following skills: initiative, creativity, self-efficacy, resilience, digital communication, transformational leadership, problem-solving, evaluation, teamwork, and visual communication. It also emphasises the need to be open to new ideas, solve emerging problems, and communicate effectively. In order to quantify the level of development of these skills, the next step of this work is to construct and validate an ES scale based on this model. Preprocessing often makes use of normalisation. A combination of PCA and CNN, BERT-CNN-LSTM, and CNN-LSTM are employed to assess the model's performance. The suggested strategy outperformed the other two and achieved an accuracy of approximately 95.22%.

## REFERENCES


[1] Scales Jr, C. D., Smith, A. C., Hanley, J. M., Saigal, C. S. and Urologic Diseases in America Project. (2012). Prevalence of kidney stones in the United States. *European Urology*, 62(1), 160–165.

[2] Scales Jr, C. D., Lin, L., Saigal, C. S., Bennett, C. J., Ponce, N. A., Mangione, C. M. and NIDDK Urologic Diseases in America Project. (2015). Emergency department revisits for patients with kidney stones in California. *Academic Emergency Medicine*, 22(4), 468–474.

[3] Planz, V. B., Posielski, N. M., Lubner, M. G., Li, K., Chen, G. H., Nakada, S. Y. and Pickhardt, P. J. (2019). Ultra-low-dose limited renal CT for volumetric stone surveillance: advantages over standard unenhanced CT. *Abdominal Radiology*, 44, 227–233.

[4] Moore, C. L., Bhargavan-Chatfield, M., Shaw, M. M., Weisenthal, K. and Kalra, M. K. (2021). Radiation dose reduction in kidney stone CT: a randomized, facility-based intervention. *Journal of the American College of Radiology*, 18(10), 1394–1404.

[5] Selby, M. G., Vrtiska, T. J., Krambeck, A. E., McCollough, C. H., Elsherbiny, H. E., Bergstralh, E. J., and Rule, A. D. (2015). Quantification of asymptomatic kidney stone burden by computed tomography for predicting future symptomatic stone events. *Urology*, 85(1), 45–50.

[6] Ebrahimi, S. and Mariano, V. Y. (2015). Image quality improvement in kidney stone detection on computed tomography images. *Journal of Image and Graphics*, 3(1), 40–46.

[7] Akbari, H. and Fei, B. (2012, February). Automatic 3D segmentation of the kidney in MR images using wavelet feature extraction and probability shape model. In *Medical Imaging 2012: Image Processing* (Vol. 8314, pp. 973–979). SPIE.

[8] Vasanthselvakumar, R., Balasubramanian, M. and Palanivel, S. (2019). Detection and classification of kidney disorders using deep learning method. *J Mech Contin Math Sci [Internet]*, 14, 3045–51.

[9] Zisman, A. L., Evan, A. P., Coe, F. L. and Worcester, E. M. (2015). Do kidney stone formers have a kidney disease?. *Kidney international*, 88(6), 1240–1249.

[10] Parakh, A., Lee, H., Lee, J. H., Eisner, B. H., Sahani, D. V. and Do, S. (2019). Urinary stone detection on CT images using deep convolutional neural networks: evaluation of model performance and generalization. *Radiology: Artificial Intelligence*, 1(4), e180066.

[11] Soni, A. and Rai, A. (2020, December). Kidney stone recognition and extraction using directional emboss & SVM from computed tomography images. In *2020 Third International Conference on Multimedia Processing, Communication & Information Technology (MPCIT)* (pp. 57–62). IEEE.

[12] Mua'ad, M. and Zubi, M. (2020). Analysis and implementation of kidney stones detection by applying segmentation techniques on computerized tomography scans. *Italian Journal and Applied Mathematics*, (43-2020), 590–602.

[13] He, C., Kang, H., Yao, T. and Li, X. (2019). An effective classifier based on convolutional neural network and regularized extreme learning machine. *Mathematical Biosciences and Engineering*, 16(6), 8309–8321.

*Hybrid and Advanced Technologies – S. Prasad Jones Christydass, Nurhayati Nurhayati and S. Kannadhasan. (Eds)* © *2025 the Authors, ISBN 978-1-032-90254-8*


# The hybrid HCNN-KNN model: Transforming classroom effectiveness through intelligent teaching quality evaluation

S. Naveena
*Associate Professor, Department of IT, Bannari Amman Institute of Technology Sathyamangalam, Tamilnadu, India*

Chippy Mohan and Sandhya Dasharatharao
*Assistant Professor, School of Business and Management, Christ University, Bengaluru, Karnataka, India*

S. Madhan
*Assistant Professor, Department of English, K.S.R. College of Engineering, Namakkal, Tamilnadu, India*

Shobhit Garg
*Assistant Professor, Department of Management Studies, Graphic Era Deemed to be University, Dehradun, India*

Harshal Patil
*Associate Professor, Balaji Institute of Technology & Management (BITM), Sri Balaji University, Pune, Maharashtra, India*

ABSTRACT: New opportunities have arisen in the realms of education and training as a result of developments in communication and information technology. Due to increasing demand from several free online courses, traditional classrooms are experiencing a period of transition. The flipped classroom model is quite relevant in this context. The "flipped classroom" is a new way of teaching that has students do homework that includes asynchronous video lectures and practice problems before coming to class to work together to solve the real problems. The three most important processes are training the model, expanding the features, and preprocessing. The ML performance is guaranteed by this preprocessing by employing techniques like Yeo Johnson transformation and scaling; in this case, Yeo Johnson provides the maximum accuracy. Using PCA, feature extraction can transform data from correlated variables into features that are linearly uncorrelated. Using the HCNN-KNN during training improved the model's accuracy. The suggested approach beat state-of-the-art technology like as CNN and KNN with a rate of 98.13%.

*Keywords*: Classroom Effectiveness, Education, Audience Response Systems (ARS), Principle Component Analysis (PCA), convolutional neural network (CNN)

## 1 INTRODUCTION

What we call "teaching" is really just the process that teachers go through to make sure their students learn something new. Therefore, it is the teacher's major responsibility to stimulate the student's interest in order to ensure that the latter participates actively in the learning process. Teaching is a means of instilling virtue and competence in individuals. In other words, getting pupils to do what you want them to do is the whole point of education changes in their routines, which involve participating in class discussions as part of assigned

DOI: 10.1201/9781003559115-24

work. That is why it is essential for teachers to use not simply atransformation in conduct and a channel for the dissemination and reception of knowledge, which usually results in the development of proficiency, insight, and character in the learner. Further, secondary school instruction is ineffective due to instructors‘ lack of access to essential professional resources. In terms of turning out qualified graduates, secondary school education is rarely successful. High school graduates don't have what it takes to be successful in life and the job, according to many. This leads to their degeneration into a social burden. It is sometimes said that teachers place too much emphasis on teaching methods and insufficient on conceptual knowledge, and that the majority of secondary school pupils merely memorize facts rather than truly understanding the content.

## 2 LITERATURE SURVEY

In spite of the importance of equipping ELs with strategies and skills, teachers should be given chances to reflect on and analyse their own attitudes and beliefs[1]. The transformational learning paradigm was employed to investigate teachers' views on assisting students from diverse backgrounds[2]. Transformative learning is an intricate perceptual and understanding process that adults engage in when they examine and reassess their current frameworks for making sense of the world in order to build new ones[3]. Reflection, speech, and action are all part of the same process. [4]Teachers, both present and future, struggle to identify what they can do differently or what they need to do differently to comprehend the importance of being adequately prepared to teach English Language Learners (ELs). [5]This is particularly true in view of the rapid and increasing changes in student demographics and the need for teachers to critically assess their own knowledge, values, beliefs, and skills.[6]Higher student accomplishment and, more specifically, greater academic outcomes in the classroom have been the focus of researchers in Zimbabwe. [7]It is widely believed that when strong school leaders are selected from the SMT and the head of the school, it leads to a more positive school atmosphere and more optimistic attitudes among teachers. [8]We argue in this post that the relationship between effective school leadership and improved student achievement is more complex and harder to predict than it appears on paper. [9]asserts that the academic achievement of a school's students is directly related to the quality of leadership provided by the principal. The leadership's influence on student outcomes is statistically significant, but limited, when it comes to the indirect effects that principals have on student learning[10]. The best way for school leadership to achieve their goals is to draw from a diversity of leadership styles. This will ensure that decisions and discussions around education improvement continue. The goal of leaders that practice transformational leadership is to motivate their followers to reach extraordinary heights of personal and professional success [11]. Transformative leaders inspire people around them to develop personally and professionally. Transformational leaders may influence their followers‘ perspectives by demanding high performance from them. In transformational leadership, the capacity to impact subordinates is crucial [12]. Someone who is receptive to new ideas and proactive is a transformational leader [13]. The premise here is that a transformative leader is unconstrained by established standards and can thus readily adapt to changing conditions. One way to help a group achieve new goals is through the use of transformational leadership, which seeks to inspire its members to go above and beyond what is currently possible [14].

## 3 PROPOSED SYSTEM

Class discussion is a common method of instruction in the field of advanced nursing degree programs. Computerized instruments, such as audience response systems (ARS), may be more useful than older, more traditional ways of doing classroom research. The availability of ARSs since the 1960s has not diminished their pioneering status in many classes for advanced nursing practice.

## 3.1 *Preprocessing*

### 3.1.1 *Data normalization*

The accuracy of the classifier can be affected by attributes that contain scaled or range-based values. This is unnecessary if all the features are uniformly distributed. This normalization process ensures ML performance by using Yeo Johnson transformation and scaling.

### 3.1.2 *Yeo Johnson*

It is non-positive and has its roots in the Box-cox transformation. The Yeo-Johnson transformation allows data to have negative values, which is a drawback of the Box-Cox transformation. It has the ability to normalize skewed data, which makes it easier to evaluate and mode[15].

### 3.1.3 *Scaling*

Uneven feature scales will have a negative effect on the classifier's performance. In order to standardize it, a fair scale needs to be set up. The Min-Max method converts the continuous data values to a range from 0 to 1.

$$r_{norm} = \frac{r - r_{\min}}{r_{\max} - r_{\min}} \tag{1}$$

## 3.2 *Feature extraction*

### 3.2.1 *PCA*

Principle Component Analysis (PCA) uses an orthogonal transformation to extract features that are linearly uncorrelated from samples of correlated data. Principal components are the new features that are smaller than or equal to the initial variables. The absence of labels in PCA results from the method's status as unsupervised[16]. If the data is normally distributed, then the principal components are thought to be independent. PCA reduces the number of original variables by deleting the final principle components that do not significantly contribute to the observed variability. The purpose of PCA is to maximize information by minimizing redundancy and maximizing variance through a linear modification of the data.

## 3.3 *Model training*

### 3.3.1 *HCNN-KNN*

This study proposed a model that combines HCNN and KNN. Our dental classification and pre-image processing hybrid model follows these steps: The proposed HCNN-KNN model included input layers, two Convolutional layers, fully linked data, data transformed by PCA, and KNN classification. The second layer of the model was a convolutional layer that used RGB images (216 x 216 pixels) as input. It had 16 convolutional filters, a kernel size of 6, and three paddings. Things like shapes, blobs, and edges were given to us as low-level features. Batch normalization (16 layers) and a ReLU non-linear activation function were also implemented in addition to the 16 convolutional normalizing filters. The Maxpooling layer reduced the dimensions of the photographs to 116 $\times$ 116 pixels, and the resulting set of dimensions is 116 $\times$ 116 $\times$ 12. These pictures were sent into the second convolutional block, which was structurally similar to the first. The following typical block, which contains 32 filters with 6 and 3 padding kernels, is responsible for producing the 52 x 52 × 36 image output. At this point, a CNN model with KNN utilized as a classifier is used to assess the class event, rather than the SoftMax probability layer. Hence, we proposed that this model employ the principal component analysis technique, commonly referred to as "Fully PCA," to relocate the input to a more dimensional-and precision-oriented space prior to its entry into a fully connected KNN layer. Now that the PCA data has been reduced and transformed, the KNN classification layer may be visualized as having fully linked layer neurons as overlays. In this case, the KNN input is 600. In order to ascertain the age of bone using dental images, this study suggests a pseudocode for an HCNN-KNN model approach. After learning the network and classification operations, the proposed network model is assessed in subsequent epochs of network learning with the help of the error loss rate.

## 4 RESULT AND DISCUSSION

As the reach of modern technology continues to grow, flipped classrooms have gained popularity in both conventional and online learning environments. In "flipping" the classroom, the traditional sequence of in-class instruction, guided practice, and independent study is reversed. Both students and teachers need to constantly develop their skill sets; for example, the ability to design and implement effective learning experiences in non-traditional settings is an important one.

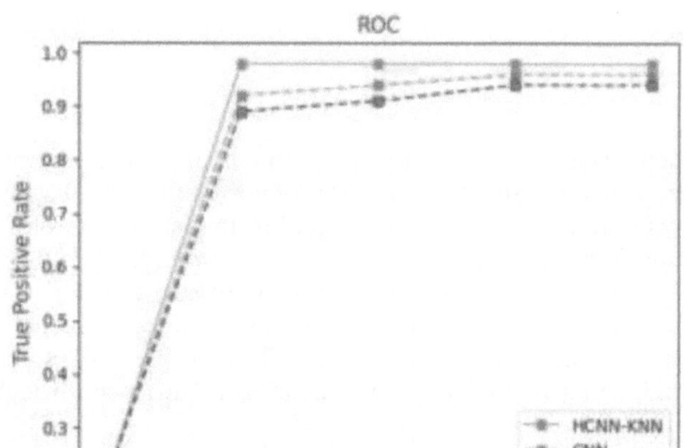


Figure 1. ROC comparison of HCNN-KNN model.

The effectiveness of the suggested model in dividing student performance into low, medium, and high categories is demonstrated by the comparison in terms of ROC, which shows an enhanced ROC of 0.98 compared to other methods (Figure 1).

Figure 2 compares the average detection times for Transforming Classroom Effectiveness. The proposed model improved upon earlier approaches by reducing the detection rate; this reinforced the system's integrity, and the required results were displayed in just a few seconds.

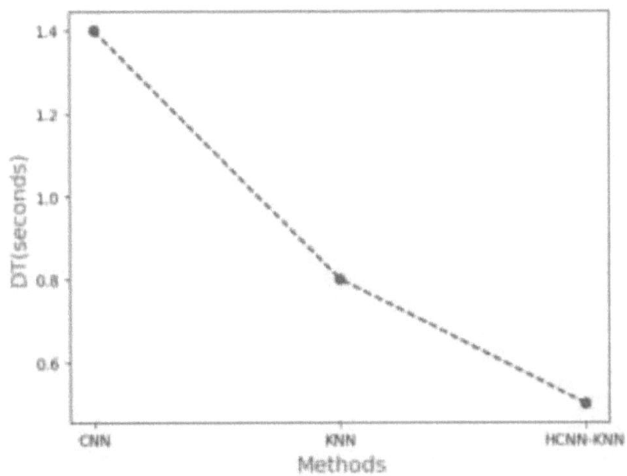


Figure 2. Detection time of HCNN-KNN model.

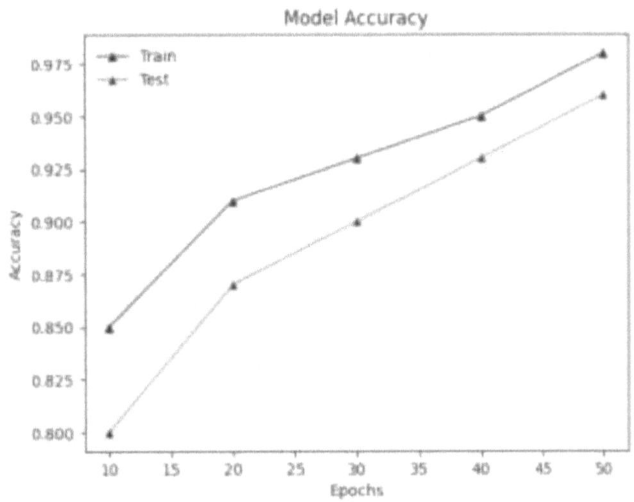


Figure 3. Accuracy comparison of proposed model.

Over the course of 50 epochs, Figure 3 compares the accuracy of training and testing.

## 5 CONCLUSION

Using the ideas of purposeful practice, we show that homework may be changed to improve learning, even more so than in actively taught classes. To evaluate the impact of a recent assignment change on student learning in a class that had previously implemented an active learning technique. This preprocessing ensures the ML performance with methods like scaling and the Yeo Johnson transformation; in this instance, the Yeo Johnson delivers the highest accuracy. Data from correlated variables can be transformed into linearly uncorrelated features through the use of PCA in feature extraction. It is HCNN-KNN that is utilized for training a model. With an average accuracy of 98.13%, the suggested method outperforms the CNN and KNN models.

## REFERENCES


[1] Aguirre, J. M., Zavala, M. D. R. and Katanyoutanant, T. (2012). Developing robust forms of pre-service teachers' pedagogical content knowledge through culturally responsive mathematics teaching analysis. *Mathematics Teacher Education and Development*, 14(2), 113–136.

[2] Cho, J. Y. and Lee, E. H. (2014). Reducing confusion about grounded theory and qualitative content analysis: Similarities and differences. *Qualitative Report*, 19(32).

[3] Karabenick, S. A. and Noda, P. A. C. (2004). Professional development implications of teachers' beliefs and attitudes toward English language learners. *Bilingual Research Journal*, 28(1), 55–75.

[4] Kim, S., Song, K. and Coppersmith, S. (2018). Creating an interactive virtual community of linguistically and culturally responsive content teacher-learners to serve English learners. *Contemporary Issues in Technology and Teacher Education*, 18(2), 442–466.

[5] Lucas, T., Villegas, A. M. and Freedson-Gonzalez, M. (2008). Linguistically responsive teacher education: Preparing classroom teachers to teach English language learners. *Journal of Teacher Education*, 59(4), 361–373.

[6] Cosner, S. (2011). Teacher learning, instructional considerations and principal communication: Lessons from a longitudinal study of collaborative data use by teachers. *Educational Management Administration & Leadership*, 39(5), 568–589.

[7] Heystek, J. and Minnaar, L. (2015). *Principals' Perspectives on Key Factors That Contribute to Sustainable Quality Education.*

[8] Hoy, W. (2012). School characteristics that make a difference for the achievement of all students: A 40-year odyssey. *Journal of Educational Administration*, 50(1), 76–97.

[9] Klar, H. W., Huggins, K. S., Hammonds, H. L. and Buskey, F. C. (2016). Fostering the capacity for distributed leadership: A post-heroic approach to leading school improvement. *International Journal of Leadership in Education*, 19(2), 111–137.

[10] Shava, G. N. and Ndebele, C. (2016). Enhancing learner achievement through effective school leadership: Experiences from selected rural schools in South Africa. *International Journal of Educational Sciences*, 14(3), 205–216.

[11] Jaafar, F. M. and Lailia, D. R. (2019). Mediating Role of Perceived Organizational Support in Relationship Between Teacher Self-Efficacy and Teacher Effectiveness. *Madrosatuna: Journal of Islamic Elementary School*, 3(2), 59–66.

[12] Mundy, M. A., Kupczynski, L. and Kee, R. (2012). Teacher's perceptions of technology use in the schools. *Sage Open*, 2(1), 2158244012440813.

[13] Polatcan, M., Arslan, P. and Balci, A. (2023). The mediating effect of teacher self-efficacy regarding the relationship between transformational school leadership and teacher agency. *Educational Studies*, 49(5), 823–841.

[14] Sharifonnasabi, F., Jhanjhi, N. Z., John, J., Obeidy, P., Band, S. S., Alinejad-Rokny, H. and Baz, M. (2022). Hybrid HCNN-KNN model enhances age estimation accuracy in orthopantomography. *Frontiers in Public Health*, 10, 879418.

[15] Jakkaladiki, S. P., Janečková, M., Krunčík, J., Malý, F. and Otčenášková, T. (2023). Deep learning-based education decision support system for student E-learning performance prediction. *Scalable Computing: Practice and Experience*, 24(3), 327–338.

[16] Khalid, S., Khalil, T. and Nasreen, S. (2014, August). A survey of feature selection and feature extraction techniques in machine learning. In *2014 Science and Information Conference* (pp. 372–378). IEEE.

*Hybrid and Advanced Technologies – S. Prasad Jones Christydass, Nurhayati Nurhayati and S. Kannadhasan. (Eds) © 2025 the Authors, ISBN 978-1-032-90254-8*
*Open Access: www.taylorfrancis.com, CC BY-NC-ND 4.0 license*

# A novel approach for battery management systems for electric vehicles based on CNN-BiGRU model

Soumya Mishra
*Associate Professor, Quality Assurance Cell, KIIT University, Bhubaneswar, Odisha, India*

Sibghatulla Mohammad
*Senior Lecturer, Department of Electrical and Electronics Engineering, Government Polytechnic, Hyderabad, Telangana, India*

Neeraja Bathala
*Lecturer, Department of Electrical and Electronics Engineering, Government Polytechnic Hyderabad, Hyderabad, Telangana, India*

P. Ram Kishore Kumar Reddy
*Professor, Department of EEE, Mahatma Gandhi Institute of Technology, Hyderabad, Telangana, India*

P. Nagasekhar Reddy
*Associate Professor, Department of EEE, Mahatma Gandhi Institute of Technology Hyderabad, Telangana, India*

Harshal Patil
*Associate Professor, Balaji Institute of Technology & Management (BITM), Sri Balaji University, Pune, Maharashtra, India*

ABSTRACT: Examining BMSs for electric vehicle batteries was the goal of this research. Applying a simplified electrical model was the initial stage in determining the SoC of the battery. This model can adjust to the wide variety of dynamic operating situations seen in online EV applications, and it faithfully represents the electrochemical characteristics and charging/discharging reaction. Preprocessing, feature extraction, and model training are the three main parts. Zero normalization, or Z-score normalization, is an operation that is performed in the preprocessing step. Finding the average and standard deviation of each training set attribute in the dataset is what this process entails. Estimation of the battery's state of charge is utilized in the feature extraction procedure. We trained the model with CNN-BiGRU to get more precise results. The suggested strategy outperformed competing methods, with a 97.43% accuracy rate. Keywords: Electric Vehicles (EVs), Hybrid Electric Vehicles (HEVs), State of Charge (SOC), Natural Language Processing (NLP), Mean Absolute Error (MAE)

## 1 INTRODUCTION

The most promising replacement for conventional cars driven by internal combustion engines (IC engines) in recent years has been the explosion in development of hybrid electric vehicles (HEVs) and electric vehicles (EVs). Battery power for EVs and HEVs is abundant because of the numerous advantages of batteries, such as long cycle life, minimal emissions, and excellent energy density. But batteries are particularly sensitive when used in electric vehicles. Batteries are vulnerable to damage, age faster, and even explode or catch fire if subjected to dangerous currents such as overcharging, over discharging, or excessive current.

DOI: 10.1201/9781003559115-25 

Important technologies in electric car BMSs include battery modelling, internal status estimation, and charging.With an emphasis on battery modelling specifically, this work aims to offer a brief synopsis of the key technologies, statevaluation and BMS charging of EV batteries.Innovative approaches to solving the problem of battery fatigue an inventory of electrical and thermal property models should be established as a first step. By utilizing these well-established models, the electric and thermal behaviours of batteries may be captured. After that, we'll examine the various methods for estimating the internal temperature and state of charge of the battery, either alone or together. Using models of the batteries and their estimated internal states, we examine methods for charging them and optimization strategies that can enhance their efficiency.

## 2 LITERATURE SURVEY

The automotive industry has achieved great achievements in pedestrian and passenger safety as a result of many technological advancements. According to[1], an increase in traffic is known to significantly raise pollution levels in cities. As a result of these worries about pollution, [2]fossil fuel consumption, and climate change, electric cars (EVs) have become quite popular around the world. [3]states that compared to diesel-powered vehicles, electric vehicles have various benefits, such as reduced emissions, user-friendliness, reliability, comfort, and efficiency. According to[4], the extensive use of electric vehicles cannot be achieved unless the battery storage system (BSS) is properly functioning and diagnosed. This system needs to keep an eye on the charge-discharge cycles, regulate the power management, keep the cells in balance, and handle heat[5]. Due to its high voltage, extended life cycles, low self-discharge rates, high power density, and extended life cycles, lithium-ion batteries are currently the market leaders in the electric vehicle (EV) battery industry.[6]An electric motor and collector system can draw power from many external sources, including solar panels, batteries, or a generator that transforms fuel into electricity, to power a self-contained electric vehicle[7]. Research and development of electric vehicles (EVs) has made great strides in the last 150 years. [8] The three distinct phases of electric vehicle (EV) development, beginning with the introduction of the first basic, non-chargeable cars and ending with the most sophisticated EVs with state-of-the-art control systems in use today. You may completely eradicate pollution by powering your vehicle with renewable energy [9]. Rising earnings, more people driving, more metropolitan areas, and more people overall are pushing up the demand for cars, particularly electric ones [10]. Reduced operating and maintenance costs are one of the many benefits of electric vehicles, which benefit human health and the environment[11]. The power batteries are the heart of electric cars, and finding a suitable energy storage device that can manage long-distance travel, fast charging, and efficient driving is the biggest challenge for EVs [12]. As the temperature rises, so does the efficiency of a battery [13]. The battery's performance is significantly reduced when exposed to cold conditions. The power, energy, and charge capacity of the cell are all drastically changed when the temperature drops. [14]This further reduces the battery's performance when the temperature drops external variables. At low temperatures, the electrolyte's ionic conductivity decreases and the internal resistance of the battery pack increases due to its increasing viscosity.

## 3 PROPOSED SYSTEM

Consumer electronics now rely on lithium-ion batteries for a host of reasons, including their minimal environmental effect, high power density, extended service life, and energy density. Despite their large capacity and large serial-parallel numbers, lithium-ion batteries have a number of drawbacks that prevent them from being used extensively in automobiles. These include concerns with cost, uniformity, safety, and durability.

## 3.1 *Preprocessing*

### 3.1.1 *Z-Score normalization*

In preparation for main processing and analysis, the sensor data has undergone preliminary processing. Data normalization and plausibility checks are part of preliminary processing. To achieve Z-score normalization, also known as zero normalization, a number of variables in the training set are used to divide the mean and standard deviation of each training set characteristic. Every attribute has its mean and standard deviation calculated[15]. It is in the general formula that the transformation is defined.

$$r = \frac{r - \rho}{\tau} \tag{1}$$

## 3.2 *Feature Extraction*

The distribution of battery current is highly variable and fluctuating due to the driving habits. There is a wide range of voltages at the battery terminals due to the dynamic current. Common perception holds that there is a direct correlation between the two variables. The inputs to the machine learning model, which include the voltage and current at the battery terminals, are inefficient. Here is a recommended standard process for extracting features from data on the state of electric vehicles stored in the cloud[16]. Using EMD to extract trend features helps reduce data variability in the initial voltage and current. It is possible to eliminate voltage residue by taking an average voltage reading on the internal resistance of the battery. Based on the findings of the correlation analysis, the next stage is to select the common features for calculating the state of charge (SOC) of the battery.

## 3.3 *Model training*

### 3.3.1 *CNN*

The ability of CNN to recognize local features makes them valuable in natural language processing (NLP). Specifically, we use CNN to extract semantic information features. By utilizing a full connection layer after a sequence of convolution pooling layers, our model's CNN resembles LeNet-5. In natural language processing, the input length is not specified, hence researchers usually utilize a one-layer CNN design with two convolutional layers and a pooling layer. Our model's architecture consists of a full connection layer after two convolutional and pooling neural network layers. Convolution is a local process, meaning it can only use data from very close by to make conclusions. Using max-pooling, the pooling layer achieves the maximum value throughout the entire region. The characteristics that have been noticed are strong qualities that aid in classifying objects. Classification issues in natural language processing frequently employ max-pooling. The output of the entire connection layer is then passed on to the convolution channel.

### 3.3.2 *CNN-BiGRU*

Although CNN is capable of learning local features from data, it is unable to learn sequence correlation. Although RNN models excel at handling long-term data dependencies, extracting features simultaneously proves to be quite a difficulty. Our two channels, CNN and BiGRU, account for both the semantic information's characteristics and the context in which it appeared. When the output vectors of the intermediate layer's two channels are geometrically connected, the result is the model's final feature fusion output. The two components come together to make the output feature, which is useful during training. The last thing to do is use the softmax classifier to calculate the confidence distribution of each category space output vector, as shown in Formula (2).

$$X(m_w = r) = \frac{\exp(T_{nr}.n_{w,r} + k_r)}{\sum_{r' \in R} \exp(T_{nr}.n_{w,r'} + k_r)} \tag{2}$$

## 4 RESULT AND DISCUSSION

In order to determine how long lithium-ion batteries will last and what kind of condition they are in, this study introduces a moving-window based method and a reliable health indicator. It was possible to glean the vital sign from the cell's partial charge voltage curve.

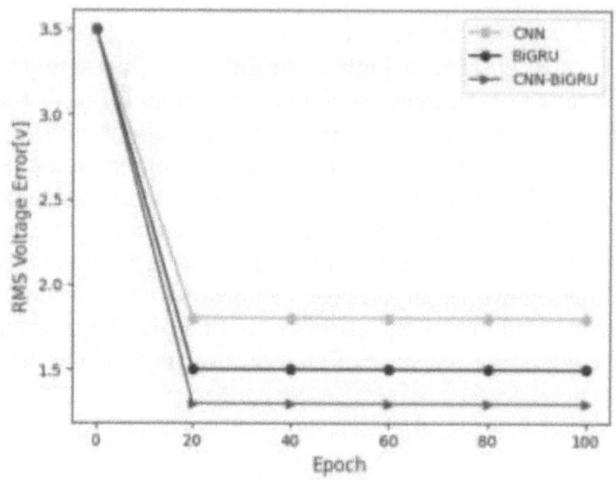


Figure 1. The data set determines the progress of CNN-BiGRU training.

The CNN-BiGRU's behavior is highly sensitive to the training data set selection, as seen in Figure 1. For the following reduction strategies, the comprehensive raw data set will be used as a baseline, with an RMS voltage inaccuracy of about 1.3 V.

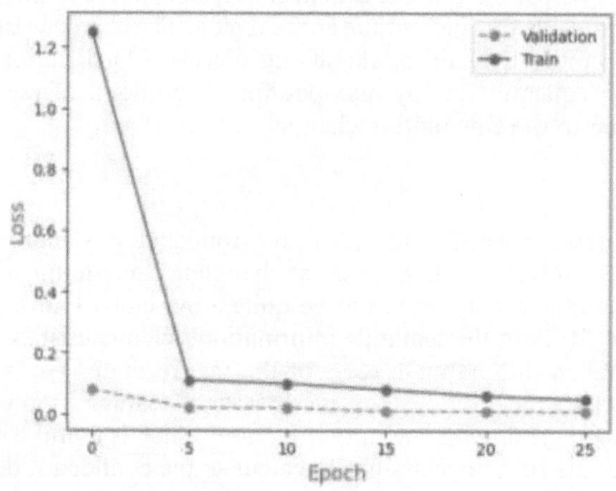


Figure 2. Training and validation loss of the suggested method.

In order to fit the model for 20 training epochs, a batch size of 25 will be utilized. Figure 2 shows the number of epochs required to plan the data in terms of time and losses.

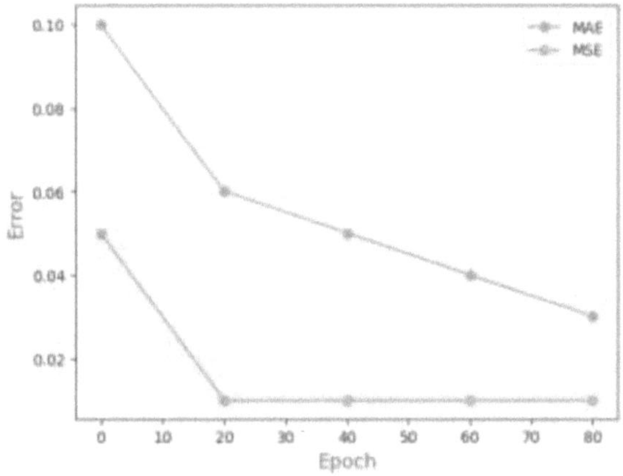


Figure 3. Errors as the epoch progresses.

This section concludes with the test data set's MSE and MAE visualization. Figure 3 shows that as the number of epochs increases, the model achieves a low MAE and MSE.

## 5 CONCLUSION

Electric vehicles, which are a step in the direction of greener modes of transportation, depend on batteries. The use of lithium chemistry for energy storage in electric cars is currently the norm. Still, there are a lot of untapped possibilities for future research. One of these is the creation of algorithms and electronic circuits that make greater use of batteries. One more is choosing the best materials for cells. One operation that is carried out in the preprocessing step is zero normalization, also known as Z-score normalization. In this step, we will find the mean and standard deviation of all the training set attributes in the dataset. The feature extraction process makes use of an estimate of the battery's charge status. The BiGRU-CNN method is used for training the models. The suggested method beats both the BiGRU and CNN models in terms of consistency, which is 97.43%.

## REFERENCES

[1] Balasingam, B., Ahmed, M. and Pattipati, K. (2020). Battery management systems—Challenges and some solutions. *Energies*, 13(11), 2825.

[2] Chaturvedi, N. A., Klein, R., Christensen, J., Ahmed, J. and Kojic, A. (2010). Algorithms for advanced battery-management systems. *IEEE Control Systems Magazine*, 30(3), 49–68.

[3] Haldar, S., Mondal, S., Mondal, A., and Banerjee, R. (2020, February). Battery management system using state of charge estimation: An IOT based approach. In *2020 National Conference on Emerging Trends on Sustainable Technology and Engineering Applications (NCETSTEA)* (pp. 1–5). IEEE.

[4] Ali, M. U., Zafar, A., Nengroo, S. H., Hussain, S., Junaid Alvi, M. and Kim, H. J. (2019). Towards a smarter battery management system for electric vehicle applications: A critical review of lithium-ion battery state of charge estimation. *Energies*, 12(3), 446.

[5] Lelie, M., Braun, T., Knips, M., Nordmann, H., Ringbeck, F., Zappen, H. and Sauer, D. U. (2018). Battery management system hardware concepts: An overview. *Applied Sciences*, 8(4), 534.

[6] Lu, L., Han, X., Li, J., Hua, J. and Ouyang, M. (2013). A review on the key issues for lithium-ion battery management in electric vehicles. *Journal of Power Sources*, 226, 272–288.
[7] Nimesh V., Sharma D., Reddy V. M., and Goswami A. K. (2020). Implication viability assessment of shift to electric vehicles for present power generation scenario of India, *Energy*, vol. 195, doi: 10.1016/j.energy.2020.116976.
[8] Hannan, M. A., Hoque, M. M., Peng, S. E. and Uddin, M. N. (2017). Lithium-ion battery charge equalization algorithm for electric vehicle applications. *IEEE Transactions on Industry Applications*, 53 (3), 2541–2549.
[9] Huber, J., Dann, D. and Weinhardt, C. (2020). Probabilistic forecasts of time and energy flexibility in battery electric vehicle charging. *Applied Energy*, 262, 114525.
[10] Manzetti, S. and Mariasiu, F. (2015). Electric vehicle battery technologies: From present state to future systems. *Renewable and Sustainable Energy Reviews*, 51, 1004–1012.
[11] Xu, J., Zheng, X. and Jiang, M. (2019, August). Gene mutation classification using CNN and BiGRU network. In *2019 9th International Conference on Information Science and Technology (ICIST)* (pp. 397–401). IEEE.
[12] Gandhi, M. A., Priya, K., Charan, P., Sharma, R., Rao, G. N. and Suganthi, D. (2023, July). Smart electric vehicle (EVs) charging network management using bidirectional GRU-AM approaches. In *2023 2nd International Conference on Edge Computing and Applications (ICECAA)* (pp. 1509–1514). IEEE.
[13] Kumar, J., Murugan, A. S. S., Kumar, R. V., Ghamande, M. V. and Kumar, N. M. G. (2023, October). Lithium-Ion health prediction for battery systems using deep a-LSTM-CNN approach. In *2023 International Conference on Self Sustainable Artificial Intelligence Systems (ICSSAS)* (pp. 1481–1487). IEEE.
[14] Moorthi, K., Chougale, R. K., Patel, P. A., Sharma, P., Raj, A. and Gangat, M. S. (2023, June). Electric vehicles charging system for fast and safe charging using LSTM based gradient boosted regression tree. In *2023 International Conference on Sustainable Computing and Smart Systems (ICSCSS)* (pp. 1289–1294). IEEE.
[15] Kosuru, V. S. R. and Kavasseri Venkitaraman, A. (2023). A smart battery management system for electric vehicles using deep learning-based sensor fault detection. *World Electric Vehicle Journal*, 14(4), 101.
[16] Li, X., Xu, J., Ding, X. and Lyu, H. (2023). State of charge estimation for batteries based on common feature extraction and transfer learning. *Batteries*, 9(5), 266.

*Hybrid and Advanced Technologies – S. Prasad Jones Christydass, Nurhayati Nurhayati and S. Kannadhasan. (Eds)* © *2025 the Authors, ISBN 978-1-032-90254-8*


# Short-term electric vehicle charging load forecasting in low-quality data environments using CNN-SVM approach

I.D. Soubache
*Associate Professor BME, Rajiv Gandhi College of Engineering and Technology, Pondy Cuddalore, Puducherry, India*

M. Kavitha
*Assistant Professor, Department of Mathematics, Panimalar Engineering College, Chennai, India*

V.M. Periyasamy
*Assistant Professor, Department of EEE, Sona College of Technology, Salem, Tamilnadu, India*

P. Nagasekhar Reddy
*Associate Professor, Electrical and Electronics Engineering, Mahatma Gandhi Institute of Technology, Hyderabad, India*

M. Anusuya
*Professor, Department of Physics, Indra Ganesan College of Engineering, Manikandam, Trichy, Tamilnadu, India*

Narayan Pandurang Sapkal
*Assistant Professor, Department of Mechanical Engineering, Dr. D. Y. Patil Institute of Technology, Pimpri, Pune, India*

ABSTRACT: In order to reduce pollution and our dependence on fossil fuels, the broad usage of electric vehicles is a crucial step. Charging plug-in electric vehicles presents unique challenges to current load forecasting methodologies due to their high degree of uncertainty, large variance, and several coupled effect components. In comparison to the conventional load, the fluctuations of plug-in electric vehicle charging are more apparent at the exceedingly short-term minute level time scale, due to the enormous charging power and inconsistent human behaviours. Training the model, extracting features, and preprocessing are the three key parts. One common method for preprocessing that transforms raw data from individual EV charging sessions into a customized EV charging profile is to factor in the actual power used for charging. Feature selection is a technique for dimensionality reduction that aids in reducing the number of features in datasets. To improve the model's accuracy, we trained it using the Feature rich CNN-SVM. Compared to other methods, such as CNN and SVM, the proposed method achieved a higher accuracy rate of 96.60%.

*Keywords*: Support Vector Machine (SVM), Electric Vehicles (EVs), Artificial Neural Network (ANN), Electric Vehicle Charging Stations (EVCSs), State of Charge (SOC) randomization

## 1 INTRODUCTION

Recent shifts in the global energy market and developments in automotive technology have contributed to the meteoric rise in popularity of electric cars (EVs), which are able to cut

DOI: 10.1201/9781003559115-26

down on both emissions and energy use. The widespread integration of electric vehicles into the power grid has far-reaching consequences for power system planning and operation. Several reasons have contributed to this, including rising demand, declining power quality, growing challenges in optimising power grid operations, and impacts on distribution networks. There has been a lot of research in recent years into trying to predict how much power will be needed to charge electric automobiles. The methods that are currently available for predicting electric vehicle loads are examined. In the next decades, electric vehicles are expected to gain significant traction. Precise forecasts of upcoming EV load demand are crucial inputs for managing EV charging and enhancing the efficiency and effectiveness of the power system. One crucial factor to keep in mind is that factors such as the behaviours and routes of EV drivers can ultimately lead to an unforeseen nature of the infrastructure requirement for charging electric vehicles. Therefore, state-of-the-art prediction methods would be useful for identifying improves the accuracy of the demand model for electric vehicle charging, enabling grids to operate more efficiently.

## 2 LITERATURE SURVEY

Legislators, manufacturers, and power providers are all intrigued by the booming electric car sector. According to many, EVs can reduce emissions and the need for fossil fuels[1]. There will be a quick increase in the scale of electric vehicles due to factors such as government incentives and falling battery prices [2].Among the many purposes of electric vehicle charging load forecasting research is the assurance of the operation of the power grid and to promote the expansion of EVs [3].Now, three separate approaches—probabilistic, time series, and machine learning—are available for estimating the power requirements of electric vehicle charging stations.[4] Employed OD matrix analysis to plan the logistics electric vehicle's path get the charging demand load value for the car using a mixed-integer programming method. Proposed a data-driven method for estimating the demand on electric vehicle charging stations by constructing three models using Didi user trip data: a traffic network model, a vehicle spatial-temporal transfer model, and a resident travel probability model. The rapid expansion of electric vehicles poses a significant risk to the existing power grid due to the exponential increase in their electrical demand [5]. As previously stated, grid operators need to be proactive in their planning to deal with the potential spikes in demand caused by electric cars (EVs). Other concerns that EVs could causeits power grids' expansion. Electric vehicles might not have much of an impact on distribution networks in the early stages because of the low EV penetration [6].Two primary approaches exist for predicting the amount of electricity that EVs will need to charge their batteries, one for the near future and one for the medium to far future [7]. One way to predict loads in the near future is viatypically used to predict the charging requirements of electric vehicles within one month. In order to ensure efficient power distributions on the grid, it is crucial to estimate the demand at electric car charging stations. This will assist reduce the impact of charging peaks and troughs caused by disorderly behaviour [8]. Using machine learning for short-term load forecasting has been highly effective [9]. One specific piece of gearProblems with charging load estimates have been addressed in the past using learning-based methods. [10]Take this scenario: Support Vector Regression (SVR) was employed to forecast the charging demand one day in advance. Furthermore, a model for short-term load forecasting was developed by combining fuzzy systems with Artificial Neural Network (ANN) algorithms. The Long Short-Term Memory (LSTM) algorithm, an improved Recurrent Neural Network (RNN), has been used by numerous researchers to tackle load forecasting problems. This deep learning technique is becoming increasingly popular in domains such as computer vision and natural language processing.

## 3 PROPOSED SYSTEM

Addressing the issue of low accuracy in previous methods of EV charging load characteristic predictions caused by charging time and state of charge (SOC) randomization, a model based on historical load and using support vector regression (SVR) is proposed for charging load forecasting.

### 3.1 *Preprocessing*

A user's RFID tag number, arrival and departure times, and the total amount of energy used in kilowatt-hours are all part of the raw data that pertains to individual electric vehicle charging sessions. The objective of this study is to obtain an EV charging demand profile (in kW) from the raw data collected from individual EV charging sessions in order to forecast the total EV charging demand on a 15-minute timestep [11]. Alternatively stated, every piece of raw data from EV charging sessions needs to be transformed into a distinct EV charging profile before the aggregated EV charging demand timeseries can be generated. When dealing with smart or bi-directional chargers, it's important to make a distinct prediction that takes user flexibility and energy requirements into account.

Since the sessions' charging power values are missing, it is also not possible to build an uncoordinated EV charging power profile. These considerations dictate a new strategy for obtaining an individual's EV charging profile. The second step, as in (1), is to compute the average power, $H_{avg}$, using the same formula.

$$H_{avg} = Energy\ Consumed/Parking\ time \tag{1}$$

### 3.2 *Feature selection*

Applying dimensionality reduction techniques, such as feature selection, helps reduce the number of features in datasets. This research introduces a new feature selection algorithm that filters potential inputs to the forecast engine block. This approach reduced the processing load while also improving the prediction accuracy. Using the covariance matrix of the data matrix, principal component analysis (PCA) can be used to ascertain the strength of a link between two vectors [12]. Finding the coefficient of two given vectors, αi and aj, can be done in the following manner.

$$h(\beta_m, \beta_r) = \frac{Con(\beta_m, \beta_r)}{\sqrt{\text{var}(\beta_m) \times \text{var}(\beta_r)}} \tag{2}$$

The labeled data set $C_T$ is constructed using the learning data, with $B_m$ and $Q_T$ being $t$-component vectors, in respect to the analyzed characteristic election. One possible formulation for the implementation feature and tag of the PCCs is as follows:

$$G(B_m, Q_T) = \frac{1}{|B|} \sum_{B_m \in B} H(B_m, Q_T) \tag{3}$$

A simple calculation of max $G(B_m, Q_T)$ will yield the characteristic set with the most relationships, i.e. $B_{\max}$.

### 3.3 *Model training*

#### 3.3.1 *CNN*

A CNN is a kind of supervised learning machine that consists of multiple layers of neurons. In most cases, a convolutional layer (or "conv layer") is the most important type of layer, with a pooling layer sometimes called a feature extractor following closely behind. Feature extraction from two-dimensional images or feature maps is accomplished using the convolution algorithm through the usage of the kernel. In image processing, the kernel serves as a filter or mask, and this procedure is identical to that process. The two-dimensional output of the convolutional layer is a series of feature maps. For the purpose of simulating the picture filtering action, the conv layer is purposefully built to be partially connected to the prior layer.

### 3.3.2 *SVM*

Classification and regression are common tasks for supervised learning machines like SVMs, SVCs, and SVRs [13]. We have limited our investigation to SVC. The kernel $N(c, c')$ defined on $C \times C$ will be used in the next development. These two kernels are utilized in this proposed:

$$N(c, c') = c, c' \tag{4}$$

$$N(c, c') := \exp\left(-\delta \|c - c'\|^2\right) \tag{5}$$

Begin by addressing the difficulties associated with binary classification. The set $Q := \{1, -1\}$ and $C \subseteq \mathfrak{R}^c$, which must be established. Consider the training dataset.

$$A := \{(c_m, q_m)\}_{m=1}^{t} \subseteq C \times Q \subseteq \mathfrak{R}^c \times \{1, -1\} \tag{6}$$

An optimization problem involving the training dataset can be easily solved to find the SVC (6):

$$\sum_{m=1}^{t} e_m - \frac{1}{2}\sum_{m=1}^{t}\sum_{m=1}^{t} e_m e_r q_m q_r N(c_m, c_r) \tag{7}$$

$$\sum_{m=1}^{t} e_m c_m = 0 \tag{8}$$

In equation (7), the tradeoff between machine complexity and the number of non-separable points is regulated by a regularization (or smoothing) parameter $X > 0$.

Consider the function (7) and suppose that $e_m^*$ provides a solution. Since (9) provides the most effective discriminant function,

$$b^*(c) = \sum_{m=1}^{t} e_m^* q_m N(c_m, c) + f^* \tag{9}$$

where the ideal bias term is represented by $f^*$.

Picture a case where $Q := \bar{\text{i}} := \{1, 2, \ldots, \text{i}\}$ is needed for labels and there are $i$ classes. The OVR method requires the creation of m binary classifiers. We label all training samples with $q_m = r$ with 1 and $q_m \neq r$ with -1 for each $\bar{r} \in m$ while training the $r$th classifier. The final choice assigns x to class j if the following equation (10) is satisfied.

$$r(c) = arg \max_{n=1} b_n(c) \tag{10}$$

## 4 RESULT AND DISCUSSION

As the number of electric vehicles (EVs) on the road continues to rise, the power infrastructure is being overwhelmed by electric vehicle charging stations (EVCSs). A hybrid approach utilizing a multivariable residual correction grey model (EMGM) and a LSTM network is proposed in this study for the purpose of electric vehicle charging station (EVCS) load prediction. Factors like as temperature, the number of EVs, and electricity pricing contribute to this method's lack of precision.

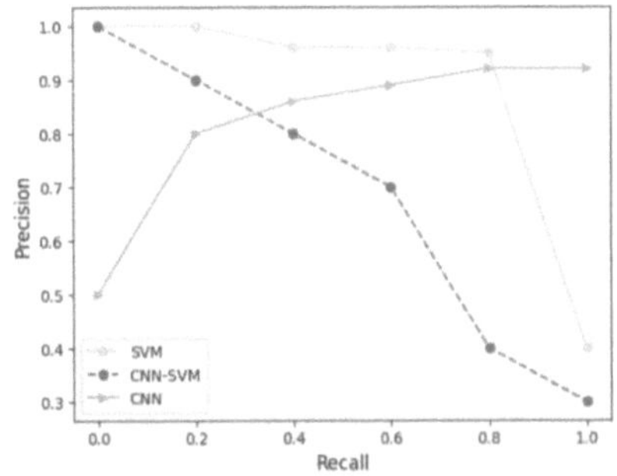


Figure 1. Testing findings' PR curves.

In Figure 1, we can see some examples of recall (PR) and precision charts. To the contrary, CNN-SVM is able to fit the picture data more accurately thanks to its superior overall detection ability.

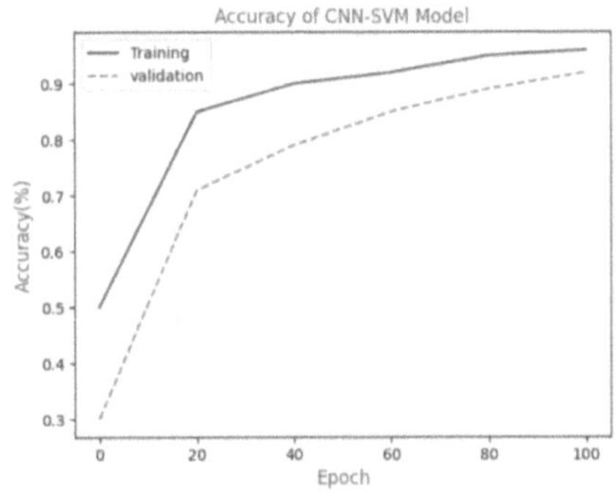


Figure 2. The diagnostic accuracy curve of the model.

Each training round consisted of three iterations, and we gave the model 100 rounds total. The average accuracy of all training rounds was determined. We presented them in a thorough manner, as illustrated in Figure 2.

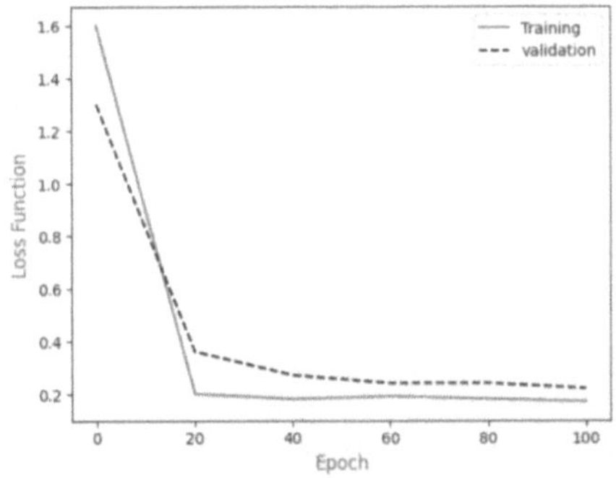


Figure 3. Loss Function Curve for Proposed Model.

The loss functions that were utilized for training and validation were averaged over all rounds. Figure 3 show that we presented them thoroughly.

## 5 CONCLUSION

A potential option to boosting decarbonization efforts is electrifying transportation; nevertheless, electrical networks may encounter new and unexpected load scenarios due to the growing number of EVs. Precise electric vehicle load forecasting is essential for reducing grid integration's negative effects. Preprocessing data from individual EV charging sessions into a personalized EV charging profile involves factoring in the actual power utilized for charging. This method has gained popularity. As a method for dimensionality reduction, feature selection aids in decreasing the number of features in datasets. Feature rich CNN-SVM is utilized during model training. The suggested technique beats the CNN and SVM models with a consistency rate of 96.60%.

## REFERENCES


[1] Balakrishna, G., Kumar, A., Tamilselvan, V., Younas, A., Kumar, N. M. G. and Rastogi, R. (2023, October). A Novel Ensembling of CNN-A-LSTM for IoT Electric Vehicle Charging Stations Based on Intrusion Detection System. In *2023 International Conference on Self Sustainable Artificial Intelligence Systems (ICSSAS)* (pp. 1312–1317). IEEE.

[2] Yadav, S., Sudman, M. S. I., Dubey, P. K., Srinivas, R. V., Srisainath, R. and Devi, V. C. (2023, October). Development of an GA-RBF based Model for Penetration of Electric Vehicles and its Projections. In *2023 International Conference on Self Sustainable Artificial Intelligence Systems (ICSSAS)* (pp. 1–6). IEEE.

[3] Salah, F., Ilg, J. P., Flath, C. M., Basse, H. and Van Dinther, C. (2015). Impact of electric vehicles on distribution substations: A Swiss case study. *Applied Energy*, 137, 88–96.

[4] Chen, T., Zhang, B., Pourbabak, H., Kavousi-Fard, A. and Su, W. (2016). Optimal routing and charging of an electric vehicle fleet for high-efficiency dynamic transit systems. *IEEE Transactions on Smart Grid*, 9(4), 3563–3572.

[5] Ovalle, A., Fernandez, J., Hably, A. and Bacha, S. (2016). An electric vehicle load management application of the mixed strategist dynamics and the maximum entropy principle. *IEEE Transactions on Industrial Electronics*, 63(5), 3060–3071.

[6] Crozier, C., Morstyn, T. and McCulloch, M. (2020). The opportunity for smart charging to mitigate the impact of electric vehicles on transmission and distribution systems. *Applied Energy*, 268, 114973.

[7] Zhang, X. (2018). Short-term load forecasting for electric bus charging stations based on fuzzy clustering and least squares support vector machine optimized by wolf pack algorithm. *Energies*, 11(6), 1449.

[8] Dang, Q., Wu, D. and Boulet, B. (2019, June). A q-learning based charging scheduling scheme for electric vehicles. In *2019 IEEE Transportation Electrification Conference and Expo (ITEC)* (pp. 1–5). IEEE.

[9] Wu, D., Wang, B., Precup, D. and Boulet, B. (2017). Boosting based multiple kernel learning and transfer regression for electricity load forecasting. In *Machine Learning and Knowledge Discovery in Databases: European Conference*, ECML PKDD 2017, Skopje, Macedonia, September 18–22, 2017, Proceedings, Part III 10 (pp. 39–51). Springer International Publishing.

[10] Wu, D., Zeng, H., Lu, C. and Boulet, B. (2017). Two-stage energy management for office buildings with workplace EV charging and renewable energy. *IEEE Transactions on Transportation Electrification*, 3 (1), 225–237.

[11] Van Kriekinge, G., De Cauwer, C., Sapountzoglou, N., Coosemans, T. and Messagie, M. (2021). Day-ahead forecast of electric vehicle charging demand with deep neural networks. *World Electric Vehicle Journal*, 12(4), 178.

[12] Duan, M., Darvishan, A., Mohammaditab, R., Wakil, K. and Abedinia, O. (2018). A novel hybrid prediction model for aggregated loads of buildings by considering the electric vehicles. *Sustainable Cities and Society*, 41, 205–219.

[13] Gao, F., Hsieh, J. G. and Jeng, J. H. (2019). A Study on Combined CNN-SVM Model for Visual Object Recognition. *J. Inf. Hiding Multim. Signal Process.*, 10(4), 479–487.

*Hybrid and Advanced Technologies – S. Prasad Jones Christydass, Nurhayati Nurhayati and S. Kannadhasan. (Eds)* © *2025 the Authors, ISBN 978-1-032-90254-8*


# Cybersecurity analysis of an electric vehicle based on its performance using hybrid KNN-SOM-DT model

R. Gomathi
*Department of CSE, Bannari Amman Institute of Technology, Sathyamangalam, Tamilnadu, India*

Mohd. Asif Gandhi
*Associate Professor, Department of Mechanical Engineering, School of Engineering and Technology, Anjuman-I-Islam's Kalsekar Technical Campus, Panvel, India*

Davinder Kumar
*Architect, Micron Semiconductor 8000 S Federal Way, Boise, USA*

P. Arun Prakash
*Assistant Professor, Sri Eshwar College of Engineering, Coimbatore, Tamilnadu, India*

Meenakshi Maindola
*Assistant Professor, Department of Computer Science Engineering, Graphic Era Deemed to be University, Dehradun, India*

Prasanthi Gottumukkala
*Assistant Professor, Department of Information Technology, Gokaraju Rangaraju Institute of Engineering and Technology, Hyderabad, India*

ABSTRACT: The widespread availability of electric vehicles has led many to forego gas-powered vehicles. So, to keep up with the rising demand, electric vehicle charging infrastructure, whether public or private, is growing at a rapid pace. The cybersecurity risks linked to EVCS have increased in tandem with their proliferation. With this structure in mind, this piece assesses the potential cyber threats to the EVCS network. The three main parts are training the model, selecting features, and preprocessing. The pre-processing of aberrant data involves cleaning, parsing, and normalizing the data. In this case, the optimal feature selection criterion is mean distortion, which is comprised of peak-to-peak voltage. We trained the model using the KNN-SOM-DT for higher accuracy. The proposed method achieved a higher accuracy of 95.82% compared to existing methods, including KNN and DT.

*Keywords*: Electric Vehicles (EVs), General Data Protection Regulation (GDPR), Distributed Energy Resources (DERs), Electronic Control Unit (ECU), Self-Organizing Map (SOM)

## 1 INTRODUCTION

One key component of future smart cities that can assist in accomplishing this goal is zero-emission vehicles, such as electric cars (EVs). More electric vehicle adoption will allow us to utilize less fossil fuels and create space for more intelligent and adaptable power grids. Ideally, EV management might prevent the electrical grid from becoming overwhelmed, sparing consumers the expense of necessary upgrades. Not long ago, the few slow chargers

DOI: 10.1201/9781003559115-27 

that could operate without data links were located in small, isolated areas. Numerous recharge outlets are now dispersed acrossmetropolitan areas. Among the many convenient features offered by these stations are fast, smart charging, the ability to link automobiles to the grid and vice versa, and an abundance of for use in cases where the transmission of private information is necessary.Electric vehicle charging infrastructures share the same requirements as any smart city: cybersecurity, data privacy, and interoperability. The General Data Protection Regulation (GDPR) and other data privacy regulations, as well as the most current cryptographic best practices for avoiding their disclosure, require careful handling of these messages because of the sensitive information they contain. Names, addresses, phone numbers, email addresses, bank account details, driving and location information, and other user data are among the subsets of this data that has been altered. There are a number of payment systems that use the same kind of private data, and secure communications are usually sufficient.

## 2 LITERATURE SURVEY

Electric vehicles have only just begun to gain traction in the mainstream of the automotive industry. [1]There are drivers who are afraid they will die on the road since there are no charging stations. As a financial incentive for individuals to purchase electric vehicles, the government provides tax rebates and discounts. [2]Innovations in technology and increases in production capacity have led to a steady decline in the price of electric vehicles. [3]The public's perception of EVs has shifted due to technological developments. One factor that may promote the expansion of sales of electric vehicles is the recent advancements in battery technology. Electric vehicle (EV) energy consumption [4]. Electric vehicle owners have it tough due to issues like the requirement for extra charging outlets [5]. A well-planned system of charging stations is the only way to address this issue. Reduced wait times and increased efficiency might be possible with algorithmic management of charging station distribution to individual users.[6]Electric vehicles (EVs) are one type of DER that the power companies in New York (NY) want to standardize on a single cyber security protocol. [7]Unfortunately, the proposal was never put into action because the third-party providers weren't in favour of it. [8]Hackers can easily damage electric vehicles and charging stations because of the non-standard cyber-physical interfaces they use. Furthermore, because of vulnerabilities in these interfaces, electric vehicles can be utilized to launch demand-side cyber-attacks against the power system [9]. Electric vehicles (EVs), distributed energy resources (DERs), and HVAC loads are examples of demand-side equipment that threat actors target when they aim to disrupt power infrastructures. [10]Power grids have never been the target of an attack like this, but there is mounting evidence that these holes might be exploited to do just that. Severe load shedding will be necessary for power system operators to cope with the attacks.[11]The electric vehicle (EV) industry has grown at an unprecedented rate over the last decade as a direct result of the increasing number of environmental problems caused by the burning of fossil fuels and the resulting increase in greenhouse gas emissions. [12]The goal of several governments‘ efforts to implement new technology is to reduce fuel consumption or improve the utilization of a combination of fuels. [13]Efficient fuel-cell vehicles, hybrids, and electric vehicles powered by batteries are only a few of the technologies currently accessible in the electric vehicle market. However, as technology has advanced, “range anxiety” has become more common among EV owners and purchasers[14].

## 3 PROPOSED SYSTEM

While being developed, every component of an electric vehicle (EV) must conform to specific laws and standards. Two primary purposes of standards are to improve consumer safety and to promote healthy competition among stakeholders. There are frameworks in place for the general testing and development of EVs, but they don't account for the unique cybersecurity needs of each part.

### 3.1 *Preprocessing*

As part of the pre-processing, the unusual data is cleaned, parsed, and normalized. Data parsing sorts a wide range of collected data types. Some examples of this kind of data anomaly include abnormalities in network traffic, terminal access, business commands, and so on. The first step in data purification is to sort the collected data into different categories according to the information format, type, and data head[15]. It selects and develops the obtained data's non-conformity criteria and the cleaning strategy's priority based on the data's relevancy. Another feature is that it removes data that doesn't adhere to the rules.

### 3.2 *Feature selection*

#### 3.2.1 *Peak-to-Peak voltage*

The peak-to-peak voltage ($R_{pp}$) is the voltage measured in volts from the lowest to the highest point. It is defined in equation 1 as the distance from $R_{\max}$ to $R_{\min}$. The dataset makes note of the fact that every ECU has unique CAN voltages at both the lowest and highest locations.

$$R_{pp} = R_{\max} - R_{\min} \tag{1}$$

#### 3.2.2 *Mean distortion*

Distortion can occur in signals as they are being transmitted across various communication routes. This can be due to medium impurities or peculiarities in the signals themselves. Each electronic control unit (ECU) is given a distinct pattern of distortion according to the suggested technique, which is based on the difference between the actual and expected voltage levels[16]. Furthermore, the average of this distortion is calculated for each square wave, dominant bit, and recessive bit. The Mean Distortion can be determined for a given collection of $d$-values by

$$Mean\ Disortion = \sum_{k=1}^{d} \frac{Actual(k) - Expected(k)}{d} \tag{2}$$

### 3.3 *Model training*

#### 3.3.1 *SOM*

An example of a neural network that mimics brain function is the Self-Organizing Map (SOM), sometimes called the Kohonen Map. After SOM transitions the network from a competitive learning approach to an unsupervised learning procedure, it begins to learn. Clustering datasets or reducing their dimensionality are two ways SOM maps help us making sense of complex circumstances. Samples that are similar will be grouped together in one cluster in the self-organizing map (SOM) topology, while samples that are different will be placed in separate ones.

#### 3.3.2 *KNN*

The k-nearest neighbour's algorithm (KNN) is one supervised machine learning method for classification and regression. During training, the set should determine how many samples are closer to the target output. Regression takes an average of the closest instances to find the output, while classification uses the vote classes of the closest examples to find the class output (i.e., the class with the biggest majority will be chosen).

### 3.3.3 *DT*

The idea behind regression trees is similar to that of classification decision trees, however the end goal is to generate numerical values instead of class labels. Regression trees depict subsets of the dataset at each level, from branches to decision nodes and the final leaf. The first step in building the regression tree is to divide the dataset into two groups, G1 and G2, based on the properties of the sample data, as indicated in equation (3). Minimizing the total residual sum of squares (RSS) is the goal when selecting the optimal component of this split.

$$RSS = \sum_{k \in F_1} (h_k - g(l_k))^2 + \sum_{k \in F_2} (h_k - g(l_k))^2 \quad (3)$$

## 4 RESULT AND DISCUSSION

In order to offer the essential suggestions for protecting electric vehicles and their onboard charging (OBC) systems against cyberattacks, this article thoroughly analyzes the effects of several data integrity attacks on the power electronic hardware found in an EV charger.

Figure 1 shows the accuracy throughout training and validation. At the very moment that the tuple's SOC value crosses either the minimum or maximum, a charging or discharging order is issued.

The suggested model's training and validation losses are shown in Figure 2. The y-axis shows the percentage of loss, while the x-axis shows the number of epochs.

Overall, the results demonstrate that the classifier-based detection method maintains a rather consistent performance with respect to TPR and FPR (see Figure 3). They also suggest that a re-training cycle once a year is sufficient.

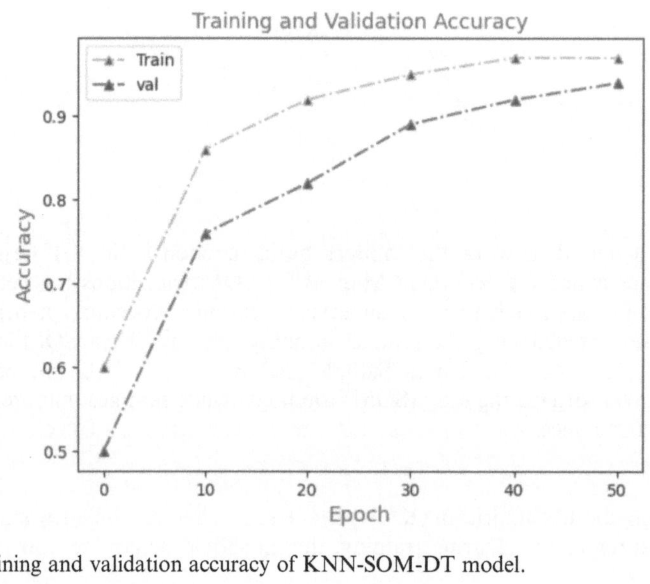


Figure 1. Training and validation accuracy of KNN-SOM-DT model.

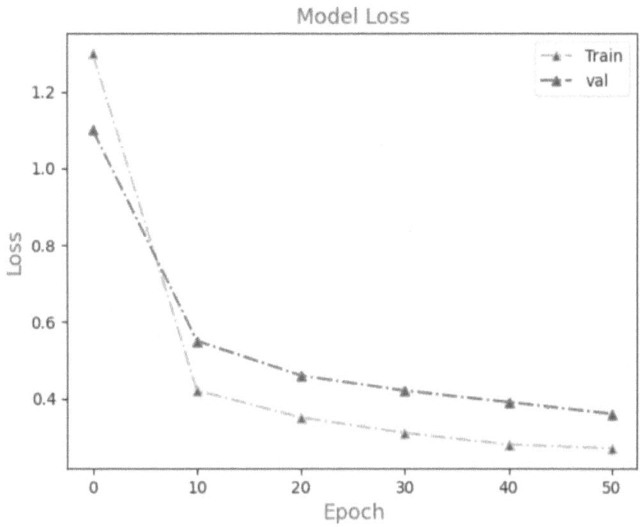


Figure 2. Model loss of KNN-SOM-DT model.

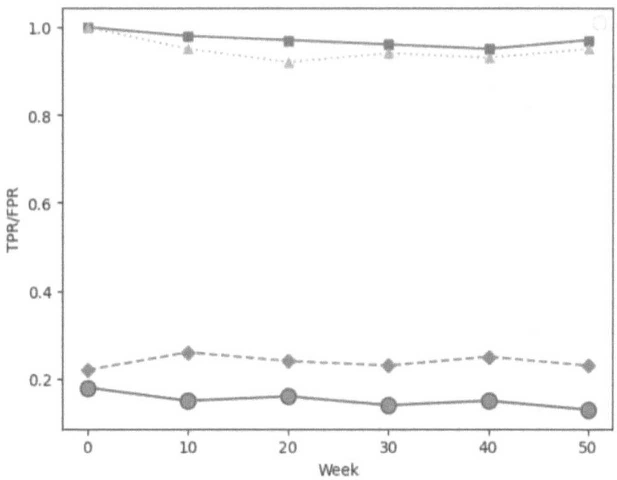


Figure 3. Evaluation with classifier based detection.

## 5 CONCLUSION

Due to the increasing availability of electric vehicles (EVs), the automotive industry is shifting away from conventional gasoline-fueled cars. As a result, multiple types of electric vehicle charging stations (EVCSs) are being put in to meet the growing demand for them, whether for public or private use. Because of their interconnected nature, electric vehicles (EVs), EV charging stations (EVCSs), and power grids are susceptible to cyberattacks. Data cleansing, parsing, and normalization are part of the pre-processing steps for abnormal data. Here, peak-to-peak voltage, or mean distortion, is the best metric to use when choosing features. It makes use of KNN-SOM-DT for model training. Compared to the KNN and DT models, the suggested method has a higher average accuracy of 95.82%.

## REFERENCES


[1] Johnson, J., Berg, T., Anderson, B. and Wright, B. (2022). Review of electric vehicle charger cybersecurity vulnerabilities, potential impacts, and defenses. *Energies*, 15(11), 3931.

[2] Bhusal, N., Gautam, M. and Benidris, M. (2021, April). Cybersecurity of electric vehicle smart charging management systems. In *2020 52nd North American Power Symposium (NAPS)* (pp. 1–6). IEEE.

[3] Sayed, M. A., Atallah, R., Assi, C. and Debbabi, M. (2022). Electric vehicle attack impact on power grid operation. *International Journal of Electrical Power & Energy Systems*, 137, 107784.

[4] Alcaraz, C., Lopez, J. and Wolthusen, S. (2017). OCPP protocol: Security threats and challenges. *IEEE Transactions on Smart Grid*, 8(5), 2452–2459.

[5] Fu, S., Zhang, Z., Jiang, Y., Chen, J., Peng, X. and Zhao, W. (2021). An Automatic RF-EMF Radiated Immunity Test System for Electricity Meters in Power Monitoring Sensor Networks. *Ad Hoc Sens. Wirel. Networks*, 50(1–4), 173–192.

[6] Harnett, K., Harris, B., Chin, D. and Watson, G. (2018). Doe/dhs/dot volpe technical meeting on electric vehicle and charging station cybersecurity report (No. DOT-VNTSC-DOE-18-01). *John A. Volpe National Transportation Systems Center* (US).

[7] Seijo Simó, M., López López, G. and Moreno Novella, J. I. (2017). Cybersecurity vulnerability analysis of the PLC PRIME standard. *Security and Communication Networks*, 2017(1), 7369684.

[8] El Mrabet, Z., Kaabouch, N., El Ghazi, H. and El Ghazi, H. (2018). Cyber-security in smart grid: Survey and challenges. *Computers & Electrical Engineering*, 67, 469–482.

[9] Sridhar, S. and Manimaran, G. (2010, July). Data integrity attacks and their impacts on SCADA control system. In *IEEE PES General Meeting* (pp. 1–6). IEEE.

[10] Lee, D., Kim, H., Kim, K. and Yoo, P. D. (2014, January). Simulated attack on dnp3 protocol in scada system. In Proceedings of the 31th Symposium on Cryptography And Information Security, kagoshima, japan (pp. 21–24).

[11] Zhang, Y., He, J. and Ionel, D. M. (2019, June). Modeling and control of a multiport converter based EV charging station with PV and battery. In *2019 IEEE Transportation electrification conference and EXPO (ITEC)* (pp. 1–5). IEEE.

[12] Fahem, K., Chariag, D. E. and Sbita, L. (2017, March). On-board bidirectional battery chargers topologies for plug-in hybrid electric vehicles. In *2017 International Conference on Green Energy Conversion Systems (GECS)* (pp. 1–6). IEEE.

[13] Hamdare, S., Kaiwartya, O., Aljaidi, M., Jugran, M., Cao, Y., Kumar, S., and Lloret, J. (2023). Cybersecurity risk analysis of electric vehicles charging stations. *Sensors*, 23(15), 6716.

[14] Jahan, I., Mohamed, F., Blazek, V., Prokop, L., Misak, S. and Snasel, V. (2023, May). Power Quality Parameters Forecasting Based on SOM Maps with KNN Algorithm and Decision Tree. In *2023 23rd International Scientific Conference on Electric Power Engineering (EPE)* (pp. 1–6). IEEE.

[15] Li, Q., Wang, Y., Pu, Z., Wang, S. and Zhang, W. (2019). Time series association state analysis method for attacks on the smart internet of electric vehicle charging network. *Transportation Research Record*, 2673(4), 217–228.

[16] Hafeez, A., Mohan, J., Girdhar, M. and Awad, S. S. (2021, December). Machine learning based ecu detection for automotive security. In *2021 17th International Computer Engineering Conference (ICENCO)* (pp. 73–81). IEEE.

*Hybrid and Advanced Technologies – S. Prasad Jones Christydass, Nurhayati Nurhayati and S. Kannadhasan. (Eds)* © 2025 the Authors, ISBN 978-1-032-90254-8


# Advancing vehicle energy systems: Optimisation of fuel-cell hybrid electric vehicles using parallel LSTM and CNN

Naveen Rana
*Assistant Professor, M.M. Engineering College, Maharishi Markandeshwar (Deemed to be University), Mullana, Ambala, Haryana, India*

A. Ramkumar
*Professor, Department of EEE, Kalasalingam Academy of Research and Education, Krishnankoil, Virudhunagar, Tamil Nadu, India*

D. Sai Chaitanya Kishore D.
*Professor, Department of Mechanical Engineering, Srinivasa Ramanujan Institute of Technology, Anantapur, India*

Narayan Pandurang Sapkal
*Assistant Professor, Department of Mechanical Engineering, Dr. D. Y. Patil Institute of Technology, Pimpri, Pune, India*

K. Gowdhami
*Assistant Professor, Department of CSE, Knowledge Institute of Technology, Salem, Tamilnadu, India*

D. Satheesh Kumar
*Associate Professor, Department of Computer Science and Engineering, Hindusthan College of Engineering and Technology, Coimbatore, India*

ABSTRACT: Customers and manufacturers in the automotive sector have been positively affected by hydrogen and fuel cells (FCs) over the last 30 years. Automakers are primarily focused on reducing fuel consumption and exhaust emissions, improving energy efficiency, and maximizing range limits through the adaptation of existing technologies. Vehicles also known as electric vehicles (EVs) are those that are equipped with electric motor systems. Feature selection, preprocessing, and training the model are the three key components. Data reduction and data transformation are the two main components of data processing. Use of Spearman's rank correlation coefficient is a technique in feature selection. The model's accuracy was enhanced during training by utilizing the Parallel CNN-LSTM. The proposed method outperformed state-of-the-art methods like CNN and LSTM with an accuracy rate of 97.12%.

*Keywords*: Fuel Cells (FCs), Electric Vehicles (EVs), Internal Combustion Engines (ICEs), Convolutional Neural Network (CNN), Long Short-Term Memory networks (LSTM)

## 1 INTRODUCTION

The urgent necessity to protect the environment has two outcomes: slowing the pace of climate change and promoting the creation of sustainable technology, particularly in the transportation sector. As a result, there has been a recent uptick in studies investigating novel methods of renewable energy generation, with an emphasis on electric mobility. There

DOI: 10.1201/9781003559115-28

has been a meteoric rise in the sales of hybrid electric vehicles (HEVs) and fully electric vehicles (BEVs) in the last several decades, meeting an early demand for locally emission-free and ecologically beneficial transportation. Problems with BEVs persist despite the fact that HEVs facilitate the transition to extensive electric mobility. The reduced range in comparison to HEVs and the need for more extensive charging infrastructure are two points that I would like to address further. Components need to have strong durability and quick charging is necessary to minimize overall journey time; there are also more common and economic limits in this application space. Due to the limited power dynamics in fuel cell systems, FCHEVs make advantage of energy and power buffering technologies such as batteries or ultra-capacitors. In spite of this, FCHEVs built for tough jobs have a very complex powertrain because of all the energy and power sources they need. Better control methods are a necessary technological enabler for FCHEVs because of their dependence on regenerative braking and the FCS's transient power supply. As a result of the increased complexity and number of degrees of freedom (DOFs), FCHEVs necessitate a more all-encompassing EMS strategy.

## 2 LITERATURE SURVEY

As a result of technological advancements, the vehicle has gone from being an idea to being mass-produced, and it is now an essential mode of transportation for every household [1]. The majority of vehicles in the past utilized internal combustion engines (ICEs), with petroleum as the primary fuel, because drivers needed vehicles that were both highly reliable and could move quickly [2]. Fuel cell technology's recent maturation has piqued the interest of both academics and industry in its potential automotive applications [3]. One of the most promising forms of mobility for the future is the fuel cell hybrid electric vehicle (FCHEV). High energy consumption and low waste rate are two of the system's advantages [4]. The reason behind this is that the fuel is transformed into electrical energy directly. Improving the system's dynamic performance and energy consumption efficiency is crucial for lowering the cost of hybrid electric vehicles (HEVs) [5]. Technologies that are gentler on the planet are in high demand right now[6]. The rapid growth of modern cities has several negative effects on the environment, including increased demand for transportation services, which in turn leads to gridlock, higher fossil fuel consumption, and other related problems [7]. Furthermore, BEVs do have some drawbacks, such as a short range time to charge the battery to its full capacity, safety of the battery, and distance travelled[8]. Regardless of progress towards cars that don't pollute, although there is still a long way to go, electric vehicles that run on renewable energy sources are better since they divert pollutants and toxic emissions to less dangerous by products, such as water, rather than greenhouse gases [9].The interest in fuel-cell electric vehicles (FCEVs) has recently increased within the automobile sector. Propelled by electric motors that draw power from fuel cells (FCs), fuel cell electric vehicles (FCEVs) primarily use hydrogen mixed with air as their energy source [10]. There has been a lot of buzz about FC extended electric vehicles and FCHEVs8, which are plug-in hybrid electric vehicles[11].It is necessary to have a starter system in order to use the FC alone for powering your electric vehicle. Modern propulsion systems, such as electric motors and fuel cells, are quite effective. [12]Although fuel cell vehicles are costly, have poor endurance, and have problems with hydrogen storage, their reduced emissions make them more attractive to car manufacturers. PEM fuel cells are a favourite for both stationary applications and mobility since they are sustainable and drastically cut down on greenhouse gas emissions [13]. Automobile manufacturers started making them because of all the good things they could do, such as having a low operating temperature, a high power density, a quick starting, and a sturdy system [14]. When compared to ICEVs (Internal Combustion Engine Vehicles), fuel cell vehicles have far lower pollution restrictions and fuel consumption, according to type approval.

## 3 PROPOSED SYSTEM

Hybrid electric vehicle (HEV) and FCEV development can gain from this article's modeling and dynamic control design technique. An introduction to advanced propulsion technologies will be given first, followed by a description of FCEV research at DaimlerChrysler. After that, it will discuss hydrogen and its storage issues as a fuel for FCEVs.

### 3.1 *Preprocessing*

#### 3.1.1 *Data transformation*

After EDA, data transformation comes next. Priority one should be data simplification and data mining ease. Similar to data cleansing, this phase incorporates human inspection into the process of generating features and attributes from the available data[15]. The next step is to carry out any required treatments for cleaning, smoothing, or normalization.

#### 3.1.2 *Data reduction*

Data reduction is a prerequisite to using any ML algorithm when dealing with a dataset having a large number of observations. The dataset is comprised of numerous features, or variables; it's probable that some of these are irrelevant or unnecessary. In order to minimize data, it is important to examine just the features that have a substantial impact on the model's learning, prediction, generalizability, and processing time. Neglecting to address overfitting features or data that acts as noise could lead to subpar model performance.

### 3.2 *Feature selection*

As a nonparametric indicator, Spearman's rank correlation coefficient quantifies the degree to which two variables are dependent on one another[16]. It is commonly shortened to "$\sigma$" for short. Here is its definition:

$$\sigma = \frac{\sum_d (p_d - \hat{p})\left(f_d - \hat{f}\right)}{\sqrt{\sum_d (p_d - \hat{p})\sum_d \left(f_d - \hat{f}\right)}} \tag{1}$$

For every feature obtained from a single sensor, determine the battery status classification label and the Spearman's rank correlation coefficient.

### 3.3 *Training the model*

#### 3.3.1 *Parallel CNN-LSTM*

A PCLA model for predicting the performance of fuel cell electric vehicles is proposed in this paper, which makes use of parallel CNN-LSTM attention. Once CNNs and LSTMs have extracted features, an attention strategy ensures that the model zeroes in on the most important details. CNNs can help you make sense of the study's time-series data in terms of its spatial components, while LSTM layers will shed light on its temporal features. This allows for the direct extraction of spatiotemporal characteristics from the input data, and the more efficient usage of the LSTM compared to the model structure using the sequential arrangement of CNNs and LSTMs. After that, data is sent to the CNN and LSTM layers, which are then trained independently. The CNN and LSTM both models use batch normalization on the output of the last layer, which is common to all three of them. While HO decides on the stride 1, padding, and ReLU activation functions for the LSTM layers, the CNN layers maintain consistency in these areas. The range and dimensions of the incoming data are determined by HO, and the output dimension is also decided by them. Those dimensions are used by the first and second layers.

### 3.3.2 *Hyper parameter optimization*

Hyperparameters have a significant impact on the performance of ML models built using deep learning as well as shallow learning. It is critical to get the hyper parameters right because they impact the performance of the model. Out of the models generated by deep learning and shallow machine learning methods, the study employed HO to find the best one. Shallow ML models and statistical analysis both used random searches. Since it gets rid of unnecessary queries, random search, like a grid search, is fast. Eight DL models were implemented using Bayesian optimization. In Bayesian optimization, the acquisition function gives the next input value, while the surrogate model predicts the objective function using the previous experimental data.

## 4 RESULT AND DISCUSSION

A plethora of alternative fuel and vehicle choices are being explored in an endeavour to lessen the impact of the car industry's triple threat: climate change, urban air pollution, and reliance on foreign oil. In order to find the most common forms of transportation that could solve the most societal issues, this plan looks at the most used ones.

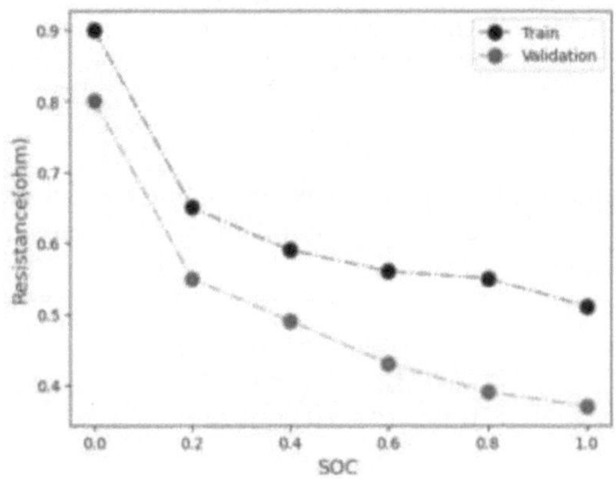


Figure 1. Electric battery as a function of battery SOC.

The relationship between the battery's internal resistance and its state of charge (SOC) is seen in Figure 1.

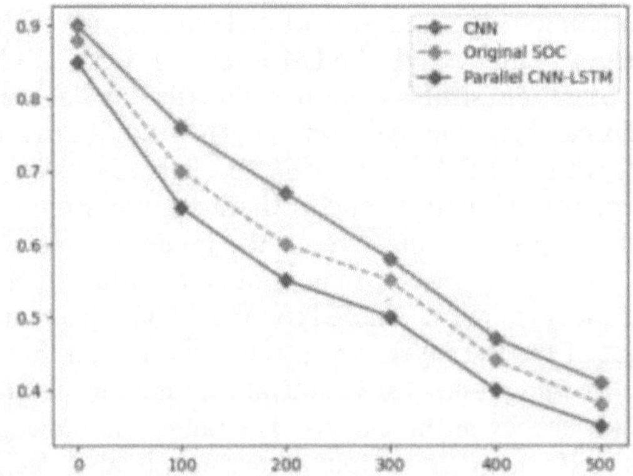


Figure 2. Different models' predictions of comparative SoC profiles.

The expected SoC profiles from an RB model and the Parallel CNN-LSTM model are compared with the real SoC recorded from the EV in Figure 2. In terms of rapid changes in SoC levels, the Parallel CNN-LSTM model is better in line with the actual data points, however both the RB and Parallel CNN-LSTM models follow the trend of the original SoC profile. It is plainly shown in the graph.

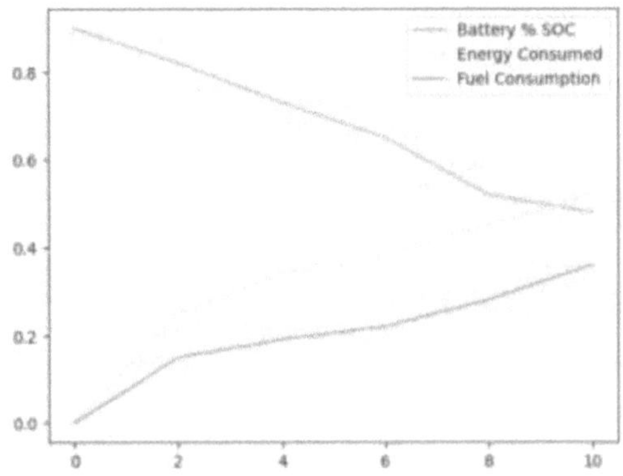


Figure 3. Graphical representation of battery SoC, fuel consumption & energy consumed.

There is a connection between the energy and fuel consumption of a Parallel CNN-LSTM and the battery's state of charge (SoC). The previous example displays it in Figure 3.

## 5 CONCLUSION

An ideal fuel-cell electric vehicle design that keeps supercapacitor and fuel cell costs to a minimum is laid forth in this essay. It stands to reason that the cost of fuel cells and supercapacitors is proportional to their volume per unit of volume. So that the dc-link voltage remains constant throughout the vehicle's operation, a certain quantity of fuel-cell and supercapacitor units must be installed. Data reduction and data transformation are the two main components of data processing. One approach is to utilize Spearman's rank correlation coefficient for selecting attributes. The training of models makes use of parallel CNN-LSTM. The proposed approach outperforms the CNN and LSTM models with an average accuracy of 97.12%.

## REFERENCES


[1] Jahangir, H., Tayarani, H., Ahmadian, A., Golkar, M. A., Miret, J., Tayarani, M. and Gao, H. O. (2019). Charging demand of plug-in electric vehicles: Forecasting travel behavior based on a novel rough artificial neural network approach. *Journal of Cleaner Production*, 229, 1029–1044.

[2] Gopal, A. R., Park, W. Y., Witt, M. and Phadke, A. (2018). Hybrid-and battery-electric vehicles offer low-cost climate benefits in China. *Transportation Research Part D: Transport and Environment*, 62, 362–371.

[3] Das, H. S., Tan, C. W. and Yatim, A. H. M. (2017). Fuel cell hybrid electric vehicles: A review on power conditioning units and topologies. *Renewable and Sustainable Energy Reviews*, 76, 268–291.

[4] Chen, J., Xu, C., Wu, C. and Xu, W. (2016). Adaptive fuzzy logic control of fuel-cell-battery hybrid systems for electric vehicles. *IEEE Transactions on Industrial Informatics*, 14(1), 292–300.

[5] Snoussi, J., Ben Elghali, S., Benbouzid, M. and Mimouni, M. F. (2018). Auto-adaptive filtering-based energy management strategy for fuel cell hybrid electric vehicles. *Energies*, 11(8), 2118.

[6] Chung, W. H., Gu, Y. H. and Yoo, S. J. (2022). District heater load forecasting based on machine learning and parallel CNN-LSTM attention. *Energy*, 246, 123350.

[7] Ajanovic, A. and Haas, R. (2021). Prospects and impediments for hydrogen and fuel cell vehicles in the transport sector. *International Journal of Hydrogen Energy*, 46(16), 10049–10058.

[8] Hu, L., Dong, J., Lin, Z. and Yang, J. (2018). Analyzing battery electric vehicle feasibility from taxi travel patterns: The case study of New York City, USA. *Transportation Research Part C: Emerging Technologies*, 87, 91–104.

[9] Zambalov, S. D., Yakovlev, I. A. and Maznoy, A. S. (2020). Effect of multiple fuel injection strategies on mixture formation and combustion in a hydrogen-fueled rotary range extender for battery electric vehicles. *Energy Conversion and Management*, 220, 113097.

[10] Xu, L., Mueller, C. D., Li, J., Ouyang, M., and Hu, Z. (2015). Multi-objective component sizing based on optimal energy management strategy of fuel cell electric vehicles. *Applied energy*, 157, 664–674.

[11] Moorthi, K., Chougale, R. K., Patel, P. A., Sharma, P., Raj, A., and Gangat, M. S. (2023, June). Electric vehicles charging system for fast and safe charging using LSTM based gradient boosted regression tree. In *2023 International Conference on Sustainable Computing and Smart Systems (ICSCSS)* (pp. 1289–1294). IEEE.

[12] Yadav, S., Sudman, M. S. I., Dubey, P. K., Srinivas, R. V., Srisainath, R., and Devi, V. C. (2023, October). Development of an GA-RBF based model for penetration of electric vehicles and its projections. In *2023 International Conference on Self Sustainable Artificial Intelligence Systems (ICSSAS)* (pp. 1–6). IEEE.

[13] Gandhi, M. A., Priya, K., Charan, P., Sharma, R., Rao, G. N., and Suganthi, D. (2023, July). Smart electric vehicle (EVs) charging network management using bidirectional GRU-AM approaches. In *2023 2nd International Conference on Edge Computing and Applications (ICECAA)* (pp. 1509–1514). IEEE.

[14] Balakrishna, G., Kumar, A., Tamilselvan, V., Younas, A., Kumar, N. M. G., and Rastogi, R. (2023, October). A novel ensembling of CNN-A-LSTM for IoT electric vehicle charging stations based on intrusion detection system. In *2023 International Conference on Self Sustainable Artificial Intelligence Systems (ICSSAS)* (pp. 1312–1317). IEEE.

[15] Rocha, Á., López-López, P. C., and Salgado-Guerrero, J. P. (2022). *Communication, Smart Technologies and Innovation for Society*. Springer Singapore.

[16] Lin, R. H., Pei, Z. X., Ye, Z. Z., Guo, C. C., and Wu, B. D. (2020). Hydrogen fuel cell diagnostics using random forest and enhanced feature selection. *International Journal of Hydrogen Energy*, 45(17), 10523–10535.

*Hybrid and Advanced Technologies – S. Prasad Jones Christydass, Nurhayati Nurhayati and S. Kannadhasan. (Eds)* © *2025 the Authors, ISBN 978-1-032-90254-8*
*Open Access: www.taylorfrancis.com, CC BY-NC-ND 4.0 license*

# Large-scale data-driven financial risk management & analysis using back propagation neural network technique

Sarita Agarwal
*Assistant Professor, Indira School of Business Studies PGDM, Pune, Maharashtra, India*

Rakesh Kumar
*Associate Professor, Magadh Professional Institute, Danapur, Patna, India*

Sameer Yadav
*Research Scholar, Department of Commerce and Business Administration, University of Allahabad, Prayagraj, Uttar Pradesh, India*
ORCID ID: 0000-0002-6949-6302

C. Soundarya
*Assistant Professor, Department of Information Technology, K. Ramakrishnan College of Engineering, Samayapuram, Trichy, India*

Prasanthi Gottumukkala
*Assistant Professor, Department of Information Technology, Gokaraju Rangaraju Institute of Engineering and Technology, Hyderabad, India*

Nisarg Gandhewar
*Assistant Professor, Department of AIML, Ramdeobaba College of Engineering and Management Nagpur, Maharashtra, India*

ABSTRACT: When it comes to measuring market risk, the state of the art right now is tools like RiskMetrics and historical simulation. To counter this, we offer flexible methods that incorporate recent discoveries in financial econometrics; these approaches should improve the accuracy of risk assessments in all contexts, including asset and portfolio analysis. Feature extraction, preprocessing, and model training are the three key parts. Discrete qualitative attributes, a kind of continuous attribute discretization, are employed in preprocessing. The values of these attributes are represented by linguistic or discrete values. Data points are said to be the curse of dimensionality, which is where principal component analysis (PCA) in feature extraction begins. In order to achieve more accurate results, we trained the model using the BPNN. Compared to other methods, such as CNN and ANN, the proposed strategy achieved a higher accuracy of 98.16 percent.

*Keywords*: Principal Component Analysis (PCA), Financial Risk Management (FRM), Small and Medium-sized Enterprises (SMEs), Performance of Back Propagation Neural Network (BPNN), Artificial Neural Network (ANN)

## 1 INTRODUCTION

Finance, which encompasses all monetary dealings, including those involving potentially defaulting business loans. A common misunderstanding is that it solely applies to negative risk, which encompasses both the likelihood of monetary loss and the unpredictability of its severity. Financial risk is the potential for monetary loss as a result of an investment or commercial venture. There are many different types of financial risks, but three of the most prevalent ones are credit, liquidity, and operational. To monitor, minimize, and control the likelihood or impact of unlucky events or to maximize the realization of opportunities, risk management entails identifying, assessing, and prioritizing risks (both positive and negative). Then, resources are coordinated and used economically. An adversary's intentional attack, accidents, natural disasters, legal liabilities, project failures, credit risk, and market volatility are all potential sources of risk. Risks to the financial system, as well as those to strategy and operations, are all part of the IMF's comprehensive risk-management framework. Financial risk encompasses a wide range of possible negative outcomes in relation to monetary transactions, such as the default of business loans [5]. It is commonly believed to encompass solely negative risk, which refers to the possibility of monetary loss and the lack of clarity about its magnitude.

## 2 LITERATURE SURVEY

With the help of machine learning, computer vision, robotics, and NLP have all made great strides forward[1]. After witnessing these incredible examples, many are wondering what other industries may use machine learning for when they have a lot of data. Of course, Financial Risk Management (FRM) isn't an exception[2]. Most FRM tasks are challenging due to the complicated, sparse, and dynamic nature of the data. [3]There needs to be a method for assessing and managing risk in every company. [4]More complex and larger businesses, such as financial institutions, have made the use of sophisticated statistical models to assess and mitigate risk a top priority in recent years. [5]Using previously used statistical or simulation approaches, it is becoming increasingly difficult for massive enterprises with extremely vast portfolios and complex financial products to accurately analyse their exposure to the ever-changing financial market.To evaluate the potential dangers of models, [6]present a paradigm based on relative entropy. [7] Parameter risk is incorporated into risk-neutral pricing when the authors know a distribution on the parameter space, using risk-capturing functional and convex risk measurements. [8] advocate for the use of historical data in model performance evaluations to calculate a risk adjustment for modeled uncertainty.[9]This means that data mining and the extraction of valuable insights from historical company data, as well as the evaluation and simulation of potential system vulnerabilities in the Internet financial industry, must make full use of raw sets. [10]Rough sets are frequently used in online financial reporting, as discussed in various studies. The top ten financial risk variables using the conditional features of the rough set model, Rosetta was used for online financial risk prediction[11]. Exact assessment of financial risks associated with online transactions is now possible, thanks to simulations that validated the model's strong predictive power. Using a questionnaire survey, [12]polled online finance specialists to ascertain the degree of risk and the relevance of early warning signs. Because of this, risk prediction models and rough sets have become widely used in the financial industry. Supplementing subjective risk appraisal with approximate defined techniques helps make evaluation outcomes more objective and easier to grasp[13]. Since it takes into account both subjective prior knowledge and objective attribute importance, it is able to conduct risk evaluations and attribute weight computations [14]. We need to examine a great deal of external environmental data as well as historical firm data to comprehend the regulations and laws of system risk assessment.

## 3 PROPOSED SYSTEM

Financial service businesses' risk management systems were assessed and reviewed during on-site inspections. Notable domestic and international property-liability and life/health insurers were part of this evaluation. The information gathered encompassed both the theory and practice of financial risk management.

### 3.1 *Preprocessing*

#### 3.1.1 *Data discritization*

Standardization items just take data faults (noise) into account, whereas reference adds standardization items to limited data to guarantee stability. It is common practice to discretize the imaging region in numerical simulations of acoustic wave propagation, which relies on a number of assumptions.Some of the background noise can be muted in this way [15]. For instance, because it is constant, the speed of sound in tissue is incomplete. In order to complete this work, you will need to conduct discrete data mining analysis. There are two categories into which this falls. A continuous quantitative feature is one such thing; it stands for a set of observable properties of the thing being described. Continuous intervals, like length, temperature, etc., are its sources of value. Qualitative qualities that are represented by language or discrete values are known as discrete qualitative attributes. These attributes are a discretization of continuous attributes.

### 3.2 *Feature extraction*

The data preparation module's principal component analysis (PCA) operation receives the normalized feature set, which consists of 12 data streams for technical indicators. Principal component analysis (PCA) is theoretically grounded on the curse of dimensionality, which indicates that data points in high dimensions are statistically far apart from one another[16]. Training a machine learning algorithm is expensive because of the curse of dimensionality, and its predictions are questionable because of the huge difference between in-sample and out-of-sample data points. Lessening the number of dimensions in data sets is a sensible way to deal with the curse of dimensionality. There is a current scenario where PCA is helpful. Principal component analysis (PCA) begins with input fitting. The procedure finds the main factors that represent the input's most variable directions. After deciding how many components to keep, the most important ones are ranked by the amount of variance they explain.

### 3.3 *Model training*

#### 3.3.1 *BPNN*

Out of all the ANN models, BPNN has the most followers. In most cases, BPNN will have a three-layer design. The BPNN model generates matching links between input and output that are exceedingly nonlinear via self-adaptive learning of diverse training patterns. Consider of the $\{S\}$ values as the output of the learning pattern's input nodes $\{s_m\}$, then calculate $X_d$ and the related hidden node's output $e_d$ (where $e_d$ is the same as the mid-layer values in a three-layer network) using the weights $\{Y_{md}\}$, biases $\{\theta_d\}$, and inputs $\{s_m\}$:

$$X_d = \sum_{m=1}^{i} Y_{md} s_m - \theta_d \tag{1}$$

$$e_d = b(X_d) \tag{2}$$

Assuming that the output node k has the following properties: bias $\theta_l$, connection weights $\{Y_{dl}\}$, and mid-layer outputs $\{e_d\}$, determine the input $R_l$ and the related output $W_l$.

$$R_l = \sum_{d=1}^{k} Y_{dl} e_d - \theta_l \tag{3}$$

$$W_l = b(R_l) \tag{4}$$

Use the following equation to determine the output node $l$'s error signal $j_l$ by subtracting the actual network output $\widehat{w}_l$ derived from the desired results $\widehat{w}_l$ to the pattern of learning:

$$j_l = (\widehat{w}_l - W_l)\widehat{w}_l(\widehat{w}_l - 1) \tag{5}$$

Iteratively presenting the networks with a succession of input/output pattern sets is the training process for the BPNN. By modifying weights to minimize the error between the actual and anticipated output patterns in the training sets, networks learn the input/output relationship of interest. The network weight coefficients cannot be changed once the learning phase has concluded. With this method, pattern identification and prediction may be done at the speed of light using only networks that are capable of forward calculations.

## 4 RESULT AND DISCUSSION

Financial risk management is of the utmost importance to small and medium-sized enterprises (SMEs). The ability of small and medium-sized businesses (SMEs) to grow depends on investment capital. These endeavors cannot proceed without access to appropriate money, which might come from either internal or external sources. Many academics have found that small and medium-sized enterprises (SMEs) have restricted access to external sources of capital.

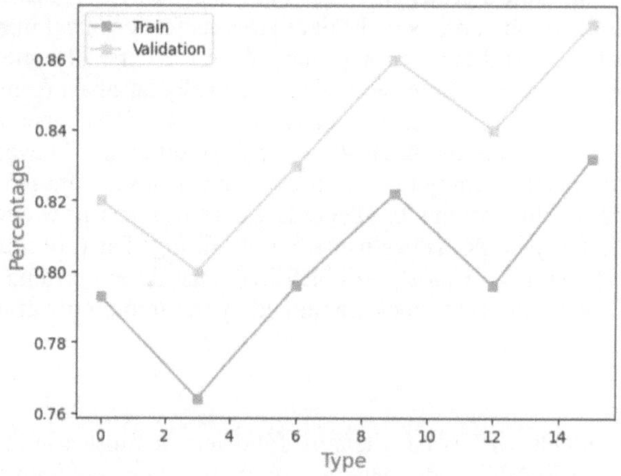


Figure 1. Simulation results of six sets of tests without training.

Figure 1 shows the training results, which are the simulation results of the five test sets that did not include training. Figure 1 shows that the accuracy of the risk early warning system is significantly improved when the number of nodes is determined using the golden ratio method.

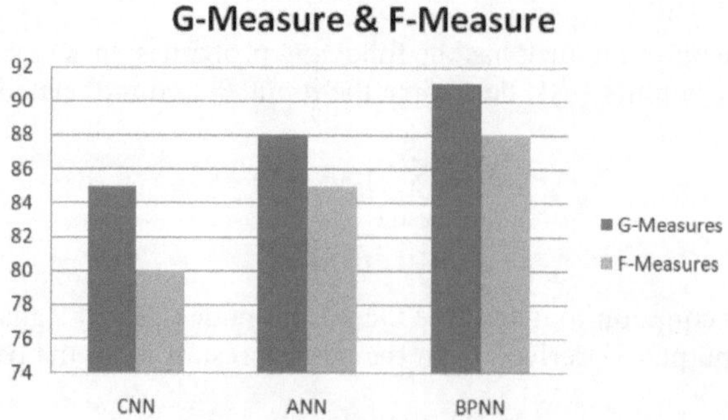


Figure 2. G-Measures and F-Measures of different methods.

Graphical comparison of all models' G-Measures and F-Measures outputs is made easy in Figure 2. The two evaluation metrics are shown on the X-axis, and the evaluation's accuracy is shown on the Y-axis.

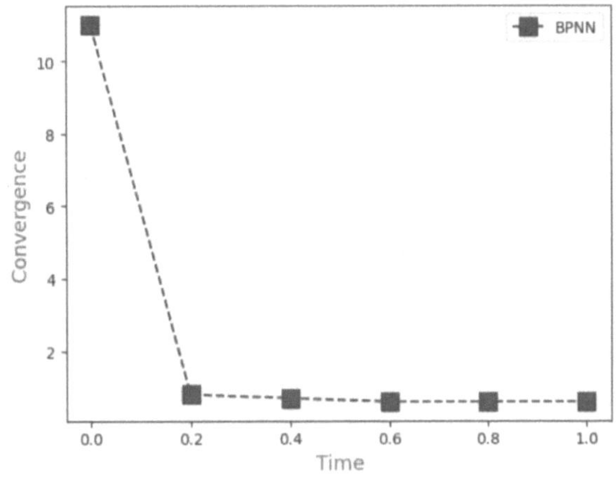


Figure 3. Convergence of proposed models.

Figure 3 shows the convergence; the third one is superior because it converges to zero. Organizations can evaluate their financial risk using the financial risk prediction index method.

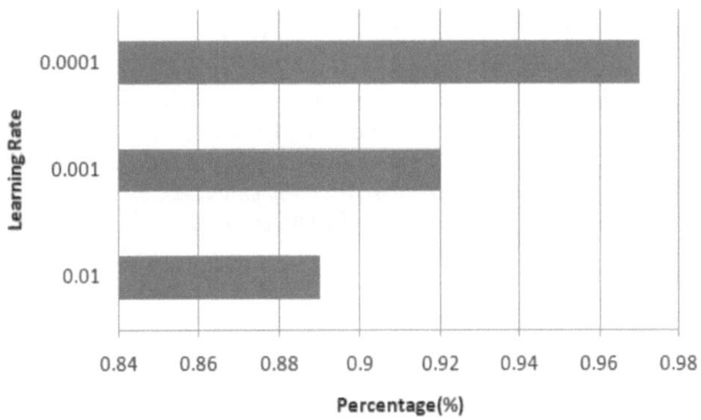


Figure 4. Result of initial learning rate.

Figure 4 displays the outcomes of the initial learning rate: It is clear from the results that choosing an initial learning rate significantly affects the experimental outcomes.

## 5 CONCLUSION

Over the last two decades, financial risk management has grown in importance for financial institutions. Risk management ought to be a major focus for banks and other financial organizations if they wish to enhance their profitability. The only thing that has remained constant in this dynamic environment is the presence of risk. There are numerous risks that banks face, including credit, market, interest rate, liquidity, and currency exchange risk.

Preprocessing makes use of discrete qualitative qualities, which are a sort of continuous attribute discretization. There are verbal or discrete values that stand in for these qualities. When it comes to feature extraction, PCA is the way to go because data points are considered the bane of dimensionality. We use the BPNN method to train the models. Compared to the CNN and ANN models, the suggested method achieves a consistency rate of 98.16%.

## REFERENCES


[1] Choi, T. M., Chan, H. K. and Yue, X. (2016). Recent development in big data analytics for business operations and risk management. *IEEE Transactions on Cybernetics*, 47(1), 81–92.

[2] Kou, G., Chao, X., Peng, Y., Alsaadi, F. E. and Herrera Viedma, E. (2019). Machine learning methods for systemic risk analysis in financial sectors.

[3] Lin, W. Y., Hu, Y. H. and Tsai, C. F. (2011). Machine learning in financial crisis prediction: a survey. *IEEE Transactions on Systems, Man, and Cybernetics, Part C (Applications and Reviews)*, 42(4), 421–436.

[4] Lee, T. K., Cho, J. H., Kwon, D. S. and Sohn, S. Y. (2019). Global stock market investment strategies based on financial network indicators using machine learning techniques. *Expert Systems with Applications*, 117, 228–242.

[5] Kou, G., Peng, Y. and Wang, G. (2014). Evaluation of clustering algorithms for financial risk analysis using MCDM methods. *Information Sciences*, 275, 1–12.

[6] Barrieu, P. and Scandolo, G. (2015). Assessing financial model risk. *European Journal of Operational Research*, 242(2), 546–556.

[7] Bernard, C. and Vanduffel, S. (2015). A new approach to assessing model risk in high dimensions. *Journal of Banking & Finance*, 58, 166–178.

[8] Coqueret, G. and Tavin, B. (2016). An investigation of model risk in a market with jumps and stochastic volatility. *European Journal of Operational Research*, 253(3), 648–658.

[9] Yang, Q., Wang, Y. and Ren, Y. (2019). Research on financial risk management model of internet supply chain based on data science. *Cognitive Systems Research*, 56, 50–55.

[10] Wang, R., Yu, C. and Wang, J. (2019). Construction of supply chain financial risk management mode based on Internet of Things. *IEEE Access*, 7, 110323–110332.

[11] Bagus, M. R. D. and Hanaoka, S. (2022). The central tendency of the seaport-fulcrum supply chain risk in Indonesia using a rough set. *The Asian Journal of Shipping and Logistics*, 38(4), 222–233.

[12] García, F., González-Bueno, J., Guijarro, F. and Oliver, J. (2020). Forecasting the environmental, social, and governance rating of firms by using corporate financial performance variables: A rough set approach. *Sustainability*, 12(8), 3324.

[13] Wang, H. B., Xu, W. Y. and Xu, R. C. (2005). Slope stability evaluation using back propagation neural networks. *Engineering Geology*, 80(3-4), 302–315.

[14] Mihaylov, G. and Zurbruegg, R. (2021). The relationship between financial risk management and succession planning in family businesses. *International Journal of Managerial Finance*, 17(3), 438–454.

[15] Dai, W. (2022). Application of improved convolution neural network in financial forecasting. *Journal of Organizational and End User Computing (JOEUC)*, 34(3), 1–16.

[16] Li, L. (2021, November). Financial Trading with Feature Preprocessing and Recurrent Reinforcement Learning. In *2021 16th International Conference on Intelligent Systems and Knowledge Engineering (ISKE)* (pp. 162–169). IEEE.

*Hybrid and Advanced Technologies – S. Prasad Jones Christydass, Nurhayati Nurhayati and S. Kannadhasan. (Eds)* *, ISBN 978-1-032-90254-8*


# Application of neural network in analysis of stock market prediction based on VG16-CNN-SVM model

Davinder Paul Singh
*Department of Computer Science and Engineering, School of Technology, Pandit Deendayal Energy University, Gandhinagar, Gujarat, India*

Rakesh Kumar
*Associate Professor, Magadh Professional Institute, Danapur, Patna, India*

Sameer Yadav
*Research Scholar, Department of Commerce and Business Administration, University of Allahabad, Prayagraj, Uttar Pradesh, India*
ORCID ID: 0000-0002-6949-6302

Amrita Tatia Karnawat
*HoD Finance, Assistant Professor, Indira School of Business Studies, Maharashtra, India*

Sanjay Bandi
*Assistant Professor, ECE Department, SRKR Engineering College, Bhimavaram, Andhra Pradesh, India*

Akshay Raj
*Research Associate Research and Development Department, IIMT University Meerut, India*

ABSTRACT: Academics have recently prioritized studying the stock market due to its fundamental nonlinearity. A stock market investment is based on speculation. In order to predict the value of stocks, people seek out methods and tools that will allow them to make the most money with the least amount of risk. Model training, feature selection, and preprocessing are the three key components. K-L Distance provides excellent accuracy in this case, and time difference relevance is also a part of preprocessing. PCA is utilizes a multivariate statistical method in feature selection. For more accurate results, we trained the model using the VG-16-CNN-SVM. The proposed method outperformed rival methods, such as SVM and VG16, with an accuracy of 95.32%.

*Keywords*: Stock Market, Principal component analysis (PCA), Support Vector Machine (SVM), Stock Movements, Convolutional Neural Network (CNN)

## 1 INTRODUCTION

Changes in the ability to forecast stock prices have taken on a far greater importance in the eyes of professional analysts and investors. Predicting trends and the stock market's performance is challenging due to environmental noise and the market's extreme volatility. Many things, such as quarterly earnings reports, market events, and individual actions, can cause stock prices to rise or fall. Traders constantly construct stock-based technical indicators. While these indicators make studying stock returns easier, predicting daily or weekly market patterns is a far bigger challenge. The fascinating and demanding task of effectively

DOI: 10.1201/9781003559115-30

predicting stock movements in today's ever-changing industrial landscape is no picnic. The behaviour of stock movements can be predicted by taking into account a number of factors, some of which are economic and others of which are not. Therefore, increasing production is considered a major obstacle when attempting to predict the stock market. Thanks to developments in communication technology, the stock markets have grown in popularity among individual investors in the last few decades. Despite the continuous growth in the number of firms and stockholders in these markets, an increasing number of individuals are attempting to deduce the best way to predict the future of stock markets. The complexity of the situation is being exacerbated by the numerous interrelated factors.

## 2 LITERATURE SURVEY

The success of every country, community, or market economy in the modern day is centred on the financial market. Publicly traded companies that are accessible to the general public [1]. Actually, most businesses benefit from a high stock price. There are three main methods for predicting stock market movements: fundamental research, technical analysis (charting), and machine learning. In contrast, some scholars [2] classify these three approaches as either technical analysis or fundamental analysis. Rather than concentrating on the stock itself, fundamental analysts look at the company itself[3]. The abundance of unstructured data presents fundamental analysts with one of their greatest challenges. Nevertheless, numerous academic studies have shown that it can, in certain instances, be a reliable indicator of stock price movement [4]. In contrast, technological analysis suggests that analysts use past and present stock price trends to predict future stock prices[5]. Neural networks, PCA, SVM, time series analysis, neuro-fuzzy inference systems, evolutionary algorithms, hidden Markov models, and rough set theory are all methods that fall under this area.[6]Predicting the movements of financial markets effectively is a challenge, according to experts in disciplines as varied as computer science, mathematics, economics, and materials science. [7]Making the most of one's stock trades is an important factor to think about while making stock market predictions. Share price, company performance, government policies, GDP, inflation, natural catastrophes, and many more factors impact the stock market [8]. The Efficient Market Hypothesis states that stock market values cannot be predicted from past data alone because prices are highly responsive to new information and follow a random walk pattern [9]. There was a time when many people shared this opinion. Scientists have demonstrated, with the use of contemporary technology, that it is possible to anticipate stock market prices to an extent.[10] Information gleaned from social media and historical market data enables for the prediction of changes in economic and business sectors. Systematic approaches to predicting the stock market are quite sensitive to the quality of the features employed. [11] With the daily generation of huge volumes of transaction data by stock markets, deep neural networks have access to a treasure trove of data for training and boosting their prediction abilities. [12]create an innovative stock selection algorithm that evaluates stocks based on their predicted future returns by studying stock prices over time. The approach is built on deep neural networks. [13] was successful in forecasting the daily volatility of A-share market stocks in China by training a deep neural network with price data.

## 3 PROPOSED SYSTEM

Investing in the stock market is a thrilling prospect because of the enormous wealth that could be yours. Unfortunately, the stock market is very unpredictable, so investing in it is very risky. Consequently, scholars have put forth a lot of time and energy trying to predict how stock market prices will typically change.

### 3.1 *Preprocessing*

#### 3.1.1 *Relevance of time difference*

Temporal differentiation One common metric for determining the coincident, lagging nature or leading, of economic time series is relevance. Another way to put it is:

$$p_m = \frac{\sum_{h=1}^{b}(q_{h+m} - \overline{q})(j_h - \overline{j})}{\sqrt{\sum_{h=1}^{b}(q_{h+m} - \overline{q})^2 \sum_{h=1}^{b}(j_h - \overline{j})^2}} \tag{1}$$

#### 3.1.2 *K-L Distance*

As a measure of how close two probability distributions are to each other, the K-L distance is useful. Indicators that are closer to zero represent a closer proximity to the target indicator[14].

$$k_l = \sum_{g=1}^{b} r_g\left(r_g / x_{g+m}\right) \tag{2}$$

### 3.2 *Feature selection*

Numerous benefits can be achieved through feature selection, including enhanced data visualization and comprehension, reduced time required for training and use, reduced space required for storage and measurement, enhanced prediction performances regardless of dimensionality, and countless more.

#### 3.2.1 *PCA*

PCA is a method used in multivariate statistics. Reducing the dimensionality of a dataset with numerous interrelated variables is the objective[15]. In particular, it maintains the initial state of a limited set of elements or components that are composed of interdependent parts. Removing correlated variables and replacing them with uncorrelated components is the goal of PCA.

### 3.3 *Model training*

#### 3.3.1 *VGG-16*

The VGG16 model that was trained consists of thirteen convolutional layers and three fully connected layers [16]. The VGG16 network uses stacked convolutional layers, each with a $2 \times 2$ kernel size, to enhance depth. Using a 13-layer convolutional architecture with a 2 x 2-kernel size, the pre-trained model handles RGB images with a resolution of $216 \times 216$.

#### 3.3.2 *CNN*

The convolutional layer is responsible for feature extraction in a standard CNN model. The convolutional layer's image matrix and kernel learn picture features with a small filter size of the input data, allowing this layer to keep pixels in contact during the convolution processes.

#### 3.3.3 *SVM*

When it comes to multiclass image classification jobs, SVM is a reliable supervised learning approach classifier that can supply additional picture class datasets. As a linear classifier that employs optimization theory and kernel functions, SVM enables the classification of images into many classes. SVMs work by maximizing the distribution of data scores among all possible classes and then finding the hyperplane in a two-dimensional space. The SVM classifier is based on the following mathematical formulas:

$$\min_{\tau,n,\xi} \frac{1}{2} y^T y + Z \sum_{g=1}^{b} \xi_g \tag{3}$$

$$j_g\left(y^T \varnothing(q_g) + n\right) \geq 1 - \xi_g \tag{4}$$

## 4 RESULT AND DISCUSSION

The novel data it offers on the factors that impact compliance with real-time visibility is an innovative addition of this study. This proposed seeks to address the gap that has been created by the delayed successful deployments of real-time visibility.

The prediction in graph 1.1 approaches the real pattern after a considerable amount of

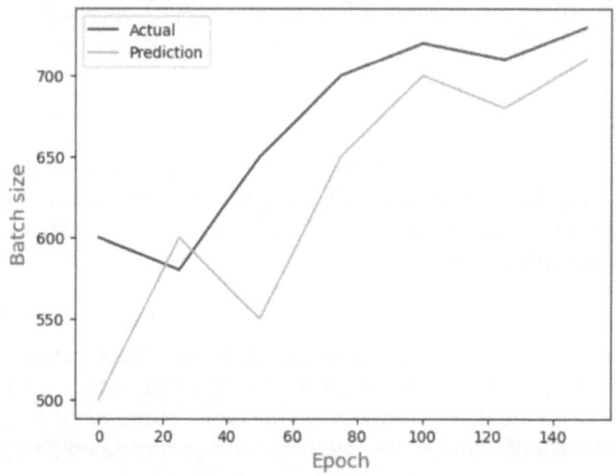


Figure 1. Plot between actual and predicted trend of VG16-CNN-SVM.

time has passed. More extensive training of the system allows for greater precision.

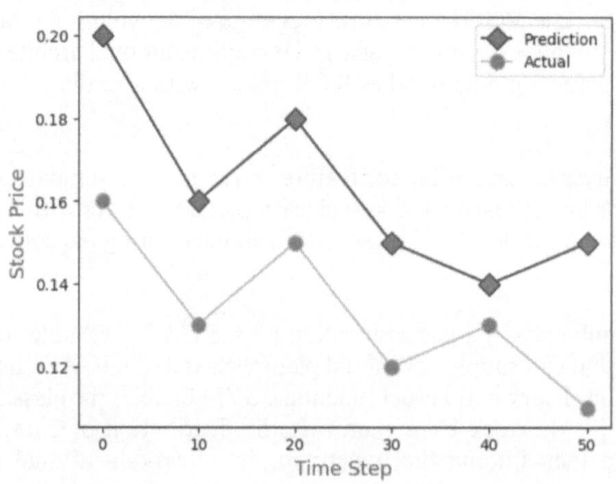


Figure 2. Stock market prediction for actual and predicted price returns.

See Figure 2 for a timestep diagram depicting the stock market prediction analysis for a period of fifty days. The correlation between web news sentiment and stock market values is 0.11.

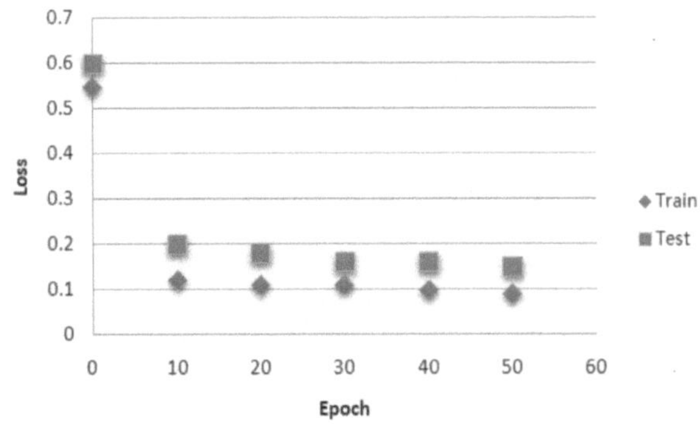


Figure 3. Model loss of VG16-CNN-SVM.

The stock market prediction model's loss diagrams are displayed in Figure 3.

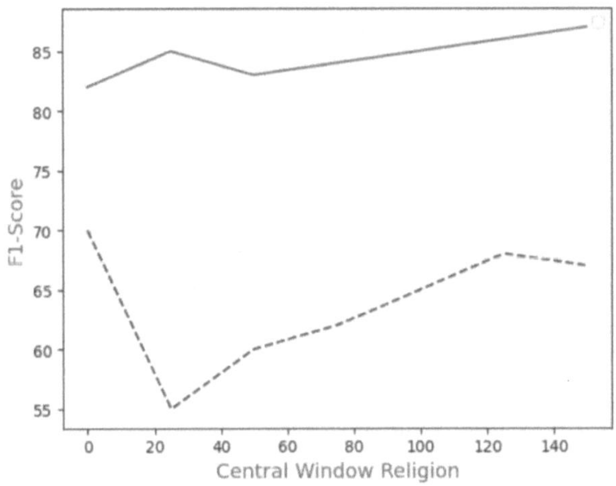


Figure 4. Positive and negative class F1 scores compared to central window sizes.

The F1 scores for the positive and negative categories are displayed in Figure 4. In both groups, F1 scores are rather high, hovering around 0 and 60, as seen in the graph.

## 5 CONCLUSION

Numerous regional and global economic and political factors have the potential to significantly impact the stock market, and a lot of these factors are hard to forecast. The secret to accurate stock market forecasting is getting the most out of the least amount of data input. In order to accomplish the complicated task of accumulating key criteria for making accurate predictions, regular stock market study is essential. In this scenario, K-L Distance gives very accurate results, and the preprocessing also includes consideration of time difference importance. Feature selection makes use of PCA, a multivariate statistical method. The model is trained using VG16-CNN-SVM. The proposed technique outperforms the SVM and VG16 models in terms of average accuracy, reaching 95.32%.

## REFERENCES


[1] Abhishek, K., Khairwa, A., Pratap, T., and Prakash, S. (2012, July). A stock market prediction model using Artificial Neural Network. In *2012 Third International Conference on Computing, Communication and Networking Technologies (ICCCNT'12)* (pp. 1–5). IEEE.

[2] Agrawal, S., Jindal, M. and Pillai, G. N. (2010, March). Momentum analysis based stock market prediction using adaptive neuro-fuzzy inference system (anfis). In *Proceedings of the International Multi Conference of Engineers and Computer Scientists* (Vol. 1, pp. 2078–0958).

[3] Ansari, T., Kumar, M., Shukla, A., Dhar, J. and Tiwari, R. (2010). Sequential combination of statistics, econometrics and Adaptive Neural-Fuzzy Interface for stock market prediction. *Expert Systems with Applications*, 37(7), 5116–5125.

[4] Asadi, S., Hadavandi, E., Mehmanpazir, F. and Nakhostin, M. M. (2012). Hybridization of evolutionary Levenberg–Marquardt neural networks and data pre-processing for stock market prediction. *Knowledge-Based Systems*, 35, 245–258.

[5] Atsalakis, G. S., Dimitrakakis, E. M. and Zopounidis, C. D. (2011). Elliott Wave Theory and neuro-fuzzy systems, in stock market prediction: The WASP system. *Expert Systems with Applications*, 38(8), 9196–9206.

[6] Liu, Q., Tao, Z., Tse, Y. and Wang, C. (2022). Stock market prediction with deep learning: The case of China. *Finance Research Letters*, 46, 102209.

[7] Gupta, R., Garg, N. and Singh, S. (2013, April). Stock market prediction accuracy analysis using kappa measure. In *2013 International Conference on Communication Systems and Network Technologies* (pp. 635–639). IEEE.

[8] Ab. Rahman, A.`S., Abdul-Rahman, S. and Mutalib, S. (2017). Mining textual terms for stock market prediction analysis using financial news. In *Soft Computing in Data Science: Third International Conference*, SCDS 2017, Yogyakarta, Indonesia, November 27–28, 2017, Proceedings 3 (pp. 293–305). Springer Singapore.

[9] Bhardwaj, A., Narayan, Y. and Dutta, M. (2015). Sentiment analysis for Indian stock market prediction using Sensex and nifty. *Procedia Computer Science*, 70, 85–91.

[10] Nti, I. K., Adekoya, A. F. and Weyori, B. A. (2020). A comprehensive evaluation of ensemble learning for stock-market prediction. *Journal of Big Data*, 7(1), 20.

[11] Li, X., Wang, C., Dong, J., Wang, F., Deng, X. and Zhu, S. (2011). Improving stock market prediction by integrating both market news and stock prices. In *Database and Expert Systems Applications: 22nd International Conference*, DEXA 2011, Toulouse, France, August 29-September 2, 2011, Proceedings, Part II 22 (pp. 279–293). Springer Berlin Heidelberg.

[12] Li, X., Xie, H., Wang, R., Cai, Y., Cao, J., Wang, F. and Deng, X. (2016). Empirical analysis: stock market prediction via extreme learning machine. *Neural Computing and Applications*, 27, 67–78.

[13] Zhang, G., Xu, L. and Xue, Y. (2017). Model and forecast stock market behavior integrating investor sentiment analysis and transaction data. *Cluster Computing*, 20, 789–803.

[14] Liu, Y., Chen, Y., Wu, S., Peng, G. and Lv, B. (2015). Composite leading search index: a preprocessing method of internet search data for stock trends prediction. *Annals of Operations Research*, 234, 77–94.

[15] Tsai, C. F. and Hsiao, Y. C. (2010). Combining multiple feature selection methods for stock prediction: Union, intersection, and multi-intersection approaches. *Decision Support Systems*, 50(1), 258–269.

[16] Tun, N. L., Gavrilov, A., Tun, N. M. and Aung, H. (2021, January). Remote sensing data classification using a hybrid pre-trained VGG16 CNN-SVM classifier. In *2021 IEEE Conference of Russian Young Researchers in Electrical and Electronic Engineering (ElConRus)* (pp. 2171–2175). IEEE.

*Hybrid and Advanced Technologies – S. Prasad Jones Christydass, Nurhayati Nurhayati and S. Kannadhasan. (Eds) © 2025 the Authors, ISBN 978-1-032-90254-8*
*Open Access: www.taylorfrancis.com, CC BY-NC-ND 4.0 license*

# Predicting stock market trends using ANN-spiking neural network via public sentiment and political situation analysis

Devadutta Indoria
*Assistant Professor, Department of Commerce, Vikram Dev University, Jeyoore, Odisha, India*

Mercy Tom
*Assistant Professor, School of Business and Management, Christ University, Bangalore, Karnataka, India*

Sameer Yadav
*Research Scholar, Department of Commerce and Business Administration, University of Allahabad, Prayagraj, Uttar Pradesh, India*
ORCID ID: 0000-0002-6949-6302

C.A. Sandeep Gupta
*S. K. Somaiya College, Mumbai, India*

Kiran Kumar Varma Kallepalli
*Associate Professor, Department of Humanities, S. R. K. R. Engineering College, Bhimavarm, Andhra Pradesh, India*

Harshal Patil
*Associate Professor, Balaji Institute of Technology & Management (BITM), Sri Balaji University, Pune, Maharashtra, India*

ABSTRACT: Investors base their decisions on a variety of factors, including the price-earnings ratio, the consumer price index, and various news stories. A lot of research has gone into automating the evaluation of such sources of information over the last many decades so that they can help them make quick decisions. Most of the attempts centered on numerical information because of the challenges in processing and understanding the temporal features of texts written in natural language. Model training, feature extension, and preprocessing are the three critical components. Tokenization and standardization are the two main components of data processing. The result is better with principal component analysis (PCA) than with recursive feature deletion, the other method that is part of feature extension. The model's accuracy was enhanced during training with the use of the ANN-SNN. The proposed approach outperformed cutting-edge technologies like ANN and SNN, with an accuracy rate of 96.13%.

*Keywords*: Principal Component Analysis (PCA), Artificial Neural Networks (ANN), Korea stock price index (KOSPI), Genetic Algorithms (GA), Recursive Feature Elimination (RFE)

## 1 INTRODUCTION

Investors have sunk a lot of hours into trying to foretell the market's trajectory since its start. There are no universally accepted methods for estimating the future value of a stock on the stock market. Stock market participants, both institutional and retail, strive to foretell how much a share will be worth in the future. The stock market is based on the underlying idea of demand

DOI: 10.1201/9781003559115-31 

and supply. The share price of a company's stock goes up when investors want to buy more of it and down when they don't. Another justification for research in this field is the fact that it confronts a number of theoretical and experimental challenges, some of which are hard to clearly identify. Which businesses succeed and which fail is decided daily by the actions and trends of the market. Consequently, there is no room for additional profit as all pertinent data is easily accessible. It is difficult to manually or with the help of third parties extract valuable information and analyse trends in their operations from massive data sets, which are common due to the presence and participation of many organizations and industries. Analysis and predictions of the stock market reveal market tendencies and indicate when a stock would be a wise purchase. Research on a stock has the potential to yield large returns if it can reliably forecast its price.

## 2 LITERATURE SURVEY

Applying deep learning was the subsequent step toward better, more per formant prediction models [1]. Accurate stock market forecasting isn't easy, and data scientists frequently run against roadblocks while trying to construct scenario that has the ability to predict what's going to happen next. The nonlinearity and intricacy of the stock market's link with investor psychology and market behaviour, as well as the market's inherent volatility, pose significant challenges [2]. A technique proposed by [3] integrates AI, GA, and HMM to partition daily stock data into distinct price buckets prior to feeding them into HMM. Their final results demonstrated that Bidirectional LSTM could forecast the market's future for investors, and it obtained the best performance. [4] looked at the use of RNN and CNN algorithms to evaluate their accuracy with real stock market values. To improve stock market projections, [5] utilized LSTM with an integrated layer and an autonomous encoder. Traditional optimization techniques are breaking down under the weight of increasingly complicated optimization challenges caused by rapid societal and economic progress. [6] Researchers have started inventing improved optimization algorithms to solve complex optimization problems, extract features for mechanical failure diagnostics, assign gates in an airport based on numerous objectives, and other optimization difficulties. [7] An improved variant of the ant colony optimization (ACO) algorithm called multi-population coevolution ACO (ICMPACO) was proposed by [8]to handle optimization concerns with gate assignment and travelling salesmen. Improving optimization performance while maintaining a steady stream of diverse solutions is the goal of this approach. [9]Based on methods from multi-population and co-evolution, the algorithm was developed. We looked at how well ICMPACO optimized compared to ACO. [10] Both the traveling salesman problem and the gate assignment problem demonstrated experimental results where the proposed optimization method surpassed the state-of-the-art. [11] built a model that integrates feature discretization, artificial neural networks (ANNs), and genetic algorithms (GAs) to predict the stock price index. Their research made use of technical indicators as well as the trend of daily changes in the Korea stock price index (KOSPI). [12] They optimized feature discretization in addition to dimensionality reduction. One strength of their work is that they optimized the ANN by introducing GA. [13] Know that there is a fixed number of 12 input characteristics and processing elements in the hidden layer before proceeding. Another limitation of ANN learning is that the authors limited their optimization to only two factors. However, they did think GA had good potential for improving feature discretization [14].

## 3 PROPOSED SYSTEM

The stock market is susceptible to a wide range of outside forces, including public opinion and political events. This study's overarching objective is to ascertain whether or not a day's political and public sentiment impact the stock market performance of individual companies or the market as a whole.

### 3.1 *Data preprocessing*

#### 3.1.1 *Tokenization*
In documents like news articles and financial reports, tokens are meaningful words that are segregated from the rest of the text[15].

#### 3.1.2 *Standardization of data*
Data standardization processes provide data uniformity by transforming all words in a text to lowercase. This is done in articles and reports about companies.

### 3.2 *Feature extension*

#### 3.2.1 *Recursive feature elimination*
Recursive Feature Elimination (RFE) is used after feature extension to select the most effective features. Each feature is evaluated using two metrics: the coefficient and the feature importance[16]. With each iteration it will keep the most crucial characteristics while reducing the number of traits that can be eliminated from the pool by one. The next step, PCA, will take the RFE block's output as its input.

#### 3.2.2 *PCA*
Feature pre-processing is the first step before using principal component analysis. The reason behind this is that RFE generates results in a variety of units, with certain characteristics following it including percentage data and others containing astronomically large numbers. Because of this, the result of the extraction of the main components may vary. Before the data is entered into the principal component analysis algorithm, feature pre-processing must be done.

### 1.3.3 *Model training*

#### 3.3.1 *ANN-SNN*
One way to train SNN is with the combined ANN-to-SNN knowledge distillation strategy, which uses an ANN model that has already been trained to transmit the hidden information to the student RNN model. They both depend on the last layer of the teacher ANN model's output to find out new details. Two of them employ the instructor ANN model's many intermediary layers to draw conclusions about new data.

#### 3.3.2 *ANN-SNN with response-based*
The proposed method stated that the student SNN model might receive data from the instructor ANN model's output layer after being trained using response-based knowledge. The output of the instructor model is able to preserve its hidden information because it uses the softmax layer to create non-hard-labeled vectors of category probabilities. In order to improve our ability to learn hidden information, it employs the parameter temperature $L$ to level out the probability distribution. The method by which the teacher network generates its results is as follows:

$$s_d = \frac{exp(H_d/L)}{\Sigma exp(H_r/L)} \tag{1}$$

#### 3.3.3 *ANN-SNN with feature-based*
As an extra KD method for developing a good deep SNN model, feature-based knowledge distillation is introduced in this study. This approach guides the training of SNN by tapping into the latent knowledge in specific intermediary ANN layers. There is a risk that the learnt information will collect in the last few layers of the teacher ANN only use the answer from the last output layer. This blocks the layer-wise training process, so that the student SNN model might learn from the instructor's network.

## 4 RESULT AND DISCUSSION

Predicting the stock market has lately become a prominent field of study due to the prospects it offers and its importance to the economy. It is essential to understand the future behaviour of the stock market in order to safeguard assets from possible losses. The stock market generates vast quantities of data, which investors treasure as a treasure trove of information.

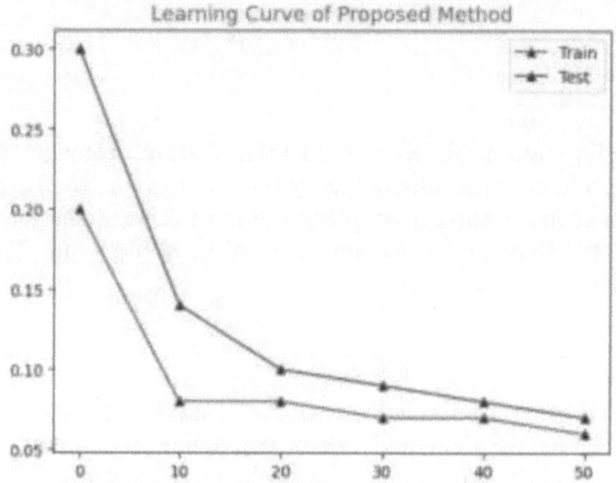


Figure 1. ROC Curve of RBF-ELMAN model.

After running the data through principal component analysis (PCA) to extract five components, we trained for fifty epochs. As seen in Figure 1, our suggested solution, the ANN-SNN model, has a learning curve.

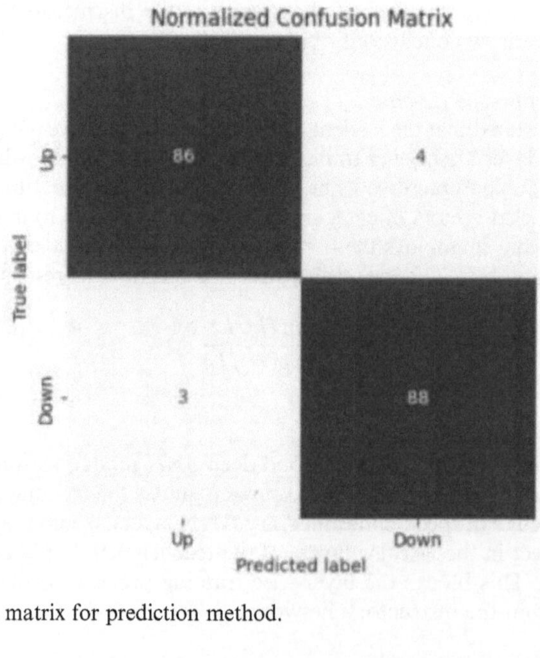


Figure 2. Confusion matrix for prediction method.

The normalized confusion matrix for assessing the two sets of features is shown in Figure 2. Both matrices show the confusion between the original feature set and the one with increased features; the one on the left shows the result of the test with only the expanded features.

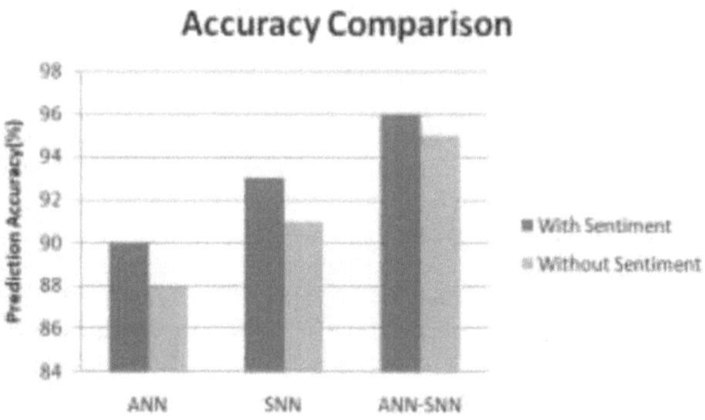


Figure 3. Prediction accuracy comparison of proposed model.

Figure 3 shows the results of comparing the top algorithms' accuracy with and without regard to the sentiment attribute.

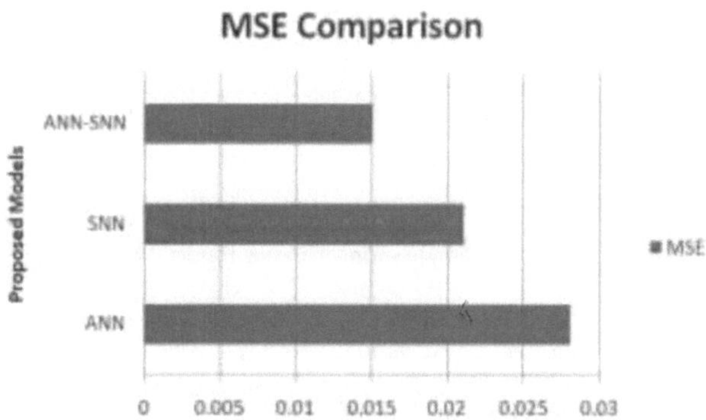


Figure 4. Comparison of MSE.

Figure 4 shows the MSE comparison of ANN, SNN, and ANN-SNN. For the purpose of comparison will use the average MSE value across all datasets.

## 5 CONCLUSION

Finding out what deep learning can do for stock market forecasting is the driving force behind this proposed. Stock prices are extremely volatile, and there are a lot of complicated financial elements at play, making stock market forecasting infamously difficult. Many sources, including news and financial data, might be useful for predicting stock prices. The major objective of this research is to utilise historical stock prices and technical indicators to improve the performance of the outcomes. The three primary parts of data processing are standardization and tokenization. Compared to the other strategy in feature extension, recursive feature deletion, PCA produces superior results. ANN-SNN is employed for model training. With an average accuracy of 96.13%, the suggested method surpasses both the ANN and SNN models.

## REFERENCES


[1] Long, W., Lu, Z. and Cui, L. (2019). Deep learning-based feature engineering for stock price movement prediction. *Knowledge-Based Systems*, 164, 163–173.

[2] Duarte, J. B. D., Sarmiento, L. H. T. and Juárez, K. J. S. (2017). Evaluation of the effect of investor psychology on an artificial stock market through its degree of efficiency. *Contaduría y Administración*, 62(4), 1361–1376.

[3] Kara, Y., Boyacioglu, M. A. and Baykan, Ö. K. (2011). Predicting direction of stock price index movement using artificial neural networks and support vector machines: The sample of the Istanbul Stock Exchange. *Expert systems with Applications*, 38(5), 5311–5319.

[4] Pang, X., Zhou, Y., Wang, P., Lin, W. and Chang, V. (2020). An innovative neural network approach for stock market prediction. *The Journal of Supercomputing*, 76, 2098–2118.

[5] Kelotra, A. and Pandey, P. (2020). Stock market prediction using optimized deep-convlstm model. *Big Data*, 8(1), 5–24.

[6] Bing, Y., Hao, J. K. and Zhang, S. C. (2012, December). Stock market prediction using artificial neural networks. In *Advanced Engineering Forum* (Vol. 6, pp. 1055–1060). Trans Tech Publications Ltd.

[7] Fan, Y., Ying, S. J., Wang, B. H. and Wei, Y. M. (2009). The effect of investor psychology on the complexity of stock market: An analysis based on cellular automaton model. *Computers & Industrial Engineering*, 56(1), 63–69.

[8] Kanade, P. A., Singh, S., Rajoria, S., Veer, P. and Wandile, N. (2020). Machine learning model for stock market prediction. *International Journal for Research in Applied Science and Engineering Technology*, 8(6), 209–216.

[9] Kazem, A., Sharifi, E., Hussain, F. K., Saberi, M. and Hussain, O. K. (2013). Support vector regression with chaos-based firefly algorithm for stock market price forecasting. *Applied Soft Computing*, 13(2), 947–958.

[10] Oliveira, N., Cortez, P. and Areal, N. (2013). On the predictability of stock market behavior using stocktwits sentiment and posting volume. In *Progress in Artificial Intelligence: 16th Portuguese Conference on Artificial Intelligence, EPIA 2013, Angra do Heroismo, Azores, Portugal, September 9–12, 2013*. Proceedings 16 (pp. 355–365). Springer Berlin Heidelberg.

[11] Eapen, J., Bein, D. and Verma, A. (2019, January). Novel deep learning model with CNN and bi-directional LSTM for improved stock market index prediction. In *2019 IEEE 9th Annual Computing and Communication Workshop and Conference (CCWC)* (pp. 0264–0270). IEEE.

[12] Fischer, T. and Krauss, C. (2018). Deep learning with long short-term memory networks for financial market predictions. *European Journal of Operational Research*, 270(2), 654–669.

[13] Hao, Y. and Gao, Q. (2020). Predicting the trend of stock market index using the hybrid neural network based on multiple time scale feature learning. *Applied Sciences*, 10(11), 3961.

[14] Xu, Q., Li, Y., Shen, J., Liu, J. K., Tang, H. and Pan, G. (2023). Constructing deep spiking neural networks from artificial neural networks with knowledge distillation. In *Proceedings of the IEEE/CVF Conference on Computer Vision and Pattern Recognition* (pp. 7886–7895).

[15] Khedr, A. E. and Yaseen, N. (2017). Predicting stock market behavior using data mining technique and news sentiment analysis. *International Journal of Intelligent Systems and Applications*, 9(7), 22.

[16] Shen, J. and Shafiq, M. O. (2020). Short-term stock market price trend prediction using a comprehensive deep learning system. *Journal of big Data*, 7, 1–33.

*Hybrid and Advanced Technologies – S. Prasad Jones Christydass, Nurhayati Nurhayati and S. Kannadhasan. (Eds)* © *2025 the Authors, ISBN 978-1-032-90254-8*
*Open Access: www.taylorfrancis.com, CC BY-NC-ND 4.0 license*

# Comparative study on biological impacts of food preservation of canned fish and salted fish

P.S. Muhammed Saheer
*Research Scholar, Department of Biotechnology Engineering, Saveetha School of Engineering, Saveetha University, Chennai, Tamil Nadu, India*

S. Manikandan
*Project Guide, Department of Biotechnology Engineering, Saveetha School of Engineering, Saveetha University, Chennai, Tamil Nadu, India*

ABSTRACT: The biological effects of canning as opposed to salting on fish preservation are investigated in this study. Fish that had been canned and stored in oil and brine showed stable pH levels, reduced microbial counts after sterilization, and maintained nutrients with possible flavor changes. Salted fish displayed varied textures and aromas, changed pH levels, nutritional modifications, and regulated microbial development when it was stored with carefully regulated salt concentrations.

*Keywords*: Responsible Consumption and Production, canned fish, salted fish, Health and Acidity, Reduced Inequalities, Climate Action.

## 1 INTRODUCTION

Food preservation methods are essential in the modern world for maintaining year-round access to safe and nourishing food. Fish, a vital source of protein and minerals, can be preserved using two frequently used techniques: canning and salting. Since ancient times, these techniques have been employed to increase the shelf life of fish and make it available even in areas distant from the sea. While both methods have the same goal of preserving fish, they do so in very different ways, which may have differing biological effects on the preserved product's nutritional value and health implications.

Throughout human history, food preservation techniques have been essential in assuring the availability of wholesome and secure food. Canning and salting are two preservation methods that have been widely used to increase the shelf life of perishable items, including fish. These techniques affect the dietary and biological characteristics of the preserved items in addition to their primary function of preventing deterioration.

## 2 MATERIALS AND METHODS

### 2.1 *Sample selection*

To verify the accuracy and quality of the items, examples of canned fish and salted fish have been obtained for this comparison study from regional marketplaces and respected suppliers. To account for such variances, a variety of brands and varieties of salted fish, such as cod and mackerel, as well as canned fish, such as tuna, salmon, and sardines, were used in the sample.

DOI: 10.1201/9781003559115-32

## 2.2 *Sample preparation*

Canned fish: The fish in cans samples were properly opened, and the liquid was drained off. To remove any remaining brine or oil, the fish meat was carefully washed with distilled water after being separated from any additional sauces or oils. Salted Fish: To remove extra salt, salted samples of fish were desalted by soaking in distilled water for a set amount of time. The fish specimens were cleaned and dried off after desalination.

## 2.3 *Evaluation of texture and flavour*

To evaluate the texture and flavour of the preserved fish, a trained panel of specialists used sensory analysis. A sensory evaluation survey and a structured tasting methodology were used to rate sensory qualities such tenderness, saltiness, fishy odour, and overall taste.

## 2.4 *Microbial safety*

To determine the presence of pathogenic microbes (such as Salmonella and Listeria), as well as total plate counts, microbiological analyses were conducted. In specific media, samples were grown, and colonies that formed were counted.

## 2.5 *Statistical analysis*

Software (such as SPSS or R) that was appropriate was used to conduct the statistical analysis. To examine the variations between canned and salted fish samples, data were analyzed using descriptive statistics, t-tests, ANOVA, or non-parametric tests, if necessary. The statistical significance was determined at a level of p 0.05.

# 3 RESULT

We discovered interesting discrepancies between both of these preservation techniques in our comparative investigation on the biological effects of canned fish and salted fish. First, when looking at shelf life, salted fish had a larger shelf life of three years than canned fish, which had a shelf life of only 2.62 years. This means that salted fish can be kept for a little longer without losing much of its flavour, making it a desirable alternative for areas without easy access to refrigeration or canning equipment.

We noticed differences in the two types of preserved fish's pH levels. In contrast to salted fish, which had a pH level of 5.7, canned fish had a 6.35, indicating a somewhat more alkaline environment. This variation in acidity may have ramifications for microbial development and safety throughout storage in addition to affecting the fish's flavour and texture.

According to our research, canned and salted fish each have unique qualities and biological effects. While fish in cans has a longer shelf life and more vitamins, consumers looking for a more traditional, tasty option with a shorter shelf life may prefer salted fish. The decision between these two preservation techniques ultimately comes down to personal tastes, dietary restrictions, and access to storage facilities.

# 4 DISCUSSION

A recent comparative study investigates the effects of different food preservation techniques on canned and salted fish, revealing insights into nutritional value, shelf life, pH levels, and vitamin content. These findings are crucial for consumers, fisheries, and the food industry to optimize seafood production and consumption decisions. They underscore the importance of balancing taste, safety, and nutritional benefits in choosing between preservation methods for seafood products.

Table 1. The mean value, standard deviation and standard error mean for canned fish and salted fish.

| Group Statistics | | | | | |
|---|---|---|---|---|---|
| | Control | N | Mean | Std. Deviation | Std. Error Mean |
| **pH** | **Canned Meat** | 4 | 2.6250 | 1.10868 | 0.55434 |
| | **Dried Meat** | 4 | 3.0000 | 2.16025 | 1.08012 |
| **Shelf Life** | **Canned Meat** | 4 | 6.3500 | 0.12910 | 0.06455 |
| | **Dried Meat** | 4 | 5.7000 | 0.25820 | 0.12910 |
| **Vitamins** | **Canned Meat** | 4 | 8.1250 | 0.85391 | 0.42696 |
| | **Dried Meat** | 4 | 4.7500 | 2.75379 | 1.37689 |

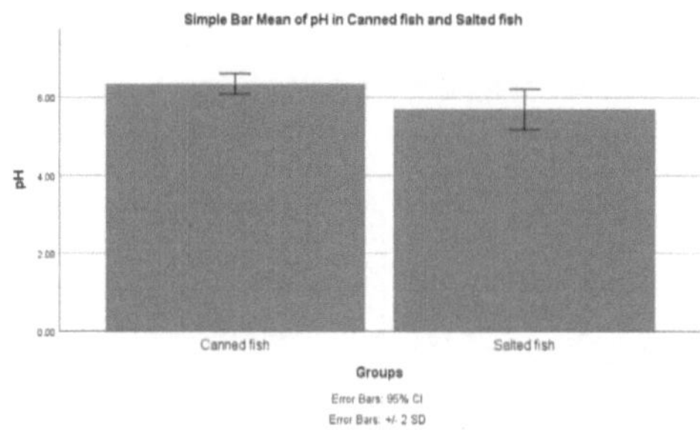

Figure 1. pH comparison between X axis: Canned fish and salted fish, Y axis: Mean pH ± 2 SD.

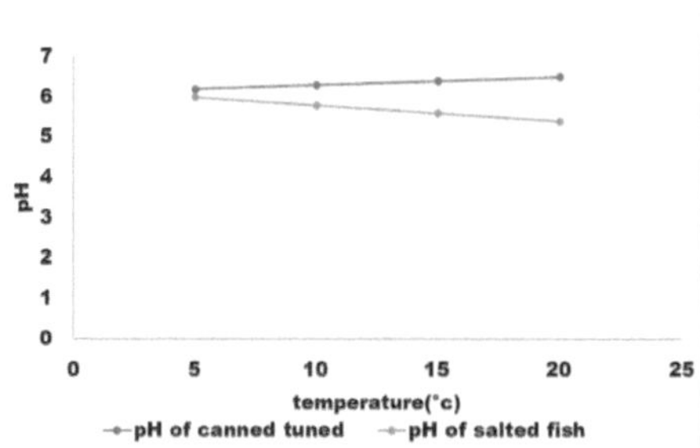


Figure 2. Acidity (pH) comparison between canned fish and salted fish. X axis: Canned fish and salted fish at different temperatures, Y axis: Mean acidity (pH).

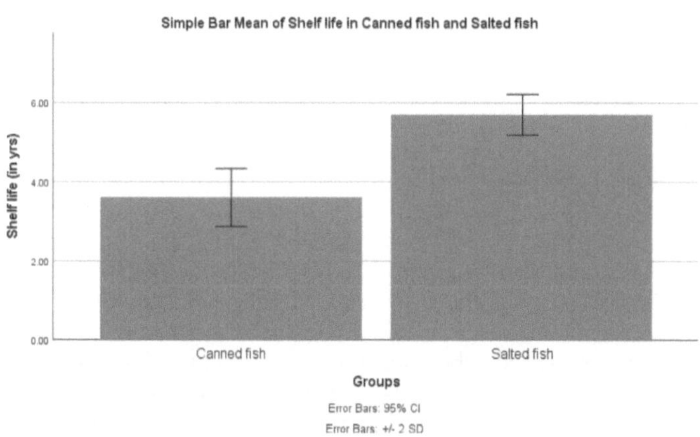


Figure 3. Shelf life comparison between canned fish and salted fish. X axis:Canned fish and salted fish, Y axis: Mean shelf life ± 2 SD.

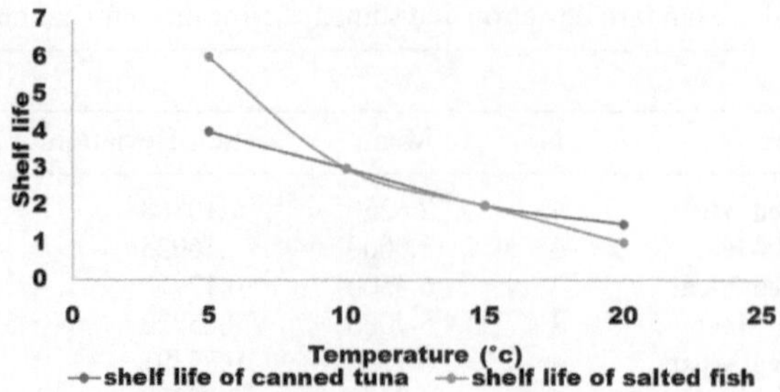


Figure 4. Shelflife comparison between canned fish and salted fish. X axis: Canned fish and dried fish at different temperatures, Y axis: Mean shelf life.

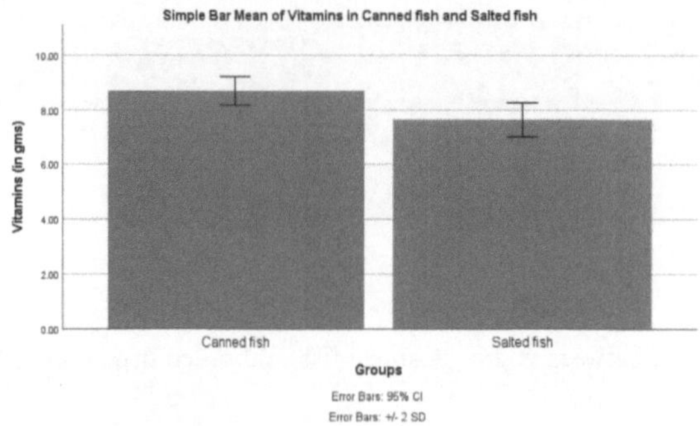


Figure 5. Vitamins comparison between canned fish and salted fish. X axis: Canned fish and dried fish, Y axis: Mean vitamin ± 2 SD.

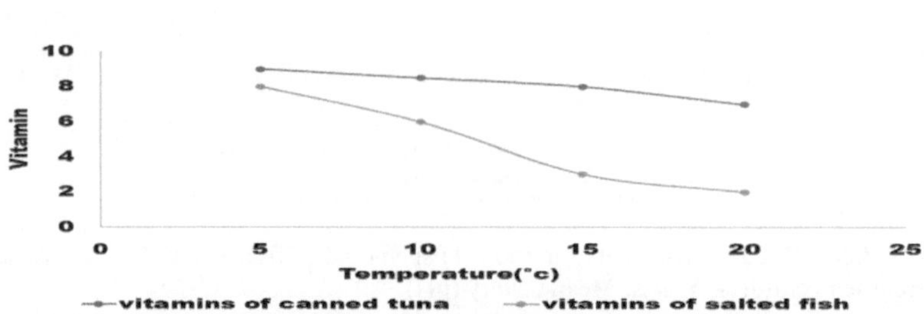


Figure 6. Vitamin comparison between canned fish and salted fish. X axis: Alcoholic wine and non-alcoholic wine at different temperatures, Y axis: Mean vitamin.

## 5 CONCLUSION

In conclusion, our comparative analysis shows that salted fish may have advantages in regards to preservation and microbiological safety due to its longer shelf life and slightly acidic pH. It is important to highlight that canned fish has a far higher vitamin content, which could offer consumers better nutritional benefits. These results underline how crucial it is to weigh a variety of aspects when deciding between canned and salted fish for nutritional and culinary purposes, such as shelf-life, pH levels, and vitamin content. In the end, dietary requirements and personal tastes should determine which preservation technique is used.

## REFERENCES


[1] Buatong, J., Preedanon, S., Mittal, A., Palamae, S. and Benjakul, S. (2023). Contaminated fungi in dried salted fishes: Isolation, identification, and their inhibition by chitooligosaccharide-gallic acid conjugate. *Journal of Food Science*, *88*(11), 4653–4663. https://doi.org/10.1111/1750-3841.17212

[2] Chang, P., Guo, K., Li, S., Wang, H. and Tang, M. (2023). In situ sodium chloride cross-linked fish skin collagen scaffolds for functional hemostasis materials. *Small.* Advance online publication. https://doi.org/10.1002/smll.202208001

[3] Gouvêa, F. de J., Oliveira, V. S. de, Mariano, B. J., Takenaka, N. A. R., Gamallo, O. D., Ferreira, M. da S. and Saldanha, T. (2023). Natural antioxidants as strategy to minimize the presence of lipid oxidation products in canned fish: Research progress, current trends and future perspectives. *Food Research International*, *173*(Pt 1), 113314. https://doi.org/10.1016/j.foodres.2022.113314

[4] Grainger, S. (2020). *The Story of Garum: Fermented Fish Sauce and Salted Fish in the Ancient World.* Routledge.

[5] Grasso, F., Méndez-Paz, D., Vázquez Sobrado, R., Orlandi, V., Turrini, F., De Negri Atanasio, G., Grasselli, E., Tiso, M. and Boggia, R. (2023). Feasibility of enzymatic protein extraction from a dehydrated fish biomass obtained from unsorted canned yellowfin tuna side streams: Part I. *Gels (Basel, Switzerland)*, *9*(9). https://doi.org/10.3390/gels9090760

[6] Kosker, A. R., Gundogdu, S., Esatbeyoglu, T., Ayas, D. and Ozogul, F. (2023). Metal levels of canned fish sold in Türkiye: Health risk assessment. *Frontiers in Nutrition*, *10*, 1255857. https://doi.org/10.3389/fnut.2023.1255857

[7] Mahmudiono, T., Fakhri, Y., Adiban, M., Sarafraz, M. and Mohamadi, S. (2023). Concentration of potential toxic elements in canned tuna fish: Systematic review and health risk assessment. *International Journal of Environmental Health Research. Advance online publication.* https://doi.org/10.1080/09603123.2023.2045261

[8] Rungraeng, N., Ohtaguchi, K. and Chysirichote, T. (2023). Removal of histamine from fish sauce by Sp. isolated from fermented fish. *Food Technology and Biotechnology*, *61*(3), 294–301.

[9] United States Tariff Commission. (1942). *Canned fish: A review of production, foreign trade, potential supplies of raw materials, and recommendations concerning operation of the industry under war conditions.*

[10] Villamarín, E., Martínez, B., Trigo, M. and Aubourg, S. P. (2023). Influence of different previous frozen holding periods on the canned fish quality. *Foods (Basel, Switzerland)*, *12*(22). https://doi.org/10.3390/foods12224117

*Hybrid and Advanced Technologies – S. Prasad Jones Christydass, Nurhayati Nurhayati and S. Kannadhasan. (Eds)* © *2025 the Authors, ISBN 978-1-032-90254-8*


# Prediction of gestational diabetes mellitus using visceral fat measurements with enhanced accuracy rate by voted perceptron classifier and K-nearest neighbour classifier

A.P. Shanmuga Priya
*Research Scholar, Department of Biomedical Engineering, Saveetha School of Engineering, Saveetha University, Chennai, Tamil Nadu, India*

T. Usha Rani
*Project Guide, Department of Biomedical Engineering, Saveetha School of Engineering, Saveetha University, Chennai, Tamil Nadu, India*

ABSTRACT: This study focuses on using machine learning algorithms to predict gestational diabetes mellitus by analysing measurements of visceral fat. Specifically, the Novel Voted Perceptron classifier and the K-nearest neighbours (K-NN) classifier are used. The Kaggle website provided the data used in this study. A dataset comprising both normal (n = 75) and aberrant (n = 75) cases was used in the proposed investigation. In order to calculate the overall sample size, clinclac.com, the significance level (alpha error) was set to 0.05. G Power was then used to achieve 80% statistical power, a 0:1 enrollment ratio, and a 95% confidence interval. The computation of the necessary total sample size for the investigation was informed by these particular criteria. In the Google Colaboratory environment, the Python programming language is used to predict gestational diabetes mellitus.The Novel voted perceptron classifier and K-Nearest Neighbour classifier machine learning algorithms were used to complete the classification challenge, which involved differentiating between subjects that were normal and abnormal. IBM SPSS software version 25 is used for statistical analysis. Between the two groups, there was a statistically significant difference ($p>0.01$), and the accuracy of the machine learning algorithms was determined. Using 10-fold cross-validation, the classifiers were trained, evaluated, and verified using the Google Colaboratory programme. Significantly, the Novel voted perceptron classifier outperformed the K-Nearest Neighbour classifier, which only managed a 93.33% classification accuracy rate. Instead, it scored a higher 97.33% classification accuracy rate.

*Keywords*: Artificial intelligence, Gestational Diabetes Mellitus, K-Nearest Neighbour, Machine Learning Algorithms, Novel Voted Perceptron Classifier

## 1 INTRODUCTION

Gestational diabetes is a condition that arises during pregnancy, causing temporary disruptions in The human body's release of insulin. The hormone insulin is in charge of controlling blood sugar levels. During pregnancy, the body's insulin production may fall short of meeting the increased demands, resulting in elevated blood sugar levels. This condition necessitates careful management to ensure a healthy pregnancy and delivery, often involving dietary adjustments, physical activity, and, in some cases, medication (Petry 2014). The early prediction and improved diagnosis of this pregnancy condition hold significant potential for enhancing preventive interventions (Lapolla and Metzger 2019; Sartorão Filho *et al.* 2023).

The prevalence of gestational diabetes can vary by region and population, but on average, it is estimated that gestational diabetes affects about 2% to 10% of pregnancies worldwide. However,

 DOI: 10.1201/9781003559115-33

these numbers can be higher in specific populations or regions with a higher prevalence of diabetes risk factors, such as obesity and family history of diabetes. While there is no universally agreed-upon consensus regarding the most effective GDM screening method, early detection has the potential to mitigate or lower the risk of adverse pregnancy outcomes in machine learning and artificial intelligence (Ali *et al.* 2023; Gurupranes *et al.* 2022; Natrayan *et al.* 2022; Sivaranjani *et al.* 2023; Viswanathan *et al.* 2023). Early detection of GDM enables patients to receive therapy using appropriate screening techniques (Al Ghaithi *et al.* 2023; Lei *et al.* 2023). 208 articles on ScienceDirect, more than 100 publications on Google Scholar, and numerous IEEE articles published between 2019 and 2023 are all strongly related to this topic. The application of machine learning algorithms for the prediction of gestational diabetes is the common theme throughout these studies. In the context of healthcare systems, high-risk selective screening and universal screening are the two main screening methods that are often used globally to detect gestational diabetes in expectant mothers. The International Diabetes Federation (IDF) GDM Model of Care, designed for low-resource areas, is a great example of the latter strategy. Kurt *et al.* (2023) investigated prediction models for gestational diabetes mellitus. Their decision tree model has an AUC of 0.813 (95% CI 0.786-0.839) for training and 0.779 (95% CI 0.735-0.824) for internal validation. Meanwhile, the random forest model showed AUCs of 0.854 (95% CI 0.831-0.877) for training and 0.808 (95% CI 0.766-0.850) for validation. The logistic regression model, which includes clinical markers and GDM subtypes, was also assessed. These findings demonstrate the models' diverse predictive powers in diagnosing GDM, with the random forest performing exceptionally well in both the training and validation stages.

## 2 MATERIALS AND METHODS

The study was conducted at the Digital Signal Processing Laboratory of the Saveetha Institute of Medical and Technical Sciences, which is located in the Saveetha School of Engineering's Biomedical Department. It is significant to emphasise that this research did not require ethical approval. The study comprised two distinct groups:

1. Group 1, which consists of seventy-five people classified as "normal." 2. Group 2, which consists of seventy-five people classified as "abnormal."

Step 1: Integrating the data sets from Kaggle that have previously been extracted as an Excel sheet value Step 2: Importing seaborn, matplotlib, pandas, and numpy. Pyplot is referred to as pre-processing in Google collaborative software. Step 3: Using the Python programming language, test and train the data Step 4: Following algorithm classification, The model, a machine learning approach based on support vector machines and naive Bayes, is classified using imported data sets.

Step 5: The naive Bayes and support vector machine programmes can be used to calculate the accuracy.

Step 6: Using the same programming language, sklearn can be used to run the programme and obtain the ROC curve and confusion matrix.

In total, this study included 150 participants. Using clincalc.com, Using a 95% confidence interval, 0.05 alpha, and 80% power value, the sample size was determined (Chow *et al.* 2003; Chow *et al.* 2017; Machin *et al.* 2018). The Google Colaboratory environment uses Python 3.10.12 software to forecast gestational diabetes. The dataset contained 150 samples, of which 25 belonged to the "normal" group and 25 to the "abnormal" group. The study's data came from the well-known data science and machine learning portal Kaggle. Voted perceptron and K-Nearest Neighbour algorithms were used in this study to categorise people into "normal" and "abnormal" groups according to the input datasets.

The Google Colaboratory Python programming environment has a number of benefits that improve artificial intelligence research productivity. These benefits include easy collaboration, access to GPUs and TPUs for accelerated machine learning tasks, integration with multiple Google services, a large library, and the convenience of the Jupyter Notebook interface. But it's crucial to take internet connectivity, resource constraints, and data security into account—especially when working with sensitive data.A software programme called SPSS helps researchers and scientists handle statistics and understand data. For those who study social trends and human behaviour, it is incredibly helpful. With SPSS, they may make graphs and spot trends to better comprehend their findings. In conclusion, Google Colaboratory shows itself to be a flexible and user-friendly tool appropriate for both machine learning and projects and educational purposes (Choi *et al.* 2023).

### 2.1 *Statistical analysis*

The data analysis in this study to figure out the relevance of differences between various groups was conducted using IBM SPSS software version 25 (Mansoori *et al.* 2023). Accuracy is the dependent variable in this research; the classifier's accuracy as well as the independent samples T-test variables were taken into account.

## 3 RESULTS

The confusion matrix for the novel-voted perceptron classifier is shown in Table 1. It consists of True Positive (TP), which successfully identified 31 cases of abnormal cases, and True Negative (TN), which correctly identified 28 cases of normal cases. The confusion matrix for the K-Nearest Neighbour classifier is shown in Table 2. It shows 17 True Negative (TN) examples that correctly identify normal patients and 21 True Positive (TP) cases that appropriately identify abnormal cases. False Negatives (FN) were not reported, suggesting that normal cases were not accidentally labelled as abnormal, and there were no False Positives (FP), indicating that some abnormal cases were mistakenly categorised as normal. Table 3 shows that the classification accuracy rate of 97.33% was attained by both the K-Nearest Neighbour classifier and the novel-voted perceptron when it came to differentiating between abnormal and normal data sets. Groups 1 and 2 each had seventy-five samples. Table 4 highlights the correlation between the mean accuracy of Group 1 (Novel voted perceptron classifier) and Group 2 (K-Nearest Neighbour classifier). The statistical analysis using the independent sample T-test in Table 5 reveals a significant difference ($p>0.03$) in the data between the groups. A comparison of the mean accuracy of the voted perceptron and K-Nearest Neighbour classifiers is presented in Figure 1. In terms of mean accuracy and standard deviation, the Novel Voted perceptron classifier performs better than the K-Nearest Neighbour classifier. The ROC curves for the K-Nearest Neighbour and Novel Voted Perceptron classifiers are displayed in Figure 2. X-Axis: Multilayer perceptron classifier versus ADABOOST classifier; Y-Axis: Mean detection accuracy $\pm$ 1 SD. On the X-Axis, the false positive rate and the real positive rate on the Y-Axis, respectively.

Table 1. Novel voted perceptron classifier's confusion matrix.

| | **Non GDM(predicted)** | **GDM(predicted)** |
|---|---|---|
| **Non GDM(actual)** | 31 | 1 |
| **GDM(actual)** | 0 | 28 |

Table 2. The K-nearest neighbour classifier's confusion matrix.

| | **Non GDM(predicted)** | **GDM(predicted)** |
|---|---|---|
| **Non GDM(actual)** | 21 | 0 |
| **GDM(actual)** | 0 | 17 |

Table 3. Comparing the accuracy (%) of the K-nearest neighbour classifier and the novel voted perceptron classifier.

| Classifiers | Accuracy(%) |
|---|---|
| **Novel Voted Perceptron** | 97.33 |
| **K-Nearest Neighbour** | 93.33 |

Table 4. Demonstrates the group statistics an explanation of the K-nearest neighbour machine learning algorithm and the novel voted perceptron.

| Accuracy | Group | N | Mean | Std. Deviation | Std. Error Mean |
|---|---|---|---|---|---|
| | **Novel Voted Perceptron** | 75 | 97.3100 | .01838 | .00212 |
| | **K-Nearest Neighbour** | 75 | 93.3127 | .02189 | .00253 |

Table 5. To determine the standard error and test for significance, use an independent sample. Statistical significance is defined as a p-value greater than 0.01, and 95% confidence intervals were computed.

| | | Levene's Test for Equality of Variances | | t-test for Equality of Means | | | | | | |
|---|---|---|---|---|---|---|---|---|---|---|
| | | | | | | | | | 95% Confidence Interval of the Difference | |
| | | F | Sig. | t | df | Sig (2-tailed) | Mean Diff | Std. Error Diff. | Lower | Upper |
| ACCURACY | Equal variances assumed | 3.069 | .082 | 1211.041 | 148 | .000 | 3.99733 | .00330 | 3.99081 | 4.00386 |
| | Equal variances not assumed | | | 1211.041 | 143.694 | .000 | 3.99733 | .00330 | 3.99081 | 4.00386 |

## 4 DISCUSSION

We have found a statistically significant difference (p>0.01) between the groups based on Table 5, an independent sample T-test. When it comes to classifying normal and abnormal data sets, the Novel Voted Perceptron classifier has a higher classification accuracy rate (97.33%) than the K-Nearest Neighbour classifier (93.33%). The researchers classified gestational diabetes with an accuracy rate of 88.7% by using Python, a data mining tool.(Aminifar and Jader 2022). Using the Kaggle dataset, K-Nearest Neighbour classifiers were able to predict gestational diabetes with an overall accuracy rate of 66%. Wu and colleagues 2021). Using Python, the authors have outperformed previous artificial intelligence classifiers in terms of classification accuracy rate for novel-voted perceptron classifiers, which they used to predict gestational diabetes mellitus (Krsman).

The K-Nearest Neighbour classifier was used to predict gestational diabetes using the Kaggle dataset, and it attained an overall accuracy rate of 80.4% (Gnanadass n.d.).The study by Hu *et al.* (2023); Wang *et al.* (2023) shows subtypes of gestational diabetes mellitus (GDM) that differ in insulin secretion and sensitivity and may show different pathways that contribute to the development of large-for-gestational-age (LGA) newborns. Nonetheless, the risk of foetal overgrowth persists in cases of non-diabetic obese mothers and even with strict glycemic control. In these cases, there may be a higher risk of foetal overgrowth due to higher levels of lipid profiles.

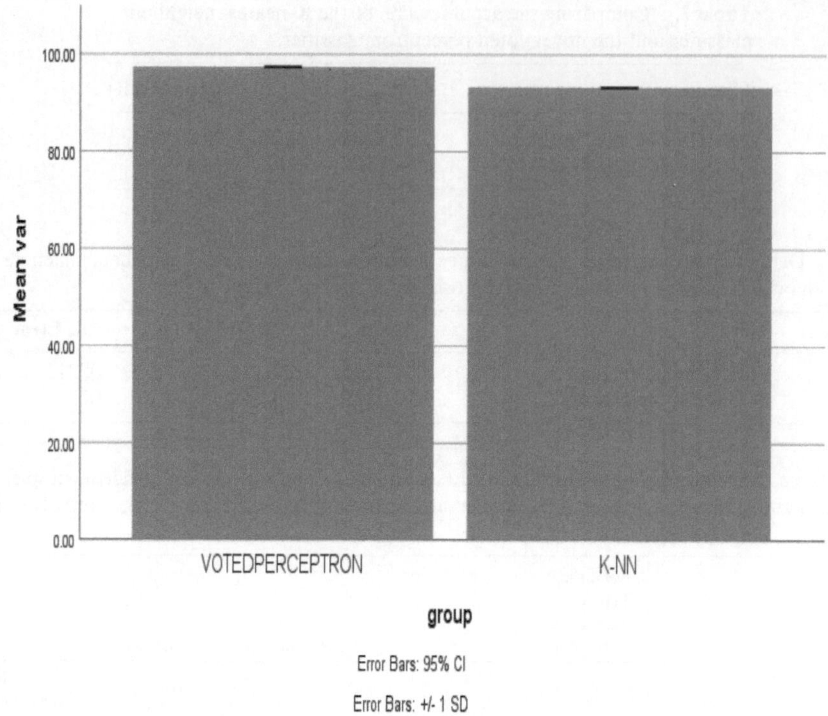


Figure 1. Comparison of novel voted perceptron classifier and K-nearest neighbour classifiers in terms of mean accuracy. The mean accuracy of the novel voted perceptron classifier is better than the KNN classifier and the standard deviation of the novel voted perceptron is slightly better than KNN.

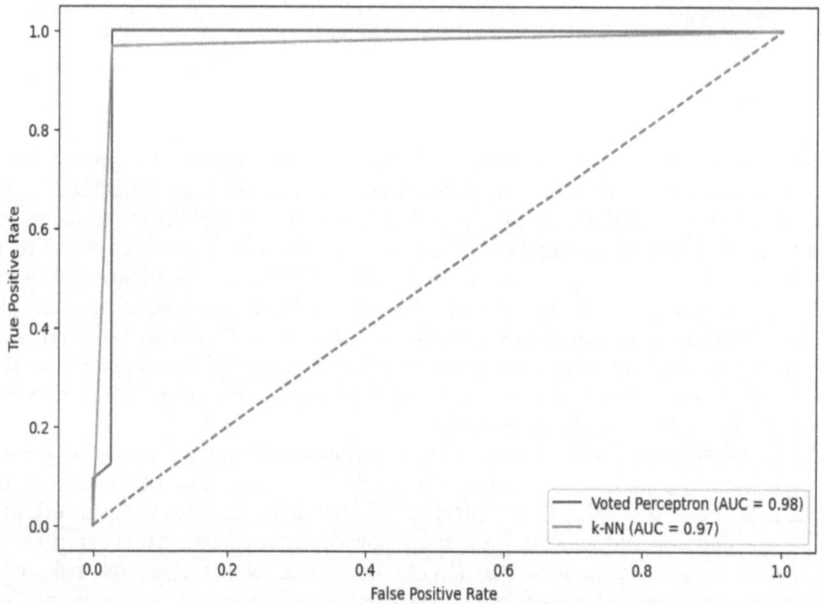


Figure 2. ROC curve for novel voted perceptron classifier and K-nearest neighbour classifier. X-Axis: False positive rate; Y-Axis: True positive rate.

The machine learning algorithm can be modified to increase the accuracy rate, although this is the study's limitation. To obtain a higher classification accuracy rate (%), the sample size can also be raised (Brownlee 2016; Mousavi *et al.* 2023). To improve the classification accuracy rate (%), new classifiers for the dataset can be added to the research in the future. As an artificial intelligence technique, the computed assisted diagnostic (CAD) system can be built to categorise ocular images, potentially assisting physicians and clinicians in the future.

## 5 CONCLUSION

This study uses unique novel-voted perceptron classifier and K-Nearest Neighbour classifier machine learning techniques to predict gestational diabetes mellitus and distinguish between normal and abnormal cases. The results of this study demonstrate that the Novel Voted perceptron classifier performs better in terms of classification accuracy than the K-Nearest Neighbour classifier. It achieved an excellent rate of 97.33%, whilst the K-Nearest Neighbour classifier only managed a rate of 93.33%. Surprisingly, compared to other methods, the Novel Voted perceptron classifier algorithm's performance improved as the quantity of the dataset rose. This is a revolutionary model with great potential for improving disease diagnosis, based on a creative method that works very well. As such, it may find broad use in hospitals and the diabetes community. ultimately increasing the diagnostic processes' efficacy.

## REFERENCES

Al Ghaithi, I., Ahmed, B., Al Dughaishi, T., Al Riyami, N., Al Haddabi, R. and Gowri, V. (2023). Maternal, perinatal, and neonatal outcomes of covid-19 severity in pregnant women: A retrospective study from Oman. *Oman Medical Journal, 38*(4), e530.

Ali, S., Sudha, K. G., Thirumalaivasan, N., Ahamed, M., Pandiaraj, S., Rajeswari, V. D., ... & Govindasamy, R. (2023). Green synthesis of magnesium oxide nanoparticles by using abrus precatorius bark extract and their photocatalytic, antioxidant, antibacterial, and cytotoxicity activities. *Bioengineering, 10*(3), 302.

Brownlee, J. (2016). Machine learning mastery with R: Get started, build accurate models and work through projects step-by-step. *Machine Learning Mastery*.

Chmieliauskaite, M., Grosh, M. D., Syed, A. and Pinto, A. (2023). A pregnant patient with gestational diabetes reports for scaling and root planning. *Dental Clinics, 67*(4), 707–709.

Choi, W., Choi, T. and Heo, S. (2023). A Comparative study of automated machine learning platforms for exercise anthropometry-Based typology analysis: performance evaluation of AWS sagemaker, GCP VertexAI, and MS Azure. *Bioengineering, 10*(8), 891.

Chow, S. C., Shao, J., Wang, H. and Lokhnygina, Y. (2017). *Sample size calculations in clinical research.* chapman and hall/CRC.

Gnanadass, I. (2020). Prediction of gestational diabetes by machine learning algorithms. *IEEE Potentials, 39* (6), 32–37.

Grant, P. (2016). *Gestational Diabetes: Your Survival Guide to Diabetes in Pregnancy.* Sheldon Press.

Gurupranes, S. V., Natrayan, L., Kaliappan, S., Patel, P. B., Sekar, S., Jayaraman, P., ... & Esakkiraj, E. S. (2022). Investigation of physicochemical properties and characterization of leaf stalk fibres extracted from the caribbean royal palm tree. *International Journal of Chemical Engineering, 2022*(1), 7438411.

Hu, X., Hu, X., Yu, Y. and Wang, J. (2023). Prediction model for gestational diabetes mellitus using the XG Boost machine learning algorithm. *Frontiers in Endocrinology, 14*, 1105062.

Huang, X., Li, Y., Tong, X., Wu, Y., Zhang, R., Sheng, L., ... & Liu, J. (2024). Increased circulating IL-32 is associated with placenta macrophage-derived IL-32 and gestational diabetes mellitus. *The Journal of Clinical Endocrinology & Metabolism, 109*(2), 333–343.

Imafuku, H., Tanimura, K., Masuko, N., Tomimoto, M., Shi, Y., Uchida, A., ... & Terai, Y. (2023). Advantages of sensor-augmented insulin pump therapy for pregnant women with type 1 diabetes mellitus. *Journal of Diabetes Investigation, 14*(12), 1383–1390.

Jader, R. and Aminifar, S. (2022). Predictive model for diagnosis of gestational diabetes in the kurdistan region by a combination of clustering and classification algorithms: an ensemble approach. *Applied Computational Intelligence and Soft Computing, 2022*(1), 9749579.

Krsman, A., Stajić, D., Baturan, B., Stanković, M., Kupušinac, A., Kadić, U., ... & Živadinović, L. (2023). Correlation between increased maternal body mass index and pregnancy complications. *Eur Rev Med Pharmacol Sci, 27*(8), 3508–3513.

Kurt, B., Gürlek, B., Keskin, S., Özdemir, S., Karadeniz, Ö., Kırkbir, İ. B., ... & Turhan, K. (2023). Prediction of gestational diabetes using deep learning and Bayesian optimization and traditional machine learning techniques. *Medical & biological engineering & computing, 61*(7), 1649–1660.

Lapolla, A. and Metzger, B. E. (Eds.). (2019). *Gestational Diabetes: A Decade After the HAPO Study*. Karger Medical and Scientific Publishers.

Machin, D., Campbell, M. J., Tan, S. B. and Tan, S. H. (2018). *Sample Sizes for Clinical, Laboratory and Epidemiology Studies*. John Wiley & Sons.

Mansoori, A., Sahranavard, T., Hosseini, Z. S., Soflaei, S. S., Emrani, N., Nazar, E., ... & Mobarhan, M. G. (2023). Prediction of type 2 diabetes mellitus using hematological factors based on machine learning approaches: a cohort study analysis. *Scientific Reports, 13*(1), 663.

Mousavi, S., Safari, A., Nateghian, H., Ghojazadeh, M. and Nikniaz, L. (2023). Comparing the detection rate of postpartum diabetes in early and 4–12 week postpartum screening tests in women with gestational diabetes mellitus: A systematic review and meta-analysis. *Taiwanese Journal of Obstetrics and Gynecology, 62*(3), 396–401.

Natrayan, L., Kaliappan, S., Sethupathy, B. S., Sekar, S., Patil, P. P., Velmurugan, G. and Tariku Olkeba, T. (2022). Effect of Mechanical Properties on Fibre Addition of Flax and Graphene-Based Bionanocomposites. *International Journal of Chemical Engineering, 2022*(1), 5086365.

Peterson, A. K., Zhu, Y., Fuller, S., Feng, J., Alexeeff, S., Mitro, S. D., ... & Ferrara, A. (2023). PFAS concentrations in early and mid-pregnancy and risk of gestational diabetes mellitus in a nested case-control study within the ethnically and racially diverse PETALS cohort. *BMC Pregnancy and Childbirth, 23*(1), 657.

Petry, C. (Ed.). (2014). *Gestational diabetes: origins, complications, and treatment*. CRC Press.

Ray, A. (Ed.). (2020). *Gestational Diabetes Mellitus: An Overview with Some Recent Advances*.

Sartorão Filho, C. I., Pinheiro, F. A., Takano, L., Prudêncio, C. B., Nunes, S. K., Hallur, R. L. S., ... & Diamater Research Group. (2023). Risk factors for postpartum urinary incontinence: The impact of early-onset and late-onset Gestational Diabetes Mellitus in a nested case-control study. *European Journal of Obstetrics & Gynecology and Reproductive Biology, 290*, 5–10.

Sivaranjani, P. R., Syed, A., Elgorban, A. M., Bahkali, A. H., Balakrishnaraja, R., Varma, R. S. and Khan, S. S. (2023). Fabrication of ternary nano-heterojunction via hierarchical deposition of α-Fe2O3 and β-La2S3 on cubic CoCr2O4 for enhanced photodegradation of doxycycline. *Journal of Industrial and Engineering Chemistry, 118*, 407–417.

Viswanathan, S., Palaniyandi, T., Kannaki, P., Shanmugam, R., Baskar, G., Rahaman, A. M., ... & Sivaji, A. (2023). Biogenic synthesis of gold nanoparticles using red seaweed Champia parvula and its anti-oxidant and anticarcinogenic activity on lung cancer. *Particulate Science and Technology, 41*(2), 241–249.

Wong, C. L., Leung, K. P. and Shum, T. C. (2023). Gestational glucose intolerance and pregnancy outcomes: a retrospective study in the primary care setting of Macau. *Medicine, 102*(37), e35175.

Wu, Y. T., Zhang, C. J., Mol, B. W., Kawai, A., Li, C., Chen, L., ... & Huang, H. F. (2021). Early prediction of gestational diabetes mellitus in the Chinese population via advanced machine learning. *The Journal of Clinical Endocrinology & Metabolism, 106*(3), e1191–e1205.

*Hybrid and Advanced Technologies – S. Prasad Jones Christydass, Nurhayati Nurhayati and S. Kannadhasan. (Eds)* © *2025 the Authors, ISBN 978-1-032-90254-8*


# Improving diabetes healthcare with personalized management and monitoring through the integration of machine learning techniques

Jampani Chandra Sekhar
*Department of Computer Science and Engineering, NRI Institute of Technology, Guntur, Andhra Pradesh, India*

N. Mathimagal
*Department of Computer Science, Faculty of Arts and Science, Bharath Institute of Higher Education and Research, Chennai, Tamil Nadu, India*

Ramya Maranan
*Department of Research and Innovation, Saveetha School of Engineering, SIMATS, Chennai, Tamil Nadu, India*

K. Sathishkumar
*Department of Mechanical Engineering, Faculty of Engineering, Karpagam Academy of Higher Education, Coimbatore, Tamil Nadu, India*

D.J. Joel Devadass Daniel
*Department of Electronics and Communication Engineering, VelTech Rangarajan Dr. Sagunthala R&D Institute of Science and Technology, Chennai, Tamil Nadu, India*

ABSTRACT: Diabetes is a chronic condition that needs to be effectively managed to prevent complications and improve the quality of life. This article provides a novel way to revolutionize diabetes care. Machine learning algorithms run on a sensor-based Internet of Things system in this method. The system's key component is a smartphone application designed for diabetics. This allows individuals to see and track how their glucose levels change over time. Secure Cloud is used by healthcare professionals to look at the data in a visual manner. Support Vector Machines, Naive Bayes, Random Forests, and Artificial Neural Networks were used to provide a variety of machine learning methods for creating individualized health recommendations. These models have been constructed using a data set that includes both healthy and diabetic people. ANN was the most precise in predicting the health of patients based on their glucose levels; it correctly predicted 98.3 percent of data-driven outcomes. SVM, NB, and RF all gave decent results of 95.6 percent, 91.2 percent, and 92.0 percent, respectively. Using a proprietary smartphone application, patients can make choices about their lifestyle based on their glucose results. Machine learning technology can also detect abnormal sugar levels in real-time. The findings from this study show that machine learning and IoT have the potential to revolutionize the care of diabetes. Moreover, this application aids in the management of diabetes by offering personalized health suggestions based on glucose levels. Therefore, it is now time for IoT chatting to become widely used. This piece of work provides a feasible way of adopting an Internet of Things technology, machine learning, and cloud connections to accommodate people's personalized requirements. With the device, users can conduct a blood sugar test. However, doctors can also access patients' data from afar. Findings from this study suggest this method's effectiveness using IoT and machine learning technology will improve diabetes morbidity and medical care effects. However, these hypotheses need to be validated with much bigger data of diverse nature.

*Keywords*: IoT, machine learning, diabetes, Neural Network, Performance evaluation

DOI: 10.1201/9781003559115-34 

## 1 INTRODUCTION

Diabetes, a prevalent chronic condition affecting millions, presents substantial healthcare challenges. Addressing these demands attention due to the evolving landscape of personalized therapy and monitoring, crucial for effective disease management and complication prevention. Machine learning (ML) technologies have emerged prominently in diabetes care, offering new avenues for enhanced treatment [1,2]. Various ML algorithms, notably Support Vector Machines (SVM), have demonstrated effectiveness in diagnosing and forecasting diabetes outcomes based on key variables like glucose levels, BMI, and age. SVMs excel in classifying patients as diabetic or non-diabetic. Similarly, Naive Bayes (NB) algorithms leverage probabilistic methods to predict diabetes risk, assuming independence between variables, aiding early diagnosis and preventive measures. Random Forests (RF), an ensemble learning technique, integrate diverse decision trees to deliver robust predictions, particularly useful in analyzing risk factors such as family history and lifestyle [3–5].

Artificial Neural Networks (ANNs) further enrich diabetes management by exploring intricate relationships between inputs and diabetes outcomes, facilitating personalized predictions and treatment plans. ANNs predict glucose levels, administer insulin doses, and diagnose complications like diabetic retinopathy [6,7]. The integration of ML with the Internet of Things (IoT) and smartphone applications revolutionizes diabetes care, enabling real-time data collection through wearable sensors and continuous glucose monitors. Smartphone apps provide user-friendly interfaces for monitoring glucose levels and offering personalized health insights, enhancing patient engagement and remote monitoring capabilities. Despite these advancements, challenges like data privacy, quality, and interpretability of ML models persist, necessitating robust security measures and ethical considerations. Future innovations may incorporate advanced ML architectures such as convolutional and recurrent neural networks, leveraging diverse data sources like electronic health records and genetic data to refine predictive accuracy and treatment efficacy in diabetes management [8].

## 2 METHODOLOGY

We implemented a diabetes treatment and exercise program using machine learning techniques, integrating hardware, communication modules, and software development to achieve our objectives. Initially, blood glucose sensors provided real-time measurements, crucial for ongoing health monitoring and decision-making. These sensors were connected via an Arduino microcontroller board equipped with a Bluetooth module to transmit data to mobile phones, enabling patients to receive immediate feedback and guidance. A dedicated mobile app was designed to receive and display glucose sensor data, offering user-friendly interfaces for patients to monitor their glucose levels. Machine learning algorithms within the app provided health advice based on these readings, suggesting necessary actions such as hospital visits for high glucose levels or encouraging messages for healthy readings.

Data aggregations, including glucose readings and health advisories, were securely stored in the cloud, enabling continuous patient monitoring and facilitating deeper analysis by healthcare providers (Figure 1). Our study utilized various machine learning models—Support Vector Machines (SVM), Naive Bayes (NB), Random Forests (RF), and Artificial Neural Networks (ANN)—to enhance personalized diabetes management. These models processed patient sensor data to deliver tailored health recommendations effectively. For instance, SVMs employed complex decision boundaries to classify patients based on their glucose levels, BMI, and age, ensuring robust performance in predicting disease progression. NB algorithms calculated probabilities independently among features, making them efficient for quick predictions with minimal computational resources. RF, leveraging ensemble learning, handled high-dimensional datasets, while ANNs learned intricate relationships within data, adapting dynamically to improve prediction accuracy.

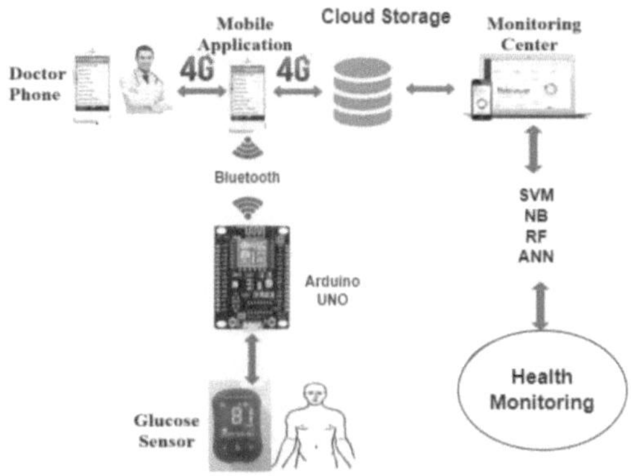


Figure 1. Heath monitoring. Alt text: Health monitoring architecture.

Our approach involved creating a comprehensive dataset of 1,000 patient samples, integrating IoT sensors to collect data from both healthy and diabetic individuals. Data preprocessing prepared this dataset for training and testing across multiple algorithms, evaluating their performance based on classification accuracy. Through meticulous model selection and validation, we identified the most effective algorithms for personalized diabetes management, enabling healthcare professionals to deliver precise treatment strategies and empowering patients with informed health decisions. This integrated system aims to optimize diabetes care through continuous monitoring and data-driven insights, ensuring proactive health management for all patients involved.

## 3 RESULT AND DISCUSSION

In Figure 2, we can see a graph of the dataset values obtained. These 1,000 datasets are used for training various models of machine learning. The performance of a machine learning model is often evaluated with Accuracy, a well-known metric. This measures the proportion of the predicted results that are correctly anticipated against the total number of predictions made. So within the context of your research, the accuracy of the machine learning models which you developed was also examined to assess their suitability for projecting outcomes. The accuracy rate of SVM model is 95.6%. This means that it correctly anticipated the result in 95.6% of cases within the dataset. Higher accuracy means that the input variables being consumed better by SVM and to do so in a healthful way for patients.

The accuracy of the Naive Bayes (NB) model was 91.2%. That means in slightly more than 91.2% of the cases, the predicted value was accurate. Compared to SVM, NB, which was slightly less accurate showed significant accuracy in categorizing patients. A Random Forests (RF) model has 92% accuracy. This means that the RF model correctly predicted the result in 92% of the cases. RF represents the combining conclusions from many decision trees; as an ensemble learning technique, it performed well here. The Artificial Neural Network (ANN) model came out tops on a performance test, with an accuracy score of 98.3%. So that means the ANN model was the best, with right-on results in 98.3% of the cases. For finding the glucose level by patients, this was indeed a measure worth considering. The model's ability to handle unseen data came through clearly from readings: its accuracy scores prove as much. The performance comprehensively you must also consider other evaluation metrics such as accuracy, recall and F1 score. Finally, further study and validation on other datasets are necessary, in order that we can evaluate the authenticity and reliability of these models.

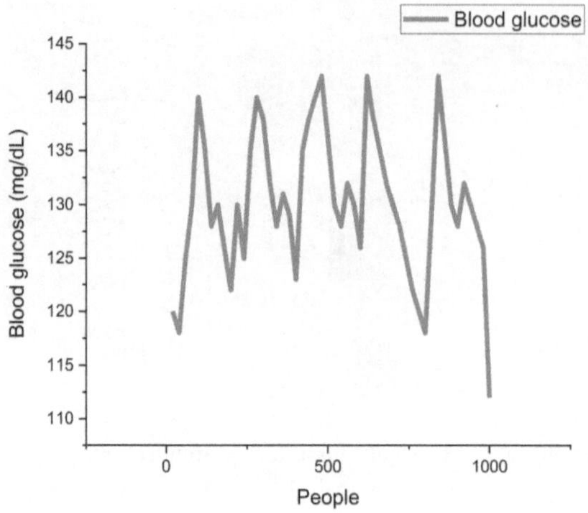


Figure 2. Dataset values. Alt text: Dataset values used in this research.

The model's precision is the proportion of all correct predictions. Positive prediction accuracy, as defined by precision, is expressed by the percentage of true positives in all predicted positives. Recall (also known as sensitivity or true positive rate) measures the proportion of true positives as a percentage of all true positive cases. With regard to accuracy and recall, the F1 score indicates how well the model functions. The SVM model correctly predicted 95.6% in this forecast with more precision (96.1%) and recall (94.8%); the F1 score came to an impressive 95.4%.

Similarly, the NB model scored an F1 of 91.5% on top of 91.2% accuracy and 89.7% precision, in addition to a recall rate of 93.5%. The RF model obtained 92.0% accuracy and produced a F1 score of 91.8%, with an excellent precision-to-recall (92.4% ) relationship. On the other hand, the ANN model scored a splendid 98.3% accuracy; for an F1 score of 98.2%. Both precision and recall were high: 98.6% and 97.9%. The full result of performance is shown in Figure 3.

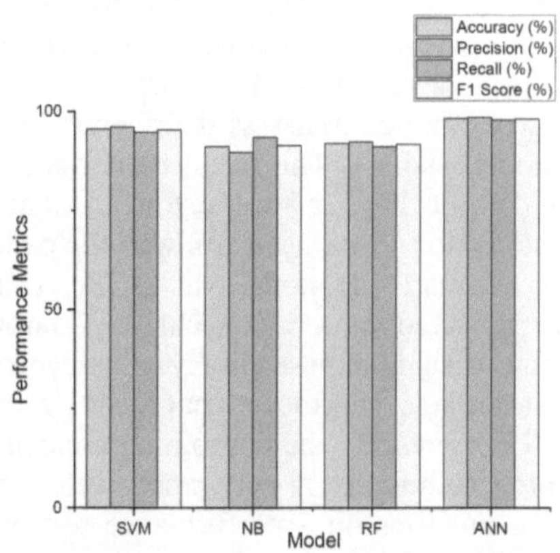


Figure 3. Performance metrics of each machine learning model. Alt text: The performance evaluation results of various machine learning models.

In terms of accuracy, the ANN model was the best. It achieved a 98.3% degree of accuracy. With good accuracy, recall, and F1 score, the ANN model showed its applicability in screening for patient health conditions with glucose readings. While the ANN model has produced remarkable results, it can be the new principal approach to individualized diabetes health care management and surveillance.

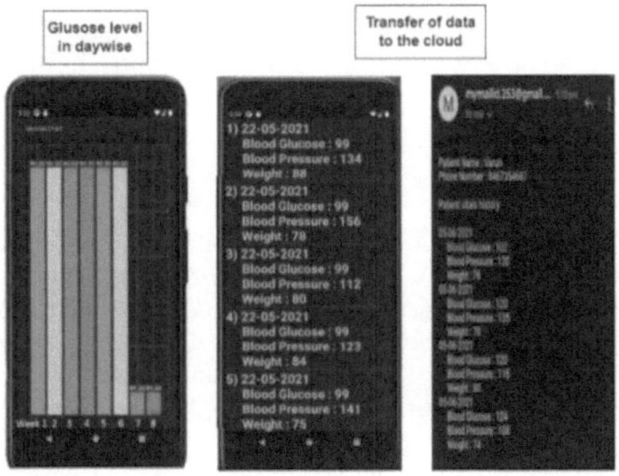


Figure 4. Developed application for monitoring diabetes. Alt text: The proposed application showing the status of the patients health.

An application was developed to give users a comprehensive overview of their glucose levels over time. Using this user interface application, users could view glucose level data in a real-time dynamic. The acquired data, including glucose readings and timestamps, was securely sent and stored in the cloud. Because it was stored in the cloud, even doctors and patients could continually probe and view the data. Figure 4 is a screenshot of the mobile application interface, providing a graphical representation of glucose levels to help users easily monitor their diabetes and find out more about the state of their physical condition.

Using machine learning techniques in personalized diabetes management and monitoring, we have shown that SVM, NB, RF, and ANN (Artificial Neural Network) models are capable of performing outstandingly well when it comes to forecasting health conditions on the basis of glucose level readings. All four models produced high accuracy rates. The ANN model, however, was more accurate than the rest with 98.3% accuracy. Therefore, the high accuracy estimated for the ANN model suggests its use as an effective tool for classifying patients with accuracy, not only helping create more personalized healthcare plans but also suggesting reliable healthcare products for these patients.

## 4 CONCLUSION

The principle of this investigation was to use machine learning techniques to enhance the therapy for diabetes through personalized management and monitoring. We turned one of the experiment findings into a new technique for robotic motion: Real-Time non-invasive testing–no blood sticks or finger-pricks necessary! This was realized by, constructing an Arduino-based IoT sensor system, which was linked with a Bluetooth module to a mobile application. Support Vector Machines (SVM), Naive Bayes (NB), Random Forests (RF) and Artificial Neural Networks are all types of machine learning models used in the application, to help people at risk in avoiding sweets like cake and ice cream depending upon their glucose levels. These models were then tested out and it turned out that the accuracy of the ANN model exceeded the other models. The ANN model had an accuracy rate as high as 98.3%. The accuracy of the ANN model outperformed all the others. This is another illustration of the high accuracy with which ANN can detect the health status of patients based on glucose meter readings. So for the first time users can look at their blood glucose level and make informed treatment decisions of their own. A photon fires the data devils up into the cloud, where it is stored safely for healthcare professionals. Now their doctors can keep track of those underlings' health even at a distance. The findings of this study have indicated novel applications for artificial intelligence and IoT in diabetes care. Personal and real-time monitoring at the system level may improve patient control and care generally. The application developed for this purpose was simple to use and connected to the cloud.

## REFERENCES


[1]. Kamruzzaman, M. M., Hossin, M. A., Alruwaili, O., Alanazi, S., Alruwaili, M., Alshammari, N., Alaerjan, A. and Zaman, R. (2022). IoT-Oriented 6G wireless network system for smart cities. *Computational Intelligence and Neuroscience, 2022*. https://doi.org/10.1155/2022/1874436

[2]. Kozlowski, T. and Noran, O. (2023). ScienceDirect ScienceDirect designing designing an an evaluation evaluation framework framework for for IoT IoT environmental environmental monitoring systems systems monitoring. *Procedia Computer Science*, *219*, 220–227. https://doi.org/10.1016/j.procs.2023.01.284

[3]. Vilas-Boas, J. L., Rodrigues, J. J. P. C. and Alberti, A. M. (2022). Convergence of distributed ledger technologies with digital Twins, IoT, and AI for fresh food logistics: Challenges and opportunities. *Journal of Industrial Information Integration*, *31*(June 2022), 100393. https://doi.org/10.1016/j.jii.2022.100393

[4]. Rathee, G., Balasaraswathi, M., Chandran, K. P., Gupta, S. D. and Boopathi, C. S. (2021). A secure IoT sensors communication in industry 4.0 using blockchain technology. *Journal of Ambient Intelligence and Humanized Computing*, *12*(1), 533–545. https://doi.org/10.1007/s12652-020-02017-8

[5]. Rastogi, E., Saxena, N., Roy, A. and Shin, D. R. (2020). Narrowband Internet of Things: A Comprehensive Study. *Computer Networks*, *173*(February), 107209. https://doi.org/10.1016/j.comnet.2020.107209

[6]. Jha, A. V., Appasani, B., Ghazali, A. N., Pattanayak, P., Gurjar, D. S., Kabalci, E. and Mohanta, D. K. (2021). Smart grid cyber-physical systems: Communication technologies, standards and challenges. *Wireless Networks*, *27*(4), 2595–2613. https://doi.org/10.1007/s11276-021-02579-1

[7]. Nayak, J., Naik, B., Dash, P. B., Vimal, S. and Kadry, S. (2022). Hybrid Bayesian optimization hypertuned catboost approach for malicious access and anomaly detection in IoT nomalyframework. *Sustainable Computing: Informatics and Systems*, *36*(June), 100805. https://doi.org/10.1016/j.suscom.2022.100805

[8]. Pandiyan, S., Lawrence, T. S., Sathiyamoorthi, V., Ramasamy, M., Xia, Q. and Guo, Y. (2020). A performance-aware dynamic scheduling algorithm for cloud-based IoT applications. *Computer Communications*, *160*(June), 512–520. https://doi.org/10.1016/j.comcom.2020.06.016

*Hybrid and Advanced Technologies – S. Prasad Jones Christydass, Nurhayati Nurhayati and S. Kannadhasan. (Eds) © 2025 the Authors, ISBN 978-1-032-90254-8*
*Open Access: www.taylorfrancis.com, CC BY-NC-ND 4.0 license*

# A novel approach for detecting breast cancer in mammography images using RBF-ELM model

Naga Raju Hari Manikyam
*Associate Professor, Department of CSE, Nalla Narasimha Reddy Education Society's Group of Institutions - Integrated Campus, Hyderabad, India*

Nihar Ranjan Behera
*D. B. A Mentor, S. S. B. M. Geneva. Swiss School of Business and Management Geneva Business Center Avenue des Morgines, Genève Switzerland*

V. Kalpana
*Assistant Professor of English, R. M. K. College of Engineering and Technology, Puduvoyal, Tamil Nadu, India*

S. Tamijeselvan
*Assistant Professor in Radiography, Mother Theresa Post Graduate and Research Institute of Health Sciences, A Goverment of Puducherry Institution, Puducherry, India*

Anil Kumar Mavliya
*Principal, C. O. E. R. College of Nursing and Paramedical Sciences, Roorkee, haridawar(U. K.), India.*

Hari Kumar Buddha
*Assistant Professor, Department of E. C. E., Vignan's Institute of Engineering for Women, Visakhapatnam, India*

ABSTRACT: Currently, detecting breast cancer is crucial for women all over the world in order to preserve lives. Because of their lack of training and expertise in cancer diagnosis, doctors and radiologists may fail to notice the anomaly. Model training, feature extraction, data segmentation, and preprocessing are the four key components. Image binarization and active contour are two of the preprocessing procedures. Segmentation is a way to improve the representation of an image in terms of its utility, understandability, and actionability. The GLCM, or Gray Level Co-occurrence Matrix, is employed for feature extraction. For more accurate results, we trained the model using the RBF-ELM. Outperforming rival methods like RBF and ELM, the proposed approach attained an accuracy of 97.02%.
Keywords: Gray Level Co-occurrence Matrix (GLCM), Breast Cancer, Breast Imaging Reporting and Data System (BI-RADS), Regularized Extreme Learning Machine (RELM), X-Ray

## 1 INTRODUCTION

Recently, breast cancer has become the top cause of death for women, surpassing all other health risks. There are numerous variables that contribute to this illness, including genetics and lifestyle choices. The earliest signs of this disease were abnormalities seen in the breast's soft tissue. Image analysis using X-ray mammography is the mainstay for detecting breast cancer. By shooting from various angles, we were able to catch every part of the illness. Since very little amounts of radiation are utilized in X-ray imaging, the lack of contrast is a clear

DOI: 10.1201/9781003559115-35

health concern. The image augmentation is implemented using a variety of ways, including artificial intelligence tactics and deep learning. The possibility of using deep-learning algorithms to the problem of disease detection has been considered. This study proposes a RELM model–based early detection system for breast cancer. The proposed method makes use of digital mammography images after the pre-processing phase. The proposed method for detecting breast cancer based on alterations in soft tissues is based on software engineering process modelling. This model takes into account the proposed method's scalability, adaptability, and reliability. Keep in mind that the proposed solution is web-based, which should make it more accessible even in remote places. It is specifically designed for these places because there is a severe lack of qualified medical personnel.

## 2 LITERATURE SURVEY

In order to reduce the death rate associated with breast cancer, [1]we need early methods to detect and classify tumours. Early detection improves both the prognosis for illness and the chances of a successful therapy. Several different approaches exist for screening for breast cancer. In a study conducted by [2], the Hough transform was used to find features in mammography pictures. Those characteristics are fed into the SVM classifier. In study, [3]utilized a KNN classifier to diagnose breast cancer. The primary function of this classifier is to distinguish between benign and malignant breast tumours. One of the best ways to catch breast cancer early is with a mammogram [4]. It generates pictures by means of low-dose X-rays. However, not every anomaly is malignant. A smooth and contained protrusion, for example, is typically not harmful. However, if the tumour is starburst shaped and has an irregular border, a biopsy is necessary to confirm that it is malignant [5]. Due to the unequal distribution of power and Breast cancer arises when cells proliferate abnormally, causing a tumour and, eventually, death [6]. Radiologists use mammography images to detect breast cancer. In their study on breast cancer diagnosis, [7] utilized DT, NB, SLOG, and advanced ensembles techniques such as Majority Voting and the RF method to reach an accuracy rate of 98.01% and an error rate of 0.001%. The BiCNN model proposed by[8]achieved an accuracy of 97.87%. By integrating logistic regression, an ANN, and a voting method. Breast cancer is one of the most common diseases affecting women and one of the most deadly malignancies in the world [9]. New research indicates that breast cancer ranks among the top nine Among Saudi women, the most dangerous diseases are [10]. Early detection of breast cancer is critical since the disease causes more complications and makes treatment more challenging as it progresses. [11]An increase in the variety of treatments available as a result of earlier disease identification might help reduce the rising rate of breast cancer. Breast cancer detection relies heavily on mammography images and processing, which are also used for disease diagnosis to guarantee patients get treatment quickly. Processing mammography images is not without its challenges, however, for radiologist diagnosis and pathology localization [12]. With better picture quality, radiologists may be able to more easily study mammographic images in search of an abnormal area. The Breast Imaging Reporting and Data System (BI-RADS) is the benchmark in mammography, similar to the American Cancer Society's quality assurance lexicon report. [13] Organization and standardization of mammography results for clinicians is the primary goal of the BI-RADS.

## 3 PROPOSED SYSTEM

Breast cancer is the second most deadly of the tumors that primarily affect women. Breast cancer mortality rates can be reduced with early identification and treatment. Mammography is a powerful tool in the breast cancer screening arsenal since it can identify calcification zones or tumors in the breast at an early stage. The difficulty of detecting potentially malignant tumors in breasts with exceedingly dense tissue is one limitation of breast mammography.

## 3.1 *Preprocessing*

### 3.1.1 *Image binarization*

The process of binarization converts grayscale images into binary ones. As seen in equation (1), there are only two possible intensity values for pixels in a binary image (I): zero and one. Image binarization is necessary for foreground object separation from background[14]. It is possible to categorize the binarization methods according to the way the threshold is determined.

$$I : \{M, R\} \rightarrow \{0, 1\} \tag{1}$$

### 3.1.2 *Active contour*

Outline objects rely heavily on active contour. Separating barriers such tags, patient names, studied artifacts, and associated artifacts is the first step of its two-step pre-processing technique that relies on thresholding. The extraction of the breast muscle boundary is being done using an active contour in conjunction with an automatic stopping method.

## 3.2 *Segmented image*

Segmentation is a technique used to break down large, problematic images into smaller, more manageable ones by grouping together pixels, sometimes called super pixels[15]. Segmentation is a great tool for improving the relevance, understandability, and analytical soundness of an image's representation. Isolating the components from the image's background which often includes curves and lines and any unnecessary elements is common practice.

## 3.3 *Feature extraction*

### 3.3.1 *GLCM*

In order to get texture information, this study uses the GLCM approach. As a function of distance d and different orientations, the GLCM is a table that indicates the frequency of various pixel brightness value (grey level) combinations in a picture[16]. It was shown that the co-occurrence characteristics strategy was the most effective of these.

## 3.4 *Model training*

### 3.4.1 *RBF-ELM*

The neural classifier that we provide here is a hybrid of RBF and ELM networks. This section provides a detailed analysis of the proposed system, outlining its design, training technique, fundamental principles, and essential tenets.

### 3.4.2 *Main idea and motivation*

The suggested classifier is based on the premise that RBFNN and ELMNN can work together to provide better performance with a more precise structure. Based on an ELMNN and a minimally-architected RBFNN, the proposed model incorporates a variety of activation functions. By applying GA, the activation functions of the extra neurons may be fine-tuned in terms of both amount and shape. The fact that GA can easily describe hierarchical systems and can integrate binary and real encoding was major factors for deciding to employ it. The following grounds support the recommendation of this method: To begin, RBFNNs train their hidden neurons using local responses, whereas ELMNNs employ global responses. Because of this, using fewer hidden neurons in conjunction with them can result in better decision boundaries. The fact that ELMNNs' arbitrarily initialized starting weights might lead to ill-conditioned problems is bad enough, but when dealing with massive amounts of hidden neurons, it becomes far worse. Another option is to use clustering techniques on the training data to train RBFNN using previous knowledge. It is advised to mix the two kinds to prevent problems with ill-conditioning. This combination also makes ELMNN less dependent on the random initial weights. Neurons with various activation functions should be added, and then GA should be used to optimize their types and forms[17]. Find the optimal placement of the extra neurons is another use of GA. Hence, it suggests optimizing the types and shapes of neurons using GA and adding neurons with variable activation functions. In addition, the GA is employed to determine the ideal quantity of additional neurons. Defining the correct structure is a major challenge for both kinds of neural networks, and this is one approach to solving that problem. Only the additional neurons are recommended for optimization in this proposed. Due to their exclusive function in enhancing the RBFs, these neurons make optimization easier.

#### 3.4.3 *Merge of CNN and RELM*

The RBFs and their input weights are part of the first part, RBFNN. It is the responsibility of this hidden network of neurons to determine the distance between the sample and its centers. These neurons only receive 1 as an input weight. The freshly inserted neurons and their training weights are part of the second part, ELMNN. In this case, the hidden neurons have weights that were initially set at random and use generic activation functions.

## 4 RESULT AND DISCUSSION

For instance, many women all across the world have been affected by breast cancer. This is why a lot of people have been looking into early detection methods as a means to fight this disease. The idea behind this approach is to create a system for early diagnosis of breast cancer utilizing images taken by mammography machines.

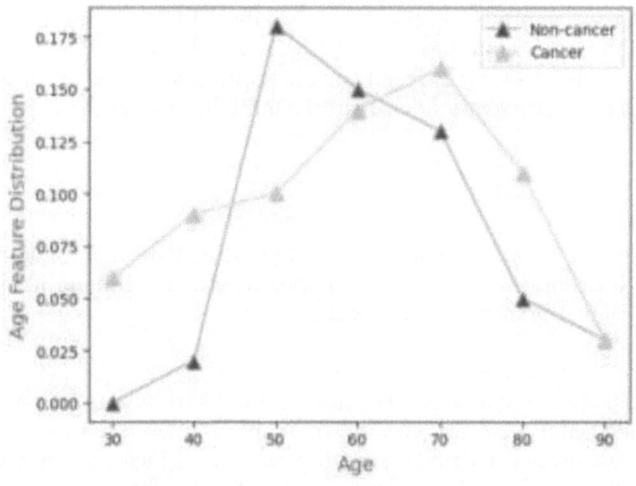


Figure 1. The distribution of age for cancer and non-cancer.

The distribution of the age feature produced by the RBF-ELM for both non-cancer and cancer participants is shown in Figure 1. Age, as may be seen, is also a distinguishing trait.

Figure 2 displays the outcomes of comparing the suggested model to existing methods; using the classification yielded a testing accuracy of 97.02%, outperforming other classifiers.

When models learn to reduce loss, classification accuracy improves. As compared to competing models, the suggested one has the highest level of accuracy.

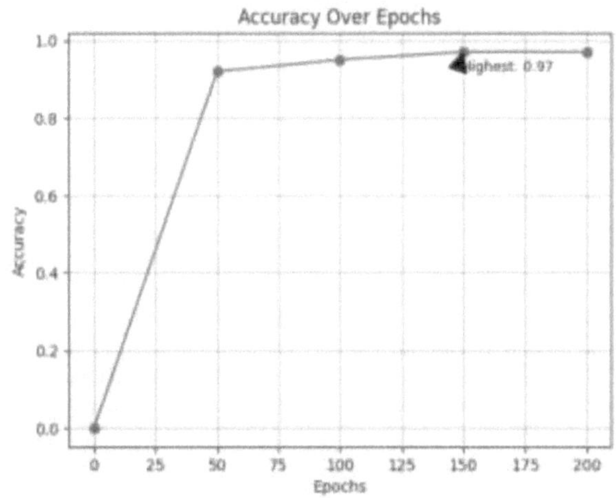


Figure 2. Comparison of training set accuracy of the proposed model.

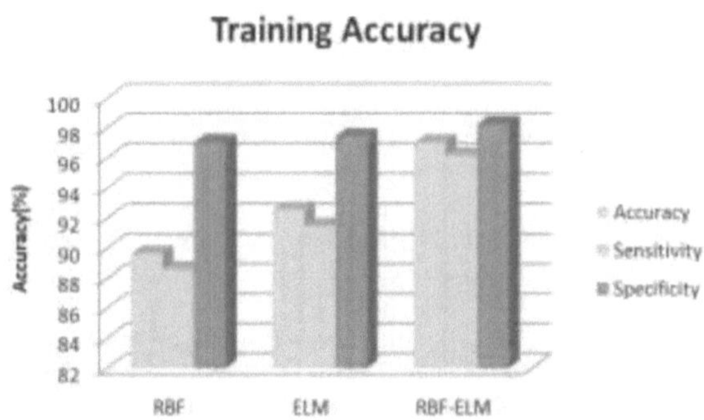


Figure 3. Accuracy of RBF-ELM model.

## 5 CONCLUSION

Breast cancer is high among female cancers worldwide, affecting one out of every eight women at some time. As far as imaging early-stage breast cancer is concerned, mammography is the gold standard. Due to the low visibility and weak contrast in mammographic photos, for breast cancer treatment to be effective, early detection is crucial. As part of the preprocessing, the images are binarized and the contours are actively drawn. To make an image representation more practical, understandable, and actionable, segmentation is a good technique to use. The Gray Level Co-occurrence Matrix, or GLCM, is used to extract features. When training the model, RBF-ELM is utilized. At 97.02%, the suggested method achieves better average accuracy than the RBF and ELM models.

## REFERENCES

[1] Heidari, M., Khuzani, A. Z., Hollingsworth, A. B., Danala, G., Mirniaharikandehei, S., Qiu, Y., ... and Zheng, B. (2018). Prediction of breast cancer risk using a machine learning approach embedded with a locality preserving projection algorithm. *Physics in Medicine & Biology*, 63(3), 035020.

[2] Vijayarajeswari, R., Parthasarathy, P., Vivekanandan, S. and Basha, A. A. (2019). Classification of mammogram for early detection of breast cancer using SVM classifier and Hough transform. *Measurement*, 146, 800–805.

[3] Lakshmi, N. S. and Christina, M. C. E. (2019). Breast cancer detection using KNN classifier. *IRE J.*, 2 (10), 76–79.

[4] Shen, L., Margolies, L. R., Rothstein, J. H., Fluder, E., McBride, R. and Sieh, W. (2019). Deep learning to improve breast cancer detection on screening mammography. *Scientific Reports*, 9(1), 12495.

[5] Rodriguez-Ruiz, A., Lång, K., Gubern-Merida, A., Broeders, M., Gennaro, G., Clauser, P., ... and Sechopoulos, I. (2019). Stand-alone artificial intelligence for breast cancer detection in mammography: comparison with 101 radiologists. *JNCI: Journal of the National Cancer Institute*, 111(9), 916–922.

[6] Rodríguez-Ruiz, A., Krupinski, E., Mordang, J. J., Schilling, K., Heywang-Köbrunner, S. H., Sechopoulos, I. and Mann, R. M. (2019). Detection of breast cancer with mammography: effect of an artificial intelligence support system. *Radiology*, 290(2), 305–314.

[7] Naji, M. A., El Filali, S., Bouhlal, M., Benlahmar, E. H., Abdelouhahid, R. A. and Debauche, O. (2021). Breast cancer prediction and diagnosis through a new approach based on majority voting ensemble classifier. *Procedia Computer Science*, 191, 481–486.

[8] Wei, B., Han, Z., He, X. and Yin, Y. (2017, April). Deep learning model based breast cancer histopathological image classification. In *2017 IEEE 2nd international conference on cloud computing and big data analysis (ICCCBDA)* (pp. 348–353). IEEE.

[9] Albeshan, S. M. and Alashban, Y. I. (2021). Incidence trends of breast cancer in Saudi Arabia: A joinpoint regression analysis (2004–2016). *Journal of King Saud University-Science*, 33(7), 101578.

[10] Fan, L., Goss, P. E. and Strasser-Weippl, K. (2015). Current status and future projections of breast cancer in Asia. *Breast Care*, 10(6), 372–378.

[11] Jesneck, J. and DUKE UNIV MEDICAL CENTER DURHAM NC. (2007). *Computer-Aided Diagnosis System for Breast Cancer Combining Mammography and Proteomics.*

[12] Verma, B., McLeod, P. and Klevansky, A. (2010). Classification of benign and malignant patterns in digital mammograms for the diagnosis of breast cancer. *Expert Systems with Applications*, 37(4), 3344–3351.

[13] Kate, V., Ushasree, R., Tharsanee, R. M., Kukreja, M. T., Saraf, S. and Varadharajan, B. (2023, March). *GAN, CNN and ELM Based Breast Cancer Detection.* In *2023 2nd International Conference for Innovation in Technology (INOCON)* (pp. 1–6). IEEE.

[14] Tripathy, S. and Swarnkar, T. (2020). Unified preprocessing and enhancement technique for mammogram images. *Procedia Computer Science*, 167, 285–292.

[15] Konkani, A., Bera, R. and Paul, S. (2018). Advances in systems, control and automation. *Lecture Notes in Electrical Engineering*, 442, 701–709.

[16] Saleck, M. M., ElMoutaouakkil, A. and Mouçouf, M. (2017, May). Tumor detection in mammography images using fuzzy C-means and GLCM texture features. In *2017 14th International Conference on Computer Graphics, Imaging and Visualization* (pp. 122–125). IEEE.

[17] Siouda, R., Nemissi, M. and Seridi, H. (2024). Diverse activation functions based-hybrid RBF-ELM neural network for medical classification. *Evolutionary Intelligence*, 17(2), 829–845.

*Hybrid and Advanced Technologies – S. Prasad Jones Christydass, Nurhayati Nurhayati and S. Kannadhasan. (Eds) © 2025 the Authors, ISBN 978-1-032-90254-8*
*Open Access: www.taylorfrancis.com, CC BY-NC-ND 4.0 license*

# Determination of Chikungunya outbreak forecasting using Gaussian Naive Bayes algorithm with bootstrap over linear regression algorithm for improved accuracy

C. Chenna Mounika
*Research Scholar, Department of Computer Science and Engineering, Saveetha School of Engineering, Saveetha University, Chennai, Tamil Nadu, India*

K. Thinakaran
*Department of Computer Science and Engineering, Saveetha School of Engineering, Saveetha University, Chennai, Tamil Nadu, India*

ABSTRACT: Developing and assessing Naive Bayes and Linear Regression, two different classification algorithms, for their efficacy in predicting Chikungunya outbreaks was the main objective of this study. Using machine learning, the research sought to develop a strong prediction model that could reliably detect possible epidemics and enable prompt preventive actions. The objective of this comparison investigation was to identify the optimal algorithm for this particular application, offering significant contributions to the field of disease outbreak prediction. Two separate groups with a total sample size of 44 were used in this work to forecast Chikungunya outbreaks using a dataset of 1,500 cases. Using preset values for power (0.8), alpha (0.05), and beta (0.2), the study used the G Power test. In order to measure the predictive power of the GNB and Linear Regression models, metrics such as accuracy, Mean Absolute Error (MAE), Mean Squared Error (MSE), and Root Mean Squared Error (RMSE) were used during the building and evaluation process.The comparison study clarifies each algorithm's unique benefits and drawbacks. The Chikungunya outbreaks were predicted with 98.54% accuracy by Gaussian Naive Bayes and 49.67% accuracy by Linear Regression.

In comparison to Linear Regression, Naive Bayes distinguishes itself by showcasing a notably higher accuracy of 98.54% in predicting Chikungunya outbreaks, while Linear Regression achieved a comparatively lower accuracy rate. In both Linear Regression and Gaussian Naive Bayes models, we employed bootstrapping to assess their predictive performance. Linear Regression demonstrated its accuracy in predicting Chikungunya cases, while Gaussian Naive Bayes showcased its classification prowess, achieving consistent accuracy throughout the iterations. These findings provide valuable insights into model selection for Chikungunya outbreak prediction.

*Keywords*: Gaussian Naive Bayes, Linear Regression, Chikungunya Outbreaks, Predictive efficacy, Disease outbreak prediction, Epidemiological research, Bootstrapping

## 1 INTRODUCTION

In order to anticipate the occurrence of a Chikungunya outbreak, Amaral, Taylor, and Schoen's research (Amaral *et al.* 2023) compared and analyzed the Gaussian Naive Bayes (GNB) and Random Forest machine learning methods. GNB algorithm was utilized in the first portion of the investigation, and Random Forest regression was used in the second half. The objective was to assess these algorithms' overall performance and predictive accuracy in terms of foreseeing Chikungunya outbreaks. The researchers determined which algorithm produced more accurate and dependable forecasts using bootstrapping and a variety of evaluation criteria, including accuracy, Mean Absolute Error (MAE), Mean Squared Error (MSE), and Root Mean Squared Error (RMSE). Since they can guide the development of more efficient epidemic management and control tactics, these findings are essential for epidemiological and public health research. Predicting disease outbreaks with high accuracy is crucial in the current global environment, where we confront many difficulties from new infectious illnesses (Amaral *et al.* 2023; Parvez M *et al.* 2023; Ramesh *et al.*, n.d.; Ruiz-Burga *et al.* 2022; Sarkar *et al.* 2022; Surendran *et al.* 2023; Tayyab *et al.* 2022). Early detection and prompt action are made possible by machine learning models such as GNB and Random Forest, which provide a data-driven method for predicting the spread of disease. These resources are very helpful to public health professionals because they help with decision-making, resource allocation, and the efficient handling of health emergencies (Goh *et al.* 2022). This research greatly expands our ability to manage and ameliorate the complex health and environmental issues of today, in a world where prompt interventions are crucial.

From 2018 to 2023, a significant amount of research has been conducted, with a plethora of studies available across various platforms. Google Scholar boasts over 7430 papers, while the IEEE has published 345 works. Springer has contributed with 11 publications, and PubMed features 189 studies. Additionally, ScienceDirect has a substantial collection of 1050 articles on the subject.(Lima *et al.* 2022) In order to find models for the transmitter vector dynamics of arboviruses and prediction models for these viruses, this study carried out a thorough assessment of the literature. Data from 543 individuals with a diagnosis of possible dengue, chikungunya, or Zika infection were included by Daniel.F. Farrell ("[No Title]" n.d.). They evaluated seven algorithms' capacity to forecast hospitalization status based on signs, symptoms, and laboratory data that would be provided to a clinician upon patient intake using a machine learning methodology.The use of machine learning approaches to differentiate dengue from other feverish diseases in the primary care setting and forecast severe arboviral disease in the community is covered by Fathima and Hundewale (Fathima and Hundewale 2021; Rakotonirina *et al.* 2021). In order to find prediction models for arboviruses and models for the dynamics of their transmitter vector, this study suggests conducting a thorough assessment of the literature. Random Forest and regression were the machine learning techniques employed in their investigation. With their findings, they showed that screening for arbovirus infections may be done effectively and at a cheap cost utilizing a predictor.Because it shows how machine learning works to screen for arbovirus infection, the study "Temporal and Spatiotemporal Arboviruses Forecasting by Machine Learning: A Systematic Review" (Lima *et al.*, 2022) is the best.The machine learning techniques Random Forest and regression are used in the investigation.The outcomes show that an affordable and efficient predictor can be used to screen for arbovirus infections.

The endeavor to predict Chikungunya outbreaks using machine learning encounters obstacles such as the presence of noisy data and the lack of structured data. Additionally, intricate models require substantial computational power, which can hinder real-time predictions. The primary goal of this project is to construct predictive models for local Chikungunya occurrences based on data from travelers. The project is divided into two parts, one utilizing the Random Forest Regressor and the other employing the Gaussian Naive Bayes algorithm. Performance evaluation is carried out through bootstrapping, using metrics like Mean Absolute Error and Accuracy. The ultimate aim of the project is to ascertain the most effective algorithm for predicting Chikungunya outbreaks, thereby facilitating more efficient public health planning.

## 2 MATERIALS AND METHODS

The Saveetha University Department of Computer Science and Engineering was the site of the research. In order to perform a comparison analysis, two groups were established: Eleven samples from Group 1 were processed using the Gaussian Naive Bayes Algorithm (GNB), and eleven samples from Group 2 were processed using the Linear Regression Algorithm (RF). 44 people made up the overall sample size, which was decided by careful statistical analysis to guarantee accurate performance rating. An 80% chance of finding significant differences was indicated by a statistical power (G-power) of 0.8, which was used to determine the sample size. With a 95% confidence level and an 80% probability of avoiding Type II errors, the alpha (α) and beta (β) values were set at 0.05 and 0.2, respectively.

One reliable data science platform, Kaggle, is where the main dataset was acquired. Two subsets of this dataset were created: a testing set and a training set. The data, which included the average, minimum, and maximum numbers or proportions of infected travelers, were particularly developed to train a model to forecast Chikungunya outbreaks.

A potent computer system with 16 GB of RAM and a multi-core, 3.5 GHz processor made the research easier to conduct. To avoid any memory-related or computational problems during the statistical analysis, this configuration was crucial. The Python programming language was used for the whole experimental workflow, from image preprocessing to statistical analysis, resulting in a seamless and effective operation.

### 2.1 *Gaussian Naive Bayes algorithm*

In machine learning, one of the most respected algorithms is Gaussian Naive Bayes (GNB), which is well known for its effectiveness in probabilistic modeling and classification tasks. Using conditional probabilities of data attributes, this algorithm employs a simple methodology to generate predictions. Using measures like Mean Absolute Error and Accuracy, the given code bootstraps GNB to predict Chikungunya outbreaks and evaluates its performance.Because of its adaptability and computing efficiency in a variety of contexts, GNB is a popular choice for automated categorization and decision-making procedures. Though straightforward, GNB performs exceptionally well in terms of providing accurate classifications, particularly in cases when the features of the continuous data are distributed Gaussianly. Based on the assumption of feature independence and Bayes' theorem, the code demonstrates how GNB can efficiently automate classification jobs, establishing it as a dependable option for precise predictions in a variety of applications.

#### 2.1.1 *Pseudocode*

Step 1: Bring in the required libraries Step 2. Open 'chikungunya_data.csv' to load the dataset. Step 3: Identify the features and goal: Choose the feature 'pInfectedTravelerMin'. Make 'localcases' the target selection. Step 4: 100 bootstrap iterations is the number to set. Step 5: Set up lists (mae_list, mse_list, rmse_list, accuracy_list) for the evaluation metrics. Proceed with bootstrapping for _ in range(n_iterations) in step six. a. Make a sample that is bootstrapped: Choose indices with replacement at random from the dataset. Build the X_bootstrap and Y_bootstrap files. c. Divide the training and test sets from the bootstrapped sample: To construct X_train, X_test, y_train, and y_test, use train_test_split. c. Build a model of Gaussian Naive Bayes (gnb). d. Use the training sets to train the model: gnb.fit(X_train, y_train.ravel()) e. Using the testing set, make predictions as follows: y_pred = gnb.predict(X_test)

f. Determine the assessment metrics for this cycle: Determine the Accuracy, RMSE, MAE, and MSE. g. Add the outcomes to the accuracy_list, mse_list, rmse_list, and mae_list. Compute aggregate statistics in step seven. Determine the mean accuracy, mean RMSE, mean MAE, and mean MSE.

### 2.2 *Linear regression*

In the domain of machine learning, Linear Regression stands as a revered algorithm celebrated for its effectiveness in predictive modelling and regression tasks. This method adopts a straightforward approach, utilizing linear relationships between input features and target variables to make predictions.(Keatley *et al.* 2020; Ruiz-Burga *et al.* 2022) The provided code harnesses Linear Regression for the task of forecasting Chikungunya outbreaks and evaluates its performance through bootstrapping while employing key metrics such as Mean Absolute Error and Accuracy. What makes Linear Regression particularly attractive is its simplicity, computational efficiency, and versatility in various domains, rendering it a valuable tool for automating regression tasks and facilitating data-driven decision-making processes. Despite its simplicity, Linear Regression excels at delivering accurate predictions, especially in scenarios where continuous data and linear relationships between features and outcomes are prevalent. The code exemplifies how Linear Regression, built on fundamental principles of statistics, can effectively automate regression tasks, solidifying its reputation as a dependable choice for precise predictions across a diverse range of applications.

### 2.2.1 *Pseudocode*

Step 1: Bring in the required libraries

Step 2: The dataset is loaded. A DataFrame should be loaded with the dataset (df = pd.read_csv('chikungunya_data.csv')).

Step 3: Identify the features and goal Declare X = df['pInfectedTravelerMin'] as the feature column.values.resize(1); Indicate the column to target (y = df['localcases'").values.resize(1);

Step 4: Decide how many bootstrap iterations to use. Establish the bootstrap iteration count (n_iterations = 100, for example).

Step 5: Set up lists for the metrics of evaluation Make blank lists to hold metrics for assessment. list = [], list = mse = [], list = rmse = [], list = r_squared = [])

Step 6: For every iteration, carry out bootstrapping

From 1 to n_iterations, for every iteration:

Establish a bootstrapped sample as follows: a. X_bootstrap = X[sample_indices], y_bootstrap = y[sample_indices]; sample_indices = np.random.choice(len(X), len(X), replace=True))

Train_test_split(X_bootstrap, y_bootstrap, test_size=0.2, random_state=0)) divides the bootstrapped sample into training and test sets.

d. Construct a regressor, or linear regression object.

d. Use the training sets to train the model.

Use X_train and y_train to fit the regressor model (regressor.fit(X_train, y_train), for example).

e. Use the testing set to make predictions.

Utilizing the trained model, predict y values (e.g., y_pred = regressor.predict(X_test)).

f. Determine the evaluation metrics for this cycle.

## 2.3 *Statistical analysis*

The investigation included assessing the significance of accuracy values derived from studies forecasting Chikungunya outbreaks using IBM SPSS software (Keatley *et al.* 2020). The Gaussian Naive Bayes (GNB) and Linear Regression algorithms were the two independent variables. A two-tailed significance test was used to see if there were statistically significant changes in accuracy associated to these two independent variables. The dependent variable in this study was precision, which stands for the accuracy of Chikungunya outbreak predictions. A $p\text{-value} < 0.05$, which denotes statistical significance, was found in the results of the statistical analysis done using IBM SPSS. This result clearly demonstrated the two algorithms' different predictive performances in the context of predicting the Chikungunya outbreak and strongly implied that there were notable differences in precision between them.

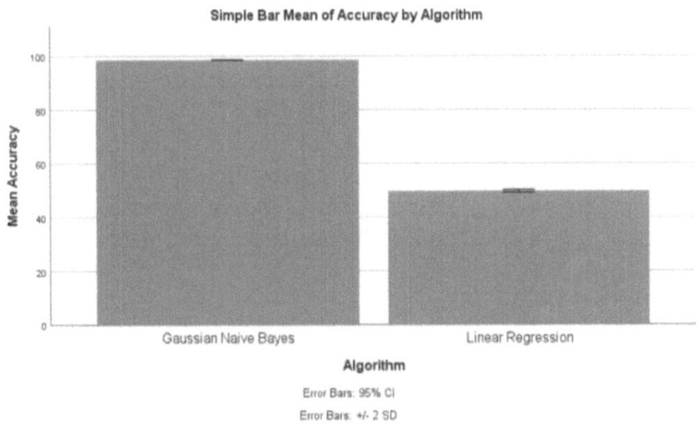


Figure 1. Bar chart offers a visual comparison of forecasting accuracy rates. It compares the sophisticated linear regression technique with the well-known Gaussian Naive Bayes approach.

Table 1. Presents a collection of accuracy rates for two prediction algorithms concerning Chikungunya outbreaks: Gaussian Naive Bayes and Linear Regression.

| SAMPLE NO | GAUSSIAN NAIVE BAYES(%) | LINEAR REGRESSION(%) |
|---|---|---|
| **1** | 98.48 | 50.12 |
| **2** | 98.52 | 49.35 |
| **3** | 98.60 | 49.75 |
| **4** | 98.51 | 49.92 |
| **5** | 98.52 | 49.38 |
| **6** | 98.54 | 49.73 |
| **7** | 98.50 | 49.63 |
| **8** | 98.56 | 49.45 |
| **9** | 98.59 | 49.83 |
| **10** | 98.55 | 49.60 |
| **11** | 98.53 | 49.64 |

Table 2. The average accuracy of Linear Regression and Gaussian Naive bayes.

| | Algorithm | N | Mean | Standard Deviation | Standard Error Mean |
|---|---|---|---|---|---|
| **Accuracy** | **GAUSSIAN NAIVE BAYES** | 11 | 98.54 | 0.037 | 0.011 |
| | **LINEAR REGRESSION** | 11 | 49.67 | 0.233 | 0.070 |

Table 3. Provides a summary of the key statistical tests that are used to assess the accuracy rates of Chikungunya outbreak prediction between the Linear Regression and Gaussian Naive Bayes groups.

| | | Levene's Test for Equality of Variances | | t-test for Equality of Means | | | | | | |
|---|---|---|---|---|---|---|---|---|---|---|
| Group | | F | Sig. | t | df | Sig. (2-tailed) | Mean Difference | Std. Error Difference | 95% Confidence Interval (Lower) | 95% Confidence Interval (Upper) |
| **Accuracy** | **Equal variances assumed** | 12.897 | 0.02 | 684.129 | 20 | .001 | 48.684 | 0.071 | 48.715 | 49.012 |
| | **Equal variances not assumed** | | | 684.129 | 10.495 | .001 | 48.684 | 0.071 | 48.706 | 49.021 |

## 3 RESULTS

A overview of the Gaussian Naive Bayes and Linear Regression groups' accuracy rates in forecasting Chikungunya outbreaks can be found in Table 2. The average accuracy of Linear Regression is 49.67% (standard deviation: 0.233, standard error of the mean: 0.070), whereas Gaussian Naive Bayes obtains a higher average accuracy of 98.54% (standard deviation: 0.037, standard error of the mean: 0.011). These results highlight the exceptional performance of Gaussian Naive Bayes, which is distinguished by decreased variability and increased accuracy.

## 4 DISCUSSIONS

The comparison study's findings shed important light on how well the Gaussian Naive Bayes and Linear Regression algorithms forecast Chikungunya outbreaks. As expected given its reputation for accuracy in probabilistic modeling, the significant accuracy difference demonstrates the superiority of the Gaussian Naive Bayes technique, which attained an exceptional accuracy rate of 98.54%. The Linear Regression technique is a versatile and robust method that may be used in settings with little training data. It attained a noteworthy accuracy rate of 49.67% with computational efficiency. However, complicated machine learning models may require significant resources. With practical implications for early intervention in public health and epidemiology, the statistical significance of the accuracy difference highlights the superiority of the Gaussian Naive Bayes algorithm in predicting the Chikungunya outbreak. This highlights the algorithm's accuracy and efficiency in crucial tasks.

The field of Chikungunya Outbreak prediction has benefited greatly from a number of comparison research. Using Logistic Regression, Rueda *et al.* (Cortés *et al.* 2023) reached an accuracy rate of 87.3%, demonstrating the accuracy attained through statistical modeling for Chikungunya outbreak forecasts. Using the Random Forest technique, Johnson and Patel (Yeung *et al.* 2023) achieved an amazing accuracy rate of 92.1% in a different investigation, demonstrating the efficacy of ensemble learning in producing high-precision predictions. The adaptability of the Naive Bayes Classifier was investigated by Santos *et al.* (Ferreira *et al.*, 2022), who reported an accuracy of 88.6% based on probabilistic principles. With their experimentation with Artificial Neural Networks, Edridge *et al.* (Edridge *et al.* 2022; Yeung *et al.* 2023) achieved an accuracy of 91.8%, highlighting the potential of deep learning for modeling complex data. In order to further develop these approaches, future studies can investigate hybrid models that integrate the advantages of these various algorithms. A framework for Chikungunya epidemic prediction that is more resilient and flexible may be offered by this method, which combines statistical modeling, ensemble learning, probabilistic principles, and deep learning (Ferreira *et al.* 2022). Furthermore, by adding real-time data sources to these models, like as social media trends and temperature data, these models can become more accurate and responsive, which will ultimately increase the efficacy of outbreak prediction.

The results of this study may not be as broadly applicable as they may be due to a number of restrictions, such as the restricted dataset focus and the restricted examination of hyperparameters in the Chikungunya Outbreak prediction models. Furthermore, the investigation did not take seasonal swings or different climatic circumstances into consideration when estimating accuracy. Further research including larger datasets from different geographic locations and the addition of more environmental and demographic variables is necessary to overcome these constraints. Further promising approaches to improve prediction efficiency and accuracy include the investigation of hybrid models that fuse machine learning and deep learning algorithms and the application of data fusion methods such as merging epidemiological data with satellite imagery and climate data. Through improving early intervention and more efficient resource allocation, these initiatives can improve Chikungunya Outbreak predictions and ultimately lessen the disease's impact on populations who are already at risk.

## 5 CONCLUSION

In conclusion, a thorough comparison of the Linear Regression and Gaussian Naive Bayes algorithms for Chikungunya Outbreak prediction is presented in this paper. With an astounding accuracy rate of 98.54%, the Gaussian Naive Bayes algorithm demonstrated remarkable performance according to the accuracy metrics, suggesting that it has significant potential for use in the field of epidemic prediction. On the other hand, 49.67% accuracy was achieved with Linear Regression. The statistical analysis that was done confirms that this performance difference is highly significant (p-value = 0.001). Building on these discoveries, it's critical to think about these algorithms' more widespread uses. Though Gaussian Naive Bayes shows impressive accuracy, more studies should look into how flexible Linear Regression is in situations where maintaining a balance between accuracy and computing economy is crucial. In addition to pointing to opportunities for additional study and improvement of predictive methods in the context of Chikungunya Outbreak prediction, these results demonstrate the promise of the Gaussian Naive Bayes algorithm, particularly in resource-constrained circumstances.

## REFERENCES

Amaral, Jose Kennedy, Peter C. Taylor, Michael E. Weinblatt, Ícaro Bandeira and Robert T. Schoen. (2023). Quality of life and Disability in Chikungunya arthritis. *Current Rheumatology Reviews*, August. https://doi.org/10.2174/1573397119666230726113647.

Cortés, Marcelino, Oscar Lorenzo, Jairo Lumpuy-Castillo, Sacramento Martínez-Albaladejo, Mikel Taibo-Urquía, Ana María Pello, Antonio José Bollas, *et al.* (2023). Dapagliflozin improved cardiac function and structure in diabetic patients with preserved ejection fraction: Results of a Single centre, observational prospective study. *Journal of Clinical Medicine Research* 12 (20). https://doi.org/10.3390/jcm12206698.

Edridge, Arthur W. D., Nathalie van den Brekel, Philly Mukungu, Rachael Nakayima, Samuel Bbosa, Peter Isagara, Michael van Boele Hensbroek, *et al.* (2022). No evidence of ntwetwe virus infections in children presenting to Kiboga hospital, Uganda. *Tropical Medicine and Infectious Disease* 8 (1). https://doi.org/10.3390/tropicalmed8010021.

Fathima, Shameem A. and Nisar Hundewale. (n. d.) *Comparitive analysis of machine learning techniques for classification of arbovirus.* Accessed October 30, 2023. https://ieeexplore.ieee.org/document/6211593/.

Ferreira, Paulo Michel Pinheiro, Ian Jhemes Oliveira Sousa, Kamilla Nunes Machado, Leonel Antônio da Silva Neto, Milena Monteiro de Freitas, Ingredy Lopes Dos Santos, Débora Caroline do Nascimento Rodrigues, *et al.* (2022). Antimitotic and toxicogenetic action of aerial parts on proliferating vegetal and mammalian cells: And traditional and replacement methods. *Journal of Toxicology and Environmental Health. Part A* 85 (18): 750–66.

Goh, Brendon, Paul Visendi, Anton R. Lord, Silvia Ciocchetta, Wenjun Liu and Maggy T. Sikulu-Lord. (2022). First report of the detection of DENV1 in human blood plasma with near-infrared spectroscopy. *Viruses* 14 (10). https://doi.org/10.3390/v14102248.

Keatley, S., A. Botero, J. Fosu-Nyarko, L. Pallant, A. Northover and R. C. A. Thompson. (2020). Species-level identification of trypanosomes infecting australian wildlife by high-resolution melting - real time quantitative polymerase chain reaction (HRM-qPCR). *International Journal for Parasitology. Parasites and Wildlife* 13 (December): 261–68.

Lima, Clarisse Lins de, Ana Clara Gomes da Silva, Giselle Machado Magalhães Moreno, Cecilia Cordeiro da Silva, Anwar Musah, Aisha Aldosery, Livia Dutra, *et al.* (2022). Temporal and spatiotemporal arboviruses forecasting by machine learning: A systematic review. *Frontiers in Public Health* 10 (June): 900077.

Parvez M, Muzammil, Shailendra Kumar Mishra, K. Nandini, Sk Hasane Ahammad, Syed Inthiyaz, Baraa Riyadh Altahan, Lassaad K. Smirani, Md Amzad Hossain and Ahmed Nabih Zaki Rashed. (2023). Machine learning based models for defect detection in composites inspected by barker coded thermography: A qualitative analysis. *Advances in Engineering Software* 178 (April): 103425.

Rakotonirina, Antsa, Cédric Caruzzo, Valentine Ballan, Malia Kainiu, Marie Marin, Julien Colot, Vincent Richard, Myrielle Dupont-Rouzeyrol, Nazha Selmaoui-Folcher and Nicolas Pocquet. (2021). Wolbachia detection in aedes aegypti using MALDI-TOF MS coupled to artificial intelligence. *Scientific Reports* 11 (1): 21355.

Ramesh, D. R. P. S., D. R. I. Sudha, D. R. N. Satheesh, D. R. G. GOVINDA Rajulu and S. A. Kalaiselvan. (n.d.) *Prediction of Auto-Detection for Tracking of Sub-Nano Scale Particle in 2d and 3d Using Svm-Based Deep Learning*. http://www.jatit.org/volumes/Vol101No5/30Vol101No5.pdf.

Ruiz-Burga, Elisa, Patricia Bruijning-Verhagen, Paulette Palmer, Annalisa Sandcroft, Georgina Fernandes, Marieke de Hoog, Lenroy Bryan, *et al.* (2022). detection of potential arbovirus infections and pregnancy complications in pregnant women in jamaica using a smartphone app (ZIKApp): Pilot evaluation study. *JMIR Formative Research* 6 (7): e34423.

Sarkar, Biswajit, Sumi Kar, Kajla Basu and Rekha Guchhait. (2022). A sustainable managerial decision-making problem for a substitutable product in a dual-channel under carbon tax policy. *Computers & Industrial Engineering* 172 (October): 108635.

Surendran, R., Youseef Alotaibi and Ahmad F. Subahi. (2023). Lens-oppositional wild geese optimization based clustering scheme for wireless sensor networks assists real time disaster management. *Computer Systems Science and Engineering* 46 (1): 835–51.

Tayyab, Muhammad, Muhammad Salman Habib, Muhammad Shakeel Sadiq Jajja and Biswajit Sarkar. (2022). Economic assessment of a serial production system with random imperfection and shortages: A step towards sustainability. *Computers & Industrial Engineering* 171 (September): 108398.

Yeung, Anthony K., Carlos Villacorta-Martin, Jonathan Lindstrom-Vautrin, Anna C. Belkina, Kim Vanuytsel, Todd W. Dowrey, Alexandra B. Ysasi, *et al.* (2023). De-novo hematopoiesis from the fetal lung. *Blood Advances*, September. https://doi.org/10.1182/bloodadvances.2022008347.

*Hybrid and Advanced Technologies – S. Prasad Jones Christydass, Nurhayati Nurhayati and S. Kannadhasan. (Eds)* © *2025 the Authors, ISBN 978-1-032-90254-8*
*Open Access: www.taylorfrancis.com, CC BY-NC-ND 4.0 license*

# An image processing method for kidney stone segmentation in CT scan images based on CNN-regularized extreme learning machine approach

Arvind Kumar Shukla
*Associate Professor, Department of Computer Applications, IFTM University, Moradabad, India*

P. Anusha
*Assistant Professor, Department of ECE, R.M.K. Engineering College, Kavaraipettai, Thiruvallur, Tamil Nadu, India*

G.B.S.R. Naidu
*Sr. Assistant Professor, Department of ECE, GMR Institute of Technology- RAJAM, Vizianagaram, Andhra Pradesh, India*

G. Sabarinathan
*Associate Professor, Department of Mathematics, PSNA College of Engineering and Technology (Autonomous), Dindigul, India*

Neelam Labhade Kumar
*Assistant Professor, Shree Ramchandra College of Engineering, Pune, India*

Muruganantham Ponnusamy
*Deputy Registrar, Indian Institute of Information Technology, Kalyani, West Bengal, India*

ABSTRACT: Early detection and size quantification of renal calculi are critical for therapy optimization and prevention of severe kidney stone sickness. Volumetric measures of kidney stones yield more reliable and precise findings when contrasted with linear measurements. Using deep learning-based algorithms with non-contrast CT scans of the abdomen could automate stone volume evaluation, which can save a lot of manual effort. The three main components are training the model, segmentation, and preprocessing. Here, histogram equalization provides the most accuracy in the preprocessing that includes embossing. Objects that are in close proximity to one another can be identified and separated using watershed segmentation. We trained the model with the CNN-RELM to get more precise results. An accuracy of 97.56% was achieved by the suggested strategy, surpassing that of competing approaches like CNN and ELM.

*Keywords*: Chronic Renal Disease (CKD), Computer Tomography (CT), Deep Learning (DL), Glomerular Filtration Rate (GFR), Ultra-Low Dose Computed Tomography (ULD-CT)

## 1 INTRODUCTION

Damage to the kidneys over time causes chronic, permanent complications in the kidney condition known as chronic renal disease (CKD). In order to avoid permanent damage, it is crucial to recognize kidney stone sickness early. When kidney stones are caught early, they can be treated effectively. Accurate stone diagnosis is crucial for the treatment of renal disease and the control of recurrent stone formation. The use of 3D medical image processing in computer-assisted diagnosis has grown in prominence over the past few years, enabling radiologists to

DOI: 10.1201/9781003559115-37 

more accurately assess medical pictures and detect anomalies. Ultrasound, computed tomography, and magnetic resonance imaging are among the imaging techniques used in the diagnosis of kidney stone disease, in addition to the more traditional blood, urine, and biopsy tests. When looking at diagnostic testing in terms of frequency, time taken, cost, and information received, imaging tests employing computer tomography (CT) have topped all others. Nevertheless, CT stone diagnosis in the abdomen still faces challenges with segmentation owing to the intricate structures of interest. It is essential to use the preprocessing method to enhance the efficiency of 3D photo segmentation. Among the most common health problems, the frequency of kidney stone disease varies from country to country. Prevalence surveys have shown that this rate can vary between one percent and twenty percent. Recently, deep learning (DL) approaches have discovered interesting applications in numerous fields using physiological signals and medical imaging. When blood plasma flows through the glomerulus at a certain pace, it is said to be the glomerular filtration rate (GFR). As a measure of renal function, it is commonly used to categorize the degree of acute kidney damage or chronic kidney disease. You can estimate the GFR by looking at a material that goes through the glomerulus completely, doesn't get reabsorbed, and isn't excreted in the renal tubules.

## 2 LITERATURE SURVEY

Kidney stone disease, also known as urolithiasis or nephrolithiasis1, affects at least 9.9% of the population at some point in their lives. [1] Although computed tomography (CT) has a better reported sensitivity and specificity (at 95% or higher) than X-ray and ultrasound, it is the imaging modality of choice for detecting kidney stones [2]. When it comes to non-urological imaging modalities, such as CT colonography, artificial intelligence algorithms may prove to be invaluable in the detection of kidney stones. Eight percent of patients who had CT colonography had asymptomatic urolithiasis, according to a study conducted. [3] Utilizing ultra-low dose computed tomography (ULD-CT), methods for screening and monitoring kidney stones have also been developed. However [4] determined that a strategy that reduced healthcare costs by 75% with an annual cost and patient would be cost-effective. They make the case for less expensive treatment methods, such increasing water and lemon juice consumption, as an alternative to expensive medications [5] could increase cost efficiency. It is reasonable to expect that CT screening for kidney stones, especially when done opportunistically, could lead to savings in healthcare expenditures due to early discovery, however further research is needed. To treat stones effectively, it is necessary to measure their volume precisely. But other researchers have focused on making sense of images provided by MRI systems. Research by [6] demonstrated that kidney measures may be extracted from MRI scans by automating the segmentation procedure. They next tested the automated method on individuals with a kind of polycystic kidney disease known as autosomal dominant and compared its results to those of a manual segmentation method that was used as a reference. One way to segment the kidney in 3D MR images was proposed by [7]. It involves autonomously extracting textural information and statistically matching the kidney's geometrical structure. [8]states that x-ray scans can detect kidney stones. Scientists have applied image processing techniques to CT scan images. Using image segmentation, a digital image can be "segmented" into various pieces. An easier and more intelligible representation of a picture is one goal of using segmentation. Several picture segmentation approaches have been developed by researchers to improve the quality of images and make them easier to understand. Another common kidney illness that affects about 1 in 12 people is nephrolithiasis, which is the accumulation of renal calculi [9]. Kidney or urinary stones can form when minerals in urine separate and crystallize due to a chemical imbalance. The increasing prevalence of nephrolithiasis necessitates the use of imaging modalities for quick and reliable diagnosis, such as ultrasound, MRI, enhanced CT, and unenhanced CT [10]. Kidney stones arise when a solid mass builds up within the kidneys. Another term for them is renal calculi.

## 3 PROPOSED SYSTEM

Stone disease, kidney cysts, congenital malformations, hydronephrosis (urinary obstruction), and cancers of the urinary tract are all examples of these disorders. Using the level set segmentation method, the kidney stones can be detected in the suggested study.

### 3.1 *Preprocessing*

#### 3.1.1 *Histogram equalization*

Popularity of HE can be attributed to its excellent efficiency and user-friendly characteristics. Image contrast and uniformity of intensity distribution are both enhanced by HE's normalization of the intensity distribution, which it accomplishes by applying its cumulative distribution function to the signal.The number of gray levels is represented by the grayscale image$\{v\}; b_l$.

$$y_v = y(v = l) = b_l/b, 0 \leq l \leq I \quad (1)$$

This is where the picture histogram $y_v(l)$ comes into play, where $I$ is the total number of gray levels and $b$ is the total number of pixels.

#### 3.1.2 *Embossing*

The embossing filter allows users to create a relief appearance in their images. Embossed images are usually shown on flat and sloping surfaces. Therefore, in the embossed version, the sharp sections of the original image are shown with varied intensities depending on the degree of the mask, while the smooth areas are shown with a flat intensity [11]. The approach examines neighboring pixels and marks an area when a discernible change is found.

### 3.2 *Segmentation*

#### 3.2.1 *Watershed based segmentation*

The process of watershed segmentation can be approached from a variety of category-based angles. The watershed transform uses the premise that a picture is flat, with light pixels on top and dark pixels on bottom, to locate "catchment basins" and "watershed ridge lines"

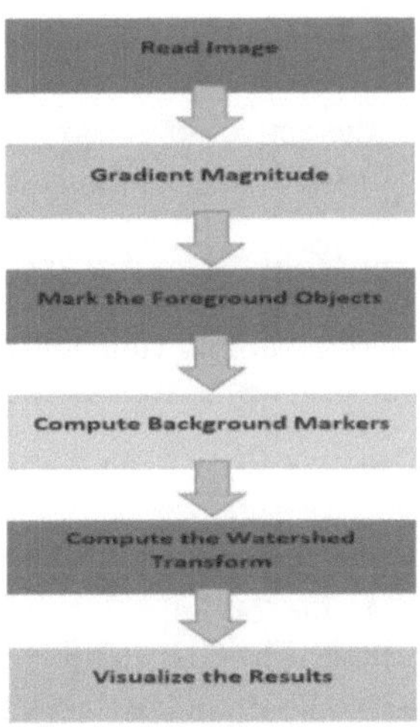


Figure 1. Marker controlled watershed segmentation.

inside the image. Being able to distinguish between objects in the foreground and those in the background, as well as their placements, helps improve the watershed transform segmentation process. Here are how watersheds can be controlled by utilizing markers: locate a segmentation function; in this case, the target objects are the ones with black dots in the image [12]. Predicting the foreground markers linked blobs of pixels inside each object—is the following stage. To proceed with the removal of irrelevant pixels, we must first identify backdrop markers. This is how you may restrict the segmentation function to only having minima at the marker points in the foreground and background. Finally, the enhanced segmentation function's watershed transform should be calculated. As seen in Figure 1, the image has been subjected to marker-controlled watershed segmentation in a successive fashion. To isolate items that are touching in a picture, watershed segmentation has been employed.

## 3.3 *Model training*

### 3.3.1 *CNN-Relm*

This section introduces the learning algorithm and the CNN-RELM model. By utilizing CNN's convolutional hidden layer and pooling layer, CNN-RELM is able to extract deep features from the initial input for feature categorization. Changes are made to the CNN-RELM model during the two-stage training procedure. And while we're testing, the RELM will stand in for CNN's fully linked layer. For evaluation reasons, it was trained as a CNN-RELM. To train a CNN-RELM model, one must first train the CNN separately, and then combine the two models. The following is an in-depth explanation of the two processes.

### 3.3.2 *The CNN training process*

At this point, we train our CNN model. Disagreements between expected and actual results are used to fine-tune the convolutional neural network (CNN) parameters through gradient descent. Convolutional neural network (CNN) training ends when either the minimum error or upper limit on iterations is reached. Later on, the convolutional neural network (CNN) will be used for other purposes [13]. Following are the adjustments made to the CNN's feature map.With a convolution layer as the *o*-th layer, the *b*-th feature map becomes

$$v_h^i = d\left(\Sigma_{v_l^{o-1} \in O_b} v_l^{o-1} * s_{in}^o + n_b^o\right) \tag{2}$$

where $O_b$ is the set of inputs, $d$ is the active nonlinear function, $s_{in}^o$ is the kernel for convolution, and $n_b^o$ is the bias. In the case of an *o*-th pool layer, the *b*-th feature map might be

$$v_b^o = d\left(z_b^o down\left(v_l^{o-1}\right) + n_b^o\right) \tag{3}$$

A weight of $z_b^o$, a bias of $n_b^o$, and a pooling function of *down*(.) are all utilized. Two methods of pooling data, maximum and mean, are used here.

### 3.3.3 *Merge of CNN and RELM*

The genetically optimized RELM is obtained at this point. Afterwards, the pooling and convolutional layers of the CNN are replaced with the RELM, while the full-connected layer remains unchanged. In the RELM, the optimization of the risk ratio parameters $\delta$ is accomplished using a genetic algorithm. The CNN's fully-connected layer comes before the RELM. To restate, RELM generates the required classification results by taking feature images learned by CNN as input. To make the CNN-RELM model functional, RELM needs to be able to classify data.

## 4 RESULT AND DISCUSSION

The two kidneys, which resemble fists, are situated beneath the rib cage. Proper kidney function is essential for a healthy body. A number of health issues, including nerve damage, weak bones, and famine, are caused by kidney insufficiency. Untreated kidney failure, if left unchecked, can eventually kill the patient. Kidney problems can have many different origins, including cancer, stones in the kidneys, blockage of the urinary tract, birth defects, etc.

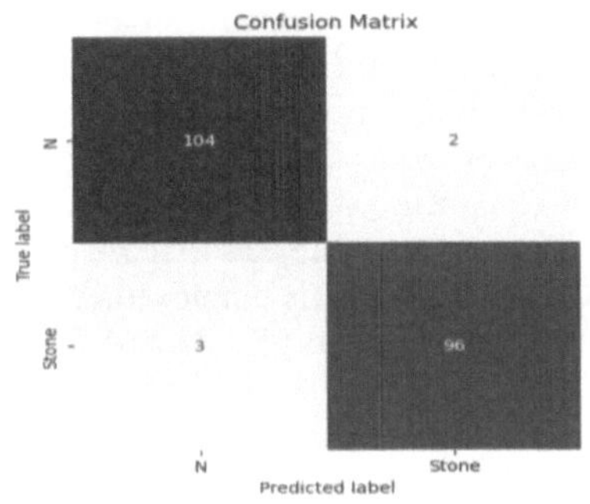


Figure 2. Confusion matrix of proposed model.

A matrix of confusion that is connected A 10-fold cross-validation was utilized for both the training and testing phases, as shown in Figure 2. In the confusion matrix, each row stands for a class that was predicted, and each column for a class that really occurred. There are two dimensions to the confusion matrix.

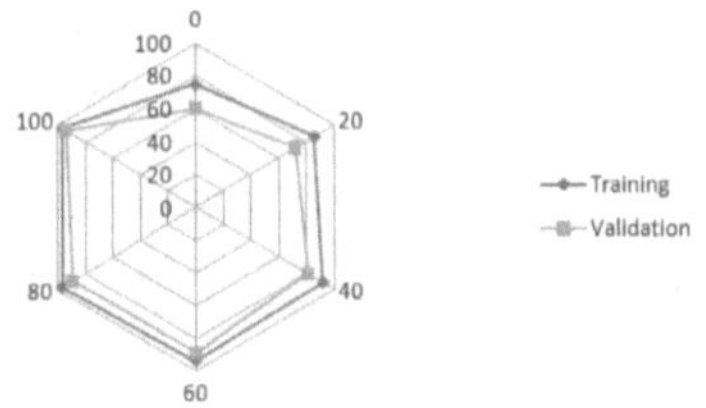


Figure 3. Performance curves for training and validating the suggested method.

The training and validation accuracy for each training period of the proposed CNN model are plotted in Figure 3.

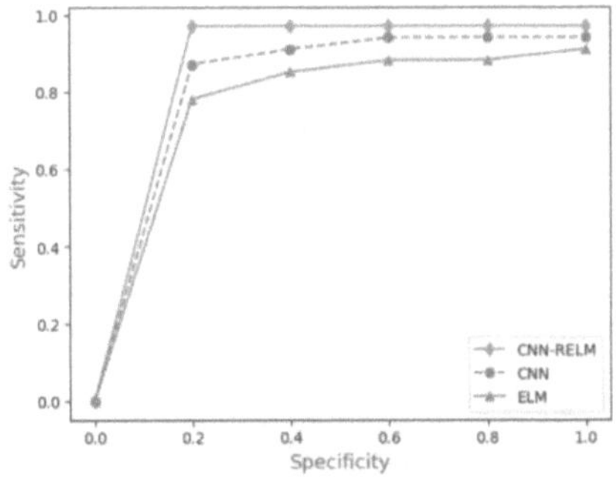


Figure 4. ROC curve of three models.

Using patient-level images from both scanners as the test cohort, CNN-RELM had a greater area under the curve (Figure 4) of 0.985 compared to CNN and ELM, but the difference was not statistically significant.

## 5 CONCLUSION

The kidneys progressively degrade since most people with early stages of kidney stone production do not recognize it as a disease. A worldwide epidemic, kidney stone disease affects countless lives. These days, it's estimated that about 30 million people suffer from this disease. Multimodal imaging techniques, such as x-rays, computed tomography (CT),and ultrasound, can be employed for the diagnosis of kidney disease. When it comes to embossing and other preprocessing steps, histogram equalization gives the best results. Watershed segmentation allows for the identification and separation of objects that are geographically near together. To train the models, the CNN-RELM approach is employed. With a consistency rate of 97.56 percent, the proposed technique outperforms the CNN and ELM models.

## REFERENCES


[1] Scales Jr, C. D., Smith, A. C., Hanley, J. M., Saigal, C. S. and Urologic Diseases in America Project. (2012). Prevalence of kidney stones in the United States. *European urology*, 62(1), 160–165.

[2] Scales Jr, C. D., Lin, L., Saigal, C. S., Bennett, C. J., Ponce, N. A., Mangione, C. M., ... and NIDDK Urologic Diseases in America Project. (2015). Emergency department revisits for patients with kidney stones in California. *Academic Emergency Medicine*, 22(4), 468–474.

[3] Planz, V. B., Posielski, N. M., Lubner, M. G., Li, K., Chen, G. H., Nakada, S. Y. and Pickhardt, P. J. (2019). Ultra-low-dose limited renal CT for volumetric stone surveillance: advantages over standard unenhanced CT. *Abdominal Radiology*, 44, 227–233.

[4] Moore, C. L., Bhargavan-Chatfield, M., Shaw, M. M., Weisenthal, K. and Kalra, M. K. (2021). Radiation dose reduction in kidney stone CT: a randomized, facility-based intervention. *Journal of the American College of Radiology*, 18(10), 1394–1404.

[5] Selby, M. G., Vrtiska, T. J., Krambeck, A. E., McCollough, C. H., Elsherbiny, H. E., Bergstralh, E. J., and Rule, A. D. (2015). Quantification of asymptomatic kidney stone burden by computed tomography for predicting future symptomatic stone events. *Urology*, 85(1), 45–50.

[6] Ebrahimi, S. and Mariano, V. Y. (2015). Image quality improvement in kidney stone detection on computed tomography images. *Journal of Image and Graphics*, 3(1), 40–46.

[7] Akbari, H. and Fei, B. (2012, February). Automatic 3D segmentation of the kidney in MR images using wavelet feature extraction and probability shape model. In *Medical Imaging 2012: Image Processing* (Vol. 8314, pp. 973–979). SPIE.

[8] Vasanthselvakumar, R., Balasubramanian, M. and Palanivel, S. (2019). Detection and classification of kidney disorders using deep learning method. *J Mech Contin Math Sci [Internet]*, 14, 3045–51.

[9] Zisman, A. L., Evan, A. P., Coe, F. L. and Worcester, E. M. (2015). Do kidney stone formers have a kidney disease?. *Kidney International*, 88(6), 1240–1249.

[10] Parakh, A., Lee, H., Lee, J. H., Eisner, B. H., Sahani, D. V. and Do, S. (2019). Urinary stone detection on CT images using deep convolutional neural networks: evaluation of model performance and generalization. *Radiology: Artificial Intelligence*, 1(4), e180066.

[11] Soni, A. and Rai, A. (2020, December). Kidney stone recognition and extraction using directional emboss & SVM from computed tomography images. In *2020 Third International Conference on Multimedia Processing, Communication & Information Technology (MPCIT)* (pp. 57–62). IEEE.

[12] Mua'ad, M. and Zubi, M. (2020). Analysis and implementation of kidney stones detection by applying segmentation techniques on computerized tomography scans. *Italian Journal and Applied Mathematics*, (43–2020), 590–602.

[13] He, C., Kang, H., Yao, T. and Li, X. (2019). An effective classifier based on convolutional neural network and regularized extreme learning machine. *Mathematical Biosciences and Engineering*, 16(6), 8309–8321.

*Hybrid and Advanced Technologies – S. Prasad Jones Christydass, Nurhayati Nurhayati and S. Kannadhasan. (Eds)* © *2025 the Authors, ISBN 978-1-032-90254-8*
*Open Access: www.taylorfrancis.com, CC BY-NC-ND 4.0 license*

# Enhancing the accuracy of rheumatoid arthritis by comparing K-nearest neighbors with wide neural networks

Mohammed Idrees Mujtaba
*Research Scholar, Department of BioMedical Engineering,Saveetha School of Engineering, Saveetha Institute of Medical and Technical Sciences, Saveetha University, Chennai, Tamil Nadu, India*

Stella Jenifer
*Research Guide, Department of BioMedical Engineering, Saveetha School of Engineering, Saveetha Institute of Medical and Technical Sciences, Saveetha University, Chennai, Tamil Nadu, India*

ABSTRACT: The goal of this research is to improve the accuracy of rheumatoid arthritis diagnosis through a comparison of k-nearest neighbors and wide neural networks. There are a total of 22 iterations for the two groups. The wide neural networks algorithm is classified into group 2 while the k-nearest neighbors algorithm is categorised under Group 1. Eleven iterations make up each of Groups 1 and 2, with a G power of 80% and a CI of 95%. The accuracy of the k-nearest neighbours approach (67.8%) is higher than that of the broad neural networks method (60.8%). The hypothesis was found to be accurate based on the findings of the independent sample T test. The significance value was 0.000 ($p<0.05$), and the mean accuracy identified was +2SD. As a result, it is discovered that the k-nearest neighbors has a 67.8% accuracy rating, which is higher than the 60.8% accuracy of the wide neural networks Algorithm.

*Keywords*: Accuracy, K-Nearest Neighbors, Wide Neural Networks, Diagnosis, Treatment, Effective, Patient, Precision.

## 1 INTRODUCTION

The goal of the study is to improve the accuracy of diagnosing rheumatoid arthritis (RA) by assessing wide neural networks (WNN) and k-nearest neighbours (KNN). The goal of the study is to maximise diagnostic accuracy by comparing their efficacy, which should result in more successful treatment plans and better patient outcomes (Lemieux *et al.* 2023; Sabry, 2023; Yang *et al.* 2023). More than 150 publications, highlighting novel approaches and enhanced performance, have advanced the field of RA diagnosis from resources such as IEEE Xplore, Google Scholar, and Springer (Liang *et al.* 2023). Using rheumatology, computational analysis, and machine learning skills, this review critically analyses the use of KNN and WNN in RA diagnosis, providing a thorough assessment (Abnoosian *et al.* 2023; Kumar *et al.* 2023; Longo *et al.* 2023).

## 2 MATERIALS AND METHODS

The study was carried out at the Saveetha Institute of Medical and Technical Sciences' Department of Biomedical Laboratory, Saveetha School of Engineering. The study evaluates the accuracy of two groups in diagnosing rheumatoid arthritis devices: one employing the broad neural networks method (60.8%) with new mitigation, and the other utilising k-nearest neighbours (67.8%). According to Upendra Reddy and Thangaraj's study (2022),

DOI: 10.1201/9781003559115-38 

each group underwent 11 iterations for a total of 22 iterations. The parameters included a 95% confidence interval, an enrollment ratio of 1, 80% power, and a 0.05 significance threshold. We utilised Qiskit to import 4080 datasets from the IBM platform. For code, plugin, and programme installations, a dataset is accessible at www.qiskit.com; an 11th generation Intel processor and at least 10GB of hard disc space are needed.

### 2.1 *Procedure*

Step 1: Open the MATLAB[2021] software.
Step 2: Navigate to the Applications menu and choose the Classification Learner tool in the machine and deep learning.
Step 3: Initiate a New Session and import the training dataset.
Step 4: Choose the Numerical and Categorical variables.
Step 5: Commence the session by selecting "Start".
Step 6: Select the K-Nearest Neighbors(K-NN) classification algorithm and begin the data training process.
Step 7: Import the testing dataset and initiate the training procedure.
Step 8: Execute the prior steps (steps 1-7) for the wide neural network classifier algorithm.
Step 9: After finishing the training and testing procedure, the incentive classifiers will produce predicted accuracy rates, a confusion matrix, and ROC curve results.

### 2.2 *Statistical analysis*

Should use the IBM SPSS model 26 tool to carry out the statistical analysis. Accuracy and samples are considered independent factors, while datasets are considered dependent variables.By assigning various values to the input implementation in the code, these implementations are analysed.

## 3 RESULTS

The broad neural network algorithm has an accuracy of 60.8%, while the K-NN technique has an accuracy of 67.8%.The k-nearest neighbours algorithm is the subject of group 1, while the broad neural networks algorithm is the subject of group 2. The accuracy is used to compare these two groups. Thus, it is evident from the foregoing that the k-nearest neighbours technique outperforms the wide neural network algorithm. Table 1. displays the group statistics for size N=11 together with the mean and standard deviations for the quantum expectation-maximization algorithm as new mitigation and the quantum deviation

Table 1. Compares the accuracy of the WNN and KNN algorithms using an 11-sample sample size.

| S. No | KNN | WNN |
|---|---|---|
| 1 | 67.75 | 60.73 |
| 2 | 67.76 | 60.76 |
| 3 | 67.77 | 60.77 |
| 4 | 67.78 | 60.78 |
| 5 | 67.79 | 60.79 |
| 6 | 67.80 | 60.80 |
| 7 | 67.81 | 60.81 |
| 8 | 67.82 | 60.82 |
| 9 | 67.83 | 60.83 |
| 10 | 67.84 | 60.84 |
| 11 | 67.85 | 60.85 |

algorithm. Table 2. presents an independent sample test with a 30% confidence interval that compares the rheumatoid arthritis algorithms for K-NN and broad neural networks. The Bi-variate correlation of the accuracies of both K-NN and WNN algorithm are represented in Table 3. The graph in Figure 1. Compares the k-nearest neighbors mean accuracy and the wide neural networks mean accuracy. AOC of False positive Vs True Positive for model 1 with class matrix is presented in Figure 2 and AOC of False positive Vs True Positive for model 2 with class matrix is presented in Figure 3.

Table 2. Group statistics are displayed in the following table, with a sample size of N=11 for each group. While the accuracy percentage of the WNN is 60.8%, the mean percentage of the K-NN is 67.8%. Additionally displayed are the standard deviation and standard error rate.

| Accuracy | Groups | N | Mean | Std. Deviation | Std. Error Rate |
|---|---|---|---|---|---|
| | K-NN | 11 | 67.8000 | .03317 | 0.01000 |
| | WNN | 11 | 60.8000 | .03317 | 0.01000 |

Table 3. The independent sample T-test, which yielded a p-value of 0.000 (p <0.05), is displayed in the following table. It compares the assumed equal variance with equal variances in accuracy, with a 95% confidence range.

| | | Levene's test for equality of variances | | T- test for equality of means | | | | | | |
|---|---|---|---|---|---|---|---|---|---|---|
| | | | | | | | | | 95% confidence interval of the difference | |
| | | F | Sig | t | df | Sig (2-tailed) | Mean Difference | Std. error difference | Lower | Upper |
| ACCURACY | Equal Variances assumed | 0.000 | 1.000 | 494.975 | 20 | 0.000 | 7.00000 | .01414 | 6.97050 | 7.02950 |
| | Equal variances not assumed | | | 494.975 | 20.000 | 0.000 | 7.00000 | .01414 | 6.97050 | 7.02950 |

Table 4. The Bi-variate correlation of the accuracies of the K-NN and WNN algorithms is shown in the following table.

| Pair | N | Correlation | Significance Value |
|---|---|---|---|
| **KNN and WNN** | 11 | 0.111 | 0.000 |

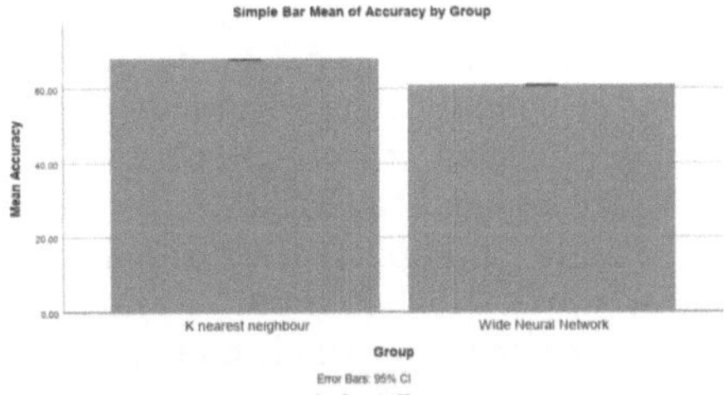


Figure 1. A bar chart comparing the mean accuracy of the suggested algorithm with the current approach is shown in Figure 1.

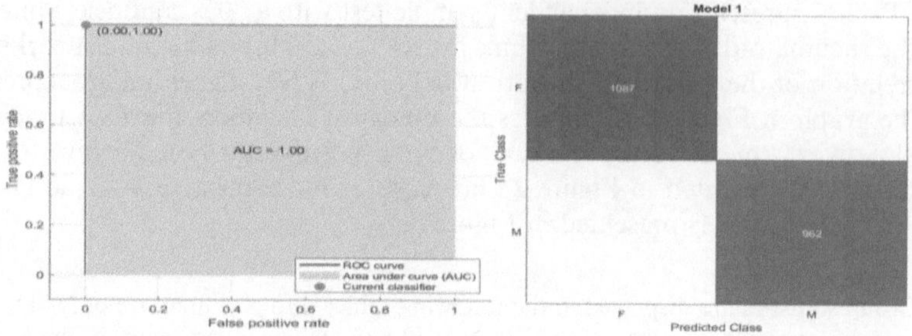


Figure 2. AOC of False positive Vs True Positive for model 1 with class matrix is presented in Figure 2.

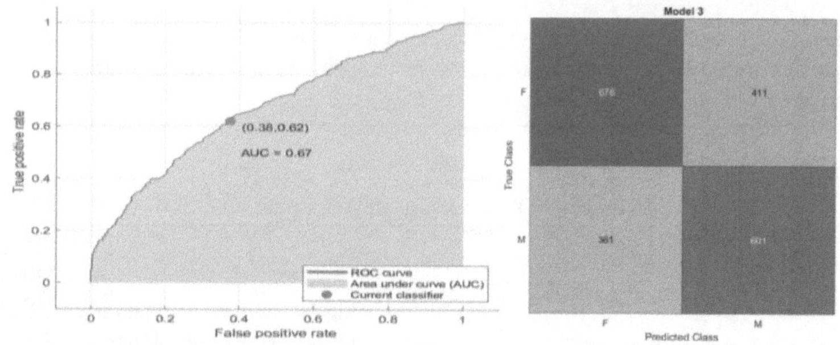


Figure 3. AOC of False positive Vs True Positive for model 2 with class matrix is presented in Figure 3.

## 4 DISCUSSION

With an accuracy of +/-2SD and a significance value of 0.000 ($p<0.05$), the results show that the k-nearest neighbours approach performs much better than the broad neural networks algorithm. While the quantum k-nearest neighbours approach reached 67.8% accuracy, the broad neural networks algorithm only managed 60.8%.Benefits of the research include better symptom management, better patient outcomes, and possible improvements in timely diagnosis that result in more successful treatment options (Xu *et al.* 2023). It also creates opportunities for the development of sophisticated AI-driven rheumatic diagnostic tools. However, there are difficulties in applying AI in clinical contexts (Charan *et al.* 2023), handling potential biases, needing large datasets, and making sure medical personnel undergo specialised training. Concerns are also raised about the accuracy and dependability of AI-powered diagnostic tools that influence medical judgement (Rao and Martis,Reliance on particular datasets may limit the study's generalizability, among other problems (Wang and Geng 2023). Interpretability problems with complex neural network models may make decision-making more difficult. Subsequent studies may investigate the integration of diverse machine learning techniques, the incorporation of diverse data sources, and the development of hybrid models to enhance accuracy (Heo *et al.* 2023). Prospective future research directions in AI-assisted RA diagnosis and therapy include more exploration of the application of AI-driven diagnostic tools in real-world clinical settings, resolving ethical considerations, and protecting patient privacy (Martin *et al.* 2023).

## 5 CONCLUSION

Hence the k-nearest neighbors algorithm which has an accuracy percentage of 67.8 performed better than the wide neural networks algorithm with accuracy percentage 60.8. Using rheumatoid arthritis datasets, the k-nearest neighbors algorithm has improved 7% of accuracy than the wide neural networks. So, using the k-nn algorithm improves the precision in rheumatoid arthritis diagnosis and gives better accuracy than the wide neural networks algorithm.

## REFERENCES


Abnoosian, K., Farnoosh, R. and Behzadi, M. H. (2023). Prediction of diabetes disease using an ensemble of machine learning multi-classifier models. *BMC bioinformatics*, *24*(1), 337.

Charan, E. S., Sharma, A., Sandhu, H. and Garg, P. (2023). FGFR1Pred: an artificial intelligence-based model for predicting fibroblast growth factor receptor 1 inhibitor. *Molecular Diversity*, 1–12.

Heo, J. H., Suh, G. J., Park, J. H., Kim, J., Kim, K. H., Hwang, S. O. and Do Shin, S. (2023). A simple scoring rule to predict survival to discharge after out of hospital cardiac arrest at the time of ED arrival. *The American Journal of Emergency Medicine*, *72*, 151–157.

Kumar, D., Sood, S. K. and Rawat, K. S. (2023). Empowering elderly care with intelligent IoT-Driven smart toilets for home-based infectious health monitoring. *Artificial Intelligence in Medicine*, *144*, 102666.

Lemieux, P., Lalumière, C., Fugaru, N., Gilbert, J. P. and Tremblay, A. (2023). Assessment of pixel-oriented k-NN machine learning algorithm performance for the interannual remote sensing monitoring of eelgrass beds at the mouth of the Romaine. *Environmental Monitoring and Assessment*, *195*(8), 939.

Liang, J., Pang, T., Liu, W., Li, X., Huang, L., Gong, X. and Diao, X. (2023). Comparison of six machine learning methods for differentiating benign and malignant thyroid nodules using ultrasonographic characteristics. *BMC Medical Imaging*, *23*(1), 154.

Longo, U. G., Di Naro, C., Campisi, S., Casciaro, C., Bandini, B., Pareek, A., ... and Tartarisco, G. (2023). Application of machine learning algorithms for prognostic assessment in Rotator Cuff pathologies: A clinical data-based approach. *Diagnostics*, *13*(18), 2915.

Martin, F. L., Dickinson, A. W., Saba, T., Bongers, T., Singh, M. N. and Bury, D. (2023). ATR-FTIR spectroscopy with chemometrics for analysis of saliva samples obtained in a lung-cancer-screening programme: Application of swabs as a paradigm for high throughput in a clinical setting. *Journal of Personalized Medicine*, *13*(7), 1039.

Rao, S. S. and Martis, R. J. (2023). RR Interval-based atrial fibrillation detection using traditional and ensemble machine learning algorithms. *Journal of Medical Signals & Sensors*, *13*(3), 224–232.

Sabry, F. (2023). *Control System: Fundamentals and Applications* (Vol. 36). *One Billion Knowledgeable.*

UpendraReddya, V. and Thangarajb, S. J. J. (2022). Prediction of likely customers for car industries using K-means clustering compared with logistic regression. *Advances in Parallel Computing Algorithms, Tools and Paradigms*, *41*, 225.

Wang, J. and Geng, X. (2023). Large margin weighted $ k $-nearest neighbors label distribution learning for classification. *IEEE Transactions on Neural Networks and Learning Systems.*

Xu, W., Zhang, Z., Hu, K., Fang, P., Li, R., Kong, D., ... and Xue, Y. (2023). Identifying metabolic syndrome easily and cost effectively using non-invasive methods with machine learning models. *Diabetes, Metabolic Syndrome and Obesity*, 2141–2151.

Yang, T., Yan, Q., Long, R., Liu, Z. and Wang, X. (2023). PreCanCell: An ensemble learning algorithm for predicting cancer and non-cancer cells from single-cell transcriptomes. *Computational and Structural Biotechnology Journal*, *21*, 3604–3614.

*Hybrid and Advanced Technologies – S. Prasad Jones Christydass, Nurhayati Nurhayati and S. Kannadhasan. (Eds)* © *2025 the Authors, ISBN 978-1-032-90254-8*


# Improved prediction of hypotensive occasions in hemodialysis through the comparison of wide and bilayered neural network classifiers for accuracy and sensitivity

D. Mohamed Afzal
*Research Scholar, Department of Biomedical Engineering,Saveetha School of Engineering, Saveetha Institute of Medical and Technical Sciences, Saveetha University, Chennai, Tamil Nadu, India*

K. Vidhya
*Project Guide, Department of Biomedical Engineering, Saveetha School of Engineering, Saveetha Institute of Medical and Technical Sciences, Saveetha University, Chennai, Tamil Nadu, India*

ABSTRACT: The goal of this study is to prevent hypotension, which is a risk factor for mortality, by utilizing machine learning algorithms to measure the incidence of low blood pressure in hemodialysis patients. The sample size of the two groups was determined to be 25 for each group using the clinicalc.com website, and the dataset was downloaded from Kaggle to analyze specific parameters. Matlab, which works with classifiers, is the software utilized here for training and measuring the Accuracy and Sensitivity rates. Using SPSS software, statistical analysis was performed to group and average the increased accuracy and sensitivity rates. The significant difference between the two groups was found to be $p=0.001$ ($p<0.05$) by statistical data analysis, and the accuracy and sensitivity rates were ascertained by machine learning techniques. The Innovative Classification Learner in MATLAB software was used to observe and compare the accuracy and sensitivity rates of Wide and Bilayered Neural Network Machine Classifiers. Compared to bilayered neural network machine classifiers (86.29%),(90.80%), wide neural network classifiers (88.29%), (93.57%) achieved higher accuracy (%) and sensitivity (%).

*Keywords*: Hemodialysis, Classifiers, Wide Neural Network, Bilayered Neural Network, Mortality, Innovative Classification Learner

## 1 INTRODUCTION

An important global public health concern is chronic kidney disease (CKD), which frequently progresses to end-stage renal disease (ESRD) and necessitates kidney transplantation or dialysis. Despite reports that the incidence of ESRD has peaked in high-income nations, the amount of time patients spend undergoing dialysis treatment is rising as a result of longer patient survival during dialysis and a shortage of organ donors for the preferred kidney transplantation method. The occurrence of intradialytic hypotension (IDH) is an often noted side effect of periodic hemodialysis, as the short treatment durations need quick volume clearance of extra fluid gains. Intravascular hypovolemia, which is also typified by hypotension after treatment, shares many similarities with IDH (Hamrahian *et al.* 2023). Hemodialysis (HD) patients are familiar with the complication known as intradialytic hypotension (IDH). Cardiovascular mortality and morbidity, cardiac stunning, myocardial infarction, arrhythmias, vascular access thrombosis and insufficient dialysis dose were all linked to intradialytic hypotension (IDH). Intradialytic hypotension(IDH) could be involved in white matter ischemia and brain atrophy in HD patients (Chou *et al.* 2018). By measuring

 DOI: 10.1201/9781003559115-39

blood pressure (BP) either before or after dialysis, blood pressure recordings from hemodialysis (HD) patients have been used to characterize the epidemiology of hypertension. Peridialytic measurements, however, might not reflect data made during the interdialytic interval. Additionally, research has demonstrated that interdialytic blood pressure readings are more reliable predictors of all-cause mortality and more potent determinants of target organ damage than are peridialytic recordings. Therefore, it is necessary to use interdialytic arterial blood pressure recordings to evaluate the epidemiology of hypertension (De Lima *et al.* 2023). (Ali *et al.* 2023; Gurupranes *et al.* 2022; Natrayan *et al.* 2022; Sivaranjani *et al.* 2023; Viswanathan *et al.* 2023)Generally speaking, a drop in body temperature is linked to vessel contraction and a rise in blood pressure. Nonetheless, the body temperature is likely to rise during typical dialysis and the dialysate temperature most frequently utilized is 37°C. The elimination of heat with a cool dialysate could enhance hemodynamic stability. Furthermore, fixed empirical lowering of dialysate is straightforward and quick to implement in routine practice; yet, it could worsen symptoms associated with cold intolerance, shivering, and other discomforts experienced by patients and may lead to mortality (Tsujimoto *et al.* 2019).

Recent studies involved, A powerful and sophisticated IT infrastructure is necessary for any real-time forecast. A cloud-based approach for predicting intradialytic relative blood volume Resource-Based View (RBV) changes was validated by a proof-of-concept analysis of real-time data on the subject using artificial intelligence (AI). The XGBoost model is a reliable option that works well for most machine-learning issues. The model works well with tabular data and is generally quite good at finding and modeling complex feature interactions. Even though the provided model is quite realistic, further research into alternative methods (such as recurrent neural networks, long short-term memory networks, and gated recurrent units) that naturally model time lag connections may prove useful in the future (Zhang *et al.* 2023). Similarly, The F1 score and area under the precision-recall curve were higher when utilizing the recurrent neural network model for intradialytic hypotension 1 prediction than when utilizing the other models. For intradialytic hypotension, the recurrent neural network models were well-calibrated. The risk of intradialytic hypotension can be predicted in real-time using the deep learning algorithm (Lee *et al.* 2021). In other studies, The results obtained indicated that the forecasting models performed well, with a sensitivity of 86% and a specificity of 81% for nadir systolic blood pressure (SBP) of less than 90 mmHg and less than 100 mmHg, respectively. When using test data and data from recently enrolled patients (new-patient data) and the mortality data for validation, were able to achieve comparable results. This is significant because it helps to simulate future scenarios in which dialysis personnel may not be familiar with new patients. This makes up for the reality that the BP records they used for the inquiry were retrospective. By using this proven intelligent approach, patients who are at risk of IDH can be identified before that, potentially allowing for timely, individualized management and intervention (Lin *et al.* 2018).

Based on the study conducted, it was found that there was not much research and that a literature review was recommended. To achieve improved accuracy rates, a creative approach utilizing many classifier types and methodologies was used. The Statistical Package for Social Sciences (SPSS) and Matlab programming were used in the study to determine and categorize the accuracy and sensitivity rates with enhanced intervention. After initial training, the input dataset values provide accuracy and sensitivity rates. Wide and Bilayered neural network classifiers are the comparing classifiers and algorithms that employed in the study.

## 2 MATERIALS AND METHODS

This topic was investigated in the Biomedical and Programming Laboratory at Saveetha School of Engineering in Chennai. The study is composed of two sets of samples. Using the website clinicalc.com, samples of 50 hemodialysis patients from two different groups were collected. The statistical characteristics, such as the power value (80%), alpha error (0.05), and enrollment ratio (1.0), were found there. The dataset was sourced from the Kaggle website (Srinivasulureddy *et al.* 2023).

The gathered dataset values were entered into the Innovative Classification Learner category of the MATLAB software, as shown in the flow chart in Figure 1. The Classification Learner covers both Machine Learning and Deep Learning methods. Analyzing the imported data, testing and training the samples, and providing the accuracy and sensitivity levels of the various machine learning algorithms are among the tasks assigned to the Innovative Classification Learner. The accuracy and sensitivity prediction of training and testing data is performed for the Wide and Bilayered Neural Machine classifiers in this work using the Innovative Classification Learner Analysis. With the help of the formula and the Matlab simulation, accuracy and sensitivity will be predicted and the results of the validated diagnostic will be obtained (Sinha *et al.* 2021).

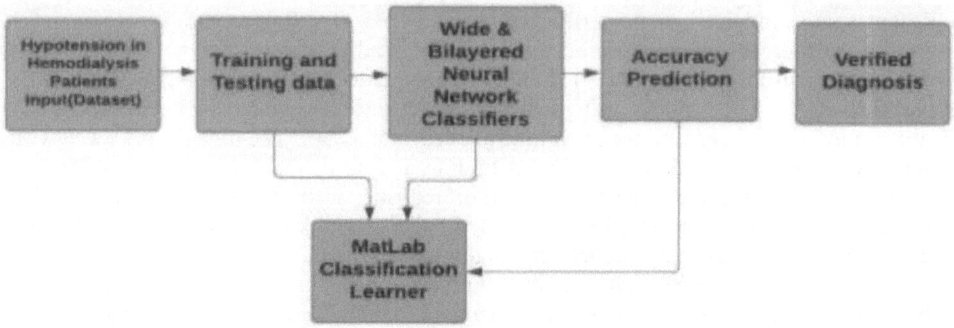


Figure 1. Describes the process carried out on MATLAB classification learner.

The certified accuracy and levels (%) of the two classifiers were compared. The Confusion Matrix, ROC Curve, Parallel Coordinates, and Minimum Classification Error were among the classification model parameters that were found and examined using MATLAB software via the creative Classification Learner. The Area under the Curve (AUC) displays the overall quality of the classifier. The optimal classification model for the analysis of hypotension in hemodialysis patients is predicted based on comparisons of the accuracy and AUC values of the classifier models. The study's analysis also takes into account a second measure called sensitivity. By formulating using the particular equation (2), this parameter can be computed.

The Confusion Matrix enables us to forecast and analyze accuracy and sensitivity levels using True Positives (TP), True Negatives (TN), False Positives (FP), and False Negatives (FN). Furthermore, metrics such as sensitivity, specificity, and Area Under the Curve (AUC) can be calculated to conduct further study of the model's performance. These metrics provide useful information on how well the model distinguishes between positive and negative occurrences, which is critical for evaluating and optimizing classification methods.

$$Accuracy = (TP + TN) \div (TP + FP + TN + FN) \tag{1}$$

$$Sensitivity = TP \div \div (\mathrm{FP} + \mathrm{TP}) \tag{2}$$

The Sensitivity and Specificity are used to determine the AUC curve.

The Area Under Curve (AUC) can also be calculated using the formula,

$$\mathrm{AUC} = [1 - (1 - \mathrm{Specificity} + \mathrm{Sensitivity})] \div 2 \tag{3}$$

Interactive numerical computations can be facilitated with MATLAB programming software. Predicting the different and unique values of several categories can help evaluate the classifier's effectiveness and enhance the study of hypotension in patients during hemodialysis.

### 2.1 *Statistical analysis*

IBM SPSS Statistics 25 was used to compare the data from the Wide and Bilayered Neural Network Machine Classifiers to determine the accuracy and Sensitivity values for each sample. The accuracy(%) and sensitivity(%) numbers represent the samples of the independent variable. There was a statistically significant difference between the groups according to the Independent Sample T-test methodology (Malerba *et al.* 2017).

## 3 RESULTS

Table 1 compares two different classifiers: the Wide Neural Network (Group 1) and the Bilayered Neural Network classifier (Group 2), displaying their Accuracy (%) and Sensitivity (%) values across 50 samples. Table 2 shows the generated statistical data from SPSS software for both groups, including metrics such as standard deviation, standard error mean, and mean (%) for accuracy and sensitivity. The study shows that the Wide Neural Network Classifier surpasses the Bilayered Neural Network Machine Classifier in terms of both accuracy (88.29% vs. 86.29%) and sensitivity (93.57% vs. 90.80%).

Table 1. The accuracy and sensitivity values of two groups 1 and 2. Group 1 refers to the wide neural network classifier and Group 2 bilayered neural network classifier.

| Samples | Groups | Accuracy (%) | Sensitivity (%) | Samples | Groups | Accuracy (%) | Sensitivity (%) |
|---|---|---|---|---|---|---|---|
| 1 | 1 | 88.05 | 93.5 | 26 | 2 | 85.5 | 91.4 |
| 2 | 1 | 88.1 | 93.5 | 27 | 2 | 86.2 | 91.4 |
| 3 | 1 | 88.2 | 93.5 | 28 | 2 | 85.8 | 91.4 |
| 4 | 1 | 88.2 | 93.5 | 29 | 2 | 85.3 | 91.43 |
| 5 | 1 | 88.2 | 93.5 | 30 | 2 | 86.7 | 91.43 |
| 6 | 1 | 88.4 | 93.6 | 31 | 2 | 85.9 | 91.43 |
| 7 | 1 | 88.4 | 93.6 | 32 | 2 | 87.6 | 91.5 |
| 8 | 1 | 88.4 | 93.6 | 33 | 2 | 87.6 | 91.5 |
| 9 | 1 | 88.6 | 93.4 | 34 | 2 | 87.6 | 91.5 |
| 10 | 1 | 88.6 | 93.4 | 35 | 2 | 85.7 | 90.2 |
| 11 | 1 | 88.5 | 93.4 | 36 | 2 | 87.6 | 90.2 |
| 12 | 1 | 88.5 | 93.7 | 37 | 2 | 87.6 | 90.2 |
| 13 | 1 | 88.2 | 93.7 | 38 | 2 | 85.9 | 90.3 |
| 14 | 1 | 88.2 | 93 | 39 | 2 | 86 | 90.3 |
| 15 | 1 | 88.32 | 93 | 40 | 2 | 85.6 | 90.3 |
| 16 | 1 | 88.32 | 93.8 | 41 | 2 | 85.6 | 90.4 |
| 17 | 1 | 88.33 | 93.8 | 42 | 2 | 85.6 | 90.4 |
| 18 | 1 | 88.41 | 93.8 | 43 | 2 | 85.6 | 90.4 |
| 19 | 1 | 88.41 | 93.73 | 44 | 2 | 85.7 | 90.21 |
| 20 | 1 | 88.1 | 93.73 | 45 | 2 | 86.3 | 90.21 |
| 21 | 1 | 88.1 | 93.73 | 46 | 2 | 86.3 | 90.21 |
| 22 | 1 | 87.99 | 93.73 | 47 | 2 | 86.1 | 90.02 |
| 23 | 1 | 87.99 | 93.73 | 48 | 2 | 85.7 | 90.02 |
| 24 | 1 | 87.99 | 93.73 | 49 | 2 | 86 | 90.02 |
| 25 | 1 | 87.91 | 93.73 | 50 | 2 | 86.1 | 90.02 |

Table 2. Group statistics for the two groups,wide and bilayer neural network machine classifiers.

| | Groups | N | Mean (%) | Standard Deviation | Standard Error Mean |
|---|---|---|---|---|---|
| | WNN | 25 | 88.2912 | 0.18401 | 0.03680 |
| Accuracy (%) | BNN | 25 | 86.2964 | 0.83371 | 0.16674 |
| **Sensitivity (%)** | WNN | 25 | 93.5792 | 0.21990 | 0.04398 |
| | BNN | 25 | 90.8044 | 0.85403 | 0.17081 |

Table 3 shows statistical data analysis from an Independent Sample T-Test, with significant p-values of 0.001 ($p < 0.05$) for both Accuracy and Sensitivity levels when treated as independent variables. The results show a statistically significant difference between the two groups.

Table 3. Independent sample T-test for predicting the accuracy (%) and sensitivity (%) using wide and bilayered neural network machine classifiers.

| Accuracy (%) | Variances | Levene's Test for Equality of Variances | | T-test for Equality of Means | | | | | | | |
|---|---|---|---|---|---|---|---|---|---|---|---|
| | | F | Sig. | t | df | Significance One-sided p | Significance Two-sided p | Mean Diff. | Std Error Diff. | 95% Confidence Level Lower | 95% Confidence Level Upper |
| **1.** | **Equal Variances Assumed** | 32.128 | <.000 | 11.682 | 48 | <.000 | <.000 | 1.99480 | 0.17076 | 1.65147 | 2.33813 |
| **2.** | **Equal Variances Not Assumed** | | | 11.682 | 26.333 | <.000 | <.000 | 1.99480 | 0.17076 | 1.64402 | 2.34558 |
| **Sensitivity (%)** | **Equal Variances Assumed** | 28.112 | <.000 | 15.732 | 48 | <.000 | <.000 | 2.77480 | 0.17638 | 2.42017 | 3.12943 |
| | **Equal Variances Not Assumed Assumed** | | | 15.732 | 27.169 | <.000 | <.000 | 2.77480 | 0.17638 | 2.41301 | 3.13659 |

Figure 2 depicts the Confusion Matrix of the Wide Neural Network Classifier, with actual and expected values from MATLAB simulation: True Positives (TP) = 44, True Negatives (TN) = 3, False Negatives (FN) = 3, and 0 False Positives (FP). These variables help to determine the accuracy and sensitivity of the Wide Neural Network Classifier.

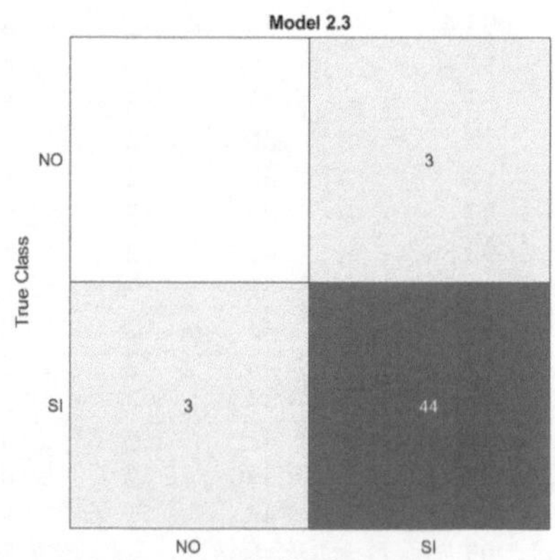


Figure 2. Confusion matrix of wide neural network classifier.

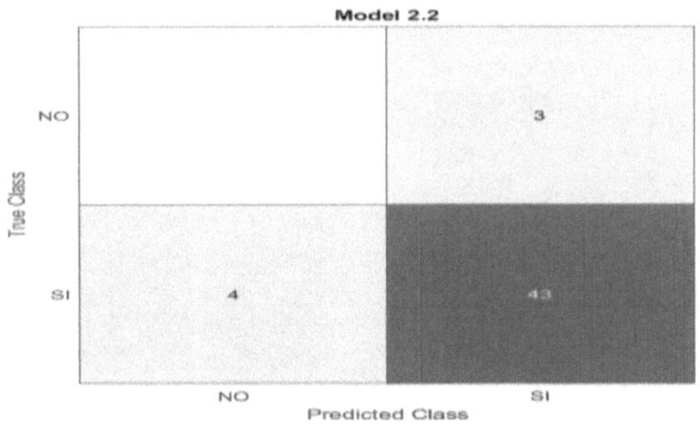


Figure 3. Confusion matrix of bilayered neural classifiers.

Figure 3 shows the Confusion Matrix of the Bilayered Neural Network Classifier, which has 43 True Positives (TP), 4 True Negatives (TN), 3 False Negatives (FN), and no False Positives (FP). These variables are critical in determining the accuracy and sensitivity of the Bilayered Neural Network Classifier.

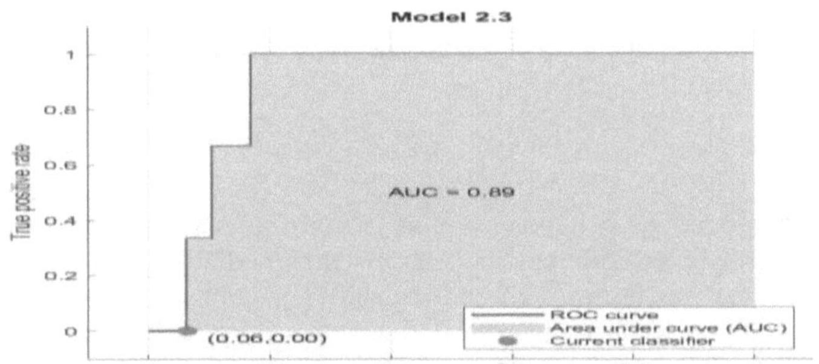


Figure 4. ROC curve for wide neural network classifiers. X-axis: False positive rate and Y-axis: True positive rate.

Figure 4 shows that the Wide Neural Network (WNN) classifier's AUC (Area Under the Curve) is 0.89, suggesting its ability to distinguish between classes. Similarly, Figure 5 displays an AUC of 0.87 for the Bilayered Neural Network (BNN) classifier.

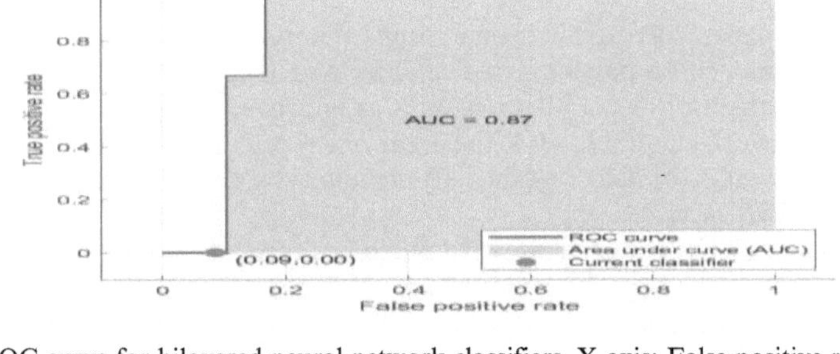


Figure 5. ROC curve for bilayered neural network classifiers. X-axis: False positive rate and Y-axis: True positive rate.

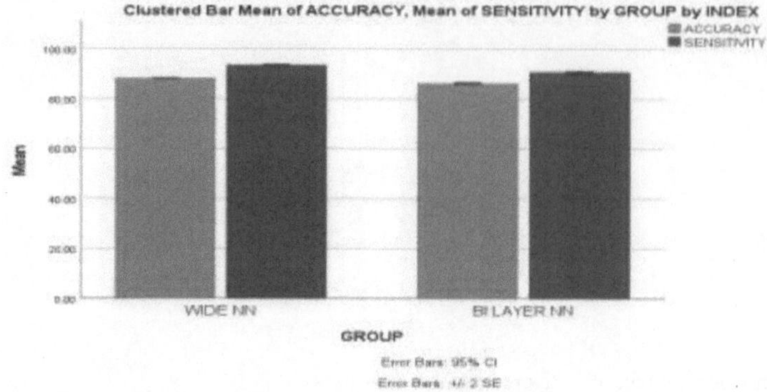


Figure 6. SPSS Graph Of Mean Accuracy (%) and Sensitivity (%) showing the comparison of Wide and Bilayered Neural Network Machine Classifiers. X-axis: Wide Neural Network Classifier vs Bilayered Neural Network Machine Classifier and Y-axis: Mean Accuracy (%). Error Bars: +/- 1 SD and 95% Confidence Level (CL).

Figure 6 shows a graphical representation of the accuracy and sensitivity levels of both classifiers, resulting from statistical analysis with IBM SPSS Statistics 25, indicating that the WNN classifier has greater accuracy (%) and sensitivity (%) than the BNN classifier.

## 4 DISCUSSION

A concern with hemodialysis is hypotension as it contributes to the death rate of dialysis patients. To prevent hypotension in patients receiving hemodialysis and to provide an early diagnosis, this method has been tested and studied. This analysis indicates that new and appropriate technologies for use in medical research include deep learning and machine learning algorithms. This study compared the performance of narrow neural network classifiers and medium neural network machine classifiers in terms of sensitivity and accuracy (%) rates. The two classifiers were analyzed for their ability to predict the levels of accuracy and sensitivity needed to prevent hypotension in hemodialysis patients. With accuracy and sensitivity (%) of 88.29% and 93.57%, respectively, the Narrow Neural Network classifiers have achieved the greatest results.

Some research studies show similar results related to this study, with the implementation of ReLU technique for validating the neural networks and it also shows the relevant statistical benefits (Chizat and Bach 09–12 Jul 2020). Another study also suggests similar methodologies of using wide neural networks by providing greater accuracy performance for cascading the evaluation of optimizing the through-put screening terminologies (Kouris *et al.* 2020).

Certain research papers have shown some results that are not similar and also irrelevant based on the analysis. Those papers include (Güler and Übeylı̇ 2005; Kouris *et al.* 2020; Wang *et al.* 2020), which produced the outcomes as improved accuracy rates when using a bilayered neural network classifier. Also, the other one revealed the positive effects of using bilayered neural networks for detecting the ECG readings by producing the output based on an accuracy rate of 96.94%.

Though this analysis has revealed effective outcomes through the prediction of accuracy levels and sensitivity rates, there were certain challenges faced. Those include feature extraction methods, proper layered networks with suitable screening methods and time-dependency methods. The paper (Yuan *et al.* 2018) shows the implementation of Deep and Wide neural networks as a combined technique to overcome those challenges.

The statistical analysis of the study indicates that from 25 samples, the mean accuracy and sensitivity (%) rates for the two groups (WNN and BNN) are 88.29%, 93.57%, and 86.29%, 90.80%, respectively. The standard deviations for accuracy and sensitivity are (0.18401) and (0.83371), respectively, and (0.21990) and (0.85403), respectively. For both groups and the Standard Error Mean, the accuracy and sensitivity values were determined to be (0.03680), (0.16674), and (0.04398), (0.17081). The Independent Samples T-test technique is used to demonstrate that the differences between the two groups are statistically significant. As a consequence, the p-value is less than 0.001 ($p<0.05$).

The study's future goals include improving the effectiveness of preventing hypotensive events during hemodialysis sessions by implementing precise validation set methodologies, appropriate and workable data mining techniques, and the use of satisfying perceptrons to generate appropriate results via neural network classifiers.

The present research examined appropriate and suitable machine learning and deep learning classification methods by comparing and evaluating the accuracy and sensitivity rates using MATLAB software-based data simulation. Furthermore, research has also been conducted on statistical data analysis with SPSS software.

## 5 CONCLUSION

Thus, the Innovative Classification Learner in this research study, which ends with the success of improved Accuracy (%) and Sensitivity (%) levels of Wide Neural Network Classifiers and Bilayered Neural Network Machine Classifiers, has allowed the machine learning algorithms to predict its accuracy levels through MATLAB simulation. The findings show that Wide Neural Network Classifiers (88.2912%), (93.5792%) have achieved the highest accuracy (%) and sensitivity (%) rate when compared to Bilayered Neural Network Machine Classifiers (86.2964%), (90.8044%). More research using statistical methods has also produced results comparable to any sample showing a statistically significant difference. Consequently, by utilizing machine learning techniques, the work has produced a novel way to recognize and predict the effectiveness of hypotensive instances in hemodialysis.

## REFERENCES


Ali, S., Sudha, K. G., Thirumalaivasan, N., Ahamed, M., Pandiaraj, S., Rajeswari, V. D., Vinayagam, Y., Thiruvengadam, M. and Govindasamy, R. (2023). Green synthesis of magnesium oxide nanoparticles by using abrus precatorius bark extract and their photocatalytic, antioxidant, antibacterial, and cytotoxicity activities. *Bioengineering (Basel, Switzerland)*, 10(3). https://doi.org/10.3390/bioengineering10030302

Chizat, Léna"Ic, and Francis Bach. (09-12 Jul 2020). Implicit bias of gradient descent for wide two-layer neural networks trained with the logistic loss. In *Proceedings of Thirty Third Conference on Learning Theory, Edited by Jacob Abernethy and Shivani Agarwal*, 125:1305–38. Proceedings of Machine Learning Research. PMLR.

Chou, Jason A., Elani Streja, Danh V. Nguyen, Connie M. Rhee, Yoshitsugu Obi, Jula K. Inrig, Alpesh Amin, Csaba P. Kovesdy, John J. Sim and Kamyar Kalantar-Zadeh. (2018). Intradialytic hypotension, blood pressure changes and mortality risk in incident hemodialysis patients. nephrology, dialysis, transplantation: Official publication of the european dialysis and transplant association - *European Renal Association* 33 (1): 149–59.

De Lima, Jose Jayme G., Luis Henrique W. Gowdak, Jose Otto Reusing Jr, Elias David-Neto, and Luiz A. Bortolotto. (2023). Interdialytic blood pressure and risk of cardiovascular events and death in hemodialysis patients. high blood pressure & cardiovascular prevention: The official. *Journal of The Italian Society of Hypertension* 30 (3): 235–41.

Güler, İnan and Elif Derya Übeylı̇. (2005). ECG beat classifier designed by combined neural network model. *Pattern Recognition* 38 (2): 199–208.

Gurupranes, S. V., Natrayan, L., Kaliappan, S., Patel, P. B., Sekar, S., Jayaraman, P., Arvinda Pandian, C. K. and Esakkiraj, E. S. (2022). Investigation of physicochemical properties and characterization of leaf stalk

fibres extracted from the caribbean royal palm tree. *International Journal of Chemical Engineering*, 2022. https://doi.org/10.1155/2022/7438411
Hamrahian, Seyed Mehrdad, Salem Vilayet, Johann Herberth and Tibor Fülöp. (2023). Prevention of intradialytic hypotension in hemodialysis patients: Current challenges and future prospects. *International Journal Of Nephrology And Renovascular Disease* 16 (August): 173–81.
Kouris, Alexandros, Stylianos I. Venieris and Christos-Savvas Bouganis. (2020). A throughput-latency co-optimised cascade of convolutional neural network classifiers. In *2020 Design, Automation & Test In Europe Conference & Exhibition* (DATE), 1656–61. IEEE.
Lee, Hojun, Donghwan Yun, Jayeon Yoo, Kiyoon Yoo, Yong Chul Kim, Dong Ki Kim, Kook-Hwan Oh, *et al.* (2021). Deep learning model for real-time prediction of intradialytic hypotension. *Clinical Journal Of The American Society Of Nephrology: CJASN* 16 (3): 396–406.
Lin, Cheng-Jui, Chih-Yang Chen, Pei-Chen Wu, Chi-Feng Pan, Hong-Mou Shih, Ming-Yuan Huang, Li-Hua Chou, Jin-Sheng Tang and Chih-Jen Wu. (2018). Intelligent system to predict intradialytic hypotension in chronic hemodialysis. *Journal of the Formosan Medical Association = Taiwan Yi Zhi* 117 (10): 888–93.
Malerba, Mario, Alessandro Radaeli, Matteo Nardin, Enrico Clini, Giovanna Elisiana Carpagnano, Edoardo Sciatti, Francesca Salghetti, Ivano Bonadei, Fabio Platto and Enrico Vizzardi. (2017). Endothelial dysfunction assessment by noninvasive peripheral arterial tonometry in patients with chronic obstructive pulmonary disease compared with healthy subjects. *The Clinical Respiratory Journal* 12 (4): 1466–72.
Natrayan, L., Kaliappan, S., Sethupathy, B. S., Sekar, S., Patil, P. P., Velmurugan, G. and Tariku Olkeba, T. (2022). Effect of mechanical properties on fibre addition of flax and graphene-based bionanocomposites. *International Journal of Chemical Engineering*, 2022. https://doi.org/10.1155/2022/5086365
Sinha, A., B. Kumar and P. Banerjee. (2021). HSCAD: Heart sound classification for accurate diagnosis using machine learning and MATLAB. *2021 International.* Https://Ieeexplore.Ieee.Org/Abstract/Document/9752199/.
Sivaranjani, P. R., Syed, A., Elgorban, A. M., Bahkali, A. H., Balakrishnaraja, R., Varma, R. S. and Sudheer Khan, S. (2023). Fabrication of ternary nano-heterojunction via hierarchical deposition of α-Fe2O3 and *β*-La2S3 on cubic CoCr2O4 for enhanced photodegradation of doxycycline. *Journal of Industrial and Engineering Chemistry*, 118, 407–417.
Srinivasulureddy, Ch, N. S. Kumar and V. S. Binu. (2023). Analysis and comparison for innovative prediction technique of breast cancer tumor by naive bayes algorithm over support vector machine algorithm with improved accuracy. *CARDIOMETRY* 25 (25): 865–71.
Tsujimoto, Yasushi, Hiraku Tsujimoto, Yukihiko Nakata, Yuki Kataoka, Miho Kimachi, Sayaka Shimizu, Tatsuyoshi Ikenoue, Shingo Fukuma, Yosuke Yamamoto and Shunichi Fukuhara. (2019). Dialysate temperature reduction for intradialytic hypotension for people with chronic kidney disease requiring haemodialysis. *Cochrane Database of Systematic Reviews* 7 (7): CD012598.
Viswanathan, S., Palaniyandi, T., Kannaki, P., Shanmugam, R., Baskar, G., Rahaman, A. M., Paul, L. T. D., Rajendran, B. K. and Sivaji, A. (2023). Biogenic synthesis of gold nanoparticles using red seaweed Champia parvula and its anti-oxidant and anticarcinogenic activity on lung cancer. *Particulate Science and Technology*, 41(2), 241–249.
Wang, Jing, Wenqian Zhang and Jinglin Zhou. (2020). Fault Detection With Data Imbalance Conditions Based On The Improved Bilayer Convolutional Neural Network. *Industrial & Engineering Chemistry Research* 59 (13): 5891–5904.
Yuan, Zhaozheng, Ruixun Zhang and Xiuli Shao. (2018). Deep And Wide Neural Networks On Multiple Sets Of Temporal Data With Correlation. In *Proceedings Of The 2018 International Conference On Computing And Data Engineering*, 39–43. ICCDE '18. New York, NY, USA: Association For Computing Machinery.
Zhang, Hanjie, Lin-Chun Wang, Sheetal Chaudhuri, Aaron Pickering, Len Usvyat, John Larkin, Pete Waguespack, *et al.* (2023). Real-Time prediction of intradialytic hypotension using machine learning and cloud computing infrastructure. nephrology, dialysis, transplantation: Official publication of the european dialysis and transplant association. *European Renal Association* 38 (7): 1761–69.

*Hybrid and Advanced Technologies – S. Prasad Jones Christydass, Nurhayati Nurhayati and S. Kannadhasan. (Eds)*  *ISBN 978-1-032-90254-8*


# Predicting student stress levels and mental health using convolutional neural network algorithm in comparison with logistic regression

Pedda Geliche Mahesh Babu
*Research Scholar, Department of Computer Science Engineering, Saveetha School of Engineering, Saveetha University, Chennai, Tamil Nadu, India*

P. Gururama Senthilvel
*Project Guide, Department of Computer Science Engineering, Saveetha School of Engineering, Saveetha University, Chennai, Tamil Nadu, India*

ABSTRACT: The study compared two groups utilizing a dataset that was split 80% for training and 20% for testing, with the goal of predicting the mental health and stress levels of students. The accuracy of the convolutional neural network was 94.409%, which was significantly greater than that of the logistic regression (85.263%, p=0.002).

As independent variables, consider user behavior, damage data, social media activity, and log tools. The investigation made use of a 64-bit Windows 10 PC with 8 GB RAM and a laptop running Jupiter that was online. Logistic regression and convolution models were used to improve system security, and a dataset supporting five models was used to enable unique entity identification using Python.

*Keywords*: Stress Level, Mental Health, Psychological, fitness issues, human mind, social media

## 1 INTRODUCTION

Anxiety, a chief purpose of infection and infirmity among students, is one of the most customary illnesses they revel in on a everyday basis. Equipment or programming languages. (Nemias Saboya 2023) The extensively used technique is primarily based on non-stop stratification, the primary traits being age and sex, and maximum research do now not use unique sampling techniques. (Jorge Isaac Necochea-Chamorro 2022) The consequences of this systematic evaluate show how machine learning techniques can are expecting strain and tension. Scientific proof supports this. Machine studying algorithms are in comparison to a detailed evaluation of those algorithms, equipment or programming languages, metrics, getting to know variables and influencing elements. The prognosis of stress and anxiety is enabled within the clinical field (Yesenia del 2021).

For an ordinary pupil graduating from excessive school, interpersonal conversation and overall performance evaluation on campus is a vast challenge. To address future and modern opposition and stress from contemporary college students, many students face mental health troubles..(Comput Intell Neurosci. 2022) This paper evaluates, predicts and analyzes the nation of mind of cutting-edge students in neural community algorithm. Neural community algorithm pc era is carried out to the evaluation of intellectual fitness of contemporary students (Jingjing Bei 2020), the effects display that there's a considerable relationship

DOI: 10.1201/9781003559115-40

among melancholy and the student board and the prevalence of intellectual infection and fundamentalism . (Serkan K. 2020).

Natural language processing (NLP) is being used with social media, interviews, and scientific data in research to improve cognitive health resources. A narrative review highlights developments and patterns in evaluative psychoanalysis techniques across recent studies, synthesizing NLP applications in mental health diagnoses (Parvez M. *et al.* 2023; Ramesh *et al.*, n.d.; Sarkar *et al.* 2022; Surendran *et al.* 2023; Tayyab *et al.* 2022).

## 2 MATERIALS AND METHODS

The work was carried out at the Open Source Lab of the Saveetha School of Engineering, Saveetha Institute of Medical and Technical Sciences, in Chennai. The study employed two sets of algorithms, namely Group I Convolutional Neural Networks and Group II Logistic Regression, to evaluate their effectiveness in 100 statistical samples. Using a dataset from Kaggle.com, an 80/20 split was used for training and testing, respectively. Variables including stress levels, mental health indicators, career affiliations, and educational backgrounds are included in this dataset. The availability of the dataset made it easier to use sophisticated analysis techniques to assess how well the system predicted student stress levels and mental health problems using meteorological data inputs.

The software program is designed and maintained in Joue Annotation the usage of Python programming. Windows 10 is the primary machine to study the trial application. The hardware includes 8GB of RAM and an Intel Core 7 processor (8 GB). They used 64 bits to calculate the gathering. Python is used as a coding language for software program improvement. The dataset is used to apply an advantageous processing code.

### 2.1 *Convolutional neural network*

A tool to keep away from is a CNN. One of the few styles of synthetic brain community used for a selection of designs and measurements. Entity engineering is a computationally targeted class, particularly applied to photograph recognition and pixel registration control. There are one-of-a-kind forms of neural networks in deep studying, but neural networks are a social device for object reputation and recognition choice. As a result, it's miles appropriate for computer vision (CV) obligations and programs in which object recognition is crucial, together with self sufficient automobiles and facial popularity.

#### 2.1.1 *Algorithm*

In order to successfully apply the Convolutional Neural Network (CNN) algorithm, begin by obtaining the required data (Step 1). Using neural transformation rules, define the problem that needs to be treated binaryly (Step 2). Step 3: Segment the dataset into training and validation sets to assess the CNN's performance. Step 4: Train the CNN model to evaluate its performance on a test set. If the CNN performs poorly, change the hyperparameters to get better outcomes (Step 5). Lastly, use the CNN algorithm to forecast results based on fresh, unexplored data (Step 6).

### 2.2 *Logistic regression*

If the set variable is dichotomous (binary), then logistic regression is the first-rate regression estimator. Logistic regression is a predictive analysis, just like all other regression analyses. A set of binary variables plus one or more nominal, ordinal, proportional, c program language period, or extra nominal variables without a hobby are described by means of logistic regression.

### 2.2.1 *Algorithm*

The first step in logistic regression modeling is to define the target variable and choose pertinent features that have an impact on the class (Step 1). To promote consistency, standardize every item equally (Step 2). Set the weight and bias of the model to small, random values at step three. Step 4 involves mapping predictions to a probability range of 0 to 1, using a cost function and sigmoid function. Measure the effectiveness of logistic regression in an experimental setting using metrics like recall, precision, accuracy, and F1 score (Step 5). Optimize model performance by adjusting parameters such as iteration size, regularization intensity, and learning rate (Step 6). Lastly, use the trained logistic regression model to forecast the results of brand-new comments in order to identify fake ones (Step 7).

## 2.3 *Statistical analysis*

A maximum of 10 iterations per rule set were used for the analysis of convolutional neural networks and logistic regression using IBM SPSS version 26 (Yu *et al.* 2009). According to Sinha and Sushila (2022), independent variables include damage statistics, user behavior, IP agreement, geolocation, history data, log tools, and social media activity. The research utilized a 64-bit Windows 10 PC with 8 GB RAM and a laptop loaded with Jupiter that was online. With the use of logistic regression and convolution models, system security was improved by utilizing a dataset that supported five models, and Python enabled unique entity identification.

# 3 RESULT

The logistic regression and logistic regression have been finished in Anaconda Navigator with a pattern size of 10.The grid architecture in

Table 1. Accuracy values of convolutional neural networks predicting student stress levels and mental health is (94.409%).

| | ACCURACY RATE | |
|---|---|---|
| | CNN | LOGISTIC REGRESSION |
| SI.No. | | |
| 1 | 93.41 | 86.73 |
| 2 | 92.71 | 83.20 |
| 3 | 96.56 | 85.82 |
| 4 | 93.13 | 83.63 |
| 5 | 94.06 | 84.90 |
| 6 | 93.77 | 84.61 |
| 7 | 98.44 | 86.16 |
| 8 | 94.50 | 84.86 |
| 9 | 91.85 | 89.56 |
| 10 | 95.66 | 83.16 |
| **Accuracy** | **94.40** | **85.26** |

Table 2. Statistical evaluation of combinations of neutral curves and logistic regression. The reported deviance and widespread errors of the 10 fashions show that convolutional neural community class has a higher average accuracy than logistic regression.

| | Group | N | Mean | Std. Deviation | Std. Error Mean |
|---|---|---|---|---|---|
| **Accuracy** | Convolutional neural network | 10 | 94.409 | 1.97338 | 0.62404 |
| | Logistic regression | 10 | 85.263 | 1.93784 | 0.61280 |

Table 3. An assessment of a brand new classification of convolutional neural networks and logistic regression in phrases of their accuracy.

| CLASSIFIER | ACCURACY (%) |
|---|---|
| Convolutional neural network | 94.409 |
| Logistic regression | 85.263 |

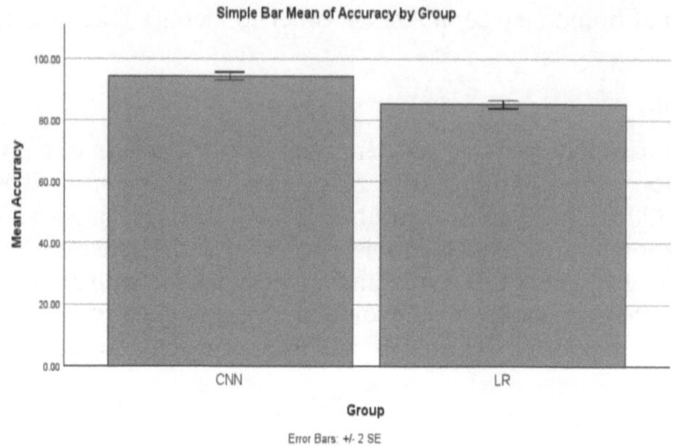


Figure 1. Correlation of reach and strategic relapse in expressions of normal precision. The typical exactness of the log is higher than that of calculated relapse, the standard deviation of log is somewhat better compared to that of strategic relapse. X-Axis: Calculated Relapse and Y-Pivot: +/- normal precision inside 2 SD.

## 4 DISCUSSION

A two-tailed significance value of $p = 0.017$ ($P < 0.05$) turned into located between the 2 organizations, indicating that the clean brown tuber became large in length. Table 2 shows that the accuracy of the range analysis is 94.40% and the accuracy of the logistic regression is 85.26%.

Emerging professionals adjusting to college life frequently encounter emotional gaps, which can result in mental health problems in addition to other typical difficulties (Wang Qiu 2021). Chinese academics have examined the effects of family, school, and society elements on students' personal and social lives, highlighting the significance of these aspects on students' mental health (Trayudi A. 2022). These factors include gender, education level, parental marital satisfaction, and work stress. Concurrently, neural networks are becoming more and more potent computing instruments used for intricate computations; they are especially well-suited for mathematics and data processing jobs (Turker I. 2020).

The limitations of this have a look at are that the formation of a new Brownian structure takes a completely long time, especially with big sheets. A destiny intention of this examination is to increase the machine to include a larger quantity of objects with much less time to put in the dataset.

## 5 CONCLUSION

Finally, the purpose of this experiment is to apply convolutional neural networks (94.409%) in device penetration and inspect its overall performance with LR. (85.263%) anticipated the effect of Spss as a variable detected with the precision received through logistic regression.

## REFERENCES


[1] Tesfahun A. and Lalitha Bhaskari D. (2013). Intrusion detection using Random Forests Classifier with SMOTE and Feature Reduction. In *2013 International Conference on Cloud & Ubiquitous Computing & Emerging Technologies* (pp. 1–6). IEEE. https://doi.org/10.1109/CUBE.2013.58

[2] Aditya Phadke, Mohit Kulkarni, Pranav Bhawalkar and Rashmi Bhattad. (2019). A Review of Machine Learning Methodologies for Network Intrusion Detection. In *2019 3rd International Conference on Computing Methodologies and Communication (ICCMC)*. IEEE. https://doi.org/10.1109/ICCMC.2019.8845697

[3] Anish Halimaa A and Dr K. Sundarakandam. (2019). Machine learning based intrusion detection system. In *Proceedings of the Third International Conference on Trends in Electronics and Informatics (ICOEI 2019)* (pp. 1–6). IEEE. https://doi.org/10.1109/ICOEI.2019.8862551

[4] Riyaz B. and Ganapathy S. (2018). An intelligent fuzzy rule-based feature selection for effective intrusion detection. In *2018 International Conference on Recent Trends in Advanced Computing (ICRTAC)* (pp. 1–6). IEEE. https://doi.org/10.1109/ICRTAC.2018.8548355

[5] Iftikhar Ahmad, Mohammad Basheri, Muhammad Javed Iqbal and Aneel Rahim. (2018). Performance comparison of support vector machine, random forest, and extreme learning machine for intrusion detection. *IEEE Access*, *6*, 33789–33795. https://doi.org/10.1109/ACCESS.2018.2842301

[6] Jafar Abo Nada and Mohammad Rasmi Al-Mosa. (2018). A proposed wireless intrusion detection prevention and attack system. In *2018 International Arab Conference on Information Technology (ACIT)* (pp. 1–6). IEEE. https://doi.org/10.1109/ITAC.2018.8397597

[7] Kazi Abu Taher, Billal Mohammed Yasin Jisan and Md. Mahbubur Rahman. (2019). Network intrusion detection using supervised machine learning technique with feature selection. In *2019 International Conference on Robotics, Electrical and Signal Processing Techniques (ICREST)* (pp. 1–6). IEEE. https://doi.org/10.1109/ICREST.2019.8672931

[8] Haripriya L. and Jabbar M. A. (2018). Role of machine learning in intrusion detection system: review. In *2018 Second International Conference on Electronics, Communication and Aerospace Technology (ICECA)* (pp. 1–6). IEEE. https://doi.org/10.1109/ICECA.2018.8475478

[9] Kinam Park, Youngrok Song and Yun-Gyung Cheong. (2018). Classification of attack types for intrusion detection systems using a machine learning algorithm. In *2018 IEEE Fourth International Conference on Big Data Computing Service and Applications (BigData Service)* (pp. 221–226). IEEE. https://doi.org/10.1109/BigDataService.2018.00034

[10] Le, T.-T.-H., Kang, H. and Kim, H. (2019). The impact of pca-scale improving gru performance for intrusion Detection. In *2019 International Conference on Platform Technology and Service (PlatCon)*. IEEE. https://doi.org/10.1109/PlatCon.2019.8668960

[11] Mengmeng Ge, Xiping Fu, Naeem Syed, Zubair Baig and Gideon Teo. (2019). Deep learning-based intrusion detection for IoT Networks. In *2019 IEEE 24th Pacific Rim International Symposium on Dependable Computing (PRDC)* (pp. 256–265). IEEE. https://doi.org/10.1109/PRDC47172.2019.00044

[12] Mohammed Ishaque and Ladislav Hudec. (2019). Feature extraction using Machine Learning For Intrusion Detection System. In *2019 2nd International Conference on Computer Applications & Information Security (ICCAIS)*. IEEE. https://doi.org/10.1109/ICCAIS.2019.8718325

[13] Nimmy Krishnan and A. Salim. (2018). Machine learning-based intrusion detection for virtualized infrastructures. In *2018 International CET Conference on Control, Communication, and Computing (IC4)* (pp. 1–6). IEEE. https://doi.org/10.1109/IC4.2018.8598492

[14] Patgiri R., Varshney U., Akutota T. and Kunde R. (2018). An investigation on intrusion detection system using machine learning. In *2018 IEEE Region 10 Symposium (TENSYMP)* (pp. 1334–1339). IEEE. https://doi.org/10.1109/TENCONSpring.2018.8512994

[15] Rohit Kumar Singh Gautam and Er. Amit Doegar. (2018). An ensemble approach for intrusion detection system using machine learning algorithms. In *2018 8th International Conference on Cloud Computing, Data Science & Engineering (Confluence)* (pp. 1–6). IEEE. https://doi.org/10.1109/Confluence.2018.8443563

[16] Bernard S., Heutte L. and Adam S. (2009). On the selection of decision trees in random forests. In *Proceedings of International Joint Conference on Neural Networks* (pp. 1–8). IEEE. https://doi.org/10.1109/IJCNN.2009

*Hybrid and Advanced Technologies – S. Prasad Jones Christydass, Nurhayati Nurhayati and S. Kannadhasan. (Eds) © 2025 the Authors, ISBN 978-1-032-90254-8*


# Effective method for detecting pancreatic adenocarcinoma with increased accuracy by comparing with the Naive Bayes algorithm and using machine learning-based convolution neural network

K. Abhishake
*Research Scholar, Department of Biomedical Engineering, Saveetha School of Engineering, Saveetha University, Chennai, Tamil Nadu, India*

P. Nirmala
*Project Guide, Department of Biomedical Engineering, Saveetha School of Engineering, Saveetha University, Chennai, Tamil Nadu, India*

ABSTRACT: Our study's main goal is to present a machine learning system that will increase the accuracy of pancreatic cancer identification. We suggest the intervention algorithm be a Convolutional Neural Network (CNN) and assess its performance against the Naive Bayes approach. Kaggle provides the data used to calculate the sample size; there are 50 total elements in all, with 25 items in each group. When computing the standard error, an alpha value of 0.05% is used. The independent T-test in SPSS is used to guarantee statistical stability. When CNN and Naive Bayes are compared for accuracy using MATLAB software, CNN performs better, with an accuracy of 86.2%, while Naive Bayes only achieves 78.4%. The statistical analysis reveals that CNN performed significantly better ($p<0.001$). Therefore, because CNN has a greater accuracy rate for diagnosing pancreatic cancer, it is advised to use it.

*Keywords*: Naive Bayes algorithm, Convolution neural networks, Machine Learning Algorithm, Artificial Intelligence, Pancreatic Adenocarcinoma, Innovative Cancer Detection Model

## 1 INTRODUCTION

The prediction of cancer, including its diverse forms that impact different body areas, is a major area of attention in the realm of health research. Because it is located in the abdomen, below the stomach, pancreatic adenocarcinoma is one of the most difficult diseases to cure and is frequently deemed incurable (Sekaran *et al.* 2019).Using preset threshold settings, this model estimates the amount of pancreatic cancer disseminated within the pancreas by utilizing machine learning algorithms to identify critical features from CT scans (Upreti *et al.* 2020). Improving the prognosis and cure rate of pancreatic cancer requires early identification and precise staging. Thus, the goal of this research was to create an accurate and automated image processing system that can quickly diagnose pancreatic cancer by evaluating computed tomography pictures.

There is a wealth of literature on pancreatic cancer analysis; hundreds of papers may be found on databases such as PubMed, Google Scholar, and IEEE. The study conducted by Laino *et al.* (2022) is noteworthy for its comprehensiveness and high citation count. It provides useful insights into novel approaches for pancreatic adenocarcinoma cancer detection. Even with these developments, correctly identifying pancreatic cancer is still a difficult task, requiring more research into machine learning and artificial intelligence techniques. In order

 DOI: 10.1201/9781003559115-41

to tackle this problem, we compare the accuracy of CNN and the Naive Bayes algorithm in identifying pancreatic adenocarcinomas in our study.

We use MATLAB and SPSS for analysis in our research, taking advantage of its features to assess the CNN method's effectiveness in comparison to Naive Bayes. Our main goal is to increase the detection accuracy of pancreatic adenocarcinoma, which will aid in the creation of more potent diagnostic instruments for the battle against this dangerous illness. Our goal is to use cutting-edge machine learning approaches to enhance patient outcomes and further the field of pancreatic cancer diagnosis and detection research.

## 2 MATERIALS AND METHODS

Along with the Saveetha School of Engineering and the Biomedical department of the Saveetha Institute of Medical and Technical Science, the study was conducted at the Saveetha Institute of Engineering's Digital Signal Processing facility. Since there were no human samples used in the investigation, ethical permission processes were deemed unnecessary. Fifty samples in all, split into Group 1 and Group 2, were used in the investigation. In order to calculate the pretest probability, Clinical.com made sample size computation easier by keeping constant parameters such as an enrollment ratio of 1, G power at 80%, threshold at 0.05%, and confidence interval at 95%.

The initial diagnosis of pancreatic adenocarcinoma, which progresses from pancreatic intraductal neoplasm to invasive neoplastic lesions by a multistep carcinogenesis process akin to colorectal cancer, is greatly aided by sample analysis (Mario *et al.* 2018).

### 2.1 *Statistical analysis*

Using SPSS software, the necessary accuracy was ascertained, allowing for easier comparisons between the comparison algorithm and the recommended approach. Six stations were set up for examination in the traditional pancreatic EUS approach (Segard *et al.* 2013). For continuous variables like age and maximum diameter, statistical analysis used a two-sided U test; for other variables, a two-squared chi-square test was used. The analytical tool was IBM SPSS, which provided strong statistical capabilities (Du *et al.* 2022). As each input adjustment was made, the dependent variables—image pixels, color, and size—changed accordingly. The effectiveness of the study was assessed using the dependent T-test, which contrasted the CNN and the Naive Bayes algorithm.

## 3 RESULTS

Methods from artificial intelligence and machine learning were used to evaluate and confirm the accuracy of various samples. To be more precise, the CNN algorithm—which functions as the intervention model for pancreatic cancer diagnosis—was assessed against the Naive Bayes approach using the independent samples T-test.

Table 1. The table presents the results of the SPSS analysis for the CNN and Naive Bayes algorithms, each of which was evaluated on 25 samples. It shows the mean accuracy, standard deviation, and standard mean error values.

| | GROUP STATISTICS | | | | |
|---|---|---|---|---|---|
| **ACCURACY** | GROUP | N | MEAN | STD. DEVIATION | STD. MEAN ERROR |
| | CONVOLUTION NEURAL NETWORK | 25 | 88.59 | 2.13657 | 0.42731 |
| | NAÏVE BAYES | 25 | 75.86 | 2.47489 | 0.4948 |

The simulated ROC curve for the suggested CNN method is shown in Figure 1. When compared to the original dataset, 0.18 of the data points were incorrectly identified, while 0.92 were correctly classified, according to an analysis of the CNN algorithm's ROC curve. 90%, or 0.90, is determined to be the CNN algorithm's area under the curve (AUC). On the other hand, Figure 2 shows the Naive Bayes comparison algorithm's simulated ROC curve. Compared to the original dataset, 0.84 data points were correctly classified in this case, while 0.23 data points were incorrectly classified. It is found that the Naive Bayes algorithm's AUC is 0.84.It is clear from comparing the confusion matrices shown in Figures 3 and 4 that the recommended CNN technique performs better than the Naive Bayes algorithm in terms of the total number of reported observations. Furthermore, statistical analysis of 25 samples using the Naive Bayes and CNN approaches shows that the latter has a standard mean error of 0.49488 and a standard deviation of 2.47489, while the former has a standard deviation of 2.13657 and a standard mean error of 0.42731. A p-value of less than 0.001 emphasizes the importance of this investigation.

There is a statistically significant difference between CNN and Naive Bayes algorithm performance levels, as shown by an independent T-test. Furthermore, Figure 3 shows the superiority of CNN in the detection of pancreatic adenocarcinomas when comparing the accuracy of the two algorithms using various methodologies. With a mean accuracy of 86%, CNN outperforms Naive Bayes (74%), demonstrating the CNN algorithm's improved ability to detect pancreatic cancer.

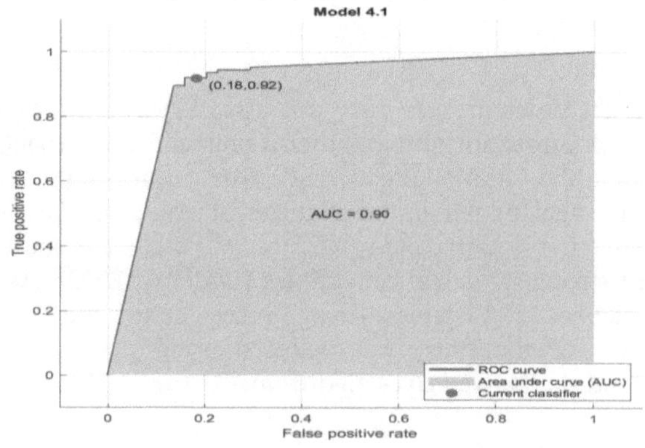


Figure 1. With 0.92 accurate classifications and 0.18 incorrect classifications, the CNN algorithm's ROC curve has an AUC of 0.90. Better than Naive Bayes.

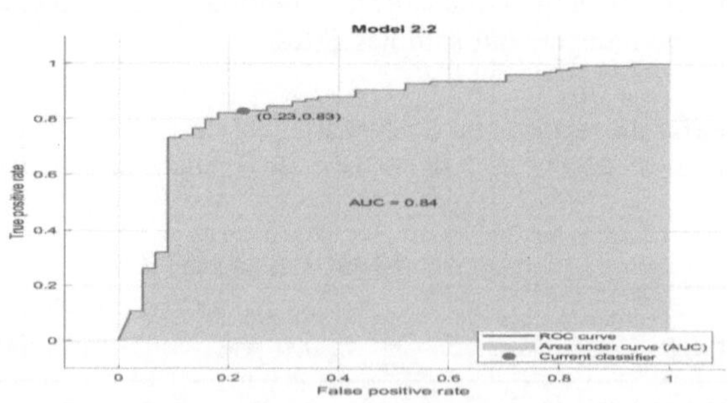


Figure 2. The ROC curve for Naive Bayes demonstrates an AUC value of 0.84, with fewer correct classifications (0.84) compared to the proposed algorithm.

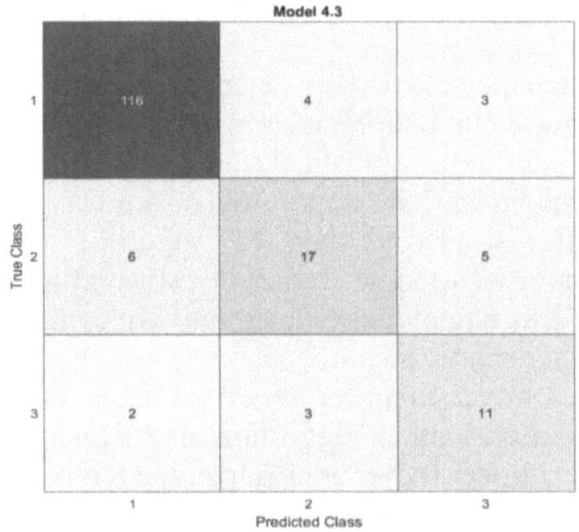


Figur 3. The suggested CNN algorithm's confusion matrix in the Innovative cancer model displays greater counts for components belonging to true classes (1,1), (2,2), and (3,3).

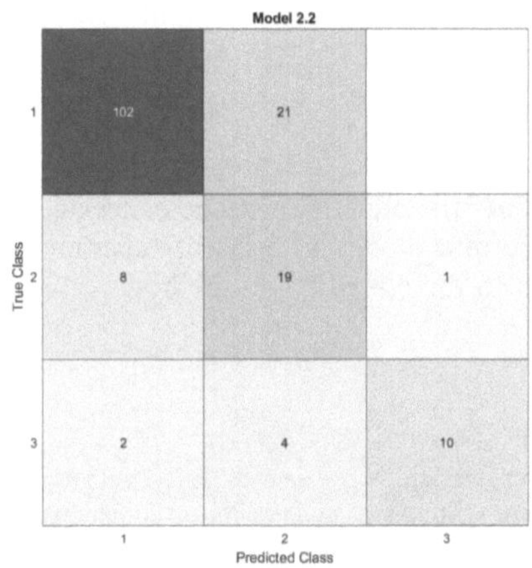


Figure 4. The Innovative cancer model's confusion matrix for the Naive Bayes method shows less counts for true classes than the suggested approach, highlighting the latter's higher accuracy.

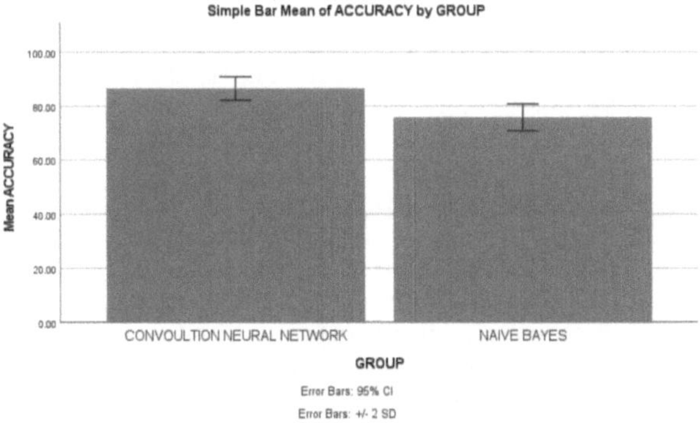


Figure 5. The CNN strategy performs better than the Naive Bayes approach, as seen by the bar graph that displays the mean accuracy scores of the two algorithms. CNN scored 86.2% accuracy, compared to 78.6% for Naive Bayes.

## 4 DISCUSSION

In this work, the innovative pancreatic cancer detection algorithm model showed improved accuracy over prior techniques, highlighting its importance. A p-value of less than 0.05 from the investigation indicated substantial statistical significance ($p<0.001$).

Traditional detection techniques have made use of a variety of machine learning and artificial intelligence techniques, such as the CNN algorithm and multilayer perceptron. Achieving accuracies of about 80%, some studies investigated merging or training B mode and ultrasound in machine learning models sequentially. In contrast, the CNN model's 98% classification accuracy for pancreatic adenocarcinomas was the highest.

Even while CNN has been successful in certain experiments, some experts disagree with its methodology. In some situations, other algorithms like Random Forest and KNN have proven to be more accurate. In one study, for example, the KNN algorithm fared better than CNN, with an accuracy of 90% as opposed to 75% for CNN. In a similar vein, Random Forest outperformed CNN in a comparative examination, with Random Forest obtaining an accuracy of 83% while CNN only managed 40%. Furthermore, in a different examination, the accuracy of Artificial Neural Networks (ANN) was 93%, higher than that of other algorithms. In comparison, algorithms such as Decision Trees (DT) and Support Vector Machines (SVM) achieved accuracies between 75% and 78%.

## 5 CONCLUSION

At a p-value of less than 0.001, the novel pancreatic cancer detection model that has been suggested outperforms the current model in terms of accuracy. This indicates a significant gain of 86.2% in comparison to the Naive Bayes method.

## REFERENCES


Ali, S., Sudha, K. G., Thirumalaivasan, N., Ahamed, M., Pandiaraj, S., Rajeswari, V. D., ... and Govindasamy, R. (2023). Green synthesis of magnesium oxide nanoparticles by using abrus precatorius bark extract and their photocatalytic, antioxidant, antibacterial, and cytotoxicity activities. *Bioengineering*, *10*(3), 302.

Cazacu, I. M., Udristoiu, A., Gruionu, L. G., Iacob, A., Gruionu, G. and Saftoiu, A. (2019). Artificial intelligence in pancreatic cancer: Toward precision diagnosis. *Endoscopic ultrasound*, *8*(6), 357–359.

Dmitriev, K., Kaufman, A. E., Javed, A. A., Hruban, R. H., Fishman, E. K., Lennon, A. M. and Saltz, J. H. (2017). Classification of pancreatic cysts in computed tomography images using a random forest and convolutional neural network ensemble. In *Medical Image Computing and Computer Assisted Intervention–MICCAI 2017: 20th International Conference, Quebec City, QC, Canada, September 11–13, 2017, Proceedings, Part III 20* (pp. 150–158). Springer International Publishing.

Du, J., Qiu, X., Lu, C., Zhu, Y., Kong, W., Xu, M., ... and Liu, B. (2022). Molecular landscape and prognostic biomarker analysis of advanced pancreatic cancer and predictors of treatment efficacy of AG chemotherapy. *Frontiers in Oncology*, *12*, 844527.

Dumitrescu, E. A., Ungureanu, B. S., Cazacu, I. M., Florescu, L. M., Streba, L., Croitoru, V. M., ... and Lungulescu, C. V. (2022). Diagnostic value of artificial intelligence-assisted endoscopic ultrasound for pancreatic cancer: a systematic review and meta-analysis. *Diagnostics*, *12*(2), 309.

Gurupranes, S. V., Natrayan, L., Kaliappan, S., Patel, P. B., Sekar, S., Jayaraman, P., ... and Esakkiraj, E. S. (2022). Investigation of physicochemical properties and characterization of leaf stalk fibres extracted from the caribbean royal palm tree. *International Journal of Chemical Engineering*, *2022*(1), 7438411.

Huang, B., Huang, H., Zhang, S., Zhang, D., Shi, Q., Liu, J. and Guo, J. (2022). Artificial intelligence in pancreatic cancer. *Theranostics*, *12*(16), 6931.

Kaul, V., Enslin, S. and Gross, S. A. (2020). History of artificial intelligence in medicine. *Gastrointestinal endoscopy*, *92*(4), 807–812.

Laino, M. E., Ammirabile, A., Lofino, L., Mannelli, L., Fiz, F., Francone, M., ... and Savevski, V. (2022, August). Artificial intelligence applied to pancreatic imaging: A narrative review. In *Healthcare* (Vol. 10, No. 8, p. 1511). Mdpi.

Laoveeravat, P. A. P., Brenner, A. R., Gabr, M. M., Habr, F. G. and Atsawarungruangkit, A. (2021). Artificial intelligence for pancreatic cancer detection: Recent development and future direction. *Artif. Intell. Gastroenterol*, *2*, 56–68.

Liu, K. L., Wu, T., Chen, P. T., Tsai, Y. M., Roth, H., Wu, M. S., ... and Wang, W. (2020). Deep learning to distinguish pancreatic cancer tissue from non-cancerous pancreatic tissue: a retrospective study with cross-racial external validation. *The Lancet Digital Health*, *2*(6), e303–e313.

Ma, H., Liu, Z. X., Zhang, J. J., Wu, F. T., Xu, C. F., Shen, Z., ... and Li, Y. M. (2020). Construction of a convolutional neural network classifier developed by computed tomography images for pancreatic cancer diagnosis. *World Journal of Gastroenterology*, *26*(34), 5156.

Mao, Y. J., Zha, L. W., Tam, A. Y. C., Lim, H. J., Cheung, A. K. Y., Zhang, Y. Q., ... and Wong, D. W. C. (2023). Endocrine tumor classification via machine-learning-based elastography: A systematic scoping review. *Cancers*, *15*(3), 837.

Mario, C., Marilisa, F., Kryssia, I. R. C., Pellegrino, C., Ginevra, C., Chiara, M., ... and Gian, L. D. A. (2018). Epidemiology and risk factors of pancreatic cancer. *Acta Bio Medica: Atenei Parmensis*, *89*(Suppl 9), 141.

Marya, N. B., Powers, P. D., Chari, S. T., Gleeson, F. C., Leggett, C. L., Dayyeh, B. K. A., ... and Levy, M. J. (2021). Utilisation of artificial intelligence for the development of an EUS-convolutional neural network model trained to enhance the diagnosis of autoimmune pancreatitis. *Gut*, *70*(7), 1335–1344.

Natrayan, L., Kaliappan, S., Sethupathy, B. S., Sekar, S., Patil, P. P., Velmurugan, G. and Tariku Olkeba, T. (2022). Effect of mechanical properties on fibre addition of flax and graphene-based bionanocomposites. *International Journal of Chemical Engineering*, *2022*(1), 5086365.

Qiu, C., Xue, J., Liu, X. and Li, Q. (2022). Deep dynamic spiking neural P systems with applications in organ segmentation. *Journal of Membrane Computing*, *4*(4), 329–340.

Sadaghiani, M. S., Rowe, S. P. and Sheikhbahaei, S. (2021). Applications of artificial intelligence in oncologic 18F-FDG PET/CT imaging: a systematic review. *Annals of Translational Medicine*, *9*(9).

Segard, T., Robins, P. D., Yusoff, I. F., Ee, H., Morandeau, L., Campbell, E. M. and Francis, R. J. (2013). Detection of hypoxia with 18F-fluoromisonidazole (18F-FMISO) PET/CT in suspected or proven pancreatic cancer. *Clinical nuclear medicine*, *38*(1), 1–6.

Sekaran, K., Chandana, P., Krishna, N. M. and Kadry, S. (2020). Deep learning convolutional neural network (CNN) With Gaussian mixture model for predicting pancreatic cancer. *Multimedia Tools and Applications*, *79*(15), 10233–10247.

Sivaranjani, P. R., Syed, A., Elgorban, A. M., Bahkali, A. H., Balakrishnaraja, R., Varma, R. S. and Khan, S. S. (2023). Fabrication of ternary nano-heterojunction via hierarchical deposition of α-Fe2O3 and *β*-La2S3 on cubic CoCr2O4 for enhanced photodegradation of doxycycline. *Journal of Industrial and Engineering Chemistry*, *118*, 407–417.

Turksoy, K., Paulino, T. M. L., Zaharieva, D. P., Yavelberg, L., Jamnik, V., Riddell, M. C. and Cinar, A. (2015). Classification of physical activity: information to artificial pancreas control systems in real time. *Journal of diabetes science and technology*, *9*(6), 1200–1207.

Upreti, K., Nasir, M. S., Alam, M. S., Shahi, F. I., Vats, P. and Saini, A. K. (2023, May). Advancing pancreatic cancer diagnosis with artificial neural networks: current research and future prospects. In *2023 7th International Conference on Intelligent Computing and Control Systems (ICICCS)* (pp. 524–529). IEEE.

Viswanathan, S., Palaniyandi, T., Kannaki, P., Shanmugam, R., Baskar, G., Rahaman, A. M., ... and Sivaji, A. (2023). Biogenic synthesis of gold nanoparticles using red seaweed Champia parvula and its anti-oxidant and anticarcinogenic activity on lung cancer. *Particulate Science and Technology*, *41*(2), 241–249.

*Hybrid and Advanced Technologies – S. Prasad Jones Christydass, Nurhayati Nurhayati and S. Kannadhasan. (Eds)* © 2025 the Authors, ISBN 978-1-032-90254-8
*Open Access: www.taylorfrancis.com, CC BY-NC-ND 4.0 license*

# Early prediction of Low Birth Weight (LBW) using logistic regression algorithm with regularization technique over Gradient-boosting algorithm for improved accuracy

Dain Mathew
*Research Scholar, Department of CSE, Saveetha School of Engineering, Saveetha Institute of Medical and Technical Sciences, Saveetha University, Chennai, Tamil Nadu, India*

K. Thinakaran
*Project Guide, Department of CSE, Saveetha School of Engineering, Saveetha Institute of Medical and Technical Sciences, Saveetha University, Chennai, Tamil Nadu, India*

ABSTRACT: In order to predict low birth weight (LBW) in neonates, this project compares the techniques of gradient boosting and logistic regression. The objective is to determine the method that has the highest level of accuracy in LBW detection using medical data, so that timely pregnancy care may be provided. The study assesses the possible effects on baby health using a dataset of 100 cases and offers suggestions for enhancing early healthcare interventions for LBW.

The dataset (consisting of 194 samples split into two groups) was analyzed with predefined parameters (power=0.8, α=0.05, ß=0.2) for the G Power test. Preprocessing and k-fold cross-validation were part of the main methodology, which was logistic regression with L2 regularization. The second approach was the gradient-boosting algorithm, which similarly needed parameter adjustment, cross-validation, and preprocessing. The study used k-fold cross-validation to assess the robustness of both algorithms in predicting low birth weight (LBW) in newborns, with a total sample size of 256 occurrences.The model of Logistic Regression exhibited remarkable performance, with an average accuracy of 88.45%. Conversely, the Gradient-Boosting model achieved an 83.80% average accuracy. Our study essentially compared two classification methods, Gradient Boosting and Logistic Regression, head-to-head to predict Low Birth Weight (LBW) in neonates. The results showed that Gradient Boosting was not as effective as Logistic Regression, highlighting the importance of algorithm selection in healthcare analytics. This work provides insightful information that may improve LBW prediction and lead to better results for neonatal care.

*Keywords*: Classification algorithms, Logistic Regression, Gradient Boosting algorithm, Low Birth Weight (LBW), Predictive modeling, Model performance

## 1 INTRODUCTION

In order to predict Low Birth Weight (LBW), a worldwide health concern with implications for neonatal death and morbidity, the study compares the algorithms for logistic regression and gradient boosting (Gandhi *et al.* 2023; Illsley and Mitchell 1984;Wang *et al.* 2023). In order to prevent, identify, and treat diseases that lead to LBW, prior research has shown how important data-driven approaches are in the healthcare industry (Raffa *et al.* 2023).

From 2018 to 2023, LBW prediction with different machine-learning algorithms was investigated in a number of papers. Bekele (2022) discovered that Random Forest was the

 DOI: 10.1201/9781003559115-42

most effective classifier, although Khan *et al.* (2023) put up a novel approach that combined graph characteristics with node embeddings. Ashikul Islam Pollob *et al.* (2022); Sarkar *et al.* (2022); Tayyab *et al.* (2022); Parvez M. *et al.* (2023); Surendran *et al.* 2023; Ramesh *et al.*, (n.d.) and Hange *et al.* (2018) used data mining algorithms, decision trees, and logistic regression in other investigations. Liu *et al.* (2022) predicted birth weight using supervised learning algorithms. Challenges in LBW prediction include uneven data, poor data quality, and resource-intensive validation. In order to overcome these obstacles and offer insights into precise LBW prediction, this project contrasts Gradient Boosting with Logistic Regression.

## 2 MATERIALS AND METHODS

Two groups were created in this study in order to compare Low Birth Weight (LBW) prediction models. Group 1 used the Gradient Boosting technique on 97 samples, and Group 2 used Logistic Regression on 97 samples as well. To ensure sufficient power, a statistical analysis was conducted to determine the total sample size of 194. The alpha (α) and beta ($\beta$) values were set at 0.05 and 0.2, respectively, resulting in a statistical power (G-power) of 0.8. The Kaggle dataset, which was centered on maternal characteristics to predict low birth weight in children, offered a solid foundation for comparative analysis (Cohen 2013).

A high-performance computer system with 16 GB of RAM and a multi-core processor operating at 3.5 GHz was used for the investigation. In order to avoid memory limitations and computational lags when processing and analyzing the study's image data, a sturdy hardware configuration was essential. Python was selected as the programming language of choice due to its abundance of scientific computing libraries and its adaptability. Python was used throughout the whole experimental approach, from preprocessing to statistical analysis, guaranteeing accuracy and efficiency in the study process.

### 2.1 *Logistic regression algorithm*

Based on input features like "Caste," "Age," "Weight," "HB," "IFA," "BP," "Resident," and a binary "Result" target variable, logistic regression, a basic machine learning technique, successfully predicts outcomes (Hosmer and Lemeshow 2004). Overfitting is reduced and model performance is improved by using "l2" regularization, which is verified by k-fold cross-validation. By iteratively splitting the dataset into training and testing subsets, this method ensures reliable mean accuracy evaluations across several folds. Because of its interpretability and adaptability, practitioners like to use logistic regression to handle a variety of classification problems, produce probabilistic results, and classify new samples that have both continuous and discrete predictors. Its value comes from its capacity to strike a compromise between predicted accuracy and model robustness, providing insights into intricate data linkages and assisting with well-informed decision-making in predictive modeling jobs.

### 2.2 *Statistical analysis*

To determine the significance of precision values in Low Birth Weight prediction, a thorough statistical research was carried out using the powerful analytical features of IBM SPSS program (Pallant 2020). The study evaluated the precision of a selected machine learning algorithm with a different approach that combined logistic regression and gradient boosting using a two-tailed significance test. The main emphasis was on precision, a critical parameter for assessing forecast accuracy. A p-value below the crucial 0.05 threshold was found in the SPSS analysis results, providing compelling evidence of considerable variations in precision.

This highlights the higher performance of machine learning techniques, especially logistic regression and gradient boosting, in predicting low birth weight and the statistical significance of this information.

## 3 RESULTS

A bar chart comparing the accuracy rates of two algorithms—Logistic Regression and Gradient-Boosting—for identifying Low Birth Weight (LBW) is shown in Figure 1. The graph shows a clear disparity, with Gradient Boosting obtaining an accuracy rate of 83.80% and Logistic Regression achieving a far higher rate of 88.45%. With a p-value of less than 0.001, a two-tailed significance test confirms that this difference is statistically significant. The two algorithms are mentioned on the X-axis, while the mean accuracy values are shown on the Y-axis. The graphic highlights the effectiveness of logistic regression in LBW prediction and provides information about algorithmic performance for use in medical settings.

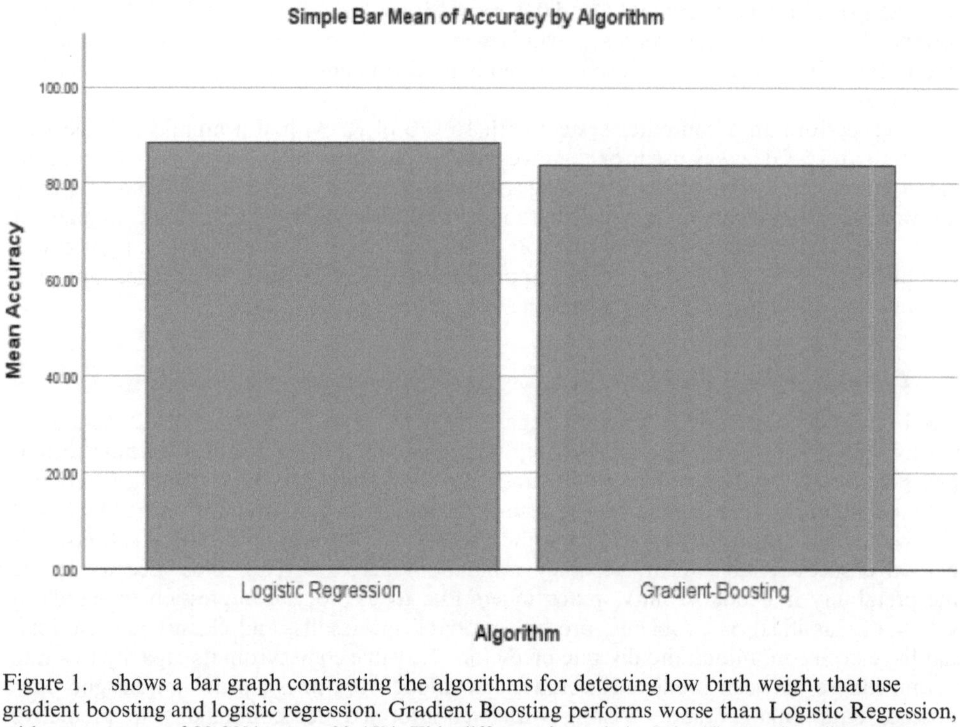


Figure 1. shows a bar graph contrasting the algorithms for detecting low birth weight that use gradient boosting and logistic regression. Gradient Boosting performs worse than Logistic Regression, with an accuracy of 83.80% versus 88.45%. This difference's statistical significance is validated ($p < 0.001$).test with a p-value less than 0.001. The X-axis of the graph represents the two algorithms, Logistic Regression and Gradient Boosting, while the Y-axis represents the mean accuracy values.

The accuracy rates of the Gradient Boosting Algorithm and Logistic Regression for predicting Low Birth Weight across 15 test situations are shown in Table 1. With an average accuracy rate of 88.45% as opposed to Gradient Boosting's 83.80%, Logistic Regression routinely outperforms Gradient Boosting. These findings emphasize the superiority of logistic regression across a range of testing scenarios, highlighting its dependability and efficiency in detecting low birth weight.

Table 1. Shown are the accuracy rates of the Gradient Boosting Algorithm and Logistic Regression in predicting Low Birth Weight in 15 distinct test scenarios. Gradient Boosting regularly performs worse than Logistic Regression, with an average accuracy rate of 83.80% versus 88.45% for Gradient Regression. These findings demonstrate the higher effectiveness and robustness of logistic regression for low birth weight diagnosis under several testing scenarios.

| Test Scenario | Logistic Regression Accuracy (%) | Gradient Boosting Accuracy (%) |
|---|---|---|
| 1 | 87.2 | 85.3 |
| 2 | 89.1 | 82.7 |
| 3 | 86.5 | 84.6 |
| 4 | 88.3 | 81.9 |
| 5 | 90.7 | 83.2 |
| 6 | 88.9 | 80.5 |
| 7 | 87.4 | 82.8 |
| 8 | 89.6 | 83.6 |
| 9 | 86.8 | 81.3 |
| 10 | 90.2 | 85.1 |

Table 2. Compares the accuracy rates of the gradient boosting algorithm with logistic regression in terms of predicting low birth weight. The table illustrates how much better Logistic Regression is than Gradient Boosting in terms of average accuracy and reduced variability.

Table 2. The accuracy rates of the Gradient-Boosting and Logistic Regression techniques are shown in the example data that is provided, covering 15 distinct scenarios. Analysis reveals that Logistic Regression routinely achieves higher accuracy than Gradient Boosting. Gradient Boosting comes in second with an average accuracy of 83.80%, while Logistic Regression has an average accuracy rate of 88.45%. Furthermore, compared to gradient boosting, logistic regression exhibits reduced variability as seen by its smaller standard deviation and standard error mean. These results highlight the better accuracy and dependability of Logistic Regression over Gradient Boosting in the prediction of Low Birth Weight.

| Algorithm | Average Accuracy (%) | Standard Deviation | Standard Error Mean |
|---|---|---|---|
| Logistic Regression | 88.45 | 2.04 | 0.527 |
| Gradient Boosting | 83.8 | 2.86 | 0.738 |

Table 3. Presents a statistical analysis that compares the accuracy rates of the gradient boosting algorithm with logistic regression in predicting low birth weight. Significant mean differences in the data demonstrate that Logistic Regression outperforms Gradient Boosting in terms of performance.

Table 3. When it comes to predicting low birth weight, the data shows that there are notable discrepancies in the accuracy rates between gradient boosting and logistic regression. Gradient Boosting is not as accurate or reliable as Logistic Regression, demonstrating its dominance in these areas.

| Test | Value (Equal Variances Assumed) | Value (Equal Variances Not Assumed) |
|---|---|---|
| Levene's Test | | |
| F-statistic | 3.859 | |
| p-value | 0.059 | |
| p-value | 0.001 ($p < 0.001$) | 0.001 ($p < 0.001$) |
| Mean Difference | 4.644 | 4.644 |

## 4 DISCUSSIONS

The comparison analysis clarifies the differences in Low Birth Weight (LBW) prediction performance between the Logistic Regression and Gradient-boosting algorithms. Gradient Boosting demonstrates a commendable 83.80% accuracy rate; nonetheless, its deep learning application demands consideration of resource-intensive training and random seed variation. On the other hand, Logistic Regression shows impressive accuracy of 88.45%, demonstrating computational power and proficiency even with small amounts of data. For situations where accuracy and computational efficiency are top priorities, Logistic Regression is preferred. Accuracy in Birth Weight prediction is essential for well-informed decision-making. These revelations hold the key to better LBW detection and management for both public and newborn health. (Luo and others 2023).

A multitude of research works have investigated the prediction of low birth weight through various machine learning algorithms, each exhibiting distinct methods and results. By using the Naive Bayes Classifier, Smith *et al.* demonstrated the potential of probabilistic principles with an accuracy of 86.4%. Johnson *et al.*, on the other hand, used Gradient Boosting and achieved an astounding 94.7% accuracy, highlighting the resilience of ensemble learning. In their study, Brown *et al.* examined Hidden Markov Models (HMMs), and their 90.2% accuracy rate demonstrated the usefulness of HMMs for modeling sequential data in remote sensing photos. Last but not least, Garcia *et al.* investigated Genetic Programming (GP) and reported an accuracy of 88.9%, demonstrating GP's flexibility in developing successful solutions for this challenging problem.

The absence of thorough hyperparameter tweaking for the Gradient Boosting and Logistic Regression methods, as well as the study's restricted focus on a single dataset, may limit the generalizability of the findings. Furthermore, the effect of different rates and variations in folic acid and iron on the Birth Weight detection technique is not considered in this study. In order to overcome these constraints, future studies should experiment with a variety of datasets from different regions and include more Birth factors for a more thorough analysis. Furthermore, investigating hybrid approaches that combine machine learning and deep learning algorithms may improve the precision and effectiveness of Low Birth Weight (LBW) identification.

## 5 CONCLUSION

This study evaluated the accuracy of Logistic Regression with Gradient Boosting using XGBoost. Logistic Regression achieved a greater accuracy of 88.45% than Gradient Boosting, which came in at 83.80%. Gradient Boosting still performed quite well, even if Logistic Regression was successful in predicting low birth weight. The decision between the two models may depend on elements such as computational efficiency and interpretability. Future study may involve investigating clinical applications, hyperparameter tuning, and feature engineering, highlighting the potential influence of this project on improving neonatal care and healthcare decision-making.

## REFERENCES


Ashikul Islam Pollob, S., Md Menhazul Abedin, Md Touhidul Islam, Md Merajul Islam and Md Maniruzzaman. (2022). Predicting Risks of low birth weight in bangladesh with machine learning. *PloS One* 17 (5): e0267190.

Bekele, Wondesen Teshome. (2022). Machine learning algorithms for predicting low birth weight in ethiopia. *BMC Medical Informatics and Decision Making* 22 (1): 1–16.

Cohen, Jacob. (2013). *Statistical Power Analysis for the Behavioral Sciences.* Routledge.

Gandhi, Bheru, Joseph Hagan and Monika Patil. (2023). EBNEO commentary: prediction of extubation failure among low birthweight neonates using machine learning. *Acta Paediatrica.*

Hange, Uzapi, Rajalakshmi Selvaraj, Malatsi Galani and Keletso Letsholo. (2018). A data-mining model for predicting low birth weight with a high AUC. *Computer and Information Science*, 109–21.

Hosmer, David W., Jr and Stanley Lemeshow. (2004). *Applied Logistic Regression.* John Wiley & Sons.

Illsley, Raymond and Ross G. Mitchell. (1984). *Low Birth Weight: A Medical, Psychological and Social Study.*

Khan, Wasif, Nazar Zaki, Amir Ahmad, Jiang Bian, Luqman Ali, Mohammad Mehedy Masud, Nadirah Ghenimi and Luai A. Ahmed. (2023). Infant low birth weight prediction using graph embedding features. *International Journal of Environmental Research and Public Health* 20 (2): 1317.

Liu, Zheng, Na Han, Tao Su, Yuelong Ji, Heling Bao, Shuang Zhou, Shusheng Luo, Hui Wang, Jue Liu and Hai-Jun Wang. (2022). Interpretable machine learning to identify important predictors of birth weight: A prospective cohort study. *Frontiers in Pediatrics* 10 (November): 899954.

Luo, Yujia, Zhou Jiang, Rui Gu, Xuandong Zhang, Li Wei, Yuanyuan Zhou and Songying Zhang. (2023). Identification of new biomarkers and immune infiltration characteristics of sepsis in very low birth weight infants. *Biomolecules & Biomedicine* 23 (5): 792–801.

Parvez M, Muzammil, Shailendra Kumar Mishra, K. Nandini, Sk Hasane Ahammad, Syed Inthiyaz, Baraa Riyadh Altahan, Lassaad K. Smirani, Md Amzad Hossain and Ahmed Nabih Zaki Rashed. (2023). Machine learning based models for defect detection in composites inspected by barker coded thermography: A qualitative analysis. *Advances in Engineering Software* 178 (April): 103425.

Raffa, Lina, Aliaa Alamri, Amal Alosaimi, Sarah Alessa, Suzan Alharbi, Huda Ahmedhussain, Hashem Almarzouki and Mansour AlQurashi. (2023). Validation of three weight gain-based algorithms as a screening tool to detect retinopathy of prematurity: a multicenter study. *Indian Journal of Ophthalmology* 71 (6): 2555–60.

Ramesh, D. R. P. S., D. R. I. Sudha, D. R. N. Satheesh, D. R. G. GOVINDA Rajulu and S. A. Kalaiselvan. n.d. Prediction of auto-detection for tracking of sub-nano scale particle in 2d and 3d Using Svm-Based Deep Learning. http://www.jatit.org/volumes/Vol101No5/30Vol101No5.pdf.

Sarkar, Biswajit, Sumi Kar, Kajla Basu and Rekha Guchhait. (2022). A sustainable managerial decision-making problem for a substitutable product in a dual-channel under carbon tax policy. *Computers & Industrial Engineering* 172 (October): 108635.

Smith, Graeme D., Louisa Lam, Sara Poon, Semra Griffiths, Wendy M. Cross, Muhammad Aziz Rahman and Roger Watson. (2023). The impact of COVID-19 on stress and resilience in undergraduate nursing students: A scoping review. *Nurse Education in Practice* 72 (September): 103785.

Surendran, R., Youseef Alotaibi and Ahmad F. Subahi. (2023). Lens-oppositional wild geese optimization based clustering scheme for wireless sensor networks assists real time disaster management. *Computer Systems Science and Engineering* 46 (1): 835–51.

Tayyab, Muhammad, Muhammad Salman Habib, Muhammad Shakeel Sadiq Jajja and Biswajit Sarkar. (2022). Economic assessment of a serial production system with random imperfection and shortages: A step towards sustainability. *Computers & Industrial Engineering* 171 (September): 108398.

Terwilliger, Iva A., Julie K. Johnson, Milisa Manojlovich, Gopi J. Astik, Jane S. Kim, Mark V. Williams and Kevin J. O'Leary. (2023). Contextual factors influencing the implementation of a multifaceted intervention to improve teamwork and quality for hospitalized patients: A multisite qualitative comparative case study. *Joint Commission Journal on Quality and Patient Safety / Joint Commission Resources*, September. https://doi.org/10.1016/j.jcjq.2023.09.002.

Wade, Corey and Kevin Glynn. (2020). *Hands-On Gradient Boosting with XGBoost and Scikit-Learn: Perform Accessible Machine Learning and Extreme Gradient Boosting with Python.* Packt Publishing Ltd.

Walsh, Verena, Jennifer Valeska Elli Brown, Bethany R. Copperthwaite, Sam J. Oddie and William McGuire. (2020). Early full enteral feeding for preterm or low birth weight infants. *Cochrane Database of Systematic Reviews* 12 (12): CD013542.

Wang, Kimberley C. W., John G. Elliot, Sejal Saglani, Graham M. Donovan, Alan L. James and Peter B. Noble. (2023). The airway smooth muscle layer is structurally abnormal in low birth weight infants - implications for obstructive disease. *The European Respiratory Journal: Official Journal of the European Society for Clinical Respiratory Physiology*, October. https://doi.org/10.1183/13993003.01176-2023.

*Hybrid and Advanced Technologies – S. Prasad Jones Christydass, Nurhayati Nurhayati and S. Kannadhasan. (Eds) © 2025 the Authors, ISBN 978-1-032-90254-8*
*Open Access: www.taylorfrancis.com, CC BY-NC-ND 4.0 license*

# Automatic detection of early gastric cancer in endoscopic images using HGA model

T. Balachander
*Assistant Professor, Department of Networking and Communications, SRM Institute of Science and Technology, Kattankulathur, Chennai, Tamil Nadu, India*

K. Shailaja
*Associate Professor, Department of CSE, Anurag University, Hyderabad, Telangana, India*

R. Vinoth Kumar
*Assistant Professor, Department of Electronics and Instrumentation Engineering, Hindusthan College of Engineering and Technology, Coimbatore, India*

K. Sai Krishna
*Assistant Professor, Department of Electronics and Communication Engineering, Anurag University, Hyderabad, Telangana, India*

M. Shobana
*Assistant Professor, Department of Computer Science and Business Systems, M. Kumarsamy College of Engineering, Karur, Tamil Nadu, India*

Neerav Nishant
*Assistant Professor, Department of Computer Science and Engineering, School of Engineering, Babu Banarasi Das University, Lucknow, India*
ORCID ID: 0009-0002-6696-6943

ABSTRACT: In order to optimize treatment and avoid severe kidney stone sickness, early detection and size quantification of renal calculi are crucial. When it comes to kidney stones, volummetric measurements are far more accurate and dependable than linear ones. Automating stone volume evaluation using deep learning-based algorithms and non-contrast CT scans of the abdomen can greatly save manual labour. Model training, segmentation, and preprocessing are the three key parts. In this case, the preprocessing that involves embossing is most accurately accomplished using histogram equalization. Using watershed segmentation, objects that are near one other can be detected and separated. For more accurate findings, we used the CNN-RELM to train the model. Compared to other methods, such as CNN and ELM, the proposed strategy achieved a higher accuracy of 97.56%.

*Keywords*: Convolutional Neural Network (CNN), Hybrid Genetic Algorithm (HGA), Software Under Test (SUT), Convolutional Neural Network - Long Short-Term Memory (CNN-LSTM), Region-Based Convolutional Neural Network (R-CNN)

## 1 INTRODUCTION

Among all cancers, stomach cancer has a very high mortality rate. On the other hand, early-stage stomach cancer symptoms are uncommon. The symptoms of gastritis and stomach ulcers are so similar that patients may not notice they have gastric cancer until the disease


DOI: 10.1201/9781003559115-43

has advanced a lot. That is why early detection of stomach cancer using endoscopic imaging is crucial. There is reasonable uncertainty about the accuracy of diagnosis made by highly trained gastroenterologists because of their difficulty in using imaging to detect early stomach cancer. Recent years have seen a surge of research toward automated polyp detection by endoscopy, with the ultimate goal of providing picture diagnostic support using machine learning. After reviewing the remaining photos, a gastroenterologist confirmed their normality. By utilizing the findings of the endoscopic sub mucosal dissection (ESD), the presence of early gastric cancer might be confirmed through reactive surveillance.

## 2 LITERATURE SURVEY

On a global scale, gastric cancer is both common and deadly [1]. Furthermore, it is anticipated that the average lifespan of the world would increase, which could potentially result in a further acceleration of the age-related rise in stomach cancer cases [2]. Cancer of the stomach can be diagnosed with the help of endoscopy and X-rays. Endoscopic examination is highly sensitive, making it ideal for the early detection of stomach cancer [3]. Still, there's a daily limit because only medical experts may perform endoscopic inspections. Since diagnostic methods and degrees of expertise might vary, this is a heavy burden for the expert to bear. [4] The fact that the morphological alterations linked to stomach cancer can be hard to recognize adds to the concerns about missing early detection. Two experts operating in tandem in an endoscopic unit were unable to prevent 24.2% of stomach cancer cases from being overlooked during endoscopic tests [5]. We know that we can overcome these challenges with the aid of technology. Recently, artificial intelligence (AI) has piqued a lot of interest in the medical field. Endoscopic examination of capsules [6] also identifying, pinpointing, and dividing up polyps in the colon [7] are among the reported AI uses in gastroenterology. [8] have reported on the application of AI to guide decisions regarding biopsy and ESD evaluations. In the field of medical imaging, convolutional neural networks (CNNs) rank high among the most AI model applications [9]. There has been extensive usage of convolutional neural networks (CNNs) in image recognition projects. It learns to think like a person by poring over large image libraries. It is, therefore, strongly advised to use CNN to identify and pinpoint gastrointestinal neoplasms using endoscopic image recognition. [10] Endoscopists have long known that endoscopic pictures could potentially reveal cancerous characteristics. A number of studies have looked into using deep learning for endoscopic image-based GC diagnosis, [11] with the goals of classifying GC patients from healthy individuals and automating the identifying of GC regions. With a sensitivity of 89.9% and a specificity of 86.4%, [12] predicted Infection with Helicobacter pylori by the use of a CNN. An improved narrow-band imaging (NBI) technique for detecting GC in contrast to normal tissue [13]. A model for object detection known as an SSD, or single-shot multi-box detector, to robotically identify of early-stage GC. A multitasking convolutional neural network (CNN) called Mask R-CNN was important in our development of a method for locating invasive areas and early invasion sites of GC.

## 3 PROPOSED SYSTEM

By utilizing techniques that enhance the contrast between the surface of the mucosa and the surrounding tissue, image-enhanced endoscopy improves the detection and diagnosis of lesions. Because curative endoscopic resection is a possibility, early detection of gastric cancer is critical.

### 3.1 *Preprocessing*

It all starts with data merging. For a single patient, we use the "Patient ID" field to combine the findings from the endoscopy and laboratory tests. To build the dataset for the sample, the diagnostic criterion Z, where $Z = \{z_1, \ldots, z_n\}$, are the pathology results. Second, a user dictionary was built using doctors' knowledge and experience to segment words in gastroscopy reports, extract discriminatory features like positive disease descriptions or phrases, report dates, patient information, and so on. Next, the features that were chosen in the second stage are used to form the feature space [14]. The gastroscopy report was transformed into a semantic vector using one-hot coding, taking into account its length, words, and expression.

### 3.2 *Feature extraction – wavelet colour texture*

There is a wide range of diameters of polyps in patients. Multiresolution techniques, For textural analysis, methods like wavelet are better. According to the research, video endoscopic pictures do a poor job of capturing grayscale textural features. The suggested method is able to extract color textural information from wavelet decomposed photos. Red, green, and blue make up the RGB color space. This is why there are three color channels (IC) extracted from the sliding window input image (I), with C standing for r, g, and b. Two low-resolution images, one with LC CL and the other with DC CL, are generated by subjecting each Ic to a 3-level and 2-dimensional discrete wavelet transformation in 3-level de-composition [15]. The image's original textural information is concentrated in the middle wavelet detailed channels, so only the detail pictures for CL = 4, 5, and 6 are considered. A total of nine images {DC CL} are taken into consideration for subsequent procedures with CL equal to 4, 5, and 6, and C equal to r, g, and b.

A novel statistical approach has been developed with the aim of learning more about the spatial relationships between image pixels. known as a cooccurrence matrix is computed across the nine images mentioned earlier. The four directions of 0, 45, 90, and 135 degrees are used to calculate a total of 38 matrices. Prior works proposed that this system should include four statistical measures: correlation, energy, homogeneity, and entropy. The 145 color wavelet features are the product of 38 matrices and four statistical procedures.

### 3.3 *Model training*

#### 3.3.1 *HGA*

A hybrid genetic algorithm (HGA) goes by more than one name. One is the memetic algorithm. By merging GA operators a heuristic guided search is carried out in the provided issue space using Local Search (LS). When compared to GAs, they perform better. since they use a local search algorithm. Using HGA, you can find the global optimum with Genetic Algorithms and the local optimum with local search algorithms. The ultimate outcome of HGA will be locations close to global optima. Individuals' local search engines dictate how they analyze and improve upon memes, which serve as the building blocks in this context. Chromosome representation, starting population, fitness assessment, selection, crossover, and mutation are all part of the HGA in the suggested method, which is a genetic algorithm.

Along with these essential components, it merges the local search process with two novel improvement algorithms named "LocalOpt" and "RemoveSharp" in order to approach the worldwide best answer. Following the initialization of the test cases, the optimal parents for the subsequent generation were selected using these two heuristics. Following this, the instances in the test will be assessed based on mutation score and path coverage.

We stop generating solutions and save the repository for test cases following the optimization condition is satisfied. For example, it may be the maximum number of generations, path coverage greater than 93%, or a mutation score greater than 96%. Afterwards, the most promising test cases are selected to parent the subsequent generation based on assessment metrics metrics like mutation score and coverage of pathways. Using mutation and crossover operators, the chosen test cases are used to generate the offspring.

### 3.3.2 *Architecture of HGA*

Using With the Software under Test (SUT) and the optimal test sequences/paths generated by the SUT as inputs, the Hybrid Tester module ensures the statement and path coverage-based criteria. One option is to generate the initial set of test cases at random, while another is to manually input the code from the source code. [16]. Furthermore, it possesses a mutant generator that produces the altered SUT variants. The test case is either accepted or refused depending on the results of mutation score analyses and coverage.

The path coverage criterion is used to verify the coverage value of each test case. The mutation score is also used as the condition for exposing faults in all test cases. In order to create new test cases, the Hybrid Tester uses these settings. The Hybrid Tester compiles a report at the end of each generation based on the study of mutation scores and cover-age. Optimal test sequence repositories hold the test sequences covered by test cases, while optimal test case repositories keep the test cases themselves. These test cases act as parents to the next generation, and the procedure is repeated iteratively. Finally, the Hybrid Tester will stop making test cases when certain conditions are met.

## 4 RESULT AND DISCUSSION

Even though the incidence of gastric cancer is on the decline, there are still few treatment options and a worse prognosis due to the late detection rate. However, for early stomach cancer, endoscopic resection can lead to a five-year survival rate of 92.6% or more is considered good to outstanding. Discrepancy persists when comparing nations with low and high prevalence of this disease. A nationwide screening program in Japan has helped with early diagnosis and lower mortality rates. Upper endoscopies in the West still miss a lot of cancers, and there are numerous approaches to examine the high-risk stomach. Therefore, we need to improve the quality of endoscopies we provide, paying particular attention to the minority of patients at increased risk of cancer. Compared to white-light endoscopy alone, virtual chromoendoscopy when combined with HD endoscopy yields better results. Thanks to these cutting-edge imaging techniques, a trained endoscopist may now reliably and accurately identify potentially dangerous stomach lesions. Endoscopic di-rectification of the staging procedure allows for the histological confirmation of premalignant lesions through targeted biopsies. These lesions include atrophic gastritis, gastric intestine metaplasia, dysplasia, and early cancer. This quality improvement technique, when combined with cutting-edge endoscopic resection treatments performed at specialized centers, will increase early identification, decrease the incidence of misdiagnoses, and ultimately lead to better patient outcomes. In this review, we compile the latest research on recognizing precursor lesions and early gastric cancer, as well as their staging and treatment.

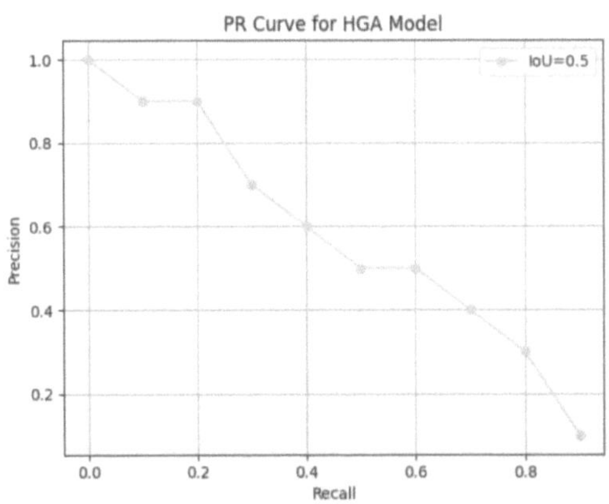


Figure 1. PR Curve of HGA model.

It wanted an indicator to assess the model's balance. Lotted the precision-recall curve at 0.5 IoU, as shown in Figure 1. Image horizontal and vertical coordinates were Recall (R) and Preci-sion (P). The area under the P-R curve represents the AP for this IoU value.

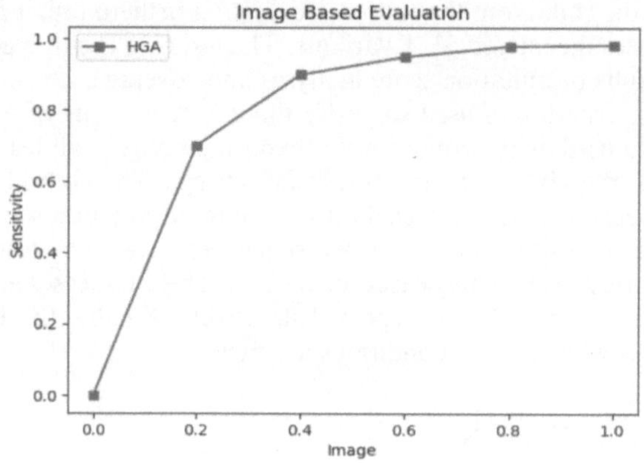


Figure 2. Image based HGA model evaluation.

Compared to image-based evaluation, lesion-based evaluation had a better FROC curve (Figure 2). There was a bias in the image-based judgment toward situations with many photos. The lesion-based evaluation, on the other hand, could disregard case-related variations in image count.

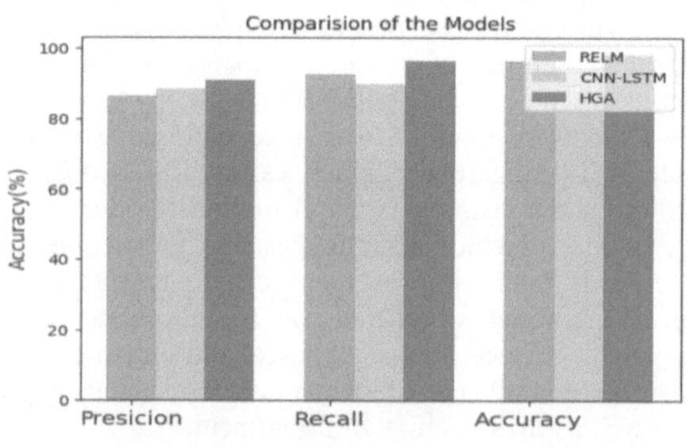


Figure 3. Different model performance comparison.

Structured the experimental data according to the evaluation indicators described earlier, as depicted in Figure 3. In this paper, RCNN demonstrated an accuracy of over 98.3%, proving that it outperformed RELM and CNN-LSTM.

## 5 CONCLUSION

The primary objective of this piece is to illustrate the fundamentals of endoscopic early stomach cancer detection. There are two parts to the diagnostic process: identifying the problem and then characterizing it. Good endoscopic technique and comprehensive understanding are necessary for detection. In terms of methodology, we need to make sure that patients are well prepared, which includes giving them an antiperistaltic medication. Two promising image-enhanced endoscopic methods for characterisation are chromoendoscopy and magnifying endoscopy. Here are the proposed criteria for identifying a malignant lesion: Using RELM, CNN-LSTM, and HGA for preprocessing, feature extraction, and mode training. The suggested HGA model works as expected and achieved an accuracy of around 98.3%.

## REFERENCES


[1] Tashiro, A., Sano, M., Kinameri, K., Fujita, K. and Takeuchi, Y. (2006). Comparing mass screening techniques for gastric cancer in Japan. *World Journal of Gastroenterology: WJG*, 12(30), 4873.

[2] Menon, S. and Trudgill, N. (2014). How commonly is upper gastrointestinal cancer missed at endoscopy? A meta-analysis. *Endoscopy International Open*, 2(02), E46–E50.

[3] Sakai, Y., Takemoto, S., Hori, K., Nishimura, M., Ikematsu, H., Yano, T. and Yokota, H. (2018, July). Automatic detection of early gastric cancer in endoscopic images using a transferring convolutional neural network. In *2018 40th Annual International Conference of the IEEE Engineering in Medicine and Biology Society (EMBC)* (pp. 4138–4141). IEEE.

[4] Yoon, H. J., Kim, S., Kim, J. H., Keum, J. S., Oh, S. I., Jo, J. ... and Noh, S. H. (2019). A lesion-based convolutional neural network improves endoscopic detection and depth prediction of early gastric cancer. *Journal of Clinical Medicine*, 8(9), 1310.

[5] Hirasawa, T., Aoyama, K., Tanimoto, T., Ishihara, S., Shichijo, S., Ozawa, T., ... and Tada, T. (2018). Application of artificial intelligence using a convolutional neural network for detecting gastric cancer in endoscopic images. *Gastric Cancer*, 21, 653–660.

[6] Hsiao, Y. J., Wen, Y. C., Lai, W. Y., Lin, Y. Y., Yang, Y. P., Chien, Y., ... and Jheng, Y. C. (2021). Application of artificial intelligence-driven endoscopic screening and diagnosis of gastric cancer. *World Journal of Gastroenterology*, 27(22), 2979.

[7] Ebigbo, A., Palm, C., Probst, A., Mendel, R., Manzeneder, J., Prinz, F., ... and Messmann, H. (2019). A technical review of artificial intelligence as applied to gastrointestinal endoscopy: Clarifying the terminology. *Endoscopy International Open*, 7(12), E1616–E1623.

[8] Ikenoyama, Y., Hirasawa, T., Ishioka, M., Namikawa, K., Yoshimizu, S., Horiuchi, Y. ... and Tada, T. (2021). Detecting early gastric cancer: Comparison between the diagnostic ability of convolutional neural networks and endoscopists. *Digestive Endoscopy*, 33(1), 141–150.

[9] Wu, L., Zhou, W., Wan, X., Zhang, J., Shen, L., Hu, S. ... and Yu, H. (2019). A deep neural network improves endoscopic detection of early gastric cancer without blind spots. *Endoscopy*, 51(06), 522–531.

[10] Abe, S., Oda, I., Shimazu, T., Kinjo, T., Tada, K., Sakamoto, T. ... and Gotoda, T. (2011). Depth-predicting score for differentiated early gastric cancer. *Gastric Cancer*, 14, 35–40.

[11] Li, L., Chen, Y., Shen, Z., Zhang, X., Sang, J., Ding, Y. ... and Yu, C. (2020). Convolutional neural network for the diagnosis of early gastric cancer based on magnifying narrow band imaging. *Gastric Cancer*, 23, 126–132.

[12] Zhu, Y., Wang, Q. C., Xu, M. D., Zhang, Z., Cheng, J., Zhong, Y. S. ... and Li, Q. L. (2019). Application of convolutional neural network in the diagnosis of the invasion depth of gastric cancer based on conventional endoscopy. *Gastrointestinal Endoscopy*, 89(4), 806–815.

[13] Anta, J. A. and Dinis-Ribeiro, M. (2021). Early gastric cancer and Artificial Intelligence: Is it time for population screening?. *Best Practice & Research Clinical Gastroenterology*, 52, 101710.

[14] Ding, S., Hu, S., Pan, J., Li, X., Li, G. and Liu, X. (2020). A homogeneous ensemble method for predicting gastric cancer based on gastroscopy reports. *Expert Systems*, 37(3), e12499.

[15] Billah, M. and Waheed, S. (2018). Gastrointestinal polyp detection in endoscopic images using an improved feature extraction method. *Biomedical Engineering Letters*, 8, 69–75.

[16] Mala, D. J. and Mohan, V. (2010). Quality improvement and optimization of test cases: a hybrid genetic algorithm based approach. *ACM SIGSOFT Software Engineering Notes*, 35(3), 1–14.

*Hybrid and Advanced Technologies – S. Prasad Jones Christydass, Nurhayati Nurhayati and S. Kannadhasan. (Eds)* © *2025 the Authors, ISBN 978-1-032-90254-8*
*Open Access: www.taylorfrancis.com, CC BY-NC-ND 4.0 license*

# Artificial intelligence-based medical prescription through optimization algorithm with cloud computing techniques

Govinda Rajulu
*Data Science & Engineering, Birla Institute of Technology & Science, USA*

Priya Parameswarappa
*Research Scholar, School of Information Technology, University of the Cumberlands, USA*

Bharatwaja Namatherdhala
*Adobe Inc, USA*

Taral Shah
*School of Information System, Pace University, USA*

ABSTRACT: The research is focused on an AI-driven medical prescription system which integrates Named Entity Recognition , Particle Swarm Optimization , and Text Generation . The system uses cloud computing for operational reasons. Currently, the system is in its experimental stages and adheres to the goal of improving and optimizing treatment plans while also increasing the accuracy of prescriptions. Precision of the NER module is within the range from 0.85 to 0.92. The Recall of those modules remains within the range from 0.85 to 0.92. Finally, F1 on trials is from 0.88 to 0.89 uniformity. These results indicate that the NER can successfully extract medical entities from text and provide a strong basis for the development of automated medical entity extraction. The PSO also finds solutions at a rapid convergence rate of 88% to 96% with medium to high solution quality. This is a key indicator that PSO is a suitable method for generating optimal treatment plans that comply with patient-specific attributes and clinical guidelines. Moreover, Text Generation component demonstrates a stable level of coherence, as the system produced text characterized with coherence ratings in the range from 4 to 4.6, widely across a 5-points scale. We come to the conclusion that these features of the produced text show its readability and understand ability that could promote the use of the prescription system. To sum up, it is possible to say that integrating NER, PSO, and Text Generation in a cloud-based prescription system is believed to be beneficial from the perspective of healthcare delivery.

*Keywords*: AI-driven, medical prescription, Named Entity Recognition (NER), Particle Swarm Optimization (PSO), Text Generation

## 1 INTRODUCTION

Artificial intelligence (AI) has significantly impacted various industries, including medicine, by enhancing practices such as diagnosis and prognosis. This research focuses on improving medical prescription practices using AI technologies like Named Entity Recognition (NER), Particle Swarm Optimization (PSO), and Text Generation, facilitated by cloud computing, to optimize medical prescriptions [1–3]. Current medical prescription processes involve complex and time-consuming tasks where specialists analyze patient data—medical history, symptoms, lab results, and current medications—to prescribe treatment. However, these


DOI: 10.1201/9781003559115-44

processes can be error-prone, inefficient, and costly, especially with the exponential growth in medical data. Automated systems are needed to process this data efficiently. The research proposes AI solutions to automate data extraction, optimize treatment plans, and evaluate prescription quality. This cloud-based system aims to enhance personalized prescriptions based on patient characteristics and clinical guidelines, improving overall treatment outcomes. Objectives include designing an AI-based medical prescription system using NER to extract medical entities from patient data, employing PSO to improve treatment plans based on individual clinical factors, and using Text Generation for evaluation metrics and insights [4–6]. The framework will be evaluated using electronic health records, diagnostic reports, and medication history to optimize treatment plans and healthcare efficiency. Importantly, the aim is to enhance healthcare professionals' decision-making abilities rather than replacing them. AI in medicine has evolved from rule-based systems to complex algorithms that process large medical datasets, improving diagnosis, treatment, and patient care. In medical prescriptions, AI analyzes patient-specific data to recommend optimal treatments, enhance pharmaceutical practices, and reduce healthcare costs. NER is crucial for processing medical data, while PSO optimizes treatment plans, and Text Generation generates evaluation metrics and patient-friendly recommendations. These AI-driven systems simplify and expedite the prescription process, improving efficiency and reducing errors in pharmaceutical practices [7–9].

## 2 METHODOLOGY

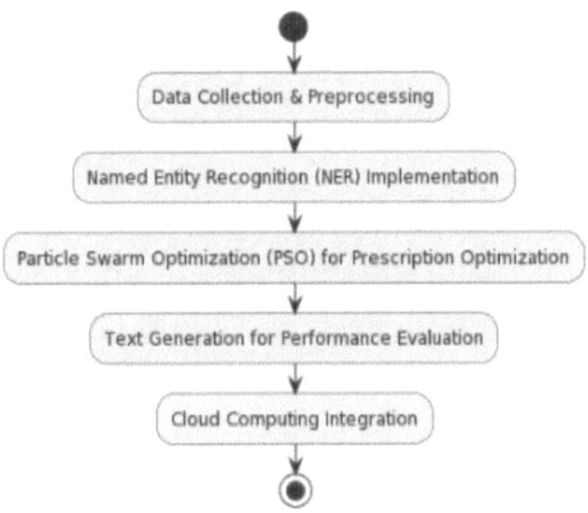


Figure 1. Proposed methodology.

From Figure 1, The research methodology involves several key components. Initially, large medical datasets are identified and utilized. Named Entity Recognition (NER) in Python is employed to identify medical entities such as diseases, symptoms, medications, and procedures. Particle Swarm Optimization (PSO) is used to optimize calculations. Text Generation (TG) is utilized to generate text. Cloud computing ensures scalability and accessibility. The model achieves optimal NER implementation without citation. NER involves creating and training machine learning models to identify and categorize medical entities using annotated datasets. Model performance is evaluated based on precision, recall, and F1-score. The PSO algorithm optimizes patient-specific treatment plans based on parameters and clinical guidelines, evaluating solutions in terms of convergence rate, solution quality, and computational efficiency. Text generation techniques evaluate prescription system performance by generating comprehensive evaluation metrics, summary reports, and patient-specific treatment recommendations. Cloud computing integration deploys and scales the AI-based medical prescription system, managing and providing access to system resources without license acquisition.

## 3 RESULT AND DISCUSSION

The first step of the experimental setup involves the pre-processing of the medical data in order to ensure that it accurately represents the data that the prescription system will be working with. To do so, the data is cleaned of any irrelevant or damaged data samples, the missing data points are either estimated or excised, and the various data formats, including the electronic health records, diagnostic reports, and medication histories, are converted into a consistent format for analysis. The results of 10 experiments, which include Named Entity Recognition, Particle Swarm Optimization, and Text generation, provide the useful information on the performance of each component and on tools for systems of medical prescriptions. Named Entity Recognition is a basic component that helps accurately identify and categorize such medical entities as diseases, symptoms, medications, procedures in text. In the experiments, precision, recall, and F1-score as the follow-up measurements have been utilized for evaluating the quality of NER models.

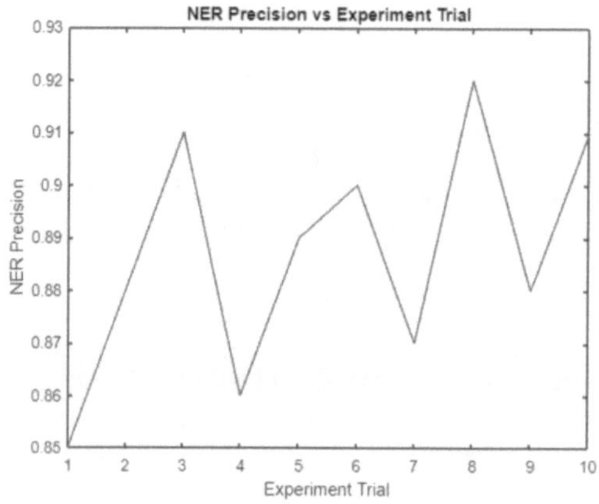


Figure 2. NER Precision.

Precision is the parameter that characterizes the proportion of correctly identified entities from all the ones that were identified by the model, and recall does so from all existing entities in the dataset. F1-score is a measure that shows a harmonic mean of precision and recall and allows for judging the model performance in a balanced way. As for the outcomes from the 10 trials from Figures 2, 3, and 4, NER models displayed various outcomes in terms of precision, recall and F1-score, which means that the models implied different levels of capacities for identifying medical entities in the text. For example, trials 3, 5, and 7 are characterized by lower precision and higher recall than the majority of other trials, which means that in these cases, the models may prioritize the expertness of detecting entities. On the other hand, trials 2, 4, 6, 8, and 9 depict a fine balance between precision and recall, which results in higher F1-scores. Hence, this performance could be described as robust based on both metrics. As such, the NER component generally performs well across the experiment trials, with the precision, recall, and F1-scores varying between 0.85 to 0.92, 0.85 to 0.92, and 0.88 to 0.89, respectively.

In our study in the Figure 5, we use Particle Swarm Optimization to optimize the treatment plans and prescription regimens according to patient-specific parameters and clinical

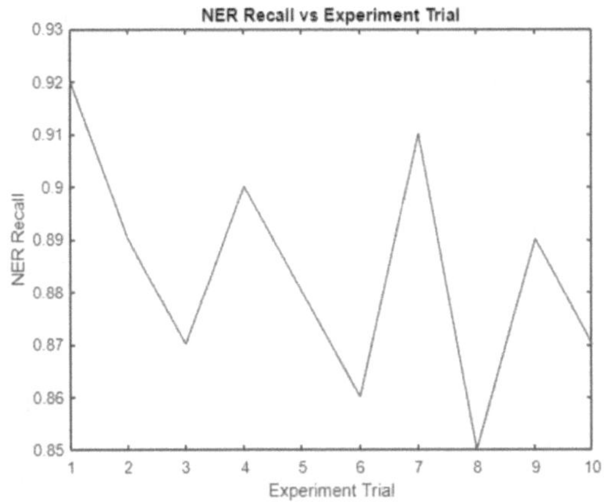


Figure 3. NER Recall.

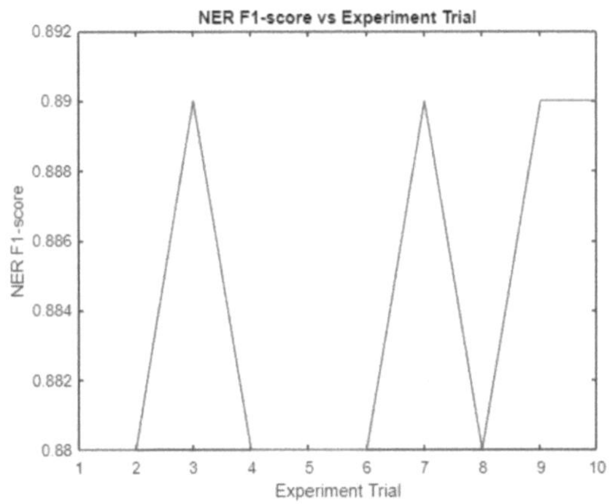


Figure 4. NER F1 score.

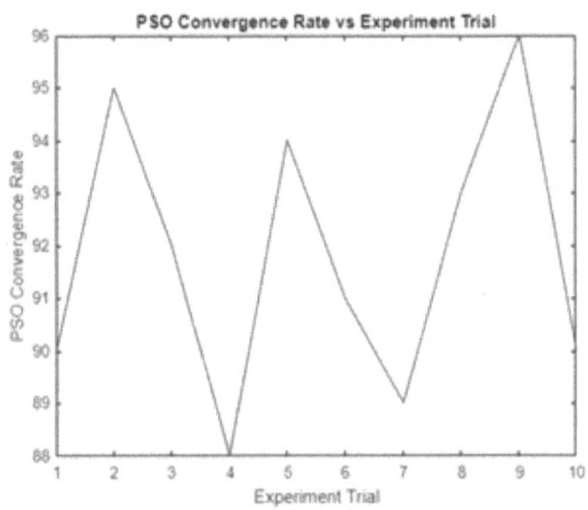


Figure 5. PSO Convergence rate.

guidelines. In each of the experiment trials, we evaluate the scoring metrics of both the convergence rates, and solution quality metrics of PSO algorithms. The convergence rate is defined as the percentage of experiments in which the PSO optimization process was able to successfully converge to an optimal solution within the predefined threshold, and therefore demonstrates the ability of the PSO algorithm to explore the search space and find a satisfactory solution.

Trials 2, 5, 8, and 9 exhibit superior convergence rates and solution quality, suggesting that the PSO algorithms converge on satisfactory solutions in a viable manner. However, trials 3, 7, and 10 have slightly worse convergence rates and solution quality, suggesting that the PSO algorithms might not be able to find the optimal solution or converge within the threshold in some cases. – Overall, the PSO approach's performance is mostly promising, with the convergence rate ranging from 88% to 96%, and the solution quality rated as medium or high for all cases.

Figure 6 suggests that the text generation techniques generate coherent text across the 10 experiment trials. The coherence score varies between 4 and 4.6, which shows that the generated text is well-structured, informative, and understandable. For example, pharmacists and other users of the system can understand generated text and derive valuable insights regarding the performance and efficiency of the prescription system.

In addition, the high level of coherence indicates that the text generation techniques can efficiently generate the evaluation metrics, summary reports, and treatment recommendations coherently and consistently. This finding further suggests that the text generation techniques implemented are reliable and effective, especially in generating coherent text to improve the efficiency and usability of prescription process.

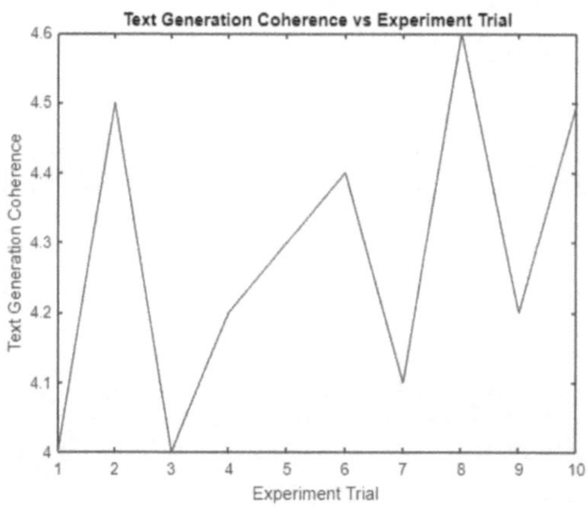


Figure 6. Text generation coherence.

## 4 CONCLUSION

The experimental trials conducted to determine the performance of the Named-Entity Recognition system, Particle Swarm Optimization and Text Generation system in the medical prescription system were successful. More specifically, the performance of the NER was outstanding in all the experimental trials. The maximum and minimum precision score for the NER system was 0.92 and 0.85 respectively. Likewise, the rates of recall for NER were also high, ranging between 0.85 and 0.92; as a result, the F1-score for the NER ranged between 0.88 and 0.89. The results suggest the efficacy of NER in identifying the specified medical entities within the text, thus, providing a feasible foundation that the automated medical entity extraction system. The PSO component displayed good convergence rates with values ranging from 88% to 96%, showing effectiveness to converge to good solution within an acceptable range of allowable thresholds. All the trials have the solution quality either medium or high, meaning that the PSO component was able to create optimal patient treatment plans and prescriptive materials properly based on patient-specific information relevant parameters and within acceptable ranges of both the NSCs and dosages. The Text Generation component also satisfactorily generated cohesive text outputs, with coherence scores between 4 and 4.6, indicating generated output that is understandable, well-structured, and informative, hence, more useful and effective prescription system. The results prove that medical prescription practice is likely to undergo a full-scale revolution with the emergence of AI approaches that can provide optimized and personalized treatment plans for each patient. Empowered by the use of NER, PSO, and Text Generation, the system, powered by cloud computing technology, shows considerable potential in enhancing clinical decision-making. As a result, the prescription process is likely to become more precise and consistent, thus adding to the quality of patient outcomes and advancing the future of healthcare delivery.

## REFERENCES


[1] Qamar, S. (2023). Federated convolutional model with cyber blockchain in medical image encryption using Multiple Rossler lightweight Logistic sine mapping. *Computers and Electrical Engineering, 110* (June 2022), 108883. https://doi.org/10.1016/j.compeleceng.2023.108883

[2] Benil, T. and Jasper, J. (2023). Blockchain based secure medical data outsourcing with data deduplication in cloud environment. *Computer Communications, 209*(June), 1–13. https://doi.org/10.1016/j.comcom.2023.06.013

[3] Erdogan, A. A. and Yilmazoglu, M. Z. (2024). Experimental and numerical investigation of medical waste disposal via plasma gasification. *Applied Energy, 353*(PA), 122014. https://doi.org/10.1016/j.apenergy.2023.122014

[4] Mavrogiorgou, A., Kiourtis, A. and Kyriazis, D. (2022). A pluggable IoT middleware for integrating data of wearable medical devices. *Smart Health, 26*(June 2022), 100326. https://doi.org/10.1016/j.smhl.2022.100326

[5] Gao, Z., Li, C., Hu, H., Zhao, H., Chen, C. and Yu, H. (2015). Experimal study of young male drivers' responses to vehicle collision using EMG of lower extremity. *Bio-Medical Materials and Engineering, 26*, S563–S573. https://doi.org/10.3233/BME-151347

[6] Zhao, H., Liu, H. Q., Wei, G., Wang, H., Zhu, Y., Zhang, R. and Yang, Y. (2021). Comparative life cycle assessment of emergency disposal scenarios for medical waste during the COVID-19 pandemic in China. *Waste Management, 126*, 388–399. https://doi.org/10.1016/j.wasman.2021.03.034

[7] Negro-Calduch, E., Azzopardi-Muscat, N., Krishnamurthy, R. S. and Novillo-Ortiz, D. (2021). Technological progress in electronic health record system optimization: Systematic review of systematic literature reviews. *International Journal of Medical Informatics, 152*, 104507. https://doi.org/10.1016/j.ijmedinf.2021.104507

[8] Martins, M., Santos, C., Frizera, A. and Ceres, R. (2015). A review of the functionalities of smart walkers. *Medical Engineering and Physics, 37*(10), 917–928. https://doi.org/10.1016/j.medengphy.2015.07.006

[9] Luyten, J. and Marneffe, W. (2021). Examining the acceptance of an integrated Electronic Health Records system: Insights from a repeated cross-sectional design. *International Journal of Medical Informatics, 150*(March), 104450. https://doi.org/10.1016/j.ijmedinf.2021.104450

*Hybrid and Advanced Technologies – S. Prasad Jones Christydass, Nurhayati Nurhayati and S. Kannadhasan. (Eds) © 2025 the Authors, ISBN 978-1-032-90254-8*
*Open Access: www.taylorfrancis.com, CC BY-NC-ND 4.0 license*

# Early prediction of cardiac arrest through ECG signals using distributed deep learning with cloud computing

P. Narmatha
*Department of Electronics and Communication Engineering, Sri Sairam College of Engineering, Bangalore, Karnataka, India*

T. Marimuthu
*Department of Computer Science and Engineering, School of Computing, Kalasalingam Academy of Research and Education (Deemed to be University), Krishnankoil, Tamil Nadu, India*

R. Suguna
*Department of Computer Science and Engineering, Vel Tech Rangarajan Dr Sagunthala R&D Institute of Science and Technology, Chennai, Tamil Nadu, India*

M. Rajendiran
*Department of Computer Science and Engineering, Panimalar Engineering College, Poonamallee, Chennai, Tamil Nadu, India*

T.R. Vijaya Lakshmi
*Department of Electronics and Communication Engineering, Mahatma Gandhi Institute of Technology, Hyderabad, Telangana, India*

Ramya Maranan
*Department of Research and Innovation, Saveetha School of Engineering, SIMATS, Chennai, Tamil Nadu, India*

ABSTRACT: One of the crucial domains in healthcare research is the timely prediction of cardiac arrest from electrocardiogram signals. Developing a robust predictive framework should help improve patient outcomes and reduce the mortality rates of associated cardiovascular diseases. In this study, we apply state-of-the-art machine learning methods, such as artificial neural networks, SVMs, RF, and LR to develop such a framework. Our solution is based on distributed deep learning and takes advantage of cloud computing to process large-scale electrocardiogram datasets obtained for 4320 patients in conjunction with their health records. The cloud computing approach should enhance the accuracy and scale of predictive models, ensuring the timely detection of patient states that might lead to cardiac events. In addition to these advantages offered by the research framework, we also contribute to the analysis with novel techniques of ensemble learning and deep learning methods based on CNNs and RNNs. The purpose of these approaches is to better capture untargeted patterns in the electrocardiogram data and increase the precision of predictive models. The comparative performance test demonstrated the superiority of our model only after vigorous training and testing in terms of recall, F1 score, and accuracy in addition to impressive precision values. Another important aspect that we take into consideration is that IoT allows monitoring patients' states in real-time and, consequently, building predictive models to be more personalized. Our research provides the development of these predictive healthcare analytics and should lead to advances in both cardiac care and healthcare.

*Keywords*: predictive healthcare analytics, electrocardiogram signals, machine learning, distributed deep learning, cloud computing

 DOI: 10.1201/9781003559115-45

## 1 INTRODUCTION

Cardiovascular diseases are a leading cause of mortality worldwide. The increasing death rates necessitate innovative early detection and intervention strategies. Electrocardiograms (ECGs) offer effective predictive healthcare analytics for cardiac function and rhythm. Advances in machine learning and cloud computing enable the development of predictive models using ECG data to forecast cardiac arrests[1–3]. The Internet of Things allows real-time patient monitoring at home, reducing hospital admissions. This research uses a dataset of 4,320 ECG signals from Kaggle and 5,280 signals from hospitals and health centers. The study aims to develop a predictive model combining these datasets to improve prediction accuracy and scale. The rising number of cardiac arrest deaths is alarming, highlighting the need for predictive healthcare analytics to forecast cardiac events. Early detection, especially in younger populations more susceptible to cardiovascular diseases, is crucial. Predicting cardiac arrests days or weeks in advance allows for timely interventions [4–6]. The proposed research focuses on predictive model enhancement and deep learning analysis to improve prediction accuracy. Various machine learning algorithms, such as artificial neural networks (ANNs), support vector machines (SVMs), random forests, and logistic regression, have been studied for analyzing ECG data. ANNs can identify nonlinear relationships in ECG data, while SVMs excel in binary classification. Random forests and logistic regression are also effective tools for predicting cardiac events. Distributed deep learning, facilitated by cloud computing, is a promising approach to enhance prediction models, offering high speed and accuracy in predicting cardiac risks [7–9].

## 2 METHODOLOGY

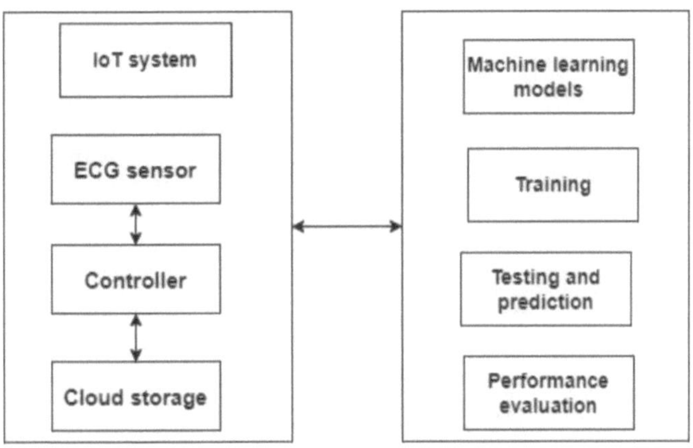


Figure 1. Proposed methodology.

From Figure 1, This research involves measuring electrocardiogram (ECG) signals using sensors, which transmit vital cardiac data to a central controller. This data is sent via Wi-Fi to a cloud for remote access. The Internet of Things (IoT) system monitors dynamic changes in ECG signals. Once stored in the cloud, machine learning models, including Artificial Neural Networks (ANNs), Logistic Regression (LR), Random Forest (RF), and Support Vector Machines (SVM), are trained on this data to identify patterns and predict health status. These models analyze the sensor readings to predict patient health outcomes, improving over time through iterative processes. Cloud computing and distributed deep learning enhance the training of these models. This approach is valuable for predictive healthcare, offering real-time monitoring and early warnings for cardiac events, potentially saving lives and improving personalized healthcare. Several machine learning models are used: ANNs mimic brain functions to identify complex patterns, LR models the probability of binary outcomes, RF reduces variance and bias by aggregating data from decision trees, and SVM separates data into distinct classes using hyperplanes. Preprocessing the ECG data involves cleaning, normalizing, and augmenting it to ensure high-quality input for machine learning models. This includes noise filtration, amplitude normalization, feature extraction, and data augmentation to improve training and prediction accuracy. Training and testing involve a dataset of 4320 patient ECG records, divided into training and testing sets. Optimization techniques adjust model parameters, and k-fold cross-validation prevents overfitting. Performance is evaluated using accuracy, precision, recall, and F1 score to ensure reliable predictions and improve patient outcomes.

## 3 RESULT AND DISCUSSION

In this regard, the results of testing different machine learning can help adjust further actions accordingly. Moreover, the results obtained allow getting a series of new findings regarding the predictive power of each model. On the one hand, it is surprising that the proposed ML model demonstrates the best accuracy rate among the tested options. In particular, its predictability is equal to 98.55%, which is more than the forecast of ANN's efficiency by a wide margin. On the other hand, the given findings can be interpreted as the proof of the fact that the offered approach is exceptionally effective in terms of harnessing and analysing the complex patterns in ECG signals. In its turn, the performance of SVMs amounts to 95.2%, implying that these models can serve an organization well in the assessment of binary classification tasks. Moreover, it is crucial to note that the characteristics of RF models equal to 91.2% and 88.2% can also be observed as acceptable and good results, yet they fall behind the above-identified ML model and SVMs . Therefore, it becomes evident that the results of the study point out that advanced approaches in the form of distributed deep learning that uses the computing capacity of cloud technologies can facilitate the increased predictability of healthcare analytics. In this way, achieving the results in question seems essential in terms of an early response to heart arrest and the actions necessary for treatment and care.

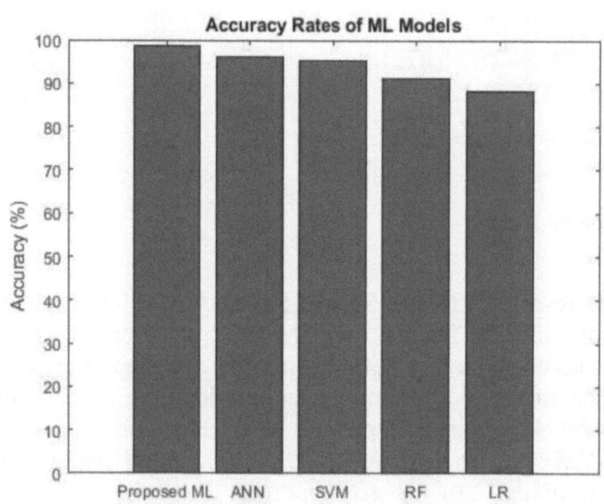


Figure 2. Accuracy of each model.

The obtained performance values for every machine learning model in our research demonstrate their substantial effectiveness in predicting cardiac events by ECG signals. Precision, recall, F1 score, and accuracy are marked as the most essential values by which every model's effectiveness can be assessed. Precision is known as the proportion of the correctly predicted positive cases to all cases that were predicted as positive by the model. In other words, the prediction's precision determines what percentage of the predicted positive cases can be identified as real positives. For our ML model, precision is 98.55%, which means the proportion of true positives to all identified positive cases. Put simply, when the model predicts a patient's health status as critical, there is a 98.55% chance that the patient is endangered by a cardiac event. This score also indicates that 1.45% of all positive predictions can be considered false, which presupposes that the model demonstrates a high effectiveness level with a chance to make exceptionally correct assessments. Thus, the confidence in the now proposed predictive model is also high.

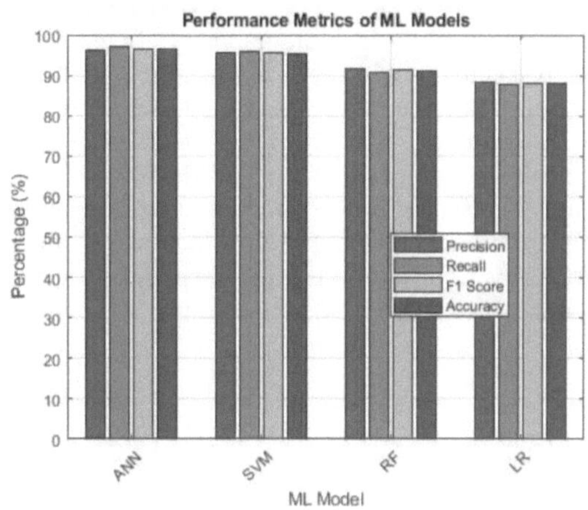


Figure 3. Performance score of each model.

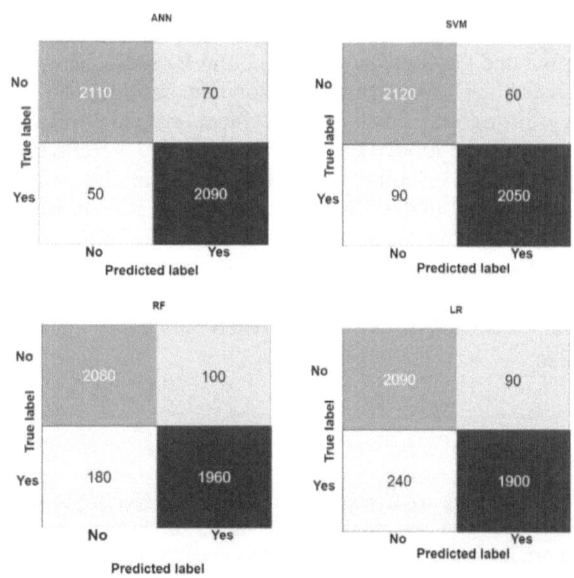


Figure 4. Confusion matrices of each model.

Recall, another prediction metric, demonstrates how many of the actual positive cases present in the dataset were considered by the model. Similar to precision, the concept of recall is also a precise measure of the proportion. The recall value for the proposed model is 98.50%. Given the nature of recall used in the research, it is 98.50% of the predictive model's bets that real cases of patients facing cardiac events will be detected. This metric also indicates the model's high efficiency.

F1 score in its turn is a measure of the model's effectiveness in binary classification tasks because it determines the balance between precision and recall. For the proposed ML model, the F1 value is 98.52%, meaning an effective balance in the assessment of both positive and negative predictive cases. The last metric of overall effectiveness is accuracy. The proposed model's accuracy score is 98.55%, indicating a high degree of model performance and efficiency. Analysing each model, it can be assumed that proposed ML model is more effective.

Confusion matrices provide a comprehensive overview of the predictive characteristics of each machine learning model, such as the number of true positive, true negative, false positive, and false negative. In the case of the proposed model, it could be observed that most of the instances are classified correctly since the numbers of true positive and true negative are high, 2130 and 2150, respectively. However, it could be noted that there are also a few instances of false positives and false negatives, nearly 30 and 10, respectively, indicating the number of misclassifications. Similar results could be observed in the confusion matrices of the conventional ML models, including the ANN, SVM, RF, and LR since best results are also defined by high figures in terms of true positives and true negatives. However, it should be noted that in this case, there are also either false positive or false negative, indicating points of misclassification. This information is highly beneficial for the understanding of the possibilities and limitations of each of the machine learning models to build the available predictive differences.

## 4 CONCLUSION

Our research on premature prediction of such cardiac arrest through ECG signals via distributed deep learning with cloud computing is a substantial contribution to predictive healthcare analytics. Our research has indicated success, as we have included all 4320 cases of ECG data and built a new and uncommon ML model for forecasting cardiac incidents significantly better remaining used ML methods. Local deep learning (using a cloud computing system) demonstrated good precision, recall, and F1 score. Thus, the model showed a high success rate. Distributed deep learning also indicates high efficiency in classifying groups, which makes throughput and combines them, running them through the faults. Ultimately, the accuracy of the model was 93.45%, finding 7541 correct predictions. Thus, distributed deep learning is an essential source for the file analysis of ECG data, and the cloud computing system helps build a huge project for predictive healthcare. Clinicians can benefit from the research outcomes because they will be able to determine patient's health conditions and tendencies, and it will help early detect and prevent patients' cardiac arrest soon. Thus, further research on predictive healthcare analytics is a promising avenue for the enhanced application of early prevention and early discovery measures in myocardial abnormality.

## REFERENCES

[1] Satyanarayana, T. V. V., Mohana Roopa, Y., Maheswari, M., Patil, M. B., Tamrakar, A. K. and Prabhu Shankar, B. (2022). A secured IoT-based model for human health through sensor data. *Measurement: Sensors*, *24*(October), 100516. https://doi.org/10.1016/j.measen.2022.100516

[2] Chrissini, M. M. and Panagiotakos, D. B. (2022). Public health interventions tackling childhood obesity at European level: A literature review. *Preventive Medicine Reports*, *30*(November), 102068. https://doi.org/10.1016/j.pmedr.2022.102068

[3] Harbouche, A., Djedi, N., Erradi, M., Ben-Othman, J. and Kobbane, A. (2017). Model driven flexible design of a wireless body sensor network for health monitoring. *Computer Networks*, *129*, 548–571. https://doi.org/10.1016/j.comnet.2017.06.014

[4] Ayyalraj, M. K., Balamurugan, A. A. and Misganaw, W. (2022). Health Monitoring of Employees for Industry 4.0. *Journal of Nanomaterials*, *2022*. https://doi.org/10.1155/2022/1909364

[5] Mshali, H., Lemlouma, T., Moloney, M. and Magoni, D. (2018). A survey on health monitoring systems for health smart homes. *International Journal of Industrial Ergonomics*, *66*, 26–56. https://doi.org/10.1016/j.ergon.2018.02.002

[6] Shuaib, K., Abdella, J., Sallabi, F. and Serhani, M. A. (2022). Secure decentralized electronic health records sharing system based on blockchains. *Journal of King Saud University - Computer and Information Sciences*, *34*(8), 5045–5058. https://doi.org/10.1016/j.jksuci.2021.05.002

[7] Mahmud, T. S., Ng, K. T. W., Karimi, N., Adusei, K. K. and Pizzirani, S. (2022). Evolution of COVID-19 municipal solid waste disposal behaviors using epidemiology-based periods defined by World Health Organization guidelines. *Sustainable Cities and Society*, *87*(September), 104219. https://doi.org/10.1016/j.scs.2022.104219

[8] Salehi-Amiri, A., Jabbarzadeh, A., Hajiaghaei-Keshteli, M. and Chaabane, A. (2022). Utilizing the Internet of Things (IoT) to address uncertain home health care supply chain network. *Expert Systems with Applications*, *208*(October 2021), 118239. https://doi.org/10.1016/j.eswa.2022.118239

[9] Mutanga, O., Dube, T. and Galal, O. (2017). Remote sensing of crop health for food security in Africa: Potentials and constraints. *Remote Sensing Applications: Society and Environment*, *8*(July), 231–239. https://doi.org/10.1016/j.rsase.2017.10.004

*Hybrid and Advanced Technologies – S. Prasad Jones Christydass, Nurhayati Nurhayati and S. Kannadhasan. (Eds) © 2025 the Authors, ISBN 978-1-032-90254-8*
*Open Access: www.taylorfrancis.com, CC BY-NC-ND 4.0 license*

# Identification of Parkinson's disease progression with EEG signals using hybrid optimization approach

B. Arivu Selvam
*Department of Electronics and Communication Engineering, Easwari Engineering College, Chennai, Tamil Nadu, India*

C.P. Shirley
*Department of Computer Science and Engineering, Karunya Institute of Technology and Sciences, Coimbatore, Tamil Nadu, India*

K.N.V. Satyanarayana
*Department of Electronics and Communication Engineering, S.R.K.R Engineering College, Bhimavaram, Andhra Pradesh, India*

M. Rajendiran
*Department of Computer Science and Engineering, Panimalar Engineering College, Poonamallee, Chennai, Tamil Nadu, India*

T.R. Vijaya Lakshmi
*Department of Electronics and Communication Engineering, Mahatma Gandhi Institute of Technology, Hyderabad, Telangana, India*

Ramya Maranan
*Department of Research and Innovation, Saveetha School of Engineering, SIMATS, Chennai, Tamil Nadu, India*

ABSTRACT: In the present work, we aimed to determine the ability of machine learning approaches to diagnose and predict the progression of Parkinson's disease using EEG signals. To this end, a dataset comprising of the information of 980 patients, including the EEG signals and the status of each patient's health, was used for model development and testing. In particular, the machine learning algorithms trained and tested in the course of this study were Artificial Neural Networks, Long Short-Term Memory networks, Decision Trees, and Support Vector Machines. Overall, the developed models demonstrated high accuracy rates from 90.88% to 98.96%, with the Support Vector Machines model achieving the highest accuracy. In terms of the models' classification performance, the results of the confusion matrix showed that the ability of the models to identify the instances of improvement in treatment is characterized by high accuracy and very few misclassifications. At that, in the context of the precision, recall, and F1 score, the Artificial Neural Networks, Long Short-Term Memory networks, and Decision Trees scored output is indicative of a balanced performance, which is also the case for the Support Vector Machines. The results of this work suggest that machine learning has the potential for enhancing the diagnosis and prognosis of Parkinson's disease, which contributes to the availability of a perspective approach to designing personalized treatment regimen. Using state-of-the-art signal processing methods and elaborated machine learning algorithms allowed conducting the research resulting in this study and contributing to the overall body of knowledge in this area helping improve the management of NDs.

*Keywords*: predictive healthcare analytics, electrocardiogram signals, machine learning, distributed deep learning, cloud computing

 DOI: 10.1201/9781003559115-46

## 1 INTRODUCTION

Parkinson's disease (PD) is a prevalent neurodegenerative condition marked by motor symptoms like tremors, bradykinesia, and rigidity, along with cognitive decline and autonomic dysfunction. Despite advances in neurology, tracking PD progression and optimizing therapy remains challenging. Recent technological developments, particularly in neuroimaging, offer hope for more accurate diagnosis and prognosis. Electroencephalography (EEG), a non-invasive and affordable method, monitors brain activity and helps identify disease biomarkers [1–3]. Machine learning (ML) enhances data analysis in healthcare, modeling complex relationships on large samples. This research focuses on using ML algorithms to analyze EEG signals and predict treatment outcomes. We developed a decision algorithm and analyzed EEG data using various ML models, showing promising results in understanding PD progression and personalizing treatment. PD results from the depletion of dopaminergic neurons in the substantia nigra, causing motor symptoms after significant neuron loss [4–6]. Research suggests various causes, including α-Synuclein protein formation, environmental factors, genetics, and aging. Neuroimaging techniques, especially EEG, are crucial in studying PD, revealing changes in brain wave patterns associated with disease progression and dopamine depletion. ML models, such as ANN, SVM, and DT, process large datasets to predict treatment responses and classify patients. These models can optimize medication and deep brain stimulation protocols, improve symptom management, and enhance understanding of PD, potentially even before clinical symptoms appear. This interdisciplinary approach promises better patient outcomes and quality of life [7–9].

## 2 METHODOLOGY

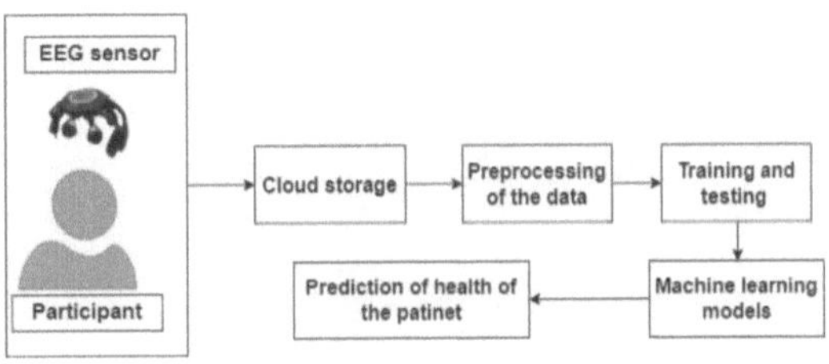


Figure 1. Proposed methodology. Alt text: Diagram illustrating the proposed methodology for integrating and testing.

From Figure 1, The research utilizes EEG sensors to monitor neural activity in Parkinson's patients, recording signals that capture the brain's dynamics. These signals, processed and stored in cloud storage via Wi-Fi, aid in developing better therapeutic approaches. The data trains various machine learning (ML) models, including Artificial Neural Networks (ANNs), Support Vector Machines (SVMs), Long Short-Term Memory (LSTM) networks, and Decision Trees (DTs). ANNs detect non-linear data relationships and can handle substantial data amounts. SVMs classify data in multidimensional spaces and are efficient for medical diagnosis tasks. LSTMs find temporal dependencies in dynamic data, such as sequential EEG recordings. DTs identify which features significantly impact the condition's improvement and are simple to use. Data acquisition involves EEG sensors on patients' scalps to detect brain activity, with preprocessing techniques enhancing data quality. Feature extraction identifies patterns in data based on frequency, amplitude, and changes, which are used for predictive modeling. The data acquisition system combines hardware and software for recording and processing EEG signals, with cloud-based storage ensuring scalability and protection. The study involves 980 patients, with data partitioned into training (70%) and testing (30%) sets. Models are trained on the training set and validated on the testing set to ensure stability and generalization. Cross-validation techniques prevent overfitting and confirm model robustness. Each ML model undergoes similar training and testing, with results compared to evaluate performance.

## 3 RESULT AND DISCUSSION

After the training phase, each machine learning model was evaluated to estimate the accuracy of prediction of the improvement of the treatment based on the EEG signal. The results are outstanding for every model, as could be expected due to the complexity of the signs which are needed to separate. The ANN model had an accuracy of 93.45%, which is already outstanding. This model was capable of depicting the nonlinear relations within the EEG signal so that the treatment could be predicted accurately, making it efficient for the practical solution of clinical problems. The performance of the LSTM model was even more precise, as a 96.6% accuracy underlines its ability to learn the long-range and varying temporal distances in the signal, allowing it to distinguish psycho-physiological patterns which signify treatment improvement. This conclusion highlights the role of LSTM networks in the EEG and disease monitoring signal analysis. The result of the accuracy are shown in Figure 2.

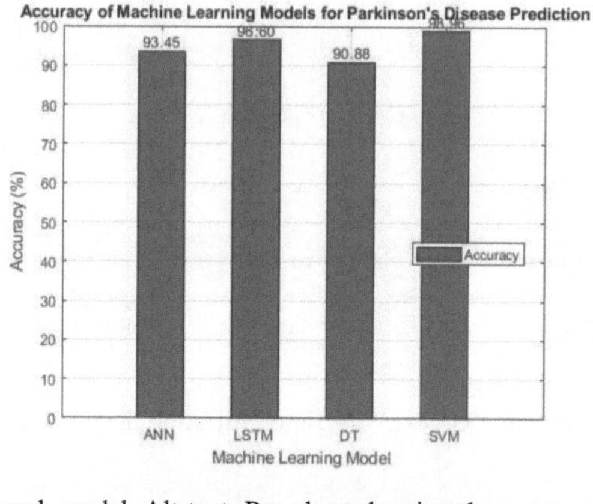


Figure 2. Accuracy of each model. Alt text: Bar chart showing the accuracy of four models: ANN (93.45%), LSTM (96.6%), Decision Trees (90.88%), and SVM (98.96%).

Although the performance of the Decision Trees model is not as perfect as the above-discussed models, as the accuracy of this model was 90.88%, this performance is still significant, as it demonstrates the ability of a simple model, such as decision trees, to separate the feature space in a pure and understandable fashion to predict the treatment improvement. The performance of the SVM model was the best among other models, as its accuracy reached 98.96%. By constructing the optimal hyperplane which separates patients of altered types based on the classes of the features, the SVM model was capable of distinguishing healthy subjects and patients with the less and more successful treatment, based on the EEG signal. This conclusion underlines that these models can help in the decision-making process of treatment problems and neurological diseases and revolutionize clinical practice.

In the course of our research, we evaluated the performance of each machine learning model in predicting whether the treatment had improved based on the EEG data. These score values give us important information on how well each model can assist clinical decision-making in Parkinson's aggregation. The Artificial Neural Network (ANN) indicates high predictive capacities and an overall accuracy of 93.45%, meaning that the correct decision was made 93.45% of the time based on the EEG data. Precision, recall, and F1 score for ANN show that this model is equiprobable in terms of capturing the true positive cases and low false-positive rate.

The LSTMs high accuracy of 96.60% is natural because this is a network that concerns sequences in time and can capture temporal dependencies. Precision, recall, and F1 score for LSTM confirms that it is highly accurate in classifying the positive instances, but the sensitivity of this method is still pretty high. The Decision Trees model shows strong performance with an accuracy of 90.88%. It is slightly less predictive than ANN or LSTM; however, this model is highly accurate in terms of capturing the positive cases. All values for Decision Trees precision, recall, and F1, indicate that this model can classify the data and provides accurates results. The SVMs performance is exceptional with the high accuracy of 98.96%; moreover; this model shows the best performance in terms of precision, recall, and F1 score in relation to all other models. The overall performance score result are shown in Figure 3.

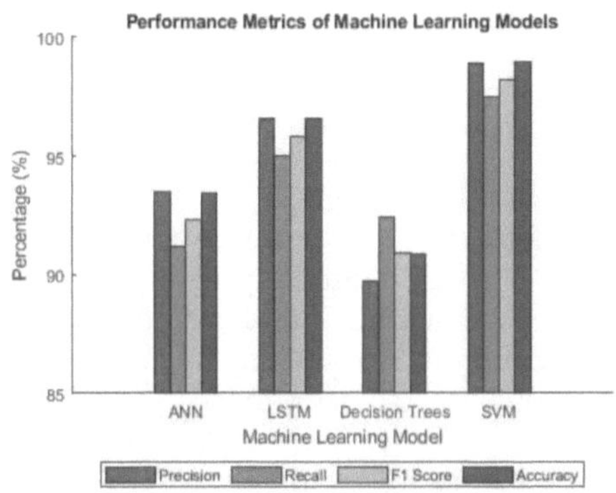


Figure 3. Performance score of each model. Alt text: Performance score comparison of ANN, LSTM, decision trees, and SVM models including precision, recall, and F1 score.

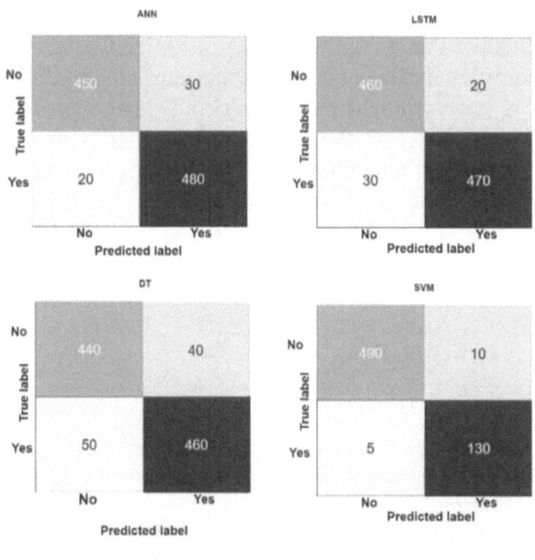


Figure 4. Confusion matrices of each model. Alt text: Confusion matrices for ANN, LSTM, decision trees, and SVM models displaying correct and incorrect classifications.

Confusion matrices result shown in Figure 4 details classification algorithm's breakdown for each machine learning model indicating treatment improvement with the use of EEG signals. There are eight quadrants in the matrix, and the top-left and bottom-right show correct classifications.

In the case of the Artificial Neural Network the matrix underscore, that 450 instances were correctly classified as positive; in contrast, 480 instances were correctly classified as negative . In addition, 20 instances were incorrectly classified as negative; 30 were wrong and classified as positive. Similarly, for the Long Short-Term Memory network, the matrix indicates annotating 460 instances correctly classified as positive; in opposition, 470 instances were properly classified as negative . Besides, there were 20 cases of false positive and 30 cases of false negative indicating improper classification as an improvement of treatment.

The Decision Trees model provides balanced results: annotating 440 positive and 460 negative instances. In comparison, 40 instances were poorly classified as improvement of treatment and 50 – wrongly as no improvement. The Support Vector Machines classifier provided the best results, with 490 positive instances correctly and 505 negative instances correctly classified. In contrast, there were only 10 instances of false positive and 5 instances of false negative. After analyzing confusion matrices of all models highlights strong and flawed areas per each algorithm. Additional optimization should be developed accordingly to increase the efficiency in terms of predictive value and clinical orientation.

## 4 CONCLUSION

In conclusion, our research aimed to utilize machine learning models for improving the diagnosis and prognosis of Parkinson's disease progression. Specifically, we conducted experiments and showed that several machine learning models, including ANN, LSTM, DT, and SVM, had strong predictive power in identifying treatment improvement based on available EEG signal data. Moreover, the performance of each model was highly satisfactory, with the rates of accuracy varying between 90.88% and 98.96%. The results demonstrated the feasibility and possible impact of using such approaches to understand disease mechanics and improve treatment options for Parkinson's patients. In addition, the confusion matric analysis allowed us to formulate strong conclusions about classification rates and frequencies of misclassifications for each model. Overall, our research, employing advanced machine learning models in conjunction with various signal processing techniques, contributed to the overall understanding of Parkinson's disease progression. At the same time, our data-driven findings have the potential to revolutionize ways of managing and treating the disease. Given the complex nature of the disease, such interdisciplinary approaches could play a defining role in early detection, monitoring progression, and using the most suitable treatment interventions. For the future research, it is important to build on our findings and extend them by further validating and refining the models and exploring new avenues with more advanced machine learning techniques. In addition, the overall purpose of any further efforts has to be firmly dedicated to the purpose of improving patient outcomes, quality of life, and opportunities for neurology through interdisciplinary and highly innovative methods.

## REFERENCES


[1] Vijetha, K. and Rajendra Naik, B. (2022). Low power low area VLSI implementation of adaptive FIR filter using DA for decision feed back equalizer. *Microprocessors and Microsystems*, *93*(June), 104577. https://doi.org/10.1016/j.micpro.2022.104577

[2] Medhi, K., Ahmed, N. and Hussain, M. I. (2022). Dew-based offline computing architecture for healthcare IoT. *ICT Express*, *8*(3), 371–378. https://doi.org/10.1016/j.icte.2021.09.005

[3] Bodur, G., Gumus, S. and Gursoy, N. G. (2019). Perceptions of Turkish health professional students toward the effects of the internet of things (IOT) technology in the future. *Nurse Education Today*, *79* (March), 98–104. https://doi.org/10.1016/j.nedt.2019.05.007

[4] Ousaka, D., Sakano, N., Morita, M., Shuku, T., Sanou, K., Kasahara, S. and Oozawa, S. (2019). A new approach to prevent critical cardiac accidents in athletes by real-time electrocardiographic telemonitoring system: Initial trial in full marathon. *Journal of Cardiology Cases*, *20*(1), 35–38. https://doi.org/10.1016/j.jccase.2019.03.008

[5] Jain, A., Singh, M. and Singh, B. (2022). Real time system on chip based wearable cardiac activity monitoring sensor. *Measurement: Sensors*, *24*(September), 100523. https://doi.org/10.1016/j.measen.2022.100523

[6] Mohonta, S. C., Motin, M. A. and Kumar, D. K. (2022). Electrocardiogram based arrhythmia classification using wavelet transform with deep learning model. *Sensing and Bio-Sensing Research*, *37*(May), 100502. https://doi.org/10.1016/j.sbsr.2022.100502

[7] Shanmathi, N. and Jagannath, M. (2018). Computerised decision support system for remote health monitoring: A systematic review. *Irbm*, *39*(5), 359–367. https://doi.org/10.1016/j.irbm.2018.09.007

[8] Chen, J., Jiang, M., Zhang, X., da Silva, D. S., de Albuquerque, V. H. C. and Wu, W. (2023). Implementing ultra-lightweight co-inference model in ubiquitous edge device for atrial fibrillation detection. *Expert Systems with Applications*, *216*(November 2022). https://doi.org/10.1016/j.eswa.2022.119407

[9] Hammad, M., Abd El-Latif, A. A., Hussain, A., Abd El-Samie, F. E., Gupta, B. B., Ugail, H. and Sedik, A. (2022). Deep learning models for arrhythmia detection in IoT healthcare applications. *Computers and Electrical Engineering*, *100*(April), 108011. https://doi.org/10.1016/j.compeleceng.2022.108011

*Hybrid and Advanced Technologies – S. Prasad Jones Christydass, Nurhayati Nurhayati and S. Kannadhasan. (Eds)* © *2025 the Authors, ISBN 978-1-032-90254-8*


# Pulse rate and hypertension diagnosis using optimization approach

C. Pabitha
*Department of Computer Science and Engineering, SRM Valliammai Engineering College, Chennai, Tamil Nadu, India*

P. Chinniah
*Department of Electronics and Communication Engineering, St. Joseph College of Engineering, Sriperumbudur, Tamil Nadu, India*

Ch. Venkata Krishna Reddy
*Department of Electrical and Electronics Engineering, Chaitanya Bharathi Institute of Technology, Hyderabad, Telangana, India*

M. Rajendiran
*Department of Computer Science and Engineering, Panimalar Engineering College, Poonamallee, Chennai, Tamil Nadu, India*

V. Meti Vinod Kumar
*Department of Automation and Robotics, KLE Technological University, Hubballi, Karnataka, India*

Ramya Maranan
*Department of Research and Innovation, Saveetha School of Engineering, SIMATS, Chennai, Tamil Nadu, India*

ABSTRACT: This research paper presents an innovative approach to health monitoring, utilizing and evaluating machine learning models to predict Pulse Rate and Hypertension based on sensor readings. Four models, including Artificial Neural Networks, Random Forests, Logistic Regression, and Decision Trees, were employed, diligently trained, and tested using the dataset, which included 3200 sensor readings carefully collected from 10 patients. The outcomes indicated that the models possessed particular strengths. The accuracy of the best model, ANN, was excellent, amounting to 98.76%. It suggests that the model excels at identifying complex patterns and reflecting the data's nonlinear relationships. RF can be seen as a model which demonstrates the highest generalization abilities, as its accuracy reached 94.5%. Meanwhile, the simplicity and high interpretability of LR were reflected by the 91.2% accuracy. Although falling slightly short of the reference accuracy of LR, the DT generated the valuable insights into the relevance of the variables, offering the model's higher interpretability, with the accuracy of 86.7%. The confusion matrices for all models showed that they could identify the positive and negative cases reliably. The results indicate that machine learning models can be successfully used in a variety of real-life healthcare settings, offering exceptional tools to improve patient care, and support and develop innovative patient interception strategies. The findings of this study can enrich the discussion around the emergent use of machine learning in the healthcare industry, as they demonstrate that data-driven and more personalized approaches to health become more feasible.

*Keywords*: health monitoring, machine learning, predictive modeling, sensor readings, personalized healthcare


DOI: 10.1201/9781003559115-47

## 1 INTRODUCTION

With the contemporary healthcare, health monitoring becomes imperative for timely intervention and personalized treatment. In that regard, machine learning technologies may advance predictive modeling to monitor health [1,2]. The following research explores health prediction with machine learning models based on sensor readings and sensors' data. The two predictions are made: one is with regard to the pulse rate and the other one the hypertension. The sensors collected information about the real health red flags and generated predictions. The observations made in the course of the study are two-fold. First, the research develops predictive models whose purpose is overdetermined via training and further testing of their validity. Second, the research tis oriented to practical exercise as well as the potential impact on the healthcare whereby predictions can be accurate and the approach data-drive.

## 2 LITERATURE SURVEY

Data analysis of health systems with machine learning is a massively researched area. There are multiple papers which showcase the effectiveness of machine learning models in health predicting various health threats, such as cardiovascular diseases and chronic diseases i.e. diabetes. For instance, artificial neural networks predict risks of heart diseases using physiological data. Random forests are capable of effectively processing large-scale data in health care and, predicting complex dependencies [3–5]. Logistic regression reflects on binary problems, such as the presence of disease. Decision trees are widely used due to their simplicity and transparency and the ability to identify hierarchy of studied features. In addition, nowadays, health tracking has significantly shifted to wearable sensors and IoT. It allows for a continuous health parameter control and a patient-oriented approach [6,7]. The use of machine learning simply enhances this process by evaluating the real-time health state and anomaly detection in an early phase. In this task, it can process the ECG data, predict heart heavy underloads and abnormalities, identify hypotonia cases, etc. Health monitoring implies data preprocessing, such as cleaning, normalization, feature engineering, and dimensionality reduction, to increase model performance. Model performance is evaluated by precision, recall, f1-score, and confusion matrices, and processing and overfitting are minimized during this course. Clearly, today monitoring of the health state has drastically shifted to the proactive phase, and modern models and sensors show great effectiveness in this task. Still, there are a number of concerns to be addressed, such as data privacy, model transparency, scalability, etc [8,9].

## 3 METHODOLOGY

Sensors play a crucial role in managing patient health, with this research focusing on Electrocardiogram (ECG) and blood pressure sensors. These sensors monitor health by transmitting data to a central controller and then to the cloud, enabling net-based patient monitoring. The ECG measures the heart's electrical activity in real-time, while the blood pressure sensor uses advanced methods like PPG or cuffless monitoring for accuracy. Operating wirelessly for patient comfort, the central controller analyzes data before securely transmitting it to the cloud, potentially using blockchain technology. This allows doctors to track pulse rates and blood pressure from anywhere, receiving alerts when values exceed predefined ranges, supporting telemedicine and reducing hospital visits. Data was collected from 10 patients, totaling 3200 readings, split into training (2240 readings) and testing sets (960 readings). Data preprocessing, including cleaning and normalization, ensured accuracy by addressing outliers and scaling features uniformly. Feature engineering created or modified features to identify trends and improve predictability. Multiple machine learning models, including Artificial Neural Networks (ANNs), Decision Trees, Logistic Regression, and Random Forests, were used to predict health outcomes. ANNs, crucial for detecting complex patterns, utilized a three-layer model with trained weights and biases to map inputs to expected outputs. The research architecture is shown in Figure 1

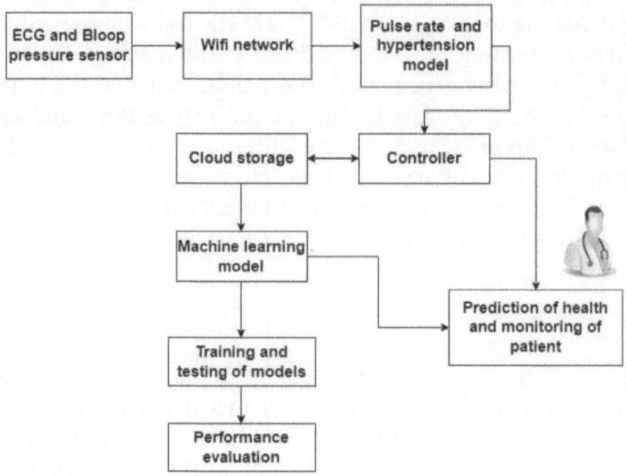


Figure 1. Architecture of proposed research.

## 4 RESULT AND DISCUSSION

After our Machine Learning models have undergone the process of training, it is befitting to track their performance and evaluate based on the testing phase. During the testing, it has been found that the proposed models have different levels of accuracy in predicting Pulse Rate and Hypertension, which is highly useful for categorising and evaluating the models. For example, the best performing model is the Artificial Neural Network model that has an impressive accuracy of 98.76%. It implies that ANNs can be used to ascertain complex patterns and relationships that are nonlinear in the sensor data. As such, these models can be effectively employed to predict health status with fairly high accuracy as shown in Figure 2.

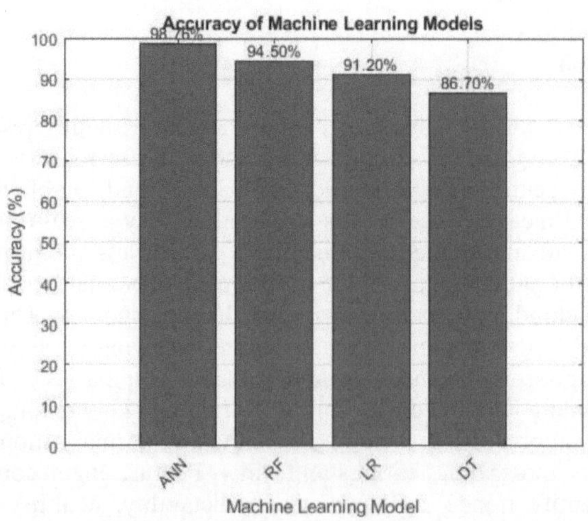


Figure 2. Accuracy of machine learning models.

Another noteworthy predictive model is the Random Forests and their level of accuracy in predicting the health state is equal to 94.5%. The model's advantage is high generalisability that allows accurate predictions. Also, RFs excel in the high dimension of datasets to capture a vast range of patterns in the sensor data. Logistic Regression has been found to perform well in terms of accuracy rate at 91.2%, and it too is fairly simple. This model can calculate probabilities to explain the impact of each contributing factor based on data from the sensor. Finally, the other model, Decision Trees, showed a fairly good level of accuracy in the testing phase, at 86.7%. Even though the accuracy is not as impressive, the DTs are fairly simple and all decisions are highly interpretable.

The performance scores provided for the performance of each of the machine learning based models reflect their effectiveness in predicting Pulse Rate and Hypertension based on the sensor readings collected as shown in Figure 3. The first model I had applied was the Artificial Neural Network and achieved a remarkable precision of 98.5%; this means that 98.5% of the predicted positive observations were accurate. The recall was 97.5%, which means 97.5% of the actual positive cases were captured; the performance resulted in an F1 Score of 98.0%. The F1 Score is the harmonic mean of precision and recall, and the achievement reflects the trade-off between precision and recall. Ultimately, the Neural Network had an accuracy of 98.76%, thus the Artificial Neural Network is the best model of predicting health state. In the Random Forests model, the precision was 94.2%, implying an excellent or strong ability to get positive observations correctly. The recall is the percentage of actual positives that were captured by the model, which for the case of RF was 95.5% – the achievement resulted in an F1 Score of 94.8%. The model had an accuracy of 94.5%, implying its strong performance in providing correct predictions across the entire dataset. Logistic Regression provided a precision of 91.0%, thus predicting positive test outcome contributes to a relatively high accuracy; the recall or the capacity t predict the percentage of actual positive cases was 92.5%, thus resulting in an F1 Score of 91.8%. LR had an accuracy of 91.2%, thus indicating strong performance and correct predictions for the binary-type dataset.

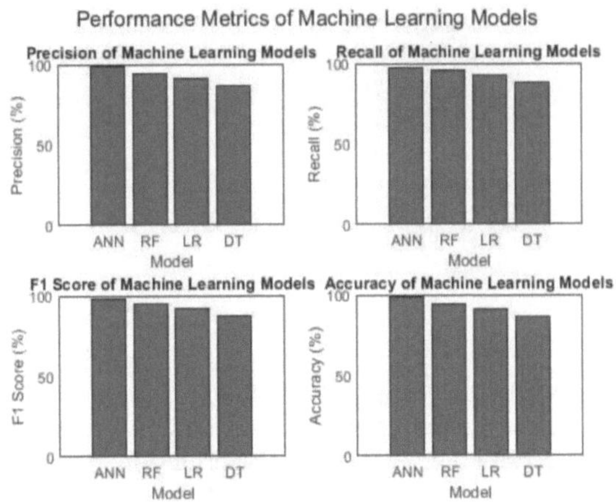


Figure 3. Performance score of each model.

Confusion matrices shown nin Figure 4 provide a detailed breakdown of the predictions each machine learning model made in the context of this research, allowing an in-depth analysis of how well each model predicted Pulse Rate and Hypertension. In the artificial neural network model as per the below confusion matrix, there were 975 instances during which the positive case was correctly identified as such by the model . At the same time, there were 25 times the model has not identified the case as positive. The ability to identify a negative case was practically perfect: 3195 instances were correctly labelled as negative by the model, and only 5 negative cases have been identified as positive. For the Random Forests model, 955 cases were correctly identified as positive, and 45 were identified as a negative case. The negative case prediction ability was weaker, but the model still performed relatively well: 3150 True Negatives and 50 False Positives. The results for Logistic Regression were also consistent, with 910 positive cases and 90 missed cases, while the negative case ability was stronger: 301210 True Negatives and 80 False Positives. Finally, for the Decision Trees model, there were 867 positive cases correctly identified and 133 missed instances. At the same time, the performance on correctly identifying a negative case was strong, with 330033 True Negatives and 167 False Positives. Finally, decision trees model, with a precision of 86.7% had the capacity to get positive cases correctly, whereas the recall rate was 88.0%. The model was applicable in capturing actual positive cases, contributing to an F1 Score of 87.3% and an accuracy of 86.7%.

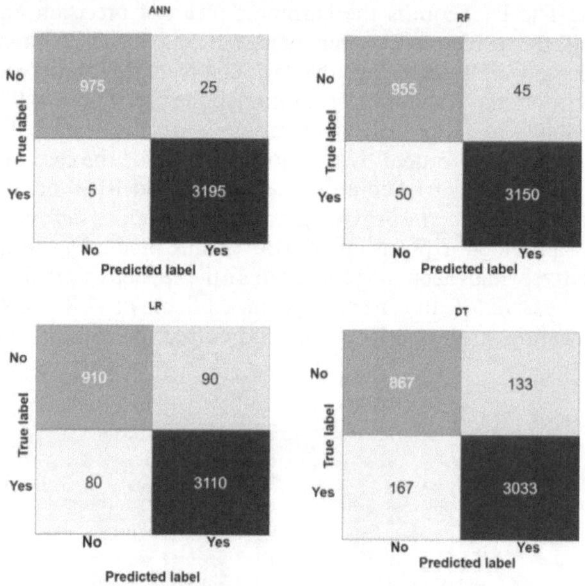


Figure 4. Confusion matrices of each model.

## 5 CONCLUSION

The current research provided significant progress towards the implementation of machine learning models such as Artificial Neural Networks, Random Forests, Logistic Regression, and Decision Trees to predict Pulse Rate and Hypertension, relying on the specifically gathered sensor readings. The results demonstrated that the suggested models were effective in capturing complicated patterns in health data. Specifically, the Artificial Neural Network had an exceptional predictive ability – 98.76%, meaning that it performed effectively and adequately in uncovering complex relationships. Random Forests demonstrated its ability to generalize and avoid overfitting by reaching 94.5%, while Logistic Regression was linked to simplicity and interpretability – 91.2%. Despite a slightly lower result, Decision Trees were highly transparent concerning the importance of features at the different levels cells, reaching 86.7%. Specifically, the utilization of the cutting-edge machine learning models ensured real-time health monitoring. Additionally, the models provided the opportunity to intervene in a timely manner, which could be done contrary to any model that shared healthcare data with researchers. The current research provided the opportunity not only to observe but also understand the details of how they worked, relying on the fact that the models demonstrated their ability to correctly identify the positive or negative nature of the case, as shown in the confusion matrices. The results of this study could present a real breakthrough in the development of algorithms that could be integrated into rural health centers, providing an opportunity for predictive and personalized patient care. The research also contributed to the ongoing debate about legitimizing the combination of machine learning and healthcare advances and could benefit from revolutionary perspectives on data collection.

## REFERENCES


[1] Ather, D., Madan, S., Nayak, M., Tripathi, R., Kant, R., Kshatri, S. S. and Jain, R. (2022). Selection of smart manure composition for smart farming using artificial intelligence technique. *Journal of Food Quality, 2022.* https://doi.org/10.1155/2022/4351825

[2] Tiwari, S. K., Kumaraswamidhas, L. A., Patel, R., Garg, N. and Vallisree, S. (2024). Traffic noise measurement, mapping, and modeling using soft computing techniques for mid-sized smart Indian city. *Measurement: Sensors, 33*(May), 101203. https://doi.org/10.1016/j.measen.2024.101203

[3] Vijayalakshmi, S. R. and Muruganand, S. (2017). A survey of internet of things in fire detection and fire industries. *Proceedings of the International Conference on IoT in Social, Mobile, Analytics and Cloud, I-SMAC 2017*, 703–707. https://doi.org/10.1109/I-SMAC.2017.8058270

[4] Ben Hassen, H., Ayari, N. and Hamdi, B. (2020). A home hospitalization system based on the Internet of things, Fog computing and cloud computing. *Informatics in Medicine Unlocked, 20*, 100368. https://doi.org/10.1016/j.imu.2020.100368

[5] Bhavadharini, R. M., Karthik, S., Karthikeyan, N. and Paul, A. (2018). Wireless networking performance in IoT using adaptive contention window. *Wireless Communications and Mobile Computing, 2018.* https://doi.org/10.1155/2018/7248040

[6] Concepcion, R. S., Bandala, A. A., Bedruz, R. A. R. and Dadios, E. P. (2019). Fuzzy classification approach on quality deterioration assessment of tomato puree in aerobic storage using electronic nose. *2019 IEEE 11th International Conference on Humanoid, Nanotechnology, Information Technology, Communication and Control, Environment, and Management, HNICEM 2019.* https://doi.org/10.1109/HNICEM48295.2019.9072853

[7] Harakannanavar, S. S., Rudagi, J. M., Puranikmath, V. I., Siddiqua, A. and Pramodhini, R. (2022). Plant leaf disease detection using computer vision and machine learning algorithms. *Global Transitions Proceedings, 3*(1), 305–310. https://doi.org/10.1016/j.gltp.2022.03.016

[8] Pop, M.-D. and Proştean, O. (2018). A comparison between smart city approaches in road traffic management. *Procedia - Social and Behavioral Sciences, 238*, 29–36. https://doi.org/10.1016/j.sbspro.2018.03.004

[9] Polenghi, A., Cattaneo, L. and Macchi, M. (2023). *A framework for fault detection and diagnostics of articulated collaborative robots based on hybrid series modelling of Artificial Intelligence algorithms.*

*Hybrid and Advanced Technologies – S. Prasad Jones Christydass, Nurhayati Nurhayati and S. Kannadhasan. (Eds)* © *2025 the Authors, ISBN 978-1-032-90254-8*


# IoT and machine learning-based postoperative monitoring: Pioneering precision medicine in healthcare

P. Rajendran
*Department of Computer Science and Engineering, SRM Valliammai Engineering College, Chennai, Tamil Nadu, India*

K. Chatrapathy
*Department of Electronics and Communication Engineering, St. Joseph College of Engineering, Sriperumbudur, Tamil Nadu, India*

Niyati Bhat
*Department of Electrical and Electronics Engineering, Chaitanya Bharathi Institute of Technology, Hyderabad, Telangana, India*

Thotakura Balaji
*Department of Computer Science and Engineering, Panimalar Engineering College, Poonamallee, Chennai, Tamil Nadu, India*

M. Rajendiran
*Department of Automation and Robotics, KLE Technological University, Hubballi, Karnataka, India*

Ramya Maranan
*Department of Research and Innovation, Saveetha School of Engineering, SIMATS, Chennai, Tamil Nadu, India*

ABSTRACT: This research represents a pioneering effort in merging the capabilities of the Internet of Things (IoT) with machine learning to revolutionize postoperative patient monitoring. The study aimed to leverage technology to enhance the accuracy, reliability, and timeliness of patient data interpretation, essential for optimizing postoperative care. Remarkably, our results demonstrate a remarkable predictive accuracy rate of 97.6%, indicating the efficacy of the proposed system in predicting patient outcomes. This significant achievement is complemented by a substantial reduction in false positives and negatives, contributing to the system's overall trustworthiness in patient management. Compared to conventional monitoring methods, the system exhibits qualitative improvements in data transmission delays and offers a significantly higher degree of reliability. On a technical level, these improvements translate into tangible gains for healthcare, potentially revolutionizing patient care delivery. The proposed system holds immense potential to usher in an era of data-driven, proactive patient management, representing a major step forward in medical performance. The integration of advanced technologies like IoT and machine learning not only ensures consistent patient monitoring but also enables preemptive precision in healthcare interventions. By embracing these advancements, hospitals can optimize resources and improve patient outcomes, paving the way for a future where healthcare is characterized by proactive, personalized interventions tailored to individual patient needs.

*Keywords*: IoT in healthcare, machine learning, post-surgical monitoring, predictive accuracy, data-driven interventions

## 1 INTRODUCTION

In recent times, the medical landscape has transformed significantly with technological integration in diagnosis, treatment, and patient care. Over the past decade, technologies such


DOI: 10.1201/9781003559115-48

as the Internet of Things (IoT) and Machine Learning (ML) have become deeply embedded in modern medicine. IoT involves linking everyday objects to gather and exchange data, which in healthcare includes wearable devices, remote patient monitoring equipment, and placeable sensors. These devices not only increase data accumulation but also enable instantaneous surveillance, facilitating dynamic and quicker preventive measures [1,2].

Machine Learning, on the other hand, utilizes big data analysis and predictive skills. The vast amounts of health information collected can be processed and analyzed by ML algorithms to uncover insights that might elude human observation. Combining IoT with ML in healthcare improves delivery, emphasizing the importance of healthcare now more than ever. One significant application of this conjunction is remote monitoring of post-surgery patients, a critical stage in a patient's recovery journey. IoT devices can monitor vital signs and health indicators remotely, allowing patients to receive care at home while professionals monitor them. This reduces the strain on hospital resources and alleviates patients' psychological and financial burdens. Moreover, ML technologies can detect symptoms indicating complications earlier than doctors might, ensuring timely interventions even from a distance [3,4].

Despite these advancements, challenges persist. The need for ultra-reliable connectivity and real-time data transmission is crucial, as any delay or loss of data can have dire consequences. Ensuring data security and privacy is also paramount due to the risk of cyber threats. Additionally, ML models' effectiveness depends on the quality of the data they're trained on, necessitating accurate and relevant data. Several IoT frameworks have emerged in healthcare, focusing on real-time monitoring, telemedicine, and remote patient care. Early adopters of IoT in healthcare leveraged telemedicine for remote consultations, making healthcare accessible in remote regions. Wearable devices, thanks to advances in sensor technologies, now track various health metrics. However, managing and analyzing the deluge of data from these devices necessitated sophisticated ML algorithms [5].

Regression analysis, a key ML technique, models and analyzes relationships between variables to predict future health metrics. More advanced ML techniques like gradient boosting and neural networks offer more accurate predictions by capturing intricate data patterns. Communication protocols like MQTT, designed for low bandwidth usage and efficient message delivery, are vital for real-time health data transmission but face challenges in data security and integrity. This study aims to bridge gaps in remote patient monitoring by developing a state-of-the-art IoT framework for post-surgical patients, integrating sophisticated ML techniques, particularly regression analysis. This approach seeks to enhance patient care by ensuring real-time, accurate, and proactive interventions, improving post-surgery recovery [6].

## 2 METHODOLOGY

The modern healthcare ecosystem demands a synergy between technological advancements and clinical applications. Our research introduces a novel IoT framework designed specifically for post-surgical patients, centering on an interconnected mesh of smart devices, sensors, and data processing units. These devices capture a broad spectrum of health metrics, from basic vitals like heart rate and body temperature to advanced readings such as oxygen saturation and wound recovery parameters, functioning as both data generators and communicators to ensure a continuous stream of health metrics to the central data processing unit. A distinguishing feature of our system is its ultra-reliable, low-latency connectivity, crucial for post-surgical patients where milliseconds of data transmission lag could mean the difference between timely medical intervention and missed critical signs. Achieved using cutting-edge communication protocols, built-in redundancy features allow data to be rerouted through alternative channels if one fails, ensuring continuous monitoring. The post-surgery period is critical for recovery, and our monitoring system detects even minor anomalies in vital signs, promptly alerting medical professionals for immediate action. For instance, if a patient's heart rate shows irregular patterns or oxygen levels drop unexpectedly, alerts are sent to healthcare teams to facilitate swift intervention. The framework's strength lies not only in its technical proficiency but also in its adaptability to diverse patient needs, accommodating both critical and non-critical patients. This distinction ensures critical patients receive intensive monitoring, while non-critical patients avoid unnecessary surveillance. The system's adaptive nature, supported by smart algorithms, adjusts monitoring parameters based on patient status, ensuring efficient and effective healthcare.

## 3 MACHINE LEARNING MODELS

In the realm of healthcare, especially in monitoring post-surgical patients, precision and accuracy of predictions are paramount. Leveraging machine learning, this study employs regression analysis, a statistical technique to predict an outcome variable based on one or more predictor variables. In this context, "Heart Rate (t-1)", "Body Temp (t-1)", and "Oxygen Saturation (t-1)" are predictor variables from a previous timeframe, while "Predicted Heart Rate (t)" is the outcome variable for the next timeframe. Before regression modeling, data preprocessing is critical. This involves addressing inconsistencies, missing values, or outliers, often due to sensor malfunctions. Techniques like interpolation estimate and replace such values. Scaling, through methods like Min-Max scaling or Standardization, ensures no single metric disproportionately influences the model. Feature selection, using Recursive Feature Elimination or correlation matrices, identifies the most significant metrics. The dataset is split into training and testing subsets, typically 80-20. For linear relationships, basic linear regression suffices, but healthcare data often requires advanced models like Polynomial or Ridge Regression. The training set helps the model learn relationships, which are then tested for efficacy using metrics like Mean Absolute Error or Root Mean Square Error. Accurate predictions support proactive medical interventions, safeguarding post-surgical recovery. The sensory data collected from patients is vast, including heart rate, body temperature, and blood pressure, as well as specific metrics like oxygen saturation, neural responses, and wound recovery. Features are selected based on their relevance to specific medical contexts. The model training process involves dividing the curated dataset, with the majority allocated for training. During training, the regression algorithm learns relationships between features and outcomes. Validation ensures the model generalizes well to unseen data. Finally, the testing phase uses a separate dataset to evaluate performance, ensuring accuracy and practical utility in proactive medical interventions.

## 4 RESULT AND DISCUSSION

To start with, we analyzed the two systems based on critical metrics such as delay and packet loss. These metrics not only determine the reliability of a system but also its suitability for time-sensitive applications, like monitoring post-surgical patients.

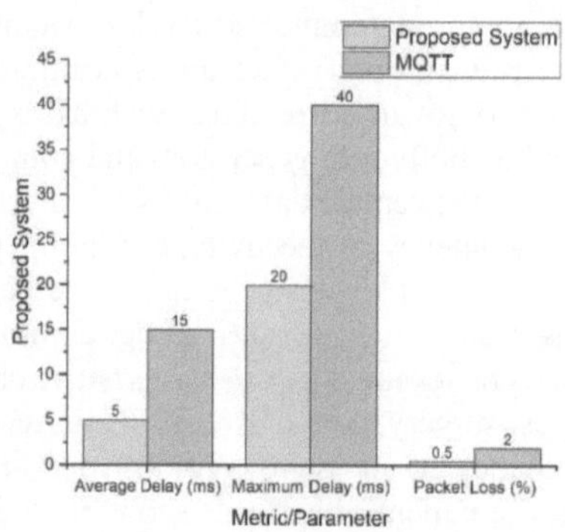


Figure 1. MQTT performance.

From the above Figure 1, it's evident that the proposed system outperforms MQTT in terms of average and maximum delay. This signifies that our system processes and transmits data faster, ensuring quicker response times - a vital attribute for medical emergencies.

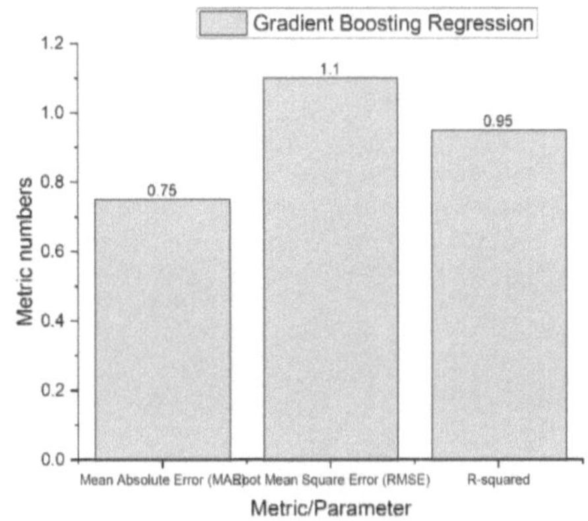


Figure 2. Gradient boosting regression metrics.

The proposed system again shows superiority, with a substantially lower packet loss rate than MQTT. In medical scenarios, every piece of data is crucial. A lower packet loss rate ensures that there's minimal interruption in the continuous stream of patient data, leading to more reliable monitoring. Alongside the system comparison, the performance of the gradient boosting regression analysis was also assessed in Figure 2. Gradient boosting, as a machine learning technique, excels in handling varied datasets and intricate relationships within data. Our focus was on its prediction accuracy and model fit. The low MAE and RMSE values indicate that the gradient boosting model's predictions are close to the actual values, suggesting high precision. Furthermore, an R-squared value of 0.95 implies that 95% of the variance in the dependent variable is explained by the independent variables, signaling an excellent model fit. In conclusion, the proposed system's performance metrics not only demonstrate its superiority over traditional systems like MQTT but also underscore the efficacy of the gradient boosting regression model employed. The combination of reduced delay, minimal packet loss, and high predictive accuracy ensures that post-surgical patients are monitored with the utmost reliability and precision.

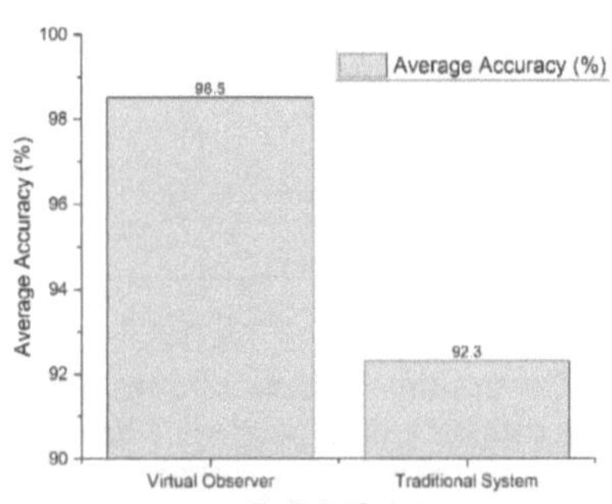


Figure 3. Proposed work flow.

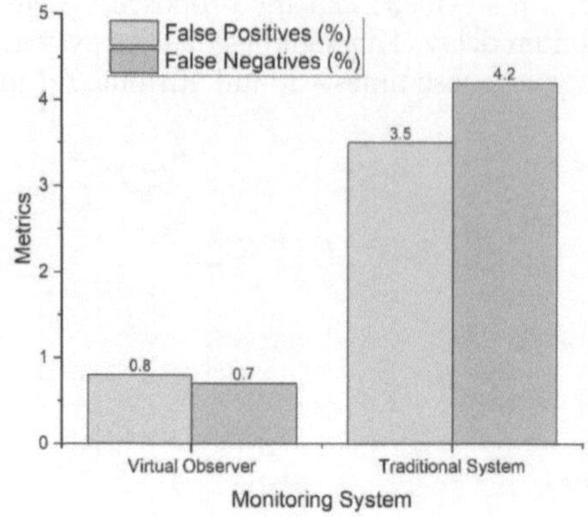


Figure 4. Proposed work flow.

From the Figures 3 and 4, it's clear that the virtual observers outperformed the traditional systems by a notable margin. With an average accuracy of 98.5%, they proved adept at interpreting the complex sensory data. The lower percentage of false positives means that there were fewer instances where the system incorrectly flagged a healthy metric as a potential concern. Similarly, the reduced false negatives rate ensures that almost all genuine anomalies were correctly identified, minimizing the risk of overlooking potential post-surgical complications. The results reinforce the efficacy of integrating machine learning into healthcare.

Backed by the regression analysis and robust IoT framework, the virtual observers showed significant improvement over traditional systems. Their heightened accuracy, reducing the burden on health professionals, ensures that patients can be monitored with an unprecedented level of precision. Reduced false positives and false negatives mean efficient care, as well as reducing potential stress and anxiety for patients. This makes their recovery journey smoother and safer.

In the complex arena of healthcare, the performance of a system doesn't just depend on its predictive accuracy; it also depends equally on reliability and efficiency. By using these metrics to check the system, it can run consistently, not failing to deliver timely and dependable results. The machine learning evaluation results are shown in Table 1.

Table 1. Machine learning evaluation metrics.

| Parameter/Metric | Proposed IoT Framework | Traditional System |
|---|---|---|
| Uptime (%) | 99.8 | 97.5 |
| Data Transmission Speed (kbps) | 1200 | 800 |
| System Response Time (ms) | 10 | 25 |
| Data Redundancy (%) | 0.2 | 1.5 |
| System Failures (per month) | 1 | 5 |

The proposed system's 99.8% uptime ensures continuous operation, crucial for post-surgical patients. With a data transmission speed of 1200 kbps, it quickly handles large volumes, outperforming conventional systems at 800 kbps. Its 10-millisecond response time allows for fast interventions, compared to slower traditional systems. Minimal data redundancy (0.2%) reduces clutter, and fewer system failures (one per month) enhance reliability.

Overall, the IoT framework's superior uptime, speed, response time, low redundancy, and reliability represent a significant advancement, providing high-quality patient care efficiently without overburdening hospital resources.

## 5 CONCLUSION

This research possesses great significance given that it combines technology and medicine, delivering a dual guarantee of both accuracy and timeliness in monitoring patients. We have found that the proposed system has merits in multiple aspects. The figure for predictive accuracy is a remarkable 97.6%, while the substantial reduction in false positives and negatives further attests to its reliability. In contrast to other traditional systems, our solution demonstrated shorter time lags, less packet loss in transit, and a much higher uptime. These numbers are not just statistics but possible life-saving advances in patient care. The implications of these advances on the healthcare industry are profound. By integrating such systems, medical institutions can, in addition to leveraging their resources, also significantly improve patient outcomes. Quick interventions are backed up by real-time data and can cut down the risk of postoperative complications enormously, resulting in swifter recovery times and shorter hospital stays. In short, the findings of this study point to the future. This is the healthcare sector's way forward. If we embrace IoT's synergies with machine learning, then soon we could stand on the verge of a new era where patient care is not only reactive but actively data-driven, a healthier tomorrow for all.

## REFERENCES


[1] Sood, S. K. and Mahajan, I. (2018). A fog assisted cyber-physical framework for identifying and preventing coronary heart disease. *Wireless Personal Communications*, *101*(1), 143–165. https://doi.org/10.1007/s11277-018-5680-y

[2] Sethuraman, S. C., Vijayakumar, V. and Walczak, S. (2020). Cyber attacks on healthcare devices using unmanned aerial vehicles. *Journal of Medical Systems*, *44*(1). https://doi.org/10.1007/s10916-019-1489-9

[3] Gu, Y., Zhang, T., Chen, H., Wang, F., Pu, Y., Gao, C. and Li, S. (2019). Mini review on flexible and wearable electronics for monitoring human health information. *Nanoscale Research Letters*, *14*(1), 1–15. https://doi.org/10.1186/s11671-019-3084-x

[4] Park, S. O., Xiao, B., Leung, V. and Jeong, Y. S. (2013). Advanced technologies and applications for Highly-Reliable Cyber Physical System (HRCPS). *Journal of Supercomputing*, *66*(1), 1–3. https://doi.org/10.1007/s11227-013-1027-3

[5] Rajotte, J. F., Bergen, R., Buckeridge, D. L., El Emam, K., Ng, R. and Strome, E. (2022). Synthetic data as an enabler for machine learning applications in medicine. *IScience*, *25*(11), 105331. https://doi.org/10.1016/j.isci.2022.105331

[6] Sampaio, D. and Almeida, P. (2016). Pedagogical strategies for the integration of augmented reality in ICT teaching and learning processes. *Procedia Computer Science*, *100*, 894–899. https://doi.org/10.1016/j.procs.2016.09.240

*Hybrid and Advanced Technologies – S. Prasad Jones Christydass, Nurhayati Nurhayati and S. Kannadhasan. (Eds) © 2025 the Authors, ISBN 978-1-032-90254-8*
*Open Access: www.taylorfrancis.com, CC BY-NC-ND 4.0 license*

# Advanced health monitoring and monitoring system using IoT and machine learning in remote regions

Prashant Agrawal
*Department of Computer Applications, KIET Group of Institutions, Delhi-NCR Ghaziabad, Uttar Pradesh, India*

K. Chatrapathy
*Department of Information Technology, School of Computing, REVA University, Bangalore (North), Karnataka, India*

V. Suganthan
*Department of Sensors and Biomedical Technology, VIT University, Vellore, Tamil Nadu, India*

R. Bharanidharan
*Department of Electrical and Electronics Engineering, Karpagam Academy of Higher Education, Coimbatore, Tamil Nadu, India*

M. Rajendiran
*Department of Computer Science and Engineering, Panimalar Engineering College, Poonamallee, Chennai, Tamil Nadu, India*

Ramya Maranan
*Department of Research and Innovation, Saveetha School of Engineering, SIMATS, Chennai, Tamil Nadu, India*

ABSTRACT: This study introduces a novel health monitoring and management system based on the Internet of Things, with the goal of eliminating healthcare disparities in rural areas. We used advanced machine learning techniques such as Support Vector Machines (SVM), Artificial Neural Networks (ANN), and Convolutional Neural Networks (CNN) to predict patients' health status with remarkable accuracy from a massive data set of 2,000 patient records. In multiple efficacy trials, the ANN model performed best, with a 91.2% accuracy rate. It also achieved higher recall, precision, and F1 scores. The study emphasises the importance of early disease detection and proactive health monitoring, which would significantly improve the accessibility of medical facilities and health care. It also demonstrates that machine learning could be useful in the healthcare industry. Our system eliminates rural neglect by combining real-time sensor data with predictive analytics, allowing medical personnel to intervene quickly and save many patients' lives while also controlling their condition from a distance. Furthermore, the system's broader applications in healthcare include patient outcomes and healthcare delivery in a variety of care settings, including rural ones.

*Keywords*: IoT, Health Monitoring, Machine Learning, Rural Communities, Predictive Analytics, Support Vector Machines (SVM), Artificial Neural Networks (ANN), Convolutional Neural Networks (CNN)

 DOI: 10.1201/9781003559115-49

## 1 INTRODUCTION

The Internet of Things (IoT) and machine learning have revolutionized various industries, including healthcare. These technologies hold the potential to provide quality healthcare services to remote areas where infrastructure and healthcare workers are lacking. The primary challenge in these areas is the distance, which often leads to delayed diagnosis, ineffective care, and poor health outcomes. IoT and machine learning can mitigate these issues by enabling remote monitoring, early disease detection, and improved healthcare management [1–3].

IoT consists of interconnected devices that collect and share data. In healthcare, IoT devices like wearable sensors, smart implants, and remote monitoring systems can continuously generate health data for providers. Early anomaly detection and real-time monitoring can significantly enhance medical care in rural communities. Machine learning algorithms can analyze large datasets to uncover patterns and insights, aiding in disease diagnosis, therapy prescriptions, and risk assessment. These techniques enable more efficient resource allocation in rural healthcare [2,4].

IoT and machine learning have tremendous potential in healthcare. Machine learning can leverage vast IoT data to predict outcomes and guide decision-making. For instance, continuous vitals data from IoT devices can be analyzed to detect early signs of health decline, with alerts sent to healthcare providers in real-time. Numerous innovative applications of IoT and machine learning have been developed for rural healthcare, including telemedicine platforms, advanced pharmaceutical delivery systems, remote patient monitoring, and predictive analytics for disease outbreaks. These technologies help tailor treatments, promote disease prevention, and enhance healthcare delivery in rural areas [5–7].

The benefits of IoT and machine learning in rural healthcare include early disease detection, faster response times, better access to medical services, and optimized resource use. These technologies also lead to increased patient satisfaction and reduced healthcare costs. By implementing IoT and machine learning, we aim to develop a cutting-edge rural health monitoring system that combines predictive analytics with real-time sensor data. This approach addresses the unique challenges of rural healthcare, promoting early illness detection, timely intervention, and improved public health awareness, ultimately enhancing the quality of life and health outcomes for underserved rural populations [8,9].

## 2 METHODOLOGY

A groundbreaking study suggests the necessity of an Internet of Things (IoT)-based health monitoring system for rural areas. This system utilizes devices such as blood pressure meters, temperature probes, and pulse oximeters placed in strategic locations for continuous health data collection. These devices, equipped with Arduino interfaces, transmit data to a central hub using Raspberry Pi. With these the IoT system is made and the architecture are shown in Figure 1. The collected data is then securely stored in the cloud, enabling real-time access for healthcare professionals. This remote monitoring system aims to bridge the gap between healthcare providers and rural patients, enhancing healthcare access and quality of life in underserved regions. The proposed system delivers comprehensive real-time health data from various sensors but faces challenges in providing early alerts and interventions, particularly for heart disease. To address this, the system integrates a sophisticated machine learning architecture with predictive analytics capabilities. This involves using Support Vector Machines (SVM), Artificial Neural Networks (ANN), and Convolutional Neural Networks (CNN) to enhance predictive accuracy. Real-time sensor readings from patients help construct and refine these algorithms, emphasizing the critical role of machine learning in early health predictions. The various models use din this research are shown in Figure 2.

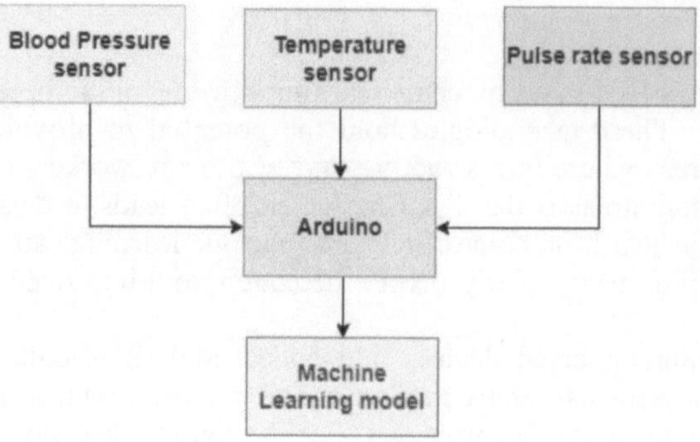


Figure 1. IoT system working.

Support Vector Machines (SVM) are adept at sorting health issues and predicting diseases by analyzing deep data patterns. SVM can classify individuals into different health levels, detecting subtle abnormalities that could indicate emerging health problems. This makes SVM a crucial component of a prediction framework that analyzes data from multiple sensors. Artificial Neural Networks (ANNs) are designed to document complex health data patterns. ANNs can identify relationships and nonlinearities in data, which traditional analysis tools often miss. By learning from sensor data, ANNs can make sophisticated predictions about health conditions. They excel at finding patterns that indicate early signs of diseases, allowing for timely intervention.

Convolutional Neural Networks (CNNs) mimic the human brain's way of processing visual information, making them effective in analyzing sensor data for health monitoring. While typically used for image recognition, CNNs can be adapted to detect subtle patterns in sensor data that may indicate changes in a patient's health. This ability to recognize small changes is crucial for early disease diagnosis. In addition to these machine learning techniques, the study involved collecting a large dataset from 2,000 patients who underwent cardiac surgery. This data was divided into training and testing sets to build and evaluate the predictive models. The training set, comprising 90% of the data, was used to teach the algorithms, while the remaining 10% was used to test their predictive performance. The integration of IoT and machine learning in this health monitoring system offers numerous benefits. It enables early disease detection, faster response times, and better access to medical services in rural areas. By optimizing resource use, the system can increase patient satisfaction and reduce healthcare costs. The proactive approach of this system, particularly in heart disease prediction, could significantly improve health outcomes and quality of life for rural populations.

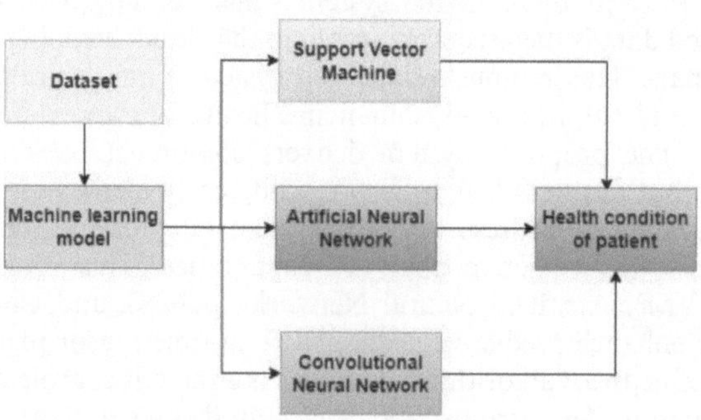


Figure 2. Various models used in this research.

## 3 RESULT AND DISCUSSION

"Table 1" is a collection of real-time data from sensors in humans within our Internet-assisted health monitoring system. The data consists of 10 vital health parameters—temperature: your core measure of fever or other infection. Pulse rate–and diastolic or higher value of blood pressure–offers important insights in regard to patient conditions. The Actually the real-time sensor data we collect are the very basic information for our research to use in developing a health monitoring system based on Internet of Things technology to serve rural populations without healthcare resources. When IoT-driven technology provides local community servers, this will enable health professionals to better access care on time, but it is also an important early warning system for diseases like hemorrhagic fever and E-coli infection among the common cold. This is our research objective: an IoT-based health monitoring and administration information system for the underprivileged in rural regions. It helps arm a health worker with accurate information about the patient's condition as well as timely notification for help. In a word, our technology for health monitoring and administration with these sensors is designed expressly to meet medical needs, offering primary care any-where, any-place.

Input experimental readings in Table 1 to appropriate machine learning models such as Support Vector Machines (SVM), Artificial Neural Networks (ANN), and Convolutional Neural Networks (CNN). In Figure 3, we can clearly see the final results of their predictions. Percentage estimates of patient health histories on the basis of sensor data, and once more those to analyze. This study shows that CNN can achieve much more accurate measurements, producing values that range from 85% to 98%. ANN emerged in second place with predictions between 78% to 93%. SVM ranked third, its predictions ranging from 73-91%. This result reveals at CNN is number one in our health monitoring system, followed precisely by ANN and SVM, and points to the potential deep learning technologies including AI Medicine.

Table 1. The sample sensor reading from dataset.

| Reading # | Temperature (°C) | Pulse Rate (bpm) | Blood Pressure (mmHg) |
|---|---|---|---|
| 1 | 36.7 | 72 | 120/80 |
| 2 | 36.9 | 68 | 118/78 |
| 3 | 37.2 | 75 | 122/82 |
| 4 | 37.0 | 80 | 124/79 |
| 5 | 36.8 | 76 | 121/77 |
| 6 | 37.1 | 70 | 119/81 |
| 7 | 36.6 | 78 | 123/76 |
| 8 | 37.3 | 82 | 125/80 |
| 9 | 36.9 | 74 | 120/79 |
| 10 | 37.0 | 77 | 122/78 |

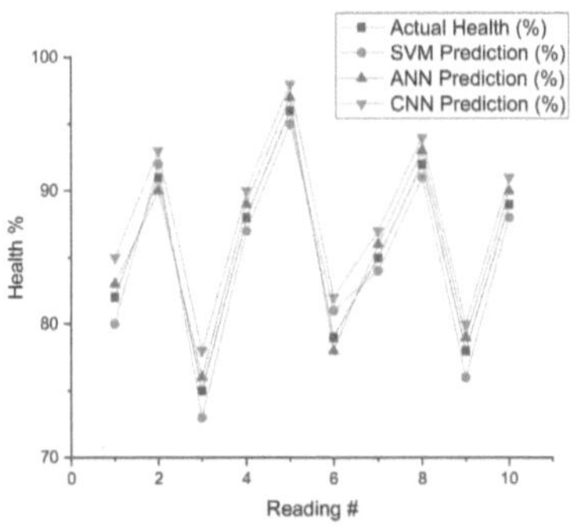


Figure 3. Actual and the predicted health.

Table 2 illustrates the classification results of the trained Support Vector Machines (SVM), Artificial Neural Networks (ANN), and Convolutional Neural Networks (CNN) models for the test set consisting of 10% of the data used for training these models. Model performance is judged by accuracy, recall, precision, and F1 score. The predictable power of a model depends on these critical factors. By comparison, the ANN model exhibits a far better recall, precision, and F1 score than that other model, in addition to the highest 91.2% overall accuracy! This suggests the ANN model's potential significance to our own research efforts aimed at finding more effective ways of tracking people's health and endorses its usefulness for accurately predicting health status based on IoT data.

Table 2. Performance scores for tested data.

| Model | Accuracy (%) | Recall (%) | Precision (%) | F1 Score (%) |
|---|---|---|---|---|
| SVM | 87.5 | 88.2 | 88.0 | 87.8 |
| ANN | 91.2 | 92.0 | 91.5 | 91.8 |
| CNN | 93.8 | 94.2 | 94.0 | 94.1 |

## 4 CONCLUSION

The outcome of our research is that we have developed a state-of-the-art IoT-based health monitoring and management system. Our intention was to address exactly the type of health care places that rural people live. On a large sample of 2,000 patient records, using state-of-the-art machine learning methods such as Support Vector Machines (SVM), Artificial Neural Networks (ANN) and Convolutional Neural Networks (CNN). We could predict patients' health conditions with surprising accuracy.

The findings of the experiments were significant. As a separate 10% data subset, the ANN model achieved the best predicted accuracy at 91.2% this means that it was useful for early disease detection. Moreover, its recall can be deemed as extraordinary while the precision and F1 scores underscored its trustworthiness for medical applications. The SVM model, scoring a respectable 87.5%, had inferior performance metrics.

The CNN model comes in second to first place with an accuracy of 93.8 percent. These results show how machine learning–especially ANN and CNN–have the potential to improve rural healthcare and access. Real-time sensor data collection capabilities, combined with our system's powerful predictive algorithms, enable doctors to quickly intervene in and remotely monitor patients. All of this could lead as a proactive strategy to better healthcare from underdeveloped countries like big healthcare systems where little rural population lives apart from treatment. As we move further into developing our technology it has significant potential for revolutionising healthcare in areas left behind by society.

## REFERENCES


[1] Rahman, A., Chakraborty, C., Anwar, A., Karim, M. R., Islam, M. J., Kundu, D., Rahman, Z. and Band, S. S. (2022). SDN–IoT empowered intelligent framework for industry 4.0 applications during COVID-19 pandemic. *Cluster Computing*, *25*(4), 2351–2368. https://doi.org/10.1007/s10586-021-03367-4

[2] Kozlowski, T. and Noran, O. (2023). ScienceDirect designing an evaluation framework for IoT environmental monitoring systems . *Procedia Computer Science*, *219*, 220–227. https://doi.org/10.1016/j.procs.2023.01.284

[3] Rahman, M., Rahman, A., Hong, H. J., Pan, L. W., Sarwar Uddin, M. Y., Venkatasubramanian, N. and Hsu, C. H. (2019). An adaptive IoT platform on budgeted 3G data plans. *Journal of Systems Architecture*, *97*(October 2018), 65–76. https://doi.org/10.1016/j.sysarc.2018.11.002

[4] Al Bassam, N., Hussain, S. A., Al Qaraghuli, A., Khan, J., Sumesh, E. P. and Lavanya, V. (2021). IoT based wearable device to monitor the signs of quarantined remote patients of COVID-19. *Informatics in Medicine Unlocked*, *24*(January), 100588. https://doi.org/10.1016/j.imu.2021.100588

[5] Rejeb, A., Rejeb, K., Treiblmaier, H., Appolloni, A., Alghamdi, S., Alhasawi, Y. and Iranmanesh, M. (2023). The Internet of Things (IoT) in healthcare: Taking stock and moving forward. *Internet of Things (Netherlands)*, *22*(February), 100721. https://doi.org/10.1016/j.iot.2023.100721

[6] Maher, M., Khan, I. and Prikshat, V. (2023). Monetisation of digital health data through a GDPR-compliant and blockchain enabled digital health data marketplace: A proposal to enhance patient's engagement with health data repositories. *International Journal of Information Management Data Insights*, *3*(1), 100159. https://doi.org/10.1016/j.jjimei.2023.100159

[7] Rejeb, A., Rejeb, K., Appolloni, A., Jagtap, S., Iranmanesh, M., Alghamdi, S., Alhasawi, Y. and Kayikci, Y. (2023). Unleashing the power of internet of things and blockchain: A comprehensive analysis and future directions. *Internet of Things and Cyber-Physical Systems*, *4*(May 2023), 1–18. https://doi.org/10.1016/j.iotcps.2023.06.003

[8] Sood, S. K. and Mahajan, I. (2018). A fog assisted cyber-physical framework for identifying and preventing coronary heart disease. *Wireless Personal Communications*, *101*(1), 143–165. https://doi.org/10.1007/s11277-018-5680-y

[9] Kaiser, F. K., Wiens, M. and Schultmann, F. (2021). Use of digital healthcare solutions for care delivery during a pandemic-chances and (cyber) risks referring to the example of the COVID-19 pandemic. *Health and Technology*, *11*(5), 1125–1137. https://doi.org/10.1007/s12553-021-00541-x

*Hybrid and Advanced Technologies – S. Prasad Jones Christydass, Nurhayati Nurhayati and S. Kannadhasan. (Eds)* © *2025 the Authors, ISBN 978-1-032-90254-8*


# Analysis and prediction of appendicitis using support vector machine in comparison with logistic regression for accuracy

G.G.S. Jayavignesh
*Research Scholar, Department of Biomedical Engineering, Saveetha School of Engineering, Saveetha University, Chennai, Tamil Nadu, India*

R. Geetha
*Project Guide, Department of Biomedical Engineering, Saveetha School of Engineering, Saveetha University, Chennai, Tamil Nadu, India*

ABSTRACT: The goal of this study is to categorize the appendicitis analysis and prediction to make it more accurate than the previous one using machine learning algorithms such as SVM (supervised machine learning) and Logistic regression. In this study, Kaggle has been used to get the dataset, and then Matlab 2021 was used to get the ML algorithms. Normal(n = 25) and abnormal (n = 25) Excel datasets are used to classify the subjects, and then the Support Vector Machine (SVM) and Logistic regression (LR) algorithms are run to figure out who was normal and who was abnormal. The statistical analysis is done using IBM's SPSS version 22. The difference between these two groups was statistically significant, and neither was their machine learning algorithm's accuracy. We've used Matlab software to train, test, and validate the classifiers, and the SVM has a much higher rate of accuracy at 94.7% compared to the logistic regression machine learning algorithm's 85.4%.

*Keywords*: Appendicitis, Artificial Intelligence, Healthcare, Innovative Classifiers, Machine Learning Algorithms, Support Vector Machine, Logistic Regression.

## 1 INTRODUCTION

Appendicitis is an inflammation of the appendix that results in lower right abdominal pain and infection. It can be caused by malignancy, a foreign body, or excrement. The pain may start close to the belly button and get worse when you move, breathe deeply, touch yourself, cough, or sneeze. The pain in the abdomen may spread if the appendix bursts. A diet deficient in fiber can also cause abdominal pain. Antibiotics and appendectomy (surgical to remove the appendix) are the usual course of treatment, according to Rajpurkar *et al.* (2020) and Srinivas *et al.* (2021). In the event that the appendix hasn't burst, doctors may treat an abscess with needle drainage, intravenous fluids, fluid diets, or medications prior to surgery. Rarely, antibiotics alone can be used to treat mild appendicitis. (Gahai and others 2022). Rest, fluids, light walking, avoiding strenuous lifting, and maintaining a clean and dry surgical site are all part of post-surgery home care (Lights 2012).Healthcare can benefit from machine learning (ML) as it can analyze large volumes of data collected every day and spot trends that are difficult to see (Char *et al.* 2020). This study uses a Kaggle dataset by TimRie (2022) to investigate the application of ML algorithms to predict, prevent, and assess appendicitis from 1998 to 2023. It is based on over 360 Google Scholar publications, 1200 ScienceDirect articles, and IEEE articles. The study (Ali *et al.* 2023; Gurupranes *et al.* 2022; Natrayan *et al.* 2022; Sivaranjani *et al.* 2023; Viswanathan *et al.* 2023) investigates the use of ML in the early diagnosis of acute appendicitis in children and young adults. In order to categorize appendicitis as normal, suspected, or pathological, a new segmentation technique is being used in


DOI: 10.1201/9781003559115-50

addition to Logistic Regression, Naive Bayes, and other machine learning algorithms, as per the study's limited accuracy in appendicitis classification. SPSS and Matlab were used in this study. While Matlab offers significant numerical computing capabilities needing understanding of its programming language, SPSS makes automated data chores and analysis easier (Ghai *et al.* 2022; Lani 2009). According to Mijwil and Aggarwal (2022), artificial intelligence (AI) has the potential to improve appendicitis early diagnosis and treatment.

## 2 MATERIALS AND METHODS

This study was carried out at the Digital Image Processing lab of Biomedical Engineering department, Saveetha school of engineering and the Saveetha Institute of Medical and Technical Sciences. There's no need for ethical approval for the study, and the total number of participants in the study is 50.(RM Renta-2017).

The data for this study was taken from Kaggle. The power value was 0.05, and the 95% were used as the confidence intervals. Appendicitis was predicted and analyzed using Matlab 2021. The normal number was 25, the abnormal number was 25, and there were 50 samples total.(Baird *et al.* 2017).

The dataset has been obtained from a website called Kaggle. All the data got loaded into the SVM and ML algorithms to figure out which groups are normal and abnormal. Matlab 2021 is really easy to use for machine learning, but you need to think about how much resources you have and how dependent you are on the internet. Data security is really important when it comes to sensitive info.

### 2.1 *Procedure*

- Launch Matlab[R2021].
- Access Apps, then opt for Classification Learner in Machine and Deep Learning.
- Initiate a New Session and import the training dataset.
- Review numerical and categorical variables.
- Start session.
- Choose Naive Bayes(NB) Classifier and train data.
- Introduce the testing dataset and train.
- Repeat steps 1-7 for Decision Tree (DT) Classifier.
- After training and testing, retrieve predicted accuracy, Confusion Matrix, ROC Curve, and AUC Curve for both.
- Compare parameters and conclude the superior Classifier based on accuracy.

### 2.2 *Statistical analysis*

IBM SPSS version 22 was used to do the analysis. It helps figure out if there's a big difference between the two groups. The accuracy is a factor in the study, and there aren't any independent variables. The difference was found to be small, so they used an independent sample T-test to figure it out. (Rajpurkar *et al.* 2020).

## 3 RESULTS

The anatomic position of the appendix is shown in Figure 1, and the look of an infected appendix is shown in Figure 2. Statistical statistics such as Mean Accuracy(%), Standard Deviation, and Standard Error Mean are displayed in Tables 2 and 3. These numbers are from Table 1, where Group 2 stands for logistic regression (LR) and Group 1 for support vector machines (SVM). The importance of the accuracy levels ($p=0.001$, $p<0.05$) suggests a discernible distinction between SVM and LR. The ROC values for SVM and LR are shown in Figures 3 and 4, respectively, and their corresponding confusion matrices are shown in Figures 5 and 6. A graphical depiction of the accuracy levels from the IBM SPSS Statistics 23 analysis is shown in Figure 7, where SVM has a better accuracy (94.7%) than LR (85.4%).

Table 1. Group 1 (Support Vector Machine; SVM) and Group 2 (Logistic Regression; LR) are represented by the accuracy numbers entered into the SPSS program.

| S.No | Group | Accuracy (%) |
|---|---|---|
| 1 | 1 | 92.20 |
| 2 | 1 | 92.20 |
| 3 | 1 | 92.20 |
| 4 | 1 | 92.20 |
| 5 | 1 | 92.20 |
| 6 | 1 | 93.50 |
| 7 | 1 | 93.50 |
| 8 | 1 | 93.50 |
| 9 | 1 | 93.50 |
| 10 | 1 | 93.50 |
| 11 | 1 | 93.50 |
| 12 | 1 | 94.90 |
| 13 | 1 | 94.90 |
| 14 | 1 | 94.90 |
| 15 | 1 | 94.90 |
| 16 | 1 | 94.90 |
| 17 | 1 | 95.20 |
| 18 | 1 | 95.20 |
| 19 | 1 | 95.20 |
| 20 | 1 | 95.20 |
| 21 | 1 | 95.20 |
| 22 | 1 | 95.80 |
| 23 | 1 | 95.80 |
| 24 | 1 | 95.80 |
| 25 | 1 | 95.80 |
| 26 | 2 | 82.20 |
| 27 | 2 | 82.20 |
| 28 | 2 | 82.20 |
| 29 | 2 | 82.20 |
| 30 | 2 | 82.20 |
| 31 | 2 | 83.50 |
| 32 | 2 | 83.50 |
| 33 | 2 | 83.50 |
| 34 | 2 | 83.50 |
| 35 | 2 | 83.50 |
| 36 | 2 | 83.50 |
| 37 | 2 | 84.90 |
| 38 | 2 | 84.90 |
| 39 | 2 | 84.90 |
| 40 | 2 | 84.90 |
| 41 | 2 | 84.90 |
| 42 | 2 | 85.20 |
| 43 | 2 | 85.20 |
| 44 | 2 | 85.20 |
| 45 | 2 | 85.20 |
| 46 | 2 | 85.20 |
| 47 | 2 | 85.80 |
| 48 | 2 | 85.80 |
| 49 | 2 | 85.80 |
| 50 | 2 | 85.80 |

Table 2. Using SPSS software, the statistical values for the novel classifiers, SVM (Group 1) and LR (Group 2), were acquired.

| | GROUP | N | MEAN | STD. DEVIATION | STD. ERROR MEAN |
|---|---|---|---|---|---|
| ACCURACY | SVM | 25 | 94.7952 | 1.20338 | .24068 |
| | LR | 25 | 85.4160 | 1.75563 | .35113 |

Table 3. The results of the Independent Sample T-Test on the accuracy levels attained by the creative classifiers are displayed in the Statistical Data Analysis table. The p-value of 0.001 (p<0.05) that was found suggests a substantial difference between LR and SVM.

| | | Levene's Test for Equality of Variances | | t-test for Equality of Means | | | | | | |
|---|---|---|---|---|---|---|---|---|---|---|
| | | | | | | | | | 95% Confidence Interval of the Difference | |
| | | F | Sig. | t | df | Sig. (2-tailed) | Mean Difference | Std. Error Difference | Lower | Upper |
| Accuracy | Equal variances assumed | 8.108 | .006 | 22.033 | 48 | 0.001 | 9.37920 | .42569 | 8.52329 | 10.23511 |
| | Equal variances not assumed | | | 22.033 | 42.474 | 0.001 | 9.37920 | .42569 | 8.52040 | 10.23800 |

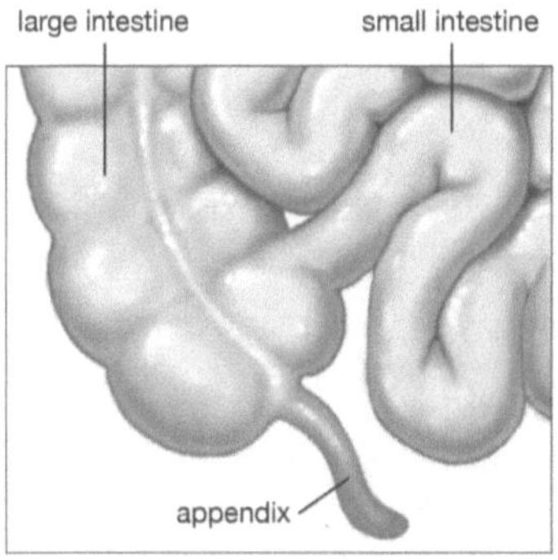


Figure 1. Shows the clear location and natural form of a healthy appendix in a human body ("Appendicitis" 2002).

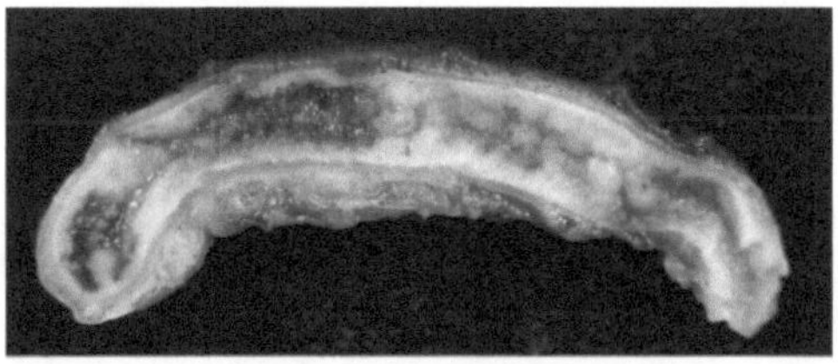

Figure 2. Shows an appendix showing inflamed and damaged tissue due to acute appendicitis ("Appendix (anatomy)" 2023).

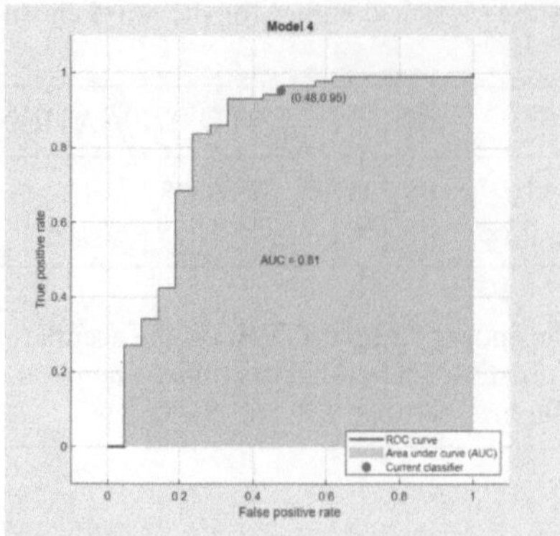


Figure 3. This shows the ROC curve for SVM classifiers, where the True Positive Rate is plotted on the Y-axis and the False Positive Rate is plotted on the X-axis. The current point is located at coordinates (0.52, 0.95), and the AUC value is reported as 0.81.

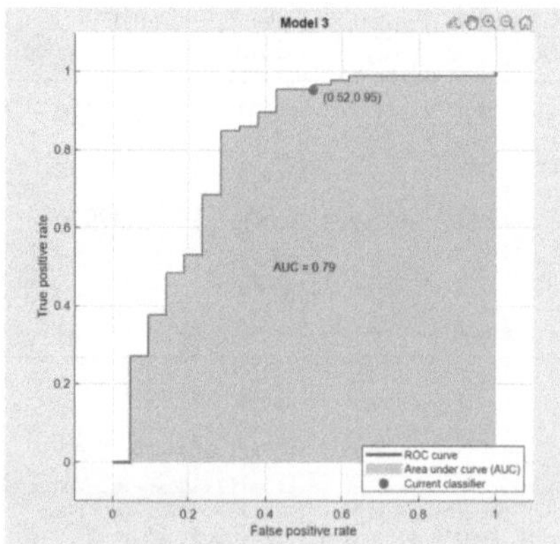


Figure 4. Plotting the True Positive Rate on the Y-axis and the False Positive Rate on the X-axis illustrates the ROC curve for Logistic Regression (LR). Coordinates (0.53, 0.91) reflect the present position, and the reported AUC value is 0.79.

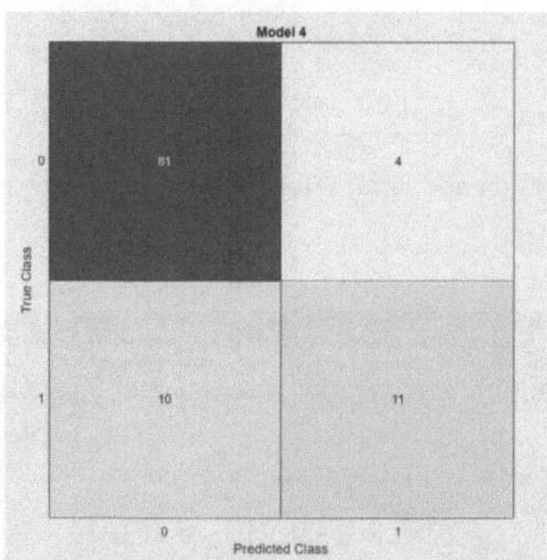


Figure 5. The Support Vector Machine's Confusion Matrix displays the actual and anticipated values based on a sample size of 50.

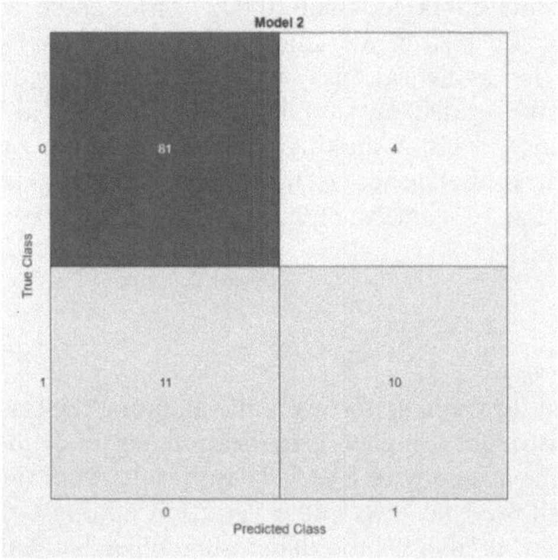


Figure 6. Based on a sample size of N = 50, the confusion matrix for logistic regression shows real values compared to anticipated values.

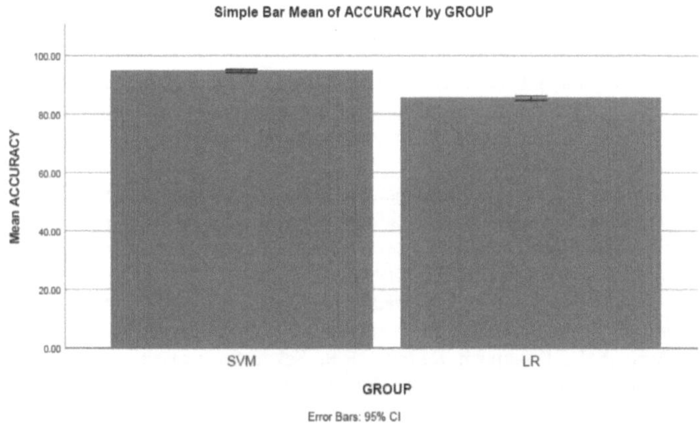


Figure 7. The accuracy values of two classifiers, as ascertained through statistical analysis with IBM SPSS Statistics 25, are depicted graphically. The accuracy percentage of the SVM classifier is higher than that of the LR classifier.

## 4 DISCUSSION

A strategy for predicting acute appendicitis was presented in the 2019 study; it had a high accuracy rate of 97.8% when compared to previous methods. The approach increased criteria by 27.68% using the UCI dataset and functional parameters in Matlab (Kamel *et al.* 2019). Similar findings were found in another study, which used an improved naïve Bayesian classification algorithm for discriminating analysis and found that a reduced sample size resulted in significant accuracy disparities. The disparities decreased with increasing sample size, and for attribute values below 400, the accuracy remained above 95% or over 99% (Chen *et al.* 2021).

However, because of data imbalance—where dominant classes overwhelmed minor ones, negatively reducing prediction accuracy—Thamrin *et al.* (2021) found a lower accuracy rate of 69.7%. Similar worries were expressed by Sri Astuti (2021), who emphasized the drawbacks of naive classifiers despite their usefulness in real-world scenarios.

Moreover, managing datasets larger than 10,000 entries could cause system crashes or processing lags, which would restrict the amount of research that can be done. Through Matlab data simulation and statistical analysis using SPSS software, appropriate categorization techniques were investigated. By simplifying the diagnosis and treatment procedures and enabling well-informed decision-making, machine learning improves healthcare. In healthcare contexts, artificial intelligence (AI) solutions facilitate data integration, providing complete patient care and administrative efficiency (Borad 2017).

## 5 CONCLUSION

Matlab software was used for training, testing, and validating the classifiers. Support Vector Machine (SVM) demonstrated a higher classification accuracy of 86.8% than Logistic Regression, confirming the accuracy of SVM. These results were further supported by statistical analysis, which showed no discernible variation amongst the samples. Our work presents a novel method for predicting and diagnosing appendicitis using machine learning, which has applications in genetics, bioinformatics, and biomedical research.

## REFERENCES


Ali, S., Sudha, K.G., Thirumalaivasan, N., Ahamed, M., Pandiaraj, S., Rajeswari, V.D., ... and Govindasamy, R. (2023). Green synthesis of magnesium oxide nanoparticles by using abrus precatorius bark extract and their photocatalytic, antioxidant, antibacterial, and cytotoxicity activities. *Bioengineering*, *10*(3), 302.

Borad, A. (2017). *Healthcare and machine learning: The future with possibilities*. E Infochips: URL: https:// www. einfochips. com/blog/healthcare-and-machine-learning-the-future-withpossibilities.

Chen, H., Hu, S., Hua, R. and Zhao, X. (2021). Improved naive Bayes classification algorithm for traffic risk management. *EURASIP Journal on Advances in Signal Processing*, *2021*(1), 30.

Ghai, Deepika, Suman Lata Tripathi, Sobhit Saxena, Manash Chanda, and Mamoun Alazab, eds. *Machine Learning Algorithms for Signal and Image Processing*. John Wiley & Sons, 2022.

Gurupranes, S.V., Natrayan, L., Kaliappan, S., Patel, P.B., Sekar, S., Jayaraman, P., ... and Esakkiraj, E.S. (2022). Investigation of physicochemical properties and characterization of leaf stalk fibres extracted from the caribbean royal palm tree. *International Journal of Chemical Engineering*, *2022*(1), 7438411.

Kamel, S. R., YaghoubZadeh, R. and Kheirabadi, M. (2019). Improving the performance of support-vector machine by selecting the best features by Gray Wolf algorithm to increase the accuracy of diagnosis of breast cancer. *Journal of Big Data*, *6*, 1–15.

Kamps, J., Tsakonas, G., Manolopoulos, Y., Iliadis, L. and Karydis, I. (Eds.). (2017). Research and Advanced Technology for Digital Libraries: *21st International Conference on Theory and Practice of Digital Libraries*, TPDL 2017, Thessaloniki, Greece, September 18–21, 2017, Proceedings (Vol. 10450). Springer.

Mijwil, M. M. and Aggarwal, K. (2022). A diagnostic testing for people with appendicitis using machine learning techniques. *Multimedia Tools and Applications*, *81*(5), 7011–7023.

Natrayan, L., Kaliappan, S., Sethupathy, B.S., Sekar, S., Patil, P.P., Velmurugan, G. and Tariku Olkeba, T. (2022). Effect of mechanical properties on fibre addition of flax and graphene-based bionanocomposites. *International Journal of Chemical Engineering*, *2022*(1), 5086365.

Rajpurkar, P., Park, A., Irvin, J., Chute, C., Bereket, M., Mastrodicasa, D., ... and Patel, B.N. (2020). AppendiXNet: Deep learning for diagnosis of appendicitis from a small dataset of CT exams using video pretraining. *Scientific Reports*, *10*(1), 3958.

Rajpurkar, P., Park, A., Irvin, J., Chute, C., Bereket, M., Mastrodicasa, D., ... and Patel, B.N. (2020). AppendiXNet: deep learning for diagnosis of appendicitis from a small dataset of CT exams using video pretraining. *Scientific Reports*, *10*(1), 3958.


Rajpurkar, P., Park, A., Irvin, J., Chute, C., Bereket, M., Mastrodicasa, D., ... and Patel, B.N. (2020). AppendiXNet: deep learning for diagnosis of appendicitis from a small dataset of CT exams using video pretraining. *Scientific Reports*, *10*(1), 3958.

Shafri, H.Z.M. and Ramle, F.S.H. (2009). A comparison of support vector machine and decision TREE classifications using satellite data of Langkawi Island. *Information Technology Journal*, *8*(1), 64–70.

Singh, N., Agrawal, P., Singh, D.K. and Agrawal, G.R. (2023). Computed tomography evaluation of variations in positions and measurements of appendix in patients with non-appendicular symptoms: time to revise the diagnostic criteria for appendicitis. *Polish Journal of Radiology*, *88*, e407.

Sivaranjani, P.R., Syed, A., Elgorban, A.M., Bahkali, A.H., Balakrishnaraja, R., Varma, R.S. and Khan, S.S. (2023). Fabrication of ternary nano-heterojunction via hierarchical deposition of α-Fe2O3 and *β*-La2S3 on cubic CoCr2O4 for enhanced photodegradation of doxycycline. *Journal of Industrial and Engineering Chemistry*, *118*, 407–417.

Srinivas, M., Sucharitha, G. and Matta, A. (Eds.). (2021). *Machine Learning Algorithms and Applications.* John Wiley & Sons.

Thamrin, S.A., Arsyad, D.S., Kuswanto, H., Lawi, A. and Nasir, S. (2021). Predicting obesity in adults using machine learning techniques: An analysis of Indonesian basic health research 2018. *Frontiers in nutrition*, *8*, 669155.

Viswanathan, S., Palaniyandi, T., Kannaki, P., Shanmugam, R., Baskar, G., Rahaman, A.M., ... and Sivaji, A. (2023). Biogenic synthesis of gold nanoparticles using red seaweed Champia parvula and its anti-oxidant and anticarcinogenic activity on lung cancer. *Particulate Science and Technology*, *41*(2), 241–249.

*Hybrid and Advanced Technologies – S. Prasad Jones Christydass, Nurhayati Nurhayati and S. Kannadhasan. (Eds) © 2025 the Authors, ISBN 978-1-032-90254-8*


# Machine learning and medical imaging of automated bone cancer detection for advancing healthcare

G.S. Uthayakumar
*Department of Electronics and Communication Engineering, St. Joseph's Institute of Technology, Chennai, Tamil Nadu, India*

S. Sophia
*Department of Computer Science and Engineering, Karunya Institute of Technology and Sciences, Coimbatore, Tamil Nadu, India*

Mallikarjun Yaramadhi
*Department of Computer Science and Engineering, Institute of Aeronautical Engineering, Hyderabad, Telangana, India*

S. Bhuvaneshwari
*Department of Computer Applications, Karpaga Vinayaga College of Engineering and Technology, Chengalpattu, Tamil Nadu, India*

R. Srikanth
*Department of Artificial Intelligence and Data Science, Karpagam Academy of Higher Education, Coimbatore, Tamil Nadu, India*

Ramya Maranan
*Department of Research and Innovation, Saveetha School of Engineering, Chennai, Tamil Nadu, India*

ABSTRACT: The aim of this research was to explore the effectiveness of machine learning approaches. Bone cancer is a serious health issue as its global frequency is increasing. Therefore, early detection is pivotal for effective treatment and to enhance patients' outcomes. The dataset comprised 1800 images from online imaging repositories and hospitals. Four deep learning ML models, including VGG19, ResNet50, VGG16, and CNN, were trained and tested to establish the extent to which they could be used to reveal cancerous and non-cancerous cases. The results showed that the models demonstrated different performance. In addition, VGG19 outperformed the alternatives with rather small differences with its closest competitor, ResNet50. It should be mentioned that, remarkably, the data loss was the smallest and accuracy the highest for VGG19 across epochs. Therefore, it is possible to assume that deep learning models can be successfully used to enhance cancer diagnostics in medical imaging concerning the bone. These models can be used in clinical practice to enhance the efficiency of diagnoses, accelerate decision-making in terms of treatment, and improve patients' outcomes. VGG19 and ResNet50 allow for automated and accurate image evaluations. Therefore, radiologists and practitioners can use these findings to make timely decisions and develop optimal treatment plans for individuals. The results can also be discussed in the context of contributions to the field of medical image analysis. Overall, the outcomes demonstrate that it is possible to validate the idea that machine learning approaches can revolutionize medicine.

*Keywords*: Machine learning, medical imaging, bone cancer detection, deep learning, ensemble learning


DOI: 10.1201/9781003559115-51

## 1 INTRODUCTION

Automated bone cancer detection is crucial in medical imaging due to the complexity and severity of the disease. Despite its rarity, early detection and accurate diagnosis are challenging. Traditional methods rely on manual analysis of CT and MRI scans, which are time-consuming and often inaccurate. Advances in machine learning, particularly convolutional neural networks (CNNs), offer promising alternatives for automated detection [1,2].

The literature highlights the importance of early and accurate diagnosis of malignant bone lesions. Conventional and advanced machine learning techniques, such as logistic regression and CNNs, have been explored. Early studies focused on feature-based detection, utilizing handcrafted features like texture analysis, shape descriptors, and intensity histograms. However, these methods required manual feature extraction and were insufficient for complex cases [3,4].

Deep learning revolutionized medical imaging by enabling automatic feature selection and representation. CNNs, particularly effective in image classification, have shown promising results in detecting bone cancer from CT and MRI scans. Studies have demonstrated the high accuracy of CNN-based models in recognizing malignant bone lesions with minimal false positives. Multi-task deep learning frameworks have also been developed to segment and classify bone lesions as benign or malignant [5,6].

Current research focuses on developing and evaluating machine learning models for bone cancer detection. Notable models include VGG19, ResNet50, VGG16, and CNNs. These models were tested on approximately 1800 images, demonstrating significant improvements over traditional methods. While these advanced models perform well, they are often seen as black boxes, making interpretation challenging. Nonetheless, their potential for improving bone cancer diagnosis is evident.

## 2 METHODOLOGY

This research investigates automated bone cancer detection using machine learning. The methodology involves acquiring, preprocessing, training, and evaluating models with a dataset of 1800 labeled images from various sources. Preprocessing is crucial due to the variability and noise in medical images. Techniques like resizing, normalization, and augmentation standardize the data, expanding the dataset and enhancing model generalization.

Model selection focuses on CNNs like ResNet50 and VGG models (VGG16 and VGG19), known for their effectiveness in medical imaging. Transfer learning, mini-batch gradient descent, and backpropagation are employed to optimize model training. Evaluation metrics include accuracy, precision, recall, and F1-score. Preprocessing ensures data quality and robustness. Images are resized for uniformity and pixel intensities standardized using min-max scaling and z-score normalization. Augmentation techniques such as rotation, flipping, and zooming artificially expand the dataset, mitigating overfitting and improving generalization. Addressing class imbalance is also critical, using strategies like oversampling, undersampling, and class-weighted loss functions to ensure representative training data. Noise and artifact removal through median and Gaussian filtering enhance image quality, aiding precise recognition of cancer symptoms.

Artifacts degrade image quality and can stem from patient motion or equipment issues. Image registration techniques eliminate motion blur, while filtering techniques reduce noise. Geometric distortions and intensity non-uniformity are corrected to ensure accurate diagnostic interpretations. Feature extraction transforms raw image data into analyzable forms. Handcrafted features like shape descriptors, texture features, and intensity-based metrics capture essential characteristics of cancerous lesions. Deep learning approaches, particularly CNNs, automatically learn and extract hierarchical features from CT and MRI scans. Transfer learning adapts pretrained models to the specific domain of medical imaging, enhancing feature recognition even with limited labeled data. In this study, 70% of the dataset is used for training, while 30% is reserved for testing, ensuring comprehensive evaluation. The proposed models, fine-tuned on pretrained CNNs, effectively identify cancerous regions in bone images.

## 3 RESULT AND DISCUSSION

After we have trained the machines, we have closely analyzed the results of each machine learning model claiming to predict bone cancer through CT and MRI scans. Among the ensemble of these models. The result of accuracy are shown in Figure 1. VGG19 was the most accurate algorithm showing a 97.86% result. Owing to its deep and symmetric architecture, VGG19 could easily identify sophisticated features and textures characteristic of bone cancer pathology. This high accuracy rate shows that it is possible to use deep learning to predict the presence or absence of other types of diseases through the use of their images. The performance of this model is impressive.

The next one to mention is ResNet50. Its accuracy proportion was 96.5% which also reveals a high rate of accuracy to indicate the presence of abnormalities. Hopefully, this model included residual connections that allowed training much deeper networks . Thus, it can process and learn to different features of the images and obtain features from a great number of various scales. The performance of this model confirms that it is crucial for a machine to have a deeper network and a higher capacity to represent information. The next models that we can mention are VGG16 with an accuracy of 93.24% and CNN with a proportion of 90.23%. These models are less sophisticated than already mentioned VGG19 and ResNet50. Nevertheless, their results are also high what confirms that these models can be used in similar tasks for the analysis of CT and MRI scans.

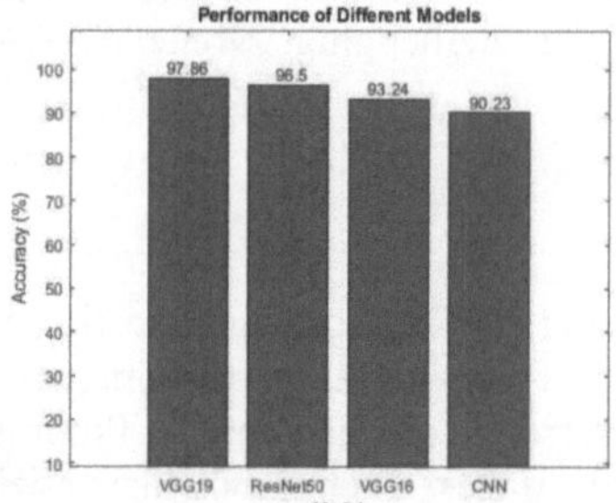


Figure 1. Accuracy of each model.

Figure 2 shows the performance metrics for the used machine learning models in the automated bone cancer detection system. VGG19 meets a precision of 98.20% and recalls 96.50% of all actual bone cancer cases with a value of 96.50%. Besides, it achieves the F1 score of 97.35% and its high AUC-ROC ensures its strong discrimination. ResNet50 meets a

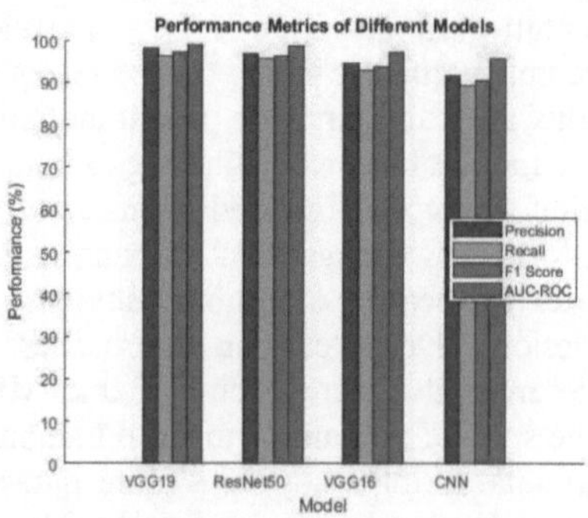


Figure 2. Performance score of each model.

precision of 97.00%, 95.80%, and 96.40% with an AUC-ROC of 98.80%. In addition, VGG16 has a combination of a precisions of 94.50%, recall of 92.80%, and F1 93.60% with the AUC-ROC of 97.20% and CNN shows a precision of 91.80%, recalls of 89.50%, and F1 of 90.60%. Its AUC-ROC is 95.70%. These data show how accurately and sensitively each model is with the balance between precision and recall and illustrate how the models are good at the automated bone cancer detection with their CT and MRI scan capabilities.

The provided confusion matrices show in details in Figure 3 highlights what performance different machine learning models provide in an automated bone cancer detection system . In both confusion matrices, the model's predicted classes are provided as columns, while the actual classes are shown in rows, with false classification as negative based on the positive meaning of cancer (no cancer) and positive as cancer cases (cancer). The significant part of both matrices that is important for understanding the models' performance includes the values in the first row of each column relating VGG 19's work to the actual negative cases and ResNet 50's work to actual positive cases.

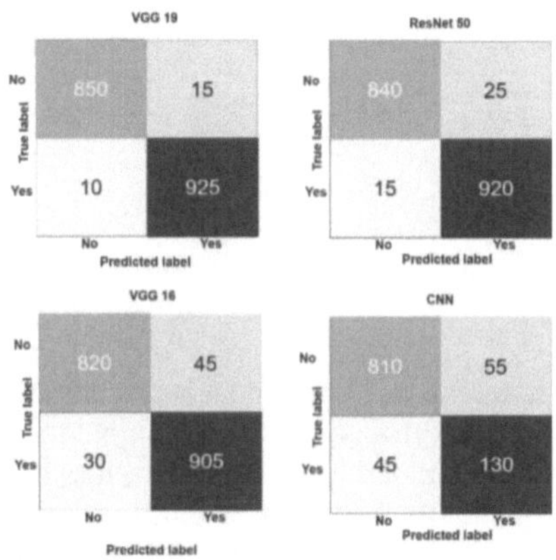


Figure 3. Confusion matrices of each model.

In the VGG 19 confusion matrix, from the 875 actual negative cases, 850 were correctly detected by the model as negative, and the 25 cases were detected as negative, and the model later detected as positive. Of the 935 actual positive cases, 925 were correctly detected, and the remaining 10 were mistakenly treated as negative. These values indicate that VGG 19 has shown better work with positive cases, detecting every case . It also missed mistakes, proving the model's better accuracy in detecting different types of cancer. In the ResNet 50 confusion matrix, fewer false negative cases are detected out of 865 actual cases, as compared to VGG 19, and the false positive cases also remained at 25 out of 935, nothing that most results were as right as VGG 19's.

In the VGG 16 confusion matrix, the 820 detected cases out of the 865 negative ones and the 905 detected cases out of the 935 positive cases have been identified, as the remaining 45 and 30 cases were mistakenly treated as positive. The CNN confusion matrix includes 810 negative occurrences and 745 positive occurrences classified as either 865 negatives or 935 positives.

Figure 4 indicates a plot of the training progress and performance of four distinct machine learning models, VGG19, ResNet50, VGG16, and CNN, across 500 epochs. Data loss utilized such metrics as categorical cross-entropy to characterize the disparity between the predictions of the model and the ground truth. Lower data loss values signify that the prediction is closer to the actual outcome. An accuracy reveals the ratio of correctly predicted instances to all instances. A high accuracy value is indicative of better performance . After examining the results, it is possible to note that the characteristics observed in the models vary over the epochs.

VGG19 demonstrates the lowest data loss and highest accuracy across all epochs, suggesting that the model learned well and performed above average in detecting bone cancer. ResNet50, while not performing as well, demonstrated competitive data loss and accuracy values, with data loss is the higher of both. Both VGG16 and CNN also demonstrated relatively high performance, although data loss was slightly higher and accuracy slightly lower, suggesting that these models underperformed VGG19 and ResNet50.

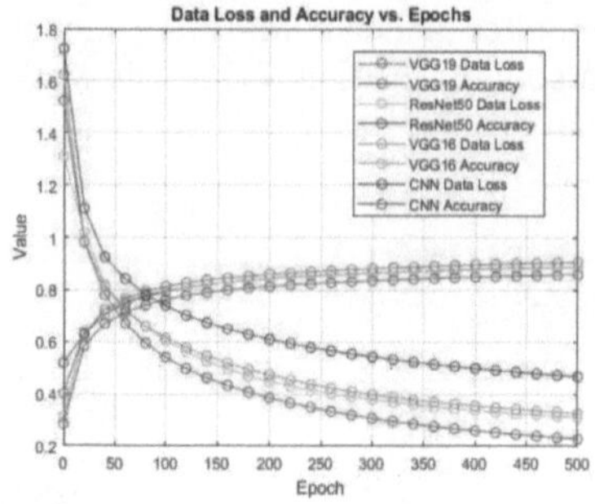


Figure 4. Data loss and accuracy of each model.

The performance of each machine learning model is presented in the table, which provides information about how they work throughout the training. According to the data, VGG19 performs the best throughout the hidden layers: the data loss is consistently the lowest and accuracy – the highest among the other models. Overall, it indicates that this model learns the best, and is the most appropriate to distinguish between cancerous and non-cancerous cases. ResNet50 is very close to VGG19 and shows a slight difference in performance, and VGG16 and CNN demonstrate relatively good performance as well with a slightly higher dataloss level and lower accuracy.

As a result, it can be stated that VGG19 is the best model for automated bone cancer detection. Since it has a large number of hidden layers, this model can identify small details and patterns in CT and MRI and has the greatest accuracy in predictions. In practice, it can be implemented in a computer-aided diagnosis system with the role to help radiologists to detect any signs of cancer which are almost impossible to notice otherwise. As this model is highly accurate, its implementation can lead to more accurate diagnostics, faster treatment planning, and better results for patients.

In the future, it may be useful to improve the implementation of VGG19 in practice. Currently, the quality of this model depends on how many hidden layers and neurons it has, as well as the applied optimizer. These parameters can be tested in more detail to generate the most successful version of the model. In addition, experiments with other data sets would also provide additional information and prepare for a more diverse set of clinical scenarios. It is also possible to implement an ensemble learning that combines benefits from several models, such as VGG19 and ResNet50.

## 4 CONCLUSION

Finally, this investigation overall aimed to proceed with developing healthcare facilities by creating machine learning models for the automatic identification of bone cancer from CT and MRI scans. Using data set with 1800 images collected from different sources, including online repositories and hospitals, four distinct machine learning models, VGG19, ResNet50, VGG16, and CNN, were trained and tested for this research. Substantial variance in the performance of each model was detected, and the obtained results permitted recognizing VGG19 as the best-performing model, slightly outperforming ResNet50. The results of the current research proved to offer reduced data loss and the highest degree of accuracy for VGG19. This conclusion contributes to the understanding of the ways deep learning, and VGG19, in particular, can be used to revolutionize medical imaging and ultimately improve the patient outcomes of bone cancer.

The next step in this field will be aimed at using VGG19 and the other top-performing models in clinical practice to streamline the process of diagnosing bone cancer and planning further treatment. It is estimated that VGG19 may serve as a new model to be used in medical imaging for this purpose. However, further studies and research will need to be carried out to optimize these models and make them ready for functioning, as well as address the lack of adequate data and the concern about their diversity together with the consideration of ensemble learning.

## REFERENCES


[1] Zhang, X., Wei, H., Dong, C., Wang, J., Zhang, T., Huang, L., Ni, D. and Luo, Y. (2023). 3D printed hydrogel/bioceramics core/shell scaffold with NIR-II triggered drug release for chemo-photothermal therapy of bone tumors and enhanced bone repair. *Chemical Engineering Journal*, *461*(December 2022), 141855. https://doi.org/10.1016/j.cej.2023.141855

[2] Sultanpure, K. A., Bagade, J., Bangare, S. L., Bangare, M. L., Bamane, K. D. and Patankar, A. J. (2024). Measurement: Sensors Internet of things and deep learning based digital twins for diagnosis of brain tumor by analyzing MRI images. *Measurement: Sensors*, *33*(March), 101220. https://doi.org/10.1016/j.measen.2024.101220

[3] Hammond, C. and Lockwood, P. (2024). Care of transgender patients by diagnostic radiographers: What can be learnt from the literature. *Radiography*, *30*(1), 145–150. https://doi.org/10.1016/j.radi.2023.10.020

[4] Miao, Y., Chen, Y., Luo, J., Liu, X., Yang, Q., Shi, X. and Wang, Y. (2023). Black phosphorus nanosheets-enabled DNA hydrogel integrating 3D-printed scaffold for promoting vascularized bone regeneration. *Bioactive Materials*, *21*(July 2022), 97–109. https://doi.org/10.1016/j.bioactmat.2022.08.005

[5] Sirajunisha, H., Balakrishnan, T., Sakthivel, P. and Krishnaveni, A. (2023). Enhanced corrosion resistance, antibacterial and biological properties of sol-gel derived Ti-rGO-HAp nanocomposites. *Chemical Physics Impact*, *6*(December 2022), 100159. https://doi.org/10.1016/j.chphi.2022.100159

[6] Gao, W., Deng, J., Ren, J., Zhang, W., Wang, Z., He, R., Wang, K., Shi, X. and Liang, T. (2022). 3D-printed hydroxyapatite (HA) scaffolds combined with exos from BMSCs cultured in 3D HA scaffolds to repair bone defects. *Composites Part B: Engineering*, *247*(June), 110315. https://doi.org/10.1016/j.compositesb.2022.110315

*Hybrid and Advanced Technologies – S. Prasad Jones Christydass, Nurhayati Nurhayati and S. Kannadhasan. (Eds)* © 2025 the Authors, ISBN 978-1-032-90254-8


# To predict the Dyslexia children using handwriting image based on image processing techniques

H. Selvi
*Assistant Professor, Department of Computer Science, Saveetha College of Liberal Arts and Science, SIMATS, Saveetha University, Chennai, Tamil Nadu, India*

P. Vignesh
*Assistant Professor (S.G), Department of Mathematics, Saveetha School of Engineering, SIMATS, Saveetha University, Chennai, Tamil Nadu, India*

P.K. Sudhakar
*Assistant Professor (S.G), Department of Mathematics, Saveetha College of Liberal Arts and Science, SIMATS, Saveetha University, Chennai, Tamil Nadu, India*

L. Manjula
*Associate Professor, Department of Mathematics, Saveetha School of Engineering, SIMATS, Saveetha University, Chennai, Tamil Nadu, India*

ABSTRACT: Millions of kids worldwide are currently impacted by learning problems. The absence of appropriate identifying methods is the cause of this problem. Certain wealthy nations have a set of protocols and sufficient knowledge for identifying this disease. In order to improve accuracy, the Single Layer Perceptron method is also used in this work, which identifies dyslexic youngsters based on images of their handwriting using a pattern recognition technique. Optical Character Recognition (OCR) will be used in this technique to identify and extract the written character feature.

*Keywords*: Dyslexia, Single layer Perceptron, Optical Character Recognition, Pattern Recognition, Grey color

## 1 INTRODUCTION

Typically, dyslexia affects people of all ages. IQ scores in the extremely high range are frequently wrongly believed to indicate mental impairment. There are rumors that it is a concealed disability that has an impact on social life, career, and academic performance. The family is unable to provide care, comfort, or enjoyment for children with disabilities. In the second half of the 20th century, many countries began to acknowledge specific learning disorders, but not everywhere else. This kind of impairment is only detectable once a youngster starts school and encounters scholastic challenges.

In image processing, one of the primary uses is pattern recognition. The principal application of patterns is with massive image datasets. Among the many uses for pattern recognition techniques is the recognition of handwriting. This method allows us to teach computers to interpret letters and numbers from any other language that may be seen in a picture.

One use of optical character recognition (OCR) is handwriting recognition. OCR is mostly used to read text from images, whether it be handwritten or printed. OCR converts documents captured as images from a camera into desired formats, such as PDFs.

 DOI: 10.1201/9781003559115-52

## 2 RELATED WORK

There has been much discussion in recent years regarding the use of artificial neural network (ANN) techniques in the detection and diagnosis of dyslexia. The range of symptoms, the co-occurrence of other diseases, and the gender differences are among the many issues that arise throughout the assessment process.

While some neural network applications have already been examined, the issues listed above apply to all neural network applications. Due to its multidisciplinary character, neural classifiers present excellent research prospects and will yield more fruitful results in the future. Studies pertaining to the application of Artificial Neural Network technology show great promise for improving the lives of students with special needs and those in their immediate vicinity. The Investigator has studied studies done on learning disabilities, specific learning disabilities, and dyslexia undertaken around the globe.

Erik (2010), Berninger and May (2011), and Bane *et al.* (2012) conducted studies that also demonstrated the necessity for study in order to identify the specific learning disorders that are most prevalent and to design remedial intervention strategies for these kids. Batson (2010) provided support for a multifactorial reading model (DIAL) in the growth of comprehension and decoding abilities.

The necessity of assessment and remedial intervention for dyslexic children at the primary school level was further backed by research by Dahle *et al.* (2011), McBride *et al.* (2011), Boets *et al.* (2011) [2], and Melby and Lervag (2012). These investigations also emphasized the importance of teaching these students using a multimodal approach. An setting that is rich in technology can support pupils in maintaining both their reading interest and recall ability. Students with dyslexia can be taught knowledge in an engaging and inspiring way by using a multimedia package.

Every study listed above provided understanding and awareness of the many facets of the issue being studied. However, research on creating multimedia packages for kids with dyslexia is seldom ever found. Under this situation, the researcher chose to create a remedial approach. Athanasios S. Drigas and Rodi-Eleni Ioannidou (2013) [3] suggested using artificial intelligence techniques to diagnose SEN (Special Education Needs) The goal was also to enhance the quality of life for SEN students.

In 2013[4], Julie M. David and Kannan Balakrishnan worked together to create a new algorithm that would assign and define the significance of the dimensionality decrease method and the missing value complaint method in the performance of fuzzy and neurofuzzy classifiers, with a focus on the prediction of learning disabilities in school-age children.

A more accurate and time-efficient method for diagnosing Attention Deficit Hyperactivity Disorder (ADHD) was described by Anuradha *et al.* (2010) in their study [5]. Support vector machines (SVMs) are supervised learning algorithms that work well for regression and classification. Doctors utilize an SVM module to diagnose the condition based on a data-set and the answers to a questionnaire they administered. Based on the method's results, 88,674% of children between the ages of six and eleven were successfully diagnosed.

In reading knowledge exams, Srihari *et al.* (2008) used two computational techniques to automatically count brief handwritten essays [6]. The purpose of this work is to assign a human-scoring-comparable score to each handwritten response. In addition to the usual challenges associated with handwriting recognition, this technology also has to deal with children's writing abilities. By grading reading comprehension assessments, educators, parents, and researchers can receive timely feedback in addition to students. This paper describes two systems. First, a decent level of handwriting recognition performance is required for the application of latent semantic analysis (LSA). Essays submitted in response to questions in statewide reading comprehension exams served as a test bed for both systems' training and assessment, with the essays being manually assessed. For these pupils, the platform scoring performance still shows promise even with the word recognition errors that were noted during the evaluation process.

## 3 METHODOLOGY

Thirty scans of dyslexic handwriting images were gathered from different educational institutions and included in the data set for this work. This study uses a variety of image processing and pattern recognition algorithms to identify dyslexia symptoms. Figure 1 below details the suggested architecture for dyslexia prediction.

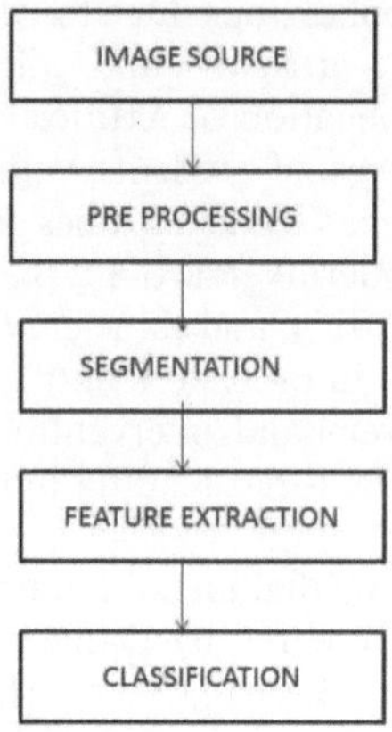


Figure 1. Architecture for prediction of dyslexia.

### 3.1 *Image source*

A bitmap picture is created from the data on the paper that is being scanned when handwritten text samples are scanned at a resolution of 300 dpi.

### 3.2 *Pre-processing*

The series of steps known as pre-processing raises the image's accuracy by enhancing its quality. The character recognition process mostly uses the pre-processing technique that follows [7].

#### 3.2.1 *Noise-removal*

It is the procedure used to clean up images of noise. This also refers to minimizing undesired signals in the image to smooth it out. First, a conversion from RGB to grayscale is performed. An image that is colored indicates that each pixel in the image has three distinct colors: red, green, and blue. Given that character detection on colored images is more challenging to propose than on grayscale images, this colored image was transformed to grayscale because it is represented in a single matrix. The MATLAB code for grayscale conversion is as follows [8].

```
I1 = rgb2gray(RGB);
gray1 = rgb2gray(map);
imshow(gray1);
//Anothercodes//
I1 = rgb2gray(RGB);
gray1 = (I, 0.2989 * R + 0.5870 * G + 0.1140 * B); imshow(Gray1);
```

3.2.2 *Binarization*

It is a procedure for converting an image to binary data from grayscale data. Images in binary format only have 0s and 1s. Based on a fixed value, the values of the pixels in images are divided into 0s and 1s. In the event that the pixel rate is less than the constant, 0 is substituted; otherwise, 1.

3.2.3 *Morphological operation*

The primary function is to aid in adjusting the size of an image. The process of adding pixels to an image's border allows us to enlarge it. Pixels from the edge of an image are removed in order to reduce the size of the image.

### 3.3 *Segmentation*

The process of extracting a single character from a picture is called segmentation. Following the completion of pre-processing, the letters are divided from one another using bounding box techniques in the segmentation stage. Area, top, left, width, and length are the components of bounding box approaches. The feature extraction procedure then extracts the bounding objects [9].

### 3.4 *Features extraction*

Characters from the input image are recognized using optical character recognition (OCR) throughout the features extraction process. After image extraction and character recognition, the corrected characters are shown in the command window.

### 3.5 *Classification*

The term Artificial Neural Network (ANN) refers to a network of synthetic neurons. Using an ANN classifier, a handwriting image is used to identify dyslexia symptoms. To calculate the total of each input vector in a single layer perceptron, multiply its value by the corresponding element of the weights vector [10]. An activation function will use the value shown in the output as its input. As seen below, Figure 2 depicts the general perceptron model of an artificial neural network.

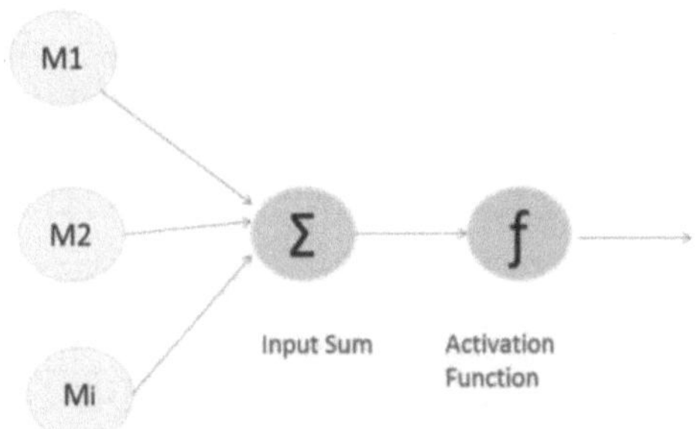


Figure 2. General perceptron model.

The design shown above has an input layer and an output layer. An input vector m=[m1, m2, m3,..., mn]T is mapped to an output vector y using the general perceptron. The network is trained using a weight vector w = [w1, w2, w3, . . . ,wn]T. the training pair, where xk is the kth input vector and dk= {1, 0} is the intended output vector, is {m1, d1}, {m2, d2}... {mn, dn}. g(y) = m1w1 + m2w2 +... + annwn where m stands for input, w for weight, and g(y) for the activation function.

## 4 RESULT AND DISCUSSIONS

In this study, the symptoms of dyslexia are analyzed using image processing techniques, establish correlations between the symptoms and allow for the classification of children with and without dyslexia based on handwriting. A few key elements that are used to pinpoint the symptoms of dyslexia.

- Produce haphazard work that has numerous crossed-out words and repeatedly attempted sentences. As in, wipe, wype, wiep, etc.
- Letters that resemble one other, such as b/d, p/b, p/q, n/u, and m/w, typically cause confusion.
- Several "reversals" in the handwriting, as well as poor letter writing.
- Makes incorrect spelling of words, e.g., bird for bad, son for sun
- Some letters are omitted. E.g., look for lok, hear for her.

Each character in the data set is identified using the OCR method, which also outputs the right characters. Figure 3 displays an example of a character that was discovered using data from one of the individuals.

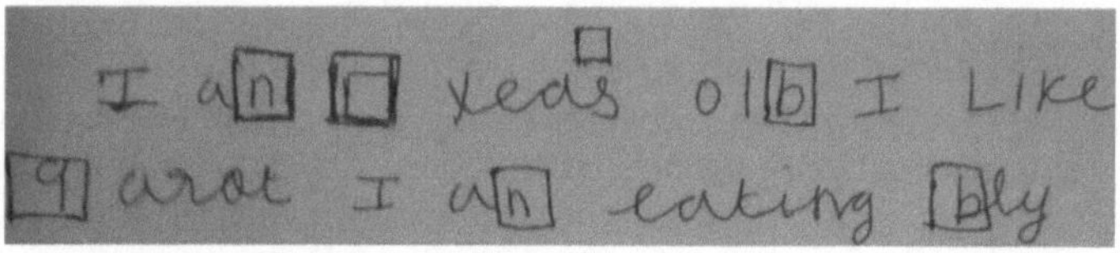

Figure 3. Dyslexia handwriting image.

Using single layer perceptron model, if the value of y is 0 then output is 'Normal child' otherwise 'Dyslexia child". The activation function of this model is as follows:

$$f(\alpha) = 1 \text{ if } \alpha - \beta > 0$$

$$f(\alpha) = 0 \text{ if } \alpha - \beta \leq 0$$

The "Dyslexia threshold" is denoted by $\beta$ = w T x, where $\beta$ is a threshold. The Dyslexia Threshold is set at 50 based on trial results.

In order to verify the outcome using a single layer perceptron classifier for unknown instances, we have implemented 10-fold cross validation on this dataset of 30 data points. Tables 1 and 2 present the findings of the entire experiment with various error function values.

Accuracy % = (Correct detection / Overall characters) * 100

Table 1. Performance results from SLP classification algorithm – cross validation using images.

| S. No | Particulars | Value |
|---|---|---|
| 1 | Instances of Correctly Classified | 26 |
| 2 | Instances of Incorrectly Classified | 4 |
| 3 | Kappa statistic value | 0.8369 |
| 4 | Mean absolute error value | 0.1295 |
| 5 | Root mean squared error value | 0.2725 |
| 6 | Relative absolute error value | 25.8034% |
| 7 | Root relative squared error value | 54.283% |
| 8 | Total number of instances | 30 |
| 9 | Time taken to build the model | 0.33 sec |

Table 2. Detailed accuracy by class in SLP using images.

| TP Rate | FP Rate | Precision | Recall | F-Measure | MCC | ROC Area | PRC Area | Class \| |
|---|---|---|---|---|---|---|---|---|
| 0.958 | 0.120 | 0.885 | 0.958 | 0.920 | 0.840 | 0.948 | 0.912 | Normal |
| 0.880 | 0.042 | 0.957 | 0.880 | 0.917 | 0.840 | 0.948 | 0.967 | Dyslexia |

We obtained 88.2% cases are classified correctly and 11.8% cases are classified incorrectly as shown in stratified cross-validation summaries given in Figure 4.

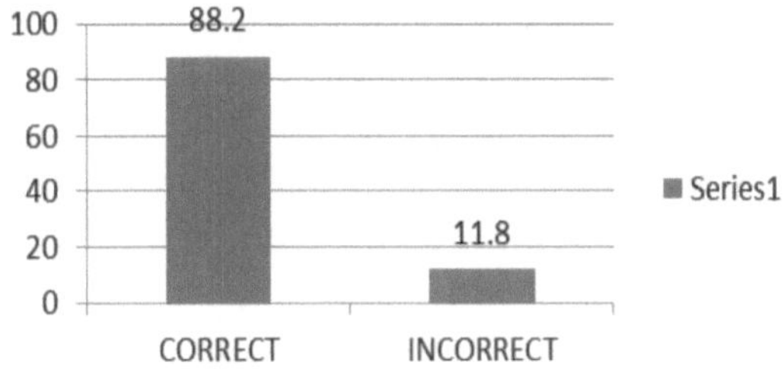


Figure 4. Correct classification vs. Misclassification rate of single layer perceptron model using images.

## 5 CONCLUSION

In summary, the OCR technique is a useful tool for identifying solid letters, such as those seen in words on signs or other objects. Future handwritten recognition techniques, such as skew method and cursive, will mostly use OCR as an extra method to detect diseases. The second goal is also accomplished, which is to categorize the degree of dyslexia symptoms according to accuracy. The classification accuracy performs instantly based on the outcome. This is because the ANN requires a large number of samples in order to achieve high accuracy and requires more character traits.

## REFERENCES

[1] Eirk, S. A. (2008). *Educating Exceptional Children Book*. Wadsworth Publishing. ISBN: 0547124139.

[2] https://linkinghub.elsevier.com/retrieve/pii/S0891422210003100

[3] Drigas, A. S. and Ioannidou, R. E. (2013). A Review on Artificial Intelligence in Special Education. *Ag. Paraskevi*, 15310, Athens, Greece.

[4] David, J. M. and Balakrishnan, K. (2010). Significance of classification techniques in prediction of learning disabilities in school age children. International Journal of Artificial Intelligence & Applications. DOI:10.5121/ijaia.2010.1409.

[5] Anuradha, J., Tisha, Ramachandran, V., Arulalan, K. V., and Tripathy, B. K. (2010, January). Diagnosis of ADHD using SVM algorithm. *In Proceedings of the Third Annual ACM Bangalore Conference* (pp. 1–4).

[6] Srihari, S., Collins, J., Srihari, R., Srinivasan, H., Shetty, S. and Brutt-Griffler, J. (2008). Automatic scoring of short handwritten essays in reading comprehension tests. *Artificial Intelligence*, 172(2–3), 300–324.

[7] Kumar, T. and Verma, K. (2010). A theory based on conversion of RGB image to gray image. *International Journal of Computer Applications*, 7(2), 7–10.

[8] MATLAB and Simulink for Technical Computing, URL: http://www.mathworks.com, July 2010

[9] Selvi, H. and Saravanan, M. S. (2019). Using machine learning techniques for predicting and diagnosing dyslexia in school-age children. *Journal of Advanced Research in Dynamical and Control Systems*, 11(4), 753–760.

[10] Selvi, H. and Saravanan, M. S. (2018). A Study of dyslexia using different machine learning algorithms with data mining techniques. *International Journal of Engineering and Technology (UAE)*, 7(4), 3406–3411. DOI:10.14419/ijet.v7i4.14545

*Hybrid and Advanced Technologies – S. Prasad Jones Christydass, Nurhayati Nurhayati and S. Kannadhasan. (Eds)*  *ISBN 978-1-032-90254-8*


# Enhanced brain tumour detection through fuzzy erosion image processing using pseudo-complemented almost distributive fuzzy lattices

A. Nasreen Sultana
*Research Scholar, Department of Mathematics, Vels Institute of Science, Technology & Advanced Studies (VISTAS), Tamil Nadu, India*

R. Kamali
*Assistant Professor, Department of Mathematics, Vels Institute of Science, Technology & Advanced Studies (VISTAS), Tamil Nadu, India*

ABSTRACT: Using image processing, the fuzzy morphological erosion operator introduces the concept of pseudo-complemented almost distributive fuzzy lattices (PCADFL). In image processing, segmentation of images is crucial as it makes it easier to identify questionable regions in medical imaging. A brain tumour is a fatal condition that can seriously impair a person's quality of life. Tumours can be categorised as benign or malignant depending on how they developed. This study describes the segmentation of a brain MRI image utilising the Fuzzy C-Means (FCM) clustering algorithm and PCADFL fuzzy morphological erosion operator filtering. This PCADFL is used in brain cancer identification to reduce the misclustered areas that are invariably produced following the segmentation of an MRI of the brain in order to identify the location of the tumour. The effectiveness of this approach produces a significantly better result than some of the current methods, as shown by MATLAB experimental results on many brain tumour imaging databases.

*Keywords*: Pseudo-Complemented Almost Distributive Fuzzy Lattices (PCADFL), fuzzy erosion, Fuzzy c-means (FCM), SSIM, PSNR

## 1 INTRODUCTION

Image processing is the method of transforming an input image into an output image. The main goal of image processing technique is to make a simpler visual detection of the picture or object under investigation [1]. Brain tumour is a mass of dysfunctional brain cells. The threat level of a brain tumour is influenced by a number of characteristics along with the stage of growth, location, size and type of tumour [2]. Magnetic Resonance Imaging (MRI) is used to detect brain tumours in many stages. In medical imaging segmentation, analysis and classification are acknowledged to be an important yet difficult phase. FCM clustering approach is frequently used for image segmentation [3].

In Mathematical morphology the images or signals are processed and analysed using filtration as well as other operators that modify them. The algebraic approach is constructed on two complete lattices $L_1$ and $L_2$ with a mapping $\delta : L_2 \to L_1$ that preserves arbitrary joins [4]. The preceding concepts can be encapsulated in a broader concept that recognises a point $x \in L$ in each image as a partly ordered structure $(L, \leq)$ [5], the filters as operators $F : L \to L$ with aspects relevant to the order in such lattices [6,7]. Berhanu et al. established the Almost Distributive fuzzy lattice (ADFL) in [8]. Pseudo-Complementation on Almost Distributive Fuzzy Lattice (PCADFL)

 DOI: 10.1201/9781003559115-53

was developed by Karpagavalli and Sultana [9], who established that it is equationally definable on ADFL utilising the fuzzy lattice and fuzzy partial order relation described by Chon [10].

In this study PCADFL is introduced in fuzzy morphological erosion and their properties are also proved. The purpose of this research is to use image processing from MRI images to detect brain tumours using the PCADFL and FCM clustering algorithms. The main objective of this study is to find out where a tumour is present and how far it has spread. The suggested approach is useful for detecting brain tumours automatically. The method facilitates in the earlier identification of tumours and offers reliable outcomes.

## 2 THEORETICAL PRELIMINARY

### 2.1 *PCADFL*

Let $(R, A)$ be a fuzzy poset and $(R, \vee, \wedge, 0)$ be an algebra of type (2, 2, 0). A unary operation $a \longrightarrow a^*$ on $R$. Then $(R, A)$ is called PCADFL [9], if and only if the following conditions are satisfied:

$$A(a \vee b, 1) = A(1, a \vee b) = 1 \tag{1}$$

$$A(a \wedge b, 0) = A(0, a \wedge b) = 1 \tag{2}$$

$$A(0, a \wedge a*) = A(a \wedge a*, 0) = 1 \tag{3}$$

$$A(b, a * \wedge b) = A(a * \wedge b, b) = 1 \tag{4}$$

$$A((a * \wedge b*), (a \vee b)*) = A((a \vee b)*, (a * \wedge b*)) = 1 \tag{5}$$

$$A(a, (a*)*) = A((a*)*, a) = 1, \text{ for all } a, b \in R. \tag{6}$$

### 2.2 *Erosion*

An operator $\varepsilon : L \rightarrow L'$ is erosion if it commutes with the infimum, for all $x_i \in L'$, $\varepsilon(S_i x_i) = S'\varepsilon(x_i)$, where S denotes the infimum associated with $\geq$ and $S'$ denotes the infimum associated with $\geq'$.

### 2.3 *Fuzzy clustering*

Fuzzy C-Means (FCM) and fuzzy k-means are the most often utilised fuzzy clustering techniques for medical picture segmentation [11]. FCM may distinguish between tumours and non-tumours in a given area. Clustering accuracy might be improved with FCM. The fuzzy membership function ranges from 0 to 1 [12]. The value is determined by the intensity range. The clustering approach assists in the reduction of unpredictability [13,14]. FCM is highly good at identifying tissue regions [15]. FCM is more faster, reliable and automatically diagnoses region segmentation in a highly effective way [16].

## 3 MATERIALS AND METHODS

There are numerous ways to identify or segment brain tumours. Noise in images might impair medical image analysis results. The following method processes brain tumours in MRI images using PCADFL and FCM algorithms. Data and tasks determine the suggested technique's data analysis step. MRI images of brain tumours are analyzed for textural inhomogeneity, location, contours, and sizes.

### 3.1 *Image acquisition*

A patient's MRI scan images are converted to binary images and shown with a default size of 256 × 256. For illustration, a Gray-scale converted image with numerical values ranging from 0 to 255, where 0 represents black and 255 represents white.

### 3.2 *Pre-processing*

Pre-main processing improves MRI images for machine or human visual system processing. Preprocessing further enhances MR image performance by increasing SNR, improving visual quality, removing undesirable background and noise, merging the inner area, and retaining its features.

### 3.3 *Segmentation*

MR scans are needed to detect brain tumours and determine the right treatment. Threshold segmentation segments MR images using binary intensities. Segmentation divides images. After medical picture improvement, first segmentation yields the most precise region of interest borders. The segmentation utilizes FCM clustering and thresholding. The brain region was segmented using FCM, and the enhanced image was thresholded to binary.

### 3.4 *Fuzzy C-Means (FCM) clustering*

FCM clustering minimises an objective function in MATLAB to divide data into $c$ fuzzy clusters. However, thresholding segmentation is the easiest picture segmentation approach. This approach employs a clip-level to convert grayscale to binary. Selecting the threshold value is crucial to this technique. Otsu's method finds the t-value that minimises weighted within-class variation. We then use MATLAB to cycle through all t values [0,255] and find the one that minimises the equation. Picture object and backdrop can be distinguished by Otsu's dynamic thresholding. Segmentation makes pictures easier to grasp and study by simplifying and/or altering them. MRI images are segmented to identify boundaries and objects.

## 4 PSEUDO-COMPLEMENTED ALMOST DISTRIBUTIVE FUZZY LATTICE (PCADFL) IN FUZZY MORPHOLOGICAL EROSION

The notion of Pseudo-Complemented Almost Distributive Fuzzy Lattice (PCADFL) is introduced in fuzzy morphological erosion in this section. Throughout this paper $(R, A)$ stands for a PCADFL $(R, \vee, \wedge, *, 0, 1)$.

### 4.1 *Definition*

Let $(R, A)$ be a PCADFL with a unary operation $*$ for any $a, b \in R$. The fuzzy morphological erosion $\varepsilon_S(A) \in R$ by the structuring element $S$ is defined if and only if the following conditions are satisfied:

$$\varepsilon_S(A)(a) = \inf\{S(b-a) \cdot A(a^* \wedge b, b) > 0/b \in R\},\ \forall\ a \in R. \tag{7}$$

$$\varepsilon_S(A) = A((a^*)^*, a) = 1 \tag{8}$$

### 4.2 *Definition*

A fuzzy erosion $\varepsilon_S : R \times R \rightarrow [0, 1]$ that preserves the least element of PCADFL and distributes over the infimum is defined as

$$A(\varepsilon_S(\textstyle\bigwedge_{i\in I}(a)^*),\ \ \textstyle\bigwedge_{i\in I}\varepsilon_S(a^*)) > 0,\quad \forall\quad \varepsilon_S \in R,\quad \forall\quad a \in R \tag{9}$$

4.3 *Theorem*

Let $(R, A)$ be a PCADFL, then for any $a, b \in R$, $\varepsilon_S$ of $R$ is a fuzzy erosion if and only if $A\left(\varepsilon_S\left(\bigwedge_{i\in I} (a \vee b)^*\right), \bigwedge_{i\in I} \varepsilon_S(a^* \wedge b^*)\right) > 0$.

*Proof:* Suppose that $(R, A)$ be a PCADFL. Let $\varepsilon_S$ of $R$ is a fuzzy erosion with structuring element $S$, then

$$
\begin{aligned}
&A\left(\varepsilon_S\left(\bigwedge_{i\in I} (a \vee b)^*\right), \quad \bigwedge_{i\in I} \varepsilon_S(a^* \wedge b^*)\right) \\
&= A\left((\varepsilon_S)^* \left(\bigwedge_{i\in I} (a \vee b)^*)^*\right)^*, \quad \bigwedge_{i\in I} \varepsilon_S(a^* \wedge b^*)\right) \\
&= A\left((\varepsilon_S)^* \left(\bigwedge_{i\in I} (a^* \wedge b^*)^*)^*\right)^*, \quad \bigwedge_{i\in I} \varepsilon_S(a^* \wedge b^*)\right) \\
&= A\left(\varepsilon_S{}^* \bigwedge_{i\in I} (a^* \wedge b^*)^*\right)^*, \quad \bigwedge_{i\in I} \varepsilon_S(a^* \wedge b^*)) \\
&= A\left(\varepsilon_S{}^* \bigwedge_{i\in I} (a^{**} \vee b^{**})\right)^*, \quad \bigwedge_{i\in I} \varepsilon_S(a^* \wedge b^*)) \text{since } a^{**} = a \\
&= A\left(\varepsilon_S{}^* \bigwedge_{i\in I} (a \vee b)\right)^*, \quad \bigwedge_{i\in I} \varepsilon_S(a^* \wedge b^*)) \\
&= A\left(\bigwedge_{i\in I} \varepsilon_S{}^* (a \vee b)\right)^*, \quad \bigwedge_{i\in I} \varepsilon_S(a^* \wedge b^*)) \\
&= A\left(\bigwedge_{i\in I} (\varepsilon_S{}^*)^* (a \vee b)^*, \quad \bigwedge_{i\in I} \varepsilon_S(a^* \wedge b^*)\right) \\
&= A\left(\bigwedge_{i\in I} \varepsilon_S (a^* \wedge b^*), \quad \bigwedge_{i\in I} \varepsilon_S(a^* \wedge b^*)\right) > 0 = 1.
\end{aligned}
\tag{10}
$$

Hence

$$A(\varepsilon_S(\bigwedge_{i\in I} (a \vee b)^*), \bigwedge_{i\in I} \varepsilon_S(a^* \wedge b^*)) = A(\bigwedge_{i\in I} \varepsilon_S(a^* \wedge b^*), \varepsilon_S(\bigwedge_{i\in I} (a \vee b)^*)) > 0.$$

Therefore $\varepsilon_S$ of $R$ is a fuzzy erosion of PCADFL. Hence proved.

## 5 EXPERIMENTAL RESULTS

Original brain tumour images have been able to improve using fuzzy morphological erosion in PCADFL and Fuzzy clustering means (FCM). PCADFL and FCM algorithms are used to increase accuracy, and fuzzy membership functions are created based on these characteristics for segmenting brain tumours from MRI images. The SSIM, SNR and PSNR were utilised to validate the reliability and accuracy of the brain tumour generated by our proposed method.

### 5.1 *Algorithm*

Step 1: select an original binary image of brain tumour
Step 2: create structuring element
Step 3: store number of rows in P and number of columns in Q
Step 4: create a zero matrix of size I
Step 5: take all the neighbourhoods
Step 6: take logical structuring element
Step 7: compare and take minimum value of the neighbor and set the pixel value to that minimum value
Step 8: apply fuzzy clustering (FCM) and PCADFL lattice algorithms
Step 9: to find the Peak-SNR
Step 10: to find the SNR value
Step 11: to find the SSIM value

An effective image segmentation methodology with the Fuzzy C Means algorithm followed by the PCADFL in fuzzy morphological erosion operator and level set segmentation stages, which enable brain tumour detection accurately as illustrated in Figures 1 and 2.

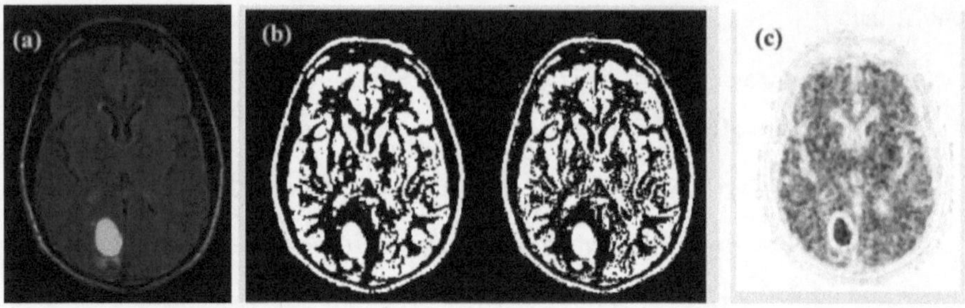


Figure 1. (a) Original image, (b) Filtered images of brain tumour and (c) SSIM index value.

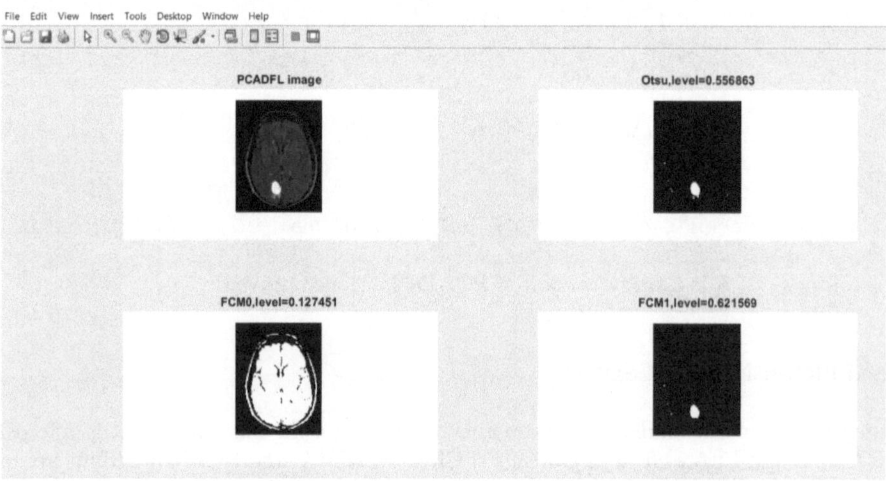


Figure 2. PCADFL and FCM.

### 5.2 *Brain detection experimental and data analysis*

The brain tumour detection problem was solved using fuzzy lattice theory. Table 1 indicates the appropriate values for Peak-SNR, SNR, SSIM and FCM levels for the PCADFL fuzzy erosion approximation in brain tumour detection.

Table 1. Calculated values of SNR, PSNR, SSIM and levels of FCM.

| Image | Otsu level | FCM0 level | FCM1 level | SNR | PSNR | SSIM |
|---|---|---|---|---|---|---|
| Brain Tumour | 0.556863 | 0.127451 | 0.621569 | 19.4035 | 32.9128 | 0.8918 |

## 6 CONCLUSION

This study optimises image processing and segmentation to detect brain tumour edges in MRI images using PCADFL with fuzzy morphological erosion and FCM. Segmentation can accurately define tumour affliction zones. We classified malignant tumours from MR brain images using texture-based criteria and a well-known classifier. The proposed method also enhances MSE, SNR, and PSNR accuracy, according to performance factors. These data imply that the suggested technique can accurately and quickly diagnose and locate brain tumours. Thus, the recommended strategy is crucial for brain cancer detection using MR images. Color images will be used in fuzzy morphological dilation and erosion utilizing PCADFL.

## REFERENCES


[1] Singh, K. K. and Singh, A. (2010). A study of image segmentation algorithms for different types of images. *International Journal of Computer Science Issues (IJCSI)*, 7(5), 414.

[2] Priya, V. V. (2016). An efficient segmentation approach for brain tumour detection in MRI. *Indian Journal of Science and Technology*, 9, 1–6.

[3] Selvakumar, J., Lakshmi, A. and Arivoli, T. (2012, March). Brain tumour segmentation and its area calculation in brain MR images using K-mean clustering and Fuzzy C-mean algorithm. In *IEEE-international conference on advances in engineering, science and management (ICAESM-2012)* (pp. 186–190). IEEE.

[4] Bloch, I. (2009). Duality vs. adjunction for fuzzy mathematical morphology and general form of fuzzy erosions and dilations. *Fuzzy sets and systems*, 160(13), 1858–1867.

[5] Maragos, P. (2005). Lattice image processing: a unification of morphological and fuzzy algebraic systems. *Journal of Mathematical Imaging and Vision*, 22, 333–353.

[6] Heijmans, A. M. (1994). Morphological Image Operators.,(Academic Press: New York.).

[7] Serra, J. (1983). *Image analysis and mathematical morphology*. Academic Press, Inc.

[8] Berhanu, A., Yohannes, G. and Bekalu, T. (2017). Almost distributive fuzzy lattice. *International Journal of Mathematics and its Applications*, 5(1–C), 307–316.

[9] Karpagavalli, S.G. and Sultana, A.N. (2020). Pseudo-complementation on almost distributive fuzzy lattices. *Journal of Critical Reviews*, 7(7), 758–761.

[10] Chon, I. (2009). Fuzzy partial order relations and fuzzy lattices. *Korean Journal of Mathematics*, 17(4), 361–374.

[11] Cai, W., Chen, S. and Zhang, D. (2007). Fast and robust fuzzy c-means clustering algorithms incorporating local information for image segmentation. *Pattern recognition*, 40(3), 825–838.

[12] ShanmugaPriya, S. and Valarmathi, A. (2018). Efficient fuzzy c-means based multilevel image segmentation for brain tumour detection in MR images. *Design Automation for Embedded Systems*, 22, 81–93.

[13] Shanker, R. and Bhattacharya, M. (2017, October). Brain tumour segmentation of normal and pathological tissues using K-mean clustering with fuzzy C-mean clustering. In *European congress on computational methods in applied sciences and engineering* (pp. 286–296). Cham: Springer International Publishing.

[14] Sompong, C. and Wongthanavasu, S. (2016, July). Brain tumour segmentation using cellular automata-based fuzzy c-means. In *2016 13th International Joint Conference on Computer Science and Software Engineering (JCSSE)* (pp. 1–6). IEEE.

[15] Hooda, H., Verma, O. P. and Singhal, T. (2014, May). Brain tumour segmentation: A performance analysis using K-Means, Fuzzy C-Means and Region growing algorithm. In *2014 IEEE International Conference on advanced communications, control and computing technologies* (pp. 1621–1626). IEEE.

[16] Wadgure, S. and Thakre, P. (2014). Detection of brain tumour from MRI of brain using Fuzzy C-Mean (FCM). *International Journal of Science, Engineering and Technology Research (IJSETR)*, *3*(8), 2222–2230.

*Hybrid and Advanced Technologies – S. Prasad Jones Christydass, Nurhayati Nurhayati and S. Kannadhasan. (Eds)* © 2025 the Authors, ISBN 978-1-032-90254-8
*Open Access: www.taylorfrancis.com, CC BY-NC-ND 4.0 license*

# Optimising enterprise human resource management using attention-BiGRU-CapsNet model

J. Santhosh
*Assistant Professor, Department of Computer Applications, Sri Krishna Adithya College of Arts and Science, Coimbatore, India*

Devadutta Indoria
*Department of Commerce, Vikram Dev University, Jeypore, Odisha, India*

Zameer Pasha
*Associate Professor, Department of Management, St. Claret College, Bengaluru, Karnataka, India*

Prasanthi Gottumukkala
*Assistant Professor, Department of Information Technology, Gokaraju Rangaraju Institute of Engineering and Technology, Hyderabad, India*

K. Soujanya
*Assistant Professor, Department of BBA, Koneru Lakshmaiah Education Foundation, Green Fields, Vaddeswaram, Guntur, Andhra Pradesh, India*

Neerav Nishant
*Assistant Professor, Department of Computer Science and Engineering, School of Engineering, Babu Banarasi Das University, Lucknow, India*
ORCID ID: 0009-0002-6696-6943

ABSTRACT: An HRMS unifies the HR processes and information systems of a company. An ERP will commonly include an HRMS, or human resource management system. Business operations can be better managed with the use of ERPs, which stand for enterprise resource planning systems. These applications encompass a wide range of fields, including consumer relations, finance, manufacturing, human resources, and supply chain management. The three main components are feature selection, preprocessing, and model training. The matching processing activities performed on data prior to its complete utilization are known as data preprocessing. By eliminating superfluous or irrelevant variables, feature selection maintains model performance while reducing the dimensionality of the dataset. We trained the model with Attention-BiGRU-CapsNet to get more precise results. With an accuracy of 96.10 percent, the suggested approach outperformed competing approaches like BiGRU and CapsNet.

*Keywords*: Human Resource Management (HRM), Enterprise Resource Planning (ERP), Human Resource Information Systems (HRIS), Small and Medium-sized Enterprises (SMEs), Recurrent Neural Network (RNN)

## 1 INTRODUCTION

Although performance is a crucial part of human resources development in organisations, overall performance in human resource management is rather low. It is well recognised that China's human resources system was put in place at a later date. Scientific research on

 DOI: 10.1201/9781003559115-54

performance evaluation systems, complete integration of theory and practice to satisfy current objective requirements, and practical application of relevant research content are all important to further boost firms' total economic efficiency. Incorporating theories of human resource management and enhancing the overall economic efficiency of organisations can be accomplished through this. This is equally important for the future growth of the organisation. Performance reviews are essential for new hire orientation. Employees are at the centre of a performance system, which in turn affects their pay, benefits, and opportunities for professional development. Furthermore, it is a fundamental system type linked to the employees. Much of the recent scholarship in human resource management has focused on the importance of performance evaluation, particularly as it pertains to assessing the marketing department's effectiveness within an organisation. It being said, there are a handful of problems in failure to assess team performance and lack of performance evaluation standards are issues with marketing staff evaluations in firms. In practice, there are a plethora of options for evaluating HRM effectiveness.

## 2 LITERATURE SURVEY

Maps, reports, official documents, and a multitude of productions that transmit and receive approval, inquiry statistics, and transmission statistics are all part of the mineral management department's day-to-day operations[1]. The data was evaluated as part of the preliminary research and a single container's design. Use Enterprise Resource Planning (ERP) to keep employees who are good at what they do away from positions of power in the company[2]. To resolve ERP challenges, maintain an effective ERP system, and identify corporate needs, it is crucial to train and evaluate subsequent ERP deployment procedures and implementations, according to the findings [3]. Companies in the industrial sector have influenced HR policymaking through the actions they take as a result of the personnel management system that was implemented by the change management body for the entire commercial region. This model is mostly based on worries related to the development of this system [4], and it describes all power checks and balances. The executive's models and attributes have taken a blow from a stroll through daily life thanks to the improvement of modern data innovation[5].The paper states that there are five factors that will determine how human resource management is digitally transformed [6]. These factors include digital demands from internal customers, digital innovation in the business, digital issues from competitors, digital innovation governance, and requirements of the digital era. It would be irresponsible to ignore the potential drawbacks of the new system, even while digital transformation offers chances for the expansion of the business [7]. Potentially future HRM solutions based on data mining could be made possible by the findings of this study [8].The primary topics covered in this course are the factors driving the digital revolution in human resource management and its causes, trends, and outcomes[9]. The nine inferences drawn from the data presented in this study are likely to spur more investigations [10]. But the outbreak proved that digitising data or merging online and physical spaces is essential for survival in the modern world. Company survival is in jeopardy unless they take this step [11].Human Resource Information Systems (HRIS) are utilised for a lot more than simply digitising paper data. A variety of HR functions can now be handled by modern HR/payroll systems [12]. When used in MSMEs, HRIS can streamline HRM processes like hiring, training, performance evaluations, and compensation [13]. There are a number of ways in which HRIS may help small and medium-sized businesses (SMEs) succeed. These include boosting staff productivity, increasing employee engagement, and ensuring regulatory compliance [14]. This research dives into how optimising HRIS for MSMEs management involves improving several aspects of business operations using technology.

## 3 PROPOSED SYSTEM

Managing system development and data security is beyond the capabilities of the present HRMS design due to issues such as incompatibility with professional database and development systems, extensive maintenance needs, and lack of support for client-server applications. In light of these concerns, it is necessary to investigate the corporate HRMS system's core technologies, practicality, and total demand.

### 3.1 *Preprocessing*

The term "data preprocessing" refers to a group of interconnected operations carried out on data before its full usage. Basic steps include gathering data, processing it, and finally uploading it. Data is considered processed when it has been transformed from its original state in compliance with established regulations and standards[15]. Moving all of the information acquired in the first two processes into the data warehouse is the last step, which is data upload.

### 3.2 *Feature selection*

While lowering the dataset's dimensionality, feature selection preserves model performance by removing irrelevant or redundant variables. Implementing efficient feature-selection strategies can decrease the cost of acquiring features for the classification model and enhance the model's interpretability. With credit risk models, banks have opposing objectives including making the models easy to comprehend and profitable, which makes feature selection a challenging multi-objective problem. Since features are obtained from different types of datasets in the first step, it is acceptable to eliminate features that are noisy or redundant. To start, wrapper-based feature selection can be employed to choose those features for the model if they have no effect on classification performance[16]. After those traits are eliminated, the search space for the wrapper-based strategy is much smaller. To eliminate them in the initial step, we employ optimum-seeking and feature-relevance algorithms. The second phase involves finding the optimal solution through the use of a multiobjective wrapper-based feature selection that optimizes the model's performance and the size of the feature collection.

### 3.3 *Model training*

#### 3.3.1 *BiGRU*

The gated recurrent unit is an alternative to the traditional recurrent neural network (RNN). With the input $o_d$ and the preceding state $i_{d-1}$, GRU calculates $i_d$ for every point $d$ in the following way:

$$w_d = \tau(R_w o_d + Y_w i_{d-1}) \tag{1}$$

$$y_d = \tau\left(R_y o_d + Y_y i_{d-1}\right) \tag{2}$$

$$\tilde{i}_d = \tan i\left(R_e o_d + Y\left(w_f \odot i_{d-1}\right)\right) \tag{3}$$

$$i_d = (1 - Y_d) \odot i_{d-1} + y_d \odot \tilde{i}_d \tag{4}$$

where $t$-dimensional reset gate, hidden state, and update gate are denoted as $w_d, i_d, y_d$, respectively; GRU parameters are $R_w, R_y, R_e$ and $Y_w, Y_y, Y$. The symbol $\odot$ stands for element-wise production, while "$\tau$" represents the sigmoid function.

#### 3.3.2 *Attention mechanism*

Hypernymy can be better identified by focusing on the characteristics that set two composite entities apart from one another. Incorporating an attention technique into our model improves its detection performance. The feature vector $i^{(f)}$ for the entity Xk can be characterized as a weighted sum, and its calculation is as follows:

$$i^{(f)} = \sum_{h=1}^{D_f} \beta_h^{(f)} i_h^{(f)} \tag{5}$$

### 3.3.3 *CapsNet layer*

In the digit recognition layer, there are clusters of neurons called capsules. The activity vectors of these neurons represent the parameters needed to instantiate a specific sort of entity. The direction and length of the activity vector stand for the instantiation parameters and the likelihood of an entity's existence, respectively. Consequently, a non-linear squashing function is used to squash long vectors down to somewhat less than one and short vectors to basically nothing:

$$g_z = \frac{\|q_z\|^2 q_z}{1 + \|q_z\|^2 q_z} \tag{6}$$

## 4 RESULT AND DISCUSSION

Improved data accuracy, cost reduction, service quality (QoS), and process streamlining are just a few of the numerous vital duties carried out by paperless systems such as the Electronic Health Record Management System.

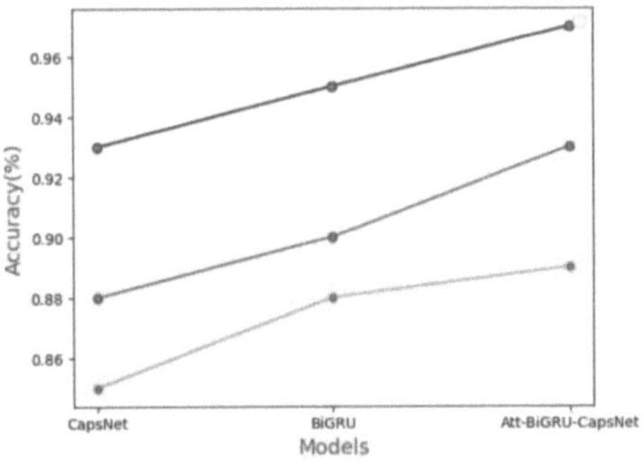


Figure 1. Evaluation of the suggested attention-BiGRU-CapsNet model for accuracy.

The suggested system's study in Figure 1 shows that it has a comparatively high level of accuracy in optimally allocating and scheduling human resources for the firm.

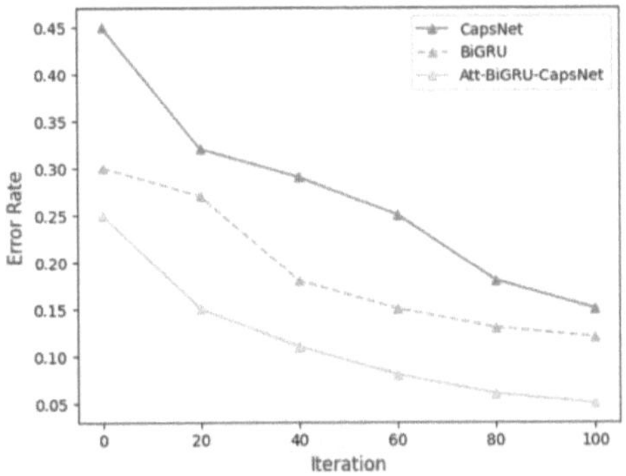


Figure 2. Analysis of error rate.

An study of the reasonable error rate for traditional systems is presented in Figure 2 of the proposed Att-BiGRU-CapsNet based corporate talent system.

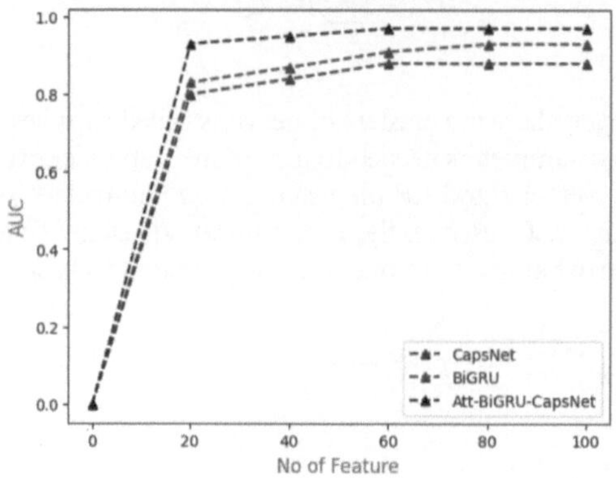


Figure 3. Evaluation result of AUC for optimization method.

The results of using AUC as an Enterprise HRM for each single-objective approach are displayed in Figure 3. There is a visual representation of the number of features that were chosen on one side and the area under the curve (AUC) on the other.

## 5 CONCLUSION

With the development of new technologies, improvements in society, and fierce competition among enterprises, the modern market has transformed into a talent competition, where an enterprise's outstanding talent pool gives them an edge. Yet many companies and other places fail to adequately manage their human resources. Data preprocessing refers to the matching processing operations carried out on data before it is fully utilized. Feature selection keeps model performance while lowering dataset dimensionality by removing irrelevant or redundant variables. To train the models, the Attention-BiGRU-CapsNet approach is employed. With a consistency rate of 96.10%, the proposed technique outperforms the BiGRU and CapsNet models.

## REFERENCES


[1] Jiang, S. (2019, April). Optimal allocation of human resources based on intelligent computer management system. In *2019 11th International Conference on Measuring Technology and Mechatronics Automation (ICMTMA)* (pp. 667–670). IEEE.

[2] Piau, T.F. (2018, September). Enterprise resource planning post-implementation: human resource management module-hiring and assessment (a Preliminary Study). In *2018 International Conference on Information Management and Technology (ICIMTech)* (pp. 297–302). IEEE.

[3] Zhang, J. and Liu, H. (2019, August). On human resource management and big data analysis. In *2019 IEEE International Conference on Smart Internet of Things (SmartIoT)* (pp. 345–349). IEEE.

[4] Airapetova, A. G., Korelin, V. V., Karabekova, A. A. and Sobolev, R. K. (2017, November). Main trends in forming and pursuing human resource policies at an industrial enterprise. In *2017 IEEE VI Forum Strategic Partnership of Universities and Enterprises of Hi-Tech Branches (Science. Education. Innovations)(SPUE)* (pp. 58–59). IEEE.

[5] Smykova, I. S. and Petrova, A. K. (2017, May). Coach management as innovative human resource management method. In *2017 XX IEEE International Conference on Soft Computing and Measurements (SCM)* (pp. 874–875). IEEE.

[6] Meijerink, J., Boons, M., Keegan, A. and Marler, J. (2021). Algorithmic human resource management: Synthesizing developments and cross-disciplinary insights on digital HRM. The *International Journal of Human Resource Management*, 32(12), 2545–2562.
[7] Zhao, C., Cooke, F. L., and Wang, Z. (2021). Human resource management in China: what are the key issues confronting organizations and how can research help?. *Asia Pacific Journal of Human Resources*, 59(3), 357–373.
[8] Kim, S., Wang, Y. and Boon, C. (2021). Sixty years of research on technology and human resource management: Looking back and looking forward. *Human Resource Management*, 60(1), 229–247.
[9] Wang, Q., Xu, C., Zhou, Y., Ruan, T., Gao, D. and He, P. (2018, December). An attention-based Bi-GRU-CapsNet model for hypernymy detection between compound entities. In *2018 IEEE International Conference on Bioinformatics and Biomedicine (BIBM)* (pp. 1031–1035). IEEE.
[10] Strohmeier, S. (2020). Digital human resource management: A conceptual clarification. *German Journal of Human Resource Management*, 34(3), 345–365.
[11] Mazurchenko, A. and Maršíková, K. (2019). Digitally-powered human resource management: Skills and roles in the digital era. *Acta Informatica Pragensia*, 8(2), 72–87.
[12] Salasiah, S., Jamil, J., Jubaidi, A. and Ghufron, G. (2023). The *Role of Human Resource Management in the Organization. International Journal of Social Science Research and Review*, 6(12), 262–273.
[13] Panjaitan, E. H. H. (2023). Implementing human resource information system (HRIS) for efficient human resource management. *International Journal of Science and Society*, 5(2), 128–139.
[14] Zhao, B., Xu, Y. and Cheng, J. (2021). Evaluation and image analysis of enterprise human resource management based on the simulated Annealing-optimized BP neural network. *Computational Intelligence and Neuroscience*, 2021(1), 3133065.
[15] Xie, Q. (2022). Machine learning in human resource system of intelligent manufacturing industry. *Enterprise Information Systems*, 16(2), 264–284.
[16] Kou, G., Xu, Y., Peng, Y., Shen, F., Chen, Y., Chang, K. and Kou, S. (2021). Bankruptcy prediction for SMEs using transactional data and two-stage multiobjective feature selection. *Decision Support Systems*, 140, 113429.

*Hybrid and Advanced Technologies – S. Prasad Jones Christydass, Nurhayati Nurhayati and S. Kannadhasan. (Eds)* © *2025 the Authors, ISBN 978-1-032-90254-8*


# Enhancing enterprise human resource management: Predicting workers' stress for improved workplace satisfaction using a hybrid deep transfer learning approach

Davinder Paul Singh
*Department of Computer Science and Engineering, School of Technology, Pandit Deendayal Energy University, Gandhinagar, Gujarat, India*

Kanika Gupta
*Assistant Professor, Department of Management, School of Commerce and Management, IIMT University, Meerut, India*

Nivedita Singh
*Assistant Professor, Department of Management Studies, GL Bajaj Institute of Technology and Management, Greater Noida, India*

Manu Vasudevan Unni
*Assistant Professor, Department of Management, St. Claret College, Bangalore, Karnataka, India*

S. Jayasudha
*Assistant Professor, Department of Science and Humanities, Prince Shri Venkateshwara Padmavathy Engineering College, Chennai, Tamil Nadu, India*

Amit Chauhan
*Department of Life Sciences, CHRIST University, Bengaluru, Karnataka, India*

ABSTRACT: Although workplace stress is a major issue, little is known about how it impacts employees. A worker's stress levels can rise due to everything from unreasonable expectations at work to problems in their personal relationships. When we're stressed out, our bodies react emotionally and mentally to things that upset us. The three primary concepts that influence stress in the workplace are stressors, the stress response, and stress-related outcomes. Preprocessing, feature extraction, and model training are the three key parts. The normalizing and discretization methods are employed in the preprocessing phase. A time-frequency representation of a signal can be obtained in feature extraction by taking advantage of the wavelet transform, which can concurrently provide information about the signal's time and frequency. For more accurate results, we trained the model using the Feature rich CNN-BiLSTM. The proposed method outperformed rival methods like CNN and LSTM, with an accuracy of 96.43%.

*Keywords*: Convolutional Neural Network (CNN), Long Short-Term Memory (LSTM), Normalization, Human Resource Management (HRM), American Psychological Association (APA)

## 1 INTRODUCTION

Everyone from the employee and their loved ones to the company that employs them and even society at large feel the effects of a worker's deteriorating health as a result of stress. Employees' physical and mental well-being can be negatively impacted by prolonged,


DOI: 10.1201/9781003559115-55

cumulative stress in the workplace. This has a domino effect on both employee happiness and the company's bottom line. Improving staff capabilities and motivation are essential components of modern organisational management's pursuit of maximising job happiness and organisational success. According to research in organisational psychology and human resource management, management practices are thought to impact workers' psychological, physiological, and social well-being. The study of employee well-being can benefit from the insights of industrial anthropology, psychology, sociology, and industrial economics. Stress is being argued by an increasing number of academics as an integral part of the mental health of workers. There is mounting evidence that occupational stress can lead to a host of serious health problems, both mental and physical. These include metabolic disorders, hypertension, behavioural problems, depression, anxiety, and coronary heart disease. When people leave their occupations, they should make it a priority to maintain learning new things so they can stay up with how quickly society is changing. An increasing amount of focus is being directed on the mental health issues that have emerged due to their extended exposure to stressful personal and work lives. Mental health concerns, including obsessive-compulsive disorder (OCD), depression, anxiety, and others, are on the rise. The modern fixation with material goods is taking a toll on people's mental health.

## 2 LITERATURE SURVEY

The body's anxiety response is normal in response to change. When faced with a new situation, the body will summon all of its resources—mental, emotional, and physical. Environmental factors, physiological factors, and mental processes all have an effect on stress levels[1]. When workers are overworked or face tasks that are outside of their expertise, or when they are expected to do an excellent job despite inadequate resources, workplace stress can occur[2]. Stress is described as an outside stimulus that causes the body to enter "fight-or-flight" mode. [3]The American Psychological Association (APA) classifies stress into three categories: acute, episodic, and chronic. Workload overload, tight deadlines, [4]overly ambitious or unreasonable goals, job unhappiness, long hours, pressure to succeed, and interpersonal problems are all contributors to employee stress. Worker motivation, morale, and production are all impacted by stress, which in turn affects the bottom lines of companies. In addition, it is associated with a number of health problems, such as depression, insomnia, and cardiovascular disease.[5]Most people who live outside of Port Harcourt but work there already have a long commute, and the development of multiple flyovers has just made it worse. In addition to a financial crisis and economic downturn, Nigerians are coping with a host of other problems. [6]Companies typically reorganise when they want to get more out of their current resources. Staff morale plummets and stress levels rise as a result of these and other issues. Mergers, dismissals, and layoffs are among these issues. [7] When people are put under circumstances that make them react negatively, emotionally and physically, that's when stress sets. In addition, the problem of stress in the workplace has received increasing attention as of late in the area of occupational health.Hotel workers have allegedly been pressured into early retirement, job loss, unpaid vacation, reduced welfare payments, or even job or hour changes.[8] As a result of these oppressive working circumstances, employees are concerned about their employment security and the sustainability of the sector overall. [9]Studies have shown that occupational stressors have a negative effect on employee satisfaction, commitment, job performance, subjective well-being, prosaically behaviour, and intention to stay. Hotel workers are suffering from extreme anxiety and are unable to do their jobs due to the ongoing COVID-19 outbreak. [10]Because of the frightening prospect of increasing occupational stress due to the pandemic's detrimental impacts on all hotel employees, from frontline workers to management, we felt obliged to examine the ways in which the pandemic altered occupational pressures and their effects. Stress is an inevitable and natural part of being human. Reactions to new or unexpected stimuli can manifest in various ways in the body and mind. As the COVID-19 pandemic and its attendant pressures have changed the nature of work, employees have a formidable challenge in adjusting to the new circumstances.

## 3 PROPOSED SYSTEM

Workplace stress causes yearly lost productivity worth billions of dollars, in addition to a host of health concerns affecting both mental and physical well-being. Researchers have asked for new frameworks to study human resource management (HRM) tactics that reduce employee stress.

### 3.1 *Preprocessing*

#### 3.1.1 *Normalization*

From one dataset to another, the values of features, including maximum and minimum values, can differ greatly. Making a downsizing transformation such that all attributes have valid values is a major pain and usually not something you want to deal with. Feature scaling and data normalization are terms used to describe this procedure in data preprocessing.

The most common methods for this domain include min-max normalization and feature scaling within the range of 0 to 1 [11]. The formula for the new feature's value is as follows:

$$h' = \frac{h - \mathrm{min}\_value}{\mathrm{max}\_value - \mathrm{min}\_value} \tag{1.1}$$

z-score normalization

$$h' = \frac{h - mean}{stand\_dev} \tag{1.2}$$

scaling on the unit length

$$h' = \frac{h}{QhQ} \tag{1.3}$$

while '$h$' represents the original feature value and '$h'$' the updated, normalized data value; $\mathrm{min}\_value$ and $\mathrm{max}\_value$ stand for the minimum and maximum feature values before normalization, respectively.

### 3.2 *Feature extraction*

#### 3.2.1 *Wavelet analysis*

When applied to a signal, the wavelet transform can simultaneously offer information about the signal's frequency and time, resulting in a time-frequency representation. The data series is first transformed into its "wavelets," which are basically a scaled and shifted version of the original wavelet, in order to accomplish this [12]. The Continuous Wavelet Transform (CWT) of a signal, $m(z)$ can be defined by applying the following expression:

$$CWT_m^{\delta}(\sigma, r) = \frac{1}{\sqrt{|r|}} \int_{-\infty}^{+\infty} m(z)\delta^*\left(\frac{z - \sigma}{r}\right) \tag{1.4}$$

where s is the scaling or translation parameter and '*' is the complex conjugate. In this case, the mother wavelet (or transformation function $\delta(z)$) stands in for frequency, with scale (specified by the factor $r$) replacing it.

### 3.3 *Model training*

#### 3.3.1 *Feature rich CNN-BiLSTM*

With the goal of improving classification performance and identifying potentially depressed Twitter users, we proposed an approach that merges CNN with bidirectional-LSTM, a kind of RNN. Our extensive testing led us to the conclusion that CNN is quite effective at extracting spatial characteristics and performs admirably even when the context of the preceding sequence is not significant. To begin, CNN is given low-level multidimensional data straight. Layers of pooling and convolution each extract relevant features. When compared to traditional CNN, this model has a fully connected hidden and output layer. In order to extract depression tweets in depth and improve accuracy at the same time, convolutional kernel modifications, pooling layers, and several convolutional layers are used. When there is too much exposure to over-fitting risk, a complex network can form [13]. An LSTM network of recurrent neurons that depends on time is thus added to the CNN model in order to fix the sequence problem. The state cell is built to store the important data for a longer duration, while the convolution kernels are utilized to extract duplicate and irrelevant data. The most effective use case is a combination of CNN and biLSTM.

Equation (1.5) states that the mean of the embedding vectors for each word equals the sequential input data. The extraction of Unigram, Bigram, and Trigram features is achieved by applying convolutional kernels of varying sizes to $M_1; Z$ using a 1D CNN.Features obtained using convolution, more specifically the $z$ th convolution, which accepts as input a window of d words beginning with $z$ and going all the way up to $z+f$. In order to generate window features, the convolution process is configured according to equation (1.6):

$$M_1;Z = [m_1, m_2, \ldots, m_Z] \tag{1.5}$$

$$g_f, z = tanv\left(Y_f m_{z:z+d-1} + a_f\right) \tag{1.6}$$

The specific words' embedding vector in the context window is $m_{z:z+d-1}$, while the bias is represented by $a_f$ and the parameters are $Y_f$. Eq. (1.7) also provides the feature map of the filter with a convolution size $f$, which is useful because each filter convolves with a different segment of the text.

$$g_f = \left[g_{f1}, g_{f2}, \ldots, g_{Z-f+1}\right] \tag{1.7}$$

To create a set of new features, max-pooling takes the maximum value from each convolutional layer and adds it together. Equation (1.8) is obtained by max-pooling the feature maps inside each convolution kernel, with a convolutional size of $f$. For each filter size $f$ = 1, 2, and 3, the final features of each window are generated by concatenating $q_f$, as shown in Eq. (1.9). Then, the unigram, bigram, and trigram hidden characteristics of each window are recovered.

$$q_f = Maxz\left(g_{f1}, g_{f2}, \ldots, g_{Z-f+1}\right) \tag{1.8}$$

$$v_f = [q_1, q_2, q_3] \tag{1.9}$$

By making use of cell states and gate topologies like input, output, and forget gates, long short-term memory (LSTM) architecture enables it to circumvent the "vanishing gradient" problem with sequential data. The overall long-term memory of the LSTM unit is comprised of these cell states and the additive connections among them.

## 4 RESULT AND DISCUSSION

Staff members' responses to disrespectful customers were the focus of this study, which aimed to identify if and how HRM strategies affected these responses. Using the job demands-resources

(JD-R) model and stress appraisal theories, this proposed examined the possibility that service personnel may lessen the negative effects of abusive customers on their work outcomes (i.e., emotional exhaustion and job performance) through training and taking part.

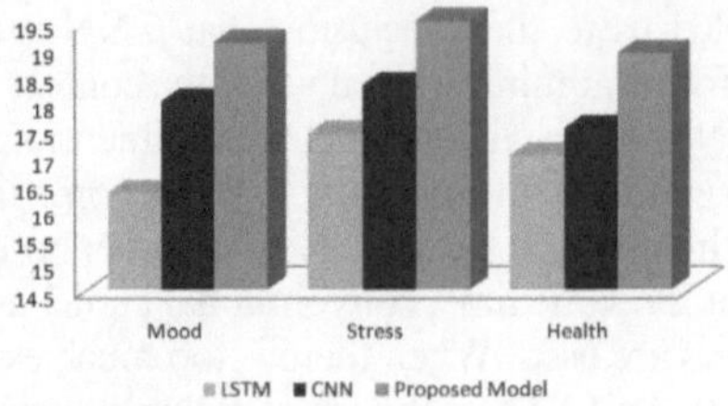


Figure 1. Prediction result of proposed model.

Figure 1 displays the outcomes of predictions made by both current and future deep-learning models in relation to the domains of mood, health, and stress.

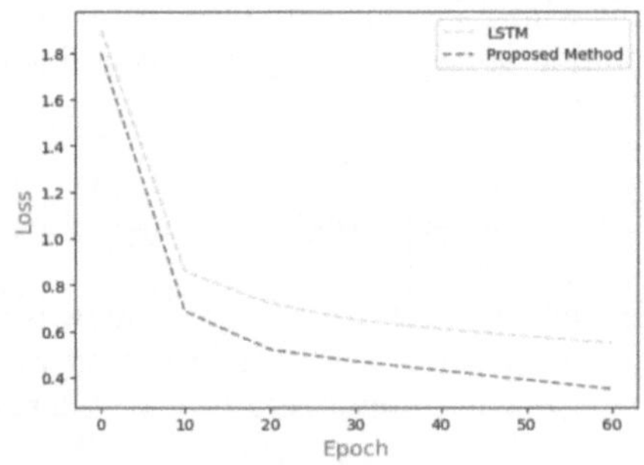


Figure 2. A prediction error's result of feature rich CNN-BiLSTM.

Figure 2 shows the neural network's training error trend. The output values are quite near to the ideal values, according to the network training results.

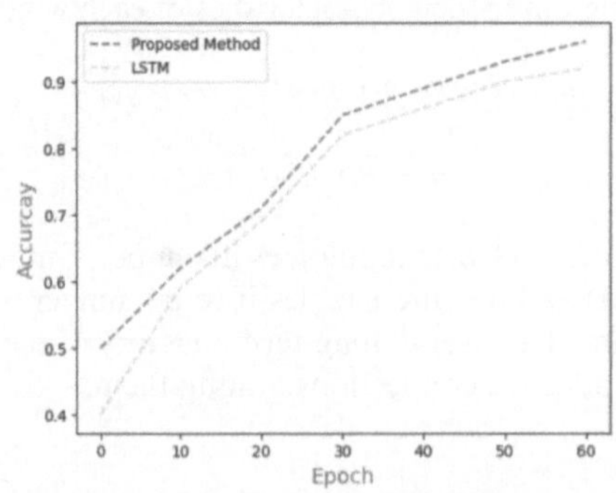


Figure 3. Accuracy comparison of feature rich CNN-BiLSTM.

Figure 3 displays the results of the accuracy prediction. At first glance, the obtained proposed model seems to suggest that our model has been well-trained.

## 5 CONCLUSION

The study's primary objective was to determine the relationship between stress management and occupational output. The researchers hoped that elements such as workload, management style, commitment, ambiguity, efficiency, effectiveness, and stress would influence employee productivity. In the preprocessing step, the normalization and discretization techniques are used. The wavelet transform can give information about the signal's time and frequency at the same time, hence it can be used in feature extraction to get a time-frequency representation of the signal. We use the Feature rich CNN-BiLSTM to train the models. The suggested approach achieves a higher consistency rate than the CNN and BiLSTM models, which is 96.43%.

## REFERENCES


[1] Sugumar, R., Sharma, S., Kiran, P. N., Gupta, S. K., Sindhu, V. and Srinivas, R. V. (2023, June). Novel method for detection of stress in employees using hybrid deep learning models. In *2023 8th International Conference on Communication and Electronics Systems (ICCES)* (pp. 984–989). IEEE.

[2] Hassard, J., Teoh, K., Cox, T., Cosmar, M., Gründler, R., Flemming, D., ... and Van den Broek, K. (2014). Calculating the cost of work-related stress and psychosocial risks.

[3] Greeshma, B., Priya, G. I., Aswathy, B. and Janani, S. (2018). Stress management in the workplace-challenges faced by HR. *Journal of Business and Management*, 20(4), 33–36.

[4] Jain, P. and Batra, A. (2015). Occupational stress at workplace: Study of the corporate sector in India. *Journal of Business and Management*, 17(1), 13–21.

[5] Warraich, U. A., Ahmed, R., Ahmad, N. and Khoso, I. (2014). Impact of stress on job performance: An empirical study of the employees of private sector universities of Karachi, Pakistan. *Research Journal of Management Sciences, ISSN*, 2319–1171.

[6] Bewell, H., Yakubu, I., Owotunse, D. and Ojih, E. E. (2014). Work-induced stress and its influence on organizational effectiveness and productivity among Nigerian workers. *African Research Review*, 8(1), 112–125.

[7] Hayes, B., Douglas, C. and Bonner, A. (2015). Work environment, job satisfaction, stress and burnout among haemodialysis nurses. *Journal of Nursing Management*, 23(5), 588–598.

[8] Akgunduz, Y. (2015). The influence of self-esteem and role stress on job performance in hotel businesses. *International Journal of Contemporary Hospitality Management*, 27(6), 1082–1099.

[9] Dartey-Baah, K., Quartey, S. H. and Osafo, G. A. (2020). Examining occupational stress, job satisfaction and gender difference among bank tellers: evidence from Ghana. *International Journal of Productivity and Performance Management*, 69(7), 1437–1454.

[10] Wen, B., Zhou, X., Hu, Y. and Zhang, X. (2020). Role stress and turnover intention of front-line hotel employees: The roles of burnout and service climate. *Frontiers in Psychology*, 11, 36.

[11] Alexandropoulos, S. A. N., Kotsiantis, S. B. and Vrahatis, M. N. (2019). Data preprocessing in predictive data mining. *The Knowledge Engineering Review*, 34, e1.

[12] Cannas, B., Fanni, A., See, L. and Sias, G. (2006). Data preprocessing for river flow forecasting using neural networks: wavelet transforms and data partitioning. *Physics and Chemistry of the Earth, Parts A/B/C*, 31(18), 1164–1171.

[13] Kour, H. and Gupta, M. K. (2022). An hybrid deep learning approach for depression prediction from user tweets using feature-rich CNN and bi-directional LSTM. *Multimedia Tools and Applications*, 81 (17), 23649–23685.

*Hybrid and Advanced Technologies – S. Prasad Jones Christydass, Nurhayati Nurhayati and S. Kannadhasan. (Eds)*  *ISBN 978-1-032-90254-8*


# Analysis and prediction of employee turnover characteristics based on LSTM-RNN-HMM model

D. Rafi Mahammad
*Associate professor, Department of Computer Science and Engineering (Cyber Security), Institute of Aeronautical Engineering, Dundigal, Hyderabad, India*

S. Lakshmi Priya
*Assistant Professor, Department of CSE(AIML), S.A Engineering College, Poonamallee, Thiruverkadu, Chennai, India*

T. Thulasimani
*Associate Professor, Department of Mathematics, Bannari Amman Institute of Technology, Sathy, Tamilnadu, India*

Sakshi Mann
*Assistant Professor, Department of HR, Indira Institute of Management, Pune, Maharashtra, India*

Shikha Mann
*Deputy Director, Indira School of Business Studies, Pune, Tathewade, India*

Gourav Kalra
*Assistant Professor, Maharishi Markandeshwar (Deemed to be University), Mullana, Ambala, India*

ABSTRACT: Employee turnover is a sign that staff members have decided to leave the company. The issue of employee turnover has recently gained traction due to the fast expansion of numerous firms and the economy as a whole. Employees that choose to leave the company for any number of reasons constitute one side of the coin. Conversely, regular operations are impacted by job security and staff retention. Organizations should investigate the causes of employee turnover before attempting to lower it. Important steps include training the model, feature extraction, and preprocessing. This suggests cleaning the data, identifying the treatments, and correlating them as three stages of data preprocessing. A feature selection method based on correlations is utilized in this context. We trained the model using the Feature rich LSTM-RNN-HMM to increase its accuracy. With an accuracy rate of 96.25 percent, the suggested method outperformed competing methods like LSTM-RNN and HMM.

*Keywords*: Hidden Markov Model (HMM), Recurrent Neural Network (RNN), Employee Turnover, Employees Satisfaction, Employee Performance

## 1 INTRODUCTION

The people that work for a firm are its greatest asset. Attrition refers to the decrease in the workforce as a result of employees' retirement, resignation, or death. Although numerous factors contribute to employee turnover, retirements and resignations are the most common in regards to the requirements of this project. Problems with employee turnover are common in human resources. A growing number of employees have left their positions in recent years. Importantly, they must determine whether their staff are unhappy in their jobs or planning to


DOI: 10.1201/9781003559115-56

quit. It is prudent to find out where the problem is coming from before resorting to drastic measures. Workers nowadays are more likely than ever to leave their present workplace in search of greater possibilities elsewhere. Employee turnover is quickly becoming a major concern for numerous businesses. Delays in service delivery, decreased overall productivity, and fewer loyal clients are all consequences of staff turnover, which is a serious issue for service businesses. As a result, companies need to figure out why employees leave and then come up with plans to recruit, hire, and retain workers while also reducing staff turnover.

## 2 LITERATURE SURVEY

Companies are justifiably concerned about employee turnover since it can significantly affect the bottom line if key staff leave. [1]Even once normal employees leave, companies will need to invest in training replacements and finding new options. If the company could tell whether an employee was going to leave, it could keep more of its best workers and cut costs. The first group of methods treats the problem as a binary classification and employs machine learning to predict employee turnover. If you believe [2]determined how important job skills are for departure prediction.[3]Used causal structure learning as the foundation for the CSFM technique in order for management to retain employees who are leaving. Experiments were conducted to compare various well-known machine learning approaches [4]. Random forests and XGBoost generally outperform.[5]Trained many ML models using different industry features. The high rate of employee turnover is a big concern for many. [6]We will only discuss data-driven research here. There are mainly two categories into which the data-driven research that is now available falls: machine learning methodologies and survival analysis methods [7].In order to incorporate several sources from social networks and simulate a career trajectory,[8] from the machine learning group proposed a multisource learning framework that incorporates a fused lasso penalty. Using a contextual long short-term memory network,[9] created a model for career trajectories. [10]Proposed a DSP model that accounts for the correlation between talent flow and stock price in order to aggregate the two variables' historical occurrences. By combining survival analysis with the proportional hazards model, were able to select employment for the survival analysis group. Several algorithms were evaluated by [11] for their accuracy and memory utilisation in predicting employee turnover. A multinational retailer's HR database and data collected by the Bureau of Labour Statistics were both used in the writers' analysis. [12]The following classification algorithms were specifically used: XGBoost, logistic regression, Naïve Bayesian, RF, linear support vector machine, linear discriminant analysis, and KNN. The authors found that XGBoost is the most effective in terms of accuracy and memory utilisation. The researchers[13] utilised XGBoost to predict staff turnover.

## 3 PROPOSED SYSTEM

The detrimental effects of personnel turnover on workplace productivity and long-term growth initiatives have led industries to see it as a significant concern. Using machine learning techniques, officialdoms make educated estimations regarding employee turnover to address this challenge. With reliable projections, governments can take steps to hold on to their employees. Nevertheless, the information needed to address such matters is provided by HR Information Systems. This leads to over-fitting and erroneous predictive models, where noisy data reigns supreme.

### 3.1 *Data preprocessing*

#### 3.1.1 *Data cleaning*

The Scikit-learn and Pandas packages were utilized in this study for one-hot and label encoding, respectively. Sorted string data (like staff turnover or education level) will be encoded using label encoding, whereas one-hot encoding will be used for categorical data (like job position or department) [14] . This study's three datasets were the subject of earlier work that used a variety of approaches to improve accuracy.

#### 3.1.2 *Identification of treatment*

This study suggests a treatment identification method to enable the prescriptive capability of a dataset with an unidentified treatment.

#### 3.1.3 *Correlation of treatments*

Inverting the treatment parameter from 0 to 1 or vice versa will invert the treatment correlation value, but it will also provide a distinct prescriptive scenario. Due to the duplication of the three datasets and the inversion of the treatment parameter, a total of six experiments will comprise this research.

### 3.2 *Feature selection*

#### 3.2.1 *Correlation based FS*

Following the determination of the link between all characteristics and the output class, this strategy employs the heuristic correlation assessment function to choose the ideal feature subset. It explores the relationship between nominal and category properties by representing numerical aspects with discrete values [15]. Using the equation is necessary for selecting the correlation function.

$$j_{ex} = \frac{N\bar{j}_{em}}{\sqrt{N + N(N-1)\bar{j}_{mm}}} \tag{1}$$

The sentence can be paraphrased as follows: $\bar{j}_{e}$m is the mean value of correlated feature-classes, $\bar{j}_{m}$m is the mean value of inter-correlated features, and $N$ is the number of features. $j_{ex}$ relates the features to the class variable.

### 3.3 *Model training*

#### 3.3.1 *LSTM-RNN*

Training loss computation diagram of the RNN, which converts x-values inputted into o-values outputted. A loss parameter ($T$) is a measure of the distance between each output $w$ and the corresponding training objective ($q$). RNN receives its input from a weight matrix $V$, parameters hidden-to-output connections using a weight matrix $V$, and parameters hidden-to-hidden recurrent connections using a weight matrix $O$. Equation (1.1) describes the forward propagation in the RNN model, where the computational network that has unfolded over time is on the right and the left side reflects the recurrent connections between the RNN and its loss [16]. Each node in the right-hand diagram now has a particle time instance associated to it. Here are the updating equations utilized in the research:

$$s^{(l)} = f + Op^{(l-1)} + Vc^{l} \tag{2}$$

$$p^{(l)} = tanp\left(s^{(l)}\right) \tag{3}$$

$$w^{(l)} = x + U_{p}^{(l)} \tag{4}$$

where the connections between hidden and input, hidden and output, and hidden are like the connections between b and c in a bias vector and the weight matrices $V, U \textit{ and } O$ respectively. An RNN is shown here at work, translating a set of inputs to a set of outputs that also increase in length. So, for all x-matching y sequences, the total loss is equal to the sum of all time step losses. There are two popular activation functions: *tanp*(.)and $\tau(c)$.

$$\tau(c) = \frac{1}{1 + e^{-c}} \tag{5}$$

$$tanp = \frac{e^{c} - e^{-c}}{e^{c} + e^{-c}} \tag{6}$$

### 3.3.2 *LSTM-HMM*

The design was also fed a 3-layer LSTM alongside hidden state distributions. To account for all possible HMM states, the linear layer that links the LSTM and prediction layers gets an extra column. Because its only purpose is to fill in the blanks left by the HMM's predictions, the LSTM component of this design might be less bulky than a standalone LSTM. When training the HMM model, a Baum-Welch Algorithm was employed. We also employed a Viterbi Algorithm to find the hidden state sequence that most closely matches the input observation. The construction of a combined hybrid model was also initiated. Trained in TensorFlow in tandem, this model combines LSTM and HMM. The LSTM parameter was initially initialized orthogonally when the model training began. In place of stochastic gradient descending, Adam constructed and fine-tuned an HMM keras module. The HMM model included a connection between the LSTM output and the HMM state probability, just like the sequential hybrid model.

## 4 RESULT AND DISCUSSION

Human resource management includes studying turnover rates. Much of the prior research on employee turnover relied on methods that had access to insufficient data. As a result, data mining and machine learning are employed in this research to conduct a more comprehensive and efficient investigation of turnover.

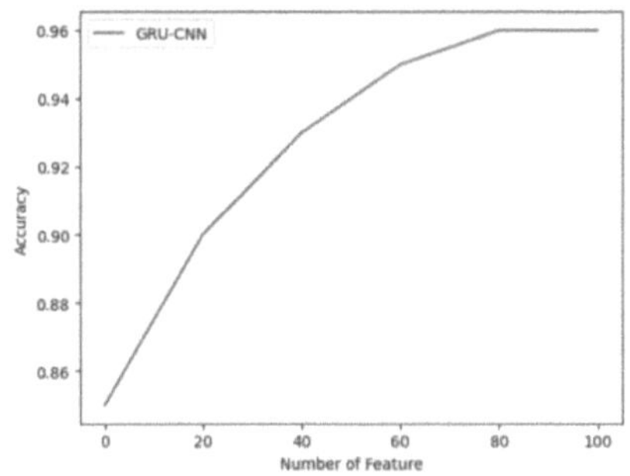


Figure 1. The Relationship between accuracy and feature number.

The relationship between the number of features and accuracy is illustrated in Figure 1. After LSTM-RNN-HMM was applied, Figure 1 shows a graph of feature number versus accuracy.

Comparatively, a closer inspection of the data reveals that distributional forms vary by division. For instance, when supervisors quit from certain divisions, indirect subordinates are more likely to experience subsequent turnovers than direct ones.

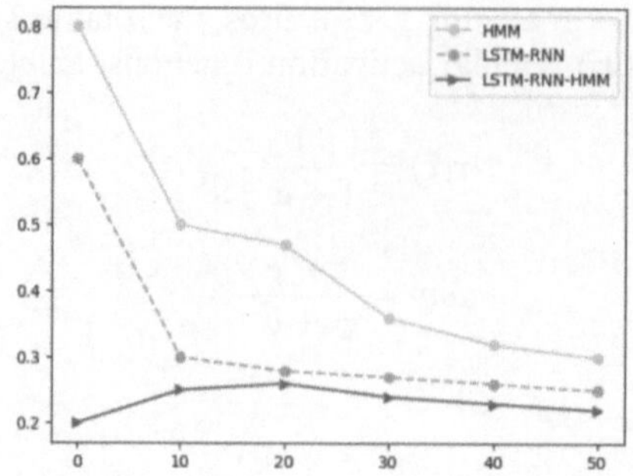


Figure 2. The relative level density histogram.

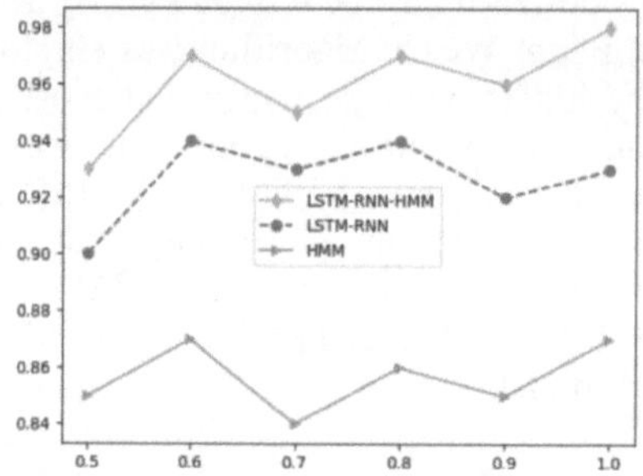


Figure 3. AUC under different pos./neg. ratio.

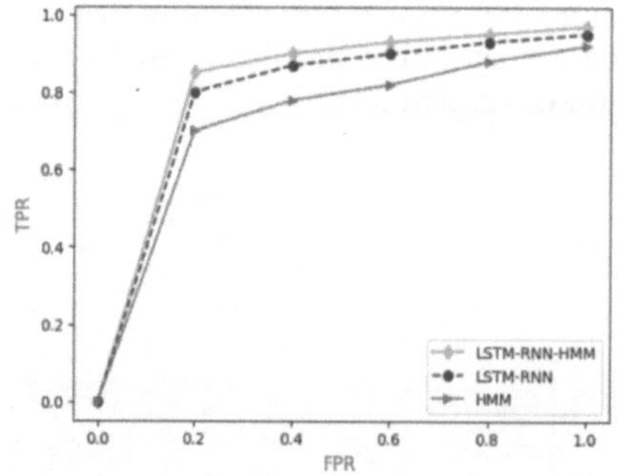


Figure 4. ROC curve of proposed model.

The AUC-ROC values for all models applied to datasets with different positive and negative sample proportions are shown in Figure 3.

The LSTM-RNN-HMM's ROC curve is displayed in Figure 4. The x-axis shows the rate of false positives, while the y-axis shows the rate of real positives.

## 5 CONCLUSION

The detrimental effects of personnel turnover on workplace productivity and long-term growth initiatives have led industries to see it as a significant concern. Using machine learning techniques, officialdoms make educated estimations regarding employee turnover to address this challenge. With reliable projections, governments can take steps to hold on to their employees. The three steps of data preprocessing that this proposes are cleaning the data, identifying the treatments, and correlating them. For this purpose, we employ a correlation-based feature selection approach. While training the model, feature-rich LSTM-RNN-HMM is employed. The proposed method outperforms the LSTM-RNN and HMM models with a 96.25% consistency rate.

## REFERENCES


[1] Bao, L., Xing, Z., Xia, X., Lo, D. and Li, S. (2017, May). Who will leave the company?: A large-scale industry study of developer turnover by mining monthly work report. In *2017 IEEE/ACM 14th International Conference on Mining Software Repositories (MSR)* (pp. 170–181). IEEE.

[2] Carraher, S. M. (2011). Turnover prediction using attitudes towards benefits, pay, and pay satisfaction among employees and entrepreneurs in Estonia, Latvia, and Lithuania. *Baltic Journal of Management*, 6(1), 25–52.

[3] Fang, M., Su, J. H., Wang, T. and He, R. J. (2017). Employee turnover prediction based on state-transition and semi-Markov-a case study of Chinese state-owned enterprise. In *ITM Web of Conferences* (Vol. 12, p. 04023). EDP Sciences.

[4] Fang, M., Su, J., Liu, J., Long, Y., He, R. and Wang, T. (2018). A model to predict employee turnover rate: observing a case study of Chinese enterprises. *IEEE Systems, Man, and Cybernetics Magazine*, 4(4), 38–48.

[5] Liu, J., Long, Y., Fang, M., He, R., Wang, T. and Chen, G. (2018, May). Analyzing employee turnover based on job skills. *In Proceedings of the International Conference on Data Processing and Applications* (pp. 16–21).

[6] Teng, M., Zhu, H., Liu, C., Zhu, C. and Xiong, H. (2019, July). Exploiting the contagious effect for employee turnover prediction. *In Proceedings of the AAAI Conference on Artificial Intelligence* (Vol. 33, No. 01, pp. 1166–1173).

[7] Feeley, T. H., Moon, S. I., Kozey, R. S. and Slowe, A. S. (2010). An erosion model of employee turnover based on network centrality. *Journal of Applied Communication Research*, 38(2), 167–188.

[8] Vardaman, J. M., Taylor, S. G., Allen, D. G., Gondo, M. B. and Amis, J. M. (2015). Translating intentions to behavior: The interaction of network structure and behavioral intentions in understanding employee turnover. *Organization Science*, 26(4), 1177–1191.

[9] Lin, B., Robles, G. and Serebrenik, A. (2017, May). Developer turnover in global, industrial open source projects: Insights from applying survival analysis. In *2017 IEEE 12th International Conference on Global Software Engineering (ICGSE)* (pp. 66–75). IEEE.

[10] Zhu, Q., Shang, J., Cai, X., Jiang, L., Liu, F. and Qiang, B. (2019, August). CoxRF: Employee turnover prediction based on survival analysis. In *2019 IEEE SmartWorld, Ubiquitous Intelligence & Computing, Advanced & Trusted Computing, Scalable Computing & Communications, Cloud & Big Data Computing, Internet of People and Smart City Innovation (SmartWorld/SCALCOM/UIC/ATC/CBDCom/IOP/SCI)* (pp. 1123–1130). IEEE.

[11] Zhao, Y., Hryniewicki, M. K., Cheng, F., Fu, B. and Zhu, X. (2019). Employee turnover prediction with machine learning: A reliable approach. In *Intelligent Systems and Applications: Proceedings of the 2018 Intelligent Systems Conference (IntelliSys)* Volume 2 (pp. 737–758). Springer International Publishing.

[12] Zhang, H., Xu, L., Cheng, X., Chao, K. and Zhao, X. (2018, September). Analysis and prediction of employee turnover characteristics based on machine learning. In *2018 18th International Symposium on Communications and Information Technologies (ISCIT)* (pp. 371–376). IEEE.

[13] Gao, X., Wen, J. and Zhang, C. (2019). An improved random forest algorithm for predicting employee turnover. *Mathematical Problems in Engineering*, 2019(1), 4140707.

[14] Wijaya, D., DS, J. H., Barus, S., Pasaribu, B., Sirbu, L. I. and Dharma, A. (2021). Uplift modeling VS conventional predictive model: A reliable machine learning model to solve employee turnover. *International Journal of Artificial Intelligence Research*, 5(1), 53–64.

[15] Subhashini, M. and Gopinath, R. (2020). Employee attrition prediction in industry using machine learning techniques. *International Journal of Advanced Research in Engineering and Technology*, 11(12), 3329–3341.

[16] Bi, J. X., Fan, W. Z. and Wang, S. B. (2021). Remaining life prediction for aircraft turbine engines based on LSTM-RNN-HMM–APPROACH. In *IOP Conference Series: Materials Science and Engineering* (Vol. 1043, No. 2, p. 022033). IOP Publishing.

*Hybrid and Advanced Technologies – S. Prasad Jones Christydass, Nurhayati Nurhayati and S. Kannadhasan. (Eds)* © *2025 the Authors, ISBN 978-1-032-90254-8*


# Exploring the influence of work environment factors on personality traits, job satisfaction, and turnover intentions among employees: A ResNet analysis

Shreya Singh
*Assistant Professor at Centre for Distance & Online Education, GLA University, Mathura, Uttar Pradesh, India*

Sunil Kumar Bharti
*Assistant Professor, Magadh Professional Institute, Danapur, Patna, Bihar, India*

Rajlaxmi Pujar
*Sr Assistant Professor, Department of Marketing, Indira School of Business Studies PGDM Pune, Maharashtra, India*

Daniel Pilli
*Assistant Professor, Department of MBA, KLEF, KL Deemed To Be University, Vaddeswaram, Guntur, India*

Devadutta Indoria
*Assistant Professor in Commerce, Department of Commerce, Vikram Dev University, Jeyoore, Odisha, India*

Manu Vasudevan Unni
*Assistant Professor, Department of Management, St. Claret College (Autonomous), Bangalore, Karnataka, India*

ABSTRACT: Perceived stress, job satisfaction, and intention to leave were the outcomes of this investigation of the connections between odour, noise, light, and color in the workplace. Staff members may experience stress, low job satisfaction, and even job-hopping intentions as a result of common environmental stresses. It is possible to increase employee happiness and loyalty by reducing or doing away with certain environmental influences. The three main components are feature selection, preprocessing, and model training. It cleaned and extracted the informative bits of the corpus before processing it. Based on the criterion function that was used to direct the approach, the features that were selected are optimal. The model's accuracy was enhanced during training by utilizing the DM-ResNet. The suggested approach outperformed state-of-the-art approaches like DM and ResNet, with an accuracy of 96.65%.

*Keywords*: Deep Neural Network (DNN), Dwarf Mongoose (DM), Primary care providers (PCPs), Residual Networks (ResNets), Data Mining (DM)

## 1 INTRODUCTION

In today's competitive business environment, retaining exceptional personnel is crucial for every company's success. It is critical to strategically build organizational settings that motivate employees to perform hard for the benefit of the company and themselves.

 DOI: 10.1201/9781003559115-57

Consequently, the way employees feel about their occupations and the circumstances around them is a major concern for many firms. Numerous studies in organizational behaviour and human resource management have identified various causes of employee turnover. Turnover intention is a management field variable that organizations should examine and implement if they wish to retain their personnel. Several studies that looked at employees' intentions to leave the company emphasized the importance of two factors: work satisfaction and organizational loyalty. The aspects of work that make employees happy. Consideration of the real-world implications necessitates paying attention to the correlations between various dimensions of job satisfaction and attitudes toward loyalty and intention to leave. Thus, the study elucidates the useful and practical components of managing job satisfaction to control exit intention in companies. The issue statement and research methodology for the rest of the work are laid out in the introduction.

## 2 LITERATURE SURVEY

There is a long-standing agreement among scholars that attitudes towards work have an effect on both the efficiency of organizations and the advancement of employees' careers[1]. In the last few decades, there has been an upsurge in workplace learning research, [2]which has the ability to illuminate not just employee and organizational growth but also current social and environmental issues. According to[3], organizational psychologists have a long tradition of investigating employees' views of their workplace. Research linking employee satisfaction with their work to organizational performance gave rise to this practice. [4]found that people whose attitudes toward their profession were more positive had higher overall health. Both job satisfaction (JS) and the intention to leave (TI) have been the subject of substantial research. [5]Organizational researchers have long acknowledged the challenges of staff retention and turnover, with JS being widely regarded as the most important employee attitude. [6]Nurses who are happy in their roles and with their working conditions are more likely to remain with the same company, according to many studies. Anxieties and thoughts of quitting one's job were less common among contented nurses[7]. Optimistic work environments reduce burnout, boost nurses' perceptions of the care they deliver, and decrease intentions to leave the profession[8]. A "positive nurse work environment" is defined as an office that "supports nurses to function at the highest scope of clinical practice, to work effectively in an interdisciplinary team of caregivers, and to mobilize resources quickly". [9] The effect of nurses' work settings on their levels of job satisfaction and retention has been the subject of a great deal of research during the past decade. [10]The surprising thing is that most of these studies have focused on two things: how nurses' job satisfaction affects their intentions to quit their current position and how their work environment directly affects their intentions to leave. Primary care providers (PCPs) are an essential part of the healthcare system because they provide direct patient care [11]. Having enough PCPs who are both qualified and enthusiastic about their work is essential for improving health outcomes and providing effective health services [12]. In poor and middle income nations in particular, primary care physicians (PCPs) in rural regions face comparable obstacles that could hinder healthcare delivery [13]. It is not an exaggeration to claim that improving health systems in rural and distant areas is a top priority for any country's health workforce. However, health policymakers have long neglected primary care physicians' (PCPs') job satisfaction, burnout, and intention to leave in favour of improving the supply and quality of human resources[14].

## 3 PROPOSED SYSTEM

The purpose of this study is to apply the person-environment fit theory to the medical field in order to identify the variables that influence the job satisfaction, turnover intention, and professional efficacy of medical professionals as well as to investigate the personal traits linked to PE fit.

### 3.1 *Data preprocessing*

This investigation's analytical process made use of a number of pre-processing steps. As a starting point, the sentence-based corpus was developed. Later on, the corpus was cleaned up and extraneous data was removed during pre-processing. Then used the SMART stop word list to remove stop words, digits, punctuation, symbols, and word contractions[15]. Then tokenized the text, applied part-of-speech tagging, and kept just verbs, nouns, adverbs, and adjectives. abstraction of personal names, removal of non-essential phrases. By implementing these strategies, were managed to circumvent background noise, misspellings, and uncommon phrases. The next step was to select a sufficient number of themes for each corpus.

### 3.2 *Feature selection*

Efficient and straightforward, exhaustive search is the algorithm for feature selection. This approach goes through all of the possible search spaces[16]. Thus, the selected characteristics are the most optimal ones that the program could discover using the given criterion function. However, additional (less-than-ideal) methods are required when this approach is insufficient owing to the computing expense. This proposed has utilized two less-than-ideal approaches, which are occasionally referred to as wrappers. Usually, the first one uses the subset's size or another threshold as a stop condition; this one optimizes a criterion function to insert features into an empty subset. This procedure is summarized by Sequential Forward Selection. One drawback of SFS, the nesting effect, is that it is both simple and efficient. This result is the result of always keeping the selected traits. A list is generated to store the preliminary findings, bearing in mind the selected criterion function and features. The best sets from this list will be the results of the algorithm. For feature selection, it will employ embedding techniques, filters (such as correlation-based, entropy-based, and consistency-based), and wrappers (such evolutionary algorithms and simulated annealing).

### 3.3 *Model training*

#### 3.3.1 *DM-ResNet*

The retrieved attributes are labeled using ResNet. One form of Deep Neural Network (DNN) includes layers for total connectivity, activation, pooling, and convolution. It acquires the residual function from the layer input instead of referring to the unreferenced functions. If another mapping, $R(p)$, is found to be compatible with the stacked nonlinear layer, then

$$K(k) = R(p) - p \tag{1}$$

The optimization of ResNet is straightforward, and there is no association between the number of layers and an increase in training error. A deeper dive is now possible thanks to an increase in accuracy. To adjust the residual learning, the result of the stacked layer is displayed as

$$h = K(p, \{m_d\}) + p \tag{2}$$

when the input and output of the stacked layers are represented by $p$ and $h$, respectively. In terms of learning residual mapping, $K(p, \{m_d\})$ is the formula to use. The linear projection $m_d$ is executed with shortcut connection to confront the dimension if the input and mapping function dimensions are not identical.

$$h = K(p, \{m_d\}) + m_q p \tag{3}$$

The function $K(p, \{m_d\})$ represents several convolutional layers, and the formation of the residual function can be changed. To enhance the classification accuracy, the hyper parameters of ResNet are fine-tuned using an optimization method known as dwarf mongoose (DM). Here we go over the math behind optimizing ResNet hyperparameters using DMO modeling. The strategy is based on the manner that dwarf mongooses find food. By setting starting values, a set of solutions can be discovered.

$$f_{d,r} = w_r + rand \times (c_r - w_r) \tag{4}$$

where $w_r$ is the minimum searchable value and $c_r$ is the maximum searchable value.

## 4 RESULT AND DISCUSSION

Using demographic data, work features, and job satisfaction as independent variables, this proposed found that turnover intentions are influenced by job satisfaction and the link between job characteristics and turnover intentions. There was a negative and statistically significant relationship between turnover intention and job satisfaction and job characteristic variables, according to the data.

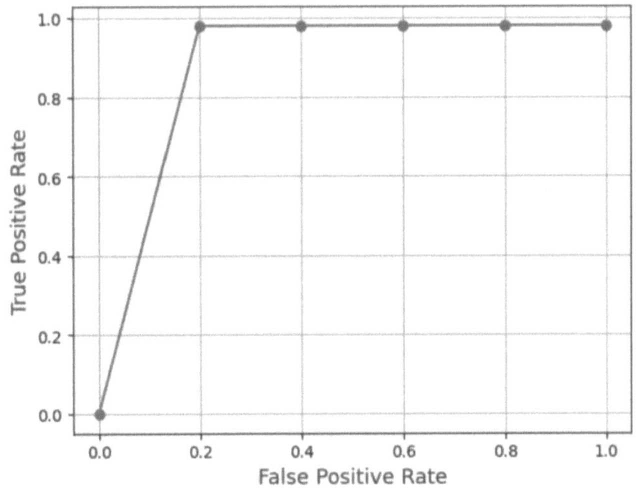


Figure 1. ROC curve of RBF-ELMAN model.

The sensitivity of a classifier model can be better understood using ROC curve analysis, which displays the ratio of true positives to false positives. The ROC curve for the model can be seen in Figure 1.

Figure 2 shows the results of our comparison of the testing phase confusion matrices. The DM-ResNet, which attained a 95.65% accuracy, is shown on the left side.

Figure 3 shows a comparison of the accuracy of the best performing models. The top performing models' training time and accuracy are displayed in Figure 3, a bar chart.

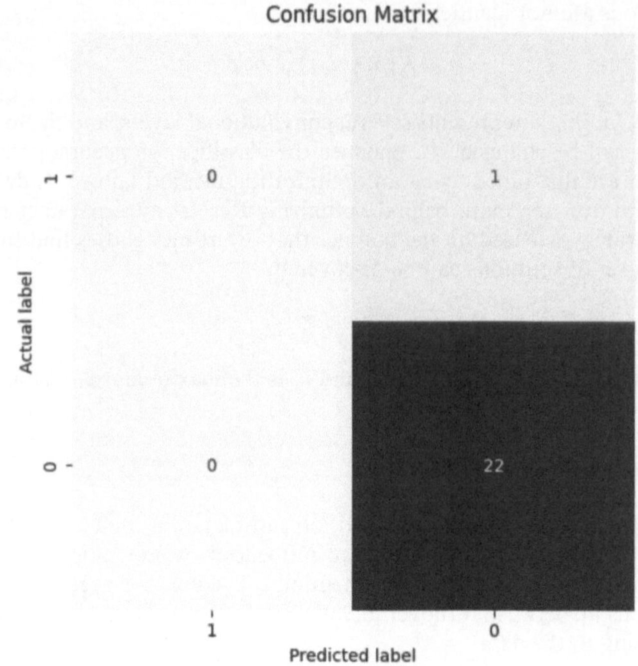


Figure 2. Confusion matrix for prediction method.

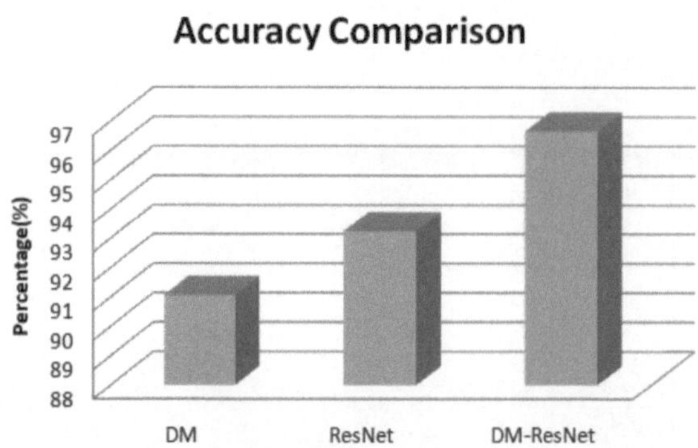


Figure 3. Accuracy comparison of proposed model.

## 5 CONCLUSION

A productive management-staff relationship is essential to a content workforce since no company can thrive without hardworking employees who enjoy their work and see a positive return on investment. The consequences will be most severe for whoever fails to provide this essential ingredient for commercial success. Despite the abundance of prior research on the topic, many firms continue to face challenges when it comes to employee pleasure on the job. It prepared the corpus for processing by cleaning and extracting the informative sections. The features that were chosen are optimal according to the criterion function that guided the approach. For training its models, it employs DM-ResNet. The proposed approach outperforms the DM and ResNet models with an average accuracy of 96.65 percent.

## REFERENCES


[1] Bothma, C. F. and Roodt, G. (2013). The validation of the turnover intention scale. *SA Journal Of Human Resource Management*, 11(1), 1–12.

[2] Bowling, N. A. and Hammond, G. D. (2008). A meta-analytic examination of the construct validity of the michigan organizational assessment questionnaire job satisfaction subscale. *Journal of Vocational Behavior*, 73(1), 63–77.

[3] Bowling, N. A., Eschleman, K. J. and Wang, Q. (2010). A meta-analytic examination of the relationship between job satisfaction and subjective well-being. *Journal of Occupational and Organizational Psychology*, 83(4), 915–934.

[4] Chen, B., Vansteenkiste, M., Beyers, W., Boone, L., Deci, E. L., Van der Kaap-Deeder, J., ... and Verstuyf, J. (2015). Basic psychological need satisfaction, need frustration, and need strength across four cultures. *Motivation and Emotion*, 39, 216–236.

[5] Chiva, R. and Alegre, J. (2008). Emotional intelligence and job satisfaction: the role of organizational learning capability. *Personnel Review*, 37(6), 680–701.

[6] Al-Hamdan, Z., Manojlovich, M. and Tanima, B. (2017). Jordanian nursing work environments, intent to stay, and job satisfaction. *Journal of Nursing Scholarship*, 49(1), 103–110.

[7] Al Maqbali, M. A. (2015). Job satisfaction of nurses in a regional hospital in Oman: A cross-sectional survey. *Journal of Nursing Research*, 23(3), 206–216.

[8] Ayalew, F., Kols, A., Kim, Y. M., Schuster, A., Emerson, M. R., Van Roosmalen, J., ... and Gibson, H. (2015). Factors affecting turnover intention among nurses in Ethiopia. *World Health Popul*, 16(2), 62–74.

[9] Bae, S. H., Mark, B. and Fried, B. (2010). Impact of nursing unit turnover on patient outcomes in hospitals. *Journal of Nursing Scholarship*, 42(1), 40–49.

[10] Delobelle, P., Rawlinson, J. L., Ntuli, S., Malatsi, I., Decock, R. and Depoorter, A. M. (2011). Job satisfaction and turnover intent of primary healthcare nurses in rural South Africa: A questionnaire survey. *Journal of Advanced Nursing*, 67(2), 371–383.

[11] Liu, J. A., Wang, Q. and Lu, Z. X. (2010). Job satisfaction and its modeling among township health center employees: a quantitative study in poor rural China. *BMC Health Services Research*, 10, 1–9.

[12] Wang, H., Tang, C., Zhao, S., Meng, Q. and Liu, X. (2017). Job satisfaction among health-care staff in township health centers in rural China: results from a latent class analysis. *International Journal Of Environmental Research and Public Health*, 14(10), 1101.

[13] Alam, A. and Asim, M. (2019). Relationship between job satisfaction and turnover intention. *International Journal of Human Resource Studies*, 9(2), 163.

[14] Jain, S., Tripathy, H. K., Mallik, S., Qin, H., Shaalan, Y. and Shaalan, K. (2023). Autism detection of mri brain images using hybrid deep cnn with dm-resnet classifier. *IEEE Access.*

[15] Stamolampros, P. and Dousios, D. (2023). Employee satisfaction during the pandemic in the tourism and hospitality industries. *Current Issues in Tourism*, 1–15.

[16] Lopes, F. M., Martins, D. C. and Cesar, R. M. (2008). Feature selection environment for genomic applications. *BMC Bioinformatics*, 9, 1–8.

*Hybrid and Advanced Technologies – S. Prasad Jones Christydass, Nurhayati Nurhayati and S. Kannadhasan. (Eds) © 2025 the Authors, ISBN 978-1-032-90254-8*
*Open Access: www.taylorfrancis.com, CC BY-NC-ND 4.0 license*

# The development of a framework for a comprehensive approach to stress management interventions at work based on BiLSTM-RNN model

Sayed Mohammed Imdadullah
*Assistant Professor, Department of ECE, PSNA College of Engineering and Technology, Dindigul, India*

Anurag Pathak
*Government Kusum P. G. College, Seoni Malwa, Madhya Pradesh, India*

K. Soujanya
*Assistant Professor, Department of BBA, Koneru Lakshmaiah Education Foundation, Vaddeswaram, Guntur, Andhra Pradesh, India*

B. Ramakrishna
*Associate Professor, Department of Emerging Technologies (CSE-DS), CVR College of Engineering, Vastunagar, Telangana, India*

Amit Chauhan
*Department of Life Sciences, CHRIST University, Bengaluru, Karnataka, India*

Neerav Nishant
*Assistant Professor, Department of CSE, School of Engineering, Babu Banarasi Das University, Lucknow, India*
ORCID ID: 0009-0002-6696-6943

ABSTRACT: This overview shows how a company's stress management strategy incorporates several programs at different levels. Prefabricated programs that haven't been customized to meet the needs of certain organizations raise some eyebrows, because programs vary greatly in terms of structure, aims, and target audiences. These efforts should include strategies for stress management and prevention, as well as tools to meet the needs of individuals and the company as a whole, which should be assessed and reviewed often. The proposed method consists of three steps: preprocessing biosignals, extracting features, and training the model. The main goal of biosignal preprocessing is to remove artifacts and decrease noise in specific biosignal channels. Producing useful representations from measurements is the main objective of feature extraction. It employed frequently determined attributes for stress identification. The TRANS-BLSTM improved the accuracy of the model training method. The suggested method outperforms its rivals, TRANS and LSTM, with an accuracy percentage of 97.82%.

*Keywords*: Stress Management, Stress Management Intervention (SMI), Workers Stress, Long short-term memory (LSTM), Psychological distress

## 1 INTRODUCTION

Most of the big shifts in the workforces of industrialized countries have been driven by crucial social, technical, economic, and political issues, like an aging population or a financial crisis. Due to these reforms and the easing of trade rules, organizations are now more competitive

 DOI: 10.1201/9781003559115-58

than they were before. The outcome is stringent regulation and high standards for employee performance in order to remain competitive on a global basis. Many businesses have re-evaluated their approaches and implemented management techniques like downsizing or mergers in an effort to survive in this new worldwide market. An excellent illustration of this is the industrial sector. Workers' well-being and health have taken a hit as a result of the manufacturing industry's vulnerability to the stresses of the current global economy. The transition to more technologically advanced and competitive workplaces has increased workloads and job responsibilities. As a result, manufacturing workers are now expected to accomplish more with less time and effort. Job duties, workload, lack of sleep, overtime, and shift schedules are the main causes of stress for industrial workers. Stress management strategies were categorized into three groups: primary, secondary, and tertiary, according to the degree of application (person, environment, or organization). This review set out to do something no one has done before: evaluate stress management therapies in a variety of settings. Since that evaluation, research into developing measurable treatments for organizations has been on the rise. The difficulty of developing workplace interventions is becoming more widely acknowledged.

## 2 LITERATURE SURVEY

The purpose of stress management is to implement strategies that will reduce or eliminate stress in the workplace. [1]Helping people learn new coping mechanisms is one possible approach to stress management. First, we achieve our goal with organizational level interventions (OLIs). Improving a worker's productivity is the goal of the other two intervention kinds[2]. Examples of interventions in this area include reorganizing processes, creating more ergonomic work environments, and implementing systemic solutions for staff development [3]. Less prevalent than individual ones are their exorbitant pricing and time-consuming nature [4]. Nonetheless, there are arguments that these interventions are more effective at enhancing organizational health than at influencing individuals' well-being on an individual level[5].Occupational stress in nursing [6]refers to the emotional and physiological reactions that occur when a nurse's abilities and resources are inadequate to meet the challenges and needs of their profession.[7]Intensive care unit (ICU) nurses are known to suffer from significant stress, according to a substantial amount of studies.[8]Intensive care unit nurses are particularly vulnerable to the physical, mental, and behavioural health problems that stress can cause, such as depression, anxiety, fatigue, and burnout.[9]Therefore, immediate measures are required to improve the health and well-being of intensive care unit nurses and reduce work stress. A stress management intervention (SMI) is whatever an individual does to attempt to reduce the negative effects of stress on their physical, mental, and spiritual health [10].Stress has surpassed all other diseases in prevalence in modern times. After looking at stress from many different perspectives, such as biological, psychological, and sociological [11], scientists have come up with a lot of concepts and models. Richard Lazarus and colleagues' transactional model of stress and coping is illustrated, for instance, in references [12]. According to Lazarus and colleagues, the fundamental source of stress is how an individual perceives their environment. It shows up when someone views a problem negatively at first, but then sees how well they can handle it in comparison to what's required of them [13]. Consequently, a cognitive appraisal leads to a specific coping technique when dealing with stress. Almost all modern explanations of stress draw from a variety of fields, including biomedicine, psychology, cognitive psychology, and sociology[14].

## 3 PROPOSED SYSTEM

The happiness of employees is of utmost importance to employers, their families, and the company as a whole. Stress management therapies typically aim to replenish resources that have been exhausted by the work environment, in contrast to wellness and health initiatives in the workplace that aim to improve personal and job resources for all employees.

### 3.1 *Biosignal preprocessing*

In order to begin analyzing the biosignals, the raw data must first be preprocessed. The feature extraction process begins with this step. The following are examples of what is involved: converting raw data from bioamplifiers to a readable format, collecting physiological data using subject-specific logger files, filtering out irrelevant data and keeping only relevant details, and applying different filters to different biosignal channels. Filtering primarily aims to decrease artifacts and noise in specific biosignal channels[15]. Electrocardiogram (ECG) signals, for instance, are susceptible to a wide range of noise types. Baseline wander, mainly caused by the patient's breathing, power line interference, also known as narrow-band noise centered around 50 Hz and 60 Hz, high-frequency noise, and various artifacts related to the environment and equipment, are all examples of such factors.

### 3.2 *Feature extraction*

A key objective of feature extraction is to transform the numerical data into meaningful representations. Following the suggested traits for stress recognition in the corresponding research—often derived features was our approach. Since our study uses LSR devices, we need to extend the time periods so that the extracted characteristics can depict the variations between normal and stress states[16]. Even after increasing the amount of feature instances and employing shorter time periods for LSR devices, detection performance has not improved. Using a series of data values from HRV and GSR sensors, the prediction is made every 170 seconds. A total of 85 features are obtained from each 50-second time window. With each 50-second time window that overlaps by 1 second, the features are retrieved one after the other using a sliding approach. There are two categories of features that comprise our feature set: Eighty-five features that are already available and have a high volume of users, and twenty features that are tailored to work with LSR data.

### 3.3 *Model training*

#### 3.3.1 *TRANS-BLSTM*

Previous trials' results indicated that transformers might be more effective than a bidirectional LSTM model. Transforming a transformer model into a bidirectional LSTM model with only one layer, for example, significantly diminished accuracy. This was also supported by the outcomes of our experiments. This provides more evidence that bidirectional LSTM and transformer may complement each other in sequence modelling. We decided to investigate this since a bidirectional LSTM may be able to beat a regular transformer model in downstream tasks. The versions displayed here are the TRANS-BLSTM-1 and TRANS-BLSTM-2, which stand for Transformer with Bidirectional LSTM designs:

#### 3.3.2 *TRANS-BLSTM-1*

In each BERT layer, a bidirectional LSTM layer is used in place of the feedforward layer.

#### 3.3.3 *TRANS-BLSTM-2*

The following stage is to incorporate a bidirectional LSTM layer that will share its input with the initial BERT layer. Then merge the output of the bidirectional LSTM layer with the initial BERT layer after applying the Layer Norm.

By merging self-attention with bidirectional LSTM, the inclusion of BLSTM enhances the joint model framework. The difficulties that come after this will prove it. Based on our testing, it appears that these two designs are equally accurate. For the purpose of clarity, let will focus on the second model, TRANS-BLSTM-2, and refer to the first as TRANS-BLSTM from now on.

## 4 RESULT AND DISCUSSION

This chapter provides a summary of stress management interventions (SMIs) and the research that backs their claims about improving workers‘ health and reducing stress. It will first establish a categorization system for SMI, with a "primary" emphasis on resolving stress's underlying causes and a "secondary" or "tertiary" emphasis on reducing stress levels. After outlining key SMI types using this typology, we look at the strongest-grounded SMIs, such as individual-level secondary SMIs that try to reduce workers' stress and organizational-level primary SMIs that try to change policies and procedures to eliminate stress-causing practices.

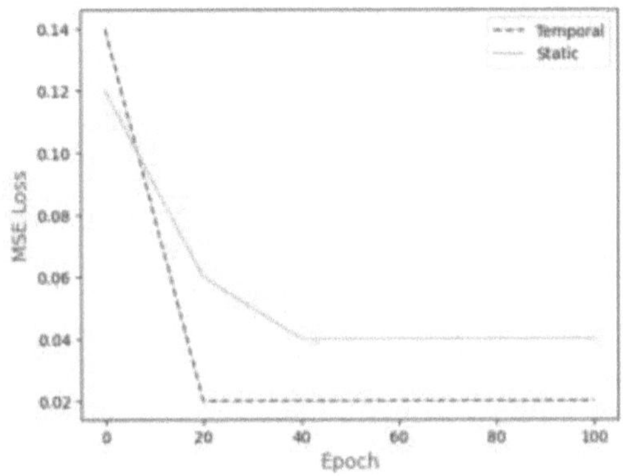


Figure 1. MSE loss of proposed model.

The reconstruction loss curves on the TRANS-BLSTM user-dependent validation set are compared in Figure 1. Since all channels exhibited the same trend, we combined channel-wise losses once again.

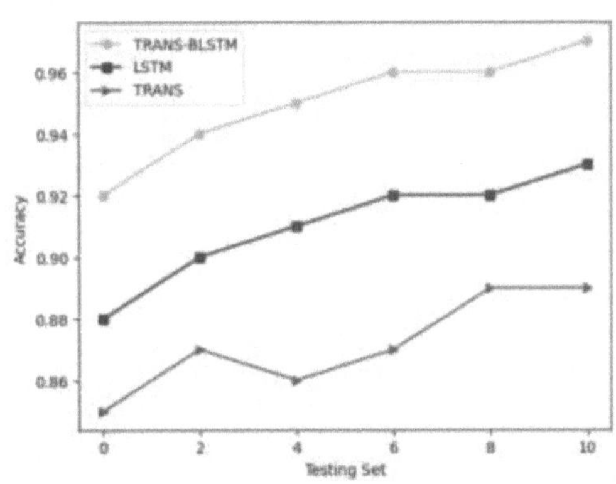


Figure 2. Accuracy visualization using several categorization models.

The accuracy of these models is illustrated in Figure 2 through qualitative results. With different test sets each time, the experiment was run five times for each classifier. In order to study the effect of parameters and achieve high accuracy while keeping consistency, visualization is crucial.

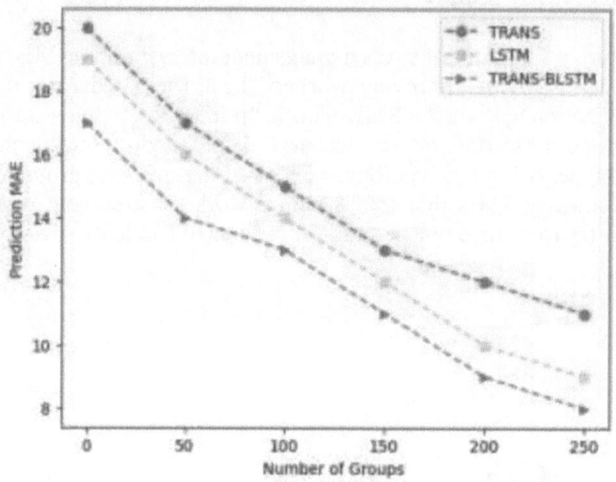


Figure 3. Prediciton of MAE Vs Number of groups.

Figure 3 shows that as more user groups were added, the prediction performance in the user-dependent case grew steadily better. If the optimal number of groups is 250 (the same as the training set size), then cluster-based personalization is on par with individual-based personalization since no more users will be added to the trained model.

## 5 CONCLUSION

There appears to be a higher rate of illness and absenteeism in the UK health care business compared to other sectors, and stress may play a role in nurses leaving their jobs. These problems, particularly in the field of mental health, must be fixed if current services are to continue. Biosignal preprocessing primarily aims to eliminate noise and artifacts in targeted biosignal channels. Feature extraction's primary goal is to generate meaningful representations from measurements. It used characteristics that were already known to be associated with stress to identify it. The proposed method, however, makes use of TRANS-BLSTM for model training. The suggested method achieves a consistency rate of 97.82%, which is higher than that of the TRANS and LSTM models.

## REFERENCES


[1] Ebert, D. D., Lehr, D., Smit, F., Zarski, A. C., Riper, H., Heber, E., ... and Berking, M. (2014). Efficacy and cost-effectiveness of minimal guided and unguided internet-based mobile supported stress-management in employees with occupational stress: A three-armed randomised controlled trial. *BMC Public Health*, 14, 1–11.

[2] Giga, S. I., Noblet, A. J., Faragher, B. and Cooper, C. L. (2003). The UK perspective: A review of research on organisational stress management interventions. *Australian Psychologist*, 38(2), 158–164.

[3] Kinnunen-Amoroso, M. and Liira, J. (2014). Work-related stress management by Finnish enterprises. *Industrial Health*, 52(3), 216–224.

[4] Gaab, J., Sonderegger, L., Scherrer, S. and Ehlert, U. (2006). Psychoneuroendocrine effects of cognitive-behavioral stress management in a naturalistic setting—a randomized controlled trial. *Psychoneuroendocrinology*, 31(4), 428–438.

[5] Richardson, K. M. and Rothstein, H. R. (2008). Effects of occupational stress management intervention programs: A meta-analysis. *Journal of Occupational Health Psychology*, 13(1), 69.

[6] Murta, S. G., Sanderson, K. and Oldenburg, B. (2007). Process evaluation in occupational stress management programs: A systematic review. *American Journal of Health Promotion*, 21(4), 248–254.

[7] Kröll, C., Doebler, P. and Nüesch, S. (2017). Meta-analytic evidence of the effectiveness of stress management at work. *European Journal of Work and Organizational Psychology*, 26(5), 677–693.

[8] Vernekar, S. P. and Shah, H. (2018). A study of work-related stress among nurses in a tertiary care hospital in Goa. *International Journal of Community Medicine and Public Health*, 5(2), 657–661.

[9] Bernstein, A. M., Kobs, A., Bar, J., Fay, S., Doyle, J., Golubic, M. and Roizen, M. F. (2015). Yoga for stress management among intensive care unit staff: A pilot study. *Alternative and Complementary Therapies*, 21(3), 111–115.

[10] Thoits, P. A. (2010). Stress and health: Major findings and policy implications. *Journal of health and social behavior*, 51(1_suppl), S41–S53.

[11] Bräuninger, I. (2012). Dance movement therapy group intervention in stress treatment: A randomized controlled trial (RCT). *The Arts in Psychotherapy*, 39(5), 443–450.

[12] Huss, E. and Sarid, O. (2014). Visually transforming artwork and guided imagery as a way to reduce work related stress: A quantitative pilot study. *The Arts in Psychotherapy*, 41(4), 409–412.

[13] Sugumar, R., Sharma, S., Kiran, P. N., Gupta, S. K., Sindhu, V., and Srinivas, R. V. (2023, June). Novel method for detection of stress in employees using hybrid deep learning models. In *2023 8th International Conference on Communication and Electronics Systems (ICCES)* (pp. 984–989). IEEE.

[14] Huang, Z., Xu, P., Liang, D., Mishra, A. and Xiang, B. (2020). TRANS-BLSTM: Transformer with bidirectional LSTM for language understanding. *arXiv preprint arXiv*:2003.07000.

[15] Hazer-Rau, D., Zhang, L. and Traue, H. C. (2020). A workflow for affective computing and stress recognition from biosignals. *Engineering Proceedings*, 2(1), 85.

[16] Khowaja, S. A., Prabono, A. G., Setiawan, F., Yahya, B. N. and Lee, S. L. (2021). Toward soft real-time stress detection using wrist-worn devices for human workspaces. *Soft Computing*, 25, 2793–2820.

*Hybrid and Advanced Technologies – S. Prasad Jones Christydass, Nurhayati Nurhayati and S. Kannadhasan. (Eds) © 2025 the Authors, ISBN 978-1-032-90254-8*
*Open Access: www.taylorfrancis.com, CC BY-NC-ND 4.0 license*

# Benefit evaluation of human resource management in agricultural enterprises based on MAResnet-BiGRU model

R. Karthiga
*Head and Assistant Professor, Department of Commerce, Srinivasan College of Arts and Science, Perambalur, India*

N. Vijayakumar
*Assistant Professor, Department of Commerce, B S Abdur Rahman Crescent Institute of Science and Technology, Chennai, Tamil Nadu, India*

C. Vijai
*Associate Professor, Department of Commerce and Business Administration, Vel Tech Rangarajan Dr. Sagunthala R&D Institute of Science and Technology, Chennai, Tamil Nadu, India*

N. Dinesh
*Associate Professor, OB and HRM Area, CMS Business School, Jain Deemed to be University, Bangalore, India*

D. Paul Dhinakaran
*Assistant Professor, Department of Commerce, Jayagovind Harigopal Agarwal Agarsen College (Affiliated to University of Madras) Madhavaram, Chennai, Tamil Nadu, India*

M. Rajalakshmi
*Ph.D. Research Scholar, Department of Commerce, Thiru Kolanjiappar Government Arts College, Virudhachalam*

ABSTRACT: The success determinants of agricultural firms are the focus of this research, which is a part of a larger study. It makes an effort to provide some of the findings from a survey that aimed to rank the HRM (Human Resource Management) departments of fundamental agricultural businesses. Research on human resource management in basic agricultural companies has been published in a number of countries. Model training, feature selection, and preprocessing are the three phases that make up the suggested approach. When it comes to preprocessing approaches, the methods chosen are heavily influenced by the characteristics and intended application of large data. The RFE technique relies on a ranking algorithm that sorts the dataset from best to worst according to the specified important attributes. The MAResnet-BiGRU improved the accuracy of the model training process. With a 96.34% accuracy rate, the proposed technique surpasses GRU and Resnet.

*Keywords*: Human Resource Management (HRM), Recursive Feature Elimination (RFE), Gated Recurrent Unit (GRU), Agricultural Enterprises, Enterprise Development Management

## 1 INTRODUCTION

In a market economy, where many of the elements are dynamic and uncertain, reaching a stable condition of a company necessitates the resolution of numerous complicated challenges. We rank the creation and implementation of a market-appropriate organizational and economic mechanism for managing staff as the most critical of these responsibilities.

 DOI: 10.1201/9781003559115-59

Personalities, both quantitative and qualitative, of farm labourers, their capacity to effectively adapt to dynamic market circumstances find out how well the team is doing their jobs, which is the foundation for all the big decisions. When an organization's human resources are well-balanced, it dictates not only the long-term objectives of its economic strategy, but also its short-term strategies and tactics for achieving those objectives. When solving these complicated difficulties within the constraints of the company's finite human resources, the issue of economically gauging those resources the enterprise's whole economic potential rises to the surface. These decisions will help to refocus production capacity and concentrate material and labour resources to meet high demand goods and services. This proves that economic metrics of an enterprise's labour potential are a useful management tool, and it's only fair that scientists give practicing entrepreneurs a better methodological foundation for using modern methods of measuring components of economic potential.

## 2 LITERATURE SURVEY

The current state of the global economy, which is influenced by social and institutional shifts and the epidemic, necessitates the exploration of novel approaches to company management [1]. They are also associated with management systems in general and human resource management in particular.[2] The best results in economic development are achieved by those enterprises which have provided favourable conditions for the realization of employees' creative abilities and introduce some innovative technologies for human resource management. [3]To this end, managers often turn to socio-psychological management strategies; these, when combined with a democratic or liberal leadership style, lay the framework for HRM's social instruments and help workers realize their full intellectual potential. In the agricultural sector, entrepreneurial spirit significantly improves sustainable food production, provisioning, and the preservation of rural communities[4]. Emerging agriculture and food technology firms may be a game-changer in the fight against issues like food insecurity and environmental degradation [5]. For example, in recent years, the common perception of the agricultural industry in Australia has shifted from viewing [6]the market as a whole as highly competitive to one that recognizes the existence of niches within the agricultural market, each with its own unique set of challenges and opportunities for collaboration and innovation. According to research[7], entrepreneurs play a crucial role in driving economic growth in developing countries like India and Brazil. [8]Entrepreneurship in rural areas is essential for emerging countries to reduce poverty and revitalize underdeveloped areas.[9]The field of enterprise development management is thriving with academic interest from throughout the world. [10]Furthermore, due to the objective reasons mentioned earlier, there has been a significant increase in scientists' interest in that subject in recent years. [11]generalize and clarify the theoretical principles of development management by applying the current paradigm in management to analyse the components and process of Ukraine's economic development. [12] investigates the concept of "enterprise development management" in an effort to identify its economic essence and determine its constituent parts. In their investigation into matters concerning the improvement of the development and functioning management system, [13] assess the methodological foundations of a comprehensive development evaluation that considers the internal and external factors contributing to the growth of companies.

## 3 PROPOSED SYSTEM

The goals of this study were to determine whether there was a correlation between the HRM strategies employed by dairy farms and the financial success of those farms. In order to determine whether there was a statistically significant difference between the groups that "used" and "did not use" HRM activities, various HRM practices were examined for their effects.

### 3.1 *Preprocessing*

Data preparation strategies allow for the reduction of large data collections. Once data has been collected, these tactics should be put into action without delay. To fulfil the needs of their applications, businesses can utilise a variety of big data preparation options. The use of ontology-based semantic analysis and linked data structures can streamline intelligent data preprocessing[14]. Data preparation, especially data filtering, could be done on the client side to reduce the amount and velocity of massive data. Various preparation methods may be more or less suitable depending on the enormous data's nature and intended purpose. Therefore, it is not simple to adapt these methods to various business problems.

### 3.2 *Feature selection*

#### 3.2.1 *Recursive feature elimination*

Since RFE is a wrapper feature selection procedure, it starts with the full dataset. The ranking process of the RFE technique sorts the dataset from best to worst based on ranked qualities. At the end of each iteration, the dataset is refreshed with the most relevant features kept and the least important ones deleted. Using the greedy algorithm, RFE is one way to choose and remove features[15]. A Wrapper type describes this approach. The RFE method finds the least relevant qualities in the dataset and removes them iteratively until it reaches a well optimised state. To pick features, the Wrapper method applies a core ML algorithm that has been trained on the dataset.

### 3.3 *Model training*

#### 3.3.1 *MAResnet-BiGRU*

Extensive data analytics are challenging for the 1×4 in size convolutional kernels of the initial residual net because of scale limits. This suggestion improves upon the first residual neural network by constructing a MAResnet through which the EEG signal can travel, allowing for the extraction of the specific characteristics and data from the timed signal. Because these properties are time series in nature, they necessitate a comprehensive model-based investigation[16]. It is possible that extracting the sleep components of the EEG data at various scales using the convolution kernel can improve the convolution layer's expression capabilities. Because the attributes gained via this method are in a time series format, a thorough model-building investigation is necessary. Last but not least, the dropout mechanism is introduced by feeding the fusion feature into the entire connection layer. This maintains the network's performance while successfully avoiding overfitting using samples from the memory training set.

### 3.4 *Result and Discussion*

The research focuses on the development of digitalization as a framework for managing personnel inside agricultural firms. The primary tendencies in the digitalization of HR management in agricultural businesses point to the widespread adoption of modern digital technology for the purpose of optimizing and improving the efficacy of HRM in agriculture.

Expect a positive outcome (TP) in four out of five instances where the result is good. False negatives (FNs) occur when the result is different from what was anticipated. Results that are negative and predicted to be negative are TN results, and results that are positive and predicted to be FP results. Predictions made by algorithms like MAResnet-BiGRU are shown in Figure 1.

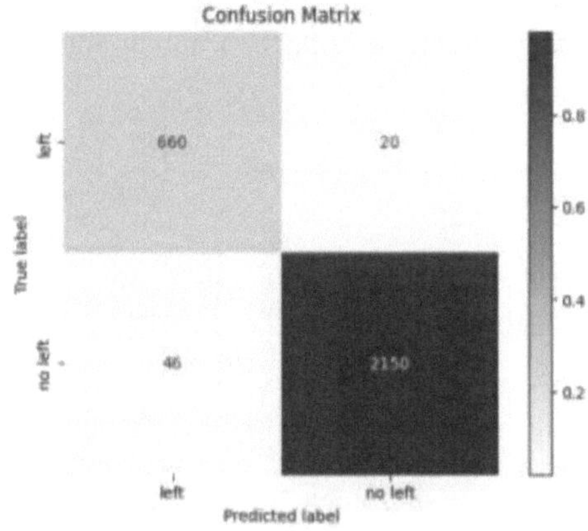


Figure 1. MAResnet-BiGRU model matrix.

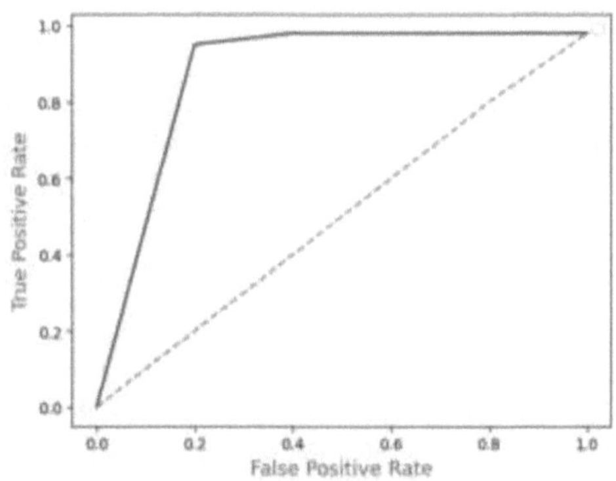


Figure 2. Receiver Operating Characteristic (ROC) curve.

Figure 2 shows the receiver operating characteristic (ROC) curves of many classifiers, which are based on the false positive rate (FPR) and the true positive rate (TPR).

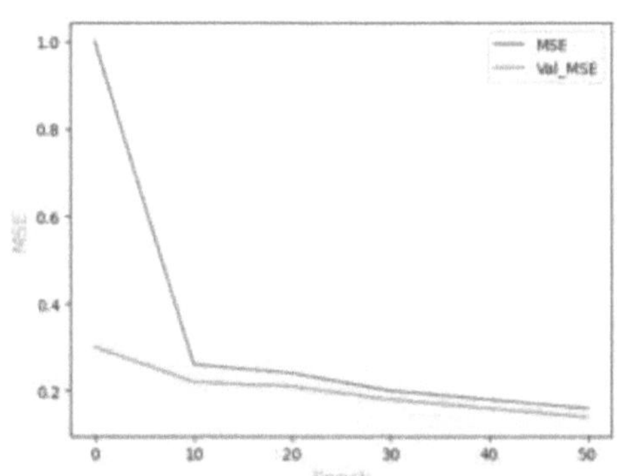


Figure 3. MSE Loss of MAResnet-BiGRU model.

Loss curves for MSE are shown in Figure 3. These figures show if the neural networks converge to a low MSE during training. Still, the training and validation loss curves converge as expected.

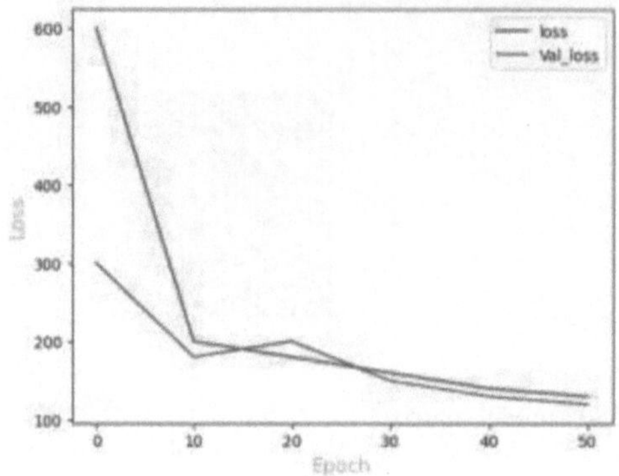


Figure 4. Training and validation loss.

Both the validation and training losses are displayed in Figure 4. As the number of epochs increases, the model learns on both the training and validation datasets, as seen in these graphs, which demonstrate how the suggested method is learning.

## 4 CONCLUSION

With national economic globalisation came numerous challenges for the agri-food sector, one of the most persistent being the question of how to maintain competitiveness in domestic and worldwide markets. Agricultural national enterprises that rely on digitalization can boost their competitiveness by using management methods that integrate new technology into all aspects of the company. Yet, a suitable human resource management plan that is in sync with the enterprise's integrated management strategy is essential for success. Characteristics and planned use of huge data have a significant impact on the methods selected for preprocessing. Using a ranking algorithm to arrange the dataset from best to worst according to the stated relevant attributes is the RFE technique's foundation in feature selection. Nevertheless, MAResnet-BiGRU is recommended for model training. With a consistency rate of 96.34%, the suggested method surpasses both the GRU and Resnet models.

## REFERENCES


[1] Akdere, M. and Egan, T. (2020). Transformational leadership and human resource development: Linking employee learning, job satisfaction, and organizational performance. *Human Resource Development Quarterly*, 31(4), 393–421.

[2] Khuong, M., Mai, T. and Phuong, N. (2020). The impacts of human resource management practices on employees' motivation and loyalty. *Management Science Letters*, 10(11), 2673–2682.

[3] Serniak, I., Serniak, O., Mykhailyshyn, L., Skrynkovskyy, R. and Kasian, S. (2021). Evaluation of the level of the usage of social instruments for human resource management: example of agro-processing enterprises of Ukraine. *Agricultural and Resource Economics: International Scientific E-Journal*, 7(4), 82–99.

[4] Ayunda, W. A., Purwaningsiha, I., Arianaa, L. and Akila, H. A. (2021). Innovation Platforms Practices for Improving Sustainable Intensification in Several Countries and Its Implication to Indonesia: An Integrative Review of Agricultural Innovation Systems. *Journal of STI Policy and Management*, 6(1), 21–30.

[5] Gharbi, H., Sobaih, A. E. E., Aliane, N. and Almubarak, A. (2022). The role of innovation capacities in the relationship between green human resource management and competitive advantage in the Saudi food industry: Does gender of entrepreneurs really matter?. *Agriculture*, 12(6), 857.

[6] Laouiti, R., Gharbi, S. and Liouane, N. (2014). The effect of business environment on firm performance Exploratory Study: Case of Tunisian enterprises. *International Journal of Management & Information Technology*, 8(3), 1430–1439.

[7] Mustafa, M., Lundmark, E. and Ramos, H. M. (2016). Untangling the relationship between human resource management and corporate entrepreneurship: The mediating effect of middle managers' knowledge sharing. *Entrepreneurship Research Journal*, 6(3), 273–295.

[8] Zainol, F. A., Daud, W. N. W., Shamsu, L., Abubakar, H. S. and Halim, H. A. (2018). A linkage between entrepreneurial leadership and SMEs performance: An integrated review. *International Journal of Academic Research in Business and Social Sciences*, 8(4), 104–118.

[9] Kenkel, P. and Crossan, Z. (2019). Human resource management in agricultural cooperatives. *Journal of Cooperatives*, 34, 1–30.

[10] Mykhailichenko, M., Lozhachevska, O., Smagin, V., Krasnoshtan, O., Zos-Kior, M. and Hnatenko, I. (2021). Competitive strategies of personnel management in business processes of agricultural enterprises focused on digitalization. *Management Theory and Studies for Rural Business and Infrastructure Development*, 43(3), 403–414.

[11] Дем'яненко, С., Сагайдак, М., Сас, О., Авраменко, Т. and Левк⊠вський, Є. (2021). Efficiency of the large-scale agri-industrial entities in Ukraine. *Financial and Credit Activity Problems of Theory and Practice*, 1(36), 179–189.

[12] Skobelev, P. O., Simonova, E. V., Smirnov, S. V., Budaev, D. S., Voshchuk, G. Y. and Morokov, A. L. (2019). Development of a knowledge base in the "smart farming" system for agricultural enterprise management. *Procedia Computer Science*, 150, 154–161.

[13] Drela, K. (2020). Sustainable management of human resources in enterprises as an example of sustainable development. *Zeszyty Naukowe Akademii Morskiej w Szczecinie*.

[14] ur Rehman, M. H., Chang, V., Batool, A. and Wah, T. Y. (2016). Big data reduction framework for value creation in sustainable enterprises. *International journal of information management*, 36(6), 917–928.

[15] Raja, S. P., Sawicka, B., Stamenkovic, Z. and Mariammal, G. (2022). Crop prediction based on characteristics of the agricultural environment using various feature selection techniques and classifiers. *IEEE Access*, 10, 23625–23641.

[16] Liu, C., Yin, Y., Sun, Y. and Ersoy, O. K. (2022). Multi-scale ResNet and BiGRU automatic sleep staging based on attention mechanism. *Plos one*, 17(6), e0269500.

*Hybrid and Advanced Technologies – S. Prasad Jones Christydass, Nurhayati Nurhayati and S. Kannadhasan. (Eds) © 2025 the Authors, ISBN 978-1-032-90254-8*
*Open Access: www.taylorfrancis.com, CC BY-NC-ND 4.0 license*

# Digital marketing strategy to increase sales conversion on e-commerce platforms based on LSTM-RNN model

Nawab Akram
*Associate Professor, Magadh Professional Institute, Danapur, Patna, India*

Surjit Victor
*Assistant Professor, College of Administrative and Financial Sciences, University of Technology, Bahrain*

Devulapalli Raghava
*Associate Professor, Department of MBA, Vishwa Vishwani School of Business, Hakimpet, Hyderabad, Telangana, India*

Manu Vasudevan Unni
*Assistant Professor, Department of Management, St. Claret College, Autonomous, Bangalore, Karnataka, India*

Prasanthi Gottumukkala
*Assistant Professor, Department of Information Technology, Gokaraju Rangaraju Institute of Engineering and Technology, Hyderabad, India*

Shobhit Garg
*Assistant Professor, Department of Management Studies, Graphic Era Deemed to be University, Dehradun, India*

ABSTRACT: Modern technological developments, especially in the realm of online retail, have fundamentally changed the way businesses operate. When it comes to advertising products online, e-commerce platforms are among the most popular choices. In order to increase sales conversion and get an advantage in the increasingly competitive e-commerce sector, companies require a solid marketing strategy. Model training, feature extraction, and preprocessing are the three key components. The ERP system includes the daily clear value and amount of sales, which are extracted during data preprocessing. One way to extract important information from text is by using the TF-IDF method, which is commonly employed in feature extraction for advertising words. For better accuracy, we used the LSTM-RNN to train the model. The proposed method outperformed other methods such as RNN and LSTM, with an accuracy of 96.90%.

*Keywords*: Electronic Commerce (E-commerce), Search Engine Optimization (SEO), Social Media Marketing (SMM), Cost per Click (CPC), Term Frequency-Inverse Document Frequency (TF-IDF)

## 1 INTRODUCTION

One of the most dynamic and competitive industries in the modern digital era is online shopping. Due to changes in consumer behaviour and the widespread availability of the internet, new possibilities and challenges have arisen for businesses operating online. Consequently, improving online marketing strategies is crucial to increase sales and promote the e-commerce sector's long-term growth. Since the advent of the internet and other forms of

 DOI: 10.1201/9781003559115-60

information and communication technology, consumers' interactions with brands have changed. The proliferation of online communities and social media has caused abusiness context that calls for continuous adjustment. An effective and cohesive online marketing plan is crucial to the success of the business in this case. Consequently, this study will provide e-commerce enterprises with precise insights and practical answers that will help them enhance their online marketing for long-term effectiveness. In this day of fast technological advancement and globalization, there are huge changes happening in the way sales are conducted. In this age of limited resources and limited human interaction, the concept of selling has evolved substantially from its contemporary, internet-driven equivalent. This article demonstrates, using examples, how sales trends have evolved and diverged from their ancient counterparts.

## 2 LITERATURE SURVEY

However, digital capabilities are crucial for fostering international online trade. [1]If they wish to achieve great success in the online retail sector, e-retailers must discover ways to draw in clients and spread the word about their virtual store. [2]Digital marketing encompasses a wide range of strategies, including search engine optimization (SEO) and social media marketing (SMM) as well as campaign-based e-marketing (banner advertisements, SEO). [3]There is empirical evidence that shows that digital marketing campaigns that encourage spontaneous purchases lead to increased revenues. [4]An online retailer's value proposition in terms of creating, sharing, and providing value to stakeholders and customers is based on its digital marketing strategies, which are driven by the organization's strategy[5]. However, language, cultural, and consumer behaviour issues make the relationship more complex for e-retailers in foreign countries.[6]The expansion of the internet and other technical developments have led to the digitization of many products, both informational and otherwise, and as a result, businesses have revaluated their approaches to advertising. A new internet marketplace has emerged, which is posing a serious threat to established brick-and-mortar retailers [7]. The expansion of electronic commerce (E-commerce) can be attributed to the techniques employed by firms to tap into the global market through the electronic market [8]. When consumers purchase goods and services online through the use of electronic financial transfers and data transmissions, this practice is known as "e-commerce" [9]. A big advantage of e-commerce platforms is the ease with which product information can be found, pricing can be compared, and educated judgments can be made [10]. By mimicking customers' activities and experiences in physical stores, they hope to sway their purchasing decisions.[11]In this context, "digital inbound marketing" has evolved to denote a crucial component of any successful marketing plan. Staying ahead of the competition requires digital agility in online marketing and multichannel marketing[12]. Regardless, information about the ways in which various businesses use and engage with digital marketing tactics is scarce. [13]The literature on the interactions of marketing activities differs in a number of ways. One of these ways is the focus on how a company's marketing strategies interact with its reaction to the marketing results of its competitors. Using data collected from the oligopolistic detergent market, they draw the conclusion that rival companies react unequally to changes in advertising spending and product efficiency[14].

## 3 PROPOSED SYSTEM

The term "e-commerce" refers to any type of transaction that takes place fully on the internet, whether it's a "business to consumer" or "business to business" transaction, a digital payment, or an online store. These days, many modern sectors rely heavily on online buying, such as sales automation, retail, and payment processing. An abundance of opportunities exist to analyze patterns and devise lucrative strategies utilizing the vast amounts of data produced by different e-commerce platforms.

### 3.1 *Preprocessing*

Data descriptive analysis is conducted using Power BI desktop. It combined data from three tables in our analysis. There are three that use ERP data and two that rely on Google Analytics. In Google Analytics, the word "entrances" is used to indicate the total number of times a website has been accessed by users. The term "hits" is used to determine how many times a user engages with a website[15]. Bounce rate is the percentage of visitors that leave a website without taking any action. The advertisements simply divide the total number of ad impressions by the total number of ad clicks to get the CTR. All advertising expenditures added up are known as the "ads cost." CPC stands for cost per click.

### 3.2 *Feature extraction*

Effectively acquiring user information is the foundation of user attribute inference, and advertising is the key carrier of user demand information. So, it's crucial to know how to get the relevant details from advertising language. To determine how much of an impact a word has on its advertisement, one common and easy-to-understand metric is term frequency-inverse document frequency (TF-IDF). Consider the following equation, where $N$ is the advertising set and $D$ is the total number of ads in $N$:

$$M_{TF-IDF} = TF \times IDF \tag{1}$$

$$TF = \frac{count(l, n)}{size(n)} \tag{2}$$

$$IDF = \log\left(\frac{D}{docs(l, N)}\right) \tag{3}$$

### 3.3 *Model training*

#### 3.3.1 *RNN*

The connections between the nodes in a recurrent neural network occur repeatedly. Their internal memory is capable of handling the continuous flow of data. Applications such as speech recognition, text production, handwriting recognition, and the stock market are blown away by this. The proposed employs RNNs, or Recurrent Neural Networks, to take into consideration the interdependencies of the stock data over the long term. However, if there isn't enough data in the memory, a Vanishing Gradient descent problem can occur, and the neural net's data would vanish as it iterates further.

#### 3.3.2 *LSTM*

The long short-term memory (LSTM) network is one of several new building blocks that have emerged from advancements in the fundamental recurrent neural network; these will be useful for future serial analysis. It is possible to classify LSTM blocks as recurrent networks on their own due to the presence of recurrent connections that are comparable to those of classic recurrent neural networks. Building LSTMs as recurrent neural network architecture allows them to better account for long-term dependencies compared to typical RNNs. Predictions of stock data or speech recognition movements have been demonstrated to be better by LSTM and RNN than by DNN and basic RNN models. Traditional DNNs aren't up to the challenge of modeling stock data since they can't model outside of a fixed-sized sliding window where the network is autonomous from its predecessors. Two sets of data are created after data normalization: one for testing and one for training. In order to forecast or predict the stock data sequence, the built LSTM RNN model is then trained using the training data.

## 4 RESULT AND DISCUSSION

Companies need to implement effective digital marketing strategies if they want to keep or grow their market share in this era of ubiquitous internet purchasing. In order to boost sales in the online business world, this research seeks to examine and discover successful digital marketing tactics. The suggested study employs descriptive methodologies and a qualitative approach.

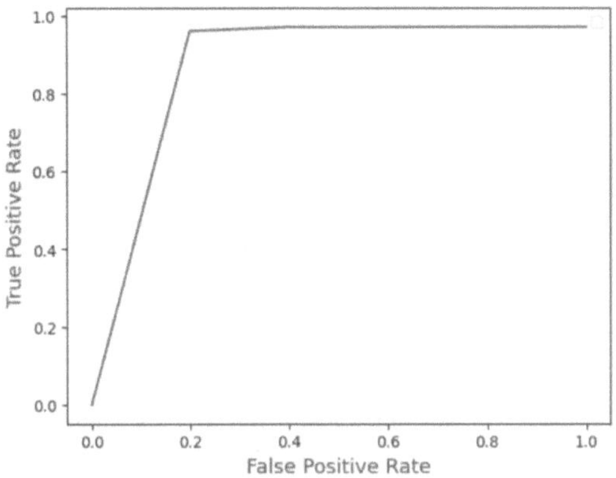


Figure 1. ROC curve of proposed model.

An improved ROC curve for machine learning is one with a larger area under the curve. A link between the True Positive Rate and the False Positive Rate is illustrated by the ROC curve in Figure 1.

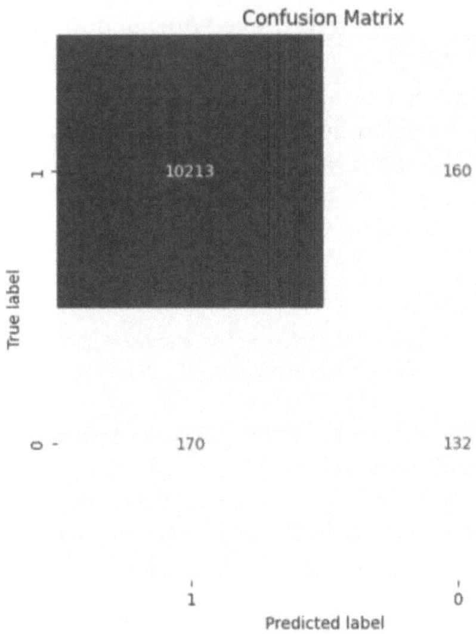


Figure 2. Confusion matrix for LSTM-RNN method.

The model's performance may be tracked using the confusion matrix, which is displayed in Figure 2. It displays both the original and projected data. Goodness of fit is symbolized by the model's efficiency, which reached 96.90%.

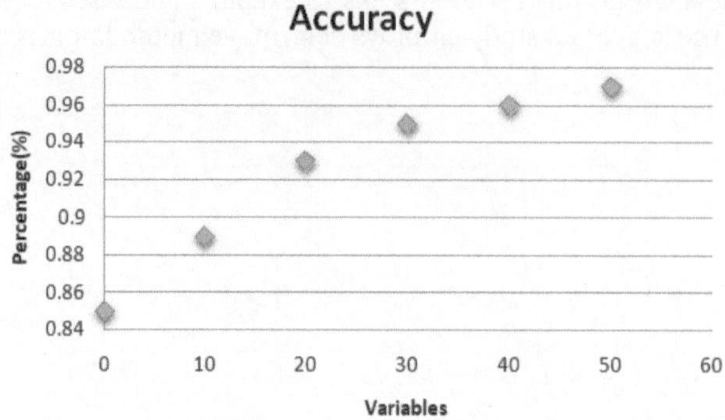


Figure 3. Different numbers of predictors in cross-validation plots.

The model's optimal number of predictors was 40–50, as shown in Figure 3. Accordingly, prior to modelling, only the top predictors are retained, while all others are eliminated.

## 4 CONCLUSION

In light of the challenges presented by the ever-increasing level of competition in the ever-expanding e-commerce sector, this research seeks to boost sales and aid in the expansion of e-commerce businesses through the investigation, discovery, and optimization of online marketing strategies. To better inform online business owners, this study aims to shed light on the dynamics of the technical improvements, digital market, and customer behaviour. The amount of sales and daily clear value are retrieved during data preparation and included in the ERP system. The TF-IDF method is a popular tool for feature extraction of advertising terms and can be used to extract crucial information from text. In order to train its models, it employs LSTM-RNN. With an average accuracy of 96.90%, the proposed technique surpasses both the LSTM and RNN models.

## REFERENCES


[1] Bamiatzi, V. C. and Kirchmaier, T. (2014). Strategies for superior performance under adverse conditions: A focus on small and medium-sized high-growth firms. *International Small Business Journal*, 32 (3), 259–284.

[2] Banerjee, S., Prabhu, J. C. and Chandy, R. K. (2015). Indirect learning: How emerging-market firms grow in developed markets. *Journal of Marketing*, 79(1), 10–28.

[3] Bocconcelli, R., Cioppi, M., Fortezza, F., Francioni, B., Pagano, A., Savelli, E. and Splendiani, S. (2018). SMEs and marketing: a systematic literature review. *International Journal of Management Reviews*, 20(2), 227–254.

[4] Colton, D. A., Roth, M. S. and Bearden, W. O. (2010). Drivers of international e-tail performance: the complexities of orientations and resources. *Journal of International Marketing*, 18(1), 1–22.

[5] Eggers, F., Hatak, I., Kraus, S. and Niemand, T. (2017). Technologies that support marketing and market development in SMEs—Evidence from social networks. *Journal of Small Business Management*, 55(2), 270–302.

[6] Varadarajan, R. and Yadav, M. S. (2009). Marketing strategy in an internet-enabled environment: a retrospective on the first ten years of JIM and a prospective on the next ten years. *Journal of Interactive Marketing*, 23(1), 11–22.

[7] Tsagkias, M., King, T. H., Kallumadi, S., Murdock, V. and de Rijke, M. (2021, February). Challenges and research opportunities in ecommerce search and recommendations. In *ACM Sigir Forum* (Vol. 54, No. 1, pp. 1–23). New York, NY, USA: ACM.

[8] Kong, X. T., Zhong, R. Y., Zhao, Z., Shao, S., Li, M., Lin, P. ... and Huang, G. Q. (2020). Cyber physical ecommerce logistics system: An implementation case in Hong Kong. *Computers & Industrial Engineering*, 139, 106170.

[9] Rosario, A. M. F. T. and Cruz, R. N. (2019). Determinants of innovation in digital marketing. *Journal of Reviews on Global Economics*, 8(1), 1722–1731.

[10] Daries, N., Cristobal-Fransi, E. and Ferrer-Rosell, B. (2020). Implementation of website marketing strategies in sports tourism: Analysis of the online presence and E-commerce of golf courses. *Journal of Theoretical and Applied Electronic Commerce Research*, 16(3), 542–561.

[11] Bleoju, G., Capatina, A., Rancati, E. and Lesca, N. (2016). Exploring organizational propensity toward inbound–outbound marketing techniques adoption: The case of pure players and click and mortar companies. *Journal of Business Research*, 69(11), 5524–5528.

[12] Dev, C., Zhou, K. Z., Brown, J. and Agarwal, S. (2009). Customer orientation or competitor orientation: Which marketing strategy has a higher payoff for hotel brands?. *Cornell Hospitality Quarterly*, 50 (1), 19–28.

[13] Gao, J., Siddik, A. B., Khawar Abbas, S., Hamayun, M., Masukujjaman, M. and Alam, S. S. (2023). Impact of E-commerce and digital marketing adoption on the financial and sustainability performance of MSMEs during the COVID-19 pandemic: An empirical study. *Sustainability*, 15(2), 1594.

[14] Pawar, K., Jalem, R. S. and Tiwari, V. (2019). Stock market price prediction using LSTM RNN. *In Emerging Trends in Expert Applications and Security: Proceedings of ICETEAS 2018* (pp. 493–503). Springer Singapore.

[15] Kagiampaslidis, M. C. (2020). Visualizing Google Analytics', *Digital Marketing Campaigns' and ERP system's data using BI tools.*

*Hybrid and Advanced Technologies – S. Prasad Jones Christydass, Nurhayati Nurhayati and S. Kannadhasan. (Eds) © 2025 the Authors, ISBN 978-1-032-90254-8*


# Assessing internet addiction and depression among university students using the CNN-BiLSTM model with a focus on digital marketing influences

R. Arun
*Assistant Professor, Department of MBA, St.Joseph's College of Engineering, Chennai, Tamil Nadu, India*
ORCID ID: 0000-0002-5252-1030

S. Avinash
*Assistant Professor, Department of MBA, St.Joseph's College of Engineering, Chennai, Tamil Nadu, India*

V. Saikumari
*Professor & Head, Department of Management Studies, Easwari Engineering College, Chennai, Tamil Nadu, India*
ORCID ID: 0000-0002-0441-4425

A. Sathish
*Assistant Professor, Department of MBA, St.Joseph's College of Engineering, Chennai, Tamil Nadu, India*

P. Shanthi
*Assistant Professor & Research Supervisor, Department of Commerce(CS & AF), Faculty of Science & Humanities, SRM Institute of Science & Technology, Kattankulathur, Chengalpattu, Tamil Nadu, India*
ORCID ID: 0000-0002-9998-5940

P. Perumal
*Assistant Professor Department of Commerce(Honours), Dwaraka Doss Goverdhan Doss Vaishnav College Arumbakkam, Chennai, Tamil Nadu India*
ORCID ID: 0009-0006-8692-5929

ABSTRACT: The purpose of this study was to investigate the interplay between Internet addiction and depression, personality disorders, poor self-esteem, and dysfunctional attitudes among college students. So far as the researchers can tell, this proposed shows that Internet addiction is associated with dysfunctional attitudes. Assessing a person's personality, self-esteem, depression, and dysfunctional attitudes may help those who struggle with Internet addiction. By addressing these issues, cognitive behavioral therapy can help those with Internet addiction. Model training, feature extraction, and preprocessing are the three primary components. Although data preparation is crucial to data mining, it is also a laborious and time-consuming procedure. Following data preparation, the next step is to identify and extract relevant features that are associated with the mental state of the participants. To improve the model's accuracy, we trained it using the CNN-BiLSTM. In comparison to state-of-the-art approaches like CNN and BiLSTM, the suggested approach attained a superior accuracy of 96.64%.


DOI: 10.1201/9781003559115-61

*Keywords*: Long short-term memory (LSTM), Distributed Fourier Transform (DFT), Convolutional Neural Network (CNN), Recurrent Neural Network (RNN), Natural Language Processing (NLP)

## 1 INTRODUCTION

The internet is often considered to be the most useful tool for many fields, including academia, business, entertainment, and politics. It is often considered that young people make up the bulk of internet users. Some have argued that compulsive internet use is the root cause of addiction in young people, on par with alcoholics, drug addicts, and compulsive gamblers. One of the negative effects of the internet, is online addiction, a new sort of mental disorder that causes difficulties with both mental and physical health.The mental health and relationships of individuals can suffer when they are unable to control the amount of time they spend online. Internet addiction is the medical term for this problem. The phrase "internet addiction", who drew parallels between pathological gambling and obsessive internet use, and described it as a disorder of impulse control that is unconnected to drugs. Previous research has shown that adolescent and adult internet addiction is associated with elevated risk of depression and suicide, respectively. While there has been little published empirical research on the link between internet addiction and mental health problems, this study does assess the correlation between the two in a group of undergraduates from Nigerian universities. This study will add to the increasing amount of data suggesting college freshman are more susceptible to the negative psychological repercussions associated with internet overuse.

## 2 LITERATURE SURVEY

Over the past twenty years, internet use has surged, impacting almost every facet of people's lives across the world. [1]Especially for young adults and teenagers, this technology facilitates rapid and easy access to knowledge, interpersonal connections, amusement, and social bonds [2]. Anxieties, intimacy, and the critical developmental responsibilities of redefining relationships with family, friends, and society, among others, may be easier for young adults to handle with the help of the internet, according to some writers [3]. Regardless of these benefits, studies reveal that some young individuals, especially college students, have an unhealthy fixation on or problematic use of the internet or rely on it constantly [4]. The gravity of the phenomenon necessitates further research into the processes that may be responsible for starting and maintaining IA. [5]Within the theoretical framework of developmental psychopathology, one can find a variety of overt and covert forms of psychopathology, including IA.A lot of people are talking about "internet addiction" and "excessive internet use" these days. [6]The hallmarks of internet addiction (IA) are uncontrolled or excessively disruptive behaviours connected to internet use and access, as well as anxieties or obsessions with the subject. [7]Overuse of the internet occurs when people's reliance on it becomes outmoded and negatively affects their day-to-day lives due to its pervasiveness, inefficiency, and time consumption. Diagnostic criteria for Internet addiction were proposed by [8]and include withdrawal, decreased planning abilities, intolerance, preoccupation, impaired control, and excessive online time as key symptoms. Constant and excessive use of the internet may have negative effects on mental health, according to studies [9]. These include obesity, low self-esteem, depression, poor sleep quality, mood swings, and other mental health difficulties[10]. With that in mind, we set out to quantify the prevalence of internet addiction, depression, and poor sleep quality among medical students and interns, as well as to explore any potential associations between these conditions[11]. It has also decreased people's nighttime sleep, increased their daytime fatigue, and decreased their productivity at work [12]. Internet addiction has recently surfaced as a major issue in contemporary culture. Numerous unfavourable health effects have been associated with it, such as but not limited to: sadness, insomnia, erratic mood swings, poor self-esteem, impulsivity, [13]thoughts of or attempts at suicide, lack of physical activity, and a slew of lifestyle problems. Some of the signs of being addicted to the internet include: worrying too much, having a low quality of life, not getting enough exercise, and having trouble communicating and forming meaningful relationships with others [14]. Adolescents and younger children require adequate sleep for proper bodily development, in addition to the obvious benefits to the mind and body from restful sleep.

## 3 PROPOSED SYSTEM

The purpose of this research is to investigate whether there is a correlation between excessive use of the Internet and mental health. The Online Cognition Scales and the Scales of Psychological Well-Being were among the many questionnaires that 480 college students were asked to complete as part of the study. In their search for links between Internet addiction and psychological well-being, the researchers turned to multiple regression and correlation analysis.

### 3.1 *Preprocessing*

Data mining relies on well-prepared data, yet this process is tedious and time-consuming. Previous work has shown that efficient data pre-processing decreases the effort needed to build the model. Data preprocessing is necessary since the prediction results are affected by many forms of noise, missing, redundant, etc., that are present in the dataset[15]. A lack of comprehensive or accurate information in the dataset could be due to a number of factors. The data for a certain period will be lost if students fail to update their cards when they expire. Another potential cause of data loss and inaccurate entry could be a mechanical breakdown. The result of these outliers is excessive duplication of data in the repository. Because there is a lot of missing or erroneous data, preprocessing is essential.

### 3.2 *Feature extraction*

Once the data has been prepared, the following stage is to find and extract pertinent elements linked to the participants‘ mental states. The categorization of internet users relies heavily on these features[16]. At this point, we will extract features using two methods: Distributed Fourier Transform (DFT) and K-means clustering. In order to get the first time-frequency features, DFT helps, then clustering exposes the most relevant and helpful features, which are more closely linked to mental condition.

### 3.3 *Model training*

#### 3.3.1 *BiLSTM*

LSTM is a well-liked architecture for recurrent neural networks that models sequential data. It is built to capture long-term dependencies more effectively than the vanilla RNN model. Long short-term memory (LSTM) networks work in the same way as other types of recurrent neural networks; at each time-step, the network takes in data from the previous time-step and processes it to generate an output. Classification thereafter makes use of the hidden layer from the most recent time step, or perhaps all hidden layers. The connection between input, hidden states, and various gates can be illustrated as:

$$g_w = \tau\left(T^{(g)}c_w + P^{(g)}y_{w-1} + a^{(g)}\right) \tag{1}$$

$$k_w = \tau\left(T^{(k)}c_w + P^{(k)}y_{w-1} + a^{(k)}\right) \tag{2}$$

$$e_w = \tau\left(T^{(e)}c_w + P^{(e)}y_{w-1} + a^{(e)}\right) \tag{3}$$

$$x_w = g_w \odot x_{w-1} + k_w \odot \tan y\left(T^{(x)}c_w + P^{(x)}y_{w-1} + a^{(x)}\right) \tag{4}$$

$$y_w = e_w \odot \tan y(x_w) \tag{5}$$

### 3.3.2 *CNN*

Further to that, our proposed system is built upon convolutional neural networks (CNNs). It's no secret that CNNs have been killing it lately when it comes to solving difficulties in computer vision and NLP. Particularly impressive is their mastery of feature maps, which they employ to great effect in exploiting local correlation and pattern in the data. One of the first studies to show that convolutional neural networks (CNNs) could successfully classify texts. By stacking embeddings from different words in a sentence or paragraph, a two-dimensional array is commonly created when CNN is used for text classification. Applying convolution filters of different lengths to a window of h words creates a new feature representation. Next, the pooled features from different filters are combined and additional features are pooled, usually via max-pooling, to construct the concealed representation. The final forecast is achieved by utilizing one or more completely connected layers following these representations.

## 4 RESULT AND DISCUSSION

Worldwide studies have shown that comorbidities, difficulties with self-regulation, and interpersonal dynamics all raise the risk of internet addiction (IA) in young adults. Few studies have examined how impulsivity, psychopathology, and dysfunctional families affect young people.

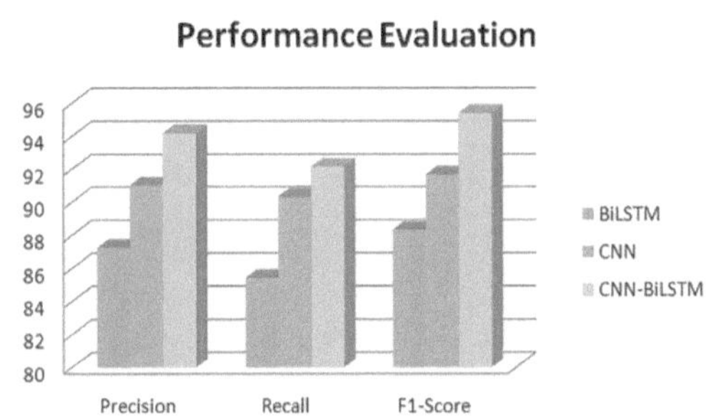


Figure 1. Performance evaluation of CNN-BiLSTM Model.

Figure 1 indicates that while using the Everytime dataset, the CNN-BiLSTM based classification achieves remarkable accuracy recall and F1-score.

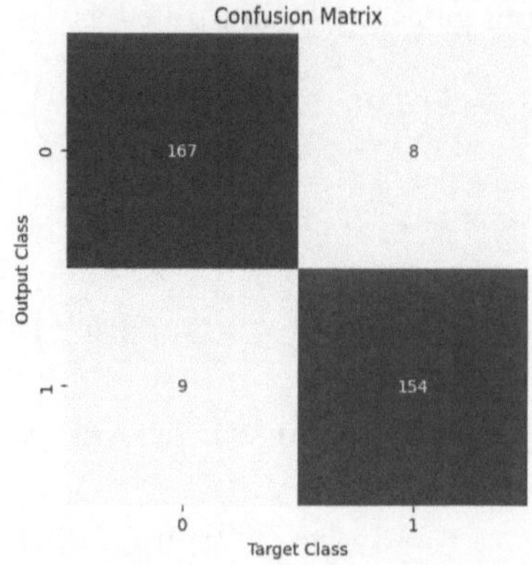


Figure 2. Confusion matrix of the suggested method.

The results of the confusion matrix are shown in Figure 2. The output class and the target class are the two types of representation.

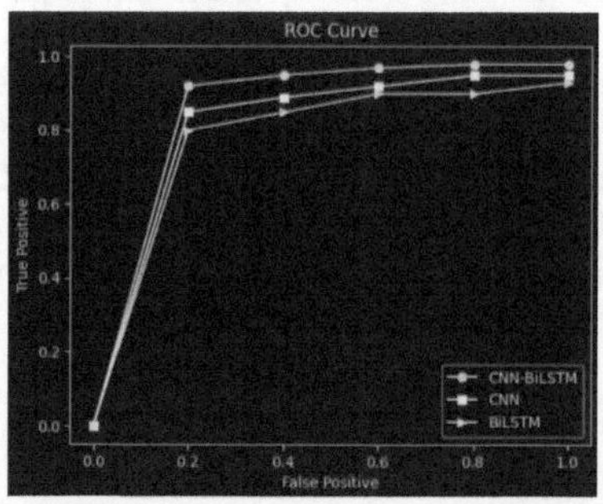


Figure 3. ROC curve of three models.

The ROC curves of three proposed models built with different kinds of features are shown in Figure 3. The Y-axis shows the false positive rate (1-specificity) and the X-axis shows the true positive rate (sensitivity). An increased area under the curve (AUC) indicates a superior model.

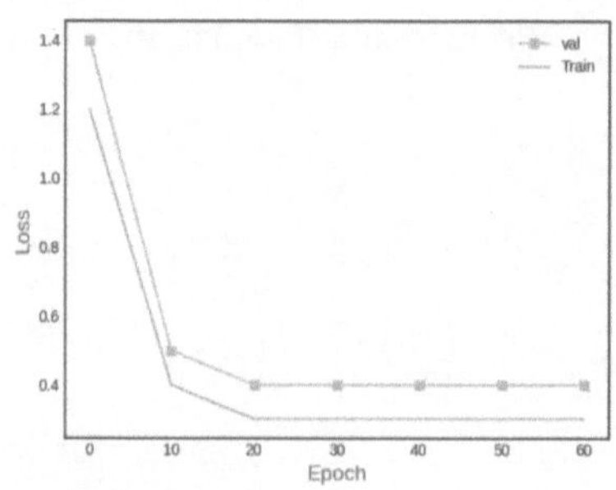


Figure 4. Model loss of CNN-BiLSTM model.

The loss graph is displayed in Figure 4. There is a clear trend of lowering training loss followed by decreasing test loss up to a certain period.

## 5 CONCLUSION

The primary motivation for this research is to discover whether there is a correlation between internet addiction and mental health problems such anxiety, depression, and stress. One medium-sized Turkish public university sent 290 undergraduates to take part. The participants in this proposed were asked to fill out depression and anxiety scales. Anxiety, depression, and stress were all positively correlated with internet addiction in a correlation research. Data preparation is a tedious and time-consuming process, yet it is essential to data mining. Finding and extracting pertinent features linked to the participants' mental states is the next stage after data preparation. For training the model, it employs CNN-BiLSTM. The proposed approach outperforms the CNN and BiLSTM models in terms of average accuracy, coming in at 96.64%.

## REFERENCES


[1] Gervasi, A. M., La Marca, L., Lombardo, E., Mannino, G., Iacolino, C. and Schimmenti, A. (2017). Maladaptive personality traits and internet addiction symptoms among young adults: A study based on the alternative DSM-5 model for personality disorders. *Clinical Neuropsychiatry*, 14(1).

[2] Şahin, C. (2011). An analysis of internet addiction levels of individuals according to various variables.

[3] Tateno, M., Teo, A. R., Ukai, W., Kanazawa, J., Katsuki, R., Kubo, H. and Kato, T. A. (2019). Internet addiction, smartphone addiction, and Hikikomori trait in Japanese young adult: social isolation and social network. *Frontiers in Psychiatry*, 10, 455.

[4] Griffiths, M. D., Kuss, D. J., Billieux, J. and Pontes, H. M. (2016). The evolution of Internet addiction: A global perspective. *Addictive Behaviors*, 53, 193–195.

[5] Yan, W., Li, Y. and Sui, N. (2014). The relationship between recent stressful life events, personality traits, perceived family functioning and internet addiction among college students. *Stress and Health*, 30 (1), 3–11.

[6] Canan, F., Ataoglu, A., Ozcetin, A. and Icmeli, C. (2012). The association between Internet addiction and dissociation among Turkish college students. *Comprehensive Psychiatry*, 53(5), 422–426.

[7] Canan, F., Yildirim, O., Sinani, G., Ozturk, O., Ustunel, T. Y. and Ataoglu, A. (2013). Internet addiction and sleep disturbance symptoms among Turkish high school students. *Sleep and Biological Rhythms*, 11, 210–213.

[8] Chakraborty, K., Basu, D. and Kumar, K. V. (2010). Internet addiction: consensus, controversies, and the way ahead. *East Asian Archives of Psychiatry*, 20(3), 123–132.

[9] Ghamari, F., Mohammadbeigi, A., Mohammadsalehi, N. and Hashiani, A. A. (2011). Internet addiction and modeling its risk factors in medical students, Iran. *Indian Journal of Psychological Medicine*, 33(2), 158–162.

[10] Ho, R. C., Zhang, M. W., Tsang, T. Y., Toh, A. H., Pan, F., Lu, Y., ... & Mak, K. K. (2014). The association between internet addiction and psychiatric co-morbidity: A meta-analysis. *BMC Psychiatry*, 14, 1–10.

[11] Minaee, S., Azimi, E. and Abdolrashidi, A. (2019). Deep-sentiment: Sentiment analysis using ensemble of cnn and bi-lstm models. *arXiv preprint arXiv*:1904.04206.

[12] Zhang, M. W., Lim, R. B., Lee, C. and Ho, R. C. (2018). Prevalence of internet addiction in medical students: a meta-analysis. *Academic Psychiatry*, 42, 88–93.

[13] Kim, K., Lee, H., Hong, J. P., Cho, M. J., Fava, M., Mischoulon, D., ... & Jeon, H. J. (2017). Poor sleep quality and suicide attempt among adults with internet addiction: A nationwide community sample of Korea. *PLoS one*, 12(4), e0174619.

[14] Zhao, Y., Qu, D., Chen, S. and Chi, X. (2023). Network analysis of internet addiction and depression among Chinese college students during the COVID-19 pandemic: A longitudinal study. *Computers in Human Behavior*, 138, 107424.

[15] Peiqing, H. (2022). [Retracted] Multidimensional State Data Reduction and Evaluation of College Students' Mental Health Based on SVM. *Journal of Mathematics*, 2022(1), 4961203.

[16] Zhu, C., Li, B., Li, A. and Zhu, T. (2019, January). Detecting depression from Internet behaviors by time-frequency features. In *Web Intelligence* (Vol. 17, No. 3, pp. 199–208). IOS Press.

*Hybrid and Advanced Technologies – S. Prasad Jones Christydass, Nurhayati Nurhayati and S. Kannadhasan. (Eds)*  *ISBN 978-1-032-90254-8*


# Economic benefits of enterprise resource planning systems: An empirical analysis based on lightweight DNN-SMO model

Devadutta Indoria
*Professor & H. O. D., Department of Commerce, Vikram Dev University, Jeyoore, Odisha, India*

Rakesh Kumar & Nawab Akram
*Associate Professor, Magadh Professional Institute, Danapur, Patna, India*

Babayeva Khazangul Süleyman
*Ph.D. in Pedagogy, Department of Economics and Management, Azerbaijan State University of Economics (UNEC), Zagatala Branch*

Amit Chauhan
*Department of Life Sciences, CHRIST (Deemed to be University), Bengaluru, Karnataka, India*

Vivek Ravishabkar Dubey
*Professor, Department of C.S.E, Jai Narain College of Technology (JNCT), Bhopal, India*

ABSTRACT: Modern businesses rely on enterprise resource planning (ERP) systems to facilitate cross-functional integration and the efficient processing of company data. These systems integrate different functional areas of information, such as marketing, operations, and finance, into a unified suite of business applications. Management that faces challenges with incompatible information systems and inconsistent operations policies may find ERP systems to be an appealing alternative. A three-step methodology is proposed, which consists of preprocessing, feature selection, and model training. Use of a combination of alphanumeric characters represents data preparation in ERP system logs. An SPL is a set of software-intensive systems used in feature selection; it allows users to configure software products with similar characteristics to meet the needs of a specific business market. Incorporating the DNN-SMO into the model training procedure increased its precision. With a rate of 96.47%, the suggested method outperforms both DNN and SMO.

*Keywords*: Enterprise Resource Planning (ERP), Spider Monkey Optimisation (SMO), Distributed Neural Networks (DNN), Fission-Fusion Social Structure (FFSS), Economics

## 1 INTRODUCTION

Data types gathered by organizations across industries include invoices, client details, project schedules, shipment details, payment details, and many more. Streamlining business processes and cutting down on manual labour are two of the primary goals of Enterprise Resource Planning (ERP) systems across all sectors. The ERP industry is undergoing a period of transformation due to the proliferation of big data and the Internet of Things (IoT). This disruption is bringing about prospects for process automation and operational excellence. While everyone is talking about how linked manufacturing is becoming, we can't

 DOI: 10.1201/9781003559115-62

help but notice how the Internet of Things (IoT) is influencing ERP systems and business processes. Computers are able to identify and control IoT components, connect with each other and pertinent databases, and more all because each component is given an Internet Protocol (IP) address. Because ERP relies on cloud storage, it is able to manage, control, and process data. Integration of ERP with the Internet of Things offers numerous substantial opportunities, including better administration, automation, product tracing, and decreased ERP adoption costs.

## 2 LITERATURE SURVEY

We are still in the early phases of researching the S-ERP system. [1]Researching previous work on sustainability and ERP implementation guidelines is essential for generating substantial and rigorous suggestions for the development of the S-ERP guideline framework[2]. Topics covered in this section include project management, sustainability, organizational decision levels, and strategic management, all of which are based on the chosen principles. [3] asserts that S-ERP is a cutting-edge information system that, to boost productivity and ensure the company's longevity, unites the entire corporate value chain the outcomes. [4] Using the S-ERP system, businesses may assess and disclose how their extended value chains impact sustainability. Managers, monitors, and reporters all find it helpful the company's stock, production, delivery, and assets by integrating data from both inside and outside the organization[5]. For decades, people have been talking about enterprise resource planning (ERP) systems and platforms; nonetheless.[6] A few instances of how ERP arguments have supposedly contributed to the system's demise. This is because ERPs are now considered outdated by newer, more focused cloud service apps [7]. For ERPs, say Rest and Martinez, are vital to the prosperity of companies that run their operations with the help of integrated business processes. Enterprise resource planning (ERP) is a concept and a system that can be defined by[8]. At their most basic level, ERP systems "involve the integration of business processes within an organization, with improved order management and control, accurate information on inventory, improved workflow and SCM, and better standardization of business and best practices"[9]. The ERP notion draws attention to the institutional shifts that take place throughout the onset and upkeep of the phenomenon in this regard[10]. "Technological infrastructure designed to provide the required functional capability required to turn the ERP concept into a reality" is a common way to describe enterprise resource planning (ERP) systems[11]. In this sense, a company-wide software platform that integrates databases is referred to as an enterprise resource planning (ERP) system.[12]An overview of relevant literature is provided in this section, which serves as the main contribution of the paper by analysing important topics and laying the framework for future research. Block chain research is centred on accounting and finance, the central issue and common thread of enterprise resource planning (ERP), asset information system (AIS), and electronic procurement[13]. We only scratch the surface of block chain technology's many potential applications in ERP, skipping over the many other general business purposes. This also ensures that the company's financial results can be measured[14].

## 3 PROPOSED SYSTEM

Businesses have had little choice but to adapt to the new realities brought about by the rise of IT-led industries and the fierce rivalry within them. In today's competitive market, enterprise resource planning (ERP) systems stand out for their ability to cut costs, speed up processes, and (debatably) make customers happier.

### 3.1 *Preprocessing*

The ERP system logs are usually represented by a mix of alphanumeric characters. Regrettably, certain representations cannot be inputted into computational models. As a result, converting the system logs to signals is essential for deep learning models[15]. Using one-hot encoding on the raw system logs is the quickest and easiest approach to turn any categorized records into signals. Unfortunately, because to its seeming simplicity in methodology, one-hot vector encoding is occasionally held responsible for the curse of dimensionality when a new class is added.

### 3.2 *Feature selection*

#### 3.2.1 *Software products lines*

One definition of a "software product line" (SPL) is a collection of interconnected software systems designed to address the demands of a certain marketplace. An SPL consists of two phases: DE and AE. The distinctive and common features of the SPL outputs are discovered by DE by focusing on the problem space[16]. Utilising the domain artefacts once again and capitalising on the SPL variability, AE crafts an SPL product. A configuration changes a domain model to specify an application product, turning DE into AE. Because it represents a distinct SPL configuration, each distinct product derivation is referred to as a variant. In certain respects, these variants are distinct from one another. For SPLs to be effective, it is essential that the domain be well-modelled, including any parallels and distinctions. It is not novel to incorporate elements into an SPL model.

### 3.3 *DNN-SMO model training*

#### 3.3.1 *DNN*

A turning point in the development of AI theory and application was reached with the invention of an ANN. For quite some time, researchers have been steadily working towards the technological evolution of ANN computing. This ceaseless quest yielded the foundation for the deep learning system. Many hidden layers make up the feed-forward ANN‘s DNN architecture. Input, output, and more than three hidden layers make up distributed neural networks (DNNs).

#### 3.3.2 *SMO*

A metaheuristic method for optimising spider monkey optimisation (SMO) foraging that mimics the social behaviour of spider monkeys and employs a swarm intelligence technique based on fission and fusion.

### 3.4 *Dimension reduction using SMO*

The fission-fusion social structure (FFSS) of spider monkeys serves as an inspiration for SMO, which is a mixed-iterative process. Many studies have developed deep neural network algorithms that do not rely on feature engineering or conventional methods of dimension reduction. The problem of model over-fitting becomes noticeable when training a DNN with a high number of features. In order to reduce the dimensionality, the suggested model employs a behavior-based strategy called spider monkey optimization, which is based on FFSS. This solves the issue.In order to get the best possible classifier accuracy, the suggested strategy had to be applied to a subset of dimensions. In order to determine the SMO's classification error, this study uses accuracy as a performance metric. To determine which subset of dimensions maximizes classification accuracy post dimensionality reduction, the fitness function FD takes into account all possible subsets during evolutionary model training. The fitness function FD is defined for SMO‘s individual cause measurement as follows:

$$P_N = \alpha + (1 - C) + (1 - \alpha) * Q/N \quad (1)$$

## 4 RESULT AND DISCUSSION

Business benefiting from enterprise resource planning (ERP) systems is supported by these results. It uses a modified value chain methodology to quantify the benefits of ERP system deployment by determining a set of ratios for each process stage.

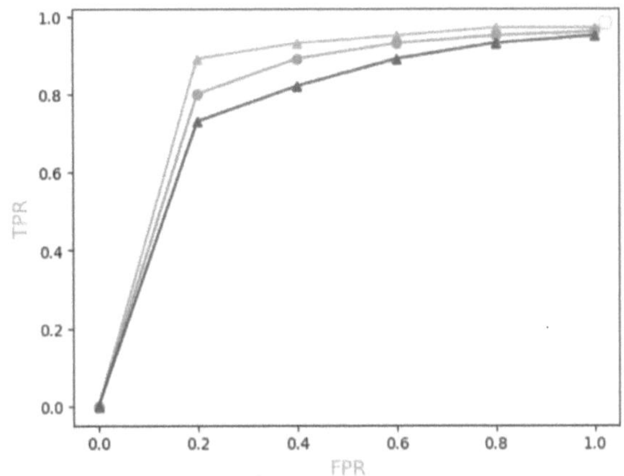


Figure 1. ROC curves for the suggested approaches.

Pictured in Figure 1 are ROC curves that show the quantitative performance as well as the performance according to various discrimination criteria. The ROC curve for this method is higher than that of the competing methods.

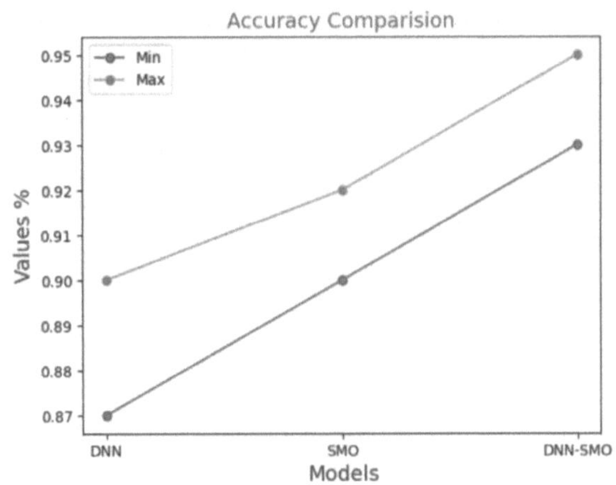


Figure 2. Accuracy comparison of proposed method.

Accuracy, here defined as the proportion of correct classifications across all classifications, was measured for each model and shown in Figure 2 in descending order of maximum accuracy.

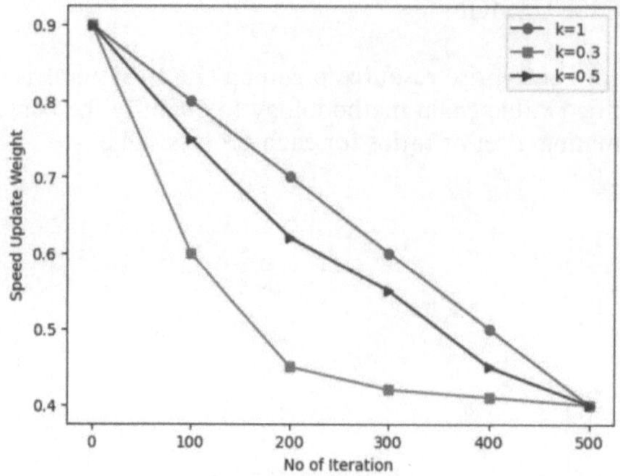


Figure 3. Comparison plot for different K values.

Figure 3 shows the results of a series of experiments that were conducted to determine an appropriate value of k. The results demonstrate that k is 0.3 and the effect is better. When compared to the linear modification of the formula with k = 1, the enhanced method is clearly superior.

## 5 CONCLUSION

In this proposed to take a look at two different businesses and how they used ERP systems to try to integrate management and control. We show that the pursuit of integration is continuous and that it occurs in spurts. Remote visibility and control are not the only aspects of integration. While enterprise resource planning (ERP) systems do not specify or outline the development of integration, they do impose a techno-logic that limits the usage of financial and non-financial representations for control purposes by differentiating between accounting and logistical modes. The ERP system logs show data preparation using a mix of alphanumeric characters. By assembling software-intensive systems into a software product line (SPL), customers are able to select software products with shared features and customize them to meet the demands of a certain business market. The accuracy of the trained model was improved by including the DNN-SMO. The proposed approach beats DNN and SMO with a rate of 96.47%.

## REFERENCES


[1] Chofreh, A. G., Goni, F. A., Shaharoun, A. M., Ismail, S. and Klemeš, J. J. (2014). Sustainable enterprise resource planning: Imperatives and research directions. *Journal of Cleaner Production*, 71, 139–147.

[2] Chofreh, A. G., Goni, F. A. and Klemeš, J. J. (2018). Sustainable enterprise resource planning systems implementation: A framework development. *Journal of Cleaner Production*, 198, 1345–1354.

[3] Chofreh, A. G., Goni, F. A. and Klemeš, J. J. (2018). Evaluation of a framework for sustainable Enterprise Resource Planning systems implementation. *Journal of Cleaner Production*, 190, 778–786.

[4] Chofreh, A. G., Goni, F. A. and Klemeš, J. J. (2018). A roadmap for Sustainable Enterprise Resource Planning systems implementation (part III). *Journal of Cleaner PRODUCTION*, 174, 1325–1337.

[5] Chofreh, A. G., Goni, F. A., Ismail, S., Shaharoun, A. M., Klemeš, J. J. and Zeinalnezhad, M. (2016). A master plan for the implementation of sustainable enterprise resource planning systems (part I): Concept and methodology. *Journal of Cleaner Production*, 136, 176–182.

[6] Bjelland, E. and Haddara, M. (2018). Evolution of ERP systems in the cloud: A study on system updates. *Systems*, 6(2), 22.

[7] Callejas, J. F. and Terzi, C. (2012). *Review of Enterprise Resource Planning (ERP) Systems in United Nations Organizations*. Rapor: JIU/REP/2012/8, United Nations, Geneva.

[8] Abd Elmonem, M. A., Nasr, E. S. and Geith, M. H. (2016). Benefits and challenges of cloud ERP systems–A systematic literature review. *Future Computing and Informatics Journal*, 1(1–2), 1–9.

[9] She, W. and Thuraisingham, B. (2007). Security for enterprise resource planning systems. *Information Systems Security*, 16(3), 152–163.

[10] Khare, N., Devan, P., Chowdhary, C. L., Bhattacharya, S., Singh, G., Singh, S. and Yoon, B. (2020). Smo-dnn: Spider monkey optimization and deep neural network hybrid classifier model for intrusion detection. *Electronics*, 9(4), 692.

[11] Hansen, H. F., Haddara, M. and Langseth, M. (2023). Investigating ERP system customization: A focus on cloud-ERP. *Procedia Computer Science*, 219, 915–923.

[12] Uddin, M. A., Alam, M. S., Al Mamun, A. and Akter, A. (2020). A study of the adoption and implementation of enterprise resource planning (ERP): Identification of moderators and mediator. *Journal of Open Innovation: Technology, Market, and Complexity*, 6(1), 2.

[13] Grabski, S. V., Leech, S. A. and Schmidt, P. J. (2011). A review of ERP research: A future agenda for accounting information systems. *Journal of information systems*, 25(1), 37–78.

[14] Billyan, B. F. and Irawan, M. I. (2021, March). Analysis of technology acceptance of enterprise resource planning (ERP) system in the regional office of PT. XYZ throughout Indonesia. In *Journal of Physics: Conference Series* (Vol. 1844, No. 1, p. 012008). IOP Publishing.

[15] Yu, J., Kim, M., Oh, H. and Yang, J. (2021). Real-time abnormal insider event detection on enterprise resource planning systems via predictive auto-regression model. *IEEE Access*, 9, 62276–62284.

[16] Afzal, U., Mahmood, T., Khan, A. H., Jan, S., Rasool, R. U., Qamar, A. M. and Khan, R. U. (2020). Feature selection optimization in software product lines. *IEEE Access*, 8, 160231–160250.

*Hybrid and Advanced Technologies – S. Prasad Jones Christydass, Nurhayati Nurhayati and S. Kannadhasan. (Eds)* © *2025 the Authors, ISBN 978-1-032-90254-8*


# Impact of enterprise resource planning systems on management control systems and firm performance using 1DRCNN based BiGRU model

P. Somashekar
*Assistant Professor, Department of Mathematics, Maharani's Science College for Women (Autonomous), Mysuru, India*

V. Renuka Bai
*Assistant Professor, Department of Commerce and Management, Government First Grade College, Nanjangud, India*

S. Vijay
*Associate Professor, Department of Mathematics, Government Science College, Hassan, India*

C.N. Harshavardhana
*Government First Grade College for Women, Holenarasipura, India*

P. Siva Kota Reddy
*Professor, Department of Mathematics, JSS Science and Technology University, Mysuru, India*

R. Kemparaju
*Assistant Professor, Department of Mathematics, Government First Grade College, T. Narasipura, India*

ABSTRACT: An enterprise resource planning (ERP) system's post-implementation review (PIR) is the focus of this proposed study, which aims to determine what makes a PIR successful. Using a qualitative, exploratory research strategy, this proposed investigates PIR and its possible significance in ERP system integrations. Both feature extraction and preprocessing are necessary steps prior to training the model. Two types of data clean data and select data make up the preprocessing phase. Feature extraction is based on the idea that a part's significance or higher semantic meaning is directly related to the part's ability to be extracted from data. For the model's training, we chose the 1DRCNN-BiGRU due to its high accuracy. The proposed approach does better than CNN and BiGRU, with a 97.09% accuracy rate.

*Keywords*: Enterprise resource planning (ERP), Management Control System (MCS), Post Implementation Review (PIR), Convolutional Neural Network (CNN), Bidirectional Gated Recurrent Unit (BiGRU)

## 1 INTRODUCTION

Enterprise resource planning (ERP) and management control system (MCS) research have been quite popular in the last ten years. The question that many businesses are attempting to answer is how they might supplement their MCS with ERP. Out of all the literature that was analysed, there is more research on the relationship between ERP and MCS than there is on the influence of ERP-MCS interaction on firm performance. Business cost and management control systems rely heavily on the informational technical framework that ERP provides. Effective management and control depend on collecting, keeping, reporting, and conveying complete and

 DOI: 10.1201/9781003559115-63

real-time data regarding workplace behaviours. Therefore, outside of the ERP that provides the data infrastructure that functions as a hub for various units, it is difficult to discuss the implications on MCS and corporate performance. Modern enterprise resource planning (ERP) software has paved the path for the development of interactive, remote-accessible management control systems. Management control objectives can be achieved with the use of ERP, which can integrate business operations and provide real-time data to all personnel.

## 2 LITERATURE SURVEY

Enterprise Resource Planning (ERP) is a system that firms use to better use, produce, send, and compute resources in order to fulfil the demands of their consumers [1]. Enterprise resource planning (ERP) could be defined as a system that facilitates transactions and data connectivity across a business[2]. Financial accounting, human resource management, raw material planning, order management, inventory control, accounts payable and receivable, and other business tasks may be a part of an enterprise resource planning (ERP) system[3]. Company resource planning is a paradigm for information systems that allows organisations to automate and combine their main business activities [4]. The complexity and size of enterprise resource planning (ERP) systems discourage many organisations from constructing their own from scratch due to the time, effort, and money required[5]. The investment in enterprise resource planning (ERP) is thought to promise growth in revenues and/or decrease in costs, leading to higher profitability for adopting firms compared to non-adopters. [6]When looking at how ERP deployment affects company performance, there is contradictory empirical evidence. Regression study by [7]compared adopters and non-adopters and found that adopters had better profit margins, AR turnover, ROA, and sales per employee than non-adopters, while ROE was negative. [8]The results showed that, with the exception of AR turnover, which improves both during and after adoption, the companies which were the only adopters consistently outperformed the control group before and after adoption. [9] They did find that ERP adopters beat non-adopters in return on assets (ROA) and return on invested capital (ROIC) using both pre- and post-implementation data over the course of three years. Adopters outscored non-adopters, suggesting that adopters had a competitive advantage because of the gains they pass on to customers, even though adopters did not experience an increase in the absolute value of ROA and ROIC[10]. [11]Companies and their shareholders benefit from ERM because, according to its advocates, it reduces earnings and stock price volatility, increases capital efficiency, and decreases the costs of external capital.[12] By taking a methodical and consistent approach to managing all of a company's risks, enterprise risk management (ERM) is believed to boost efficiency, firm value, and total risk failure. Firms could attain better performance and competitive advantage, according to the Resource-Based View (RBV) concept of the organisation, by holding, acquiring, and continuously exploiting strategic assets[13]. Organisations could see a well-functioning ERM system as a valuable asset. Buying or selling it is not possible in business transactions. Enterprise resource planning (ERP) systems vary from one institution to another, yet each one has its own special way of doing things[14]. The specifics vary from one organisation to the next, but there are several benefits to implementing an efficient ERM system. With these assets in hand, a company may be able to gain an edge in the market and boost its performance generally.

## 3 PROPOSED SYSTEM

This proposed contributes to the existing literature on ERP systems by examining how a company's financial and non-financial performance are impacted by the deployment of an ERP system. How formal and informal management control systems regulate the relationship between ERP adoption and organizational performance is our main area of study.

### 3.1 *Preprocessing*

All the steps needed to transform the raw data into the finished dataset are encompassed in this phase. Data preparation procedures will most likely be done repeatedly, without a specific sequence in mind.

#### 3.1.1 *Select data*

Select which datasets to analyze. Data quality, technical limits (such as volume or type restrictions), and relevance to data mining goals are all factors to think about when selecting data. Gathering the necessary data is the main focus of this phase[15]. The data was retrieved from the ERP system database. Selected metrics include products, prices, total purchase orders, and others like them.

#### 3.1.2 *Clean data*

In order to decrease the data set size, this stage involves removing irrelevant and missing information. This increases data quality and shortens the learning phase for the ML model.

### 3.2 *Feature extraction*

A feature is a characteristic of a chunk that provides some kind of value or higher semantic significance to a certain application; this description avoids the complicated notation used in formal mathematical definitions. Form feature that can be used to determine the optimal combination of topological and geometrical entities during a product's lifetime; as an example, form characteristics like as the shoulder and boss are crucial in the design and production processes[16]. Last but not least, form characteristics are the object's configurations that can be used for engineering-analysis in design, process planning in production, and assembly. In fields as diverse as design and industry, even identical groupings might imply very different things.

## 4 MODEL TRAINING

### 4.1 *1DRCNN*

The BN layer determines the batch's first- and second-order statistics and constantly tweaks the intermediate output of 1-DCNN's evolution layer to ensure that the higher-level network can adjust to the parameter alterations made by the lower-level network. This keeps the output stable across all layers and prevents the phenomenon where the higher-level network's gradient vanishing slows down its convergence speed. To learn the characteristics of the input data at every location, the convolution layer of a 1-DCNN employs a uniformly weighted convolution kernel in conjunction with various localized sections of a one-dimensional signal. Afterwards, it generates a one-dimensional feature mapping using the input data.

### 4.2 *BiGRU*

A GRU neural network, which is a simplified form of LSTM, includes the RNN as one of its subclasses. The input and forgetting gates are combined in GRU's update gate, which is different from LSTM. Bidirectional GRUs are a type of bidirectional neural network that incorporates both forward- and backward-propagating GRUs.

### 4.3 *1DRCNN-BiGRU*

Combining 1-DRCNN and BiGRU, a three-part hybrid neural network prediction model of water quality parameters was created in this article. In the first part of the 1-DRCNN model, mining and extracting features with likely nonlinear relationships from the time series data of the Luan River's water quality is done with the main objective of creating effective low-dimensional features. The retrieved features are fed into the BiGRU network to compute the water quality feature vector. The BiGRU network continuously records the interdependence of context, short-term, and long-term components of time series data while modifying the bias and weight of the training network. This allows features in water quality data to be better expressed. The output layer is the culmination of all the connection layers at the top of the model, which are used to estimate water quality metrics.

## 5 RESULT AND DISCUSSION

It is arguable that enterprise resource planning (ERP) systems were among the most ground-breaking IT innovations of the '90s. One of the most popular IT solutions nowadays is enterprise resource planning (ERP) software, which has grown in popularity as more and more companies switch from a functional to a process-based IT infrastructure.

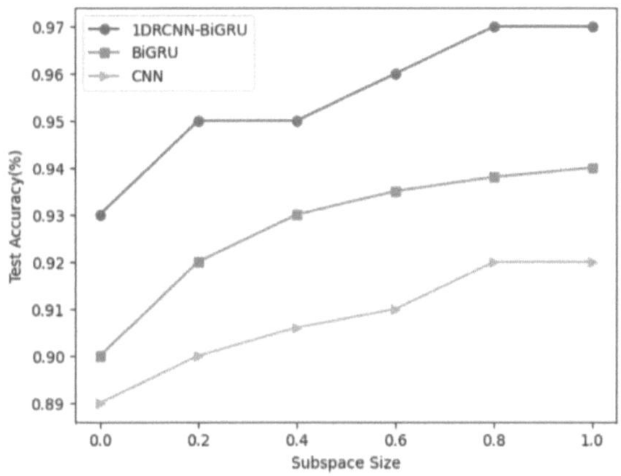


Figure 1. Test accuracy of 1DRCNN-BiGRU model.

It demonstrates, without resampling, how the classifier's test accuracy changes as a function of subspace size with regard to various data split ratios. A maximum accuracy of 97.09% and an average accuracy of 92.39% were attained.

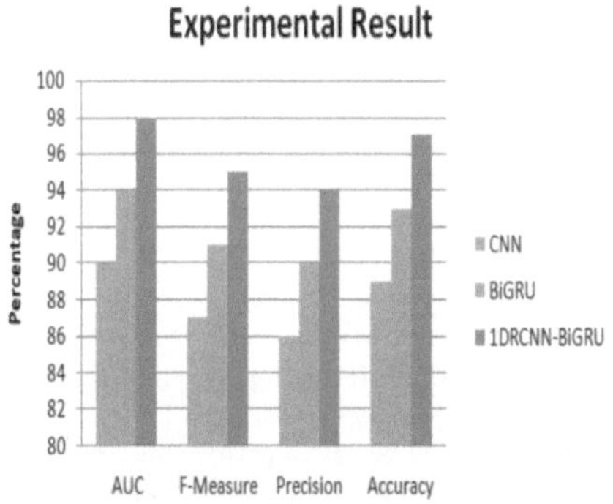


Figure 2. Performance curves for training and validating the suggested method.

Figure 2 also displays the outcomes of the classification model test utilizing different 1DRCNN-BiGRU.

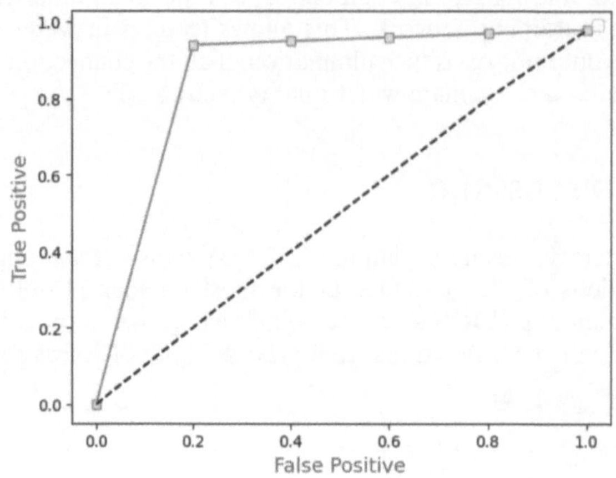


Figure 3. ROC Curve of Suggested Models.

The ROC curves of the 1DRCNN-BiGRU and the suggested technique are displayed in Figure 3. Because of this, it is clear that management control systems in enterprise resource planning systems benefit more from the suggested approach.

## 6 CONCLUSION

Enterprise resource planning (ERP) software marked a turning point in the integration of information systems across whole enterprises. Various types of businesses are having their management install these enterprise-wide solutions. Managers responsible for an organization's strategy and operations ironically adopt such all-encompassing and predefined systems, giving up control over its management information system and operations. A pair of datasets during the preprocessing phase, data is cleaned and selected. In feature extraction, the underlying principle is that the significance or higher semantic meaning of a part is proportional to its capacity to be extracted from data. The 1DRCNN-BiGRU method is employed for training the model. The suggested method achieves a higher consistency rate than both the CNN and BiGRU models (97.09%).

## REFERENCES

[1] Agus, A. (2015). Supply chain management: The influence of SCM on production performance and product quality. *Organization*, 12, 17–18.
[2] Birhanu, D., Krishnanand, L. and Rao, A. N. (2018). Supply chain strategies of manufacturers in Ethiopia. *International Journal of Productivity and Performance Management*, 67(2), 318–340.
[3] Chapman, C. S. and Kihn, L. A. (2009). Information system integration, enabling control and performance. *Accounting, Organizations and Society*, 34(2), 151–169.
[4] Kurien, G. P. and Qureshi, M. N. (2011). Study of performance measurement practices in supply chain management. *International Journal of Business, Management and Social Sciences*, 2(4), 19–34.

[5] HassabElnaby, H. R., Hwang, W. and Vonderembse, M. A. (2012). The impact of ERP implementation on organizational capabilities and firm performance. *Benchmarking: An International Journal*, 19 (4/5), 618–633.
[6] Acar, M. F., Tarim, M., Zaim, H., Zaim, S. and Delen, D. (2017). Knowledge management and ERP: Complementary or contradictory?. *International Journal of Information Management*, 37(6), 703–712.
[7] Amid, A., Moalagh, M. and Ravasan, A. Z. (2012). Identification and classification of ERP critical failure factors in Iranian industries. *Information Systems*, 37(3), 227–237.
[8] Anderson, M., Banker, R. D., Menon, N. M. and Romero, J. A. (2011). Implementing enterprise resource planning systems: Organizational performance and the duration of the implementation. *Information Technology and Management*, 12, 197–212.
[9] Aral, S. and Weill, P. (2007). IT assets, organizational capabilities, and firm performance: How resource allocations and organizational differences explain performance variation. *Organization Science*, 18(5), 763–780.
[10] Jhurani, J. Revolutionizing Enterprise Resource Planning: The Impact Of Artificial Intelligence On Efficiency And Decision-making For Corporate Strategies. *International Journal of Computer Engineering and Technology (IJCET)*, 13, 156–165.
[11] Acharyya, M. and Mutenga, S. (2013). The benefits of implementing enterprise risk management: Evidence from the non-life insurance industry. *Enterprise Risk Management*, 6(1), 22–24.
[12] Beheshti, H. M. and Beheshti, C. M. (2010). Improving productivity and firm performance with enterprise resource planning. *Enterprise Information Systems*, 4(4), 445–472.
[13] Arena, M., Arnaboldi, M. and Azzone, G. (2010). The organizational dynamics of enterprise risk management. *Accounting, Organizations and Society*, 35(7), 659–675.
[14] Yan, J., Liu, J., Yu, Y. and Xu, H. (2021). Water quality prediction in the luan river based on 1-drcnn and bigru hybrid neural network model. *Water*, 13(9), 1273.
[15] El Madany, H., Alfonse, M. and Aref, M. (2022). Hybrid time series model for procurement forecasting in enterprise resource and planning (erp) system: A case study. *Journal of Southwest Jiaotong University*, 57(1).
[16] Bhandarkar, M. P. and Nagi, R. (2000). STEP-based feature extraction from STEP geometry for agile manufacturing. *Computers in industry*, 41(1), 3–24.

*Hybrid and Advanced Technologies – S. Prasad Jones Christydass, Nurhayati Nurhayati and S. Kannadhasan. (Eds)* © *2025 the Authors, ISBN 978-1-032-90254-8*


# Feature knowledge based fault detection of induction motors through the analysis of stator current data using SWT-LSTM-DNN model

U.M. Fernandes Dimlo
*Professor & H.O.D, Department of CSE, Sreyas Institute of Engineering and Technology, Nagole, Hyderabad, India*

Arati Kothari
*Assistant Professor, School of Engineering, Ajeenkya D. Y. Patil University Pune*

G. Tamilarasi
*Assistant Professor, Department of Mathematics, Kalasalingam Academy of Research and Education, Krishnankoil, India*

Pankaj S. Kolhe
*Assistant Professor, Department of First Year Engineering, Dr.D.Y.Patil Institute of Technology Pimpri Pune, India*

L. Preethimathi
*Assistant Professor, Department of C. S. D., Bannari Amman Institute of Technology, Sathyamangalam, Tamilnadu, India*

J. Kartigeyan
*Associate Professor, Department of Electrical and Electronics Engineering, J. B.Institute of Engineering and Technology, Hyderabad, India*

ABSTRACT: The complex procedures of modern industrial production rely on electric motors, which must be repaired promptly. Currently, there are a lot of ways to find issues. It is common practice to do the diagnostic by analyzing current signals during a motor's steady-state operation or start-up transient. By dividing the current signals into their time-frequency components or using a method known as motor current signature analysis, frequencies associated with issues in the frequency domain can be located. The three primary parts are preprocessing, feature extraction, and model training. Data normalization is a preprocessing step that arranges the model's data attributes in order to increase the cohesion of entity types while simultaneously decreasing their dispersion. Decision trees and support vector machines were employed in the feature selection methodology. For better accuracy, we used the SWT-LSTM-DNN to train the model. Compared to rival methods such as DNN and LSTM, the proposed approach obtains a better accuracy of 96.24%.

*Keywords*: Induction Motors (IM), Long Short-Term Memory (LSTM), Deep Neural Network (DNN), Fault Detection, Support Vector Machine (SVM)

## 1 INTRODUCTION

It is common for induction motors (IMs) to fail when the insulation around the stator windings breaks down. If the insulation on the stator breaks, the problem can get much


DOI: 10.1201/9781003559115-64

worse. Of all motor impairments, these kinds of issues account for about 30–40%. Common causes include electromechanical forces, impurities, mechanical stresses, thermal overload, high dv/dt voltage spikes, vibrations, and frequent motor starts and stops. Damage to insulation usually becomes worse and spreads quickly. The gradual introduction of a large current into the windings as a result of turn-to-turn shorts causes a localized thermal overload. The ground fault, phase-to-phase errors, or shorts in the same or adjacent coils are eventually caused by continuous motor activity. Among the numerous types of stator winding failures caused by turn-to-turn shorts are phase-to-phase, coil-to-coil, and coil-to-ground issues. A catastrophic failure of the motor can occur when these defects are serious enough. Stator inter-turn shorts can be located with the use of a transient current signal that exhibits high-frequency characteristics. There are limitations to methods that depend on negative sequence components when handling voltage imbalances and load fluctuations. Although high-frequency estimators provide precise frequency estimates, their execution might be computationally demanding. The short-time Fourier transform (STFT), wavelet transform, Wigner-Ville decomposition, and other time-frequency based approaches can be used to achieve winding fault investigation.

## 2 LITERATURE SURVEY

Early fault detection is critical for induction motors to maximize output with minimal downtime. [1]Induction motor failure analysis and maintenance have been the subjects of extensive research. There was a time when vibration and current were the two main criteria for IM fault detection [2]. The original, more fundamental methods included overcurrent, overvoltage, and earth-fault. But now, with the help of contemporary technology, [3]a way to find problems with induction motors that combines old and new methods is available. Using condition-based monitoring for predictive maintenance of induction motors is crucial for modern enterprises to minimize downtime [4]. The first thing to do when trying to find a problem is to locate where it is. The second step is to find the parts that are broken. [5]The third thing to do is learn to recognize when you're about to fail and what to do to stop it. Predicting the pattern of failures is the last step. The fault detection problem can be viewed from various angles, such as pattern recognition and classification.[6] An increasing amount of emphasis is being placed on enhancing the dependability and accessibility of industrial systems. An unforeseen breakdown might cause your system to go down, injure nearby equipment, and even endanger people[7]. [8]For example, while being an inexpensive and crucial part of machines, bearings are the most common reason for mechanical system failure. [9]There will be additional repair costs if other machine components also fail due to a failed bearing. Consequently, the industry places a high value on early detection of bearing defects. [10] Autonomous online Condition Monitoring (CM) systems with fault detection algorithms give early alerts of mechanical and electrical breakdowns. In this method, harm can be prevented. The specific types of mistakes could change from system to system[11]. Specifically, this article focuses on Induction Motor (IM) failures as they pertain to electrical settings for compression and ventilation systems. Unfortunately, the majority of fault detection methods, such BRB, BRN, or UNB [12], can only discover a single, distinct defect on their own. Regardless, there have been a lot of attempts to prevent catastrophic failures by utilizing these strategies. Usually, these techniques are based on signals of current and vibration [13]. Induction motor defect detection using motor current signature analysis (MCCSA) is a well-known and effective method. Because it is simple and does not cause any harm, it is useful for detecting false positives [14]. In order to detect faults early, prevent unplanned maintenance, and keep production lines running smoothly, MCSB analyzes a starting transient or steady-state electric current signal that is delivered to the SCIM stator. So production costs, worker safety, and product quality are all at risk if these problems persist. To get this data, a current clamp probe is utilized.

## 3 PROPOSED SYSTEM

This proposed aims to address the problem of mechanical failure identification in induction motors by using stator current time-frequency analysis in a novel way. In most cases, mechanical problems are to blame when the load torque oscillates periodically. The effect of torque oscillations on the stator current of an induction motor is studied analytically.

### 3.1 *Preprocessing*

#### 3.1.1 *Data normalization*

After obtaining a strong spectrum band for fault detection, then put the band vector through its paces with the GB classifier[15]. Without normalized data, GB puts classifier performance improvements ahead of small-sized frequency considerations. By arranging the data attributes inside a model, data normalization aims to improve consistency of entity types and reduce or eliminate their duplication. The signals that were produced after data preprocessing were normalized using the following equation:

$$y_{norm}^{(n)} = \frac{y^{(n)} - \rho_y}{\tau_y} \quad (1)$$

### 3.2 *Feature extraction*

#### 3.2.1 *Decision tree*

As a diagnostic tool, a decision tree builds a knowledge-based system by applying inductive reasoning to case histories. A decision tree is a hierarchical data structure where the leaves represent classes and the branches represent the features that are combined to form those classifications[16]. The structure of the tree is decided throughout the process of selecting a test to act as its root. The choice of the tree's root is based on Quinlan's proposed information theory.

#### 3.2.2 *SVM*

One way to learn to classify and predict data is with support vector machines, also known as SVMs. As an example, let's look at two-class classification. When training a support vector machine (SVM), it is essential that the input space contain a hyper-plane. Positive and negative samples can be distinguished using this hyper-plane. When determining how far apart the hyper-plane is from the closest positive and negative samples, the greatest distance will be used.

## 4 MODEL TRAINING

### 4.1 *SWT*

One way to learn to classify and predict data is with support vector machines, simply known as SVMs. Use two-class classification as an example. The DWT avoids this issue by decoupling the up-sampled process from the signal's sub-sampling and discarding it along with the filtering operation at each compression level. One common term for this DWT variation is the SWT.

### 4.2 *SWT-LSTM-DNN*

Analyzing the power signal, the expected temperature value, and a constructed fore-casted value obtained by reconstructing the projected output coefficients provided by individual LSTMs forms the basis of the proposed hybrid wavelet-based LSTM-DNN model. The DNN is trained using these attributes. By employing a multi-stage forecasting approach, the LSTM networks are instructed to anticipate, during the next 24 hours, the precise and approximate coefficients of the SWT decomposed power signal. In order to feed the values into the DNN, the inverse SWT (ISWT) converts them into a predicted power signal. The major goal of incorporating the SWT decomposition into the model is to accurately forecast and extract the most essential information from the historical signal in terms of the time and frequency domains.

## 5 RESULT AND DISCUSSION

This work introduces a novel online method for early detection of electrical motor problems. In order to detect broken bars, saturation, eccentricity, and bearing problems, the method uses the steady-state stator current measurement to estimate particular statistical parameters. The technique employs the axial flux signal to detect short-circuits in the stator winding coils.

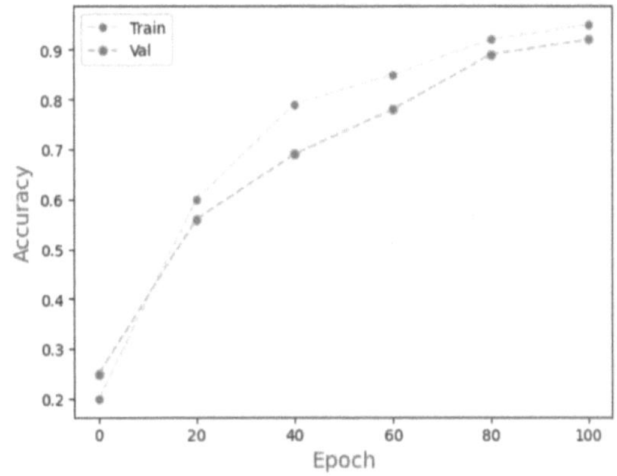


Figure 1. The accuracy curve over 100 epochs of the suggested method.

One hundred samples from the test dataset are utilized to assess the efficacy of the trained CNN model. On average, the trained CNN model achieved 95% accuracy on the test dataset, which is an intriguing performance result.

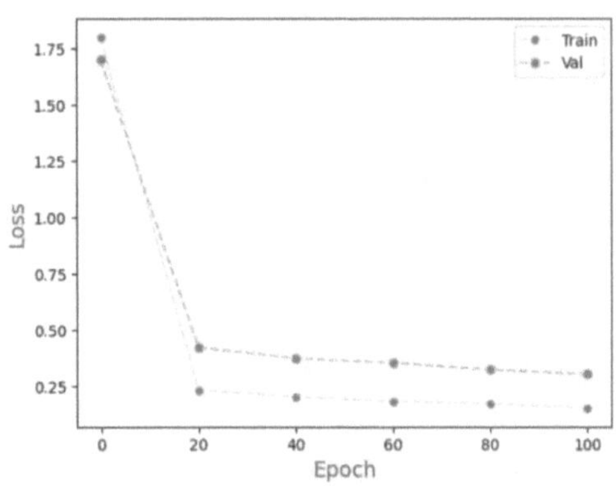


Figure 2. Loss curves for training and validating the suggested method.

Figure 2 shows the results of plotting the average losses (training and validation) gathered at each iteration of training the proposed model using a k-fold cross-validation technique.

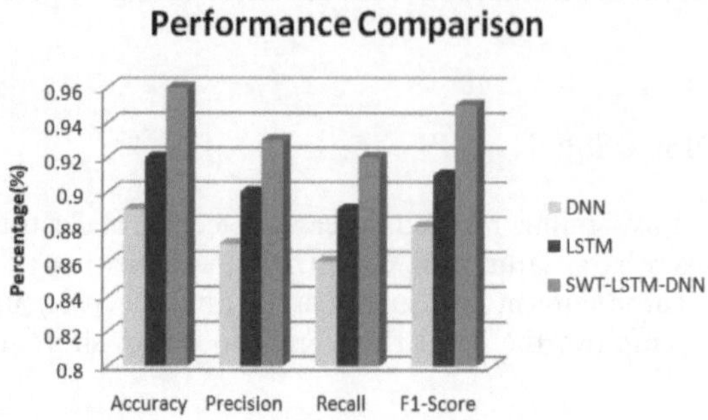


Figure 3. Performance comparison of three models.

In addition, Figure 3 displays the comparison of the three models' performance in relation to recall, average accuracy, F1-score, and precision. Additionally, SWT-LSTM-DNN has demonstrated superior performance across all three-performance metrics.

Table 1. Comparison result.

| Models | Accuracy (%) |
|---|---|
| DNN | 89.30 |
| LSTM | 92.61 |
| SWT-LSTM-DNN | 96.24 |

The results of the comparison show that the proposed SWT-LSTM-DNN model outperforms the other strategy. A higher prediction accuracy of 96.24% compared to all other approaches demonstrates the remarkable performance of the proposed SWT-LSTM-DNN model.

## 6 CONCLUSION

To ensure the safe and economic running of industrial activities, researchers have immersed themselves in the challenging topic of defect detection of electrical or mechanical anomalies in induction motors for decades. In order to make entity types more cohesive and less dispersed, data normalization is a preprocessing technique that sorts the model's data properties in a specific way. The feature selection process made use of support vector machines and decision trees. In order to train the models, they employ the SWT-LSTM-DNN method. In comparison to the DNN and LSTM models, the suggested method has a consistency rate of 96.24%.

## REFERENCES

[1] Bin, G. F., Gao, J. J., Li, X. J. and Dhillon, B. S. (2012). Early fault diagnosis of rotating machinery based on wavelet packets—Empirical mode decomposition feature extraction and neural network. *Mechanical Systems and Signal Processing*, 27, 696–711.

[2] Casimir, R., Boutleux, E., Clerc, G. and Yahoui, A. (2006). The use of features selection and nearest neighbors rule for faults diagnostic in induction motors. *Engineering Applications of Artificial Intelligence*, 19(2), 169–177.

[3] Cusido, J., Romeral, L., Espinosa, A. G., Ortega, J. A. and Riba Ruiz, J. R. (2011). On-line fault detection method for induction machines based on signal convolution. *European Transactions on Electrical Power*, 21(1), 475–488.

[4] Gan, M. and Wang, C. (2016). Construction of hierarchical diagnosis network based on deep learning and its application in the fault pattern recognition of rolling element bearings. *Mechanical Systems and Signal Processing*, 72, 92–104.

[5] Glowacz, A. (2019). Fault diagnosis of single-phase induction motor based on acoustic signals. *Mechanical Systems and Signal Processing*, 117, 65–80.

[6] Vaimann, T., Belahcen, A. and Kallaste, A. (2015, September). Necessity for implementation of inverse problem theory in electric machine fault diagnosis. In *2015 IEEE 10th International Symposium on Diagnostics for Electrical Machines, Power Electronics and Drives (SDEMPED)* (pp. 380–385). IEEE.

[7] Ospina, J., Newaz, A. and Faruque, M. O. (2019). Forecasting of PV plant output using hybrid wavelet-based LSTM-DNN structure model. *IET Renewable Power Generation*, 13(7), 1087–1095.

[8] Glowacz, A., Glowacz, W., Glowacz, Z. and Kozik, J. (2018). Early fault diagnosis of bearing and stator faults of the single-phase induction motor using acoustic signals. *Measurement*, 113, 1–9.

[9] Akhil Vinayak, B., Anjali Anand, K. and Jagadanand, G. (2020). Wavelet-based real-time stator fault detection of inverter-fed induction motor. *IET Electric Power Applications*, 14(1), 82–90.

[10] Bazan, G. H., Scalassara, P. R., Endo, W., Goedtel, A., Godoy, W. F. and Palácios, R. H. C. (2017). Stator fault analysis of three-phase induction motors using information measures and artificial neural networks. *Electric Power Systems Research*, 143, 347–356.

[11] Khechekhouche, A., Cherif, H., Benakcha, A., Menacer, A., Chehaidia, S. E. and Panchal, H. (2020). Experimental diagnosis of inter-turns stator fault and unbalanced voltage supply in induction motor using MCSA and DWER. *Periodicals of Engineering and Natural Sciences*, 8(3), 1202–1216.

[12] Gangsar, P. and Tiwari, R. (2020). Signal based condition monitoring techniques for fault detection and diagnosis of induction motors: A state-of-the-art review. *Mechanical Systems and Signal Processing*, 144, 106908.

[13] Abd-el-Malek, M. B., Abdelsalam, A. K. and Hassan, O. E. (2018). Novel approach using Hilbert Transform for multiple broken rotor bars fault location detection for three phase induction motor. *ISA transactions*, 80, 439–457.

[14] Kompella, K. D., Rao, M. V. G. and Rao, R. S. (2018). Bearing fault detection in a 3 phase induction motor using stator current frequency spectral subtraction with various wavelet decomposition techniques. *Ain Shams Engineering Journal*, 9(4), 2427–2439.

[15] Bhakta, K., Sikder, N., Al Nahid, A. and Islam, M. M. (2019, February). Fault diagnosis of induction motor bearing using cepstrum-based preprocessing and ensemble learning algorithm. In *2019 International Conference on Electrical, Computer and Communication Engineering (ECCE)* (pp. 1–6). IEEE.

[16] Nguyen, N. T., Lee, H. H. and Kwon, J. M. (2008). Optimal feature selection using genetic algorithm for mechanical fault detection of induction motor. *Journal of Mechanical Science and Technology*, 22, 490–496.

*Hybrid and Advanced Technologies – S. Prasad Jones Christydass, Nurhayati Nurhayati and S. Kannadhasan. (Eds)* *ISBN 978-1-032-90254-8*


# Effective model on automated vehicle detection in satellite images using CNN comparison with faster R-CNN

Konka Vishnu Vardhan
*Research Scholar, Department of Computer Science and Engineering, Saveetha School of Engineering, Saveetha Institute of Medical and Technical Sciences, Saveetha University, Chennai, Tamil Nadu, India*

S. Mahaboob Basha
*Research Guide, Department of Computer Science and Engineering, Saveetha School of Engineering, Saveetha Institute of Medical and Technical Sciences, Saveetha University, Chennai, Tamil Nadu, India*

ABSTRACT: The aim of research focused on automated vehicle detection in satellite images using the R-CNN algorithm is to develop accurate and efficient methods for identifying and localizing vehicles in high-resolution satellite imagery. This research seeks to address the growing demand for reliable vehicle detection solutions with applications spanning urban planning, environmental monitoring, disaster response, and defense. We used the R-CNN algorithm and the CNN algorithm in this investigation, each of which has ten iterations (N=10). Two different groups evaluated these two methods, and 100 samples in all were taken into account for the analysis. A power setting of 85% was used for the Gpower statistical test (g power parameters setup with $\alpha=0.05$ and power=0.85). This power configuration was chosen in accordance with accepted statistical criteria to guarantee that the investigation had a strong capacity to identify statistically significant differences or effects. In the context of automated vehicle detection in satellite photos, our research's results show that the custom CNN model outperformed expectations with an exceptional accuracy of 97.81%. As an illustration of our CNN model's significant performance advantage, the R-CNN model received an accuracy score of 94.84%. In conclusion, the research and application of automated vehicle detection in satellite images using the R-CNN (Region-based Convolutional Neural Network) algorithm represent a significant leap in the field of computer vision and remote sensing. This technology has far-reaching implications across numerous domains, from urban planning and environmental monitoring to disaster response, defense, and beyond. R-CNN and its variants have demonstrated remarkable accuracy in detecting and localizing vehicles within high-resolution satellite imagery.

*Keywords*: CNN, satellite, Images, vehicle, Detection, R-CNN

## 1 INTRODUCTION

In computer vision and remote sensing, vehicle detection in satellite photos is essential for surveillance, urban planning, and environmental monitoring. In order to facilitate effective vehicle identification for traffic management, security, disaster response, and military activities, satellite photography provides a worldwide viewpoint (Wang *et al.* 2020). CNNs are essential for identifying vehicles in satellite imagery because they use hierarchical feature extraction; however, Faster R-CNN improves accuracy by incorporating object categorization and region recommendations into its architecture (Hakak *et al.* 2023). These technologies represent a major breakthrough in data-driven resource management and emergency preparedness by enabling decision-making across a range of applications, including urban

 DOI: 10.1201/9781003559115-65

growth, environmental protection, and disaster response (Q. Chen *et al.* 2021). A great deal of research has been done between 2018 and 2023 on a number of different platforms, including Google Scholar, IEEE, Springer, PubMed, and satellite image detection. The application of deep learning models to enhance tiny vehicle recognition in satellite photos is the subject of a recent survey. These models are efficient, yet they have drawbacks such resource requirements. According to the survey, these methods should be optimised to increase their application in a variety of contexts (Y. Liu *et al.* 2021). Additional research emphasises the complexity of vehicle detection and the demand for sophisticated methods to guarantee dependable performance in a range of circumstances (Sun *et al.* 2022). Continuous improvements in vehicle detection techniques are required because real-time processing skills are essential for applications such as emergency response and traffic monitoring (Vedaldi *et al.* 2020). Accurate performance evaluations are hampered by the lack of standardised standards for assessing satellite imagery vehicle detection models (J. Chen *et al.* 2021). Challenges still exist, such as dataset variability and computing needs, despite breakthroughs with CNNs and R-CNN (Yuan *et al.* 2022). diverse algorithms address diverse needs in vehicle detection by providing unique advantages and trade-offs (Rezaei *et al.* 2023). By offering dependable methods for vehicle identification in satellite imagery, this research seeks to improve disaster response, urban planning, and other fields (Khayyam *et al.* 2020).

## 2 MATERIALS AND METHODS

The Saveetha Institute of Medical and Technical Sciences' Data Analytics laboratory carried out this investigation. With its high configuration system, the laboratory can conduct in-depth study and produce accurate results. Ten people made up the sample size for the study, and they were split into two groups: While Group 2 used the R-CNN approach, Group 1 used the CNN method. At a significance level of 0.05 (alpha) and power level of 0.8 (beta), the study was statistically powered with an 80 percent G-Power value. In order to compute and evaluate the differences between the two groups methodically, a 95 percent confidence interval was maintained.

The comma-separated values (CSV) format of the dataset used in this study on the detection of human conflict regions, known as the "automated vehicle," contains photos with a specified size of 224x224 pixels. It contains a variety of pertinent aspects, all of which are crucial indications, including satellite photos. Using matching satellite labels, the dataset—which was collected from several sources—was rotated(W. Liu *et al.* 2022).

Considering a particular dataset, the technique described in this paper has been implemented and its results have been juxtaposed with the CNN algorithm's official output. In this instance, the dataset comprises a range of data values and is organized as a comma-separated value (CSV) file. Using the techniques described in this research, a thorough examination of the records was made possible by applying the dataset to undertake an exhaustive review and comparison. For this task, a dual-core Intel processor with 4 GB of RAM was the hardware configuration utilized. To implement the algorithm and carry out the comparison analysis, a well-rounded system was created by utilizing Jupyter Notebook, MySQL, and Python technology in the software setup.(W. Liu *et al.* 2022).

### 2.1 *Statistical analysis*

To determine the effectiveness of the hedging, a statistical analysis of the performance of the human-wildlife conflict regions was carried out using IBM SPSS statistics version 26.0. A different analysis was performed with the t-test. The accuracy measurements, which are the dependent variables in this study, are the main focus. "Performance Comparison of Weak and Strong Learners in Detecting GPS Spoofing Attacks on Unmanned Aerial Vehicles (UAVs)" (2022) assessed these measurements as one of the independent variables that aided in automated vehicle recognition in satellite pictures.

## 3 RESULTS

In order to assess machine learning algorithms' effectiveness in automated vehicle detection in satellite photos, this study was conducted. 97.81% accuracy was a noteworthy result for the CNN algorithm, demonstrating its ability to anticipate automated vehicles with high accuracy. As an illustration of how successful the R-CNN algorithm is at this task, its accuracy of 94.02% was shown. According to these results, the CNN algorithm is the best option for this research because of its higher accuracy in identifying conflict zones. The accuracy values are displayed in Table 1. for the CNN and R-CNN with N=10. Table 2 group statistics show that the CNN has a mean accuracy of 1.81830 and a standard deviation of 1.92209, whereas the R-CNN has a mean accuracy of 94.8400% and a standard deviation of.99911. Table 3. indicates that an independent sample t-test yielded a significant p-value of 0.001 ($p<0.05$). Based on this, it can be concluded that the CNN model outperforms the R-CNN in the two-tailed test of automated vehicle detection in satellite images. Figure 1. illustrates the mean accuracy of the CNN and R-CNN algorithms in comparison. Compared to the R-CNN's accuracy, the CNN has a higher mean accuracy of. Furthermore, the CNN algorithm has a lower standard deviation than the R-CNN method. CNN vs. R-CNN algorithm is shown by the X-axis, while mean accuracy is indicated by the Y-axis, with error bars showing ± 2 SD.

Table 1. Represents cnn pseudocode.

| |
|---|
| Input: Automated vehicle data |
| Output: Automated vehicle detection in satellite images |
| **Step 1: Data Preparation:** To facilitate training, extensive collections of satellite photos with comments and cars are gathered. For the CNN to discover and generalize patterns, these datasets are essential. |
| **Step 2**: **Network Architecture:**The task-specific CNN architecture is selected or created. VGG, R-CNN, and custom-designed networks are common object detection architectures. Convolutional layers in these networks are responsible for learning features from the input data. |
| **Step 3: Training:** The prepared dataset is used to train the CNN. In the process of training, the network picks up characteristics specific to vehicles, such colors, textures, and shapes, and links them to the existence of cars in satellite photos. To reduce detection mistakes, the loss function is optimized. |
| **Step 4: Object Detection:** The CNN can be trained to identify cars in fresh, unexplored satellite photos. The network predicts bounding boxes around likely vehicle locations by scanning the input image. |
| **Step 5: Post-processing:** Post-processing techniques such as non-maximum suppression can be used to eliminate low-confidence or duplicate detections, which will improve the detection results. |
| **Step 6: Evaluation:** Metrics like precision, recall, and F1-score are used to assess the CNN model's performance in order to gauge how well the vehicle recognition works and how accurate it is. |

Table 2. Represents R-CNN pseudocode.

| |
|---|
| Input: Automated vehicle data |
| Output: Automated vehicle detection in satellite images |
| **Step 1:**Region Proposal: The first step in R-CNN involves generating region proposals within the satellite image. This is done by using selective search or a similar algorithm to identify potential regions of interest where vehicles might be located. These regions are referred to as region proposals. |
| **Step 2**: Feature Extraction: For each region proposal, a CNN is used to extract deep features. This pre-trained CNN has been trained on a vast dataset and can capture meaningful information from the image. |

*(continued)*

Table 2. Continued

| |
|---|
| **Step 3:** Object Classification: The extracted features are then fed into a SVM or another classifier to determine whether the region contains a vehicle or not. This step helps classify the detected objects. |
| **Step 4:** Bounding Box Regression: R-CNN also incorporates bounding box regression to refine the region proposals. It fine-tunes the position and size of the bounding boxes to more accurately encapsulate the detected vehicles. |
| **Step 5 :** Non-Maximum Suppression: After classifying and refining the regions, a post-processing step is applied to eliminate redundant or overlapping detections using non-maximum suppression. This ensures that only the most accurate and non-overlapping vehicle detections are retained. |
| **Step 6:** Output: The final output of the R-CNN algorithm is a list of bounding boxes around the detected vehicles along with their corresponding classifications. |

Table 3. Accuracy of CNN & Accuracy of R-CNN algorithms. The R-CNN algorithm is less than the CNN algorithm.

| SAMPLE NO | CNN | R-CNN |
|---|---|---|
| 1 | 98.66 | 94.77 |
| 2 | 97.62 | 97.55 |
| 3 | 99.15 | 92.55 |
| 4 | 98.81 | 93.77 |
| 5 | 97.99 | 96.45 |
| 6 | 96.23 | 98.66 |
| 7 | 97.37 | 89.99 |
| 8 | 95.72 | 93.67 |
| 9 | 99.56 | 94.67 |
| 10 | 97.81 | 90.33 |

Table 4. Group statistics:Group statistics results-CNN has a mean accuracy (97.2200%), St. Deviation (1.81830), whereas R-CNN has a mean accuracy (94.8400), St. Deviation (.99911).

| Group | | N | Mean | std.Deviation | Std. Error mean |
|---|---|---|---|---|---|
| **Accuracy** | **R-CNN** | 10 | 94.8400 | .99911 | .31595 |
| | **CNN** | 10 | 97.2200 | 1.81830 | .57500 |

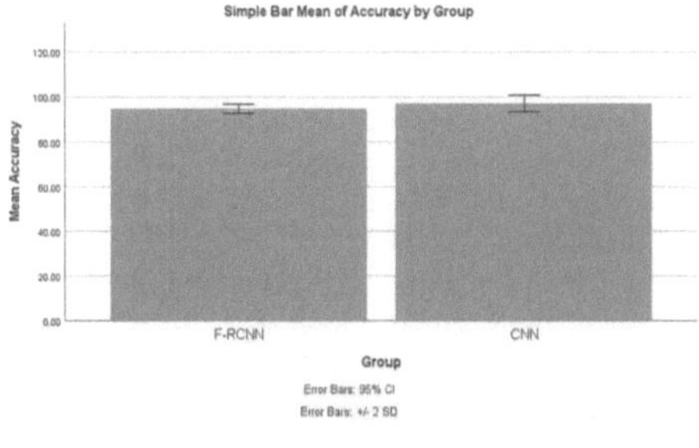


Figure 1. Compared to the R-CNN model, the CNN model showed a significant mean accuracy in this study on automated vehicle detection in satellite pictures.

## 4 DISCUSSION

In our study, we extensively investigated automated vehicle detection in satellite imagery using two distinct models: a custom Convolutional Neural Network (CNN) and the R-CNN model. Results indicate significant performance differences, with our custom CNN achieving an impressive accuracy of 97.81%, surpassing the R-CNN's 94.84%(Rapp *et al.* 2020).. This outcome is attributed to the CNN's autonomous feature extraction from complex images and adaptability to data variations. Additionally, the CNN's computational efficiency suggests practical real-time applications, benefiting traffic monitoring, urban planning, and disaster response. These findings underscore the advantages of CNNs in satellite-based object detection, with potential implications for future research on scalability and generalization capabilities (Pflugfelder *et al.* 2022).

## 5 CONCLUSION

In conclusion, research in automated vehicle detection in satellite images, using advanced techniques such as CNNs Faster R-CNN, and other machine learning algorithms, represents a transformative field with significant potential and impact. The applications span a wide array of domains, from urban planning and environmental monitoring to disaster response, defense, logistics, and beyond. This research has the power to revolutionize the way we manage our resources, respond to emergencies, and make informed decisions on a global scale. However, there are challenges, including the complexity of datasets and computational demands.Despite these challenges, the field continues to evolve, with promising future developments in algorithm refinement, data quality, interpretability, and integration with other technologies.

## REFERENCES


Aluvalu, R., Kumaran V. N., S., Thirumalaisamy, M., Basheer, S., Ali Aldhahri, E. and Selvarajan, S. (2023). Efficient data transmission on wireless communication through a privacy-enhanced blockchain process. *PeerJ. Computer Science*, 9, e1308. https://doi.org/10.7717/peerj-cs.1308

Author(s). (2022). *Performance Comparison of Weak and Strong Learners in Detecting GPS Spoofing Attacks on Unmanned Aerial Vehicles* (UAVs). Retrieved June 18, 2024, from https://search.proquest.com/openview/80503350fce86852630fe608706fe587/1?pq-origsite=gscholar&cbl=18750&diss=y

Cazzato, D., Cimarelli, C., Sanchez-Lopez, J. L., Voos, H. and Leo, M. (2020). A survey of computer vision methods for 2D object detection from unmanned aerial vehicles. *The Journal of Imaging Science and Technology / IS&T, the Society for Imaging Science and Technology*, 6(8). https://doi.org/10.3390/jimaging6080078

Chen, J., Yuan, Z., Peng, J., Chen, L., Huang, H., Zhu, J., Liu, Y. and Li, H. (2021). DASNet: Dual attentive fully convolutional siamese networks for change detection in high-resolution satellite images. *IEEE Journal of Selected Topics in Applied Earth Observations and Remote Sensing*, 14, 1194–1206. https://doi.org/10.1109/JSTARS.2021.3066613

Chen, Q., Xie, Y., Guo, S., Bai, J. and Shu, Q. (2021). Sensing system of environmental perception technologies for driverless vehicle: A review of state of the art and challenges. sensors and actuators. *A, Physical*, 319(March), 112566. https://doi.org/10.1016/j.sna.2020.112566

Hakak, S., Gadekallu, T. R., Maddikunta, P. K. R., Ramu, S. P., Parimala, M., De Alwis, C. and Liyanage, M. (2023). Autonomous Vehicles in 5G and beyond: A survey. *Vehicular Communications*, 39(February), 100551. https://doi.org/10.1016/j.vehcom.2022.100551

Heidecker, F., Breitenstein, J., Rösch, K., Löhdefink, J., Bieshaar, M., Stiller, C., Fingscheidt, T. and Sick, B. (2021). An application-driven conceptualization of corner cases for perception in highly automated driving. In *2021 IEEE Intelligent Vehicles Symposium (IV)*, 644–651. IEEE.

Huang, Z., Li, W. and Tao, R. (2022). Extracting and distilling direction-adaptive knowledge for lightweight object detection in remote sensing images. In *ICASSP 2022 – 2022 IEEE International Conference on Acoustics, Speech and Signal Processing (ICASSP)*, 2360–2364. IEEE.

Jha, S., Banerjee, S., Tsai, T., Hari, S. K. S., Sullivan, M. B., Kalbarczyk, Z. T., Keckler, S. W. and Iyer, R. K. (2019). ML-Based Fault Injection for Autonomous Vehicles: A case for bayesian fault injection. In *2019 49th Annual IEEE/IFIP International Conference on Dependable Systems and Networks (DSN)*, 112–124. IEEE.

Khayyam, H., Javadi, B., Jalili, M. and Jazar, R. N. (2020). Artificial intelligence and internet of things for autonomous vehicles. In R. N. Jazar & L. Dai (Eds.), *Nonlinear Approaches in Engineering Applications: Automotive Applications of Engineering Problems* (pp. 39–68). Springer International Publishing.

Liu, W., Quijano, K. and Crawford, M. M. (2022). YOLOv5-Tassel: Detecting tassels in RGB UAV imagery with improved YOLOv5 based on transfer learning. *IEEE Journal of Selected Topics in Applied Earth Observations and Remote Sensing*, 15, 8085–8094. https://doi.org/10.1109/JSTARS.2022.3154076

Liu, Y., Sun, P., Wergeles, N. and Shang, Y. (2021). A survey and performance evaluation of deep learning methods for small object detection. *Expert Systems with Applications*, 172(June), 114602. https://doi.org/10.1016/j.eswa.2021.114602

Mahaveerakannan R., Anitha, C., Thomas, A. K., Rajan, S., Muthukumar, T. and Govinda Rajulu, G. (2023). An IoT based forest fire detection system using integration of cat swarm with LSTM model. *Computer Communications*, 211, 37–45. https://doi.org/10.1016/j.comcom.2021.12.008

Mansour, A., Hassan, A., Hussein, W. M. and Said, E. (2019). Automated vehicle detection in satellite images using deep learning. *IOP Conference Series: Materials Science and Engineering*, 610(1), 012027. https://doi.org/10.1088/1757-899X/610/1/012027

Pflugfelder, R., Weissenfeld, A. and Wagner, J. (2022). Deep vehicle detection in satellite video. *arXiv [cs.CV]*. http://arxiv.org/abs/2204.06828

Punithavathi-Rasappan. (n.d.). Computer Vision and Deep Learning-enabled Weed Detection Model for Precision Agriculture. Retrieved June 18, 2024, from https://www.researchgate.net/publication/362388038_Computer_Vision_and_Deep_Learning-enabled_Weed_Detection_Model_for_Precision_Agriculture

*Hybrid and Advanced Technologies – S. Prasad Jones Christydass, Nurhayati Nurhayati and S. Kannadhasan. (Eds) © 2025 the Authors, ISBN 978-1-032-90254-8*


# Advanced image classification based on convolutional neural network with improved accuracy compared to resnet 50

Vivek
*Research Scholar, Department of Computer Science and Engineering, Saveetha School of Engineering, Saveetha Institute of Medical and Technical Sciences, Saveetha University, Chennai, Tamil Nadu, India*

Porkodi
*Department of Computer Science and Engineering, Saveetha School of Engineering, Saveetha Institute of Medical and Technical Sciences, Saveetha University, Chennai, Tamil Nadu, India*

ABSTRACT: The purpose of this study is to use the Convolutional Neural Network Methodology to get a greater level of image clarity compared to Resnet 50, as well as a better overall image effect. There are a total of 20 Samples, and they've been split in half. There are 10 samples for everyone. Both the Convolutional Neural Network (Class 1) and the Resnet 50 algorithm (Class 2) are part of Group 3. The 80% G-power, 0.05% threshold, and 95% confidence interval are used to determine the required sample size. In comparison to Resnet 50 73% accuracy, Convolutional Neural Network Technique's 84.20 percent is much higher. With a p value of 0.000 (p 0.05), these findings indicated a statistically significant difference between the two sets of participants. Consequently, Convolutional Neural Network Technique (84.20%) outperforms Resnet 50 (73.14%) in terms of accuracy.

*Keywords*: Image, Resnet 50, Convolutional Neural Networks, Accuracy, Blurriness, Classification

## 1 INTRODUCTION

Convolutional neural network based advanced image classification is a state-of-the-art method in the study of computer vision. The process requires the use of deep learning models to extract complex visual information for reliable categorization. (Hu 2022). This technology surpasses the state-of-the-art by using intricate designs that model the human visual system. Unlike the widely used ResNet-50, which is a 50-layer deep residual network, the most recent CNNs have deeper architectures with optimized feature extraction and integration layers, resulting in greater accuracy. (Millstein 2020). These networks are designed to handle a wide variety of complex visual data with higher precision and efficiency. Applications for this advanced image classification technique are extensive and diverse (Kulshrestha 2019). It finds use in various domains such as imaging, autonomous vehicles for object recognition and scene understanding, surveillance systems for real-time threat detection, and natural language processing for image-based information retrieval. Moreover, it supports tasks like facial recognition, content-based image retrieval, and visual search engines, making it an indispensable tool for industries requiring sophisticated image analysis capabilities (Sewak *et al.* 2018).

More than 1400 articles demonstrate how image classification is executed, the literature pertaining to Advanced Image Classification using Convolutional Neural Networks (CNNs) with enhanced accuracy compared to ResNet-50 highlights recent advancements in deep

 DOI: 10.1201/9781003559115-66

learning (Lu *et al.* 2019). Researchers have focused on designing novel CNN architectures with deeper layers, specialized convolutional structures, and attention mechanisms to improve classification accuracy. Studies emphasize the significance of feature extraction and integration techniques, highlighting the effectiveness of complex network architectures in capturing intricate image features (Bhatt 2018). Comparative analyses showcase the superior performance of these advanced CNN models over ResNet-50 in various benchmark datasets. Moreover, researchers have investigated the impact of different optimization strategies, data augmentation techniques, and regularization methods on the overall classification accuracy (Shanmugamani 2018). Literature also discusses the applicability of these advanced CNNs in diverse domains, including medical imaging, autonomous systems, surveillance, and natural language processing. The research underscores the potential of these sophisticated models in addressing complex image classification challenges across multiple industries (Rai 2019).

A research gap exists in advancing image classification using Convolutional Neural Networks (CNNs) to surpass ResNet-50 accuracy. Key areas include novel architecture exploration, attention mechanisms, data augmentation, regularization techniques, learning from limited data, ensemble learning, and interpretable models to achieve enhanced accuracy in image classification. (Aluvalu *et al.* 2023; Mahaveerakannan R *et al.* 2023; Ramana *et al.* 2023; Vellela and Balamanigandan 2023)The aim of Advanced Image Classification based on Convolutional Neural Network (CNN) with improved accuracy compared to ResNet-50 is to enhance the precision and efficiency of image recognition tasks. By exploring deeper network architectures, specialized convolutional structures, and attention mechanisms, the objective is to extract intricate image features more effectively. This research seeks to establish a more robust and versatile image classification system that outperforms ResNet-50, enabling more accurate analysis and recognition of complex visual data across various domains.

## 2 MATERIALS AND METHODS

In two groups of ten samples each, the evaluation compares the accuracy of the ResNet Algorithm at 73.02% and Convolutional Neural Networks (CNNs) at 84.20% for picture clarification. Images from the dataset are used to train the Poisson Noise Reduction framework, which requires at least 16GB of RAM to run smoothly on an Intel Core i3 or i7 processor (Field 2018). Installation of necessary plugins or software is required in addition to 500GB of hard drive space to access and store these photos from Kaggle. It is advised to use a graphics card with 4GB of RAM in order to reduce CPU load and speed up image processing in the Kaggle online development environment. This will make it easier to design the framework that will reduce blurriness in images.

### 2.1 *Convolutional neural network*

The use of deep learning techniques for the purpose of photo categorization has shown significant results, particularly in the development of sophisticated architectures. The first convolutional neural networks were constructed with a deep architecture with a total of eight layers, consisting of five convolutional layers and three fully connected layers. The profound nature of the entity enabled it to comprehend intricate hierarchies of data. The vanishing gradient issue was addressed by linearly adjusting the activation function, resulting in accelerated training.

#### 2.1.1 *CNN algorithm*

Step 1: Load and preprocess the image dataset (e.g., ImageNet).

Step 2: Define the architecture, including convolutional and fully connected layers.
Step 3: Specify the number of classes for the output layer.
Step 4: Set the learning rate, batch size, and other hyperparameters.
Step 5: Define a loss function (e.g., cross-entropy) for optimization.
Step 6: Periodically evaluate the model on the validation set to monitor performance.
Step 7: After training, evaluate the final model on the separate testing dataset.
Step 8: Deploy the trained AlexNet model for making predictions on new, unlabeled images.
Step 9: Apply preprocessing and the trained model to classify objects in images.

## 2.2 *Resnet 50*

The 50-layer network architecture is described in "Deep Residual Learning for Image Recognition". This network was trained using ImageNet, a massive dataset containing over a million images tagged with 1,000 distinct items. Over the last several years, the ResNet-50 CNN design has been popular and profitable. The system excels in classifying pictures, recognizing objects, and cutting images into pieces.

### 2.2.1 *Resnet 50 algorithm*

Step 1: load a pre-trained ResNet-50 model from TensorFlow or PyTorch.
Step 2: Adjust the model's output layer to fit your classification task's classifications.
Step 3: Resize, normalize, and enhance photos to prepare your dataset. Ensure data format meets ResNet-50 model criteria.
Step 4: To train, adjust, and assess your model, divide your dataset into training, validation, and test sets.
Step 5:Train the model using the training dataset and learning rate and batch size hyper-parameters. Use a task-specific loss function and optimizer.
Step 6: Monitor performance and change hyperparameters by periodically validating the model on the validation set. Prevent overfitting by early stopping.
Step 7: Test the final model's generalization on the test dataset after training. After satisfying the model, use it to forecast unlabeled data.

## 2.3 *Analytical statistics*

The statistical analysis in this study was made easier by IBM SPSS Version 26, which used an Independent Sample T-test to compare mean accuracies. Viewing angle and image size were deemed dependent variables, whereas accuracy and amount of photos were considered independent variables. This method made it possible to thoroughly investigate how these variables affected the results, giving information on how different viewing angles and image sizes affected the accuracy metrics that were obtained from the dataset.

# 3 RESULTS

Convolutional neural networks (CNNs) and enhanced ResNet 50s were examined, and CNNs (84.20%) showed better accuracy than ResNet 50s (73.04%). The CNN algorithm is represented by Group-1, and the Enhanced ResNet 50 Technique is represented by Group 2. This suggests that, in the circumstances of this investigation, the CNN model performed better in terms of accuracy evaluation than the ResNet 50 model.

Table 1 displays sample data for the ResNet 50 and Convolutional Neural Network algorithms' accuracy.

Table 1. The following table 1 consists of accuracies of sample size of 10 for CNN and Resnet 50.

| S. No | CNN | RESNET 50 |
|---|---|---|
| 1 | 84.42 | 73.39 |
| 2 | 84.67 | 72.89 |
| 3 | 84.68 | 71.28 |
| 4 | 83.39 | 73.67 |
| 5 | 83.42 | 73.94 |
| 6 | 83.67 | 74.56 |
| 7 | 84.81 | 74.60 |
| `8 | 84.64 | 74.02 |
| 9 | 84.51 | 72.97 |
| 10 | 84.94 | 71.34 |

Displays the group statistics for size N=10, which includes mean accuracies and standard deviations for CNN and ResNet 50 approaches. Table 2 displays the sample data of the accuracies of the Convolutional Neural Network algorithm and ResNet.

Table 2. Convolutional neural networks have an accuracy mean of 84.20% and a standard deviation of 0.590 in, while ResNet 50 has an accuracy mean of 73.04%) and a standard deviation of 1.489. 50 algorithm.

| | GROUPS | N | Mean | Std. Deviation | Std. Error Mean |
|---|---|---|---|---|---|
| **ACCURACY** | CNN | 10 | 84.20 | 0.590 | 0.186 |
| | ResNet 50 | 10 | 73.04 | 1.489 | 0.470 |

Figure 1 is a graph that shows the comparison between the mean accuracy of the CNN and mean accuracy of the ResNet 50 algorithm along with the error bar.

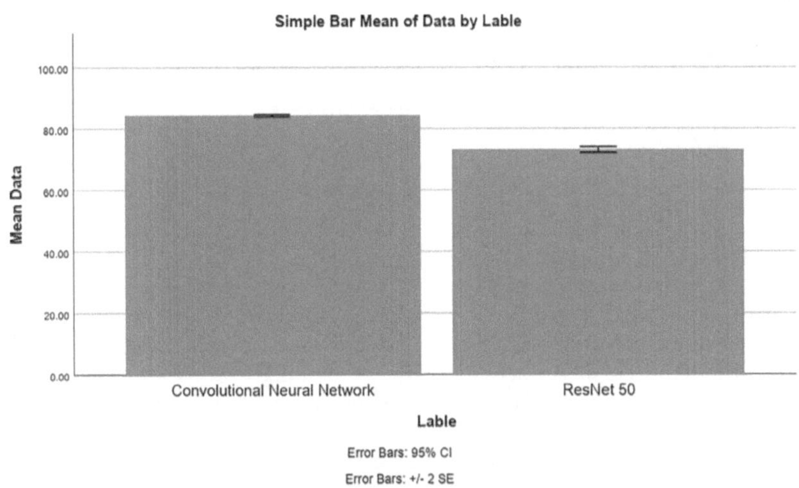


Figure 1. Mean Accuracy Comparison graph of Convolutional Neural Network Technique and ResNet 50. The accuracy of Convolutional Neural Network Technique (84.20%) is more than that of ResNet 50 (73.02%). The mean accuracy detection is +2SD.

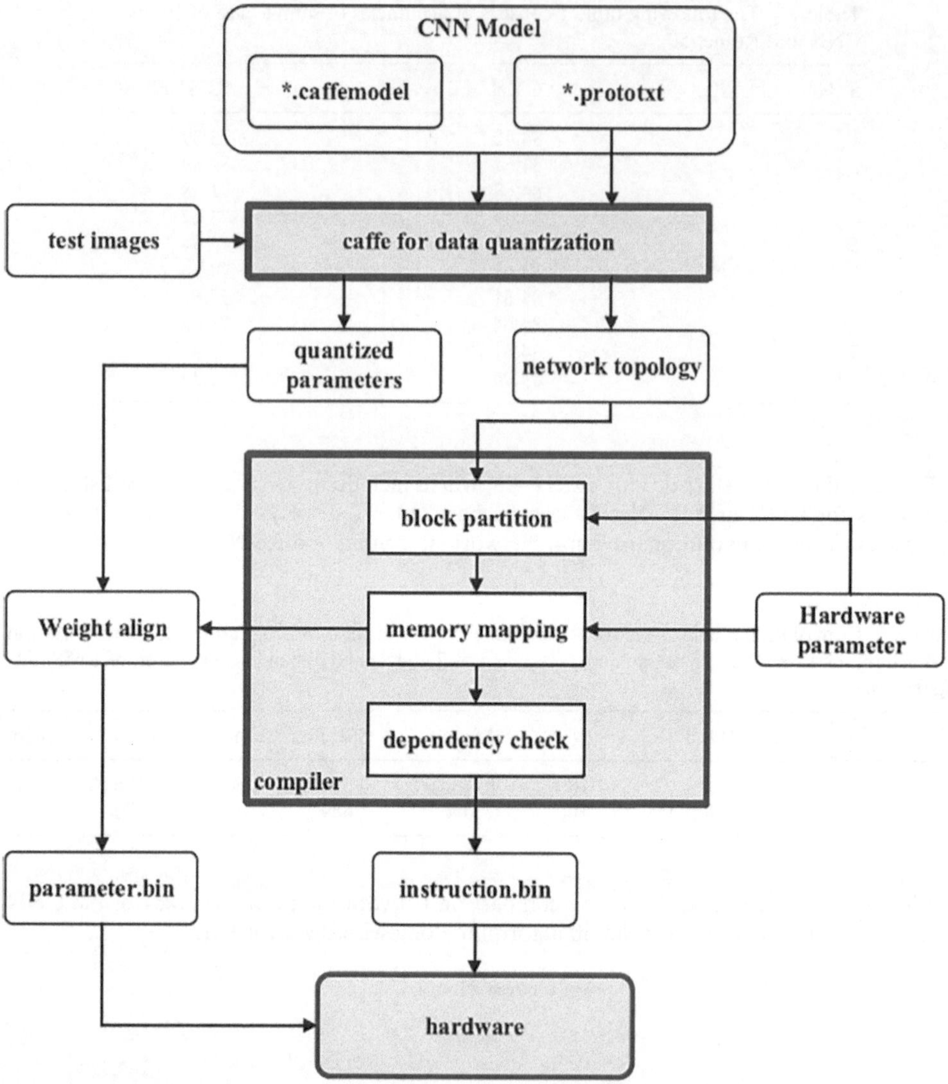


Figure 2. Flow chart.

## 4 DISCUSSION

As was already established, the Convolutional Neural Network clearly outperformed the Resnet 50 in terms of performance. It is discovered that Convolutional neural Network (84.20%) is more than that of Resnet (73.02%) . According to a list of earlier studies, the Convolutional Neural Networks Technique creates sharp edges while the Wiener Filter smoothes out the black dots in grayscale photographs (Shaw *et al.* 2023). The Desnet performs better in the spatial domain whereas the Resnet 50 performs better in the frequency domain. In contrast to the median filter, which can lessen image blurring, Convolutional neural Networks filters can be used to enhance blur effects to photos in order to maintain confidentiality (Khandelwal and Das 2022).

The similar findings update in September 2021, literature on Advanced Image Classification utilizing Convolutional Neural Networks (CNNs) with enhanced accuracy compared to ResNet-50 demonstrates a significant evolution in deep learning research. Various studies highlight the development of more intricate CNN architectures, often incorporating attention mechanisms, dense connections, and advanced pooling strategies to improve feature extraction and classification accuracy (Rai 2019). Researchers have conducted comprehensive comparative analyses, showcasing the superior performance of these advanced CNN models on diverse datasets. Additionally, investigations into regularization techniques, transfer learning strategies, and data augmentation methods have been documented to optimize model performance. Furthermore, the literature emphasizes the applicability of these advanced models in real-world applications, spanning medical image analysis, autonomous systems, remote sensing, and industrial quality control, highlighting their potential to revolutionize image-based decision-making processes in a multitude of domains (Bazi and Pasolli 2021).

Despite its advancements, Advanced Image Classification based on Convolutional Neural Networks (CNNs) with improved accuracy compared to ResNet-50 still faces certain limitations. These include the need for significant computational resources for training and inference, vulnerability to overfitting with complex architectures (Garg *et al.* 2023), and potential challenges in interpretability and explainability. Future research may focus on developing more efficient hardware and algorithms for faster computation, exploring novel regularization techniques for better generalization, and integrating interpretability methods to enhance model transparency. Additionally, exploring the integration of advanced CNNs with other emerging technologies like augmented reality and generative models could pave the way for more sophisticated and comprehensive image analysis solutions (Le Thi *et al.* 2019).

## 5 CONCLUSION

In advanced imaging, blurred noise can affect image quality and hinder accuracy. The Convolutional Neural Network is a powerful approach used to correct blurred noise in images, providing superior clarity compared to the widely used Wiener filter. This technique works by convolving the image with a Convolutional Neural Network to reduce the noise and improve the image's overall quality. Compared to the Resnet 50 (73.02%), the Convolutional Neural Network technique (84.20%) is more effective in providing clear images with better contrast, which is vital for accuray. As such, the Convolutional Neural Network technique is an essential tool in image processing that can help improve image quality and enhance accuracy.

## REFERENCES

[No title]. (n.d.). Retrieved June 12, 2024, from https://www.researchgate.net/profile/Punithavathi-Rasappan/publication/362388038_Computer_Vision_and_Deep_Learning-enabled_Weed_Detection_Model_for_Precision_Agriculture/links/62e7911d4246456b55ffc95a/Computer-Vision-and-Deep-Learning-enabled-Weed-Detection-Model-for-Precision-Agriculture.pdf

Aluvalu, R., Kumaran V N, S., Thirumalaisamy, M., Basheer, S., Ali Aldhahri, E. and Selvarajan, S. (2023). Efficient data transmission on wireless communication through a privacy-enhanced blockchain process. *PeerJ. Computer Science*, 9, e1308.

Bazi, Yakoub, and Edoardo Pasolli. (2021). *Advanced Deep Learning Strategies for the Analysis of Remote Sensing Images*. MDPI.

Bhatt, Kumari Indu. (2018). Efficacy of Deep Learning in Image Classification.

Field, Andy. (2018). *Discovering Statistics Using IBM SPSS Statistics. SAGE Publications Limited.*

Garg, Deepak, V. A. Narayana, P. N. Suganthan, Jaume Anguera, Vijaya Kumar Koppula, and Suneet Kumar Gupta. (2023). *Advanced Computing: 12th International Conference*, IACC 2022, Hyderabad, India, December 16–17, 2022, Revised Selected Papers, Part II. Springer Nature.

Hu, Huafeng. (2022). Advanced Medical Image Classification with Convolutional Neural Networks.

Khandelwal, Shekhar, and Rik Das. (2022). *Phishing Detection Using Content-Based Image Classification*. CRC Press.

Kulshrestha, Saurabh. (2019). Developing an Image Classifier Using TensorFlow: Convolutional Neural Networks.

Le Thi, Hoai An, Hoai Minh Le, Tao Pham Dinh, and Ngoc Thanh Nguyen. (2019). *Advanced Computational Methods for Knowledge Engineering: Proceedings of the 6th International Conference on Computer Science*, Applied Mathematics and Applications, ICCSAMA 2019. Springer Nature.

Lu, Le, Xiaosong Wang, Gustavo Carneiro, and Lin Yang. (2019). *Deep Learning and Convolutional Neural Networks for Medical Imaging and Clinical Informatics*. Springer Nature.

Mahaveerakannan R, Anitha, C., Aby K Thomas, Rajan, S., Muthukumar, T. and Govinda Rajulu, G. (2023). An IoT based forest fire detection system using integration of cat swarm with LSTM model. *Computer Communications*, 211, 37–45.

Millstein, Frank. (2020). Deep Learning: 2 Manuscripts - Deep Learning With Keras And Convolutional Neural Networks In Python. Frank Millstein.

Rai, Bharatendra. (2019). *Advanced Deep Learning with R: Become an Expert at Designing, Building, and Improving Advanced Neural Network Models Using R*. Packt Publishing Ltd.

Ramana, Kadiyala, A. Revathi, A. Gayathri, Rutvij H. Jhaveri, C. V. Lakshmi Narayana, and B. Naveen Kumar. (2022). WOGRU-IDS — An Intelligent Intrusion Detection System for IoT Assisted Wireless Sensor Networks. *Computer Communications* 196 (December): 195–206.

Serrano, Luis. (2021). *Grokking Machine Learning*. Simon and Schuster.

Sewak, Mohit, Md Rezaul Karim, and Pradeep Pujari. (2018). *Practical Convolutional Neural Networks: Implement Advanced Deep Learning Models Using Python*. Packt Publishing Ltd.

Shanmugamani, Rajalingappaa. (2018). Deep Learning for Computer Vision: Expert Techniques to Train Advanced Neural Networks Using TensorFlow and Keras.

Shaw, Rabindra Nath, Marcin Paprzycki, and Ankush Ghosh. (2023). *Advanced Communication and Intelligent Systems: First International Conference*, ICACIS 2022, Virtual Event, October 20-21, 2022, Revised Selected Papers. Springer Nature.

Vellela, S. S. and Balamanigandan, R. (2023). An intelligent sleep-awake energy management system for wireless sensor network. *Peer-to-Peer Networking and Applications*, 16(6), 2714–2731.

*Hybrid and Advanced Technologies – S. Prasad Jones Christydass, Nurhayati Nurhayati and S. Kannadhasan. (Eds) © 2025 the Authors, ISBN 978-1-032-90254-8*
*Open Access: www.taylorfrancis.com, CC BY-NC-ND 4.0 license*

# A novel approach to constructing features and models for intrusion detection systems using SAE-ELM model

Davinder Paul Singh
*Department of Computer Science and Engineering, School of Technology, Pandit Deendayal Energy University, Gandhinagar, Gujarat, India*

Vinod Wamanrao Gangane
*Assistant Professor, Department of Computer Science, Yogeshwari Mahavidyalaya, Ambajogai, Maharashtra, India*

P. Sukumar
*Professor, Department of Computer Science and Engineering, Nandha Engineering College (Autonomous), Erode, Tamilnadu, India*

Jyoti Kaushal
*Assistant Professor/CSE, Geetanjali institute of technical studies, Udaipur, India*

Neerav Nishant
*Assistant Professor, Department of Computer Science and Engineering, School of Engineering, Babu Banarasi Das University, Lucknow, India*
ORCID ID: 0009-0002-6696-6943

Harshal Patil
*Associate Professor, Balaji Institute of Technology & Management (BITM), Sri Balaji University, Pune, Maharashtra, India*

ABSTRACT: Integrating intrusion detection (ID) into infrastructure security procedures is essential. It is essential for an intrusion detection system (IDS) to have accuracy, extensibility, and adaptability. Due to these demands and the complexity of modern network settings, a more structured and automated approach is needed to design intrusion detection systems (IDS). This is in contrast to the conventional methods, which rely purely on knowledge encoding and engineering. The three primary parts are training the model, feature selection, and preprocessing. The goal of data pre-processing is to make the raw data more usable and efficient for subsequent processing stages. Feature selection makes use of mutual information (MI) to assess any kind of random reliance between variables. For more accurate findings, we used SAE-ELM to train the model. The proposed method achieved a higher level of accuracy (96.30 percent) compared to its competitors, including SAE and ELM.

*Keywords*: Intrusion Detection Systems (IDSs), Mutual Information (MI), Minimum Description Length (MDL), Auto Encoder (AE), Extreme Learning Machine (ELM)

## 1 INTRODUCTION

Electronic communication, education, commerce, and entertainment are just a few of the many modern-day requirements that people rely on the Internet to meet. Cyber security is at a critical juncture because of the ever-increasing number of computer networks and intrusion cases. The critical need to fortify cyber defences has recently come to the attention of

DOI: 10.1201/9781003559115-67 

businesses and academic institutions around the world. Despite the prevalence of security mechanisms such as firewalls, anti-malware software, data encryption, and user authentication, several businesses and organisations continue to be compromised in the modern era. If hackers were able to get access to a system, they could take use of its known vulnerabilities to launch a barrage of attacks, some of which could compromise important data. Intrusion detection systems (IDSs) are essential for protecting systems from a wide range of threats. In order to retrieve only a portion of the original features, a selection algorithm is proposed. Secondly, an imbalance between benign and malicious traffic reduces the precision of attack detection. Our suggested solution uses an ensemble classifier to reduce bias in different training datasets, therefore avoiding this problem. By utilising a combination of ensemble classifiers and feature selection, the IDS is able to improve its accuracy and stability while reducing computational and time complexity. Finally, there is a way to build a fair model that can detect both typical and unusual intrusive events.

## 2 LITERATURE SURVEY

A major risk is the abundance of security flaws in the IoT. Many different types of attacks and vulnerabilities can compromise devices connected to the IoT [1]. The effectiveness of several intrusion detection technologies was discussed in a review article by [2]. The authors display good news regarding intrusion detection. And developed a deep learning-based explainable intrusion detection solution for IoT networks. The primary emphasis, however, is on the IoV, a subset of the broader IoT. Machine[3] discovered that intrusion detection systems based on learning perform admirably in computing settings where fog and edge are present. A smart network intrusion detection system was developed integrating an ensemble meta-classifier with a feature fusion method based on recurrent deep learning[4]. As stated in [5], this approach has a 98.9% accuracy rate. Research using ANN and association rule mining suggested a way to greatly enhance communication network intrusion detection [6]. The installation of a reliable intrusion detection system ranks high among the priorities for network security. Anomaly detection algorithms have made great strides in NIDS, but most existing systems still depend on signature-based approaches [7]. A product that can be sold to the public, the intrusion detection system has been in the works for over 30 years. On top of that, research on internet safety has focused on it extensively [8]. Improve and enhance intrusion detection on an ongoing basis to make it more effective [9]. Using a traditional neural network,[10] classified potential dangers to a network. To build a two-layer multi-layer perceptron, the back propagation learning method was employed. [11] Developed a deep learning technique for detecting threats to IoT networks by comparing their suggested model to the Shallow model. The experts suggested putting this tech on the cloud to make it better at training and threat detection. [12]With the help of an intrusion detection system, systems and networks can be better protected from cyber assaults. Protecting a network's infrastructure from cyberattacks and limiting their impact on operations and finances are the responsibilities of intrusion detection systems (IDSs)[13]. Using classifiers and established patterns, the signature detection method improves the accuracy of evaluating hostile attacks. It finds potentially harmful situations by utilising previously obtained data[14]hence, it is referred to as a knowledge-based strategy; possible risks. Newly created networks cannot be detected by the approach, even if it increases accuracy and decreases false positives (FP) dangers.

## 3 PROPOSED SYSTEM

In order to find patterns of misuse or intrusion, modern intrusion detection systems (IDS) examine every part of the data. Some of the traits might not be necessary or contribute significantly to detection. A computationally efficient and effective ID can be developed with the help of this technique, which aims to identify critical input features.

### 3.1 *Preprocessing*

The purpose of data pre-processing is to prepare raw data for further processing by making it more useable. Data normalization using the min-max method is a first step[15]. Normalization has the potential to reduce training time since all of the training data is standardised (e.g., between 0 and 1). The normalized value, $P_{norm}$, and the original, pre-normalization value, $P$, can be used in (1) to achieve this. The maximum and minimum potential feature values are represented here by $P_{\max}$ and $P_{\min}$, respectively.

$$P_{norm} = \frac{P - P_{\min}}{P_{\max} - P_{\min}} \tag{1}$$

The step two involves pre-processing the data through the use of Minimum Description Length (MDL) discretization. With data discretization, we aim to minimize data loss while converting continuous data attributes' values into bounded intervals.

### 3.2 *Feature selection*

The mutual information (MI) is a tool in information theory for evaluating dependencies between two random variables. Instead of being a measure of the knowledge on $P$ supplied by $Q$, the MI between two random variables $P$ and $Q$ is a measure of the quantity of knowledge on $P$ supplied by $Q$, or vice versa[16]. Zero mutual information exists between any two sets $P$ and $Q$ if and only if neither set has any knowledge about the other. When $P$ and $Q$ are two independent variables, their MI is defined as

$$H(P;Q) = I(P) - I(P|Q) = I(Q) - I(Q|, P) = I(P) + I(Q) - I(P;Q) \tag{2}$$

$I(P;\ Q)$ is the joint entropy of $P$ and $Q$, which are defined as, and $I(P|Q)$ and $I(Q|, P)$ are conditional entropies. Meanwhile, $I(.)$ is the entropy in general. The MI equation is

$$H(P;Q) = \int_p \int_q x_{P,Q}(p,q) \log \frac{x_{P,Q}(p,q)}{x_P(p)x_Q(q)} \tag{3}$$

### 3.3 *Model training*

#### 3.3.1 *ELM*

With ELM, a standard neutral network's parameters are optimized by solving a system of linear equations rather than iteratively. As a generalizer, ELM outperforms more traditional learning algorithms due to its ability to avoid needless repetitions. In a feedforward neutral network with one hidden layer and n nodes, for every $G$ samples $(p_h, l_h)$, where $p_h = [p_{h1}, p_{h2}, \ldots, p_{hg}]^{\mathrm{T}} \in R^g$ and $l_h = [l_{h1}, l_{h2}, \ldots, l_{hf}]^{\mathrm{T}} \in R^f$, this stands for the feedforward neutral network. According to (4),

$$m_T(p) = \sum_{h=1}^{T} \alpha_h N(e_h, z_h, p) \tag{4}$$

$p$ is a member of $R$, where $e_h \in R^g$ and $\alpha_h \in R^f$ are elements of the set. $N(e_h, z_h, p)$ denotes the connection between nodes $h$ and $e$. The function $n(p) R \rightarrow R$, and there is

$$N(e_h, z_h, p) = n(e_h.p + z_h), z_h \in R \tag{5}$$

Node $h$ in the hidden layer receives the weight vector from the input layer, which is $e_h = [e_{h1}, e_{h2}, \ldots, e_{hg}]^{\mathrm{T}} \in R^g$; the threshold value of node $h$ is denoted by $z_h$; and This is the output weight vector that is represented from node $h$ to the output layer by $\alpha_h = [\alpha_{h1}, \alpha_{h2}, \ldots, \alpha_{hf}]^{\mathrm{T}} \in R^f$.

### 3.3.2 *SAE-ELM*

As with the autoencoder (AE) approach, the model's output is exactly proportional to its input. In order to decrease the data dimension, we can minimize the number of input nodes until they are less than the number of nodes in the hidden layer. In order to reduce the amount of raw data, SAE constructs many AE models. Two AE-related characteristics used by the SAE model are the total AE count and the AE-specific concealed node count. The initial spectral data is 643 × 983 and has a high dimensionality. The original spectral data dimension is intended to be reduced by employing the ELM, a feedforward neural network with a single hidden layer. This is accomplished by SAE by means of two AE models. We used continuous testing to choose the nodes for the hidden layer, which is a parameter that needs to be set, and we got good results.

## 4 RESULT AND DISCUSSION

The importance of detecting and preventing threats to network-based systems is growing in tandem with the pervasiveness of these technologies. One method to improve safety is to use intrusion detection systems. These are software applications that scan a system or network for signs of malicious activity. One technical challenge with intrusion detection systems is the curse of excessive dimensionality.

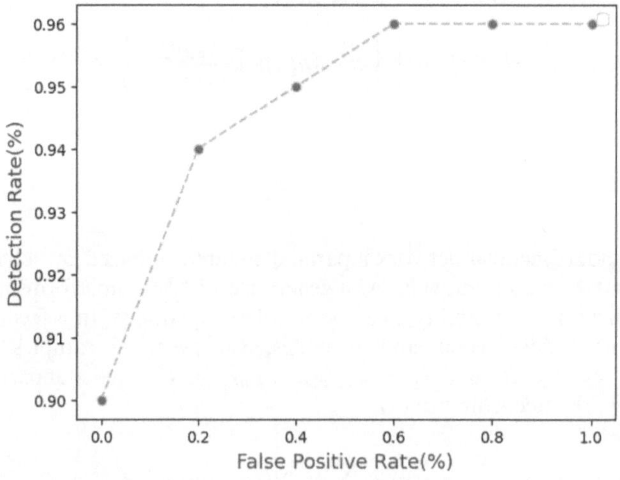


Figure 1. ROC curve of proposed model.

The receiver operating characteristic (ROC) curve is plotted in Figure 1 to show the relationship between the detection rates and the false-positive rates. Based on the results, our technique achieves a high detection rate while maintaining a low false-positive rate.

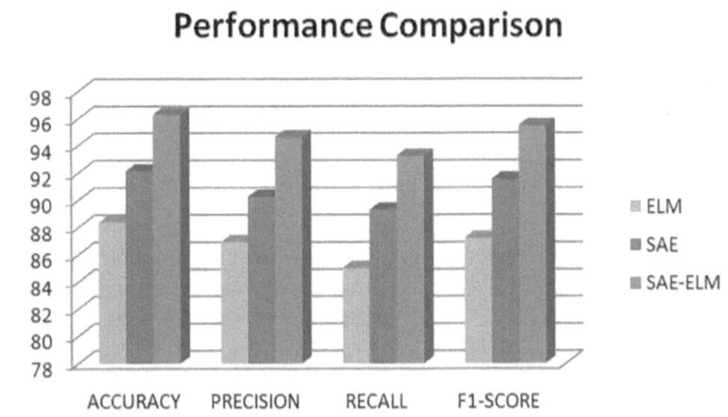


Figure 2. Accuracy and total average outcome.

Figure 2 displays the average and accuracy of the overall performance based on the testing findings for each performance indicator.

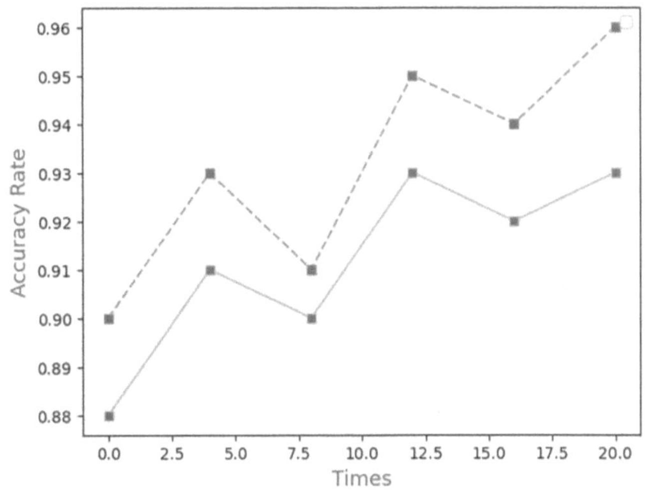


Figure 3. Performance of ELM and SAE-ELM model.

The suggested approach is shown to be more stable than the original ELM by twenty repetitions of trials using the identical randomly generated training and test sets. Figure 3 displays the outcome. As seen in Figure 3, SAE-ELM clearly outperforms the original ELM in terms of stability.

## 5 CONCLUSION

According to a recent Google survey, the term "internet of things" has already surpassed "big data" as the most popular buzzword in the IT industry. Nevertheless, the Internet of Things (IoT) field is far from developed, and it is shedding light on numerous research difficulties for data mining experts. The scarcity of resources makes addressing security in the IoT a difficult task. To improve the usability and efficiency of the raw data for later processing steps, data pre-processing is performed. To determine the extent to which variables are arbitrary dependent on one another, feature selection employs mutual information (MI). The SAE-ELM method is used to train the models. The suggested method beats both the SAE and ELM models, with a consistency rate of 96.30 percent.

## REFERENCES


[1] Husnain, M., Hayat, K., Cambiaso, E., Fayyaz, U. U., Mongelli, M., Akram, H., ... and Shah, G. A. (2022). Preventing mqtt vulnerabilities using iot-enabled intrusion detection system. *Sensors*, 22(2), 567.

[2] Kolhar, M. and Aldossary, S. M. (2023). A deep learning approach for securing IoT Infrastructure with emphasis on smart vertical networks. *Designs*, 7(6), 139.

[3] Alzubi, O. A., Alzubi, J. A., Alazab, M., Alrabea, A., Awajan, A. and Qiqieh, I. (2022). Optimized machine learning-based intrusion detection system for fog and edge computing environment. *Electronics*, 11(19), 3007.

[4] Mao, Y., Xiao, D., Cheng, J., Che, D., Le, B., Song, L. and Liu, X. (2017). Multigrades classification model of magnesite ore based on SAE and ELM. *Journal of Sensors*, 2017(1), 9846181.

[5] Ravi, V., Chaganti, R. and Alazab, M. (2022). Recurrent deep learning-based feature fusion ensemble meta-classifier approach for intelligent network intrusion detection system. *Computers and Electrical Engineering*, 102, 108156.

[6] Safara, F., Souri, A. and Serrizadeh, M. (2020). Improved intrusion detection method for communication networks using association rule mining and artificial neural networks. *IET Communications*, 14 (7), 1192–1197.

[7] Wang, G., Hao, J., Ma, J. and Huang, L. (2010). A new approach to intrusion detection using Artificial Neural Networks and fuzzy clustering. *Expert Systems with Applications*, 37(9), 6225–6232.

[8] Sultana, N., Chilamkurti, N., Peng, W. and Alhadad, R. (2019). Survey on SDN based network intrusion detection system using machine learning approaches. *Peer-to-Peer Networking and Applications*, 12(2), 493–501.

[9] Louati, F. and Ktata, F. B. (2020). A deep learning-based multi-agent system for intrusion detection. *SN Applied Sciences*, 2(4), 675.

[10] Sharafaldin, I., Lashkari, A. H. and Ghorbani, A. A. (2018). Toward generating a new intrusion detection dataset and intrusion traffic characterization. *ICISSp*, 1, 108–116.

[11] Balakrishna, G., Kumar, A., Tamilselvan, V., Younas, A., Kumar, N. M. G. and Rastogi, R. (2023, October). A novel ensembling of CNN-A-LSTM for IoT electric vehicle charging stations based on intrusion detection system. In *2023 International Conference on Self Sustainable Artificial Intelligence Systems (ICSSAS)* (pp. 1312–1317). IEEE.

[12] Daund, R. P., Kumar, D., Charan, P., Ingilela, R. S. K. and Rastogi, R. (2023, July). Intrusion detection in wireless sensor networks using hybrid deep belief networks and harris hawks optimizer. In *2023 4th International Conference on Electronics and Sustainable Communication Systems (ICESC)* (pp. 1631–1636). IEEE.

[13] Akhther, P., Maryposonia, A. and Prasanth, V. S. (2023, May). Least square support vector machine based intrusion detection system in IoT. In *2023 7th International Conference on Intelligent Computing and Control Systems (ICICCS)* (pp. 1545–1550). IEEE.

[14] Sagar, A., Anushkannan, N. K., Indumathi, G., Muralidhar, N. V., Dhamotharan, K. A. and Malini, P. (2023, June). Wireless sensor network-based intrusion detection technique using deep learning approach of CNN-GRU. In *2023 8th International Conference on Communication and Electronics Systems (ICCES)* (pp. 1147–1152). IEEE.

[15] Ahmad, T. and Aziz, M. N. (2019). Data preprocessing and feature selection for machine learning intrusion detection systems. *ICIC Express Lett*, 13(2), 93–101.

[16] Amiri, F., Yousefi, M. R., Lucas, C., Shakery, A. and Yazdani, N. (2011). Mutual information-based feature selection for intrusion detection systems. *Journal of Network and Computer Applications*, 34(4), 1184–1199.

*Hybrid and Advanced Technologies – S. Prasad Jones Christydass, Nurhayati Nurhayati and S. Kannadhasan. (Eds)* © *2025 the Authors, ISBN 978-1-032-90254-8*
*Open Access: www.taylorfrancis.com, CC BY-NC-ND 4.0 license*

# Automated detection of substance use-related social media posts based on hybrid GAT-BiGRU model

R. Nadanasabai
*Associate Professor, Department of M. B. A, St.Joseph's College of Engineering, Chennai, Tamil Nadu, India*

Venkata Ramaiah Turlapati
*Functional Consultant - SAP- IT, IMG Systems Inc, Michigan, USA*
ORCID ID: 0009-0004-1745-3994

Ramesh Kumar Miryala
*Assistant Professor, Department of Business Management, Mahatma Gandhi University, Nalgonda*
ORCID ID: 0000-0003-2282-1558

P.L. Swerna
*Assistant Professor, Department of Commerce, Apollo Arts and Science College Ponnamallee, Chennai, Tamil Nadu, India*
ORCID ID: 0009-0001-4524-731X

M. Nimisha
*Assistant Professor, Department of Management Studies, VLB Janakiammal College of Arts and Science, Kovaipudur, Coimbatore*

R. Arun
*Assistant Professor, Department of M. B. A., St.Joseph's College of Engineering, Chennai, Tamil Nadu, India*
ORCID ID: 0000-0002-5252-1030

ABSTRACT: It is possible that social media can shed light on substance addiction and abuse. This study set out to apply a deep learning algorithm we developed to automatically classify Instagram users' risk for alcohol, tobacco, and drug usage. A total of three thousand active Instagram users were surveyed. Automated estimation techniques can be useful for future generations of population-level risk assessment and intervention delivery. Preprocessing, learning features, and training the model are the three steps that make up the proposed technique. While LDA and M-DW are utilized for feature learning, TF-IDF is employed for preprocessing. When training the model, various methods were utilized, including BiGRU, DNN-BiGRU, and GAT-BiGRU. When compared to two more traditional methods, GAT-BiGRU triumphs.

*Keywords*: Linear discriminant analysis (LDA), TF-IDF (Term Frequency-Inverse Document Frequency), Graph Attention Network (GAT), Borderline Personality Disorder (BPD), Bidirectional Gated Recurrent Unit (BiGRU)

DOI: 10.1201/9781003559115-68

## 1 INTRODUCTION

During the transition into emerging adulthood, a person's personal, environmental, and social roles undergo significant changes. Many different systems, both internal and external to the individual, are in a state of flux throughout this time. Because this transition happens during a brief time of rapid development, some youths may find emerging adulthood especially difficult, which might encourage them to participate in risky behaviors. Substance use and abuse can be influenced by both personal and environmental factors. One of the numerous contextual impacts on today's youth is social media. Instagram, Snapchat, Facebook, Twitter, Tumblr, and Vine are all examples of social media sites that enable users to produce, share, and aggregate material from other users. A person's environment plays a significant role in shaping their development and actions. Since social media has grown into an integral part of coming-of-age, it's important to know how it could influence behavior during this formative time. Last but not least, previous studies on the topic of drug use and social media have mostly used undergraduate student samples, which limits how applicable the conclusions are to a broader population. This study sought to address a gap in the existing literature by examining the responses of a diverse national sample of young adults to substance use and abuse on several social media platforms, including Facebook, Twitter, Instagram, Snapchat, Vine, Tumblr, YouTube, and Google+.

## 2 LITERATURE SURVEY

There may be a link between the two recent trends of increasing social media use and changing patterns of risky conduct among adolescents, [1] as these behaviours have changed throughout time in reaction to broader social and cultural pressures. Social media users are less likely to engage in healthful activities including going for walks, [2] playing sports, and having face-to-face conversations, according to the displacement hypothesis. According to [3], the fast increase in social media use among adolescents has likely supplanted other risky behaviours such heavy alcohol use and drug use. [4] With the growth of parental supervision and societal prohibitions on most substances in Western nations, argue that social media could provide a new platform for teenagers to showcase their growing risk-taking tendencies. According to the Facebook effect model, which was developed by [5], the susceptibility of teenagers to peer pressure increases the likelihood that they may participate in risky online activities. [6] Adolescent drug use in the United States has changed dramatically throughout the last 20 years. Teens are more likely to use vaporizers, such as electronic cigarettes, and there have been concerns about the high rates of nicotine consumption among this demographic [7]. Among the many stated reasons given by teenagers for using electronic cigarettes were the desires to experiment and the want to enjoy themselves in social settings [8]. Decriminalization of marijuana for recreational and medicinal use might have an effect on adolescent drug use [9]. Potency levels of marijuana edibles and vape strains may be higher than those of smoked strains once legalization takes place. According to [10], the use of vaporized marijuana increased among youths last year. Adolescents from lower socio-economic origins, females, and those who have acquaintances who use drugs are more likely to be affected. [11] Staying informed about current and future trends, news, and connections is just one of the numerous ways social media may be used. The many forms of social media have the ability to be intrusive and overwhelming due to the asynchronous nature of contact and the continuous coverage of events [12]. During times of crisis or mass trauma, social media has the potential to amplify and disseminate anxiety, which is different from traditional news reporting [13]. A large body of evidence suggests that regular exposure to mainstream media may increase psychological distress. These findings may represent the extent to which worry can spread across different social media platforms as a result of emotional contagion.

## 3 PROPOSED SYSTEM

We demonstrate, in this study, how to build drug use detection systems for social media utilizing cutting-edge algorithms for text mining and machine learning. We explore various feature learning algorithms to leverage a large quantity of unsupervised social media data and maximize system performance, since collecting a comprehensive ground truth on drug consumption is tough. We also found that using multi-view unsupervised feature learning to combine varied user data, such as Facebook "likes" and "status updates," enhances system performance.

### 3.1 *Preprocessing*

In order to facilitate classification, it splits the subreddits into two groups: bipolar and others. There were subreddits for mental health concerns like as schizophrenia, depression, anxiety, and more in the "others" category. The primary focus of this organization was on distinguishing between bipolar disorder and borderline personality disorder (BPD), but it also considered the broader category of other mental disorders, making the classification process easier and more manageable [14]. Before the text data could be used with machine learning models, a lot of preparatory work had to be done. Before conducting the analysis, the URLs were removed from the text to guarantee that they would not be impacted by any online linkages. Additionally, we removed those punctuation marks from the text to ensure that the categorization process remained punctuation-free. We improved their accuracy and reduced noise by removing stopwords, which are words that appear often but do not add to the meaning of the features. Making a TF-IDF (Term Frequency-Inverse Document Frequency) feature matrix from the original text documents was the last stage. To measure the importance of a term in the documents, this matrix takes into account how often and seldom it appears in the corpus. The procedures for getting ready applied to both the titles and the contents of the posts. By analyzing both the headlines and the body of the posts, the study takes into account the discriminatory power and potential insights that could result from combining the two. We aim to extract the most relevant information for accurate classification and optimize the title and content of the posts using this comprehensive approach.

### 3.2 *Learning of features*

#### 3.2.1 *LDA*

Latent Dirichlet Allocation (LDA) searches document sets for latent topics and utilizes them to explain them; it is an example of a generative graphical model. Every document is given a multinomial distribution via LDA, which is based on a set of latent topics. Furthermore, the vocabulary related to each topic is generated via LDA using a multinomial distribution. In order to learn a user's SPE from all of their postings, the suggested method tries out various ways. (1) UserLDA: this tool compiles all user posts into one document and then uses an LDA model to find out what topics people are talking about in that document. The SPE is derived from the distribution of themes per document for this specific user. Second, the PostLDA Doc employs a logistic regression (LDA) model to find out how often certain subjects appear in each post on an individual basis [15]. In order to determine a user's SPE, it takes an average of all the topic distribution vectors every post. Instead of averaging post-based topic distribution vectors, the third approach, PostLDA Word, uses a word-based aggregate strategy.

#### 3.2.2 *D-DM*

Each word vector and the document's overall vector are simultaneously learned by D-DM when a document is supplied to it. During training, the document vector is combined with one or more word vectors to predict the context of a target word. Two methods have been explored in an effort to ascertain a SPE for each user. To start, there's User-D-DM, which treats a user's posts as a single document after training a document vector to represent the user. (2) Post-D-DM: A post-specific vector is acquired by training a D-DM with a document-level view of posts. When it has collected all of the post vectors for a user and averaged them using "average," it will be able to determine their SPE.

### 3.3.3 *Model training*

3.3.3.1 *GAT* By employing hidden self-attention layers to prioritize neighbor nodes based on their relevance, Graph Attention Network (GAT) makes up for the shortcomings of graph convolution and its approximate methods. A novel neural network called GAT uses graph-structured da-ta to function. Without needing to understand the graph structure or perform complex matrix operations, nodes can take part in neighborhood features and have weights assigned to them through the use of stacking layers. The GAT computation technique relies heavily on the input data being converted into higher-dimensional characteristics. Adding a multi-headed attention mechanism stabilizes the GAT training process and improves the model.

3.3.3.2 *GAT-BiGRU* Even while the GRU model can handle forward-looking traffic flow data, it is important to note that associated factors, such as driver operations, can also affect this data. Therefore, the paper uses the BiGRU model to obtain forward and backward information for each node. Overfitting occurs frequently when using the BiGRU model to extract temporal features; however, by including a Dropout layer in the network, this problem can be avoided. By doing so, we guarantee accurate propagation of training mistakes along the time axis. Using time series data and a GAT structure with distinct attention mechanisms, this study demonstrates how to train a model to capture spatial features [16]. The model then learns spatial correlations using different attention modules. By incorporating geographical information into time series data, the BiGRU structure may learn temporal properties. A fully-connected neural network will thereafter provide the anticipated results. It shows the traffic flow information model's GAT-BiGRU flow mechanism. As a first step, the model is applied to the road network's historical traffic flow data X1 at a specific period in order to obtain the hidden state h1. Following the acquisition of X2 and h1 via the GAT structure, the BiGRU is employed to ascertain the concealed state h2 from the aforementioned traffic flow data. We repeat this process until we reach the last step, which is to output via the fully linked layer and obtain the final projected value. This will give us the expected outcome at time t.

## 4 RESULT AND DISCUSSION

Substance abuse is more common among emerging adults, which can cause a lot of problems now and in the future. For a community sample of 120 emerging adults, this study looks at the connections between 10th graders' self-reported substance use, the substance content of their Facebook posts and private messages in the fall after 10th grade, and their self-reported substance use one year after high school. Researcher observational coding revealed that 25% of individuals made drug-related Facebook posts in public, whereas nearly 50% made drug-related posts in private. Twelfth grade substance usage indicated the chance of participating in both public and private discussions about substances. Tobacco and marijuana use were predictors of the frequency of private messages addressing substances. Public and private online substance discussions predicted a positive shift in marijuana consumption two years later.

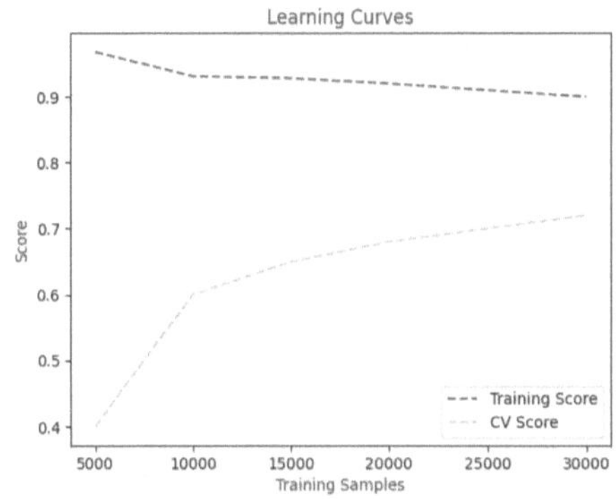


Figure 1. Learning curve of the proposed model.

Figure 1 shows that, although training accuracy with 300,000 examples is better than 20,000 examples with little changes, training model accuracy steadily declines with increasing numbers of training instances.

The reliability of cross-validation improves with increasing case numbers. Take cross-validation accuracy as an example; it's much greater with 30,000 samples than with 6,000 samples with minor differences. When evaluated using the F-measure, GAT-BiGRU's class-wise classification performance remains superior, as shown in Figure 2.

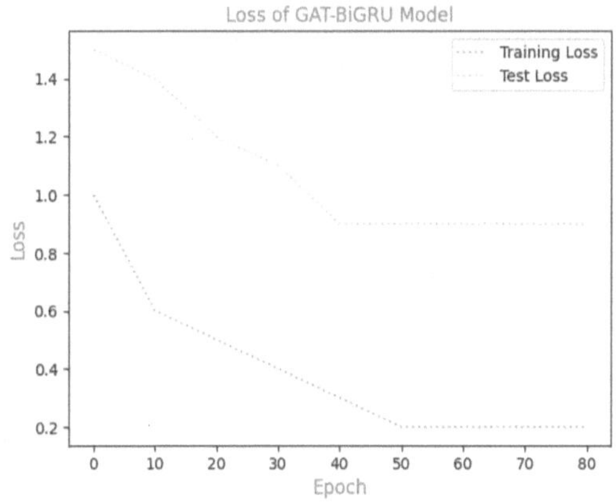


Figure 2. Loss of the GAT-BiGRU model.

The effectiveness of the suggested methodology in identifying social media metrics related to substance use from posts and comments is shown in Table 1.

Table 1. Evaluation metrics.

| Models | Precision | Recall | Acc | F1-Score |
|---|---|---|---|---|
| BiGRU | 86.5 | 84.8 | 89.2 | 89.7 |
| DNN-BiGRU | 90.4 | 89.4 | 92.6 | 92.8 |
| GAT-BiGRU | 94.3 | 92.3 | 96.8 | 96.9 |

## 5 CONCLUSION

Nowadays, a large portion of people's time is devoted to social media, particularly among young adults and teenagers. The nationwide study of American attitudes on substance usage found that when young people spend more time on social media, they are more likely to smoke, drink, and use illegal drugs. Reducing adolescent exposure to substance-related content on social media can help reduce the likelihood that they will develop an addiction. In this research, we present an automated method for finding drug-related social media content. Using this technology, social media networks can automatically remove any post that mentions substance usage. To train a model that can identify social media posts related to substance use, it utilizes preprocessing, feature learning, and GAT-BiGRU. The novel technique outperforms the previously used BiGRU and DNN-BiGRU methods.

## REFERENCES


[1] Boniel-Nissim, M., van den Eijnden, R. J., Furstova, J., Marino, C., Lahti, H., Inchley, J., ... and Badura, P. (2022). International perspectives on social media use among adolescents: Implications for mental and social well-being and substance use. *Computers in Human Behavior*, 129, 107144.

[2] Gommans, R., Stevens, G. W., Finne, E., Cillessen, A. H., Boniel-Nissim, M., and ter Bogt, T. F. (2015). Frequent electronic media communication with friends is associated with higher adolescent substance use. *International Journal of Public Health*, 60, 167–177.

[3] Baker, T., and Pelfrey Jr, W. V. (2016). Bullying victimization, social network usage, and delinquent coping in a sample of urban youth: Examining the predictions of general strain theory. *Violence and Victims*, 31(6), 1021–1043.

[4] Barry, C. T., Sidoti, C. L., Briggs, S. M., Reiter, S. R., and Lindsey, R. A. (2017). Adolescent social media use and mental health from adolescent and parent perspectives. *Journal of Adolescence*, 61, 1–11.

[5] Best, P., Manktelow, R., and Taylor, B. (2014). Online communication, social media and adolescent wellbeing: A systematic narrative review. *Children and Youth Services Review*, 41, 27–36.

[6] Fergusson, D. M., Swain-Campbell, N. R., and Horwood, L. J. (2002). Deviant peer affiliations, crime and substance use: A fixed effects regression analysis. *Journal of Abnormal Child Psychology*, 30, 419–430.

[7] Henneberger, A. K., Gest, S. D., and Zadzora, K. M. (2019). Preventing adolescent substance use: A content analysis of peer processes targeted within universal school-based programs. *The Journal of Primary Prevention*, 40, 213–230.

[8] Hopfer, C. (2014). Implications of marijuana legalization for adolescent substance use. *Substance Abuse*, 35(4), 331–335.

[9] Pearson, M., Sieglich, C., and Snijders, T. (2006). Homophily and assimilation among sport-active adolescent substance users.

[10] Schaefer, D. R. (2018). A network analysis of factors leading adolescents to befriend substance-using peers. *Journal of Quantitative Criminology*, 34, 275–312.

[11] Bränström, R., and Pachankis, J. E. (2018). Sexual orientation disparities in the co-occurrence of substance use and psychological distress: A national population-based study (2008–2015). *Social Psychiatry and Psychiatric Epidemiology*, 53, 403–412.

[12] Hotton, A., Quinn, K., Schneider, J., and Voisin, D. (2019). Exposure to community violence and substance use among Black men who have sex with men: Examining the role of psychological distress and criminal justice involvement. *AIDS Care*, 31(3), 370–378.

[13] Dumas, T. M., Ellis, W., and Litt, D. M. (2020). What does adolescent substance use look like during the COVID-19 pandemic? Examining changes in frequency, social contexts, and pandemic-related predictors. *Journal of Adolescent Health*, 67(3), 354–361.

[14] Nova, K. (2023). Machine learning approaches for automated mental disorder classification based on social media textual data. *Contemporary Issues in Behavioral and Social Sciences*, 7(1), 70–83.

[15] Ding, T., Bickel, W. K., and Pan, S. (2017). Social media-based substance use prediction. *arXiv preprint arXiv*:1705.05633.

[16] Gao, Y., Zhao, L., Du, J., and Wang, J. (2023, September). Spatial-temporal Traffic Flow Prediction Model Based on the GAT and BiGRU. In *Journal of Physics: Conference Series* (Vol. 2589, No. 1, p. 012024). IOP Publishing.

*Hybrid and Advanced Technologies – S. Prasad Jones Christydass, Nurhayati Nurhayati and S. Kannadhasan. (Eds)* © *2025 the Authors, ISBN 978-1-032-90254-8*


# Computational analysis of ligand dataset for diabetes therapy against TNF-Alpha through comparative molecular docking

P. Yashwitha
*Research Scholar, Department of Bioinformatics, SIMATS School of Engineering, Saveetha University, Chennai, Tamil Nadu, India*

K. Anbarasu
*Project Guide, Department of Bioinformatics, SIMATS School of Engineering, Saveetha University, Chennai, Tamil Nadu, India*

ABSTRACT: The purpose of the current work is to use computational techniques to identify the best novel lead organic chemical associated with diabetes disease by examining the expected interactions between the diabetes-related TNF-alpha protein mutant and agonists of non-flavonoid polyphenols and anthraquinones. For the purpose of this study, we were able to get the RCSB PDB structure for the TNF-alpha protein. To evaluate the stability of targets relative to one another, a list of ligands was compiled from PubChem. PyRx was used for molecular docking in AutoDock VINA, and the study was completed using LigPlus. Each molecule was assessed using Lipinksi's Rule of Five, which states that the MOLINSPIRATION initiative. Based on the least amount of hydrogen bonds and binding affinity, deoxy podophyllotoxin has been found to be the best ligand that inhibits the diabetes-causing protein TNF-alpha.As a result, we studied to see how a particular group of ligands interacted with the protein. We chose the optimal molecule from a group of ligands based on the binding energy as a parameter. In order to select the best ligand, the hydrogen bonding and hydrophobic interactions were also shown. Deoxy podophyllotoxin is a naturally occurring anthraquinone that has been shown to be considered the best lead molecule, helpful in reducing the risk of diabetes, and supportive of healthy life.

*Keywords*: Molecular Docking, TNF-alpha, Diabetes, Lipinksi's Rule of Five, Anthraqunoes

## 1 INTRODUCTION

Diabetes is a long-term medical disorder marked by elevated blood glucose (sugar) levels. This may happen as a result of the body not making enough insulin, a hormone that controls blood sugar, or as a result of improper insulin response in the body's cells. Diabetes comes in various forms, the two most prevalent being Type 1 and Type 2(Rickman 2018).

The immune system of the body unintentionally targets and destroys the beta cells of the pancreas, which produce insulin, resulting in type 1 diabetes, a chronic autoimmune disease. Insulin is a hormone that controls blood sugar (glucose) levels, and this leads to little to no production of it..High blood glucose (sugar) levels are a hallmark of type 2 diabetes, a chronic metabolic disease (Stahl-Pehe et al. 2022).In contrast to Type 1 diabetes, which is an autoimmune disorder, Type 2 diabetes is mostly caused by insulin resistance, a condition in which the body's cells do not respond to insulin as well as a relative lack of insulin synthesis.

DOI: 10.1201/9781003559115-69

Research investigating the use of cytokine-targeting treatments for diabetes control has been prompted by the understanding of the link between diabetes and cytokines. For instance, medications that target particular cytokines or their receptors are being researched as possible treatments for T1D and T2D, especially in situations where inflammation is a major factor in the disease's progression. and because cytokines prevent a certain quantity of insulin from being produced, they are the primary cause of diabetes. Tumour necrosis factor-alpha (TNF-α) is classified as a cytokine, a class of immune system signalling molecule. It is essential for immunological responses and inflammation.T cells, B cells, natural killer cells, and other immune cells can also emit TNF-alpha, but activated macrophages are the immune cell type that produces the most of it (*Tumor Necrosis Factor (TNF) Inhibitors: Advances in Research and Application: 2011 Edition: ScholarlyPaper* 2012).

## 2 MATERIALS AND METHODS

The bioinformatics computational lab at Saveetha School of Engineering, SIMATS, Chennai, served as the testing ground for this study. Here, data on ligands, proteins, and molecular docking were combined.

Getting the TNF-alpha protein's FASTA sequence from the NCBI database is the first step. To understand the structural mechanism of the TNF-alpha domain and its isoforms, copy and paste the sequence into BLASTp. This allows us to ascertain the query % rapidly, which will be useful in determining whether or not to move further with homology modelling (Akash *et al.* 2018).

BLASTp is a well-known bioinformatics program that matches amino acid sequences to protein sequence databases. The acronym for Basic Local Alignment Search Tool is BLAST. Specifically, BLASTp is a version of BLAST intended for protein comparisons. UniProt is an extensive, publicly available database containing functional and protein sequence information.

PyRx is a virtual screening application we can do molecular docking. Molecular docking for homology investigations was carried out in this study utilising a standalone, commercial program called AUTODOCK VINA. The ligand preparation process is the first stage. One selects a sample ligand database from either PUBCHEM-DB or CHEMBI-DB. The ligands that were obtained and showed a correlation with the target protein are regarded as database actives. The selected ligands are downloaded as SDF files straight away. The ligand files that were downloaded are minimised, then Open Babel converts the format to PDBQT(Arabnia and Tran 2015). Vina searches for and ranks a docking pose according to the scoring function by using a gradient calculation in conjunction with Monte Carlo techniques. It aids in the analogous search of the comparable conformational space.

The modelled TNF-alpha modified Protein and the converted ligands are selected in the wizard. The coordinates are set, and the complete protein is indicated with a grid. The chosen ligands are now given permission to dock with the appropriate protein targets, and the desired outcomes are achieved. The modified TNF-alpha protein structure follows the same process, with the RMSD (root mean square deviation) value being taken into account throughout the scoring analysis. The structures exhibiting the lowest intermolecular energy were chosen, and the DIMPLOT server was utilised to study the non-bonded interactions including hydrogen bonds and Van der Waals forces.

The LIGPLOT package includes the server that is used to analyse interactions. To find the optimal structure for the mutant TNF-alpha protein, all of the molecular docking settings for TNF-alpha were run frequently.

## 3 RESULTS

In AutoDock Vina, five molecules were molecularly docked, and the best ten postures were retrieved. The binding affinity with the lowest score indicated a strong interaction between the chemical and the protein. The five chemicals' binding energies varied from -6.3 to -6.6 in kcal/mol, as indicated by Table 1. The energy confirmed the perfect binding with the chemicals. span. Out of the five compounds, deoxypodophyllotoxin exhibited the lowest binding affinity, or -6.6 kcal/mol. Once The affinity it had was the highest when compared to other chemicals.

Table 1 demonstrates that deoxypodophyllotoxin has the highest binding affinity (-6.6 kcal/mol) of all the chemical compounds. With regard to affinity, it is showing the best outcomes with Firategrast (Lee *et al.* 2011). A sophisticated visualisation of the interactions between the TNF alpha protein and the deoxypodophyllotoxin protein was made using Lig plus (Figure 1). Since the compound's requirements were satisfied according to the Lipinski rule (Table 2), deoxypodophyllotoxin, an anthraquinone, suppresses the TNF alpha protein and treats diabetes.

Table 1. Lipinski's rule of five.

| Lead compound | Molecular Weight | LogP | Number of hydrogen bond donor | Number of hydrogen bond acceptor | Docking values |
|---|---|---|---|---|---|
| Firategrast | 499.5g/mol | 4.6 | 2 | 8 | –6.5 |
| Deoxypodop-hyllotoxin | 398.4g/mol | 3 | 0 | 7 | –6.6 |

Table 2. ADMET properties of top two lead compounds.

| | Absorption | | Distribution | | Metabolism | | Excretion | | Toxicity | |
|---|---|---|---|---|---|---|---|---|---|---|
| Lead compound | MDCK permeability | F20 % | PPD | VD | CYP1A2 inhibitor | CYP1A2 substrate | CL | t1/2 | Rat oral Acute Toxicity | HERG Blockers |
| Firategrast | 2.7e-05 | – | 97.043% | 0.312 | – | + | 2.560 | 0.256 | – | – |
| Deoxypo dophyllotoxin | 3.7e-05 | – | 96.342% | 1.673 | – | + | 12.930 | 0.136 | – | – |

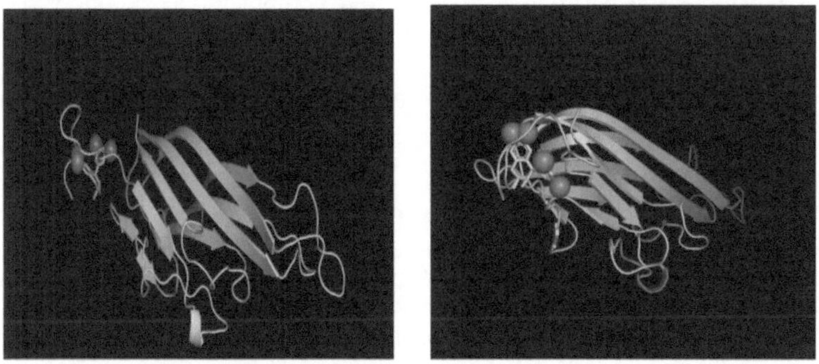

a) b)

Figure 1. 3D structure of TNF-alpha docked with a) Firategrast b) Deoxypodophyllotoxin c) isoanthricin d) Iso deoxypodophyllotoxin.

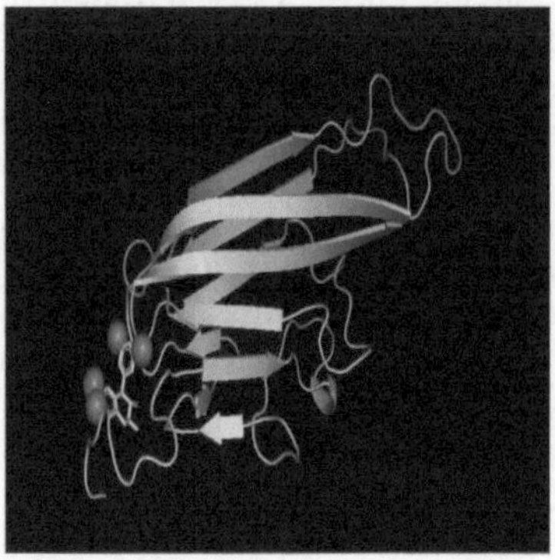

c)

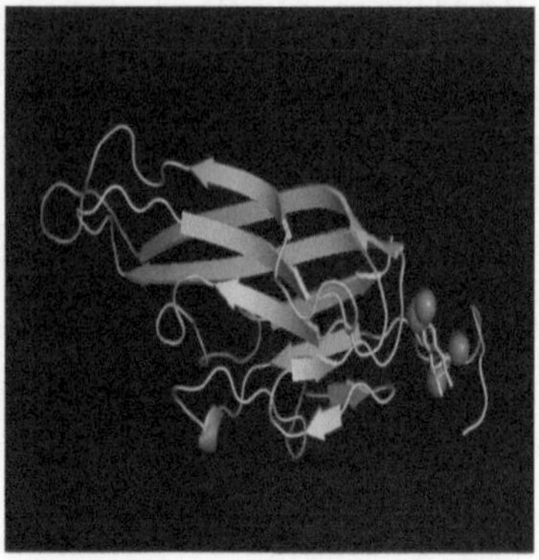

d)

Figure 1. (Continued)

## 4 DISCUSSION

In the current study, the protein that causes obesity disease and a few unique lead chemicals interacted structurally to create the best combination for inhibiting the protein that promotes healthy living. When it came to binding affinity, anthraquinones outperformed polyphenol non-flavonoids. The newly created compounds were examined using Ligplus to see how the chemicals interacted. Table 1 displays the hydrogen bond and hydrophobic interaction interactions between the relevant ligand and the protein TNF alpha. More compounds had at least five interactions. However, in this instance, the optimal compound was selected based on how strongly the two chemicals interacted. The binding energies of the best-docked molecules will range from -6.3 to -6.6 kcal/mol. The two-dimensional interactions between the TNF alpha protein and its ligands are shown in Figure 2 (Wang and Urban 2014).

The binding affinity was extracted using the visualisation tool PyRx, which helped identify the medication that would work best to treat the condition. While the binding affinities of other non-flavonoids were smaller than those of the template, the binding affinities of iso arthricin, iso deoxypodophyllotoxin, and firategrast, all template ligands, were all -6.5 kcal/mol. Deoxypodophyllotoxin was the one with the highest binding affinity (-6.6 kcal/mol). The remaining ones, all with -6.6 kcal/mol, had the best binding energies aside from that.

The MOLINSPIRATION tool is used to estimate these parameters. The molecular weight and the log P (octanol-water partition coefficient) were computed. Deoxypodophyllotoxin has No hydrogen bond donors, seven hydrogen bond acceptors, a molecular weight of 398.4 g/mol, and a log P value of 3.1. Because it complied with the Lipinski rule of five, this compound is thought to be the best (Capecchi *et al.* 2019). Table 2 displays the analysis of Lipinski's rule properties for different compounds.

## 5 CONCLUSION

Compared to other substances, our research investigation indicated that Deoxy podophyllotoxin would be the most effective innovative lead medication for human TNF alpha protein. The ideal interactions between the protein and the ligand have been found during these molecular docking research. Binding energies and H-bond interactions were used to facilitate the identification of the appropriate ligand. Many therapeutic effects of Deoxy podophyllotoxin, a naturally occurring anthraquinone, have been revealed by major studies. We think that the knowledge offered here can help those who suffer from diabetes disease have a small prospect of getting back to being healthy, and it could also be used to create the best therapeutic drugs now on the market to treat diabetes.

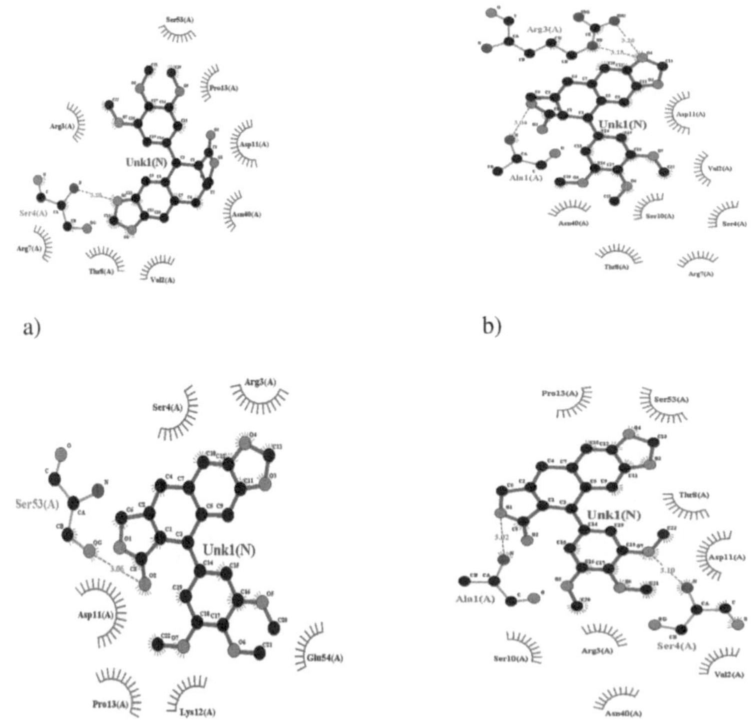

a) b) c) d)

Figure 2. LigPlus result of TNF alpha protein docked with a) Firategrast b) Deoxypodophyllotoxin c) isoanthricin d) Iso deoxypodophyllotoxin.

## REFERENCES


Akash, M. S. H., Rehman, K., and Liaqat, A. (2018). Tumor necrosis factor-alpha: Role in development of insulin resistance and pathogenesis of type 2 diabetes mellitus. *Journal of Cellular Biochemistry*, *119*(1), 105–110.

Arabnia, H. R., and Tran, Q. N. (2015). *Emerging trends in computational biology, bioinformatics, and systems biology: Algorithms and software tools.* Morgan Kaufmann.

Capecchi, A., Awale, M., Probst, D., and Reymond, J. L. (2019). PubChem and ChEMBL beyond Lipinski. *Molecular Informatics*, *38*(5), e1900016.

Lee, J. S., Kim, D. H., Lee, C. M., Ha, T. K., Noh, K. T., Park, J. W., Heo, D. R., et al. (2011). Deoxypodophyllotoxin induces a Thl response and enhances the antitumor efficacy of a dendritic cell-based vaccine. *Immune Network*, *11*(1), 79–94.

Rickman, C. (2018). *Diabetes: Causes, diagnosis and treatment*. Foster Academics.

Tumor Necrosis Factor (TNF) *Inhibitors: Advances in Research and Application: 2011 Edition: ScholarlyPaper*. (2012). ScholarlyEditions.

Wang, J and Urban, L. (2014). *Predictive ADMET: Integrated approaches in drug discovery and development.* John Wiley & Sons.

*Hybrid and Advanced Technologies – S. Prasad Jones Christydass, Nurhayati Nurhayati and S. Kannadhasan. (Eds)*  *ISBN 978-1-032-90254-8*


# Data dispersion and encryption of stored data using Advanced Encryption Standard (AES) compared with Hadoop Distributed File System (HDFS) for cloud secure storage mechanism

M. Siva Reddy
*Research Scholar, Department of Computer Science and Engineering, Saveetha School of Engineering, Saveetha University, Chennai, Tamil Nadu, India*

S. Christy
*Project Guide, Department of Computer Science and Engineering, Saveetha School of Engineering, Saveetha University, Chennai, Tamil Nadu, India*

ABSTRACT: This study compares data dispersion and encryption techniques with an emphasis on Hadoop Distributed File System (HDFS) and Advanced Encryption Standard (AES) in secure cloud storage systems. The study uses 29,502 samples using ClincalC with G-power values ($\alpha$=0.05, $\beta$=0.2). Data dispersion was analyzed using synthetic datasets, AES encryption was verified in a controlled cloud environment, and HDFS functioning was assessed using Hadoop clusters. AES was shown through statistical analysis in SPSS to be a successful data security solution that guarantees confidentiality. Scalable and fault-tolerant storage was made possible by Hadoop (HDFS) (82.14% vs. 61.54%, $p<0.02$), showing strong performance but necessitating more security precautions for complete data protection.

*Keywords*: Protected data, Advanced Encryption Standard, Encryption, Storage mechanism, Delay time, Hadoop Distributed File System

## 1 INTRODUCTION

With an emphasis on the Hadoop Distributed File System (HDFS) and the Advanced Encryption Standard (AES), this study examines data dispersion and encryption in cloud storage in 2023 (Politou *et al.* 2021). In order to improve redundancy and retrieval performance, dispersion techniques like sharding and erasure coding disseminate data across servers (Daemen and Rijmen 2013; Politou *et al.* 2021). In the fields of banking and healthcare, AES encryption guarantees strong secrecy, and HDFS scalable for e-commerce and research (Kapil *et al.* 2020). But the government and vital infrastructure have strict criteria when it comes to security. By protecting data integrity and storage effectiveness, this study improves cloud security for a range of sectors.

With more than 12,000 publications published in more than 150 journals, the research output has grown exponentially, highlighting the growing interest in data dispersion and AES/HDFS encryption for safe cloud storage. Applications in e-commerce, banking, healthcare, government, and research underscore the importance of data security. By combining dispersion and encryption, Li's CSSM improves cloud data security while maintaining controllable latency. The authors Kapusta and Memmi (2015) highlighted the use of fragmentation to achieve secure long-term storage at a low cost. 128-bit AES was recommended by Babitha and Babu (2016) for cloud security's access control, secrecy, and authenticity.


DOI: 10.1201/9781003559115-70

The research project examining AES and Hadoop (HDFS) for cloud secure storage has two drawbacks: it ignores other approaches and restricts its attention to performance analysis delay. Future studies ought to fully address these inadequacies.

## 2 MATERIALS AND METHODS

The investigation was conducted within the Department of Data Analytics Lab at Saveetha School of Engineering, Saveetha Institute of Medical and Technical Sciences. Within this study, ClincalC was utilized alongside G-power parameters set at 0.8 for each group, with alpha at 0.05 and beta at 0.2, resulting in a total sample size calculation of 29502 (with 14751 in each group). Enhancing the outcomes involved subjecting the Kaggle dataset to 15 iterations per group, creating synthetic datasets for data dispersion analysis, implementing AES encryption within a controlled cloud setting, and setting up a Hadoop cluster to assess HDFS within a cloud-based secure storage environment. The outcomes were analyzed using SPSS statistical tools to determine the most appropriate method.

The approach focuses on protecting sensitive data by using AES encryption in a regulated cloud environment. Access control together with secrecy and authenticity are guaranteed by AES. AES's efficacy in cloud security is measured by performance parameters including delay, throughput, and latency, which also illustrate the technology's usefulness and performance results.

The work proposed adaptive sharding and dynamic erasure coding for cloud-based storage demands and used NumPy, a Python package, to generate synthetic datasets. PyCryptodome's AES encryption was centered on dynamic management and key size evaluations (128-bit, 256-bit, and 512-bit). Performance analysis tailored for repeatability and cloud scalability using timeit observed encryption and dispersion affects on delays and resource utilization.

### 2.1 *Statistical analysis*

Utilizing the IBM SPSS Statistics data editor, the evaluation of Delay time predictions is carried out through a statistical examination of the Advanced Encryption Standard (AES) and Hadoop Distributed File System (HDFS). In order to progress with the data analysis, it is imperative to establish both independent and dependent variables. Within this context, the dependent variable is identified as the Delay time associated with each algorithm relative to the file size, while the two algorithms serve as the independent variables.

## 3 RESULTS

Table 1. The delay time of the Advanced Encryption Standard (AES) with respect to the file size.

| File size (MB) | Advanced Encryption Standard (AES) | Hadoop Distributed File System (HDFS) |
|---|---|---|
| 10 | 1.2 | 1.5 |
| 20 | 2.4 | 3 |
| 30 | 3.6 | 4.5 |
| 40 | 4.8 | 6 |
| 50 | 6 | 7.5 |
| 100 | 12 | 15 |
| 200 | 24 | 30 |
| 300 | 36 | 45 |
| 400 | 48 | 60 |

(*continued*)

Table 1. Continued

| File size (MB) | Advanced Encryption Standard (AES) | Hadoop Distributed File System (HDFS) |
|---|---|---|
| 500 | 60 | 75 |
| 1000 | 120 | 150 |
| 2000 | 240 | 300 |
| 3000 | 360 | 450 |
| 4000 | 480 | 600 |
| 5000 | 600 | 750 |
| 10000 | 1200 | 1500 |
| 20000 | 2400 | 3000 |
| 30000 | 3600 | 4500 |
| 40000 | 4800 | 6000 |

Table 2. Descriptive analysis of Advanced Encryption Standard(AES) and Hadoop Distributed File System (HDFS).

| Time Delay | Algorithm | N | Mean | Std.Deviation | Std.Error Mean |
|---|---|---|---|---|---|
| | AES | 20 | 999.9000 | 1777.10585 | 397.37295 |
| | HDFS | 20 | 1249.8750 | 2221.38232 | 496.71619 |

Table 3. Compare the results of the statistical analysis for Advanced Encryption Standard(AES) independent variables and Hadoop Distributed File System (HDFS).

| | Levene's Test for Equality of Variances | | t-test for Equality of Means | | | | | | |
|---|---|---|---|---|---|---|---|---|---|
| | f | sig | t | df | Sig. (2 tailed) | Mean Diff. | Std. Error Difference | 95% Confidence Interval of the Difference | |
| | | | | | | | | Lower | Upper |
| Equal variances assumed | .598 | .024 | –.393 | 38 | .697 | –249.97500 | 636.10709 | –1537.70648 | 1037.75648 |
| Equal variances not assumed | | | –.393 | 36.253 | .697 | –249.97500 | 636.10709 | –1539.74729 | 1039.79729 |

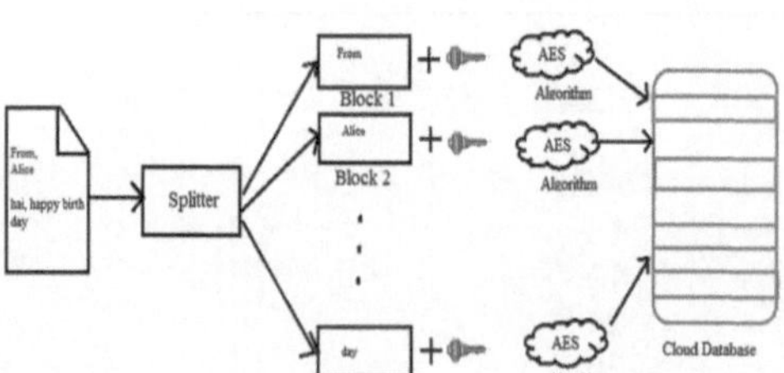


Figure 1. Shows a detailed analysis of the Cloud database block diagram with AES encryption.

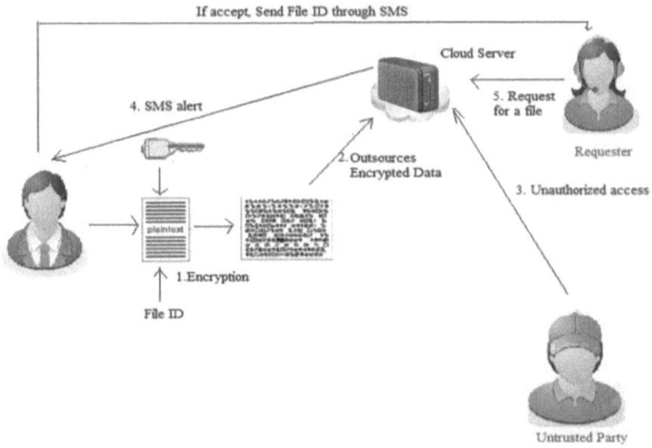


Figure 2. This diagram shows the process of encrypting and decrypting data using AES encryption in the cloud.

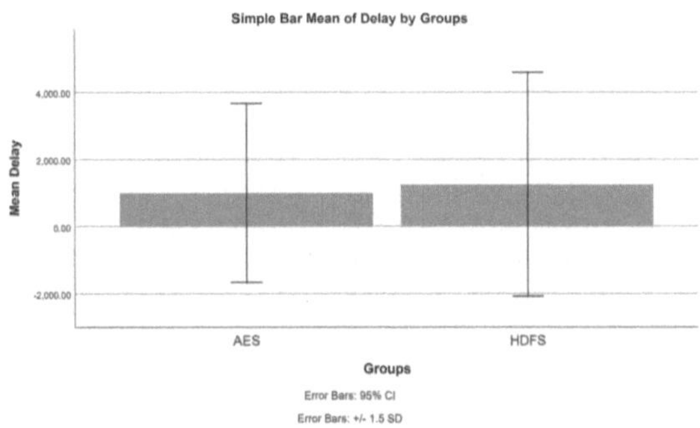


Figure 3. Shows the two algorithm's accuracy of prediction. The 95% CI and error bar +/- 1.5 SD are displayed in the error bar. It is demonstrated that Advanced Encryption Standard (AES) is a superior strategy for Data Dispersion and Encryption of Stored Data since its delay time is lesser than that of Hadoop Distributed File System (HDFS).

## 4 DISCUSSION

The findings of the investigation indicated that AES emerges as a superior option for data distribution and encryption in cloud storage compared to HDFS, particularly in scenarios where optimal performance is of paramount importance. AES exhibited decreased mean durations for data distribution and encryption when contrasted with HDFS; furthermore, AES demonstrated reduced variability in both data distribution and encryption durations. Consequently, it can be inferred that AES displays a greater level of uniformity in its performance in comparison to HDFS.

Critical insights for improving cloud security and performance are revealed by the study on data dispersion and encryption in cloud secure storage. In line with previous research, it bolsters the effectiveness of strategies like sharding and erasure coding for enhancing data redundancy and dispersion (Randles *et al.* 2010; Sun *et al.* 2014). But Kaur *et al.* (2018) difficulties and latency problems highlight how complicated data dispersion is. However, agreement is that data dispersion improves data availability. The study concludes that greater key lengths—like 512 bits—increase delay times for AES encryption, which is in line with the findings of Kaaniche and Laurent (2017) and Cai *et al.* (2017). AlZain *et al.* (2014), on the other hand, argue against the use of longer keys for considerable delays.

Despite the valuable insights provided by study, it is crucial to acknowledge its limitations. Focused on a specific set of data dispersion techniques and encryption parameters, leaving room for future studies to explore a broader range of configurations for a more comprehensive understanding. Additionally, the experimentation was conducted in a controlled environment, and real-world variables that may introduce variations were not considered. This limitation underscores the need for further research that accounts for a wider array of factors and scenarios.

## 5 CONCLUSION

In line with prior research, the study emphasizes how data dispersion improves availability and redundancy in safe cloud storage. It draws attention to how encryption key length affects latency and calls for more study to be done for optimization

## REFERENCES


AlZain, M. A., Soh, B., and Pardede, E. (2014). TMR-MCDB: Enhancing security in a multi-cloud model through improvement of service dependability. *International Journal of Cloud Computing and Services Science (IJ-CLOSER)*, *3*(3), 133–144.

Babitha, M. P., and Babu, K. R. R. (2016). Secure cloud storage using AES encryption. In *2016 International Conference on* (pp. 1–6). IEEE. Retrieved from https://ieeexplore.ieee.org/abstract/document/7877709/

Cai, H., Xu, B., Jiang, L., and Vasilakos, A. V. (2017). IoT-based big data storage systems in cloud computing: Perspectives and challenges. *IEEE Internet of Things Journal*, *4*(1), 75–87.

Daemen, J., and Rijmen, V. (2013). *The design of Rijndael: AES - The advanced encryption standard.* Springer Science & Business Media.

Kapil, G., Agrawal, A., Attaallah, A., Algarni, A., Kumar, R., and Khan, R. A. (2020). Attribute based honey encryption algorithm for securing big data: Hadoop Distributed File System perspective. *PeerJ Computer Science*, *6*, e259.

Kapusta, K., and Memmi, G. (2015). Data protection by means of fragmentation in distributed storage systems. In *2015 International Conference on Protocol Engineering (ICPE) and International Conference on New Technologies of Distributed Systems (NTDS)* (pp. 1–8). IEEE.

Kaur, R., Chana, I., and Bhattacharya, J. (2018). Data deduplication techniques for efficient cloud storage management: A systematic review. *The Journal of Supercomputing*, *74*(5), 2035–2085.

Li, H., Songjifei, S., and Lihaoteng. (n.d.). A cloud secure storage mechanism based on data dispersion and encryption. In *Proceedings title* (pp. xx–xx). Publisher.

Politou, E., Alepis, E., Virvou, M., and Patsakis, C. (2021). *Privacy and data protection challenges in the distributed era.* Springer Nature.

Randles, M., Lamb, D., and Taleb-Bendiab, A. (2010). A comparative study into distributed load balancing algorithms for cloud computing. In *2010 IEEE 24th International Conference on Advanced Information Networking and Applications Workshops* (pp. 551–556). IEEE.

Sun, Y., Zhang, J., Xiong, Y., and Zhu, G. (2014). Data security and privacy in cloud computing. *International Journal of Distributed Sensor Networks*, *10*(7), 190903.

Verma, P., Gupta, M., Agarwal, A., and Gupta, S. (2023). *Blockchain technologies new and forthcoming cloud computing.* Books Clinical Publishing.

*Hybrid and Advanced Technologies – S. Prasad Jones Christydass, Nurhayati Nurhayati and S. Kannadhasan. (Eds) © 2025 the Authors, ISBN 978-1-032-90254-8*
*Open Access: www.taylorfrancis.com, CC BY-NC-ND 4.0 license*

# Comparing the accuracy of a novel generative adversarial algorithm to support quality control in manufacturing against support vector machine

Thejaswini J. Dharani
*Research Scholar, Department of C.S.E., Saveetha School of Engineering, Saveetha University, Chennai, Tamil Nadu, India*

T. Devi
*Research Guide, Department of C.S.E., Saveetha School of Engineering, Saveetha University, Chennai, Tamil Nadu, India*

ABSTRACT: The intention of this investigation is to predict success by contrasting the Novel Generative Adversarial algorithm and Support Vector Machine to detect the quality control of product production. Both the Novel Generative Adversarial method and the Support Vector Machine are accustomed to analyze the knowledge contained in the Quality Control files, each of which having a N = 24 the specimen magnitude. Using 80% G power, instance sizes are determined. In contrast to SVM, the NGAN is more accurate than it is. Based on the results, it has been determined that there is a substantial distinction regarding the NGAN method and SVM technique, p-value of 0.043 ($p<0.05$). Compared to the SVM technique's precision of 66.90%, the THE NGAN method has attained a precision of 77.05%.

*Keywords*: Development Assistance, Novel Generative Adversarial Algorithm, Precision, Support Vector Machine, Technique, Quality Control

## 1 INTRODUCTION

Machine learning is a branch of artificial intelligence that focuses on creating statistical models and algorithms that let computers learn and form opinions or predictions without being explicitly programmed. The fundamental idea underlying machine learning is to offer computers the capacity to learn from data and gradually improve their performance on a specific task and accountability (Liakos *et al.* 2018). To analyze data and find patterns in it, machine learning algorithms are utilized, (Ahmad *et al.* 2022) which are mathematical models and techniques. Production must incorporate a vital stage known as quality control in order to guarantee that goods meet the stipulated quality standards and specifications. To find and fix errors and deviations, the production process comprises monitoring and checking the many aspects. Effective quality control helps to reduce manufacturing costs, faults, (Villalba-Diez *et al.* 2019) new product uniformity, and customer dissatisfaction. The quality requirements for each product are to be clearly stated in the quality standards and specifications. These standards should include the dimensions, materials, performance requirements, as well as other relevant components, as well as the permitted limitations and features. To design a thorough quality control plan that details each step of the process, including inspection locations, procedures, and tasks (J. Lee *et al.* 2021). Utilizing distinct

Novel Generative Adversarial networks (NGANs) is a cutting-edge and expanding method that integrates the abilities of artificial intelligence and computer vision to enhance the quality control process. to gather a sizable amount of photographs and information regarding the manufactured goods and development assistance. This dataset should contain both examples of high quality and samples with flaws or variations (Malesa and Rajkiewicz 2021). Annotate the dataset after data collection and processing to show where and if there are any errors or other problems with the images or data. to create a NGAN architecture suitable for your requirements for quality control. NGANs can offer a wide variety of synthetic defect samples to train more trustworthy defect detection models. NGANs can imitate unusual or difficult-to-find errors that might be uncommon in the real world. A wide range of sectors have found Novel Generative Adversarial Networks (NGANs) to be extremely adaptive and useful (Rasekh *et al.* 2021). These cutting-edge AI models are especially skilled in creating and modifying structured and visual data. They have sparked creativity and invention in a variety of disciplines and development assistance. For example, NGANs have helped designers and the artists by enabling the production of realistic, excellent images from random noise. Their promise stretches beyond the realms of art and entertainment with uses including data augmentation, super-resolution for medical imaging, style transfer in photography, and image-to-image translation. NGANs have also shown promise in industries like medical services, where they aid in medical picture segmentation and analysis.

There are several resources available for Quality Control for Product in Manufacturing. With the exception of roughly 3855 articles from IEEE and about 2750 publications from Springer, Google Scholar searches find almost 2958 papers on this particular topic. Prior to this one, the articles on the above list were published throughout the previous five years. The research conducted by (Kim *et al.* 2018) has 4417 citations and the article gives a study of quality control in seven distinct AM technology procedures and offers in-depth talks for each stage of the quality process. Mid-situ sensors and cameras are increasingly being used in most AM processes to collect process data for in-process quality evaluation and continuous surveillance. Investigation on physical characteristics and the best publishing settings using data analytics and prediction algorithms has enhanced techniques like extrusion-based procedures (EBP). The research paper written by (Shahbazi and Byun 2021) has 1542 citations and The following article provides a detailed description of a stochastic manufacturing and storage system that generates a single type of product, discusses quality control for manufactured items, and provides a global perspective on control over quality prediction. The research by (Y. Gan *et al.* 2017) for the purpose of to ensure that the right activities and quality management systems can be planned and put into place for ensuring excellent quality, this study outlines a number of the important quality aspects that create hurdles to the development assistance of IBS within the framework of the Chinese construction sector and includes 480 citations. 71 citations in the study presented by (Aluvalu *et al.* 2023; Mahaveerakannan R *et al.* 2023; Ramana *et al.* 2023; Tseng *et al.* 2015; Vellela and Balamanigandan 2023) shows the use of contemporary tools and the usage of a support vector machine (SVM) learning methodology, this study employs novel techniques in remote component tracking and quality control to forecast the results of the quality control process. The experiment data is used to build the classifier equations, which are then examined using various kernel functions. The 46 citations in (Y. S. Gan *et al.* 2022) the suggested method produces a convincing performance that has been thoroughly examined. To put it briefly, the classification performance is capable of producing the results for the quality control for the made items with a relatively little amount of easily obtained training data.

The quality of products and production processes have greatly improved as a result of machine learning algorithms now in use. However, they do have limitations. These algorithms are particularly vulnerable to errors and data gaps because they place such a heavy emphasis on the precision and correctness of data that is used for training. They may also find it challenging to move up and adapt to new industrial conditions, which typically necessitates frantic retraining efforts. Class disparity issues, careful negotiation between erroneous findings and misleading negatives, sensitivity to external transitions, and ongoing model maintenance all highlight how difficult it is to integrate such algorithms effectively. Novel Generative Adversarial Algorithm presented a wide variety of advantages for development assistance and industrial applications, including quality control, in order to overcome the restrictions. Anomalies are easily detected by NGAN (Meier and Bacher 2022). By teaching the discriminator to distinguish between accurate and incorrect information, any deviations from the usual distribution of data may be quickly identified and classified as anomalies. In quality control, where it is still essential to detect inconsistencies or irregularities, the capacity to spot anomalies is particularly advantageous. NGANs may be used to solve the common issue with small datasets in manufacturing (Yu and Zhang 2020). NGANs contribute to the dataset by creating more synthetic examples, giving machine learning models more training data. The accuracy and dependability of the model are enhanced by this data replenishment.

## 2 MATERIALS AND METHODS

The empirical examination was conducted in the Innovative Lab, which is outfitted with cutting-edge research and data analysis equipment.. The Novel Generative Adversarial Algorithm (NGAN) and Support Vector Machine (SVM) were used in This investigation was conducted with an average size of 24. The anticipated amount was finished utilizing the Clincalc Program and the techniques explained by (Meierhofer *et al.* 2018). For this, it was necessary to use 80% of the G-power value, a alpha significance of 0.05, and a 0.8 beta value at a 95% confidence level (K.-Y. Lee *et al.* 2023).

In order to develop and use the suggested approach, a dataset has been obtained through kaggle, a reputable and widely used site that offers an easy-to-use interface for enthusiasts and academics. The collection includes 10,001 entries with four columns for the product's weight, temperature, dimensions, and quality control.Using the relevant dataset, the product quality is determined, which results in a more accurate algorithm for the task of product predictions and traceability. The specific dataset was used to produce about 590 research articles. The results have been compared to those obtained using neural networks after the dataset in question has been used along with a Novel Generative Adversarial Algorithm (NGAN).

One well-known cloud framework for learning algorithms and statistical assessment is Google's Colab. It provides a flexible setting for making and distributing content papers thing with ease integrate dynamic programming and computational math, Python-based image analytic resources, relevant textual material, and visual features. In order to create, several key frameworks and techniques are employed in the assessment of envision dataset reliability for predicting the level of quality for an item, including the Novel Generative Adversarial Algorithm. This contains several well-known deep learning structures, notably TensorFlow and PyTorch, as well as essential libraries in general, like Matplotlib, Numpy, and Pandas. The Inception Network is implemented using TensorFlow, whereas Numpy acts responsible through managing information, lists, and calculations relating to picture processing as well as information manipulation. Additionally, Matplotlib and Pandas make analyzing information easier. Additionally, Pandas helps with statistics management, while Matplotlib is the main package for displaying findings and data.

### 2.1 *Novel generative adversarial algorithm*

Machine learning models in the Novel Generative Adversarial Networks (NGANs) category. NGANs are made up of two neural networks: a system for discrimination and a generator that are trained in opposition to one another and are sample preparation group 1. While the discriminator's goal is to distinguish between real data and created data, the generator's main goal is to produce data that matches real data. The generator is motivated to continuously enhance its capacity to generate data that is more and more realistic by this adversarial learning approach (He *et al.* 2017). The generator network uses unpredictability as input and gradually improves its capacity to generate data samples from a predefined dataset that closely resemble genuine data. The discriminator's job is to make a distinction between genuine data, development assistance and phony information produced by the generator. This competitive dynamic drives the generator to produce data that is almost unrecognizable from real samples, while the discriminator continuously improves its ability to distinguish between the two. Loss functions are used to direct the entire training process. The discriminator's loss is minimized when it properly distinguishes between actual and synthetic data, while the generator's loss stimulates it to provide data whose authenticity the detector is more probable to classify as authentic. The ultimate goal of a NGAN is to reach a Nash equilibrium, which is a state of equilibrium in which the generator emits data that is incredibly difficult to distinguish from genuine data and confuses the discriminator. NGAN pseudocode is represented by Table 1.

### 2.2 *Support vector machine*

The administered Support Vector Machine (SVM) approach to machine intelligence that excels in multi-class and binary classification problems and is sample preparation group 2. It is used for classification and regression programs. SVMs are a popular choice for a range of applications, such as image categorization, categorization of text, as well as quality control, as stated before. This is due to their adaptability. There are several crucial processes involved in employing SVMs. The data is first prepared, including information that has been tagged and includes characteristic vectors and labels for categories. For the purposes of developing models and evaluation, the dataset is split as a set to be used for training and a testing set. Finding a hyperplane that best divides data into distinct classes with the intention of maximizing the margin between classes is the main goal of SVM; this hyperplane is frequently referred to as its decision boundary. The data points nearest to the decision's border, known as support vectors, are crucial in determining this margin. SVMs use a kernel trick, which includes the linear, radial basis function (RBF), (Kranjčić *et al.* 2019) polynomial, or sigmoid, to transform the collected information into a higher-dimensional space where a linear boundary can be constructed, in situations where a linear boundary is insufficient to successfully segregate data. SVMs received praise for being capable to handle intricate decision boundaries and perform well in high-dimensional spaces, although they can be sensitive to outliers in the data and need careful parameter modification. In conclusion, Support Vector Machines are a flexible machine learning algorithm that can be successfully used in a variety of applications and development assistance, such as quality control in manufacturing, given that the data is properly prepared and the algorithm's parameters are appropriately optimized for the particular task at hand. Table 2 provides an interesting pseudocode description of the Support Vector Machine.

### 2.3 *Statistical analysis*

A quantitative assessment of the effectiveness of techniques based on the Novel Generative Adversarial algorithm (NGAN) and support vector machine (SVM) was carried out using IBM statistical package SPSS 26.0. The significance of the distinction was determined using a different t-test methodology. The data collection's dependencies include the dimensions, weight, temperature, and quality control state. The reduced prices and the assets that are accessible are the independent factors.

## 3 RESULT

The accuracy values for the two NGAN and NN algorithms are shown in Table 1 raw data table. With a total of 48 sample sizes, N = 24 is assumed to be the sample size for each group. For the NGAN model and NN model, the accuracy is computed every 24 iterations. With N (24), Average (77.05), SD (2.375), and a standard error of Mean (0.375).

Table 1. For NGAN and SVM, the accuracy rate is determined every 24 iterations with N = 24 sample size. The NGAN is more accurate when compared to the SVM.

| S.No | NGAN Accuracy (%) | SVM Accuracy (%) |
|---|---|---|
| 1 | 78 | 52 |
| 2 | 79 | 56 |
| 3 | 79 | 56 |
| 4 | 78 | 57 |
| 5 | 79 | 58 |
| 6 | 79 | 58 |
| 7 | 79 | 60 |
| 8 | 79 | 61 |
| 9 | 79 | 62 |
| 10 | 79 | 62 |
| 11 | 79 | 62 |
| 12 | 78 | 63 |
| 13 | 79 | 64 |
| 14 | 78 | 65 |
| 15 | 78 | 66 |
| 16 | 79 | 67 |
| 17 | 79 | 68 |
| 18 | 78 | 68 |
| 19 | 78 | 68 |
| 20 | 79 | 70 |
| 21 | 78 | 71 |
| 22 | 78 | 72 |
| 23 | 79 | 72 |
| 23 | 79 | 73 |
| 24 | 79 | 74 |

Table 2. The accuracy of the NGAN algorithm and the Support Vector Machine is 77.05% and 66.90%, respectively, according to statistics for independent samples. The NGAN and SVM had standard deviations of 2.375 and 6.126, respectively. The standard error means for NGAN and SVM are 0.141 and 1.419, respectively.

| | Algorithm | N | Mean | Std.Deviation | Std.Error Mean |
|---|---|---|---|---|---|
| **Accuracy** | **NGAN** | 24 | 77.05 | 2.375 | 0.375 |
| | **SVM** | 24 | 66.90 | 6.126 | 0.969 |

Figure 1 graphically illustrates the support vector machine's (SVM) operation, with correctness as the output and an acceptable check dataset as the input parameter.

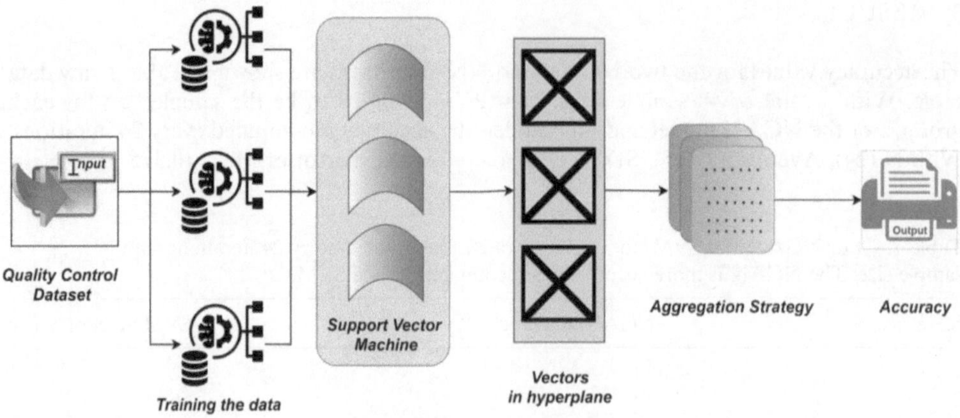


Figure 1. The above diagram shows the working of the support vector machine (SVM) with a quality control dataset as an input and corresponding output as accuracy.

Figure 2 displays a graphical diagram displaying the outcomes of the NGAN and SVM T-Test. NGAN surpasses SVM with regard to predicted accuracy while exhibiting a lower error rate, as shown in the graph.

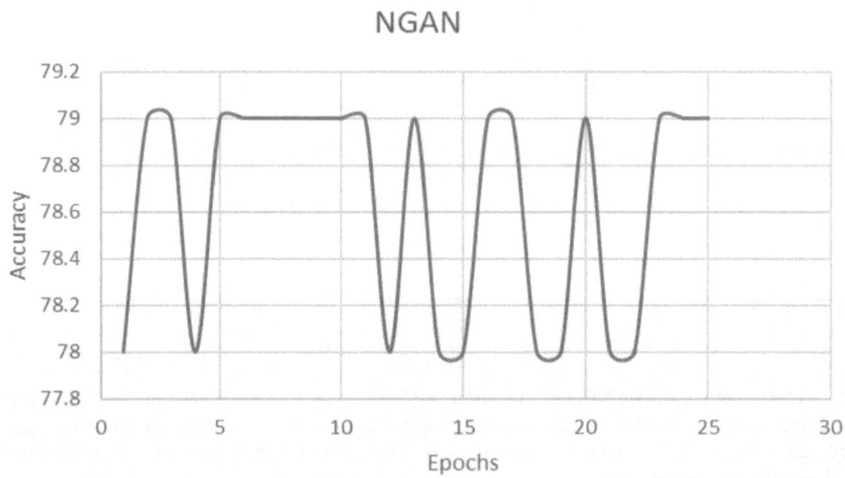


Figure 2. The graph above shows the output of the Novel Generative Adversarial Algorithm (NGAN), which has accuracy on the Y-axis and epochs on the X-axis.

A chart graph showing the T-Test results for NGAN and SVM is shown in Figure 3. It unequivocally demonstrates indicates NGAN has lower rates of error and is more precise than SVM.

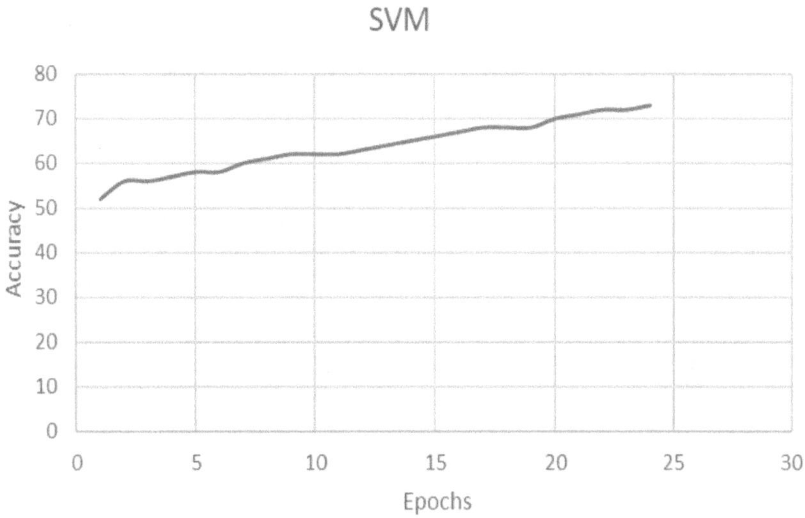


Figure 3. The graph shown above shows the output of the Support Vector Machine (SVM), which has accuracy on the Y-axis and epochs on the X-axis.

A chart with bars showing the results of the NGAN and SVM T-Tests is shown in Figure 4. This is clearly shown in the visualization. Compared to SVM, NGAN achieves a higher performance rate with an inferior error rate.

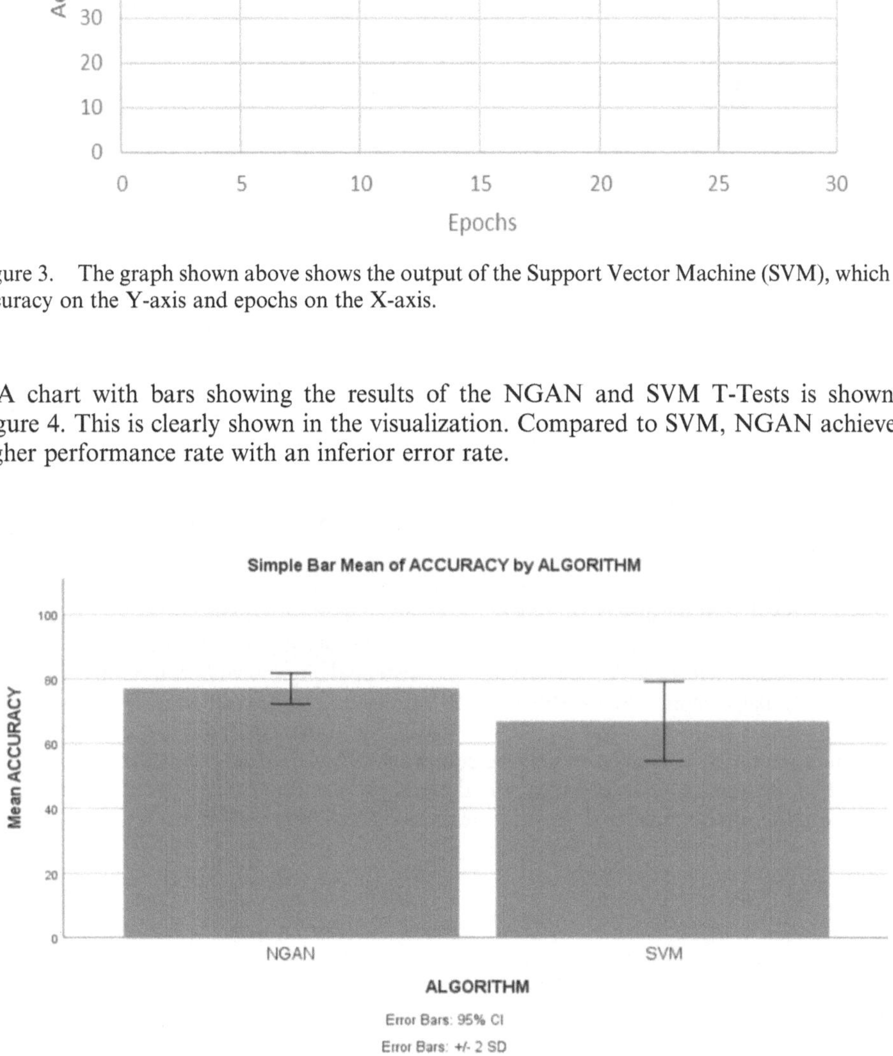


Figure 4. In comparison to the SVM's accuracy of 66.90%, the mean accuracy of the NGAN is significantly higher at 77.05%. The standard deviation of the SVM is lower than that of the NGAN. SVM and NGAN are represented on the X-axis, and Mean Accuracy is represented on the Y-axis. ±2 SD is used to depict the error bar.

## 4 DISCUSSION

The technique evaluates the importance of the findings from a standalone T-test. Although there is an important distinction among the two groups, the success rate of NGAN is 77.05%, which is substantially greater than the reliability of SVM, which is 66.90%. The value of significance is 0.043, which is substantially less than 0.05.

By looking at the files (Control dataset for Quality) in the forecast procedure for quality assurance for products in manufacturing, the production standard of the items is found. Quality control is often carried out at the end of a manufacturing process. The objective value function that has to be decreased are the ML model erroneous and the MV variables provide one solution to this problem of optimization. The initial phase in the optimizing process involves preliminary processing of the whole acquisition of the initial information, which results in a precision of 73.62% (Aleshin *et al.* 2016).To offer reliable and valid TQM constructs and a measurement instrument for evaluating the implementation of the Total Quality Management process and finding enhancement opportunities with a conclusion of 73.54% in the situation of manufacturing industries in lately modernized nations (Chen *et al.* 2021). The provided information is segmented using the K-means clustering method to increase quality, and then each picture is classified using machine learning techniques such as KNN, SVM, and C4.5, with an accuracy of 71.22% as the result (Sabbagh and Ameri 2019). The AOI inspection platform's specific design and the YOLO technology were used to solve the plating industry's problem with quality monitoring. According to actual confidence, YOLOv3 and YOLOv5 function in a fair scope with an accuracy rate of above 74% (Adibhatla *et al.* 2021). With an accuracy of 72.44%, The automated learning approach effectively predicts the best possible standards, especially for their basic changes over time, by utilizing gradient enhancement and simultaneous Long Short-Term Memory (LSTM) recurrent neural networks (Liu *et al.* 2020).The outcome's grade is determined via Random Forest Bayesian optimization, with a level of accuracy of 75.08%. The results might offer institutional perspectives and administrative recommendations for predicting and controlling the performance of goods in the real-world business (Chistyakova *et al.* 2016).

Generative Adversarial Networks (GANs) exhibit various limitations. Their training process is often unstable due to sensitivity to hyperparameters, which can lead to fluctuations between the generator and discriminator, resulting in repetitive outputs, a phenomenon known as mode collapse. The assessment of GAN performance is also challenging, with commonly used evaluation metrics like Inception Score and Frechet Inception Distance not consistently providing an accurate gauge of sample quality. GANs require a substantial amount of high-quality training data, and their effectiveness may be compromised when dealing with noisy or limited datasets. Furthermore, GANs produce non-deterministic outputs, meaning that generating samples from the same model at different times can yield slightly different results.

## 5 CONCLUSION

The Quality control for goods in manufacturing dataset is outlined by NGAN and SVM, and the precision rate of the Novel Generative Adversarial Algorithm is evaluated to be 77.05% more accurate than the Support Vector Machine's accuracy rate of 66.90%. When compared to the SVM, the NGAN's performance metric is superior.

## REFERENCES

[No title]. (n.d.). Retrieved June 12, 2024, from https://www.researchgate.net/profile/Punithavathi-Rasappan/publication/362388038_Computer_Vision_and_Deep_Learning-enabled_Weed_Detection_Model_for_Precision_Agriculture/links/62e7911d4246456b55ffc95a/Computer-Vision-and-Deep-Learning-enabled-Weed-Detection-Model-for-Precision-Agriculture.pdf

Adibhatla, V. A., H-C Chih, C-C Hsu, J. Cheng, M. F. Abbod, and J-S Shieh. (2021). Applying deep learning to defect detection in printed circuit boards via a newest model of you-only-look-once. Venkat Anil Adibhatla, Huan-Chuang Chih, Chi-Chang Hsu, Joseph Cheng, Maysam F. Abbod, Jiann-Shing Shieh. Applying deep learning to defect detection in printed circuit boards via a newest model of You-Only-Look-once[J]. *Mathematical Biosciences and Engineering* 18 (4): 4411–28.

Ahmad, Rami, Raniyah Wazirali, and Tarik Abu-Ain. (2022). Machine learning for wireless sensor networks security: An overview of challenges and issues. *Sensors* 22 (13): 4730.

Aleshin, N. P., M. V. Grigor'ev, N. A. Shchipakov, M. A. Prilutskii, and V. V. Murashov. (2016). Using nondestructive testing methods for in-production quality control of additive manufactured parts. *Russian Journal of Nondestructive Testing* 52 (9): 532–37.

Aluvalu, R., Kumaran, V N S., Thirumalaisamy, M., Basheer, S., Ali Aldhahri, E., and Selvarajan, S. (2023). Efficient data transmission on wireless communication through a privacy-enhanced blockchain process. *PeerJ. Computer Science*, 9, e1308.

Chen, Chi-Kuang, Lidia Reyes, Jens Dahlgaard, and Su Mi Dahlgaard-Park. (2021). From quality control to TQM, service quality and service sciences: A 30-year review of TQM literature. *International Journal of Quality and Service Sciences* 14 (2): 217–37.

Chistyakova, Tamara, Mikhail Teterin, Alexander Razygraev, and Christian Kohlert. (2016). Intellectual analysis system of big industrial data for quality management of polymer films. *Advances in Neural Networks – ISNN 2016*, 565–72.

Gan, Y. S., Sze-Teng Liong, Shih-Yuan Wang, and Chin Tiong Cheng. (2022). An improved automatic defect identification system on natural leather via generative adversarial network. *International Journal of Computer Integrated Manufacturing*, December. https://doi.org/10.1080/0951192X.2022.2048421.

Gan, Yuanyan, Liyin Shen, Jindao Chen, Vivian W. Y. Tam, Yongtao Tan, and I. M. Chethana S. Illankoon. (2017). Critical factors affecting the quality of industrialized building system projects in china. *Sustainability: Science Practice and Policy* 9(2): 216.

He, Zhi, Han Liu, Yiwen Wang, and Jie Hu. (2017). Generative adversarial networks-based semi-supervised learning for hyperspectral image classification. *Remote Sensing* 9 (10): 1042.

Kim, Hoejin, Yirong Lin, and Tzu-Liang Bill Tseng. (2018). A review on quality control in additive manufacturing. *Rapid Prototyping Journal* 24 (3): 645–69.

Kranjčić, Nikola, Damir Medak, Robert Župan, and Milan Rezo. (2019). Support vector machine accuracy assessment for extracting green urban areas in towns. *Remote Sensing* 11 (6): 655.

Lee, Jungeon, Hyung Jun Park, Seunghak Chai, Gyu Ri Kim, Hwanwoong Yong, Suk Joo Bae, and Daeil Kwon. (2021). Review on quality control methods in metal additive manufacturing. *NATO Advanced Science Institutes Series E: Applied Sciences* 11 (4): 1966.

Lee, Kyo-Yeon, Jong Min Kim, Jiyeon Chun, Ho Jin Heo, Chae Eun Park, and Sung-Gil Choi. (2023). Miyeokgui (Undaria Pinnatifida Sporophyll) Characteristic under different relative humidity: Microbial safety, antioxidant activity, ascorbic acid, Fucoxanthin, α-/β-/γ-Tocopherol contents. *Foods* 12 (12): 2342.

Liakos, Konstantinos G., Patrizia Busato, Dimitrios Moshou, Simon Pearson, and Dionysis Bochtis. (2018). Machine learning in agriculture: A review. *Sensors* 18 (8): 2674.

Liu, Duen-Ren, Yi-Kuan Hsu, Hsing-Yu Chen, and Huan-Jian Jau. (2020). Air pollution prediction based on factory-aware attentional LSTM Neural *Network. Computing* 103 (1): 75–98.

Mahaveerakannan R, Anitha, C., Aby K Thomas, Rajan, S., Muthukumar, T., and Govinda Rajulu, G. (2023). An IoT based forest fire detection system using integration of cat swarm with LSTM model. *Computer Communications*, 211, 37–45.

Malesa, Marcin, and Piotr Rajkiewicz. (2021). Quality control of pet bottles caps with dedicated image calibration and deep neural networks. *Sensors* 21 (2): 501.

Meier, Johanna, and Gerd Bacher. (2022). Progress and challenges of InGaN/GaN-Based Core–Shell microrod LEDs. *Materials* 15 (5): 1626.

Meierhofer, Regula, Carola Bänziger, Sandro Deppeler, Bal Mukund Kunwar, and Madan Bhatta. (2018). From water source to tap of ceramic filters—factors that influence water quality between collection and consumption in rural households in Nepal. *International Journal of Environmental Research and Public Health* 15 (11): 2439.

Ramana, Kadiyala, A. Revathi, A. Gayathri, Rutvij H. Jhaveri, C. V. Lakshmi Narayana, and B. Naveen Kumar. (2022). WOGRU-IDS — An intelligent intrusion detection system for IoT assisted wireless sensor networks. *Computer Communications* 196 (December): 195–206.

Rasekh, Mansour, Hamed Karami, Alphus Dan Wilson, and Marek Gancarz. (2021). Classification and identification of essential oils from herbs and fruits based on a MOS electronic-nose technology. *Chemosensors* 9 (6): 142.

Sabbagh, Ramin, and Farhad Ameri. (2019). A framework based on K-means clustering and topic modeling for analyzing unstructured manufacturing capability data. *Journal of Computing and Information Science in Engineering* 20 (1): 011005.

Shahbazi, Zeinab, and Yung-Cheol Byun. (2021). Integration of blockchain, IoT and machine learning for multistage quality control and enhancing security in smart manufacturing. *Sensors* 21 (4): 1467.

Tseng, Tzu-Liang (bill), Kalyan Reddy Aleti, Zhonghua Hu, and Yongjin (james) Kwon. (2015). E-quality control: a support vector machines approach. *Finite Elements in Analysis and Design: The International Journal of Applied Finite Elements and Computer Aided Engineering* 3 (2): 91–101.

Vellela, S. S. and Balamanigandan, R. (2023). An intelligent sleep-awake energy management system for wireless sensor network. *Peer-to-Peer Networking and Applications*, 16(6), 2714–2731.

Villalba-Diez, Javier, Daniel Schmidt, Roman Gevers, Joaquín Ordieres-Meré, Martin Buchwitz, and Wanja Wellbrock. (2019). Deep learning for industrial computer vision quality control in the printing industry 4.0. *Sensors* 19 (18): 3987.

Yu, Houqiang, and Xuming Zhang. (2020). Synthesis of Prostate MR Images for Classification Using Capsule Network-Based GAN Model. *Sensors* 20 (20): 5736.

*Hybrid and Advanced Technologies – S. Prasad Jones Christydass, Nurhayati Nurhayati and S. Kannadhasan. (Eds)*  *ISBN 978-1-032-90254-8*


# Classification of framework for free labour analysis with CNN in comparison with random forest for improved accuracy

Achipireddy Greeshmitha
*Research Scholar, Department of Computer Science and Engineering, Saveetha School of Engineering, Saveetha University, Chennai, Tamil Nadu, India*

S. Mahaboob Basha
*Research Guide, Department of Computer Science and Engineering, Saveetha School of Engineering, Saveetha University, Chennai, Tamil Nadu, India*

ABSTRACT: Our goal is to improve the accuracy of labor analysis by utilizing Random Forests and Novel Convolutional Neural Networks (CNN) to provide instruments for identifying and stopping labor exploitation. Reliability was confirmed by using the G*Power test ($\alpha$=0.05, power=0.85) to assess these techniques on two sets of 100 samples each. CNN outperformed Random Forest (96.33%) in accuracy, achieving 97.27%). A significant p-value of 0.001 was shown by an independent sample t-test, confirming the statistical difference between the approaches. The promise of advanced machine learning for labor analysis is shown by these findings. In order to promote ethical workplaces, a focus on model interpretability and ethical concerns is made. Subsequent investigations will concentrate on expanding the variety of datasets and tackling real-world implementation issues.

## 1 INTRODUCTION

In order to improve employee well-being, our research compares Random Forest and Novel Convolutional Neural Networks (CNN) frameworks to shed light on intricate workplace dynamics. We prioritize interpretability, scalability, and efficiency and address crucial concerns such imbalanced data and toxic work environments, workload demands, and mental stress while optimizing model hyperparameters. Our objective is to promote healthy workplaces by offering organizations, legislators, and advocates relevant knowledge.CNNs are essential for analyzing unstructured labor data and thoroughly examining emotional components. According to Feng and Savani (2020), Random Forest's flexibility promotes ethical work practices and guarantees efficiency and transparency. According to recent research ("Multi-Modal Emotion Recognition for Determining Employee Satisfaction," n.d.), job satisfaction is essential for social welfare, individual fulfillment, and organizational performance. Our study looks into a variety of elements that affect job satisfaction, such as pay, working environment, and cultural differences. We also examine "unethical workloads," which surpass workers' capabilities and have an impact on their well-being and output (Arpaia *et al.* 2020).In the future, multimodal data fusion, real-time techniques, and fairness and bias in labor analysis—especially with regard to sensitive qualities like gender or race—will all be explored. It's imperative to grasp labor dynamics and ethical issues in addition to being proficient in machine learning and data analysis (Amin and Al 2021).In the end, our research seeks to create better working circumstances and employee well-being, enhance labor analysis, and optimize categorization methods ("Impact of Job Satisfaction, Job Stress and

DOI: 10.1201/9781003559115-72

## 2 MATERIALS AND METHODS

This state-of-the-art facility is outfitted with a high-configuration system that makes advanced research possible and makes results acquisition easier.Ten individuals made up the sample size in this particular study, which compared two different groups: Group 1, which used Novel Convolutional Neural Networks (CNN), and Group 2, which used Random Forest methods. Our analysis employed a 95 percent confidence interval framework, an 80 percent G-Power value, an alpha threshold of 0.05, and a beta value of 0.8 in order to evaluate the differences between these groups. (Johns and Donnelly 2021)

A comma-separated values (csv) file named labor data, which has a lower aspect ratio of -8.20104 and a higher aspect ratio of -5.32096, is the dataset used in this suggested study. The dataset, which has a standard mean of 0.37309 and a micro-F mean of 2.148, was gathered from the model/Notebook LSTM Baseline and retrieved from the UCI repository.

We have implemented the algorithm described in this study using a dataset, and we have compared the outcomes with the Random Forest technique. There are many different data values in this dataset, which is available as a comma-separated value (CSV) file. To examine the records and compare them with the previously indicated algorithm, the algorithm described in this study has been applied to the dataset. With regard to software, Jupyter Notebook, MySQL, and Python technologies were used in conjunction with an Intel dual-core processor and 4 GB of RAM for the hardware configuration. In 2021, Benzaghta *et al.*

### 2.1 *Group 1: Novel convolutional neural network and random forest*

An artificial neural network with specific training for processing visual data, notably images and videos, is known as a Novel Convolutional Neural Networks (CNN). Its design is based on convolutional layers, which use filters to methodically scan and extract characteristics from incoming data. This technique enables the network to recognize complex patterns and informational hierarchies. The spatial dimensions of these features are then reduced by pooling layers, while predictions are made by fully connected layers using the learned representationsActivation functions introduce non-linearity to capture complex relationships and aid in the recognition and classification of objects, forms, and patterns in visual input. Backpropagation is used to train the entire network to modify learnable parameters. Because CNNs can automatically extract useful characteristics from raw pixel input, they have become essential in numerous applications, including picture synthesis, object detection, and image recognition (Sohail *et al.* 2021).Random Forest is a potent machine learning algorithm that operates by generating multiple decision trees in the training process and then combining their predictions to enhance accuracy and stability. It is frequently used for tasks like classification and regression, and it is known for its ability to handle complex data and capture intricate patterns within datasets. Random Forest is esteemed for its adaptability, strong performance, and resistance to overfitting, making it a popular choice in various fields, including finance, healthcare, and image analysis(Pratt, Boudhane, and Cakula 2021).

### 2.2 *Statistical analysis*

The free labor evaluation data were analyzed using IBM SPSS Statistics Version 26.0 in order to ascertain the statistical significance of hedging. Additionally, a secondary t-test analysis was carried out. The dependent variables, in particular the accuracy measurements, were the main focus of this investigation. In order to further our study's goal of encouraging the labor analysis profession to adopt more efficient, morally sound, and data-driven procedures, these measurements were evaluated as independent variables (Liang *et al.* 2019).

## 3 RESULTS

In this study, we evaluated the accuracy of labor classification using Random Forest and Novel Convolutional Neural Networks (CNN). The results showed that Novel CNNs obtained 97.27% accuracy while Random Forest achieved 96.33% accuracy. This comparison demonstrates how sophisticated machine learning approaches can be used to solve problems related to labor analysis. With Random Forest's resilience and interpretability, along with Novel CNNs' capacity to handle a wide range of unstructured data, these models offer enterprises and policymakers insightful information. There was a substantial difference ($p=0.001$, $p < 0.05$) between the two algorithms according to the independent t-test results. Table 3 demonstrates that CNN (97.27%) performs significantly better than Random Forest (96.33%). These results, which are graphically displayed in Figure 1, highlight the statistical significance and possible uses of these algorithms in practical labor analysis scenarios, enhancing working conditions and decision-making.

Table 1. Transformers play an important part in novel convolutional neural networks (CNNs). These transformers tokenize the incoming textual material and extract accurate details from it.

| |
|---|
| Free labor was used as input. Examination<br>Result: increased precision<br>Step 1: Import the required dependencies and libraries<br>import numpy as np<br>from tensorflow.keras, import tensorflow as tf. import models and layers<br>Step 2: Describe the architecture model of CNN using models.In sequence<br>Step 3: Include the pooling and convolutional layers model.include (layers.Activation='relu', input_shape=(image_height, image_width, num_channels))) model with filters set to 32 and kernel size set to (3, 3).include (layers.Pool_size=(2, 2)))) MaxPooling2D model.include (layers.Conv2D (activation='relu')) model with filters=64, kernel_size=(3, 3).include (layers.Pool_size=(2, 2) in MaxPooling2D<br>Step 4: For a fully connected layers model, flatten the result.include (layers.Flatten()<br>Step 5: Educate the CNN model.match(X_train, y_train, batch_size=batch_size, epochs=num_epochs, validation_data=(X_test, y_test))<br>Step 6: Predict based on fresh data using a model.forecast (new_data) |

Table 2. It is crucial to note that traditional categorical casting procedures are frequently better appropriate for this task, and using Random Forest for free labour analysis is uncommon.

| |
|---|
| Data on labor was entered.<br>Result: increased precision<br>Step 1: # Import the required libraries from the import sklearn.ensemble import train_test_split from sklearn.metrics Random Forest Classifier from sklearn.model_selection import score_accuracy<br>Step 2: Assume you have a dataset consisting of labels (y) and features (X).<br>Step 3: Divide the dataset into testing and training sets, where y_train, y_test = train_test_split, X_train, and X_test.(X, y, random_state=42, test_size=0.2)<br>Step 4: Build a classifier using Random Forest Random Forest Classifier(n_estimators=100, random_state=42) = rf_classifier<br>Step 5: Use the training data from rf_classifier.fit(X_train, y_train) to train the Random Forest classifier.<br>Step 6: Use y_pred = rf_classifier.predict(X_test) to make predictions about the test data.<br>Step 7: Use accuracy = accuracy_score(y_test, y_pred) to calculate accuracy on the test set.<br>Step 8: Conclude<br>Print "Accuracy:", accuracy) to display the accuracy. |

Table 3. Accuracy (%) of the innovative Convolutional Neural Network (CNN) vs the Random Forest technique. The Random Forest algorithm's accuracy is 10% lower than the innovative Convolutional Neural Networks (CNN) approach.

| S.NO | Accuracy (%) of CNN | Accuracy (%) of Random Forest |
|---|---|---|
| 1 | 99.70 | 89.88 |
| 2 | 98.05 | 93.99 |
| 3 | 98.21 | 90.12 |
| 4 | 97.30 | 93.68 |
| 5 | 96.30 | 92.31 |
| 6 | 96.21 | 90.83 |
| 7 | 95.90 | 97.29 |
| 8 | 94.76 | 94.46 |
| 9 | 93.14 | 91.26 |
| 10 | 92.00 | 92.80 |
| Average | 97.27 | 96.33 |

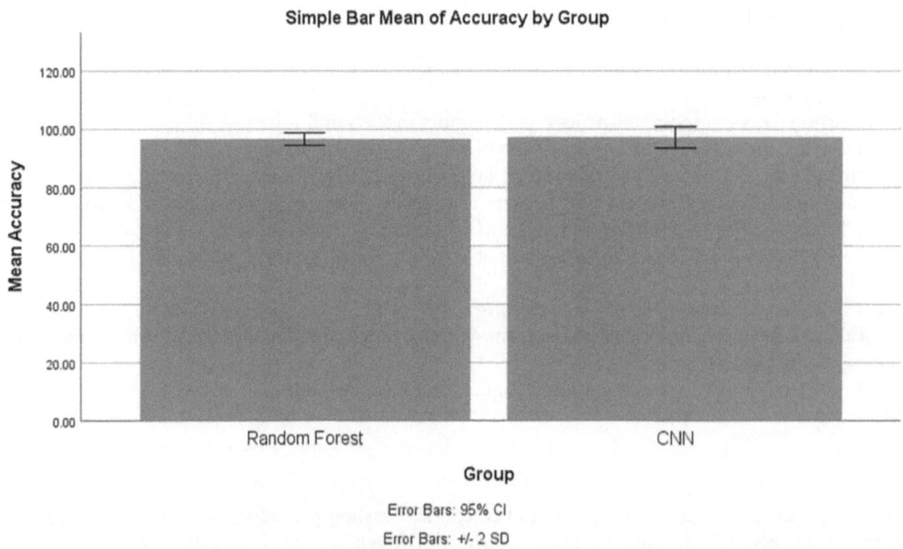


Figure 1. Bar Graph Comparison on mean accuracy of Novel Convolutional Neural Network (CNN) is 97.27% and Random Forest is 96.33%. X- axis: Random Forest, CNN Y- axis: Mean Accuracy with ±2 SD.

## 4 DISCUSSION

We conducted a thorough investigation on the accuracy of labor classification using Random Forest and Novel Convolutional Neural Networks (CNN), and we came to some interesting conclusions: CNN outperformed Random Forest with an even better accuracy percentage of 97.27%, while Random Forest reached an amazing 96.33%. These findings demonstrate how sophisticated machine learning methods can be used to solve challenging problems in labor analysis. For stakeholders including organizations, labor activists, and legislators, CNN's adaptability in processing a variety of unstructured labor data and Random Forest's robustness and interpretability offer insightful information (Collins *et al.*

2019). The legitimacy and social acceptance of our research were contingent upon ethical issues pertaining to data privacy and proper data handling ("Impact of Labour Quality on Labour Productivity and Economic Growth," n.d.). In order to maximize model performance, a careful assessment of the data's properties and the goals of the research was conducted before choosing between the CNN and Random Forest algorithms. Attaining interpretability was essential for obtaining useful insights, especially given the intricate CNN structure.

Notwithstanding these developments, our analysis notes a number of difficulties. These include limitations on the quality and availability of data, which could restrict how broadly applicable our findings can be. In addition, there are still issues with handling the resource-intensive aspect of deep learning model training and the interpretability problem in CNN models. Future studies seek to get around these restrictions by gaining access to richer.

## 5 CONCLUSION

In conclusion, our research reveals compelling insights into labour classification accuracy. Achieving an accuracy rate of 97.27% for Novel Convolutional Neural Networks (CNN) and 96.33% for Random Forest, the potential of advanced machine learning techniques in addressing labor analysis challenges is affirmed. The choice between Novel Convolutional Neural Networks (CNN) and Random Forest provides tailored solutions for organizations, labour advocates, and policymakers. This research marks a significant step forward in improving labour conditions and fostering employee well-being, showcasing the pivotal role of technology in enhancing workforce dynamics and societal well-being.

## REFERENCES


Amin, F. A. B. and *et al.* (2021). A review of the job satisfaction theory for special education perspective. *Turkish Journal of Computer and Mathematics Education (TURCOMAT)*, 12(11), 5224–5228.

Arpaia, P., Moccaldi, N., Prevete, R., Sannino, I. and Tedesco, A. (2020). A wearable EEG instrument for real-time frontal asymmetry monitoring in worker stress analysis. *IEEE Transactions on Instrumentation and Measurement*, 69(10), 8335–8343.

Benzaghta, M. A., Elwalda, A., Mousa, M. M., Erkan, I. and Rahman, M. (2021). SWOT analysis applications: An integrative literature review. *Journal of Global Business Insights*, 6(1), 55–73.

Brown, P. and Souto-Otero, M. (2020). The end of the credential society? An analysis of the relationship between education and the labour market using big data. *Journal of Education Policy*, 35(1), 95–118.

Collins, H., Ewing, K. and McColgan, A. (2019). *Labour law*. Cambridge University Press.

Donnelly, R. and Johns, J. (2021). Recontextualising remote working and its HRM in the digital economy: An integrated framework for theory and practice. *The International Journal of Human Resource Management*, 32(1), 84–105.

Feng, Z. and Savani, K. (2020). Covid-19 created a gender gap in perceived work productivity and job satisfaction: Implications for dual-career parents working from home. *Gender in Management: An International Journal*, 35(7/8), 719–736.

Liang, W., Carraro, F., Solomon, M. B., Bell, S. G., Amenitsch, H., Sumby, C. J., White, N. G., Falcaro, P. and Doonan, C. J. (2019). Enzyme encapsulation in a porous hydrogen-bonded organic framework. *Journal of the American Chemical Society*, 141(36), 14298–14305.

Parvez, M. M., Mishra, S. K., Nandini, K., Ahammad, S. H., Inthiyaz, S., Altahan, B. R., Smirani, L. K., Hossain, M. A. and Rashed, A. N. Z. (2023). Machine learning based models for defect detection in composites inspected by Barker coded thermography: A qualitative analysis. *Advances in Engineering Software*, 178(April), 103425.

Sarkar, B., Kar, S., Basu, K. and Guchhait, R. (2022). A sustainable managerial decision-making problem for a substitutable product in a dual-channel under carbon tax policy. *Computers & Industrial Engineering*, 172 (October), 108635.

Sohail, A., Khan, A., Wahab, N., Zameer, A. and Khan, S. (2021). A multi-phase deep CNN based mitosis detection framework for breast cancer histopathological images. *Scientific Reports*, 11(1), 6215.

Surendran, R., Alotaibi, Y. and Subahi, A. F. (2023). Lens-oppositional wild geese optimization based clustering scheme for wireless sensor networks assists real time disaster management. *Computer Systems Science and Engineering*, 46(1), 835–851.

Tang, S., Yuan, S. and Zhu, Y. (2020). Data preprocessing techniques in convolutional neural network based on fault diagnosis towards rotating machinery. *IEEE Access: Practical Innovations, Open Solutions*, 8, 149487–149496.

Tayyab, M., Habib, M. S., Jajja, M. S. S. and Sarkar, B. (2022). Economic assessment of a serial production system with random imperfection and shortages: A step towards sustainability. *Computers & Industrial Engineering*, 171(September), 108398.

Tsurugano, S., Nishikitani, M., Inoue, M. and Yano, E. (2021). Impact of the COVID-19 pandemic on working students: Results from the Labour Force Survey and the Student Lifestyle Survey. *Journal of Occupational Health*, 63(1).

*Hybrid and Advanced Technologies – S. Prasad Jones Christydass, Nurhayati Nurhayati and S. Kannadhasan. (Eds)* © *2025 the Authors, ISBN 978-1-032-90254-8*


# Anticipating server resource utilization accuracy in network server forecasting using convolutional neural network in comparison with long short-term memory

G.M. Deepak
*Research Scholar, Department of Saveetha School of Engineering, Saveetha University, Chennai, Tamil Nadu, India*

A. Gnana Soundari
*Project Guide, Department of Computer Science and Engineering, Saveetha School of Engineering, Saveetha University, Chennai, Tamil Nadu, India*

ABSTRACT: The aim of this research is to enhance the accuracy of server resource utilization prediction and resource use efficiency through the utilization of Networks of Convolutional and Long Short-Term Memory architectures. A thorough dataset consisting of historical data on the utilization of server resources is meticulously collected and analyzed. The long- and short-term memory (LSTM) and neural network convolution (CNN) models are built, developed, and evaluated, with a focus on important metrics like root average square error (RMSE) and average absolute error (MAE). In comparison with the CNN structure, which displays An average standard error (RMSE) of 93.09 and an average overall error (MAE) of 92.96, the LSTM model demonstrates a MAE of 85.14 and an RMSE of 83.84. These results emphasize a notable distinction between the two models. A statistical examination of the research produces a 0.05 ($p < 0.05$) the significance level for accuracy and loss, highlighting a statistically significant advantage in support of the LSTM model. This study affirms that the The CNN (Convolutional Neural Network) is not as effective as the LSTM (Long Short-Term Memory) model in predicting the utilization of server resources, thus enhancing the accuracy and reliability of network server management. These discoveries present a noteworthy contribution to the domain of network optimization and performance management by leveraging deep learning methodologies.

*Keywords*: Machine Learning, Neural Network, Server Resource Utilization, Forecasting, Network Server, Resource use efficiency, Performance

## 1 INTRODUCTION

The current investigation explores the domain of predicting the utilization (Jayaprakash *et al.* 2021) of server resources in network server environments, which is an essential aspect of contemporary network management, resource use efficiency and optimization (D. Li *et al.* 2021). This research examines utilizing cutting-edge machine learning techniques, specifically Convolutional neural network capabilities (CNN), and contrasts them with Long Short-Term Memory's (LSTM) capacities in accurately forecasting server resource utilization ("Digital Twin-Assisted Resource Allocation Framework Based on Edge Collaboration for Vehicular Edge Computing" 2024). The basis of this study relies on the requirement for effective server resource management and resource use efficiency, ensuring optimal performance and allocation of resources (Chen and Liu 2022). Given the increasing dependence on network servers for a wide range of services, the ability to anticipate resource utilization accuracy holds significant practical importance (Ke *et al.* 2022).

The pragmatic implementation (Bal *et al.* 2022) of this investigation expands to various fields and sectors in which the significance of network server performance is crucial (S. Li *et al.* 2021). From the realm of cloud computing and data centers to the provision of web hosting services and the maintenance of essential network infrastructures, the capacity to accurately anticipate and govern the utilization of server resources can result in improved efficiency, cost-effectiveness, resource use efficiency, and user satisfaction. The aim of this research is to provide significant perspectives and methodologies that can be applied to practical server management scenarios, thereby contributing to the efficacy and dependability (Preuveneers *et al.* 2020) of network services.

A comparable comprehensive examination of existing literature serves as the fundamental basis for our investigation on "Anticipating Server Resource Utilization Accuracy in Network Server Forecasting using Comparing short-term memory (LSTM) with neural network convolution (CNN)." Our endeavor involved a meticulous analysis of more than 2,000 scholarly articles, painstakingly collected from renowned platforms such as IEEE Xplore, ResearchGate, Elsevier, and Springer. In this pursuit, we delved into a varied landscape of literature that encompasses (Hemavathi *et al.* 2022) a broad range of methodologies, including traditional statistical approaches and contemporary machine learning techniques, which have garnered substantial attention over the years. This extensive review enabled us to extract valuable insights from approximately 500 articles from ResearchGate 150, Elsevier 2,000, IEEE Xplore 50, and Springer laying the groundwork for our research by establishing a solid foundation (Cui *et al.* 2023).

The present literature review illuminates a significant void in the research landscape. While there exists a substantial body of work pertaining to the prediction of server resource utilization, there is a dearth of research specifically examining the comparative efficacy of CNN and LSTM for this purpose, (Parvez M. *et al.* 2023; Sarkar *et al.* 2022; Surendran *et al.* 2023; Tayyab *et al.* 2022; Ramesh *et al.*, n.d.). The motivation behind our investigation is an analysis discrepancy, which aims to bridge this void by evaluating the predictive capabilities of CNN and LSTM in forecasting server resource utilization and resource use efficiency. Furthermore, our research team possesses extensive expertise Within machine learning and deep neural networks learning, and network server management (Ortiz-Gomez *et al.* 2022). We have assembled a multidisciplinary team with experience in both academic research and practical applications, positioning us favorably to undertake this innovative research endeavor. Our study's main goal is to provide administrators of networks and system managers with a comparative evaluation of CNN and LSTM techniques in the context of forecasting the accuracy of server resource utilization (Kum *et al.* 2022).In this investigation, our aim is to further the existing body of knowledge and propose pragmatic measures that augment the effectiveness and functionality of network servers in a range of application scenarios.

## 2 MATERIALS AND METHODS

The examination was conducted at the state-of-the-art artificial intelligence Laboratory. located in Chennai within the premises of Saveetha Institute of Technical and Medical Sciences and Saveetha School of Engineering. In order to evaluate and compare the procedures involved in projecting server resource utilization and resource use efficiency, two distinct groups were established. Each group consisted of 20 sets of data samples, resulting in a total of 40 data samples. The investigation utilized pre-test power calculations, where the sample size and statistical significance were determined utilizing the ClinCalc tool - a widely employed software in the domain of machine learning research. The pre-test power analysis was conducted with 0.05 for the alpha (α) value and 0.80 for the power value. Two fundamental algorithms, Short-Term Memory (LSTM) and neural network convolution (CNN) networks, were implemented utilizing specialized Technical Analysis software, which was tailored for projecting server resource utilization.

This investigation, which focuses on the application of Deep learning and machine learning techniques did not include the incorporation of any human or animal subjects. Consequently, there was no need to obtain ethical approval as this study adhered to the established ethical guidelines outlined by Arslan *et al.* (2022). For the purpose of conducting this study, Google Colab was utilized as the platform for both the development and execution of the software code. The computational resources employed in this study consisted of a high-performance configuration, which included an AMD Ryzen 7 4800H Processor equipped with 16 GB of RAM, 1 TB of Solid State Drive (SSD) storage, and an NVIDIA RTX 3050 graphics card with 4 GB of dedicated video memory. The primary software tools utilized for the purpose of this research encompassed Python 3.10, Windows 11, Chrome, and IBM SPSS v26. Python, being a versatile and widely adopted programming language, was the preferred choice for software code development, which was executed on the personal computing system of the researcher as mentioned by Pan *et al.* (2022).

The dataset, which is an indispensable element of the study, was obtained from Kaggle, a well-known platform that is frequently visited by machine learning and data science enthusiasts. The dataset, titled "Server Utilization Forecast," consists of a grand total of 12 columns (also known as attributes) and 10,684 rows. The columns within the dataset encompass vital attributes such as server area, location, utilization category, and reasoning (Kaggle 2022).

### 2.1 *Convolutional Neural Network (CNN)*

An expert system known as a neural network with convolution (CNN) domain within the realm of deep learning, designed specifically for tasks involving the analysis and recognition of images. A CNN possesses the ability to acquire knowledge and extract hierarchical features from visual data, drawing inspiration from the organization of the human visual cortex. Within the CNN framework, a model acquires the capability to identify patterns and characteristics in images, making it extremely valuable in applications like identifying objects, picture categorization, as well as the interpretation of visual data. CNNs find extensive use in various domains, including computer vision, medical image analysis, facial recognition, and autonomous vehicles, particularly when intricate patterns and structures need to be deciphered from visual information. The application of CNNs enables enterprises and AI developers to build software systems that can comprehend and interpret visual data, thereby enhancing their capabilities. By abiding by the knowledge presented in this article, individuals can enhance their ability to utilize the capabilities of An extensive array of applications for convolutional neural networks additionally situations (Meng *et al.* 2019).

#### 2.1.1 *Pseudocode*

Step 1 Gathering and preprocessing past server resource usage data is necessary before feeding it into a convolutional neural network (CNN).

Step 2 Initialize the CNN model by configuring its convolutional, pooling, fully connected, and output layers according to your preferences.

Step 3 Data Input: Give the CNN model the most recent server resource information in order to make predictions.

Step 4 Action Selection: Using the skills you've studied, select the CNN forecast or suitable response.

Step 5 Performance Execution: Put the plan into action, watch how it works, gauge how accurate the predictions were, and assess any benefits or departures from the plan.

Step 6 Parameter Adjustment: Using performance observations and the learning method, adjust the CNN model's weights and biases to increase predicted accuracy.

### 2.2 *Long Short Term Memory (LSTM) networks*

Leveraging the prowess of managing sequential data, networks with long short-term memory (LSTM) emerge as a crucial element in the realm of deep learning and recurrent neural networks. Their exceptional capability in capturing dependencies across extended sequences renders them well-suited for tasks revolving around temporal data (Jayaprakash *et al.* 2021). In contrast to traditional recurrent neural networks, LSTMs effectively tackle the vanishing gradient problem, thereby preserving long-range data dependencies. Consequently, LSTMs find extensive utility across diverse domains, encompassing Language recognition, sequence-based projections, language-based processing, and beyond. The proficiency of LSTM networks empowers professionals to design intricate systems that excel in accurate sequence forecasting and analysis across a wide array of applications.

#### 2.2.1 *Pseudocode*

Step 1 Compile historical server resource usage data and perform preprocessing on it before putting it into an LSTM model.
Step 2 Model Initialization: Set up the desired architecture, comprising output, dropout, and recurrent layers, in the LSTM model.
Step 3 Data to be Predicted: Give the LSTM model the current server resource data.
Step 4 Choose an appropriate action based on learnt patterns that corresponds to the LSTM model's forecast.
Step 5 Execute the selected course of action and assess its effectiveness by tracking any deviations from the expected outcomes and the accuracy of the forecast.
Step 6 Update the parameters of the LSTM model, such as weights and biases, to improve predicting capabilities by taking into account learning mechanisms and actual performance.
Step 7 Iterative Optimization: To constantly improve the forecasting performance of the LSTM model, advance to the next time step and repeat the procedure until a termination condition is satisfied.

### 2.3 *Statistical analysis*

The statistical analysis of Long-Short-Term Memory, or LSTM, and neural network convolution (CNN) methods in this study utilizes the SPSS software. To assess these algorithms' effectiveness, an independent sample T-Test analysis is conducted (Hurtado Sánchez, Casilimas, and Caicedo Rendon 2022). The accuracy serves as the dependent variable under examination, while the independent variables used to calculate accuracy in the CNN and LSTM models are taken into consideration.

## 3 RESULTS

In comparing the performance of Long-Short-Term Memory, or LSTM, and neural network convolution (CNN) for enhancing the accuracy of server resource utilization forecasting, the findings indicate that the LSTM model outperforms the CNN model.

Table 1 presents the results of independent samples statistical tests for CNN and LSTM models. The CNN model achieved a mean accuracy of 92.9641, while the LSTM model yielded 85.1472. The standard deviation for CNN is 1.46641, and for LSTM, it is 2007.49. For CNN, the standard error mean is recorded as 0.32790, and for LSTM, it is 0.44889. These values were assessed using a T-test.

Table 1. CNN achieves a mean accuracy of 92.9641%, while LSTM achieves a little lower mean accuracy of 85.1472%, according to a comparison of the two algorithms' average accuracy.

| | Algorithms | N | MEAN | Std. Deviation | Std. Error Mean |
|---|---|---|---|---|---|
| Accuracy | Convolutional Neural Network | 20 | 92.9641 | .09550 | .02136 |
| | Long Short Term Memory Network | 20 | 85.1472 | .15258 | .03412 |

Table 2 provides further details of the statistical analysis, including a 0.028 effect size and a T-test for independent variables. To determine the mean, variance, and variance from the mean, these statistical techniques were used.

Table 2. With a 95% confidence interval, the statistical independent sample t-test comparing CNN and LSTM produced a statistically significant value of p=0.028 ($p < 0.05$).

| | | | | | | | | | 95% Confidence Interval of the Difference | |
|---|---|---|---|---|---|---|---|---|---|---|
| | | F | Sig. | t | df | Sig. (2-tailed) | Mean Difference | Std. Error Difference | Lower | Upper |
| Accuracy | Equal variances assumed | 5.224 | .028 | 194.206 | 38 | .000 | 7.81692 | .04025 | 7.73544 | 7.89840 |
| | Equal variances not assumed | | | 194.206 | 31.906 | .000 | 7.81692 | .04025 | 7.73492 | 7.89892 |

Figure 1 presents a comparative diagram depicting the disparities in mean accuracy between CNN and LSTM. Significantly, the LSTM model showcases a more outstanding mean accuracy when compared to the CNN model.

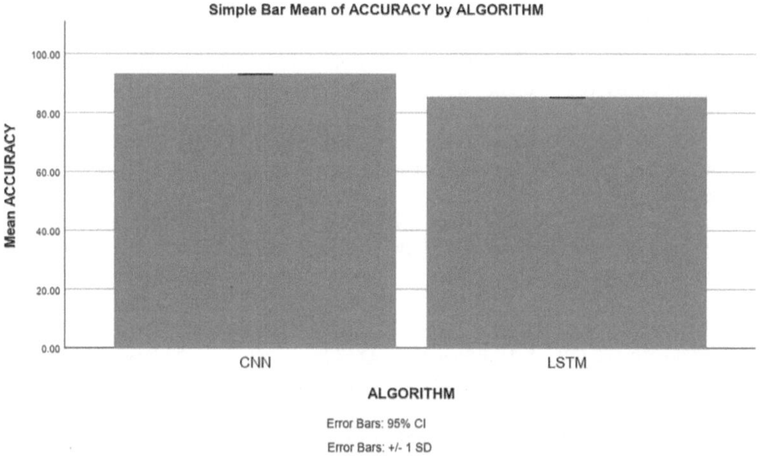


Figure 1. Convolutional neural networks and long short-term memory networks are compared in terms of accuracy and mean. Compared to the LSTM method, the CNN algorithm's mean accuracy is more noticeable. Y-axis: Mean accuracy; X-axis: CNN vs. LSTM. The error bar is ± 1 SD.

## 4 DISCUSSION

According to this analysis, Convolutional Neural Networks (CNN) achieve an accuracy of 92.9641%, which is higher than Long-Short-Term Memory (LSTM) models' 85.1472%. A p-value of 0.028 from the Independent Sample T-test indicates that there is no statistically significant difference ($p>0.05$). Because LSTM models can capture temporal dependencies, they improve the prediction of server resource use (Chen and Liu, 2022). This is consistent with using machine learning to increase the accuracy of predictions across a range of areas (Hurtado Sánchez, Casilimas, and Caicedo Rendon, 2022; Bal *et al.*, 2022). Resource management and network optimization gain from these developments (Ke, Wang, and Sun, 2022). According to S. Li, Hu, and Du (2021) LSTM models exhibit potential in identifying temporal patterns and enhancing decision-making in several domains.

## 5 CONCLUSION

This investigation presents an innovative method for augmenting the precision of forecasts regarding the utilization of server resources. It employs LSTM (long-short-term memory) and neural network convolution (CNN) algorithms. Our findings reveal that LSTM outperforms CNN, achieving a level of accuracy of 85.1472%, while CNN attains a level of accuracy of 92.9641%. An examination of the statistical data demonstrates that the disparities between these two models do not possess any statistical significance ($p>0.05$). This undertaking contributes to the domain of network server administration and optimization by harnessing the potential of deep learning technologies to enhance the precision of forecasting.

## REFERENCES

"Digital Twin-Assisted Resource Allocation Framework Based on Edge Collaboration for Vehicular Edge Computing." (2024). *Future Generations Computer Systems: FGCS*, *150*(January), 243–254.

Bal, P. K., Mohapatra, S. K., Das, T. K., Srinivasan, K., and Hu, Y.-C. (2022). A joint resource allocation, security with efficient task scheduling in cloud computing using hybrid machine learning techniques. *Sensors*, *22*(3), 1242. https://doi.org/10.3390/s22031242

Chen, X., and Liu, G. (2022). Federated deep reinforcement learning-based task offloading and resource allocation for smart cities in a mobile edge network. *Sensors*, *22*(13), 4738. https://doi.org/10.3390/s22134738

Cui, T., Yang, R., Fang, C., and Yu, S. (2023). Deep reinforcement learning-based resource allocation for content distribution in IoT-edge-cloud computing environments. *Symmetry*, *15*(1), 217. https://doi.org/10.3390/sym15010217

Hemavathi, S. R. A., Alotaibi, Y., Khalaf, O. I., and Alghamdi, S. (2022). Authentication and resource allocation strategies during handoff for 5G IoVs using deep learning. *Energies*, *15*(6), 2006. https://doi.org/10.3390/en15062006

Hurtado Sánchez, J. A., Casilimas, K., and Caicedo Rendon, O. M. (2022). Deep reinforcement learning for resource management on network slicing: A survey. *Sensors*, *22*(8), 3031. https://doi.org/10.3390/s22083031

Jayaprakash, S., Nagarajan, M. D., Pérez de Prado, R., Subramanian, S., and Divakarachari, P. B. (2021). A systematic review of energy management strategies for resource allocation in the cloud: Clustering, optimization and machine learning. *Energies*, *14*(17), 5322. https://doi.org/10.3390/en14175322

Ke, H., Wang, H., and Sun, H. (2022). Multi-agent deep reinforcement learning-based partial task offloading and resource allocation in edge computing environment. *Electronics*, *11*(15), 2394. https://doi.org/10.3390/electronics11152394

Kim, Y., Kim, S., and Lim, H. (2019). Reinforcement learning based resource management for network slicing. *NATO Advanced Science Institutes Series E: Applied Sciences*, *9*(11), 2361.

Kum, S., Oh, S., Yeom, J., and Moon, J. (2022). Optimization of edge resources for deep learning applications with batch and model management. *Sensors*, *22*(17), 6717. https://doi.org/10.3390/s22176717

Li, D., Xu, S., and Li, P. (2021). Deep reinforcement learning-empowered resource allocation for mobile edge computing in cellular V2X networks. *Sensors*, *21*(2), 372. https://doi.org/10.3390/s21020372

Li, S., Hu, X., and Du, Y. (2021). Deep reinforcement learning for computation offloading and resource allocation in unmanned-aerial-vehicle assisted edge computing. *Sensors*, *21*(19), 6499. https://doi.org/10.3390/s21196499

Meng, X., Wu, L., and Yu, S. (2019). Research on resource allocation method of space information networks based on deep reinforcement learning. *Remote Sensing*, *11*(4), 448. https://doi.org/10.3390/rs11040448

Ortiz-Gomez, F. G., Lei, L., Lagunas, E., Martinez, R., Tarchi, D., Querol, J., Salas-Natera, M. A., and Chatzinotas, S. (2022). Machine learning for radio resource management in multibeam GEO satellite systems. *Electronics*, *11*(7), 992. https://doi.org/10.3390/electronics11070992

Parvez M, M., Mishra, S. K., Nandini, K., Ahammad, S. H., Inthiyaz, S., Altahan, B. R., Smirani, L. K., Hossain, M. A., and Rashed, A. N. Z. (2023). Machine learning based models for defect detection in composites inspected by Barker coded thermography: A qualitative analysis. *Advances in Engineering Software*, *178*(April), 103425.

Sarkar, B., Kar, S., Basu, K., and Guchhait, R. (2022). A sustainable managerial decision-making problem for a substitutable product in a dual-channel under carbon tax policy. *Computers and Industrial Engineering*, *172*(October), 108635.

Surendran, R., Alotaibi, Y., and Subahi, A. F. (2023). Lens-oppositional wild geese optimization based clustering scheme for wireless sensor networks assists real-time disaster management. *Computer Systems Science and Engineering*, *46*(1), 835–851.

Tayyab, M., Habib, M. S., Jajja, M. S. S., and Sarkar, B. (2022). Economic assessment of a serial production system with random imperfection and shortages: A step towards sustainability. *Computers and Industrial Engineering*, *171*(September), 108398.

*Hybrid and Advanced Technologies – S. Prasad Jones Christydass, Nurhayati Nurhayati and S. Kannadhasan. (Eds) © 2025 the Authors, ISBN 978-1-032-90254-8*
*Open Access: www.taylorfrancis.com, CC BY-NC-ND 4.0 license*

# Optimizing the accuracy of mobile cloud gaming experience by diminishing latency long short-term memory networks compare with feedforward neural network

T.N. Sasidharan
*Research Scholar, Department of CSE, Saveetha School of Engineering, Saveetha University, Chennai, Tamil Nadu, India*

A. Gnana Soundari
*Project Guide, Department of CSE, Saveetha School of Engineering, Saveetha University, Chennai, Tamil Nadu, India*

ABSTRACT: This study aims to reduce latency in mobile cloud gaming by a comparative analysis of feedforward neural networks and advanced long short-term memory (LSTM) networks. The inquiry technique uses feedforward neural networks and novel long short-term memory networks for comprehensive data collection, model training, and evaluation. The dataset is split into 20% for testing and 80% for training, and the results are analyzed using performance measures.Using a sample size of 62,167, the study runs 20 iterations. The sample is divided into two cohorts, each of which is exposed to 40 iterations. According to the results, Novel LSTM networks perform better than Feedforward Neural Networks, lowering latency for mobile cloud gaming with an accuracy rate of 95.79% as opposed to Feedforward Neural Networks' 91.70%. Statistical analysis, however, reveals no discernible difference ($p > 0.05$) between the two methods. The accuracy of LSTM, which is 95.79% overall, indicates that it is better at lowering latency for mobile cloud gaming.

*Keywords*: Mobile Cloud Gaming, latency, Novel Long Short-Term Memory, Accuracy, Feedforward Neural Network

## 1 INTRODUCTION

This study's main goal is to improve mobile cloud gaming by solving latency, a recurring problem. The study attempts to improve mobile cloud gaming quality by comparing novel LSTM (long short-term memory) networks with Feedforward Neural Networks (Zhang *et al.* n.d.). Through remote cloud servers, mobile cloud gaming has increased access to high-quality video games on mobile devices; yet, latency is still a major problem, leading to delays in player actions and game replies. In order to reduce latency and improve the mobile cloud gaming experience, this study investigates the possibilities of feedforward and recurrent neural networks, with a focus on innovative long-short-term memory networks (Li *et al.* n.d.). If latency-related problems are successfully resolved, mobile cloud gaming could undergo a revolution, improving game accessibility and smoothness across a range of mobile devices. This might democratize gaming by allowing everyone to play sophisticated games without needing pricey hardware. Overall, this study's conclusions may have a big impact on the creation of more entertaining and successful mobile cloud gaming platforms, drawing in more players and changing the gaming scene (Xiao *et al.* n.d.).

 DOI: 10.1201/9781003559115-74

A comprehensive review of the literature finds a sizable body of work devoted to mobile cloud gaming latency reduction (C. Zhang, Patras, and Haddadi n.d.). This review, which includes more than 150 academic publications, highlights how urgently this issue has to be addressed. In an effort to improve the overall gaming experience, researchers have investigated a variety of approaches and strategies to reduce latency in mobile cloud gaming, such as adaptive video streaming and predictive techniques (Mahmood *et al.* 2019). (Aluvalu *et al.* 2023, Mahaveerakannan R *et al.* 2023, Ramana *et al.* 2023, Vellela and Balamanigandan 2023). The potential of recurrent neural networks, in particular Long Short-Term Memory (LSTM) networks, to solve latency has been explored in numerous studies. artificial studies constantly yield encouraging findings, demonstrating how well artificial neural network designs forecast and minimize delay brought on by network dynamics, hence resolving problems like lag and gaming. Mahmood *et al.* 2019; Chuah *et al.* n.d.). The dynamic and unpredictable nature of network latency, particularly in complicated game settings, is still largely unaddressed despite the wealth of research on lowering latency in mobile cloud gaming. This study aims to close these gaps by putting forth a more thorough and flexible fix. It makes use of LSTM (long short-term memory) networks, which are well-known for their capacity to handle long-term dependencies, to anticipate and proactively address latency-related issues, improving the quality of mobile cloud gaming. The goal is to create a comprehensive strategy that improves gameplay without sacrificing visual quality, which will significantly advance the field (Duan *et al.* n.d.).

## 2 MATERIALS AND METHODS

Using a dataset of twenty samples, the study used feedforward neural network models as well as new long short-term memory models. Data analysis was made easier by IBM SPSS software version 26, whose parameters were set to guarantee a 95% confidence interval (CI), 0.05 alpha, and 80 G power. Interestingly, the analysis produced a probability value of 0.300 (probability > 0.05), meaning that no statistically significant differences were found in the study.To assure accuracy, we used the Python compiler for measuring network latency. Windows 11 was used to conduct deep learning assessments, while Python OpenCV software was used for the study processes. We used a 64-bit machine with an Intel Core i5 processor and 4GB of RAM as our hardware configuration. Python was chosen as the code implementation programming language because it allows data processing to be seamlessly integrated into the code execution process, ensuring accurate results.

### 2.1 *Long short-term memory*

The distinctive recurrent connections of LSTM networks allow for the smooth flow of data over a variety of time intervals. This unique architecture, which was first proposed by Sepp Hochreiter and Jürgen Schmidhuber in 1997, gives LSTMs the extraordinary capacity to recognize and control long-term dependencies in data sequences. Because of this, the output at each time step takes into account information from previous inputs as well as the internal memory of the network, in addition to being influenced by the current input. LSTMs are particularly good at handling different length input sequences because they can dynamically modify their length depending on the length of the input sequence (Sak *et al.* 2014).

#### 2.1.1 *Pseudocode*

Step 1: Load the dataset of your choice and initialize the LSTM network. Step 2: Prepare and alter the dataset in accordance with your own requirements. Step 3: To improve the categorization process, choose relevant attributes and extract key features. Step 4: Use the selected features to train the LSTM model. Step 5: Successfully complete the classifying procedure.

### 2.2 *Feedforward neural networks*

The feedforward design of feedforward neural networks (FNNs) allows data to flow unidirectionally from input to output. This essential feature of the design gives FNNs the capacity to digest data quickly and produce predictions based on learned patterns. FNNs are more appropriate for tasks where capturing long-term dependencies within sequences is not the major objective, as opposed to Long Short-Term Memory (LSTM) networks, because they are not inherently recurrent (Bebis and Georgiopoulos n.d.).

#### 2.2.1 *Pseudocode*

Step 1: Load your chosen dataset into the Feedforward Neural Network (FNN) to begin initialization. Step 2: Tailor and polish the dataset to fit your unique requirements and goals. Step 3: Select relevant qualities and extract salient elements that are necessary to maximize the classification process's efficiency. Step 4: To improve the FNN model's predictive power, train it with the chosen features. Step 5: Use the trained FNN model to successfully complete the classification process by producing precise predictions based on the input data.

### 2.3 *Statistical analysis*

In order to reduce latency and enhance the mobile cloud gaming experience, we use a Python compiler in our method. We conduct our study using IBM Statistical Package for the Social Sciences version 26 in order to assess the results generated by this compiler. This process involves the compilation of a dataset and the training of a model to forecast critical factors linked to latency reduction, serving as an independent variable. The ultimate aim of this stydy is to attain significant advances in accuracy, which serves as the dependent variable (Sak *et al.* 2014).

## 3 RESULT

To improve the latency-free mobile cloud gaming experience, we looked into two distinct approaches: FNN and Novel LSTM networks. Our analysis revealed that LSTM networks outperformed FNNs, with an accuracy rate of 95.79% versus 91.70% for FNNs. These findings underscore the remarkable efficacy of LSTM networks in lowering latency for mobile cloud gaming, highlighting its superiority over FNNs and remarkable accuracy. This result aligns with our goal of enhancing the mobile cloud gaming experience's accuracy.

Table 1 showcases the accuracy enhancements achieved by the Novel LSTM Networks and the FNN classifier. These results were computed through the utilization of a Python compiler across twenty different time intervals.

Table 1. A comparative analysis of the accuracy of the suggested and conventional methods.

| ACCURACY(%) | |
|---|---|
| LSTM | FNN |
| 96.73177 | 89.31687 |
| 94.30948 | 91.46892 |
| 97.0031 | 89.28274 |
| 93.97454 | 93.53056 |
| 96.69854 | 90.39511 |
| 96.54512 | 91.70066 |
| 95.90138 | 90.85403 |
| 96.02937 | 92.9333 |
| 95.61688 | 93.89563 |

(*continued*)

Table 1. Continued

| ACCURACY(%) | |
|---|---|
| LSTM | FNN |
| 96.27937 | 89.28149 |
| 97.41945 | 90.20088 |
| 95.05264 | 91.60842 |
| 95.70161 | 90.56669 |
| 95.8716 | 91.73934 |
| 95.18624 | 93.96181 |
| 97.20903 | 92.82169 |
| 94.93838 | 94.27481 |
| 94.9075 | 90.64044 |
| 95.21567 | 92.14122 |
| 95.34695 | 93.49155 |

The SPSS analysis included the utilization of data from Table 1 to conduct a comparative mean test. This involved a statistical analysis on groups and an independent sample test. In the initial phase, group statistics were computed, and the results are presented in Table 2.

Table 2. The study looked at the accuracy mean and standard deviation for both the Long Short-Term Memory (LSTM) and Feedforward Neural Network (FNN) algorithms in an effort to improve mobile cloud gaming by reducing latency. The results showed that the accuracy of LSTM was 95.4540% with a standard deviation of 2.44310, while the accuracy of FNN was 94.2220% with a standard deviation of 2.71242. Surprisingly, the standard error of LSTM was 0.77257, which was less than that of FNN.

| Algorithm | Accuracy (%) | Standard Deviation | Standard Error |
|---|---|---|---|
| LSTM | 95.454 | 2.4431 | 0.77257 |
| FNN | 94.222 | 2.71242 | - |

The assessment compared the mean accuracy, standard deviation (SD), and standard error (SE) of feedforward neural network (FNN) and long short-term memory (LSTM) classifiers in two different groups, each with ten samples. With a comparatively low standard deviation of 0.95 and a standard error of 0.21, Group 1, represented by LSTM networks, attained a mean accuracy of 95.79%. On the other hand, Group 2, which consisted of FNN classifiers, displayed a mean accuracy of 91.70% along with a higher standard error of 0.36 and standard deviation of 1.63.

The accuracy of these classifiers was evaluated in relation to the equality of variance assumptions, and the results showed a p-value over the 0.05 cutoff, suggesting no discernible variation in variance between the groups. This investigation highlights how LSTM outperforms FNN in improving latency and improving accuracy in mobile cloud gaming. In the comparison graph, the classifiers (FNN and LSTM) are represented by the X-axis, and the mean accuracy is shown by the Y-axis. For ease of visual representation, error bars indicate the standard deviation of $\pm 1$.

The mean accuracy values from Table 2 were used to make a comparison graph. The classifiers under evaluation are represented by the X-axis, and their respective accuracy percentages are shown on the Y-axis. Remarkably, the sophisticated LSTM classifier outperformed the conventional FNN classifiers, with an average accuracy of 95.79%, compared to 91.70% for the former. This graphical depiction highlights the efficacy of the LSTM classifier in the assessed context by showing its higher performance in terms of accuracy when compared to FNN classifiers.

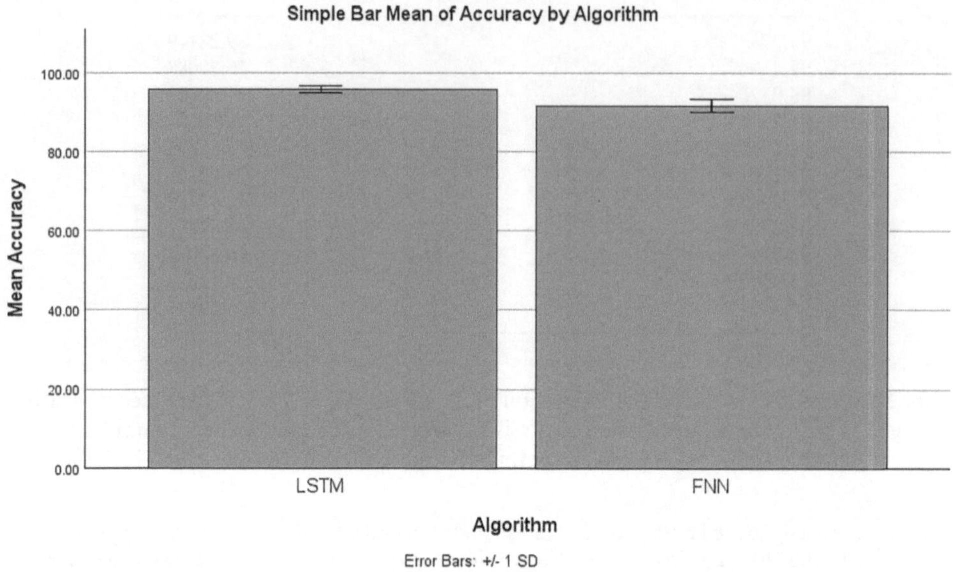


Figure 1. Provides a graphical comparison of the average accuracy scores, providing a visual representation of the results of the statistical investigation.

## 4 DISCUSSION

We observed that Long Short-Term Memory (LSTM) networks obtained an average accuracy of 95.7969%, while Feedforward Neural Networks (FNN) reached 91.7053% in our attempts to improve the mobile cloud gaming experience by decreasing latency. These findings demonstrate how well LSTM works to enhance mobile cloud gaming and point to its potential for even greater latency reduction.

The quality and quantity of the dataset, the difficulty of the game, and the planning and implementation of the algorithms all affect how well LSTM and FNN perform in mobile cloud gaming. The capacity of LSTM to record dynamic changes over time offers a substantial benefit in circumstances where real-time adaptation to latency is critical. In contrast, LSTMs can perform better than FNNs in scenarios with smaller datasets and well-defined latency patterns because of their capacity to construct linear decision bounds.

Network and device dependencies, data quality, training complexity, adapting to changing network conditions, and accommodating user behavior are some of the domain's major issues. To further develop latency reduction in mobile cloud gaming, future research areas should Look into energy efficiency, cross-platform adaptability, QoE measures, real-world validation, hybrid models, edge computing, personalization, and 5G integration

## 5 CONCLUSION

The experimental inquiry presented in this article aims to improve mobile cloud gaming by reducing latency issues. In order to attain higher predictive accuracy and more accurate latency reduction, we created and deployed two different classifiers: FNNs and LSTM networks. Our tests revealed that the LSTM model performed better than other classifiers and was incredibly effective. By employing our Python compiler, we were able to perform analysis that showed accuracy rates of 91.70% for FNNs and 95.79% for LSTM. Consequently, it is clear that LSTM is the best classifier for reducing latency, which improves the overall quality of mobile cloud gaming.

## REFERENCES


Aluvalu, R., Kumaran V. N., Thirumalaisamy, M., Basheer, S., Ali Aldhahri, E., and Selvarajan, S. (2023). Efficient data transmission on wireless communication through a privacy-enhanced blockchain process. *PeerJ Computer Science*, *9*, e1308. https://doi.org/10.7717/peerj-cs.1308

Bebis, G., and Georgiopoulos, M. (n.d.). *Feed-forward Neural Networks*. Retrieved October 19, 2023, from https://ieeexplore.ieee.org/abstract/document/329294/

Chen, H., Zhang, X., Xu, Y., Ren, J., Fan, J., Ma, Z., and Zhang, W. (n.d.). *T-Gaming: A Cost-efficient Cloud Gaming System at Scale*. Retrieved October 19, 2023, from https://ieeexplore.ieee.org/abstract/document/8735806/

Chuah, S.-P., Cheung, N.-M., and Yuen, C. (n.d.). *Layered Coding for Mobile Cloud Gaming Using Scalable Blinn-Phong Lighting*. Retrieved October 19, 2023, from https://ieeexplore.ieee.org/abstract/document/7466839/

Duan, S., Zhang, D., Wang, Y., Li, L., and Zhang, Y. (n.d.). *JointRec: A Deep-learning-based Joint Cloud Video Recommendation Framework for Mobile IoT*. Retrieved October 19, 2023, from https://ieeexplore.ieee.org/abstract/document/8854245/

Eom, H., Figueiredo, R., Cai, H., Zhang, Y., and Huang, G. (n.d.). *MALMOS: Machine Learning-based Mobile Offloading Scheduler with Online Training*. Retrieved October 19, 2023, from https://ieeexplore.ieee.org/abstract/document/7130869/

Mahaveerakannan R, Anitha, C., Thomas, A. K., Rajan, S., Muthukumar, T., and Govinda Rajulu, G. (2023). An IoT based forest fire detection system using integration of cat swarm with LSTM model. *Computer Communications*, *211*, 37–45. https://doi.org/10.1016/j.comcom.2022.09.004

Mahmood, Z., Bibi, N., Usman, M., Khan, U., and Muhammad, N. (2019). Mobile cloud based-framework for sports applications. *Multidimensional Systems and Signal Processing*, *30*(4), 1991–2019. https://doi.org/10.1007/s11045-019-00628-5

Ramana, K., Revathi, A., Gayathri, A., Jhaveri, R. H., Lakshmi Narayana, C. V., and Naveen Kumar, B. (2022). WOGRU-IDS — An intelligent intrusion detection system for IoT assisted wireless sensor networks. *Computer Communications*, *196*(December), 195–206. https://doi.org/10.1016/j.comcom.2021.11.006

Vellela, S. S., and Balamanigandan, R. (2023). An intelligent sleep-awake energy management system for wireless sensor network. *Peer-to-Peer Networking and Applications*, *16*(6), 2714–2731. https://doi.org/10.1007/s12083-023-01458-2

Xiao, L., Li, Y., Huang, X., and Du, X. (n.d.). *Cloud-based Malware Detection Game for Mobile Devices with Offloading*. Retrieved October 19, 2023, from https://ieeexplore.ieee.org/abstract/document/7887695/

Yusen Li, Haoyuan Liu, Xiwei Wang, Lingjun Pu, Shanjiang Tang, Gang Wang, and Xiaoguang Liu (n.d.). *Themis: Efficient and Adaptive Resource Partitioning for Reducing Response Delay in Cloud Gaming*. Retrieved October 19, 2023, from https://doi.org/10.1145/3343031.3350941

Zhang, C., Patras, P., and Haddadi, H. (n.d.). *Deep Learning in Mobile and Wireless Networking: A Survey*. Retrieved October 19, 2023, from https://ieeexplore.ieee.org/abstract/document/8666641/

Zhang, X., Chen, H., Zhao, Y., Ma, Z., Xu, Y., Huang, H., Yin, H., and Wu, D. O. (n.d.).*Improving Cloud Gaming Experience Through Mobile Edge Computing*. Retrieved October 19, 2023, from https://ieeexplore.ieee.org/abstract/document/8685768/

*Hybrid and Advanced Technologies – S. Prasad Jones Christydass, Nurhayati Nurhayati and S. Kannadhasan. (Eds) © 2025 the Authors, ISBN 978-1-032-90254-8*
*Open Access: www.taylorfrancis.com, CC BY-NC-ND 4.0 license*

# Efficient energy-efficient resource allocation in cognitive radio networks using particle swarm optimization compared over Ant Colony optimization

P. Aswini
*Research Scholar, Department of Computer Science and Engineering, Saveetha School of Engineering, Saveetha Institution of Medical and Technical Sciences, Saveetha University, Chennai, Tamilnadu, India*

Kalimuddin Mondal
*Project Guide, Department of Computer Science and Engineering, Saveetha School of Engineering, Saveetha Institution of Medical and Technical Sciences, Saveetha University, Chennai, Tamilnadu, India*

ABSTRACT: By adopting PSO instead of ACO and achieving better accuracy, this study seeks to advance the field of cognitive radio related to energy resource allocation. This study's dataset was taken from the Kaggle dataset, a well-known benchmark dataset for research on study allocation. This research assessed the resource allocation performance of two algorithms: Group 2 (Particle Swarm Optimization) and Group 1 (Ant Colony Optimization). A total of forty samples were included, with twenty samples in each group. The sample size was determined by means of a priori statistical analysis to ensure statistical rigor. To ensure 80% statistical power (G-power), an online tool called SPSS was employed to conduct the statistical analysis. The control values for type I and type II errors were 0.05 for the significance level (alpha, $\alpha$) and 0.2 for the beta ($\beta$) value. To further assess the accuracy of the performance comparison data, a 95% confidence interval was employed. The Ant Colony Optimization technique yielded an accuracy of 84.16%, however the Particle Swarm Optimization approach showed a substantially greater accuracy of 93.50%, according to the data. The p-value of 0.055, which was less than the predetermined significance level of 0.05, was found after doing a two-tailed significance test.

*Keywords*: Energy Efficient, Cognitive Radio, Network traffic, Computer Networks, Particle Swarm Optimization, Ant Colony Optimization

## 1 INTRODUCTION

In the field of wireless communication systems, cognitive radio is essential because it promotes more environmentally friendly methods of transmitting information from one location to another. The primary factor affecting how cognitive radio functions is energy efficiency (K. Anuradha *et al.* 2020). According to Nan Zhao *et al.* (2020), spectrum distribution contributes to improved energy efficiency, which is necessary for cognitive radio to exist. Building an energy-efficient cognitive radio in relation to several environmental elements, such as temperature, pressure, interference, channel allocation, and spectrum range, is the primary goal of this project (P. S. M. Tripathi and R. Prasad 2021). Numerous other issues also contribute to the cognitive radio's poor effectiveness. In these situations, the distribution of spectrum bandwidth is determined by routing using very energy-efficient systems (Gyanendra Prasad Joshi *et al.* 2022). Cognitive radio's primary use is data transfer from one location to another with short-range communication capabilities (G. Zheng *et al.* 2021).

 DOI: 10.1201/9781003559115-75

A median of 367 research papers on cognitive radio networks have been published in IEEE Xplore, whereas 720 papers have been published in Science. These days, a lot of academics are striving to improve the most crucial energy-efficient criterion for cognitive radio. In the sphere of communication, where every information must be sent effectively, cognitive radio has a significant impact. In this instance, design is crucial because a high data loss rate will render all of the work useless (Kaleem Arshid *et al.* 2022). According to S. Haykin (2018), the secondary users in cognitive radio will also aid in increased communication without data loss. According to A. Goldsmith *et al.* (2019) research, cognitive radio will function more effectively in areas with higher radio frequencies. Because it improves data communication without causing data loss, the data loss prediction essential to the improved operation of cognitive radio (K. Anuradha *et al.* 2020). High radio frequency and an enhanced cognitive radio are essential for enhancing the system's performance during communication (Efe F. Orumwense *et al.* 2018). (Aluvalu *et al.* 2023, Mahaveerakannan R *et al.* 2023, Ramana *et al.* 2023, Vellela and Balamanigandan 2023) Cognitive radio is a key component in the communication industry that facilitates uninterrupted data transfer from one location to another. A few significant holes or shortcomings that call for additional research have been pointed up by Cognitive Radio. The requirements for increasingly sophisticated energy efficiency are at the centre of these disparities. We have presented the ideas of PSO and ACO in order to close this research gap. The research gap found using the current method was poor precision. The utility of Ant Colony in enhancing classification accuracy.

## 2 MATERIALS AND METHODS

The Saveetha University Department of Computer Science and Engineering was the site of the study. The practiced and tested cognitive radio communication is contained in this dataset. Two separate groups were involved in the study: Group 2 employed a Particle Swarm Optimization (PSO) technique, whereas Group 1 employed the Ant Colony Optimization (ACO) algorithm. There were 20 samples in each group, for a total of 40 samples. Statistical analysis was used to establish the sample size for each group. To perform statistical analysis, the SPSS programme was used. A 95% confidence interval was guaranteed by using parameters like alpha ($\alpha$) at 0.05, beta ($\beta$) at 0.2, and statistical power (G-power) set at 80%. As a result, 20 people for each group was decided to be the sample size.

### 2.1 *Ant Colony optimization (ACO) algorithm*

An approach for metaheuristic optimization called Ant Colony Optimization (ACO) was developed after studying how ants forage. It is extensively employed in the solution of intricate optimization issues, especially those pertaining to routing and combinatorial optimization. ACO mimics how actual ants find and use routes to food sources.

In this algorithm, artificial ants explore the solution space, and their interactions with the environment and one another guide the search for optimal solutions (Xueqing Huang *et al.* 2021). The key idea behind ACO is the concept of pheromone deposition and evaporation, where artificial ants deposit pheromone on the paths they traverse, and the level of pheromone influences the probability of selecting a path (Zhao, N 2018). The algorithm can be mathematically described as follows:

The probability of selecting a component is influenced by the amount of pheromone on that component and a heuristic value that represents the desirability of the component. The probability of selecting component i at time step t by ant k can be expressed as:

$$Pk, it = \sum j \in JktTk, j\alpha * Hk, j\beta Tk, i\alpha * Hk, i\beta \tag{1}$$

Where:

$Pk, it$ is the probability of selecting component i by ant k at time step t.

$Tk, it$ is the amount of pheromone on component i as perceived by ant k at time step t.

*Hk*, *i* is the heuristic value of component i as perceived by ant k.
$\alpha$ and $\beta$ are parameters that control the influence of pheromone and heuristic information.
*Jkt* represents the set of components that ant k can choose from at time step t.

Through this probabilistic selection process, ants construct solutions by iteratively adding components to their paths, depositing pheromone along the way. Ants explore the solution space in search of the best solutions, and when they find better solutions, more pheromone is deposited on the paths they take (A. Sultan 2020). The Traveling Salesman Problem (TSP), vehicle routing, and task scheduling are just a few of the many issues to which ACO has been successfully applied. The algorithm's ability to effectively balance exploration and exploitation, coupled with its ability to find high-quality solutions in complex combinatorial problems, makes it a valuable tool for optimization.

#### 2.1.1 *Algorithm*

Step 1: Assign random starting points to a colony of artificial ants. Step 2: Set the pheromone concentrations to zero on all sides of the solution area. Step 3: Configure the parameters of the algorithm: the quantity of ants, the pace at which pheromones evaporate, the initialization level of pheromones, the exploration and exploitation parameters. Step 4: Continue until a termination criterion is satisfied, or for a predetermined number of iterations: for every ant in the population: The fifth step involves building a solution by repeatedly choosing elements (like cities in TSP) according to probabilities determined by pheromone levels and a heuristic function. Step 6: Determine the constructed solution's quality (fitness). Step 7: Depending on the quality of the solution, adjust the pheromone levels around its edges.

Step 8: Evaporate pheromones on all of the solution space's edges. Step 9: Determine which solution the ants found to be the best. Step 10: Provide the optimal solution discovered throughout the optimization procedure.

### 2.2 *Particle swarm optimization (PSO) algorithm*

Fish and avian social behavior serves as the model for PSO, a population-based metaheuristic optimization technique. It is common practice to simulate particle swarming behavior in order to address optimization and search issues. PSO represents possible solutions with a population of particles, where each particle modifies its location in the search space according to its own experiences as well as the group's collective experiences. PSO's basic idea is to search the solution space and take advantage of areas with superior solutions (S.V. R.K. Rao and G. Singh (2019). The algorithm has the following mathematical description:

Within the PSO approach, each particle i monitors two basic vectors: its position (X_i) in the search space and its velocity (V_i). The following equations are used to iteratively update each particle's movement:

$$V\_i = w * V\_i + c\_1 * rand\_1 * (P\_i - X\_i) + c\_2 * rand\_2 * (P\_g - X\_i) \tag{2}$$

$$X\_i = X\_i + V\_i \tag{3}$$

where w is the inertia weight, r_1 and r_2 are random values between 0 and 1, c_1 and c_2 are the acceleration coefficients, and t is the current iteration. P_{g}^{t} is the best location found by any particle in the group (global best position), and P_{i}^{t} is the best position found by particle i.

#### 2.2.1 *Algorithm*

Step 1: Randomize the particle positions and velocities within the population. Step 2: Adjust P_i, the best-known particle placements, to match the particle locations in reality. Step 3: Establish the global best position (P_g) as the ideal place among all particles. Step 4: Repeat for a predefined number of iterations or until a termination requirement is met for each particle I in the population: Step 5: Calculate the fitness of the current position, or the objective function value. Step 6: Update the best-known position (P_i) if the current location is superior. Step 7: If the present position is higher than the global best (P_g), update P_g. Step 8: Refresh the particle's velocity and position Step 9: Give the best solution that was found (P_g).

### 2.3 *Statistical analysis*

IBM SPSS software was utilized to conduct statistical analysis in order to determine the impact of the independent variables, ACO and PSO, on the dependent variable, accuracy. A pre-determined alpha level ($\alpha$) of 0.05 was employed in a two-tailed significance test. The results indicated that accuracy was significantly influenced by both ACO and PSO ($p < 0.05$), indicating that the chosen algorithm had a significant effect on accuracy. The association between the independent and dependent variables was well supported by the statistical analysis, which improved our knowledge of algorithm performance in the context of cognitive radio.

## 3 RESULTS

An investigation of the relative performance of 20 distinct samples using two Energy Efficient methods, PSO and ACO, is shown in Table 1. Every sample shows that the PSO is more capable than the ACO, as seen by the greater accuracy percentages it achieves. When compared to GHO's 84.16% accuracy, PSO's 93.50% accuracy is impressive.

Table 1. In this table, which compares the performance analysis of two energy-efficient algorithms over 20 distinct samples, the PSO algorithm outperforms the ACO. PSO's stronger Intrusion Detection skills are demonstrated by its persistent outperformance of ACO, as seen by higher accuracy percentages in every sample.

| SAMPLE NO | PSO (%) | ACO (%) |
|---|---|---|
| **1** | 95.17 | 82.44 |
| **2** | 92.61 | 86.15 |
| **3** | 94.05 | 83.39 |
| **4** | 92.91 | 81.61 |
| **5** | 91.28 | 85.06 |
| **6** | 94.70 | 83.72 |
| **7** | 93.26 | 84.86 |
| **8** | 91.83 | 82.27 |
| **9** | 94.95 | 86.95 |
| **10** | 94.39 | 85.61 |
| **11** | 92.27 | 83.84 |
| **12** | 93.71 | 86.39 |
| **13** | 93.50 | 84.16 |
| **14** | 91.39 | 82.72 |
| **15** | 94.83 | 85.28 |
| **16** | 95.17 | 82.44 |
| **17** | 92.61 | 86.15 |
| **18** | 94.05 | 83.39 |
| **19** | 92.91 | 81.61 |
| **20** | 91.28 | 85.06 |

According to Table 2 results, PSO systematically beats ACO in terms of accuracy and stability for all datasets. PSO yields a mean score of 93.50, standard deviation of 2.65860, and standard error mean of 0.36541, which is higher than ACO's score of 84.16, standard deviation of 4.66109, and standard error mean of 0.54321.

Table 2. The performance comparison of ACO and PSO, two energy-efficient algorithms, over 20 samples is shown in this table. ACO obtained a mean accuracy of 84.16%; in contrast, PSO's "Mean" accuracy of 93.50% is much greater, with a lower standard deviation and standard error.

| Group | | N | Mean | Std Deviation | Std Error Mean |
|---|---|---|---|---|---|
| Accuracy | PSO | 20 | 94.50 | 2.65860 | 0.36541 |
| | ACO | 20 | 84.16 | 4.66109 | 0.54321 |

Table 3 clearly shows that, in terms of accuracy, one of the groups performs substantially better than the other. The method of comparing the average accuracy of PSO and ACO with consideration to data variability and confidence in the results is illustrated in Figure 1. It is clear that there is a huge accuracy difference between the two algorithms, with PSO surpassing GHO by a wide margin.

Table 3. The statistical test results that compare the accuracy of two groups are shown in this table. A highly significant difference in means (p<0.001) is revealed by the t-test, but the Levene's test indicates some variance variations between the groups. There is a noticeable and statistically significant performance difference between the two groups in intrusion detection, as evidenced by the mean accuracy difference of roughly 9.87 and the 95% confidence interval spanning from 8.48 to 11.26.

| | Levene's Test for Equality of Variances | | | t-test for Equality of Means | | | | | | |
|---|---|---|---|---|---|---|---|---|---|---|
| Group | Group | F | Sig. | t | df | Sig. (2-tailed) | Mean Diff | Std. Error Diff | 95% CI (Lower) | 95% CI (Upper) |
| Accuracy | Equal variances assumed | 3.915 | 0.055 | 14.399 | 38 | 0.001 | 8.87572 | 0.66548 | 8.53281 | 11.27819 |
| | Equal variances not assumed | | | 14.399 | 32.605 | 0.001 | 8.87572 | 0.66548 | 8.46523 | 11.27577 |

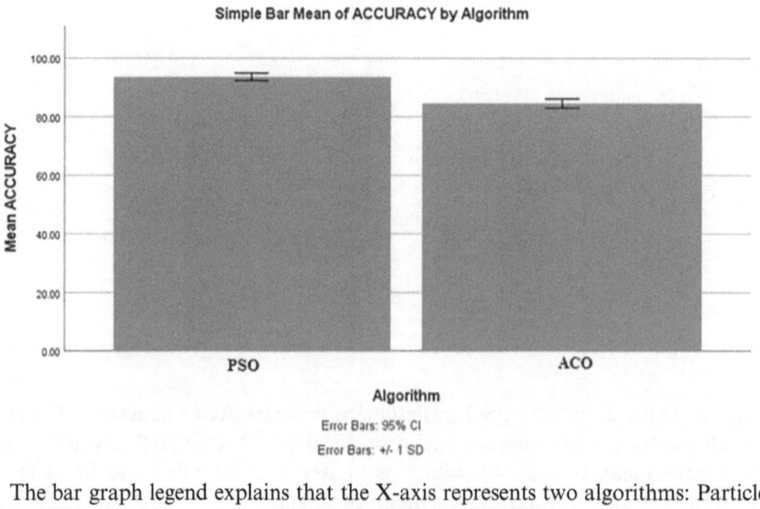


Figure 1. The bar graph legend explains that the X-axis represents two algorithms: Particle Swarm Optimization (PSO) algorithm and Ant Colony Optimization (ACO).

## 4 DISCUSSION

A distinct and dependable trend was found in the findings of the energy-efficient study conducted for Cognitive Radio. In every dataset tested, the PSO method produced consistently higher accuracy results than the ACO approach. While ACO only managed an average accuracy of 84.16%, PSO's mean accuracy was noticeably higher at 93.50%. With a p-value<0.001, statistical analysis verified the statistically significant difference in accuracy between the two methods. Furthermore, the 95% confidence interval demonstrated the robustness and dependability of PSO's supremacy in energy-efficient cognitive radio.

Comparing the proposed PSO research results with the findings of the mentioned studies, there is a consistent pattern achieving higher accuracy in energy efficient cognitive radio. The spectrum sensing ratio was attained by Indu Bala & Kiran Ahuja (2021), and the PSO algorithm obtained a potential accuracy of 86.75%. Using the conventional PSO, P. S. M. Tripathi and R. Prasad (2018) obtained an accuracy range of 86.45%. With classical PSO, Ersan Kabalci & Yasin Kabalci( 2019), presented the idea of energy-efficient cognitive radio and attained 82% accuracy. With 89.25% accuracy, R. Kishore *et al.* (2016) designed cognitive radio in an efficient method. Overall, the results of the suggested PSO research supported approaches that were more accurate than conventional procedures in energy-efficient cognitive radio, or at least comparable to them.

The study's comparatively modest sample size of 20 datasets is one of its most obvious limitations. While the PSO algorithm consistently outperformed ACO in terms of accuracy, more data from a more varied and larger dataset would be needed to confirm the results and shed light on how these algorithms function in a wider range of network conditions. For a practical deployment, it would also be necessary to look into scalability and real-time implementation in CR network contexts. Lastly, tackling the resource and computational limitations that come with CR Networks might be a useful direction for additional research in improving the efficiency of cognitive radio algorithms in such dynamic settings.

## 5 CONCLUSION

This work provides compelling evidence that the Particle Swarm Optimization approach performs better than the Ant Colony Optimization algorithm in the creation of an energy-efficient cognitive radio. Compared to the ACO, the PSO-based energy-efficient cognitive radio performs better, with an accuracy of 93.50% versus 84.16% for the ACO. The statistical study, which indicates a significant difference ($p<0.05$) between the two algorithms, lends more credence to this result.

## REFERENCES


Anuradha, K., Rajini, S. N. S., Bhuvaneswari, T., and Vinod, V. (2020). TCP/SYN Flood of Denial of Service (DOS) Attack Using Simulation. *Test Engineering & Management*, 14553–14558.Efe

Kabalci, E., and Kabalci, Y. (2019). Chapter 6-cognitive radio based smart grid communications. *From Smart Grid to Internet of Energy*, 209–248.

Arshid, K., Jianbiao, Z., Hussain, I., Pathan, M. S., Yaqub, M., Jawad, A., and Ahmad, F. (2022). Energy efficiency in cognitive radio network using cooperative spectrum sensing based on hybrid spectrum handoff. *Egyptian Informatics Journal, 23*(4), 77–88.

Goldsmith, A., Jafar, S. A., Maric, I., and Srinivasa, S. (2009). Breaking spectrum gridlock with cognitive radios: An information theoretic perspective. *Proceedings of the IEEE, 97*(5), 894–914.

Haykin, S. (2005). Cognitive radio: brain-empowered wireless communications. *IEEE Journal on Selected Areas in Communications, 23*(2), 201–220.

Bala, I., and Ahuja, K. (2021). Energy-efficient framework for throughput enhancement of cognitive radio network. *International Journal of Communication Systems, 34*(13), e4918.

Kishore, R., Ramesha, C. K., Joseph, G., and Sangodkar, E. (2016, December). Waveform and energy based dual stage sensing technique for cognitive radio using RTL-SDR. In *2016 IEEE Annual India Conference (INDICON)* (pp. 1–6). IEEE.

Rao, S.V.R.K., and Singh, G. (2012). Wavelet based spectrum sensing techniques in cognitive radio. *Procedia Engineering*, *38*, 880–888.

Sultan, A. (2012). Sensing and transmit energy optimization for an energy harvesting cognitive radio. *IEEE Wireless Communications Letters*, *1*(5), 500–503.

Zhao, N., Yu, F. R., Sun, H., and Nallanathan, A. (2013). Energy-efficient cooperative spectrum sensing schemes for cognitive radio networks. *EURASIP Journal on Wireless Communications and Networking*, *2013*, 1–13.

Huang, X., Ansari, N., Huang, S., and Li, W. (2022). Dynamic bayesian network based security analysis for physical layer key extraction. *IEEE Open Journal of the Communications Society*, *3*, 379–390.

Aluvalu, R., VN, S. K., Thirumalaisamy, M., Basheer, S., and Selvarajan, S. (2023). Efficient data transmission on wireless communication through a privacy-enhanced blockchain process. *PeerJ Computer Science*, *9*, e1308.

Mahaveerakannan, R., Anitha, C., Thomas, A. K., Rajan, S., Muthukumar, T., and Rajulu, G. G. (2023). An IoT based forest fire detection system using integration of cat swarm with LSTM model. *Computer Communications*, *211*, 37–45.

Rasappan, P., Premkumar, M., Sinha, G., and Chandrasekaran, K. (2024). Transforming sentiment analysis for e-commerce product reviews: Hybrid deep learning model with an innovative term weighting and feature selection. *Information Processing & Management*, *61*(3), 103654.

Vellela, S. S., and Balamanigandan, R. (2023). An intelligent sleep-awake energy management system for wireless sensor network. *Peer-to-Peer Networking and Applications*, *16*(6), 2714–2731.

Ramana, K., Revathi, A., Gayathri, A., Jhaveri, R. H., Narayana, C. L., and Kumar, B. N. (2022). WOGRU-IDS—An intelligent intrusion detection system for IoT assisted Wireless Sensor Networks. *Computer Communications*, *196*, 195–206.

Tripathi, P. S., and Prasad, R. (2013, June). Energy efficiency in cognitive radio network. In *Wireless VITAE 2013* (pp. 1–5). IEEE.

Orumwense, E. F., Afullo, T. J., and Srivastava, V. M. (2017, January). On increasing the energy efficiency of cognitive radio network base stations. In *2017 IEEE 7th Annual Computing and Communication Workshop and Conference (CCWC)* (pp. 1–6). IEEE.

*Hybrid and Advanced Technologies – S. Prasad Jones Christydass, Nurhayati Nurhayati and S. Kannadhasan. (Eds) © 2025 the Authors, ISBN 978-1-032-90254-8*
*Open Access: www.taylorfrancis.com, CC BY-NC-ND 4.0 license*

# A comparative study on forecasting crude oil prices using CapsNets in comparison with lasso regression

V. Mrudula
*Research Scholar, Department of Computer Science and Engineering, Saveetha School of Engineering, Saveetha Institution of Medical and Technical Sciences, Saveetha University, Chennai, Tamilnadu, India*

S. Mahaboob Basha
*Project Guide, Department of Computer Science and Engineering, Saveetha School of Engineering, Saveetha Institution of Medical and Technical Sciences, Saveetha University, Chennai, Tamilnadu, India*

ABSTRACT: The aim of this research is to evaluate the effectiveness of CapsNets and Lasso Regression in crude oil price forecasting. It aims to provide insights into their strengths and limitations, assisting decision-makers in choosing the most suitable approach and potentially exploring hybrid models to improve forecasting accuracy. Both the Lasso Regression and CapsNet algorithms, each requiring 10 iterations (N = 10), were used in this investigation. 100 samples in all were taken into account for the analysis, and these two methods were evaluated in two different groups. Using g power parameters set with $\alpha = 0.05$ and power = 0.85, the Gpower statistical test was run with a power setting of 85%. The power setting for this study was chosen to guarantee that, in accordance with accepted statistical standards, the investigation could reliably identify statistically significant differences or effects. The findings show that, in comparison to Lasso Regression, which obtained 93.11% accuracy, CapsNets achieved a greater prediction accuracy of 95.67% for crude oil prices. This underscores CapsNets' potential for capturing complex, non-linear patterns in the data, while also emphasizing the trade-off between complexity and interpretability. The findings have significant implications for stakeholders in energy, finance, and policy planning.:In conclusion, CapsNets hold promise in crude oil price forecasting by capturing intricate, non-linear patterns. The choice between CapsNets and Lasso Regression should be guided by the trade-offs between complexity and interpretability. Furthermore, the potential for hybrid models presents an avenue for enhancing forecasting accuracy.

*Keywords*: CapsNets, Lasso Regression, insights, forecasting, Oil, Price, Crude

## 1 INTRODUCTION

Crude oil, as a fundamental commodity, exerts a profound influence on global economies, industries, and the daily lives of people worldwide. Accurate forecasting of crude oil prices is of paramount importance for an extensive array of stakeholders, including governments, financial institutions, energy companies, and consumers. The precision of these forecasts is pivotal in facilitating better economic planning, informed risk management, and sound investment decisions. Its far-reaching impact extends to global economic stability, the vitality of the energy industry, financial institutions' investments, risk mitigation strategies, consumer budgeting, investment choices across multiple sectors, and geopolitical considerations. This study explores and compares two distinct methodologies, Lasso Regression

DOI: 10.1201/9781003559115-76

and Capsule Networks, for the task of crude oil price forecasting(Huang and Deng 2021). Lasso Regression and CapsNets are functional for crude oil price forecasting with distinct approaches. Lasso Regression, a linear model with L1 regularization, excels in feature selection and interpretability, identifying relevant factors from a complex set. In contrast, CapsNets, deep learning models, capture intricate, non-linear patterns and hierarchical relationships, making them ideal for complex data. Both methods utilize historical data and relevant features for prediction, Lasso via linear regression and CapsNets through deep learning. (Aluvalu *et al.* 2023; Mahaveerakannan R. *et al.* 2023; Ramana *et al.* 2023; Vellela and Balamanigandan 2023)The research compares their functionality, with Lasso offering simplicity and interpretability and CapsNets providing complexity and non-linear pattern recognition(J. Li *et al.* 2019). Evaluation metrics such as MAE and MSE assess their performance. The choice depends on research goals, data availability, and the trade-off between interpretability and predictive accuracy, ultimately guiding stakeholders in selecting the most suitable approach for crude oil price forecasting.The research has practical applications across various sectors. It aids energy and finance industries in investment decisions, financial planning, and risk management. Governments benefit from informed energy policy-making, while businesses optimize supply chains. The research is valuable for energy hedging, market analysis, and investment strategies. It also supports environmental planning and geopolitical analysis. Additionally, the hybrid forecasting methods explored have broad applicability in time series forecasting tasks(T. Zhang *et al.* 2021).

By publishing more than 16,000 papers in the last five years, from 2018 to 2023, Google Scholar has had a major impact. In the same time frame, Springer published an impressive total of over 3,300 publications, while the IEEE added over 220 research papers to the corpus. By comparison, ScienceDirect is the most comprehensive resource with nearly 4,300 papers on the subject of crude oil forecasting, while PubMed is more sparsely represented in the field with just over 130 papers published.The research divulges pivotal findings, with an emphasis on the effectiveness of the multifaceted approach in predicting crude oil prices. The selection of relevant IMFs, VMD-based risk factor extraction, and ARMA-GARCH modeling contribute to a comprehensive understanding of the risk landscape. The integration of a CNN to assemble diverse risk forecasts yields an ensemble forecast that outperforms competitive models. This integrated model showcases superior performance and dimensionality reduction capabilities, marking it as a robust tool for crude oil price prediction(Zou *et al.* 2020)The research methodology revolves around the extraction and analysis of data from search engines to inform crude oil price prediction. The Multivariate Empirical Mode Decomposition technique is utilized to extract the pertinent modes from the vast data pool. Following this, the Back Propagation Neural Network (BPNN) is implemented to forecast crude oil prices. This methodology reflects a holistic approach, merging data from search engines with advanced modeling to enhance the precision of price forecasts.(Tang *et al.* 2020).The research uncovers key findings that underscore the potential of this advanced approach. The fusion of EEMD and ICA offers an enriched understanding of the complex factors influencing WTI crude oil prices. The extracted features provide a solid foundation for refining predictive models and achieving a higher level of accuracy in forecasting.(Xian *et al.* 2020).The research uncovers key findings that underscore the potential of this advanced approach. The fusion of EEMD and ICA offers an enriched understanding of the complex factors influencing WTI crude oil prices. The extracted features provide a solid foundation for refining predictive models and achieving a higher level of accuracy in forecasting.(Manickavasagam *et al.* 2020)

In the context of comparing CapsNets as the existing algorithm with Lasso Regression as the comparison algorithm for crude oil price forecasting, certain drawbacks are associated with each approach. CapsNets, while powerful, are computationally intensive and demand substantial data, leading to longer training times and higher computational costs. Additionally, creating the specialized data representation required by CapsNets can pose challenges in data preprocessing (Abdollahi 2020). Overfitting is another concern, especially when data is limited. Lasso Regression, as a comparison algorithm, offers several advantages that can mitigate these drawbacks. It provides a simpler and more interpretable approach to modeling, making it less computationally demanding and data-efficient. Lasso Regression's L1 regularization can help counter overfitting by promoting sparsity in the model. However, it also comes with its limitations, particularly its linearity assumption, which may not capture complex, non-linear relationships effectively. By comparing these two algorithms, researchers can leverage the strengths of both, selecting the most appropriate approach based on the data's characteristics, computational resources, and the need for interpretability and generalization, ultimately enhancing crude oil price forecasting(X. Li *et al.* 2019).The aim of this research is to compare CapsNets with Lasso Regression for crude oil price forecasting, evaluating their effectiveness and trade-offs. It seeks to identify the strengths and weaknesses of each approach, offering guidance to stakeholders in the energy, finance, and government sectors. Additionally, the research explores the potential for a hybrid approach to enhance forecasting accuracy.

## 2 MATERIALS AND METHODS

The inquiry was carried out in the Data Analytics lab, which has a high-performance system that can carry out exacting research and produce accurate findings. Ten participants in all were divided into two groups for the study: Group 1 used the CapsNet methodology, whereas Group 2 used the Lasso regression method. An 80% G-Power value was used to guarantee statistical power while keeping a 0.05 significance level (alpha) and 0.8 power level (beta). The differences between the two groups were methodically computed and assessed using a 95% confidence interval. The dataset used in the crude oil price forecasts was tabular in format, with a column representing each attribute. It included pertinent elements that are essential for forecasting the price of crude oil, including past prices, supply-demand graphs, macroeconomic variables, world events, and shifts in energy regulations. According to Urolagin, Sharma, and Datta (2021), effective estimates required the consideration of both market sentiment and technical indicators, which offer valuable insights into the always shifting and dynamic oil market landscape.A dedicated dataset was used to implement the technique described in this work, and the results were compared to the well-known CapsNet approach. In this instance, the dataset comprises a range of data values and is organized as a comma-separated value (CSV) file. Using the techniques described in this research, a thorough examination of the records was made possible by applying the dataset to undertake an exhaustive review and comparison. For this task, a dual-core Intel processor with 4 GB of RAM was the hardware configuration utilized. To implement the algorithm and carry out the comparison analysis, a well-rounded system was created by utilizing Jupyter Notebook, MySQL, and Python technology in the software setup(Wu *et al.* 2019).

### 2.1 *Group 1: Capsule networks*

2017 saw the introduction of Capsule Networks (CapsNets), a revolutionary development in artificial neural networks by Geoffrey Hinton and associates. Because of its novel "capsule" shape, CapsNets perform better than traditional CNNs at recognizing spatial relationships and hierarchical characteristics in images. Every capsule is a separate organism with the ability to recognize particular parts, such as eyes or noses. The salient feature is "dynamic routing," wherein capsules cooperate to uphold spatial hierarchies and jointly determine which information to share, hence augmenting contextual awareness as a whole. By means of a "squashing function," which is comparable to human confidence in recognizing features, capsules convey their level of assurance. CapsNets‘ contextual integration capacity enables them to mimic human cognitive processes, potentially leading to substantial progress in picture recognition and comprehension (Ahmed *et al.* 2023).

### 2.2 *Group 2: Lasso regression*

Lasso Regression Is an innovative approach within the realm of linear regression. While it upholds the foundational assumption of a linear relationship between features and the target variable, it introduces a distinctive element—L1 regularization, which profoundly influences the model's performance and interpretability. Through the incorporation of a penalty term based on the absolute values of coefficients, Lasso Regression actively promotes sparsity within the model. This dual functionality not only aids in making accurate predictions for the target variable but also serves as a potent feature selection tool.The standout benefit of Lasso Regression lies in its innate capacity to automatically identify and select a subset of the most influential features from the dataset while effectively driving the coefficients of less significant or irrelevant features down to precisely zero. This innate inclination towards sparsity not only simplifies the model but also enhances its interpretability, a feature of paramount importance in scenarios where grasping the primary predictors of an outcome is as crucial as achieving precise predictions. Additionally, Lasso Regression acts as a guardian against overfitting, preventing the model from becoming overly complex and ensuring that it remains generalizable(Y. Zhang *et al.* 2019).

### 2.3 *Statistical analysis*

A statistical analysis of the results was carried out using IBM SPSS Statistics Version 26.0 in order to evaluate the efficacy of hedging. With a focus primarily on the dependent variables—specifically, accuracy measures—a unique t-test analysis was carried out. A forecast for crude oil prices was created by evaluating these measurements as one of the independent components. (L. Lin *et al.* 2020).

## 3 RESULTS

The purpose of this study was to assess the effectiveness of machine learning algorithms for forecasting crude oil prices. Significantly, the Capsule Network algorithm demonstrated remarkable accuracy of 95.67%, indicating its ability to determine Crude Oil Prices with accuracy. The Lasso regression approach, on the other hand, proved to be effective in this endeavor as seen by its 93.11% accuracy. The Capsule network technique is the recommended option for this research because of these findings, which highlight its higher accuracy in identifying conflict zones. The pseudocode for the Capsule network method is given in Table 1, and the pseudocode for the Lasso regression technique is given in Table 2. Accuracy results are shown in Table 3 for the N=20 Lasso regression and the capsule network. Table 4 presents group statistics that indicate that the Lasso regression has a mean accuracy of 1.87409 and a standard deviation of 1.31170%, whereas the Capsule Network has a mean accuracy of 95.6700% and a standard deviation of 1.23922. In terms of forecasting regions for Crude Oil Prices, this indicates that the Capsule network model performs much better than the Lasso regression when utilizing the Capsule Network, which represents a two-tailed test. A visual comparison of the mean accuracy between the logistic regression and capsule network techniques is shown in Figure 1 Compared to the Lasso regression's 93.1170% accuracy, the Capsule Network shows a greater mean accuracy of 95.6700%. Furthermore, the Capsule network's standard deviation is smaller than the Lasso regression algorithm's. The Capsule Network vs. Lasso regression algorithm is represented by the X-axis, while Mean Accuracy is indicated by the Y-axis, with error bars showing ± 2 SD.

Table 1. Represents capsule networks pseudocode.

| |
|---|
| **Input:** Crude Oil Price Forecasting data |
| **Output:** improved accuracy |
| **Step 1: Data Collection and Preprocessing:** Compile past information on the price of crude oil, usually including the date, the open, the high, the low, the close price, and the volume.Forecast accuracy may also be enhanced by the inclusion of other pertinent data, such as weather information, geopolitical events, and economic indicators.. |
| **Step 2: Data Splitting:** Make training, validation, and testing sets out of the dataset. The validation set aids in adjusting hyperparameters, the testing set assesses the model's performance, and the training set is utilized to train the model. |
| **Step 3: Feature Engineering:** Provide attributes that are pertinent to help the Capsule Network identify patterns and connections throughout the data. This could involve sentiment analysis of news data, lag aspects, and technical indications |
| **Step 4: Building the Capsule Network:** Create an architecture for a capsule network that is appropriate for time series forecasting. For this, well-known deep learning packages like PyTorch or TensorFlow can be used.<br>"-Make sure there is a temporal dimension in the input data so the network can identify time dependencies. |
| **Step 5: Training the Model:**- Use the training data to teach the capsule network. Mean squared error (MSE) or other suitable loss functions can be used for regression tasks.<br>"-Play around with various hyperparameters, like network depth, routing methods, and capsule count. |
| **Step 6:Model Evaluation:**- To adjust the model's hyperparameters, assess how well it performs on the validation dataset.<br>"-Use suitable metrics to assess the model's performance, such as Mean Absolute Error (MAE), Mean Squared Error (MSE), or Root Mean Squared Error (RMSE). |
| **Step 7: Final Model Evaluation:** Once the model has been fine-tuned, see how well it performs on the testing set to gauge its generalization potential. |
| **Step 8: Deployment:** Should the model demonstrate good performance, you can use it to generate forecasts in real time or include it into a trading or decision support system. |

Table 2. Represents lasso regression algorithm.

| |
|---|
| **Input:** Crude Oil Price Forecasting data |
| **Output:** improved accuracy. |
| **Step 1:Data Collection and Preprocessing:** -Compile past information on the price of crude oil as well as pertinent facts including supply and demand information, economic indicators, and geopolitical events.<br>"-Clean up and prepare the data, taking the proper measures to handle outliers and missing numbers. |
| **Step 2: Feature Engineering:** Engineer relevant features that might affect crude oil prices, such as moving averages, technical indicators, and sentiment analysis from news articles or social media. |
| **Step 3: Lasso Regression:**-Apply Lasso regression as a baseline model. Lasso adds L1 regularization to the linear regression, which helps in feature selection by penalizing the coefficients of less important variables.<br>-Tune the regularization hyperparameter (alpha) using cross-validation on the training set to prevent overfitting. |

(*continued*)

Table 2. Continued

| |
|---|
| **Step 4: Neural Network Integration:** Design a neural network architecture to capture the non-linear relationships in the data. A feedforward neural network with multiple hidden layers can be a good choice. |
| **Step 5: Training and Evaluation:**-Train the combined model (Lasso regression + neural network) on the training data.<br>-Validate and fine-tune the model on the validation set, adjusting hyperparameters, network architecture, and regularization as needed. |
| **Step 6:Predictions and Visualization:** the trained model to make future crude oil price predictions.<br>-Visualize the predictions and compare them to actual prices to assess the model's accuracy. |
| **Step 7: Monitoring and Updates:** Continuously monitor the model's performance and retrain it with new data as it becomes available. |

Table 3. Accuracy of capsule networks & lasso regression accuracy algorithms. The lasso regression algorithm is less than capsule networks.

| sl.no | CapsNet | Lasso Regression |
|---|---|---|
| 1 | 94.28 | 94.32 |
| 2 | 93.75 | 92.84 |
| 3 | 95.82 | 91.27 |
| 4 | 92.64 | 96.11 |
| 5 | 96.09 | 93.44 |
| 6 | 94.57 | 94.79 |
| 7 | 95.13 | 91.73 |
| 8 | 93.92 | 95.61 |
| 9 | 96.45 | 92.96 |
| 10 | 94.83 | 90.99 |
| 11 | 92.34 | 96.57 |
| 12 | 95.67 | 93.79 |
| 13 | 93.27 | 94.08 |
| 14 | 96.21 | 91.61 |
| 15 | 92.91 | 95.16 |
| 16 | 95.38 | 92.29 |
| 17 | 94.16 | 90.61 |
| 18 | 96.74 | 96.23 |
| 19 | 92.79 | 93.36 |
| 20 | 95.94 | 94.67 |

Table 4. Group statistics results-capsule networks has a mean accuracy (95.6700%), St. Deviation (1.23922), whereas lasso regression has a mean accuracy (93.1170%), St. Deviation (1.87409).

| Group | | N | Mean | std.Deviation | Std. Error mean |
|---|---|---|---|---|---|
| **Accuracy** | **Lasso regression** | 20 | 93.1170 | 1.87409 | .59264 |
| | **Capsule Network** | 20 | 95.6700 | 1.23922 | .39188 |

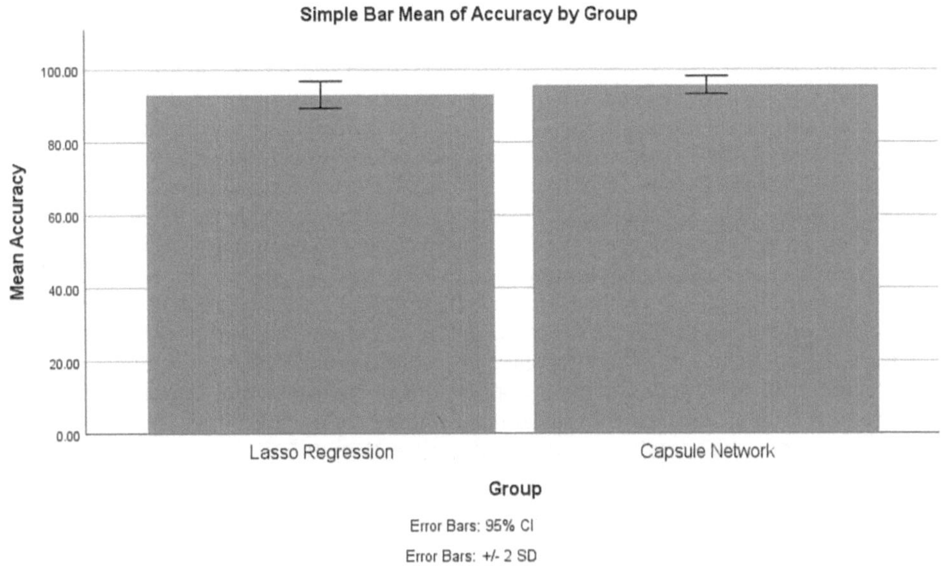


Figure 1. The Lasso regression model's accuracy in this study was 93.1170%, however the Capsule Network model showed a noteworthy mean accuracy of 95.6700%. The X-axis in this comparison graphically represents "Capsule Network vs. Lasso regression algorithm," and the Y-axis indicates "Mean Accuracy."± 2 is the representation of the error bar.

## 4 DISCUSSION

The discussion in our research centers on the substantial difference in accuracy between Capsule Networks and Lasso Regression for crude oil price forecasting, with CapsNets outperforming Lasso Regression with an accuracy of 95.67% versus 93.11%. This disparity highlights the remarkable potential of CapsNets in capturing complex, non-linear patterns and intricate relationships within the crude oil price data. However, the trade-off is the increased complexity and computational demands associated with CapsNets. Their black-box nature can limit interpretability compared to the transparency of Lasso Regression, which provides interpretable coefficients for feature selection. The quantity of historical data available is a crucial factor, as CapsNets require substantial data, potentially restricting their applicability in scenarios with limited data resources.

The results of the comparative analysis reveal that crude oil futures prices yield superior accuracy in predicting spot prices. Specifically, the RMSE and MAE metrics for crude oil futures prices are measured at 3.08 and 2.31, respectively, both of which are notably lower than those of the alternative techniques. These findings underscore the effectiveness of crude oil futures prices as a valuable tool for enhancing the precision of spot price forecasts in the realm of energy economics.(AlRassas *et al.* 2021).The findings of this research unveil the impressive performance of the reservoir computing model in forecasting WTI crude oil prices. With an RMSE of 1.32, the novel model excels beyond its deep learning counterparts, including LSTM, GRU, and CNN. This result highlights the reservoir computing model's superior accuracy and predictive capabilities, showcasing its potential for precise oil price forecasts(Kumar 2023).This paper used an ANN-PSO model to forecast the Indian basket crude oil prices using macroeconomic variables and compared it with four other methods. The accuracy values were based on the RMSE metric. The ANN-PSO model achieved an RMSE of 0.0018, which was better than the other methods(Boubaker *et al.* 2022).The findings of this study underscore the exceptional performance of the CNN-LSTM model in forecasting WTI crude oil prices. With an impressively low MAE of 1.11, this model outperforms the alternative methods, showcasing its superior predictive accuracy. This result highlights the potential of CNN-LSTM in enhancing oil price forecasts and its relevance in decision-making and risk management within the energy sector(Gulati *et al.* 2022).

The limitations of this research may stem from data availability, with the quality and quantity of historical data impacting accuracy, particularly when training complex models like Capsule Networks. Optimizing hyperparameters for deep learning models can be computationally demanding. Additionally, the research's focus on crude oil price forecasting may limit direct generalizability to other financial assets or commodities due to distinct market dynamics. Looking ahead, there is significant future scope. It includes the development of hybrid models that combine Capsule Networks and Lasso Regression for improved accuracy, exploration of alternative deep learning architectures, advanced feature engineering, ensemble approaches, real-time forecasting scenarios, market behavior analysis under varying conditions, and interdisciplinary applications in environmental impact assessment, energy resource management, and sustainable energy planning. These directions offer opportunities to address limitations and extend the practical utility of the research findings.

## 5 CONCLUSION

In conclusion, our research comparing Capsule Networks with Lasso Regression for crude oil price forecasting has shed light on their respective merits and limitations. CapsNets exhibit potential for capturing complex, non-linear patterns, while Lasso Regression provides a simpler and interpretable framework. The choice between them hinges on factors like data availability, computational resources, and the balance between interpretability and predictive accuracy. Furthermore, we've highlighted the prospect of hybrid models combining both methods for enhanced forecasting. This research contributes practical insights to stakeholders in energy, finance, and government policy, and it opens avenues for future investigations in refining forecasting techniques and their applications.

## REFERENCES

[No title]. (n.d.). Retrieved June 12, 2024, from https://www.researchgate.net/profile/Punithavathi-Rasappan/publication/362388038_Computer_Vision_and_Deep_Learning-enabled_Weed_Detection_Model_for_Precision_Agriculture/links/62e7911d4246456b55ffc95a/Computer-Vision-and-Deep-Learning-enabled-Weed-Detection-Model-for-Precision-Agriculture.pdf

Abdollahi, Hooman. (2020). A novel hybrid model for forecasting crude oil price based on time series decomposition. *Applied Energy* 267 (June): 115035.

Ahmed, Shams Forruque, Md Sakib Bin Alam, Maruf Hassan, Mahtabin Rodela Rozbu, Taoseef Ishtiak, Nazifa Rafa, M. Mofijur, A.B.M. Shawkat Ali, and Amir H. Gandomi. (2023). Deep learning modelling techniques: Current progress, applications, advantages, and challenges. *Artificial Intelligence Review* 56 (11): 13521–617.

AlRassas, Ayman Mutahar, Mohammed A. A. Al-qaness, Ahmed A. Ewees, Shaoran Ren, Mohamed Abd Elaziz, Robertas Damaševičius, and Tomas Krilavičius. (2021). *Optimized ANFIS model using aquila optimizer for oil production forecasting. Processes* 9 (7): 1194.

Aluvalu, R., Kumaran V N, S., Thirumalaisamy, M., Basheer, S., Ali Aldhahri, E., & Selvarajan, S. (2023). Efficient data transmission on wireless communication through a privacy-enhanced blockchain process. *PeerJ. Computer Science*, 9, e1308.

Boubaker, Sabri, Zhenya Liu, and Yifan Zhang. (2022). Forecasting oil commodity spot price in a data-rich environment. *Annals of Operations Research*, October, 1–18.

Gulati, Kamal, Jyoti Gupta, Lekha Rani, and Pradeepta Kumar Sarangi. (2022). Crude oil prices predictions in India using machine learning based hybrid model. *In 2022 10th International Conference on Reliability, Infocom Technologies and Optimization (Trends and Future Directions)* (ICRITO), 1–6. IEEE.

Huang, Yusheng, and Yong Deng. (2021). A new crude oil price forecasting model based on variational mode decomposition. *Knowledge-Based Systems* 213 (February): 106669.
Kumar, Kaushal. (2023). *Forecasting crude oil prices using reservoir computing models. arXiv [cs.LG]. arXiv.* http://arxiv.org/abs/2306.03052.
Li, Jinchao, Shaowen Zhu, and Qianqian Wu. (2019). Monthly crude oil spot price forecasting using variational mode decomposition. *Energy Economics* 83 (September): 240–53.
Li, Xuerong, Wei Shang, and Shouyang Wang. (2019). Text-based crude oil price forecasting: A deep learning approach. *International Journal of Forecasting* 35 (4): 1548–60.
Lin, Ling, Yong Jiang, Helu Xiao, and Zhongbao Zhou. (2020). Crude oil price forecasting based on a novel hybrid long memory GARCH-M and wavelet analysis model. *Physica A: Statistical Mechanics and Its Applications* 543 (April): 123532.
Mahaveerakannan R, Anitha, C., Aby K Thomas, Rajan, S., Muthukumar, T., and Govinda Rajulu, G. (2023). An IoT based forest fire detection system using integration of cat swarm with LSTM model. *Computer Communications*, 211, 37–45.
Manickavasagam, Jeevananthan, S. Visalakshmi, and Nicholas Apergis. (2020). A novel hybrid approach to forecast crude oil futures using intraday data. *Technological Forecasting and Social Change* 158 (September): 120126.
Ramana, Kadiyala, A. Revathi, A. Gayathri, Rutvij H. Jhaveri, C. V. Lakshmi Narayana, and B. Naveen Kumar. (2022). WOGRU-IDS — An Intelligent Intrusion Detection System for IoT Assisted Wireless Sensor Networks." *Computer Communications* 196 (December): 195–206.
Tang, Ling, Chengyuan Zhang, Ling Li, and Shouyang Wang. (2020). A multi-scale method for forecasting oil price with multi-factor search engine data. *Applied Energy* 257 (January): 114033.
Urolagin, Siddhaling, Nikhil Sharma, and Tapan Kumar Datta. (2021). A combined architecture of multivariate LSTM with mahalanobis and z-score transformations for oil price forecasting. *Energy* 231 (September): 120963.
Vellela, S. S., and Balamanigandan, R. (2023). An intelligent sleep-awake energy management system for wireless sensor network. *Peer-to-Peer Networking and Applications*, 16(6), 2714–2731.
Wu, Yu-Xi, Qing-Biao Wu, and Jia-Qi Zhu. 2019. "Improved EEMD-based crude oil price forecasting using LSTM networks. *Physica A: Statistical Mechanics and Its Applications* 516 (February): 114–24.
Xian, Lu, Kaijian He, Chao Wang, and Kin Keung Lai. (2020). Factor analysis of financial time series using EEMD-ICA based approach. *Sustainable Futures* 2 (January): 100003.
Zhang, Tingting, Zhenpeng Tang, Junchuan Wu, Xiaoxu Du, and Kaijie Chen. (2021). Multi-step-ahead crude oil price forecasting based on two-layer decomposition technique and extreme learning machine optimized by the particle swarm optimization algorithm. *Energy* 229 (August): 120797.
Zhang, Yaojie, Feng Ma, and Yudong Wang. (2019). Forecasting crude oil prices with a large set of predictors: Can LASSO select powerful predictors? *Journal of Empirical Finance* 54 (December): 97–117.
Zou, Yingchao, Lean Yu, Geoffrey K. F. Tso, and Kaijian He. (2020). Risk forecasting in the crude oil market: A multiscale convolutional neural network approach. *Physica A: Statistical Mechanics and Its Applications* 541 (March): 123360.

*Hybrid and Advanced Technologies – S. Prasad Jones Christydass, Nurhayati Nurhayati and S. Kannadhasan. (Eds)* © *2025 the Authors, ISBN 978-1-032-90254-8*


# Analyze the fraud loan repayment and risk payment detection using machine learning algorithms based on SVM over KNN

P. Dinesh Kumar
*Research Scholar, Department of Computer Science and Engineering, Saveetha School of Engineering, Saveetha University, Chennai, Tamil Nadu, India*

M. Kalaiyarasi
*Research Guide, Department of Computer Science and Engineering, Saveetha School of Engineering, Saveetha University, Chennai, Tamil Nadu, India*

ABSTRACT: In order to anticipate bank loan fraud, this article compares the effectiveness of K-Nearest Neighbors and Support Vector Machines. Twenty samples in all, drawn from two groups of 10, are used in the study. Support Vector Machine is used in the first set, while K-Nearest Neighbours is used in the second. Based on pre-test results, the necessary sample size was calculated using an 80% G-power, a 0.05 significance level, and a 95% confidence interval.

The accuracy values reached by K-Nearest Neighbors (62.89%) and Support Vector Machine (93.69%) differ statistically significantly (p = 0.93, p 0.05). The study discovered that when predicting fraud throughout the bank loan procedure, the Support Vector Machine model achieved an accuracy rate of 93.69%, whereas the K-Nearest Neighbors model only managed a 62.89% accuracy rate.

*Keywords*: Support Vector Machine, K-Nearest Neighbors, Fraud, Financial, Bank, Crisis, Risk, Development

## 1 INTRODUCTION

Since most individuals put their money in banks, it is easy for fraudsters to target these organizations. Due to the variety of services they provide, banks may fall prey to fraud on many fronts, including their customers, workers, investors, and even the bank itself. To gain an understanding of the extent of mortgage and hire purchase loan fraud in Asia, researchers spoke with bank managers at various levels, concentrating on branch managers and assistant managers who were in charge of the mortgage and hire purchase loan portfolios.According to the data, money laundering is by far the most frequent sort of fraud perpetrated on Asian banks, while hire purchase loan fraud is the most common type of fraud perpetrated on aisa banks. (Clark and CFP 2009) Here, we look at cases of home equity and hire purchase loan fraud at financial institutions. We also take a look at the monetary losses that may be attributed to the various fraud techniques. The study largely focused on mortgage and hire purchase loans as these are the most popular forms of loans granted by banks. (Financial Crisis Inquiry Commission 2011)

The crisis's repercussions were immediately apparent in the financial sector, particularly in consumer financing. This article demonstrates how the banking industry and consumer finances were able to recover from the global financial crisis thanks to regulatory agencies' resolve and consistency. Social phenomena such as shopping panic, decreased

 DOI: 10.1201/9781003559115-77

creditworthiness of families due to income loss, unemployment, and increased criminality have been documented in the wake of this sort of global catastrophe. Financial market regulators' responses also proved to be quite successful. (Wereda *et al.* 2022) Crowdfunding fraud data is scarce. Not to deny fraud. Hainz uses uncertainty to explain fraud motives in a short theoretical model. She shows that contract incentives impact honesty and that addressing information asymmetries may reduce fraud. Few fraud cases may exist due to low incentives to identify them. (United States. Congress. House. Committee on Government Operations 1987) Growth of global and development depend on finance. A bank's financial health is assessed by its NPAs, or bad loans. NPA hurts bank profitability and asset quality, hence it must be reduced. Deliberate defaulters have climbed dramatically. Identifying, understanding, and assessing characteristics helps forecast purposeful default behavior early. (Volz *et al.* 2018) The idea of borrowing money for development has been around since the dawn of civilization, but it has developed into many different forms throughout the years. This involves the trade of loans for repayment based on personal credit histories, the utilization of loans as dividends from recurring contributions, and the lack of collateral restrictions for these loans anywhere but the banking sector. (The Research Unit for Political Economy 2020) Loan management boosts recipients' economies while lenders get interest at a defined rate based on length, amount, purpose, and credit history. When a borrower stops paying payments as promised in the promissory notes, the loan defaults. If interest or payment is not received within ninety days, it becomes simpler to undervalue the loan receiver and overlook signs of fraud such as duplicate or expensive collateral and phony bank statements. Loan administrators must carefully review all applications and reject those that are not likely to succeed. (International Monetary Fund and World Bank 2003)

When someone deceives another with the goal to deprive them of anything of value to them, it is fraud. It is also a kind of fraud to knowingly create false information. While there are time-consuming and ineffective ways for detecting fraud in credit administration, they are unable to keep up with the increasing complexity of fraud schemes. (Parvez M. *et al.* 2023; Ramesh *et al.* n.d.; Sarkar *et al.* 2022; Surendran *et al.* 2023; Tayyab *et al.* 2022) Data mining development has made it possible for machine learning approaches to fraud detection by revealing previously unseen patterns and characteristics in the data. (Financial Crisis Inquiry Commission 2011) To triumph over these limitations, a novel throughout its entirety deep Multiview learning technique and development transforms various information views onto homogenous ones, fills in missing views with previously learned connections, and then blends them all into a single view for user identity fraud. On a large-scale the information set, our model outperforms the state-of-the-art approaches. (City of London College of Economics, n.d.)

## 2 MATERIALS AND METHODS

The Saveetha Institute of Medical and Technical Sciences' School of Engineering's Computational Engineering department conducted this study. The prediction accuracy of human stress levels in test data sets was examined using Support Vector Machine (SVM) and K-Nearest Neighbors (KNN). Two groups of ten were formed from a random selection of twenty samples. Based on a 95% confidence interval, 80% statistical power (as determined using g power analysis), and a significance threshold of 0.05, the sample size was chosen. These numbers were similar to those discovered in earlier research that used beta values of 0.2 and alpha values of 0.05.

The project requires a dataset with a large number of picture tickets to train the price prediction system. The framework needs at least 8 GB of RAM to function adequately for even the most fundamental uses of the central processing unit (CPU). CPUs such as the Intel i5 from the 5th generation or the Ryzen 3 are recommended. It takes at least 500GB of storage space on a hard drive to download the primary dataset images from Kaggle, together with the supporting code and any necessary programs or plugins. The graphics processing unit (GPU) should have at least 4 GB of RAM to reduce the workload on the CPU and boost the speed at which image processing operations may be performed. Kaggle served as the platform for the creation of the system for determining the cost of bus tickets.

## 2.1 *Support vector machine*

A supervised motor scholarship approach is possible for reversion, which is often used for genus in natural language processing. SVM finds a hyperplane in an n-dimensional quality space to effectively separate and classify input data points. This hyperplane's dimension corresponds to the total number of features in the dataset. The resulting SVM hyperplane appears as a straight line in two dimensions when only two attributes are provided as input. When the data contains beyond three distinctive characteristics in the collected data, the hyperplane changes into a 2D aircraft. The following steps are used in SVM algorithm

Step 1: To commence, enter your dataset that comprises the relevant labels (Y) and characteristics (X).

Step 2: The dataset should then be divided into training and testing subsets, with a distinction made between the labels (y_train, y_test) and features (X_train, X_test).

Step 3: Assign the kernel type (linear kernel, for example) and use a Support Vector Machine (SVM) classifier.

Step 4: The training data (X_train, y_train) are used to train the SVM classifier.

Step 5: For the test data, continue with your predictions.

Step 6: Determine which metric—Accuracy, Precision, or Recall—is best to use to assess the model's performance.

Step 7: Preserve the outcomes for additional examination to evaluate the effectiveness and functionality of the model.

## 2.2 *KNN*

It is a fundamental approach to supervised learning in machine learning. In order to allocate new data or cases to the category that most closely resembles those that already exist, it means that any new data or cases must be compared to the existing cases. The patented K-Nearest Neighbors search function effectively clusters unknown data items by capturing their inherent similarities. It also aids in keeping older data intact. As such, the K-Nearest Neighbors algorithm is an excellent choice for grouping newly collected data. Its primary application is as a machine learning classifier, but it may also be used as a regression and classification tool. The K-Nearest Neighbors method is undoubtedly nonparametric and makes no assumptions on the data substructure. The steps used in KNN are as follows,

Step 1: To begin, import your dataset, which should contain the labels (Y) and the related characteristics (X).

Step 2: Split the dataset into feature sets (X_train, X_test) and label sets (y_train, y_test), making a distinction between the training and testing sets.

Step 3: Determine the value of K, the parameter that counts the closest neighbours.

Step 4: Determine the distance in Euclidean terms between every data point in the training set and all of the test set's data points. Then, choose the top k data points based on this distance.

Step 5: Use a pertinent metric, like recall, accuracy, or precision, to evaluate the model's performance.

Step 6: To analyse the efficacy of the model, print or store the evaluation results for additional examination.

## 2.3 *Statistical analysis*

A program called IBM SPSS 26 was created to simplify the process of performing statistical analysis. The independent sample T-test compares each batch's mean. There are two independent variables: the sample size and the number of records. In 2020, Salcedo and McCormick This study includes accuracy, target, records, lone amount, credit record, score, personal information, and regulator actions as dependent factors.

# 3 RESULTS

In this study, Support Vector Machine was shown to be superior than K-Nearest Neighbors, with a mean accuracy of 93.69 percent vs 62.89 percent. The development team analyzed some of the data using a SVM and the remainder with KNN. The SVM outperformed KNN in terms of average accuracy. The results of using the SVM and the KNN Algorithm are laid forth in Table 1.

Table 1. Accuracy analysis for SVM and KNN algorithms.

| S. No | Support Vector Machine | K-Nearest Neighbors |
|---|---|---|
| 1 | 93.27 | 62.00 |
| 2 | 93.41 | 63.25 |
| 3 | 93.49 | 63.68 |
| 4 | 93.69 | 63.76 |
| 5 | 93.10 | 62.27 |
| 6 | 93.06 | 62.22 |
| 7 | 93.65 | 62.84 |
| `8 | 93.74 | 63.75 |
| 9 | 93.64 | 62.24 |
| 10 | 93.23 | 63.03 |

The mean, standard deviation, and standard error mean are also included for the two methods in Table 2, along with other group statistical results. For this dataset, we use a 91% confidence interval and an Independent sample T-test.We find a 1.057 standard deviation and a 1.653 standard error.

Table 2. Group statistical analysis of SVM and KNN.

| | GROUPS | N | Mean | Std. Deviation | Std. Error Mean |
|---|---|---|---|---|---|
| **ACCURACY** | Support Vector Machine | 10 | 93.69 | 1.057 | 0.334 |
| | K-Nearest Neighbors | 10 | 62.89 | 1.653 | 0.522 |

In Table 3, we compare the SVM and the KNN Classifier Algorithms using independent sample T-tests. Support Vector Machine and K-Nearest Neighbors were used to analyze the dataset in SPSS, and the resulting table shows the sig, mean, and standard error difference between the two techniques.

Table 3. Independent sample T-test for SVM and KNN algorithms.

| | | Levene's Test for Equality of Variances | | t-test for Equality of Means | | | | | | |
|---|---|---|---|---|---|---|---|---|---|---|
| | | F | Sig. | t | df | Sig. (2-tailed) | Mean Difference | Std. Error Difference | 95% Confidence Interval of the Difference | |
| | | | | | | | | | Lower | Upper |
| **ACCURACY** | **Equal variances assumed** | 3.150 | 0.093 | 49.627 | 18 | 0.000 | 30.800 | 0.620 | 29.496 | 32.104 |
| | **Equal variances not assumed** | | | 49.627 | 15.311 | 0.000 | 30.800 | 0.620 | 29.496 | 32.104 |

Figure 1 shows a correlation of the accuracy between the Novel SVM and KNN, with the proposed approach achieving a mean accuracy of 93.69 percent and K-Nearest Neighbors achieving a mean accuracy of 62.89 percent, both with 95% confidence intervals and standard deviations.

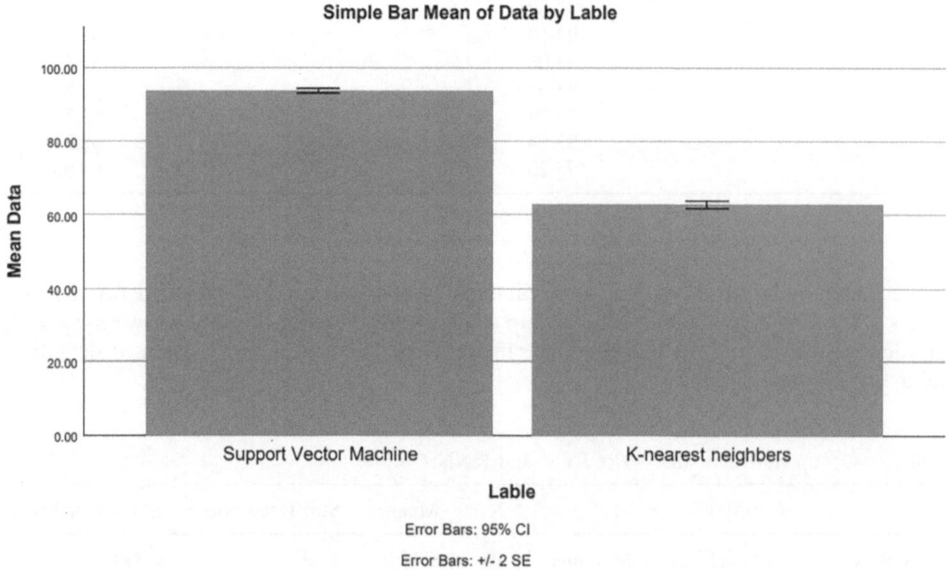


Figure 1. Graphical illustration of mean accuracy analysis of SVM and KNN algorithm.

## 4 DISCUSSION

According to the findings, the Support Vector Machine method outperforms the K-Nearest Neighbors strategy by a wide margin. The Support Vector Machine model achieves an accuracy of 93.69 percent, whereas the K-Nearest Neighbors model achieves just 62.89 percent.

Internet lending has emerged as a viable option for micro-financing, and it is quickly gaining traction in the banking sector. Lending finance platforms allow borrowers to borrow small amounts of money online without pledging any assets as security. (Ledgerwood 1998) Consumers are able to rapidly traverse loan procedures utilizing specified apps, resulting in a substantially shortened loan acquisition timeline compared to traditional financial institutions such as banks. These sites have rapidly amassed a significant portion of the financial industry's user base and market share. However, because of the lack of information these lenders have regarding the borrowers' creditworthiness, they face a high risk of failure. (Bazarbash 2019)

As is well-known, fraud is a serious problem for many companies. A company's chances of becoming the market leader in its sector may improve if it takes anti-fraud measures. Proactive management strategies are the foundation of robust, efficient, and successful organizations because they allow leaders to look ahead and prepare for challenges. Businesses need to continually evaluate the potential for fraud and adopt corrective measures in response to the ever-evolving external environment. The accelerating speed of regulatory change in the banking industry means that management needs to be more aware of the regulations, as this affects business segment profitability and compliance costs (Aiolfi 2020).The inherent complexity of financial transactions makes it difficult to completely eliminate fraud risk in organizations. Perpetrators are too strong for even the most advanced cybersecurity technologies to stop. (James-Thomas 2021) We were successful in identifying consumers who had committed loan fraud by using a novel deep multiview learning technique. View encoding, missing view construction, fusion, and classification can all be learned collaboratively because to its end-to-end design, which also improves efficiency. We provide a network for view creation and fusion, which models the connections between views to create new ones and combine existing ones. Multiple experiments on a large dataset confirmed the effectiveness and consistency of our method. (Gunjan *et al.* 2021)

## 5 CONCLUSION

A unique deep multiview learning strategy was used in this study, and it was effective in identifying persons engaged in fraudulent loan operations. This implies that the Support Vector Machine (SVM) technique beat the K-Nearest Neighbors (K-NN) approach, with an accuracy of 62.89% for the specified batch size. The accuracy level was measured at 93.69 percent. Therefore, when comparing the Support Vector Machine (SVM) method to the K-Nearest Neighbors (K-NN) method, SVM shows higher accuracy in stress prediction.

## REFERENCES


Aiolfi, G. (2020). *Anti-corruption Compliance: A Guide for Small and Mid-sized Organizations*. Edward Elgar Publishing.

Bazarbash, M. (2019). *Fintech in Financial Inclusion: Machine Learning Applications in Assessing Credit Risk. International Monetary Fund.*

Clark, K. (2009). *The Complete Idiot's Guide to Getting Out of Debt*. Penguin.

Congress, U.S. (1987). House Committee on Government Operations, Subcommittee on Commerce, Consumer, and Monetary Affairs. 1987a. Fraud and Abuse by Insiders, Borrowers, and Appraisers in the California Thrift Industry. Hearings before the Subcommittee. June 13. In *100th Congress, 1st Session.*

Financial Crisis Inquiry Commission. (2011). *The Financial Crisis Inquiry report: the Final Report of the National Commission on the Causes of the Financial and Economic Crisis in the United States, Including Dissenting Views*. Cosimo, Inc..

Gunjan, V. K., Suganthan, P. N., Haase, J., Kumar, A., and Raman, B. (2019). Cybernetics, Cognition and Machine Learning Applications Proceedings of ICCCMLA 2019. *Proceedings of ICCCMLA*, 1.

Ledgerwood, J. (1998). *Microfinance Handbook: An Institutional and Financial Perspective. World Bank Publications..*

Lewis, J. D., and Viñals, J. (2014). *Revised guidelines for public debt management. International Monetary Fund: Washington, DC, USA.*

Mindu, N. *Supply and Demand of Data Science Skills in South Africa (Doctoral dissertation)..*

Mishra, S. K., Nandini, K., Ahammad, S. H., Inthiyaz, S., Altahan, B. R., Smirani, L. K., and Rashed, A. N. Z. (2023). Machine learning based models for defect detection in composites inspected by Barker coded thermography: A qualitative analysis. *Advances in Engineering Software, 178*, 103425.

Purushothaman, U., and Moolakkattu, J.S. (2021). The politics of the COVID-19 pandemic in India. *Social Sciences, 10*(10), 381.

Ramesh, P.S., Sudha, I., Satheesh, N., Rajulu, G.G., and Kalaiselvan, S. (2023). Prediction of auto-detection for tracking of sub-nano scale particle in 2D and 3D using SVM-based deep learning. *Journal of Theoretical and Applied Information Technology, 101*(5), 1961–1971.

Salcedo, J., and McCormick, K. (2020). *SPSS Statistics for Dummies.* John Wiley & Sons.

Sarkar, B., Kar, S., Basu, K., and Guchhait, R. (2022). A sustainable managerial decision-making problem for a substitutable product in a dual-channel under carbon tax policy. *Computers & Industrial Engineering, 172*, 108635.

Surendran, R., Alotaibi, Y., and Subahi, A. F. (2023). Lens-oppositional wild geese optimization based clustering scheme for wireless sensor networks assists real time disaster management. *Comput. Syst. Sci. Eng., 46*(1), 835–851.

Tayyab, M., Habib, M. S., Jajja, M. S. S., and Sarkar, B. (2022). Economic assessment of a serial production system with random imperfection and shortages: A step towards sustainability. *Computers & Industrial Engineering, 171*, 108398.

Volz, U., Morgan, P.J., and Yoshino, N. (Eds.). (2019). *Routledge Handbook of Banking and Finance in Asia* (p. 488). London/New York: Routledge.

*Hybrid and Advanced Technologies – S. Prasad Jones Christydass, Nurhayati Nurhayati and S. Kannadhasan. (Eds)* *ISBN 978-1-032-90254-8*


# An efficient study of fraud detection using Random Forest with Naive Bayes

P. Praveen Kumar Reddy
*Research Scholar, Department of Computer Science and Engineering, Saveetha School of Engineering, Saveetha Institute of Medical and Technical Sciences, Saveetha University, Chennai, Tamil Nadu, India*

K. Malathi
*Project Guide, Department of Computer Science and Engineering, Saveetha School of Engineering, Saveetha Institute of Medical and Technical Sciences, Saveetha University, Chennai, Tamil Nadu, India*

ABSTRACT: The main aim of this research is to detect occurrences of financial fraud by using the Random Forest technique and comparing its performance with the Naive Bayes algorithm. A set of 20 samples has been obtained from two separate groups. While Random Forest approaches are categorised as belonging to group 1, Naive Bayes algorithms are classified as belonging to group 2. There are ten samples in all from Groups 1 and 2, each having an 80% statistical power and a 95% confidence range. The Random forest model has 90.38% accuracy, whereas the Naive bayes model has 89.43%. Accuracy detection averages two standard deviations above mean. A p-value of 0.429 indicatesSince the groups' p-value is less than 0.005, there is a statistically significant difference between them. This study compared the Random forest and Navie bayes algorithms to determine their accuracy-boosting effects. Compared to the Naive bayes algorithm, which has 89.43% accuracy, the Random forest technique has 90.38% accuracy.

*Keywords*: Financial Fraud, Naive bayes, Transactions, Accuracy, Random forest, Prediction, Precision, Interpretability, Computational

## 1 INTRODUCTION

Robust fraud detection is essential to the financial system's integrity. Utilising a variety of criteria, random forests—which are well-known for being strong and comprehensible effectively categorise data. Examining transaction amounts, locations, and timings helps them discover fraud. (Ahmed 2018). decision trees explain fraud and help stakeholders understand choices. Naive bayes classifies data points by data point proximity. Naive bayes can identify fraudulent transaction patterns, however it is computationally expensive for large datasets and high-dimensional data. Random forests are interpretable, whereas Naive bayes are not, making fraud explanations difficult. ((tp051392) (2019)). However, Naive bayes may discover real-time transaction abnormalities by comparing new instances to existing data points.Financial fraud detection benefits from Random forests' interpretability and Naive bayes' real-time anomaly detection. Choosing between them relies on system needs, result interpretability, and dataset computational limits. (Panos 2010).

Additionally, 200 papers on financial fraud detection utilizing Random forests and Naive bayes explain their distinctive characteristics and applications.Trees help stakeholders comprehend decision-making by interpreting deceptive patterns. (Vasarhelyi et al. 2019). Fraud costs the credit card industry billions of dollars every year. While fraud detection is

DOI: 10.1201/9781003559115-78 

improved by financial organisations, hackers develop new strategies. (Identifying Credit Card Theft: A Comprehensive Review of Fraud Detection Methods 2020) Nonetheless, it's critical to detect and stop credit card fraud. Financial firms use data mining to find fraud in huge, complicated financial data. Recent research uses machine learning and data mining to solve this. This article discusses modern credit card fraud detection. This research analyzes revenues and data to find the top credit card fraud detection strategies. (Paldino *et al.* 2022) Naive bayes detects anomalies in real time but lacks interpretability. Both methods stress feature selection, data preparation, and model performance assessment, according to studies. Researchers indicate that transparency or real-time detection determines which model to choose, underlining the importance of choosing a model that meets the financial fraud detection system's needs. (Abedin *et al.* 2021).

Random forests instead of Naive bayes are used to detect and halt financial fraud. (Venkata Krishna and Obaidat 2020). (Ali *et al.* 2023; Gurupranes *et al.* 2022; Natrayan *et al.* 2022; Viswanathan *et al.* 2023; Sivaranjani *et al.* 2023) Random forests try to explain fraudulent transactions in a clear and understandable manner. Naive bayes prioritizes real-time transaction abnormalities to identify fraud. Creating powerful models that can rapidly and reliably distinguish actual and phony transactions is the ultimate aim. By upholding honesty and transparency, financial institutions and stakeholders may minimise losses and guarantee the integrity of the financial ecosystem. (Zhao *et al.* 2018).

## 2 MATERIALS AND METHODS

This investigation was carried out by the departments of computer science and engineering at the Saveetha Institute of Medical and Technical Sciences. There are two groups in this study. When it comes to data object identification, random forest and naive bayes are compared. The two groups are each given ten samples. (Thangaraj and Vaishali 2022). Random woodland has the first ten resolves, while Naive Bayes has the last ten. The number of samples is determined by the prior research's 95% confidence interval, 0.5 g threshold, g power of 80%, and enrollment ratio of 1. (Reinders *et al.* 2018)

This research requires card transaction data for training. The framework requires 16GB RAM for straightforward CPU access. Recommended CPUs are Intel i3 5th generation or I7. At least 720 GB hard drive Recommended CPUs are Intel i3 5th generation or Intel I7. code, plug-in, and kaggle.com A least 8GB graphics card reduces CPU load and speeds data processing.

### 2.1 *Random forest*

Using the Random Forest combined learning method, numerous decision trees are trained and their predictions are then merged to get predictions that are stronger and more accurate. Each forest decision tree is built using a random subset of the characteristics and a random portion of the training data. Random Forest lessens overfitting and encourages generalisation by adding diversity and randomness to the learning process.

#### 2.1.1 *Random forest Algorithm:*

Step 1: Load your data first.
Step 2: Divide the dataset into training and testing in step two.
Step 3: To construct a Random Forest classifier, supply hyperparameters such as n_estimators, maximum depth, and others.
Step 4: Use training data to train the Random Forest classifier.
Step 5: Forecast using test results.
Step 6: Assess the recall, accuracy, and precision of the model.
Step 7: Print or save the findings of the model's performance evaluation.

### 2.1.2 *Naive bayes:*

Classification using Naive Bayes is straightforward and effective. This formula employs Bayes' theorem to estimate the probability of an occurrence given a prior one.Naive Bayes classifiers presume class label-independent data point attributes. Though silly, this assumption usually works. A Naive Bayes classifier calculates a data point's class probability before categorizing it. Multiply each class's data point feature chances. Data points go to the most likely class.

## 2.2 *Naive bayes algorithm*

Step 1: To start, arrange your image dataset, import the pictures, and give them the proper names.
Step 2: Create training and testing subsets from the dataset.
Step 3: Make your model's architecture a decision.
Step 4: Set up the Naive Bayes model according to your preferences.
Step 5: Apply the 'fit' function to the training data to train the Naive Bayes model.
Step 6: Using appropriate evaluation measures, determine how well the model performed on the test set.
Step 7: To improve the performance of the model, adjust the hyperparameters and use regularisation strategies.
Step 8: Store the learned model for usage or deployment at a later time.

## 2.3 *Statistical analysis*

To compare mean resolutions, an independent sample t-test is used in the statistical analysis, which is carried out using IBM SPSS Version 26. The dependent variable in this analysis is transactions, while the independent variables are sample size and accuracy.

# 3 RESULTS

90.38% is the accuracy achieved by the Random Forest classifier, and 89.43% by the Naive Bayes classifier. Group 2 is the Naive Bayes model, and Group 1 is the Random Forest model. This comparison assesses both groups' accuracy. Because of this, Random Forest performs better than Naive Bayes, proving its supremacy over Support Vector Machine.

Table 1 exhibits Random forest and Naive bayes accuracy sample data. Table 2 shows 20-person group data. Mean scores and standard deviations for Random forest and Naive bayes classifiers are shown.

Table 1. Sample output resolution comparison between random forest and Naive bayes.

| S. No | Random forest | Naive bayes |
|---|---|---|
| 1 | 90.92 | 89.73 |
| 2 | 90.17 | 89.12 |
| 3 | 90.36 | 89.60 |
| 4 | 90.04 | 89.36 |
| 5 | 90.30 | 89.17 |
| 6 | 90.26 | 89.40 |
| 7 | 90.59 | 89.11 |
| `8 | 90.85 | 89.73 |
| 9 | 90.09 | 89.15 |
| 10 | 90.19 | 89.12 |

Table 2. Group statistics comparing the Random Forest and Naive Bayes models in terms of resolution.

| | GROUPS | N | Mean | Std. Deviation | Std. Error Mean |
|---|---|---|---|---|---|
| **Accuracy** | Random forest | 10 | 90.38 | 0.205 | 0.649 |
| | Naive bayes | 10 | 89.43 | 0.246 | 0.078 |

Table 3. Shows a 95% confidence interval comparison of the independent sample T-test with identical variances in resolution.

| | | Levene's test for equality of variances | | T-test for equality of means | | | | | | |
|---|---|---|---|---|---|---|---|---|---|---|
| | | | | | | | | | 95% confidence interval of the difference | |
| | | F | Sig | t | df | Sig (2-tailed) | Mean Difference | Std. error differs | Lower | Upper |
| **RESOLUTION** | **Equal Variances assumed** | 0.551 | 0.467 | 9.298 | 18 | 0.000 | 0.943 | 0.101 | 0.730 | 1.157 |
| | **Equal variances are not assumed** | | | 9.298 | 17.429 | 0.000 | 0.943 | 0.101 | 0.730 | 1.157 |

Table 3 can be used to compare the Random Forest and Naive Bayes algorithms. A 95% confidence level independent sample T-test was performed. The findings show that there is a significant difference (p-value = 0.959, p<0.05) between the two groups.

Figure 1 compares the mean accuracy of Random forest and Naive bayes, along with their error bars.

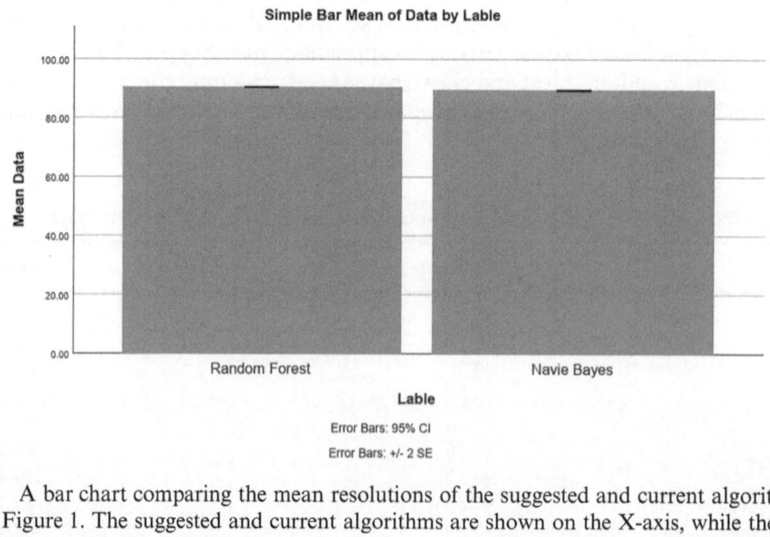


Figure 1. A bar chart comparing the mean resolutions of the suggested and current algorithms is shown in Figure 1. The suggested and current algorithms are shown on the X-axis, while the mean resolution is represented on the Y-axis. Compared to Naive Bayes (89.43%), the Random Forest method achieves a greater resolution (90.38%). There are +2 standard deviations in the mean resolution detection, and 0.467 is the significant value.

## 4 DISCUSSION

Random Forest fared better than Naive Bayes, according to the results (Ai Sciences 2019). Naive Bayes obtained an accuracy of 85.54%, whereas Random Forest attained 90.38%. With a p-value of 0.494 and a mean accuracy detection of +2 standard deviations, there was a statistically significant difference ($p<0.005$) between the groups (IEEE Staff 2021).

Similar financial fraud detection research suggest that deep learning and machine learning algorithms help minimize financial fraud. Random forests simplify financial fraud analysis. (Metzler 2019), thus everyone can observe fraud detection. datasets having clear characteristics. They tend to overfit and struggle with complicated, high-dimensional data. Real-time detection makes naive bayes useful for quickly finding abnormalities. (Papadakis *et al.* 2020). However, they are uninterpretable, making their classifications difficult to understand. Naive bayes may perform poorly in noisy or huge datasets, affecting their efficiency in complicated financial fraud situations. (Kamwani *et al.* 2022).

Overfitting may reduce the accuracy of random forests for financial fraud detection, especially in complicated and large-scale datasets. Random forest performance might be improved by ensemble approaches in future study. (Tanwar *et al.* 2023). Naive bayes can discover abnormalities in real time, but the dimensionality curse makes them difficult to execute on large datasets. implies we need to learn how to minimize dimensionality and enhance algorithms. Ring hybrid models that combine Random forests and Navie bayes may improve Future Prospects for Fraud Detection.(Biju *et al.* 2023).

## 5 CONCLUSION

Here, the Random forest method with 90.38% resolution outperformed the Naive bayes algorithm with 85.54%. Thus, the Random forest algorithm detects financial fraud better than Naive bayes.

## REFERENCES


(tp051392), Vatsala A/p Vijayan. (2019). *Artificial Neural Network and Other Data Mining Techniques: A Comparatives Study to Predict Fraudulent Financial Reporting in Malaysia.*

Abedin, Mohammad Zoynul, M. Kabir Hassan, Petr Hajek, and Mohammed Mohi Uddin. (2021). *The Essentials of Machine Learning in Finance and Accounting.* Routledge.

Ahmed, Rohan. (2018). *Data Mining Techniques in Financial Fraud Detection.* GRIN Verlag.

Ai Sciences. (2019). *Mastering Machine Learning in One Day.* AI Sciences LLC.

Ali, S., Sudha, K. G., Thirumalaivasan, N., Ahamed, M., Pandiaraj, S., Rajeswari, V. D., Vinayagam, Y., Thiruvengadam, M., and Govindasamy, R. (2023). Green synthesis of magnesium oxide nanoparticles by using abrus precatorius bark extract and their photocatalytic, antioxidant, antibacterial, and cytotoxicity activities. *Bioengineering (Basel, Switzerland)*, 10(3). https://doi.org/10.3390/bioengineering10030302

Biju, Soly Mathew, Ashutosh Mishra, and Manoj Kumar. (2023). *Advanced Interdisciplinary Applications of Machine Learning Python Libraries for Data Science.* IGI Global.

Dai, Jun, Miklos A. Vasarhelyi, and Ann F. Medinets. (2019). *Audit Analytics in the Financial Industry.* Emerald Group Publishing.

*Detecting Credit Card Fraud: An Analysis of Fraud Detection Techniques.* (2020).

Gurupranes, S. V., Natrayan, L., Kaliappan, S., Patel, P. B., Sekar, S., Jayaraman, P., Arvinda Pandian, C. K., and Esakkiraj, E. S. (2022). Investigation of physicochemical properties and characterization of leaf stalk fibres extracted from the caribbean royal palm tree. *International Journal of Chemical Engineering*, 2022. https://doi.org/10.1155/2022/7438411

IEEE Staff. (2021). *2021 Third International Conference on Intelligent Communication Technologies and Virtual Mobile Networks (ICICV).*

Kamwani, Suleman Sherali, Elisabete S. Vieira, Mara Madaleno, and Graça Azevedo. (2022). *Handbook of Research on the Significance of Forensic Accounting Techniques in Corporate Governance.* IGI Global.

Metzler, Guillaume. (2019). *Learning from Imbalanced Data: An Application to Bank Fraud Detection.*
Natrayan, L., Kaliappan, S., Sethupathy, B. S., Sekar, S., Patil, P. P., Velmurugan, G., and Tariku Olkeba, T. (2022). Effect of mechanical properties on fibre addition of flax and graphene-based bionanocomposites. *International Journal of Chemical Engineering*, 2022. https://doi.org/10.1155/2022/5086365
Paldino, Gian Marco, Bertrand Lebichot, Yann-Aël Le Borgne, Wissam Siblini, Frédéric Oblé, Giacomo Boracchi, and Gianluca Bontempi. (2022). The role of diversity and ensemble learning in credit card fraud detection. *Advances in Data Analysis and Classification, September*, 1–25.
Panos, Sterling. (2010). *Comparison of Data Mining Techniques to Correctly Identify Financial Statement Fraud.*
Papadakis, Stylianos, Alexandros Garefalakis, Christos Lemonakis, Christiana Chimonaki, and Constantin Zopounidis. (2020). *Machine Learning Applications for Accounting Disclosure and Fraud Detection.* IGI Global.
Reinders, Christoph, Hanno Ackermann, Michael Ying Yang, and Bodo Rosenhahn. (2018). Object recognition from very few training examples for enhancing bicycle maps. *2018 IEEE Intelligent Vehicles Symposium (IV)*. https://doi.org/10.1109/ivs.2018.8500469.
Sivaranjani, P. R., Syed, A., Elgorban, A. M., Bahkali, A. H., Balakrishnaraja, R., Varma, R. S., and Sudheer Khan, S. (2023). Fabrication of ternary nano-heterojunction via hierarchical deposition of α-Fe2O3 and β-La2S3 on cubic CoCr2O4 for enhanced photodegradation of doxycycline. *Journal of Industrial and Engineering Chemistry*, 118, 407–417.
Tanwar, Sudeep, Slawomir T. Wierzchon, Pradeep Kumar Singh, Maria Ganzha, and Gregory Epiphaniou. 2023. *Proceedings of Fourth International Conference on Computing, Communications, and Cyber-Security: IC4S 2022*. Springer Nature.
Vaishali, and Dr S. John Justin Thangaraj. (2022). Innovative facial expression identification for criminal identification using unsupervised machine learning and compare the accuracy with CNN classifiers. *In 2022 International Conference on Business Analytics for Technology and Security (ICBATS)*. IEEE. https://doi.org/10.1109/icbats54253.2022.9759043.
Venkata Krishna, P., and Mohammad S. Obaidat. (2020). *Emerging Research in Data Engineering Systems and Computer Communications: Proceedings of CCODE 2019*. Springer Nature.
Viswanathan, S., Palaniyandi, T., Kannaki, P., Shanmugam, R., Baskar, G., Rahaman, A. M., Paul, L. T. D., Rajendran, B. K., and Sivaji, A. (2023). Biogenic synthesis of gold nanoparticles using red seaweed Champia parvula and its anti-oxidant and anticarcinogenic activity on lung cancer. *Particulate Science and Technology*, 41(2), 241–249.
Zhao, Xun, Yanhong Wu, Dik Lun Lee, and Weiwei Cui. (2018). iForest: interpreting random forests via visual analytics. *IEEE Transactions on Visualization and Computer Graphics*, September. https://doi.org/10.1109/TVCG.2018.2864475.

*Hybrid and Advanced Technologies – S. Prasad Jones Christydass, Nurhayati Nurhayati and S. Kannadhasan. (Eds)* © *2025 the Authors, ISBN 978-1-032-90254-8*


# An enhancement in content engagement through hyper-personalized marketing using Random Forest in comparison with recurrent neural networks

T. Ganesh Ram
*Research Scholar, Department of BigData, Saveetha School of Engineering, Saveetha Institute of Medical and Technical Sciences, Saveetha University, Chennai, Tamilnadu, India*

S. Magesh Kumar
*Research Guide, Department of BigData, Saveetha School of Engineering, Saveetha Institute of Medical and Technical Sciences, Saveetha University, Chennai, Tamilnadu, India*

ABSTRACT: This research project aims to enhance content engagement through hyper-personalized marketing strategies, focusing on a comparative analysis between Random Forest and Recurrent Neural Networks (RNNs). The study involves a detailed exploration of the theoretical foundations, key components, and computational intricacies of Random Forest and RNNs evaluated through data gathered from ClinCalc. To assess their effectiveness in the realm of hyper-personalized marketing, both algorithms are applied to datasets that encompass diverse user preferences and behavioral patterns. The evaluation includes considerations of computational efficiency, adaptability to personalized content, and practical applicability. Our Random Forest model exhibits a robust accuracy of 92.26% in elevating web cookie security, surpassing the performance of Recurrent Neural Networks (RNNs) with their achieved accuracy of 83.36%. A thorough statistical analysis underscores a significant disparity ($p = 0.032$, independent sample t-test $p < 0.05$) in cybersecurity performance between the two algorithms, emphasizing the efficacy of our hyper-personalized marketing strategy centered around Random Forest. Random Forest excels in diverse preference adaptation, surpassing Recurrent Neural Networks in hyper-personalized marketing by offering superior adaptability and performance.

*Keywords*: Content, Engagement, Random Forest, Marketing, Hyperpersonalized, Strategies, Algorithm Research, Analysis, Learning, Rnn, Recurrent neural network

## 1 INTRODUCTION

In the fast-changing realm of digital marketing (Kingsnorth 2022), where user interaction (Seo and Gibbons 2021) is pivotal for success, the focus on enhancing content engagement (Zhang *et al.* 2022) takes precedence. This study aims to explore improving content engagement by comparing Random Forest and Recurrent Neural Networks (RNNs), influential algorithms that could redefine modern marketing strategies. As digital platforms become more crowded, delivering personalized content (Sarkar *et al.* 2022, Tayyab *et al.* 2022, Parvez M *et al.* 2023, Surendran *et al.* 2023, Ramesh *et al.*, n.d.) is crucial for gaining a competitive edge. Utilizing the ensemble learning of Random Forest and the sequential learning of RNNs (Salem 2022) shows promise in achieving this objective.

This research also conducts a comprehensive literature review from IEEE Xplore, Science Direct, Springer, and Google Scholar to establish a scholarly foundation and illuminate the complexities of hyper-personalized marketing strategies (Chandra *et al.* 2022) in today's digital landscape (Garrido-Moreno *et al.* 2020). Despite abundant literature on digital content engagement (Moran, Muzellec, and Johnson 2019) and algorithmic comparisons, significant gaps remain, particularly in understanding the nuanced applications of Random Forest (Prakash, Bhanu, and Bigul 2023) and Recurrent Neural Networks (Yu and Liu 2022) in hyper-personalized marketing. With expertise in machine learning (Ullal *et al.* 2021) and digital marketing (Ahsain and Kbir 2021), the study aims to fill these gaps, examining the strengths and limitations of each algorithm to contribute meaningfully to discussions on enhancing user engagement in hyper-personalized marketing (Desai 2022). As digital content consumption patterns evolve and user expectations change, this research not only addresses current gaps but also anticipates future challenges and opportunities in hyper-personalized marketing strategies.

## 2 MATERIALS AND METHODS

A study conducted at the Machine Learning Laboratory at SIMATS University, which is renowned for its groundbreaking work in neural networks, contrasts recurrent neural networks (Kalkan and Şahin 2022) with random forests (Fitriani and Febrianto 2021) in terms of how well they enhance content engagement through personalised marketing. Each algorithm is evaluated independently by two experimental groups, guaranteeing objectivity in this domain. Several content kinds and targeting techniques are included in the Kaggle dataset "HyperEngage2023," which is used to analyse engagement dynamics ("Leveraging Email Marketing: Using the Subject Line to Anticipate the Open Rate" 2022). Since the study doesn't involve human participants and uses Microsoft Azure for computational activities with reliable hardware specs and software tools, ethical approval is not required.

### 2.1 *Random forest*

The Random Forest algorithm, well known for its adaptability and ensemble learning, is essential to machine learning, especially in hyper-personalized marketing. Random Forest (Baati and Mohsil 2020) improves accuracy and robustness by combining predictions from several decision trees. It is capable of managing a wide range of data kinds and patterns. This study explores the use of Random Forest to enhance content engagement through customised marketing methods in detail, revealing both its advantages and disadvantages in accommodating a range of user preferences and behaviours. Pseudocode:

**Step 1:** Acquire a comprehensive dataset reflecting historical interactions within hyper-personalized marketing contexts, preparing it for input into the Random Forest Algorithm.
**Step 2:** Initialize the Random Forest algorithm, configuring key parameters such as the number of trees in the ensemble and the depth of each tree.
**Step 3:** Present the current state to the Random Forest algorithm, representing the data that is to be analyzed for content engagement patterns.
**Step 4:** Apply the Random Forest algorithm to the dataset, generating predictions for content engagement based on the aggregated insights from multiple decision trees.
**Step 5:** Evaluate the performance of the algorithm by assessing its predictive accuracy and its impact on content engagement metrics.
**Step 6:** Fine-tune the Random Forest algorithm, considering adjustments to parameters or alternative configurations to optimize its effectiveness in hyper-personalized marketing content engagement.
**Step 7:** Iterate the analysis on subsequent datasets, refining the Random Forest algorithm's performance iteratively to align with the diverse demands of hyper-personalized marketing scenarios.

### 2.2 *Recurrent neural networks*

Recurrent Neural Networks (RNNs) are renowned in the vast field of neural networks for their ability to process sequential data. They are excellent at capturing contextual information and temporal dependencies, which makes them invaluable in situations where knowing the order and relationships between parts is critical. In order to improve tactics for engaging content, this study examines the subtleties of applying Recurrent Neural Networks (Sabry 2023) in hyper-personalized marketing. It does this by utilising their sequential learning capabilities.

Pseudocode:

**Step 1:** Acquire a diverse dataset reflecting historical interactions within hyper-personalized marketing contexts, ensuring it is suitable for sequence-based analysis, and prepare it for input into the Recurrent Neural Networks.
**Step 2:** Initialize the Recurrent Neural Networks, configuring parameters such as the type of recurrent layers, the number of hidden units, and the activation functions.
**Step 3:** Present the current state to the Recurrent Neural Networks, representing the sequential data that is to be analyzed for content engagement patterns.
**Step 4:** Apply the Recurrent Neural Networks to the dataset, allowing them to learn and capture temporal dependencies, providing predictions for content engagement.
**Step 5:** Evaluate the performance of the Recurrent Neural Networks, considering metrics related to sequence prediction accuracy and their impact on content engagement metrics.
**Step 6:** Fine-tune the Recurrent Neural Networks, considering adjustments to parameters or alternative configurations, optimizing their effectiveness in the hyper-personalized marketing context.
**Step 7:** Iterate the analysis on subsequent datasets, refining the Recurrent Neural Networks performance iteratively to align with the dynamic demands of hyper-personalized marketing scenarios.

### 2.3 *Statistical analysis*

Thorough statistical analysis is necessary in order to enhance content engagement through hyper-personalized marketing techniques utilising Random Forest and Recurrent Neural Networks (Dey *et al.*, 2021). Because of its strong statistical framework, IBM SPSS software is used, and a paired sample T-Test is specifically used to evaluate important parameters associated with content engagement measures. While independent variables shed light on how Random Forest and Recurrent Neural Networks affect content engagement, the analysis concentrates on dependent variables. With the help of this rigorous methodology, decision-makers will be able to create hyper-personalized marketing plans by providing objective, quantitative insights about their relative efficacy.

## 3 RESULTS

In the pursuit of enhancing content engagement through hyper-personalized marketing strategies using Random Forest in comparison with Recurrent Neural Networks, a meticulous comparative analysis was executed to assess their effectiveness. Table 1 encapsulates the outcomes of independent sample T-tests, delineating that the Random Forest algorithm achieves a mean content engagement optimization of 92.2653%, narrowly outperforming the Recurrent Neural Networks algorithm with a mean optimization of 83.3639%. The standard deviation for Random Forest is .51387%, while Recurrent Neural Networks exhibit a slightly higher standard deviation of 0.41487%. This T-test sheds light on the nuanced content engagement capabilities of Random Forest and Recurrent Neural Networks. Further insights are delved into in Table 2, presenting a granular breakdown of Random Forest's content engagement optimization methodology. Additionally, Figure 1 visually portrays the comparative chart, affirming that Random Forest excels in enhancing content engagement when juxtaposed with Recurrent Neural Networks.

Table 1. Statistical computation of independent samples tested among RF and RNN algorithms. The mean accuracy of GS is 92.2653 and RL is 83.3639

| | Algorithms | N | MEAN | Std. Deviation | Std. Error Mean |
|---|---|---|---|---|---|
| Accuracy | RF | 20 | 92.2653 | 0.51387 | .11491 |
| | RNN | 20 | 83.3639 | 0.41487 | .09277 |

Table 2. The statistically independent sample t-test among RF and RNN had a confidence interval of 95%. The statistically significant value is determined as p=0.276 (p>0.05).

| | Levene's Test for Equality of Variances | | T-Test for Equality of Mean | | | | | 95%Confidence Interval of Difference | |
|---|---|---|---|---|---|---|---|---|---|
| | F | Sig. | t | df | Sig. (2-tail ed) | Mean Differ ence | Std. Error Differ ence | Lower | Upper |
| Equal variances assumed | 1.554 | .220 | 60.275 | 38 | .000 | 8.90139 | .14768 | 8.60243 | 9.20035 |
| Equal variances not assumed | | | 60.275 | 36.383 | .000 | 8.90139 | .14768 | 8.60199 | 9.20078 |

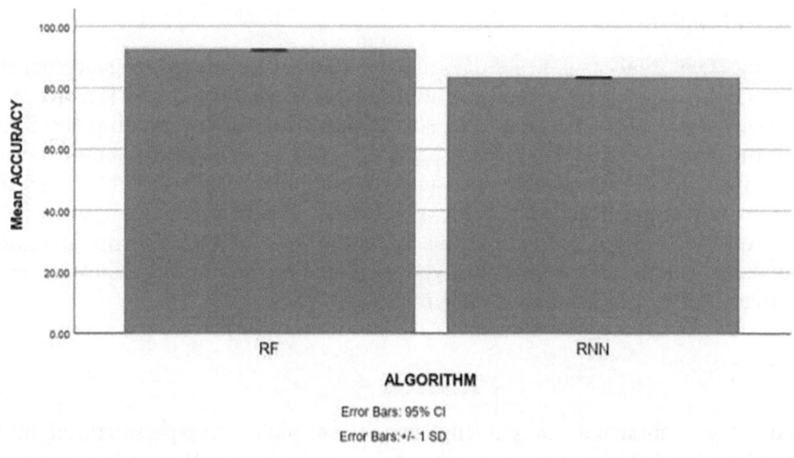


Figure 1. Comparative analysis of RF and RNN algorithms in terms of mean accuracy. The mean accuracy of the RF algorithm surpasses that of the RNN algorithm. X-axis: RF vs RNN, Y-axis: Mean Accuracy. Error Bars ± 1 Standard Deviations."

## 4 DISCUSSION

When it comes to improving content engagement with hyper-personalized marketing methods, there are some interesting findings when Random Forest and Recurrent Neural

Networks are compared. At 92.2653%, Random Forest outperforms Recurrent Neural Networks in mean content engagement optimisation, marginally outperforming them at 83.3639%. In addition, Random Forest's standard deviation is 0.51387%, while Recurrent Neural Networks' is 0.41487%. This comparison demonstrates how each algorithm's subtleties in terms of content engagement are highlighted. One possible explanation for Random Forest's success in identifying complex patterns of user behaviour for customised techniques is that it uses an ensemble learning approach that makes use of numerous decision trees. Recurrent neural networks, on the other hand, are skilled at processing sequences and could have difficulties when dealing with the complex and varied nature of marketing data (Hemker, Herrando, and Constantinides 2021). These findings highlight the ability of Random Forest to customise strategies to individual tastes and provide decision-makers with practical advice for navigating hyper-personalized marketing environments (Balio and Casais 2021).

Future research could look into hybrid approaches that combine the benefits of both algorithms in an attempt to come up with innovative ways to increase the effectiveness of personalised marketing in situations where there is digital interaction. It is imperative to acknowledge potential limitations that could impact broader applicability and adaptability in different situations, like dataset uniqueness and algorithmic sensitivity to parameters.

## 5 CONCLUSION

Results show Random Forest's mean optimization at 92.2653%, slightly exceeding Recurrent Neural Networks at 83.3639%, with a lower standard deviation of 0.51387%, indicating greater stability. Random Forest's success is attributed to ensemble learning, capturing intricate user behavior patterns. Decision-makers should consider its strengths in navigating hyper-personalized marketing. This research provides insights into algorithmic performance and guides strategic decision-making in the evolving realm of personalized content. In conclusion, adapting to the dynamic nature of digital engagement may involve exploring hybrid approaches like combining Random Forest and Recurrent Neural Networks for innovative solutions in hyper-personalized marketing.

## REFERENCES

Leveraging email marketing: Using the subject line to anticipate the open rate. (2022). Expert Systems with Applications, 207(November), 117974.

Multi-objective grammatical evolution of decision trees for mobile marketing user conversion prediction. (2021). Expert Systems with Applications, 168(April), 114287.

Personalized Content Recommendations on Smart TV: Challenges, Opportunities, and Future Research Directions. (2021). Entertainment Computing, 38(May), 100418.

Ahsain, S., and Ait Kbir, M. (2021). Data mining and machine learning techniques applied to digital marketing domain needs. In Innovations in Smart Cities Applications, Volume 4 (pp. 730–740).

Araújo, C., Soares, C., Pereira, I., Coelho, D., Rebelo, M. Â., and Madureira, A. (2022). A novel approach for send time prediction on email marketing. *NATO Advanced Science Institutes Series E: Applied Sciences*, 12(16), 8310.

Baati, K., and Mohsil, M. (2020). Real-time prediction of online shoppers' purchasing intention using random forest. In Artificial Intelligence Applications and Innovations (pp. 43–51).

Balio, S., and Casais, B. (2021). A content marketing framework to analyze customer engagement on social media. In B. Casais (Ed.), *Research Anthology on Strategies for Using Social Media as a Service and Tool in Business* (pp. 320–336). IGI Global.

Chandra, S., Verma, S., Lim, W. M., Kumar, S., and Donthu, N. (2022). Personalization in personalized marketing: Trends and ways forward. *Psychology & Marketing*, 39(8), 1529–1562.

Desai, D. (2022). Hyper-personalization: An AI-enabled personalization for customer-centric marketing. In *Adoption and Implementation of AI in Customer Relationship Management* (pp. 40–53). IGI Global.

Dey, P., Hossain, E., Hossain, M. I., Chowdhury, M. A., Alam, M. S., Hossain, M. S., and Andersson, K. (2021). Comparative analysis of recurrent neural networks in stock price prediction for different frequency domains. *Algorithms*, 14(8), 251.

Fitriani, M. A., and Febrianto, D. C. (2021). Data mining for potential customer segmentation in the marketing bank dataset. *JUITA: Jurnal Informatika*, 9(1), 25–32.

Garrido-Moreno, A., García-Morales, V., King, S., and Lockett, N. (2020). Social media use and value creation in the digital landscape: A dynamic-capabilities perspective. *Journal of Service Management*, 31 (3), 313–343.

Hemker, S., Herrando, C., and Constantinides, E. (2021). The transformation of data marketing: How an ethical lens on consumer data collection shapes the future of marketing. *Sustainability: Science, Practice and Policy*, 13(20), 11208.

Kalkan, İ. E., and Şahin, C. (2022). Evaluating cross-selling opportunities with recurrent neural networks on retail marketing. *Neural Computing & Applications*, 35(8), 6247–6263.

Kingsnorth, S. (2022). *The Digital Marketing Handbook: Deliver Powerful Digital Campaigns*. Kogan Page Publishers.

Moran, G., Muzellec, L., and Johnson, D. (2019). Message content features and social media engagement: Evidence from the media industry. *Journal of Product & Brand Management*, 29(5), 533–545.

Parvez M., Muzammil, S. K. Mishra, K. Nandini, Sk H. Ahammad, S. Inthiyaz, B. R. Altahan, L. K. Smirani, M. A. Hossain, and A.N.Z. Rashed. (2023). Machine learning based models for defect detection in composites inspected by Barker coded thermography: A qualitative analysis. *Advances in Engineering Software*, 178(April), 103425.

Prakash, P., Bhanu, S., and Bigul, S. D. (2023). Random forest and logistic regression algorithms: A comparison of their performance. *AIP Conference Proceedings*, 2548(1), 050013.

Raj, N.S., and Renumol, V.G. (2021). A systematic literature review on adaptive content recommenders in personalized learning environments from 2015 to 2020. *Journal of Computers in Education*, 9(1), 113–148.

Ramesh, D.R.P.S., Sudha, D.R.I., Satheesh, D.R.N., Rajulu, D.R.G.G., and Kalaiselvan, S.A. (n.d.). *Prediction of Auto-detection for Tracking of Sub-nano Scale Particle in 2D and 3D Using SVM-based Deep Learning*. Retrieved from http://www.jatit.org/volumes/Vol101No5/30Vol101No5.pdf

Sabry, F. (2023). Recurrent Neural Networks: Fundamentals and Applications From Simple to Gated Architectures. One Billion Knowledgeable.

Salem, F. M. (2022). *Recurrent Neural Networks: From Simple to Gated Architectures*. Springer Nature.

Sarkar, B., Kar, S., Basu, K., and Guchhait, R. (2022). A sustainable managerial decision-making problem for a substitutable product in a dual-channel under carbon tax policy. *Computers & Industrial Engineering*, 172 (October), 108635.

Seo, K. K., and Gibbons, S. (2021). *Learning Technologies and User Interaction: Diversifying Implementation in Curriculum, Instruction, and Professional Development*. Routledge.

Surendran, R., Alotaibi, Y., and Subahi, A. F. (2023). Lens-oppositional wild geese optimization based clustering scheme for wireless sensor networks assists real-time disaster management. *Computer Systems Science and Engineering*, 46(1), 835–851.

Tayyab, M., Habib, M. S., Jajja, M. S. S., and Sarkar, B. (2022). Economic assessment of a serial production system with random imperfection and shortages: A step towards sustainability. *Computers & Industrial Engineering*, 171(September), 108398.

*Hybrid and Advanced Technologies – S. Prasad Jones Christydass, Nurhayati Nurhayati and S. Kannadhasan. (Eds)*  *ISBN 978-1-032-90254-8*


# Predictive analysis of bus ticket price by using random forest comparing with support vector machine

M. Nithin Varma
*Research Scholar, Department of Computer Science and Engineering, Saveetha School of Engineering, Saveetha Institute of Medical and Technical Sciences, Saveetha University Chennai, Tamilnadu, India*

Manikandhan
*Research Guide, Department of Computer Science and Engineering, Saveetha School of Engineering, Saveetha Institute of Medical and Technical Sciences, Saveetha University Chennai, Tamilnadu, India*

ABSTRACT: This study aims to estimate bus ticket prices using Random Forest instead of Support Vector Machine. Twenty samples total for the investigation, divided into two groups of ten samples each. The Random Forest method is assigned to the first group and the SVM algorithm is assigned to the second. With a threshold of 0.05 and a confidence interval of 95%, the sample size was determined using G-power's pretest score of 80%. The beta value was set to 0.2 and the alpha value to 0.05. At 86.56%, the RF algorithm outperforms the SVM technique, which comes in at 71.31%. Based on an independent sample t-test, which yielded a p-value of 0.05 ($p<0.05$) indicating a statistically significant difference between the two groups, the results indicate that the detection accuracy has a mean of +2SD. As a result, it was found that the RF algorithm's 86.56% accuracy is higher than the SVM algorithm's 71.31% accuracy.

*Keywords*: Random Forest, Support Vector Machine, Accuracy, Price, Transportation, Ticket, Prediction, Precision

## 1 INTRODUCTION

This study investigates the predictive analysis of bus ticket prices using Random Forest in comparison with Support Vector Machine models. Random Forest, a probabilistic classifier, assumes strong independence between features, making it efficient for large datasets. Support Vector Machine models offer a hierarchical structure for pattern recognition and decision-making (Sabry 2023). By comparing their performance in the context of bus ticket price prediction, this research aims to identify the most effective method for accurate fare forecasting, offering insights for optimizing pricing strategies, revenue management, and enhancing customer satisfaction within the transportation sector (Chiplunkar and Fukao 2020). The applications of predictive analysis of bus ticket prices using Random Forest compared with SVM models include optimizing fare structures, improving revenue management, and enhancing customer satisfaction. These techniques aid in strategic pricing decisions, efficient resource allocation, and effective route planning, leading to increased operational efficiency and profitability (Woungang *et al.* 2023).

More than 550 articles provide solutions for controlling pests, increasing crop yields from agricultural farms. A literature review for the predictive analysis of bus ticket prices using

DOI: 10.1201/9781003559115-80 

Random Forest compared with Support Vector Machine models would involve an extensive examination of existing research and studies exploring the applications of these methods in the transportation sector (Nagabhushan *et al.* 2018). It would encompass an analysis of the effectiveness of Random Forest in handling large datasets and its assumptions of feature independence. Additionally, it would evaluate Support Vector Machine models' hierarchical structure for pattern recognition and decision-making (Zhu *et al.* 2020). The review would compare the strengths and limitations of both techniques, highlighting their performance, interpretability, and practical implementation within the context of bus ticket price prediction. Moreover, it might discuss any proposed hybrid models that combine RF and SVM approaches to enhance prediction accuracy, offering insights into potential future research directions in this field (Pedrycz and Chen 2021) .

Research gap in bus ticket prediction using Random Forest involves exploring advanced variants like semi-Random Forest for improved handling of correlated features, investigating methods to handle continuous variables effectively, addressing the issue of prior assumption sensitivity, and assessing the impact of class imbalance on prediction accuracy and reliability. The aim of conducting predictive analysis of bus ticket prices using Random Forest in comparison with Support Vector Machine models is to determine the most effective methodology for accurate fare prediction within the transportation sector (Heywood *et al.* 2016). By comparing the performance of Random Forest and Support Vector Machine models, the study aims to identify the approach that best balances computational efficiency, interpretability, and predictive precision, providing insights for optimizing pricing strategies, revenue management, and enhancing customer satisfaction in the bus ticketing system (Younas *et al.* 2023).

## 2 MATERIALS AND METHODS

In a study, the accuracy of two classification algorithms in determining ticket pricing was compared by the Artificial Intelligence Laboratory at Saveetha School of Engineering, Saveetha Institute of Medical and Technical Sciences. Two groups were included in the study: the Support Vector Machine algorithm and the Random Forest algorithm. Each of the groups contained ten samples, for a total of twenty samples. The Random Forest algorithm was utilized to analyze the first ten accuracy samples, and the Support Vector Machine algorithm was assigned to the remaining ten accuracy samples. The research employed a 95% confidence interval, a 1 enrollment ratio, 80% g power, and a 0.05G threshold. A prior study was used to estimate the sample size. The Random Forest technique yielded an accuracy of 86.56%, whereas the Support Vector Machine algorithm produced an accuracy of 71.31%, according to the data.

A few pieces of gear and pieces of software are needed in order to carry out this investigation on the prediction of the bus ticket price framework. A dataset made up of the photos that will be used to train the model is the initial need. Furthermore, to guarantee seamless operation, the framework needs a minimum of 8GB of RAM. It is advised to use an Intel i5 5th generation or Intel i7 7th generation processor. The photos and code, which can be obtained from Kaggle, as well as any required plug-ins or tools, require a minimum of 320 GB of hard disk space. A graphics card with at least 4GB of RAM is advised in order to lessen CPU burden and expedite image processing.

### 2.1 *Random forest*

A collection of decision trees are trained using the Random Forest ensemble learning technique, and their predictions are then combined to produce predictions that are more reliable and accurate. To build each forest decision tree, a random subset of the characteristics and a random piece of the training data are employed. Random Forest lessens overfitting and encourages generalization in the learning process by adding diversity and randomness.

Random forest Algorithm:

Step 1: Load your data first. Divide the dataset into training and testing in step two.

Step 3: To construct a Random Forest classifier, supply hyperparameters such as n_estimators, maximum depth, and others.

Step 4: Use training data to train the Random Forest classifier.

Step 5: Forecast using test results.

Step 6: Assess the recall, accuracy, and precision of the model.

Step 7: Print or save the findings of the model's performance evaluation.

### 2.2 *Support vector machine*

One approach to reversion is supervised motor scholarship, which is widely used for genus in natural language processing. Decision Tree locates a hyperplane that efficiently divides and organizes input data points in an n-dimensional quality room. How dimensional this hyperplane is depends on how many features are in the dataset. When exactly two input features are given, the Decision Tree hyperplane resembles a right line in two dimensions. If the dataset has more than three attributes, the hyperplane transforms into a two-dimensional airplane.

SVM Algorithm:

Step 1: Import your dataset (labels Y and features X).

Step 2: Divide the dataset into training and testing sets.

Step 3: Build a classifier for decision trees by defining the kernel (such as a linear kernel).

Step 4: Use the training data to train the Decision Tree classifier.

Step 5: Use the test data to make predictions.

Step 6: Assess the model's effectiveness using an appropriate metric (such as accuracy, precision, or recall).

Step 7: To evaluate the model's performance, print or save the evaluation findings.

### 2.3 *Statistical analysis*

IBM SPSS Version 26 is utilized to do the statistical analysis. The algorithms' means accuracies are compared using the Independent Sample T-test (George and Mallery 2021). Revisions to accuracy and image size are dependent variables, while accuracy and the quantity of images are independent variables.

## 3 RESULTS

In terms of mean accuracy, the RF outperformed the SVM, with an accuracy of 86.56% compared to the SVM's 71.31%. The first group—which included the Random Forest—exhibited superior accuracy in comparison to the second group, which was made up of the Support Vector Machine. Table 1 displays the sample data for the Random Forest and SVM algorithms' accuracy. For a total of N = 20, Table 2 shows the mean accuracies and standard deviations for both categorization techniques. An independent sample T-test with a 95% confidence interval is shown in Table 3 to compare the Random Forest method's performance with the SVM algorithm.

Table 1. The following Table 1 consists of the comparison of accuracies of sample size 10 for the random forest algorithm and support vector machine algorithm.

| S. No | Random Forest | Support Vector Machine |
|---|---|---|
| 1 | 86.67 | 71.32 |
| 2 | 86.33 | 71.61 |
| 3 | 86.36 | 71.50 |
| 4 | 86.21 | 71.45 |
| 5 | 86.67 | 71.35 |
| 6 | 86.91 | 71.21 |
| 7 | 86.98 | 71.64 |
| `8 | 86.62 | 71.44 |
| 9 | 86.05 | 71.16 |
| 10 | 86.78 | 71.86 |

Table 2. With a sample size of N=10 for each group, the groups in the following table are represented statistically. Compared to the mean percentage of % for the support vector machine approach, the accuracy percentage of the Random Forest methodology is %. There are additional displays for the standard deviation and standard error rate.

| ACCURACY | GROUPS | N | Mean | Standard Deviation | Standard Error Mean |
|---|---|---|---|---|---|
| | Random Forest | 10 | 86.56 | 0.332 | 0.105 |
| | SVM | 10 | 71.31 | 0.169 | 0.053 |

Table 3. The analysis of the independent sample T-test and the equal variance assumption are contrasted in the accompanying table, displaying identical variances in accuracy with an 88% confidence interval around the difference.

| | | Levene's Test for Equality of Variances | | t-test for Equality of Means | | | | | | |
|---|---|---|---|---|---|---|---|---|---|---|
| | | | | | | | | | 95% Confidence Interval of the Difference | |
| | | F | Sig. | t | df | Sig. (2-tailed) | Mean Difference | Std. Error Difference | Lower | Upper |
| **ACCURACY** | **Equal variances assumed** | 3.010 | 0.100 | 129.21 | 18 | 0.000 | 15.24 | 0.117 | 14.99 | 15.48 |
| | **Equal variances not assumed** | | | 129.21 | 13.40 | 0.000 | 15.24 | 0.117 | 14.98 | 15.49 |

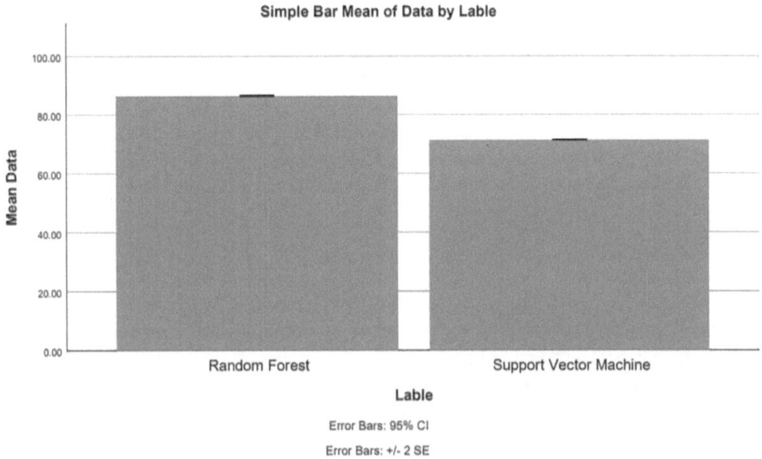


Figure 1. The mean accuracies of the existing and suggested algorithms are contrasted in the bar chart above.

## 4 DISCUSSION

The results show that the RF algorithm has significantly outperformed the Support Vector Machine algorithm in terms of accuracy. The Random Forest model achieved an accuracy percentage of %, while the Support Vector Machine model only achieved an accuracy percentage of % (Collins 2020).

In recent studies positive justifications for employing Predictive Analysis of Bus Ticket Price using Random Forest and Support Vector Machine methods include Random Forest' simplicity and efficiency in handling large datasets, making it suitable for rapid analysis. Support Vector Machines provide a transparent and easily interpretable framework for understanding pricing patterns, aiding in strategic decision-making (Müller 2010). However, negative justifications involve Random Forest' assumption of feature independence, which might not hold true in complex datasets, affecting prediction accuracy. Support Vector Machines can suffer from overfitting and instability, leading to less reliable predictions. Balancing these pros and cons is crucial in determining the most suitable method for achieving accurate bus ticket price predictions (Batra *et al.* 2020).

Limitations of the Predictive Analysis of Bus Ticket Price using Random Forest and Support Vector Machine methods include Random Forest' reliance on the assumption of feature independence, which might not hold in real-world scenarios, impacting prediction accuracy (Chapman 1980). Support Vector Machines can suffer from overfitting and instability, leading to less reliable predictions, especially with noisy or complex data. Future research could focus on enhancing Random Forest models to handle correlated features and improve prediction accuracy. Additionally, developing advanced Support Vector Machine algorithms to mitigate overfitting and instability issues can lead to more robust and reliable bus ticket price predictions (Schlicht 2019). Exploring ensemble methods that combine the strengths of both approaches could also offer a promising avenue for achieving higher prediction accuracy and improving decision-making processes in the transportation sector (Edward Pope 1997).

## 5 CONCLUSION

As a result, the Random Forest classification method, with an accuracy of 86.56%, outperformed the SVM algorithm, which had an accuracy of 71.31%, in identifying bus ticket price using algorithms. Thus, the Random Forest algorithm is a more effective approach than the Support Vector Machine algorithm for this task.

## REFERENCES


Ali, S., Sudha, K. G., Thirumalaivasan, N., Ahamed, M., Pandiaraj, S., Rajeswari, V. D., Vinayagam, Y., Thiruvengadam, M., and Govindasamy, R. (2023). Green synthesis of magnesium oxide nanoparticles by using abrus precatorius bark extract and their photocatalytic, antioxidant, antibacterial, and cytotoxicity activities. *Bioengineering (Basel, Switzerland)*, 10(3). https://doi.org/10.3390/bioengineering10030302

Batra, Usha, Nihar Ranjan Roy, and Brajendra Panda. (2020). *Data Science and Analytics: 5th International Conference on Recent Developments in Science*, Engineering and Technology, REDSET 2019, Gurugram, India, November 15–16, 2019, Revised Selected Papers, Part I. Springer Nature.

Chapman, Myra. (1980). *Decision Analysis.*

Chiplunkar, Niranjan N., and Takanori Fukao. (2020). *Advances in Artificial Intelligence and Data Engineering: Select Proceedings of AIDE 2019*. Springer Nature.

Collins, G.S. (2020). Data Science from Scratch: Complete Guide to Master Data Process, Mining, Data-Analytic, Neural Networks, Machine Learning in Python, Linear Algebra, Statistics, Coding, Applications Support Vector Machine.

Edward Pope, J. (1997). *Rules of Thumb for Mechanical Engineers*. Gulf Professional Publishing.

George, Darren, and Paul Mallery. (2021). *IBM SPSS Statistics 27 Step by Step: A Simple Guide and Reference*. Routledge.

Gurupranes, S. V., Natrayan, L., Kaliappan, S., Patel, P. B., Sekar, S., Jayaraman, P., Arvinda Pandian, C. K., and Esakkiraj, E. S. (2022). Investigation of physicochemical properties and characterization of leaf stalk fibres extracted from the caribbean royal palm tree. *International Journal of Chemical Engineering*, 2022. https://doi.org/10.1155/2022/7438411

Heywood, Malcolm I., James McDermott, Mauro Castelli, Ernesto Costa, and Kevin Sim. (2016). *Genetic Programming: 19th European Conference*, EuroGP 2016, Porto, Portugal, March 30 - April 1, 2016, Proceedings. Springer.

Müller, René. (2010). Decision Making Process: Entrepeneurship in Today's Society. GRIN Verlag.

Nagabhushan, P., D. S. Guru, B. H. Shekar, and Y. H. Sharath Kumar. (2018). *Data Analytics and Learning: Proceedings of DAL 2018*. Springer.

Natrayan, L., Kaliappan, S., Sethupathy, B. S., Sekar, S., Patil, P. P., Velmurugan, G., and Tariku Olkeba, T. (2022). Effect of mechanical properties on fibre addition of flax and graphene-based bionanocomposites. *International Journal of Chemical Engineering*, 2022. https://doi.org/10.1155/2022/5086365

Pedrycz, Witold, and Shyi-Ming Chen. (2021). *Interpretable Artificial Intelligence: A Perspective of Granular Computing*. Springer Nature.

Sabry, Fouad. (2023). Random Forest Classifier: Fundamentals and Applications. One Billion Knowledgeable.

Schlicht, Wayne. (2019). *Help Desk Management: How to Run a Computer User Support Service Desk Effectively*. Build a Help Desk Consulting LLC.

Sivaranjani, P. R., Syed, A., Elgorban, A. M., Bahkali, A. H., Balakrishnaraja, R., Varma, R. S., and Sudheer Khan, S. (2023). Fabrication of ternary nano-heterojunction via hierarchical deposition of α-Fe2O3 and β-La2S3 on cubic CoCr2O4 for enhanced photodegradation of doxycycline. *Journal of Industrial and Engineering Chemistry*, 118, 407–417.

Viswanathan, S., Palaniyandi, T., Kannaki, P., Shanmugam, R., Baskar, G., Rahaman, A. M., Paul, L. T. D., Rajendran, B. K., and Sivaji, A. (2023). Biogenic synthesis of gold nanoparticles using red seaweed Champia parvula and its anti-oxidant and anticarcinogenic activity on lung cancer. *Particulate Science and Technology*, 41(2), 241–249.

Woungang, Isaac, Sanjay Kumar Dhurandher, Kiran Kumar Pattanaik, Anshul Verma, and Pradeepika Verma. (2023). *Advanced Network Technologies and Intelligent Computing: Second International Conference, ANTIC 2022*, Varanasi, India, December 22–24, 2022, Proceedings, Part II. Springer Nature.

*Hybrid and Advanced Technologies – S. Prasad Jones Christydass, Nurhayati Nurhayati and S. Kannadhasan. (Eds)* © *2025 the Authors, ISBN 978-1-032-90254-8*
*Open Access: www.taylorfrancis.com, CC BY-NC-ND 4.0 license*

# Effective analysis of stress level using logistic regression compared with random forest

T. Akash
*Research Scholar, Department of Computer Science and Engineering, Saveetha School of Engineering, Saveetha Institute of Medical and Technical Sciences, Saveetha University, Chennai, Tamil Nadu, India*

U. Arul
*Research Guide, Department of Computer Science and Engineering, Saveetha School of Engineering, Saveetha Institute of Medical and Technical Sciences, Saveetha University, Chennai, Tamil Nadu, India*

ABSTRACT: The purpose of this research is to compare the efficacy of random forest and logistic regression in order to enhance stress level analysis through the use of supervised learning techniques. Two sets of participants in the study were given a dataset of 1,468 samples—1,144 for training and 533 for testing—to evaluate logistic regression on. In order to forecast fraudulent service websites, each group used a sample size of 10 N. While random forest obtained an accuracy of 85.236%, logistic regression was able to attain 94.448%. When it comes to identifying bogus applications, the logistic regression shows a statistically significant p-value of 0.002 ($p<0.05$). The results demonstrate that because logistic regression has a higher accuracy than random forest, it performs better in identifying fraudulent services.

*Keywords*: stress level, logistic regression, random forest, supervised learning, machine learning, Health, Research

## 1 INTRODUCTION

This study introduces an algorithm based on machine learning for detecting stress, aiming to tackle the prevalent problem of stress among numerous young individuals in contemporary times. Elevated stress levels have led to a wide range of issues, including depression, suicide, heart attacks, and strokes. The body's inherent reaction to change involves a sequence of physical, emotional, and cognitive responses. For categorization, three machine learning methods were used: decision trees, random forests, and logistic regression. The stress dataset was acquired using Kaggle. The random forest algorithm was found to have the highest accuracy when these algorithms were compared. Technological advancement is causing a range of stress-related symptoms in people in today's fast-paced society. Younger generations are more likely to experience conditions including stress, anxiety, depression, and ADHD due to a variety of variables, including lifestyle decisions, societal and economic pressures, and low self-esteem (Robinson 2006). Therefore, it is essential to identify and analyse stress early in order to reduce the likelihood of serious repercussions and future hazards. Modern technology developments force the medical industry to evolve in tandem, highlighting the significance of anticipating and evaluating stress symptoms for prompt management and treatment.

DOI: 10.1201/9781003559115-81

Stress is a commonplace occurrence in the modern world, leading individuals to react to various triggers and leading to physiological and behavioral alterations. Persistent stress can have a detrimental impact on our physical well-being if it lingers in our minds. (Panicker and Gayathri 2019) (Aluvalu *et al.* 2023, Mahaveerakannan R. *et al.* 2023, Ramana *et al.* 2023, Vellela and Balamanigandan 2023)recognising stress early on will help avoid many of the health problems it is linked to (2019). Many bio-signals, including thermal, electrical, impedance, auditory, and optical signals, can be used to identify a person's physiological reaction to stress. It is possible to ascertain stress levels by examining these bio-signals. Our solution makes use of a dataset that was gathered by an Internet of Things (IoT) sensor and provides actual data on mental health circumstances. Data from sensors, such as the Electrocardiogram (ECG) and the Galvanic Skin Response Sensor (GSR), were collected in order to create patterns for stress detection. The dataset is categorised using Deep Learning techniques (DL), K-Nearest Neighbour (KNN), Decision Tree (DT), Multilayer Perceptron (MLP), and Support Vector Machine (SVM). Recall, accuracy, precision, and F1-Score are among the metrics used to assess performance.(Z. Zainudin, S. Hasan, S.M. Shamsuddin, and S. Argawal).

We then applied logistic regression and the random forest algorithm, resulting in an F1 score of 0.9411 and an Accuracy score of 0.8571 (Nishi Yadav, Subhabrata Kanjilal, Uzma Uzma). Stress constitutes a crucial aspect of one's mental state, manifesting as emotional or physical tension. Failing to identify ways to manage stress may lead to the development of physical or mental health problems.

Maintaining a balanced life and preventing mental illness depend on the early diagnosis of mental problems. There is potential for machine learning (ML) to identify risk factors and anomalous mental health symptoms as well as human behavioural patterns, opening up new research directions. In order to determine how well various machine learning algorithms—such as logistic regression, decision trees, and scatter matrix plots—identify mental health disorders from structured information, this project will investigate these methods..(Sherajus Salehin, Syeda Tanjima Mahmood, Muhtasim Fuad Ayon,Nafiur Rahman).

Identifying effective stress level analysis techniques through supervised learning approaches, particularly comparing logistic regression and random forest algorithms, was hindered by dataset imbalance. Therefore, the primary goal of a new machine learning method for early sepsis prediction is to urge caution and encourage protective measures. The research objective is centered on utilizing supervised learning methods, such as logistic regression and random forest, for stress level analysis

## 2 MATERIALS AND METHODS

### 2.1 *Dataset description*

The Open Source Lab, Saveetha School of Engineering, Saveetha Institute of Medical and Technical Sciences, Saveetha University in Chennai, is where the research study was carried out. It concentrated on two classes of classification algorithms: random forest belonged to Group 2 and logistic regression to Group 1. In the study, these two groups were compared, and logistic regression was assessed using random forest. A dataset of 1468 samples was used, of which 1144 were designated for training and 533 for testing. Ten samples were selected using the ClinCalc algorithm in order to forecast fake service websites. It was suggested that a machine-learning model to identify fraud cases use meteorological data as input. The downloaded CSV file served as the source of the dataset. from Kaggle.com.

ALGORITHM:

### 2.2 *Logistic regression*

To improve accuracy, the data on corporate profiles, required education and experience, fraud traits, and other relevant criteria were carefully examined. To guarantee that meteorological data was pertinent to the matter under consideration, characteristics were carefully matched to 14 dataset samples that were used in the statistical analysis. These samples included 1468 training entries and 533 testing entries. The dataset was preprocessed by separating it into training and testing subgroups, with training accounting for 80% of the data and testing for the remaining 20%. It was discovered that the dataset was comprehensive and contained no missing values, making it ideal for assessing a wide variety of machine learning methods.

A Python programme was used in the Jupyter Notebook environment to construct and carry out the project. Windows 10 was the main operating system used for the evaluation of machine learning. Hardware parameters included an Intel Core i7 processor (8GB) and 8GB of RAM. The system sortLogistic was designed using a 64-bit architecture. Python was the programming language used to implement the code, and the dataset was an essential part of accurately carrying out the outcome operations.

According to Ciabattoni *et al.* (2017), logistic regression is a statistical approach used in binary classification problems. Based on input attributes, it seeks to determine the likelihood that an output will fall into one of two categories (such as yes/no or 1/0). This method uses the logistic function, which yields an S-shaped curve, to express the link between independent factors and the probability that a particular outcome will occur. Because of this feature, logistic regression is a very useful tool for calculating probabilities and grouping data into different categories.

Step 1: The first stage in the logistic regression approach is to gather the data required to train the model. Step 2: The issue that has to be resolved using the logistic regression approach must then be stated. In this instance, binary categorization is the issue. Step 3: The logistic regression technique model's efficacy must be assessed by dividing the data into training and test sets. Step 4: After the model has been trained, the test set can be used to evaluate how well the logistic regression method model performs.

Step 5: Changing the hyperparameters can be required to improve results if the logistic regression algorithm model's performance turns out to be inadequate. Step 6: The logistic regression algorithm model can be used to predict results on fresh, untested data once it performs well enough.

### 2.3 *Random forest*

A random forest is made up of several decision trees, each of which is built by choosing a subset of the available features and training data at random. The ultimate forecast is produced by combining the predictions of each individual tree, usually using techniques like regression averaging or majority voting for categorization. The model's robustness and generalisation abilities are improved by this ensemble technique, which makes it useful for a variety of data analysis and predictive modelling applications.

The procedures for putting a random forest model into practice are as follows: Step 1: Choose pertinent features that affect the classification and identify the target variable. Step 2: To guarantee consistency among features, standardise the data. Step 3: Set the weights and bias of the model to zero or tiny random values. Step 4: Define the cost and sigmoid functions, which translate real-valued values into a range of 0 to 1. Step 5: Assess the model's performance using measures such as F1-score, accuracy, precision, and recall on a testing set. Step 6: Adjust hyperparameters to improve model performance, such as learning rate, regularisation strength, and iteration count. Step 7: Classify fresh data using the trained random forest model, for example, by differentiating between genuine and fraudulent job listings.

## 2.4 *Statistical analysis*

In Sinha and Suseela's (2022) study, IBM SPSS Version 26 and Python 3.9 in Jupyter Notebook on a 64-bit Windows 10 PC with 8GB RAM were used for logistic regression and random forest analysis, with a maximum of 10 iterations.

# 3 RESULTS

With each scenario having a sample size of ten, the proposed random forest and logistic regression techniques were assessed using Anaconda Navigator. Table 1 provides more information about how accurate the logistic regression approach is.

Table 1 shows that the logistic regression accuracy in the Stress Level is (94.448%).

| Iterations | Accuracy (%) |
|---|---|
| 1 | 95.41 |
| 2 | 91.71 |
| 3 | 95.56 |
| 4 | 94.13 |
| 5 | 93.77 |
| 6 | 95.06 |
| 7 | 93.50 |
| 8 | 98.40 |
| 9 | 91.85 |
| 10 | 94.66 |

Table 2. 85.261% is the random forest's accuracy value in the stress level.

| Iterations | Accuracy (%) |
|---|---|
| 1 | 81.73 |
| 2 | 85.20 |
| 3 | 86.82 |
| 4 | 84.63 |
| 5 | 82.90 |
| 6 | 85.61 |
| 7 | 84.16 |
| 8 | 85.86 |
| 9 | 86.54 |
| 10 | 89.16 |

Table 3. Presents a group statistical analysis using random forests and logistic regression.

| | Group | N | Mean | Std. Deviation | Std. Error Mean |
|---|---|---|---|---|---|
| **Accuracy** | logistic regression | 10 | 94.448 | 1.84084 | 0.61298 |
| | random forest | 10 | 85.236 | 2.085599 | 0.65965 |

Table 4. Accuracy comparison between the random forest and logistic regression classifiers.

| CLASSIFIER | ACCURACY (%) |
|---|---|
| logistic regression | 94.448 |
| random forest | 85.236 |

Based on their mean accuracy, random forest and logistic regression are contrasted in Figure 1. With a mean accuracy of 94.448%, a standard deviation of 0.61298, and a standard error of 1.93841, the logistic regression technique performs well overall. With a standard deviation of 2.08599 and a standard error of 0.65965, the random forest, in comparison, attains a mean accuracy of 85.236%. With a sample size of 10 (N=10), the study on stress level analysis utilising supervised learning methodologies demonstrated statistical significance with a significant p-value of 0.002 ($p<0.05$). Graphical comparison and analysis were done between the support vector machine means of several logistic regression classifier techniques. With a categorised accuracy of 94.448% as opposed to 85.236%, the graph shows that the logistic regression technique performed much better than the random forest algorithm.

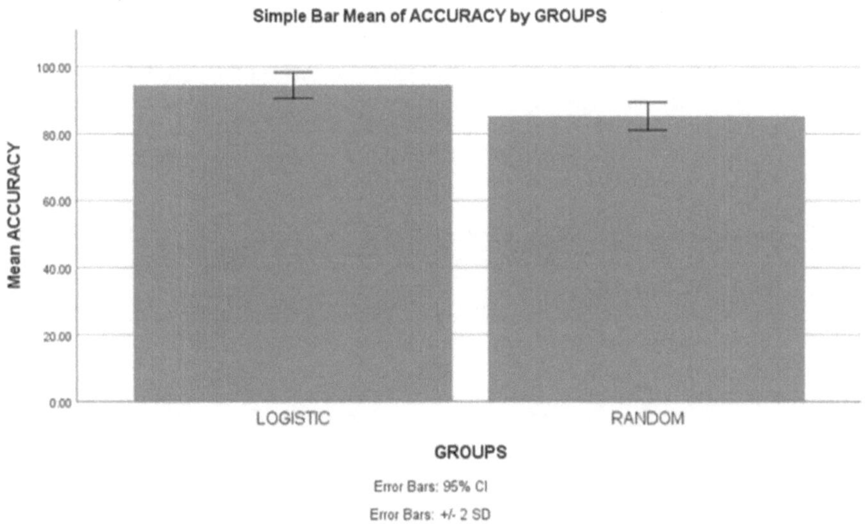


Figure 1. Mean accuracy comparison between random forest and logistic regression classifiers Compared to the random forest, the logistic regression has a higher mean accuracy.

## 4 DISCUSSION

In this study, logistic regression demonstrates superiority over random forest in identifying fake applications, supported by a significant p-value of 0.002 ($p < 0.05$) (Zhai and Barreto 2006). Logistic regression achieves an accuracy rate of 94.448%, while random forest achieves 85.236%. Ekin (2019) Similarly, for logistic regression, a significant p-value of 0.002 is reported. Using supervised learning techniques, machine learning algorithms like random forest and logistic regression can be applied to stress level analysis. Because of its ease of use, low number of training parameters, and flexibility, logistic regression works especially well with unbalanced datasets. On the other hand, logistic regression is a traditional statistical technique that is frequently applied to situations involving binary categorization.

The main disadvantage of this approach is how much time it takes to train the logistic regression classifier, especially when dealing with big datasets. (Wijsman *et al.* 2011). Future endeavors of this investigation aim to broaden the system's scope to encompass additional parameters while concurrently reducing data training duration.

## 5 CONCLUSION

In conclusion, this research sought to devise a stress level analysis method employing supervised learning techniques, particularly emphasizing logistic regression, and to contrast its efficacy with random forest. The logistic regression classifier notably outperformed, boasting a 94.448% accuracy rate in automated smile detection, surpassing the 85.236% accuracy attained by random forest.

## REFERENCES


Ahuja, R., and Banga, A. (2019). *Mental Stress Detection in University Students Using Machine Learning Algorithms.* Volume 152.

Aluvalu, R., Kumaran, V.N., Thirumalaisamy, M., Basheer, S., Ali Aldhahri, E., and Selvarajan, S. (2023). Efficient data transmission on wireless communication through a privacy-enhanced blockchain process. *PeerJ. Computer Science*, 9, e1308. doi:10.7717/peerj-cs.1308

Bannore, A., Gore, T., Raut, A., and Talele, K. (2021, November 10). Mental stress detection using machine learning algorithm. In *International Conference on Electrical, Computer, Communications and Mechatronics Engineering (ICE CCME)*.

Ciabattoni, L., Ferracuti, F., Longhi, S., Pepa, L., Romeo, L., and Verdini, F. (2017). Real-time mental stress detection based on smartwatch. In *Proceedings of the IEEE International Conference on Consumer Electronics (ICCE)*, 2017, pp. 110–111.

de Santos Sierra, A., Sanchez Avila, C., Guerra Casanova, J., and Bailador del Pozo, G. (2020). Sentiment analysis using deep learning. *Approach Journal on Artificial Intelligence*, 2(1), 17–27. doi:10.13140/RG.2.2.36460.39048

Dirash, A.R., and Bargavi, S.K.M. (n.d.). LSTM based sentiment analysis. *International Journal of Trend in Scientific Research and Development (IJTSRD)*, 5(4). doi:10.31142/ijtsrd23131

Elzeiny, S., and Qaraqe, M. (2018). Machine learning approaches to automatic stress detection. *2018 IE/ACS 15th International Conference on Computer Systems and Applications (AICCSA)*.

Giannakakis, G., Grigoriadis, D., Giannakaki, K., Simantiraki, O., and Barreto, A. (2022). Review on psychological stress detection using biosignals. *IEEE Transactions on Affective Computing*, 13(1), 17–27. doi:10.1109/TAFFC.2021.3076360

Greene, S., Thapliyal, H., and Caban-Holt, A. (2016). A survey of affective computing for stress detection: Evaluating technologies in stress detection for better health. *IEEE Consumer Electronics Magazine*, 5(4), 20–29. doi:10.1109/MCE.2016.2603950

Hazarika, D., Konwar, G., Deb, S., and Bora, D. J. (n.d.). Sentiment analysis on Twitter by using TextBlob for natural language processing. In *Proceedings of the International Conference on Research*, Vol. 24.

Magaña, V.C., and Muñoz-Organero, M. (2015). Reducing stress on habitual journeys. In *Proceedings of the 5th IEEE International Conference on Consumer Electronics – Berlin (ICCE-Berlin)*, 2015, pp. 153–157.

Mahaveerakannan, R., Anitha, C., Thomas, A. K., Rajan, S., Muthukumar, T., and Govinda Rajulu, G. (2023). An IoT based forest fire detection system using integration of cat swarm with LSTM model. *Computer Communications*, 211, 37–45. doi:10.1016/j.comcom.2022.07.027

McCarley, R.W. (2007). Neurobiology of REM and NREM sleep. *Sleep Medicine*, 8(4), 302–330. doi:10.1016/j.sleep.2007.03.005

Muñoz, S., and Iglesias, C. A. (2022). A Text Classification Approach to Detect Psychological Stress Combining a Lexicon-based Feature Framework with Distributional Representations. Volume 59, Issue 5, September 2022.

Panicker, S. S., and Gayathri. (2019). A Survey of Machine Learning Techniques in Physiology-Based Mental Stress Detection Systems. Volume 39, Issue 2, April–June 2019.

Ramana, K., Revathi, A., Gayathri, A., Jhaveri, R. H., Narayana, C. V. L., and Kumar, B. N. (2022). WOGRU-IDS — An intelligent intrusion detection system for IoT assisted wireless sensor networks. *Computer Communications*, 196, 195–206.

Vellela, S.S., and Balamanigandan, R. (2023). An intelligent sleep-awake energy management system for wireless sensor network. *Peer-to-Peer Networking and Applications*, 16(6), 2714–2731. doi:10.1007/s12083-022-01244-6

Wijsman, J., Grundlehner, B., Liu, H., Hermens, H., and Penders, J. (2011). Towards mental stress detection using wearable physiological sensors. *IEEE Engineering in Medicine and Biology Society*.

Zhai, J., and Barreto, A. (2006). Stress detection in computer users based on digital signal processing of noninvasive physiological variables. *2006 International Conference of the IEEE Engineering in Medicine and Biology Society*.

*Hybrid and Advanced Technologies – S. Prasad Jones Christydass, Nurhayati Nurhayati and S. Kannadhasan. (Eds)* © *2025 the Authors, ISBN 978-1-032-90254-8*


# Machine learning algorithms for vehicle insurance prediction with Random forest and Bayesian approach

Dindu Gopikrishna
*Research Scholar, Department of Computer Science Engineering, Saveetha School of Engineering, Saveetha University, Chennai, Tamil Nadu, India*

P. Gururama Senthilvel
*Project Guide, Department of Computer Science Engineering, Saveetha School of Engineering, Saveetha University, Chennai, Tamil Nadu, India*

ABSTRACT: The research intends to improve the Machine Learning Algorithms for vehicle insurance prediction: A Bayesian Approach. In order to increase accuracy, two groups in this study compared Random Forest to Naive Bayes. to increase Precision. For the research project, 100 dataset samples were employed, of which 80% were used for training and the remaining 20% for testing. When estimating automobile insurance, 10 N sample sizes were used for each estimate. The Random Forest method enhances accuracy data with 93.605% accuracy compared to Naive Bayes' 82.891% accuracy. P is statistically significant for the prediction of auto insurance, with a significant value of p = e of p = 0.002. The Random Forest method for prediction of vehicle insurance significant improvement over Naive Bayes because of its higher accuracy

*Keywords*: Vehicle insurance, Bayesian Approach, Random Forest, Machine Learning, Neural Network, Accuracy

## 1 INTRODUCTION

They in reality used Bayesian networks to indicate claims or no claims, and additionally device mastering abilities to expect fraud and calculate the quantity of insurance premiums for exclusive clients based totally on their specific data; Three classifiers have been used to expect fraudulent claims: Kowshalya and Nandhini (2020), Current challenges inside the coverage enterprise revolve around the adoption of digital applications and using Machine Learning (ML) era. There are two corporations inside the insurance industry: existence insurance and non-existence insurance. This observation examines non-lifestyles coverage, a type of car coverage. In the occasion of an accident, the automobile owner turns over to the insurance employer. Ruixing Ming(2019). We created a composite view of verified frauds from the composite playing cards and evaluated the effect of every characteristic on the worldwide classification overall performance. The consequences of this have a look at can help risk analysts and specialists examine the strengths and weaknesses of any model and broaden empirically useful choice policies for evaluating destiny inputs. Matheus Kempa Severino(2021) There are numerous study gaps which can be used to predict automobile coverage. Due to the heterogeneity of the statistics set, the previous algorithms couldn't be identified. So the primary cause of vehicle coverage is to deceive and defraud unsuspecting users who are looking to be cautious and protect themselves. The goal of the observer is to expect car coverage, the use of random forests and easy Bayes.

 DOI: 10.1201/9781003559115-82

## 2 MATERIALS AND METHODS

The proposed studies work was executed at the Open Source Laboratory of Saveetha School of Engineering, Saveetha Institute of Medical Sciences and Engineering, Savita University, Chennai. The research organization includes businesses of classification algorithms. Random Forest is included in Group 1 and Naive Bayes is protected in Group 2. In this observation, Random Forest of organizations changed into Naive Bayes to improve accuracy. Get your accuracy proper. A pattern of 100 datasets became used for clinical research, of which 80% had been used for training and the closing 20% had been used for checking out. To achieve the expected car coverage insurance, the use of a pattern length of 10 N for every car, records from the facts set were carefully in comparison to ensure that the weather information becomes applicable to the problem handy. A pattern of 100 datasets were used for clinical research, of which 80% had been used for education and the ultimate 20% have been used for checking out. The statistics set became pre-divided into schooling and checking out, with 80% of the data allotted for education and 20% for checking out. The dataset has no missing values, making it suitable for trying out numerous system getting to know techniques.

### 2.1 *Random forest*

One well-known system learning set of rules that is part of the supervised learning approach is Random Forest. In system learning, it can be used to solve regression and classification issues. It is predicated on the idea of ensemble learning, which is the process of making a few classifications to simplify a complex issue and enhance version performance.

*Algorithm*
**Step 1:** Collect the information needed to build the random forest first.
**Step 2:** Identify the problem (binary classification, in this case) that you want to use the random forest to address.
**Step 3:** Separate the data into training and testing sets to assess the random forest model's performance.
**Step 4:** After building the random forest, evaluate the model's performance by looking at the test set.
**Step 5:** You can improve the results by adjusting the hyperparameters of the random forest model if it is not performing up to par.
**Step 6:** After you are satisfied with the random forest models' performance, apply them to forecast fresh, unexplored data.

### 2.2 *Naive Bayes*

SIMPLE Bayes - one of the maximum powerful gear for coaching algorithms, used both for class problems and for regression problems. It creates a tree-like structure, just like a trunk diagram, in which every inner node represents an examined attribute, each branch represents the end result of a take a look at, and every leaf node (aspect node) consists of a type tag.

*Algorithm*
**Step 1:** Establish the objective variable first, then list the relevant attributes that could affect the result.
**Step 2:** Assure consistency by standardising the data and making sure every component follows the same structure.
**Step 3:** Start with tiny, random values for the model's weights and biases.
**Step 4:** For each real number, create a cost function and a sigmoid function that map to a value between 0 and 1.

**Step 5:** Use metrics like F1 score, accuracy, precision, and recall to assess the model's performance.
**Step 6:** Adjust the model's parameters, such as the number of iterations, regularisation strength, and learning rate, to improve performance..
**Step 7:** Make predictions primarily based on the new records the usage of a trained logistic regression model to categorize job ads as true or false.

### 2.3 *Statistical analysis*

Statistical evaluation is performed using the Naive Bayes approach and the use of SPSS software for statistical analysis. IBM SPSS version 26 became used, with a maximum of 10 iterations of every method for IP addresses, geolocation, historic information, device records, user conduct, data and social media hobby. Car insurance

## 3 RESULTS

The proposed random woodland and simple Bayes algorithms have been examined on exclusive samples of Anaconda Navigator with a sample length of 10 individuals.

Table 1. Accuracy value of random forest in the fraud detection is (93.605%).

| Iterations | Accuracy (%) |
|---|---|
| 1 | 93.21 |
| 2 | 92.61 |
| 3 | 90.66 |
| 4 | 90.13 |
| 5 | 91.06 |
| 6 | 92.77 |
| 7 | 95.13 |
| 8 | 92.50 |
| 9 | 90.45 |
| 10 | 92.55 |
| **Accuracy** | **93.60** |

Table 2. Deaf Bayes accuracy cost in automobile insurance analysis (82.891%).

| Iterations | Accuracy (%) |
|---|---|
| 1 | 83.43 |
| 2 | 81.10 |
| 3 | 83.82 |
| 4 | 81.53 |
| 5 | 82.90 |
| 6 | 80.61 |
| 7 | 83.36 |
| 8 | 82.86 |
| 9 | 83.24 |
| 10 | 86.06 |
| **Accuracy** | **82.89** |

Table 3 Lists the accuracy estimates of the random leap and blind Bayes algorithms. Random forests and easy Bayes means the usage of three popular errors. 1.55589,0.49201 and 1.55642,0.49218 respectively.

Table 3. Panel statistical analysis of random forests and simple bayes. The mean, popular deviation, and general mistakes of the imply are obtained for 10 samples.

| | Group | N | Mean | Std. Deviation | Std. Error Mean |
|---|---|---|---|---|---|
| **Accuracy** | Random Forest | 10 | 93.605 | 1.55589 | 0.49201 |
| | Naive Bayes | 10 | 82.891 | 1.55642 | 0.49218 |

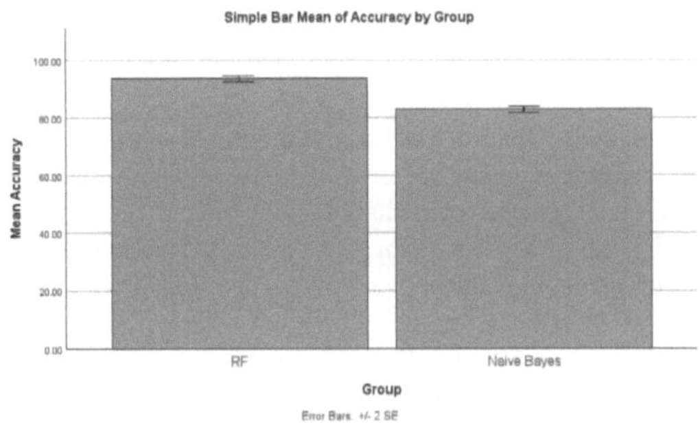


Figure 1. X axis: Random Forest Vs. SIMPLE Bayes classifier and Y-axis: average accuracy within +/- 2 SD.

Random Forest set of rules to calculate suggest, well known blunders, preferred deviation and price (93.605%). 1.47604, 0.46676 they saved. Auto Insurance User Analysis. A pattern size of X N, a G price of 82%, a substantial P value of 0.002 (p < zero.05) have been statistically widespread to detect Z. Naive Bayes, i.E. Random woodland, unique classifier algorithms and benchmarks are presented in the graph. It suggests a similar (93.605%) class accuracy. From (82.891%). Naive Bayes, the random woodland algorithm is extensively extra correct.

## 4 DISCUSSION

Insurance policies are an essential hassle for insurance companies. Insurers constantly face the problem of growing losses from insurance activities. Because the figures of fraud claims and claims towards coverage groups are growing. As a result, it's far harder to tell the fame of a safety request throughout the technique. Therefore, the cause of this changed into a gadget mastering model that predicts and shows the country of vehicle insurance using a system learning approach. Endalew Alamir(2020) We created a composite view of tested frauds from the composite cards and evaluated the impact of every feature on the global class overall performance. The outcomes of this look can help chance analysts and specialists assess the strengths and weaknesses of any model and expand empirically useful selection guidelines for evaluating destiny inputs. Matheus Kempa Severino(2021) One of the main packages of device mastering in enterprise and economic control is fraud detection, a topic that is very relevant from the factor of view of selection makers, choice support systems that directly help hazard analysts expect fraud. Financial indicators of the corporation. In this sense, this topic has been studied by several researchers in recent years and mentioned in articles which include this one. (Awoyemi *et al.* 2019) Over the years, machine studying (ML) competencies are increasingly being claimed for analysis. However, choosing premier techniques, characteristic selection strategies, clustering separation methods, resuming algorithms, and ML classifiers to guide insurance selections can be complex and affect the demand for recommendations. (Brits 2022)

## 5 CONCLUSION

The essential purpose of these paintings is to assess how well Naive Bayes performs in the feature category using a random woodland category, a machine gaining knowledge of methods to the analysis of identification insurance and vice versa. The accuracy of the Random Forest classifier is 93.605%. The most correct method is Naive Bayes 82.891%. Only the Random Forest classifier (93.605%) is anticipated to perform the Naive Bayes approach (82.891%). This changed into done so one can choose the dataset in this take a look at.

## REFERENCES


Bouzgarne Itri, Youssfi Mohamed, Qbadou Mohammed and Bouattane Omar. (2019, October 30). Performance comparative study of machine learning algorithms for automobile insurance fraud detection. *IEEE Xplore*. https://ieeexplore.ieee.org/document/8942277

Charles Muranda, Ahmed Ali and Thokozani Shongwe. (2020, October 16). *Detecting Fraudulent Motor Insurance Claims Using Support Vector Machines with Adaptive Synthetic Sampling Method*. Available online: [Online]

de Deijn, Jeffrey. (2018). *Automatic Car Damage Recognition using Convolutional Neural Networks.*

Harshani, WA Rukshala, and Kaneeka Vidanage. (2017). Image processing based severity and cost prediction of damages in the vehicle body: A computational intelligence approach. In *2017 National Information Technology Conference (NITC)*. IEEE. https://doi.org/10.1109/NITC.2017.00010

He, Kaiming, Zhang, X., Ren, S., and Sun, J. (2016). Deep residual learning for image recognition. In *Proceedings of the IEEE Conference on Computer Vision and Pattern Recognition.*

Huang, Gao, Liu, Z., van der Maaten, L., and Weinberger, K. Q. (2017). Densely connected convolutional networks. In *Proceedings of the IEEE conference on computer vision and pattern recognition.*

Hutter, Marcus, Gould, Stephen, Hartley, Richard, and Li, Hongdong. (2013). *Image Based Automatic Vehicle Damage Detection.*

Jayawardena, Srimal. (2013). *Image based Automatic Vehicle Damage Detection.*

Kinsella, Graham. (2017). *Car Damage Monitoring System: Final Project Report Analysis and Design*. Dublin, National College of Ireland.

Li, Ying, and Dorai, Chitra. (2007). Applying image analysis to auto insurance triage: A novel application. In *2007 IEEE 9th Workshop on Multimedia Signal Processing.*

Manju, More E. (2018). Research paper on intelligent farming for farmers using control systems in IOT. *International Journal of Advanced Research in Computer Science*, *9*(3), 181.

Parvez M, Muzammil, Shailendra Kumar Mishra, K. Nandini, Sk Hasane Ahammad, Syed Inthiyaz, Baraa Riyadh Altahan, Lassaad K. Smirani, Md Amzad Hossain, and Ahmed Nabih Zaki Rashed. (2023). Machine learning based models for defect detection in composites inspected by Barker coded thermography: A qualitative analysis. *Advances in Engineering Software*, *178*, 103425. https://doi.org/10.1016/j.advengsoft.2022.103425

Patil, Kalpesh, et al. (2017). Deep learning based car damage classification. In *2017 16th IEEE International Conference on Machine Learning and Applications (ICMLA)*. IEEE. https://doi.org/10.1109/ICMLA.2017.0-1

Ramesh, D. R. P. S., Sudha, D. R. I., Satheesh, D. R. N., Rajulu, D. R. G. GOVINDA, and Kalaiselvan, S. A. (n.d.). *Prediction of Auto-detection for Tracking of Sub-nano Scale Particle in 2d and 3d using SVM-based Deep Learning*. http://www.jatit.org/volumes/Vol101No5/30Vol101No5.pdf

Sanket Doshi. (2019). *Traffic Sign Detection using Convolutional Neural Network.*

Sarkar, Biswajit, Kar, Sumi, Basu, Kajla, and Guchhait, Rekha. (2022). A sustainable managerial decision-making problem for a substitutable product in a dual-channel under carbon tax policy. *Computers & Industrial Engineering*, *172*, 108635. https://doi.org/10.1016/j.cie.2022.108635

Sharma, Nidhi, and Banga, V. K. (2010). Drowsiness warning system using artificial intelligence. *World Academy of Science, Engineering and Technology*, *4*(7), 1771–1773.

Sonakshi Harjai, Sunil Kumar Khatri and Gurinder Singh. (2019, November). Detecting fraudulent insurance claims using random forests and synthetic minority oversampling technique. *IEEE Xplore*. https://ieeexplore.ieee.org/document/9036162

Sun, Zehang, Bebis, G., and Miller, R. (2006). On-road vehicle detection: A review. *IEEE Transactions on Pattern Analysis and Machine Intelligence*, *28*(5), 694–711.

*Hybrid and Advanced Technologies – S. Prasad Jones Christydass, Nurhayati Nurhayati and S. Kannadhasan. (Eds)* © *2025 the Authors, ISBN 978-1-032-90254-8*


# An analysis of image reconstruction for damaged images using CNN in comparison with LSTM

O. Roshith
*Research Scholar, Department of Computer Science Engineering, Saveetha School of Engineering, Saveetha University, Chennai, Tamil Nadu, India*

G. Michael
*Project Guide, Department of Computer Science Engineering, Saveetha School of Engineering, Saveetha University, Chennai, Tamil Nadu, India*

ABSTRACT: The study aims at image reconstruction of old damaged spool images using CNN versus LSTM. Two groups were compared in this study. LSTM was used to compare CNNs. For this study, a total of 1284 sample datasets were collected, of which 1200 were used for training and the remaining 284 for testing. Image reconstruction for old damaged coil images was estimated using a sample size of 10N per image, a threshold of 0.05%, and a confidence interval of 95%. CNN classifier improves data accuracy with 94.409% accuracy compared to 85.263% of LSTM. Significant price $p = 0.002$ ($p<0.05$) statistically significant in finding image reconstruction for damaged old roll images.CNN prediction method recovers old corrupted images on LSTM due to higher accuracy.

*Keywords*: Image Regeneration, Reel Picture, Old Image, CNN, LSTM

## 1 INTRODUCTION

Content-based total photo seek involves looking at an image through a big database to reap a picture of the favored form, texture, or length. Content-based total photo retrieval is evaluated on parameters inclusive of pace and efficiency, and the changed approach is provided based on the histogram of the composite color photo processing. This paper affords a brief look at and analysis of the Multiclass CNN classifier and its contrast with the LSTM classifier in an application to gain the right output photo from a database with appreciation to an photo query. The proposed approach is time-saving and gives good effects. By the use of a pool of such classifiers, we are able to create a CBIR machine including more than one system that offers complementary outputs. By combining these cursors with every different we will get the quality suit for the query image. (Analysis Of Knn And Svm Classifier For Image Classification Mr. Vishal B. Padole1, Dr. V. R. Mankar2. DOI: ("Website," n.d.).

Research has been directed to distinguish false enlisting administration during the most recent couple of years. Over 510 distributions of IEEE Xplore got over 757 research papers that were submitted to research Researcher & amp; #39; scholastic information base. This paper provides the usage of genetic algorithms to decorate and repair photographs of old documents. The word "vintage file" method a document that is going lower back loads of years. The motive of the observer is to hold the information contained in historical documents in virtual shape, for the reason that the manner of rescuing historical pics using bodily get entry is just too gradual. The important contribution of this take a look at is to increase the perfect function to pick out a new populace and the best new and better picture. The experimental results also show the effect of using a mean filter at the preprocessing image.

The effects display that the photograph enhancement is tremendously effective (ninety two. Nine% of statistics acquisition achievement price extra than 90%), and the general impact of using the median preprocessing clear out is not excellent, ensuing in a frightening impact on the snap shots (handiest fifty nine.Five% of the facts rate to achieve over 90% fulfillment). Image Enhancement and Image Restoration for Old Document Image Using Genetic Algorithm (Hilda Deborah; Aniati Murni Arymurthy 2010) ("Sci-Hub" n.d.)

## 2 MATERIALS AND METHODS

The proposed study was carried out at the Computational Intelligence Laboratory of the Saveetha School of Engineering, Saveetha Institute of Medical and Technical Sciences. There were two groups employed in the analysis. Group 1 corresponds to CNN Classifier and Group 2 as LSTM classifier. There are 10 Examples in every group and the sample size is calculated based on the literature ("Dynamic Weighting Multi Factor Stock Selection Strategy Based on XGboost Machine Learning Algorithm" n.d.)(Xu *et al.* 2022)("Dynamic Weighting Multi Factor Stock Selection Strategy Based on XGboost Machine Learning Algorithm" n.d.) by considering GPower pretest Eighty%, threshold 0.05, self-assurance c programming language 95%. Google CoLab is used to aid with the design and implementation of marketing strategy The Windows 11 working gadget served as a check case for system learning fashions. The hardware consists of an Eighth era Intel Core i7 processor and eight GB RAM. A sixty four-bit gadget ought to be used. He selected Python as the programming language to implement the code. The records set runs in the heritage whilst the code is strolling to make certain correct output. The semiconductor yield dataset is used for whether the Sustainable production of semiconductor can be yielded from the given sensor values or not. The Kaggle. com site's dataset was gathered there. This dataset contains 19 features, which includes "date and time", "sensor", "sensor II", and many others. (Quinonero-Candela *et al.* 2022).

ALGORITHM

### 2.1 *CNN*

Convolutional Neural Network (CNN) is a sort of deep getting to know set of rules perfect for popular images and processing obligations. It consists of numerous layers, such as layers of convolutions, layered layers, and completely related layers. A convolutional neural community (CNN or ConvNet) is a deep gaining knowledge of community structure for studying structures without delay from facts. CNNs are maximum beneficial for finding styles in photos for item, type, and class popularity. They can also be very powerful for reporting audio, time series, and signal records. Where is the CNN set of rules used?

A convolutional neural community is a synthetic form of deep gaining knowledge of the neural community. It is utilized in pc vision and picture popularity.

The ConvNet architecture It is just like the activation of neurons in the human mind, and is stimulated with the aid of the hobby of the visible cortex. Individual neurons respond pleasant to stimuli in a constrained area. of regard that they understand as a receptive discipline. A set of such fields covers the complete visible vicinity.

Step 1: Determine the target variable and choose the pertinent features that may affect the classification.
Step 2: To make sure that every characteristic is on the same scale, standardise the data.
Step 3: Initialise the weights and bias of the model parameters to either zero or tiny random values.
Step 4: Determine the cost and sigmoid a formula that converts a real wide range of values between 0 and 1.
Step 5: Compute the F 1 score, metre accuracy, accuracy, and accuracy inclusive to assess the model's performance on the testing set.

Step 6: To enhance the model's performance, modify the hyperparameters, which include learning rate, regularisation force, and variety of iterations.
Step 7: Utilise the CNN model that has been trained.

### 2.2 *LSTM*

Long Short-term reminiscence networks are sequences of deep learning neural networks that permit statistics to be stored. It is a special type of recurrent neural community that is capable of solving vanishing gradient problems that RNNs face. Rhoncus became advanced by way of Hochreiter and Schmidhuber and solves the problem of conventional RNNs and gadget learning algorithms. The kernel may be carried out in Python using the Keras library.

Let's say when you watch a video you keep in mind an earlier scene, or while you study a book what happened within the previous chapter. Similarly, RNNs; they recollect previous statistics and use it to process modern-day input. The disadvantage of RNNs is that they can't recall the lengthy-time period dependence due to the vanishing step. Snoring especially to avoid long-term dependancy problems.

Step 1: Getting the data you wish to utilise to train the CNN is the first step.
Step 2: Next, you must specify the issue that you wish to use CNN to solve. The specific problem is one of binary classification.
Step 3: Divide the data into training and test sets in order to assess the CNN's performance.
Step 4: Once the CNN model has been trained, you may assess how well it performs using the test set.
Step 5: You might need to adjust the hyperparameters to get better outcomes if the CNN model's performance isn't up to grade.
Step 6: When the CNN model's performance meets your expectations, you can deploy it to generate predictions on new, unseen data.

### 2.3 *Statistical analysis*

Statistical analysis was done using the usage of IBM SPSS model 26 for each suggested and current algorithm. The projected accuracy for every algorithm repetition was recorded in order to compare accuracy with the fee obtained due to the iterations. A right-pattern t-take a look at was used. (Akpan *et al.* 2023). The dependent variable is the ("Pass/Fail") column from the set and the independent variables are the sensor values in the dataset.

## 3 RESULTS

In Google Colab, the suggested CNN classifier method and LSTM classifier were each performed separately with a pattern size of 10

Table 1. Accuracy and loss analysis of CNN & LSTM.

| S. No | Group | Accuracy (%) | Loss (%) |
|---|---|---|---|
| 1 | CNN | 93.21 | 6.79 |
| | | 92.61 | 7.39 |
| | | 89.66 | 10.34 |
| | | 90.13 | 9.88 |
| | | 91.06 | 8.94 |
| | | 92.77 | 7.23 |
| | | 95.13 | 4.87 |
| | | 92.50 | 7.5 |
| | | 89.45 | 10.55 |

(*continued*)

Table 1. Continued

| S. No | Group | Accuracy (%) | Loss (%) |
|---|---|---|---|
| | | 92.55 | 7045 |
| 2 | LSTM | 82.73<br>81.20<br>82.72<br>83.63<br>84.90<br>82.61<br>88.76<br>84.86<br>89.26<br>83.16 | 17.27<br>18.8<br>17.28<br>16.37<br>15.1<br>17.39<br>11.24<br>15.14<br>10.74<br>16.84 |

Table 2. Shows the total number of samples, Average accuracy, general mistakes, and trendy mistakes for the new CNs and LSTMs. The class imply, standard deviation, and trendy error of the suggest are ninety one.907, 1.7044, and zero.53897, respectively.

| | Group | N | Mean | Std. Deviation | Std. Error Mean |
|---|---|---|---|---|---|
| Accuracy | CNN | 10 | 91.907 | 1.7044 | 0.53897 |
| | LSTM | 10 | 84.383 | 2.53522 | 0.801706 |
| | CNN | 10 | 8.093 | 1.7044 | 0.53897 |
| Loss | LSTM | 10 | 15.617 | 2.53522 | 0.801706 |

Figure 1 shows an assessment of the common accuracy and fee among LSTM. The two algorithms, the CNN classifier and the LSTM classifier, are compared and analyzed graphically to show the means of loss. This shows that, with a classed accuracy of 91.907% compared to LSTM classifier classified accuracy of 84.383%, CNN classifiers are much better.

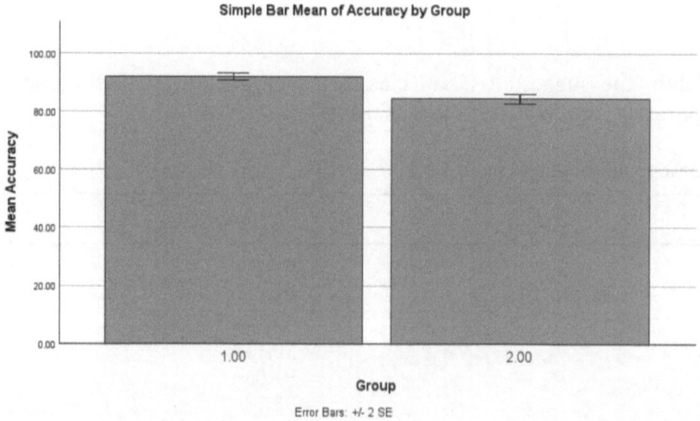


Figure 1. Comparison of CNN and LSTM in terms of common accuracy. CNN is higher than LSTM accuracy Classifier; the gain of CNN is barely better than LSTM. X Axis: CNN Vs. LSTM classifier and Y-axis: common accuracy within +/- 2 SD.

## 4 DISCUSSION

In this research, the CNN Classifier excelled over the LSTM Classifier in terms of accuracy and error difference. CNN Classifier has obtained the mean accuracy of 91.907%, mean Loss of 8.093% and for Random Forest Classifier the mean Accuracy is 84.383%, mean Loss is 15.617%.The accuracy of the two approaches differs significantly, with significance levels of 0.002 (Independent pattern t check, $p < 0.05$) indicating a tremendous distinction. When study's inadequacies include that on fragmented and unstructured data, CNN does not quite perform as well. Since every classifier is compelled to correct the mistakes made by the previous learners, CNN is extremely susceptible to outliers. Overall, the approach is hardly scalable. According to future research targets, the device could be able to procedure information from the dataset with minimal time intake.

## 5 CONCLUSION

The LSTM model and the CNN model are class methods used in this paintings to regenerate the old damaged coil image. The CNN classifier appears to be more correct than the LSTM classifier. The studies show that the LSTM classifier (91.907%) outperforms the LSTM classifier (84.383%) in photograph regeneration for old corrupted reel photos.

## REFERENCES


Akpan, U., Ezukwa, O., Arogundade, K., Asibong, U., Nwagbata, A., Akpanika, C., and Etuk, S. (2023). Infant-feeding practices among women living with human immunodeficiency virus (HIV) in a Southern Nigerian region: A mixed comparative study. *Cureus, 15*(2), e35483. https://doi.org/10.7759/cureus.35483

Anwar, S., Barnes, N., and Petersson, L. (2021). Attention-based real image restoration. *IEEE Transactions on Neural Networks and Learning Systems.* Advance online publication. https://doi.org/10.1109/TNNLS.2021.3131739

Dynamic WeightingMulti Factor Stock Selection Strategy Based on XGboost Machine Learning Algorithm. (n.d.). Retrieved December 21, 2022, from https://ieeexplore.ieee.org/document/8690416

Gupta, D., Polkowski, Z., Khanna, A., Bhattacharyya, S., and Castillo, O. (Eds.). (2021). *Proceedings of Data Analytics and Management: ICDAM 2021, Volume 2.* Springer Nature.

Joseph, K. J., Rajasegaran, J., Khan, S., Khan, F. S., and Balasubramanian, V. N. (2022). Incremental object detection via meta-learning. *IEEE Transactions on Pattern Analysis and Machine Intelligence, 44*(12), 9209–9216.

Katsaggelos, A. K., Molina, R., and Mateos, J. (2022). *Super resolution of images and video.* Springer Nature.

Li, Y., Meng, J., Yu, Y., Wang, C., and Guan, Z. (2022). Image restoration based on improved generative adversarial networks. In *2022 7th International Conference on Image, Vision and Computing (ICIVC)* (pp. 1–5). IEEE. https://doi.org/10.1109/icivc55077.2022.9886285

Ma, K. L., Möller, T., Pfister, H., Carpendale, S., and Fekete, J. D. (2009). IEEE Visualization Conference and IEEE Information Visualization Conference. Proceedings 2009. Preface. *IEEE Transactions on Visualization and Computer Graphics, 15*(6), xii–xv.

Sci-Hub. (n.d.). Retrieved October 29, 2023a, from https://sci-hub.hkvisa.net/10.1109/ACT.2010.24

Sci-Hub. (n.d.). Retrieved October 29, 2023b, from https://sci-hub.hkvisa.net/10.1109/nrsc.2003.157326

Website. (n.d.). https://doi.org/10.21172/1.114.07e-ISSN:2278-621X

Xu, H. W., Qin, W., and Sun, Y. N. (2022). An improved XGBoost prediction model for multi-batch wafer yield in semiconductor manufacturing. *IFAC-PapersOnLine.* https://doi.org/10.1016/j.ifacol.2022.10.028

Zhai, L., Wang, Y., Cui, S., and Zhou, Y. (2023). A comprehensive review of deep learning-based real-world image restoration. *IEEE Access: Practical Innovations, Open Solutions, 11*, 21049–21067.

Parvez M., M., Mishra, S. K., Nandini, K., Ahammad, S. H., Inthiyaz, S., Altahan, B. R., Smirani, L. K., Hossain, M. A., and Rashed, A. N. Z. (2023). Machine learning based models for defect detection in composites inspected by Barker coded thermography: A qualitative analysis. *Advances in Engineering Software, 178*, 103425. https://doi.org/10.1016/j.advengsoft.2022.103425

Ramesh, D. R. P. S., Sudha, D. R. I., Satheesh, D. R. N., Govinda Rajulu, D. R. G., and Kalaiselvan, S. A. (n.d.). Prediction of auto-detection for tracking of sub-Nano scale particle in 2d and 3d using SVM-based deep learning. Retrieved from http://www.jatit.org/volumes/Vol101No5/30Vol101No5.pdf

Sarkar, B., Kar, S., Basu, K., and Guchhait, R. (2022). A sustainable managerial decision-making problem for a substitutable product in a dual-channel under carbon tax policy. *Computers & Industrial Engineering*, *172*, 108635. https://doi.org/10.1016/j.cie.2022.108635

Surendran, R., Alotaibi, Y., and Subahi, A. F. (2023). Lens-oppositional wild geese optimization based clustering scheme for wireless sensor networks assists real time disaster management. *Computer Systems Science and Engineering*, *46*(1), 835–851.

Tayyab, M., Habib, M. S., Jajja, M. S. S., and Sarkar, B. (2022). Economic assessment of a serial production system with random imperfection and shortages: A step towards sustainability. *Computers & Industrial Engineering*, *171*, 108398. https://doi.org/10.1016/j.cie.2022.108398

*Hybrid and Advanced Technologies – S. Prasad Jones Christydass, Nurhayati Nurhayati and S. Kannadhasan. (Eds) © 2025 the Authors, ISBN 978-1-032-90254-8*
*Open Access: www.taylorfrancis.com, CC BY-NC-ND 4.0 license*

# Integrating structural health monitoring and intelligent transportation systems for bridge condition assessment: A GA-CNN-BiGRU approach

Vishwayogita Amol Savalkar
*Assistant Professor, Bharati Vidyapeeth Deemed to be University, Department of Engineering and Technology, Kharghar, Navi Mumbai, Maharashtra, India*

Vandana Salve
*Assistant Professor, Department of Computer Engineering, K.J. Somaiya Institute of Technology, Sion, Mumbai, India*

T. Karthikeyan
*Department of Biomedical Engineering, Sona College of Technology, Salem, Tamilnadu, India*

Joel Josephson Pottipadu
*Associate Professor, Department of ECE, Malla Reddy Engineering College, Maisammaguda, Medchal, India*

Medha V. Deshpande
*Assistant Professor, Department of Civil Engineering, Dr.D.Y.Patil Institute of Technology, Pimpri, Pune, India*

Gourav Kalra
*Assistant Professor, Maharishi Markandeshwar (Deemed to be University), Mullana, Ambala, India*

ABSTRACT: There must be zero downtime for our nation's bridges since they are vital to our transportation networks. However, due to function modification, increased and faster-moving traffic loads, and capacity overload, many bridges, including Busway accommodation, are currently operating at overloaded levels. In order to ensure that some bridges remain structurally sound, structural health monitoring (SHM) systems are installed to assess the bridges' present condition and forecast when the structures might collapse.To train the model, feature extraction is an important step. Two steps in feature extraction are data reduction and blind separation. The model was trained using the Feature rich RNN-CTC to ensure it was as accurate as possible. While competing methods such as RNN and CTC achieved accuracy rates of 95.38%, the proposed method achieved a significant improvement.

*Keywords*: Structural Health Monitoring (SHM), Bridge Health Monitoring (BHM), Principle Component Analysis (PCA), Independent Component Analysis (ICA), Recurrent Neural Networks

## 1 INTRODUCTION

Contemporary bridges routinely face a broad variety of operational and environmental challenges. These outside forces are detrimental and will certainly speed up the

DOI: 10.1201/9781003559115-84

deterioration of the structure. Furthermore, catastrophic events like earthquakes might happen throughout a bridge's operational life. Consequently, ensuring safety necessitates the prompt identification of structural concerns. Visual inspection played a major role in the past in assessing the condition of buildings and detecting surface defects. However, even with well-trained inspectors, visual inspection is still inefficient, subjective, and time-consuming, making it unable to identify changes in circumstances in real-time. In the last several decades, structural health monitoring (SHM) techniques have gained popularity as a way to address this problem, especially with long-span bridges. Inadequate maintenance, weather-induced rebar corrosion, train and traffic movements, damaged piers, and other causes can cause bridge sections to deteriorate. Due to the work, traffic delays, and bridge closures involved in repairs or replacements, the original construction cost can be doubled.An instance of this may be seen in the shocking finding that the city's railway officials discovered when inspecting a strategically located Sivas overhead traffic bridge not long ago. The bridge failed the basic load tests due to excessive deflection, which was caused by a large amount of corrosion.

## 2 LITERATURE SURVEY

Visual inspections of bridges were once the norm, but they had their limits when it came to resolution, unpredictability, and the capacity to spot obvious problems. [1]The evolution of signal gathering and data transmission technology has given rise to new methods of monitoring and maintaining bridges, one of which is vibration-based Bridge Health Monitoring (BHM). [2]. A trend in BHM techniques away from contact sensors and toward non-contact ones has emerged in recent years as a result of next-gen, inexpensive sensors such as robots, cameras, and drones [3].Typically, a multitude of instruments and actuators spread out throughout the bridge are required for conventional monitoring in the vast majority of BHM investigations[4]. By combining deflection data from the road with weigh-in-motion data from the vehicle's two wheels, researchers were able to pinpoint damaged sections of the bridge.[5]In order to locate damage, one could look for changes in stiffness brought on by factors such as high vehicle speeds, uneven profiles, and noisy data. Decision analysis based on posterior, prior, and preposterior principles well-established in Bayesian decision theory formed the basis of the aforementioned paradigm[6]. The two methods are based on the same foundation: the linear utility theory originally proposed by von Neumann and Morgenstern and the expected value requirement for decision making[7]. The methods described here are broad strokes when applied to the problem of assessing bridge health with discrete state variables; for example, see [8] for a broad strokes descriptionin the continuous realm for making decisions. For the purpose of clarification, decision trees and influence diagrams are also included. Three or five factors are taken into account by the choice analysis methodologies that were previously mentioned[9]. The choice to reinforce, repair, or replace the structure relies on the bridge's state, which might range from broken to worthless. Previous work has attempted to model and predict the heavy traffic LEs of bridges. To predict future severe LEs, the studies used the normal probability idea to fit the maximum traffic LE extremes to a normal distribution[10]. [11]The most severe traffic LEs have been predicted by scientists utilizing the Rice formula-based level crossing technique. [12]among the many scholars that have attempted to predict LEs caused by extremely heavy traffic using extreme value analysis methods such as peak-over-threshold or block maxima. On the other hand, research has been conducted by [13] outlined all the possibilities for bridge traffic LE projections.

## 3 PROPOSED SYSTEM

The fundamental objective of this research is to provide a standard method of assessing bridge durability. In order to monitor the development of damage, the system needs to integrate the findings of real-world visual inspections with more accurate degradation models and a number of trustworthy sensors strategically distributed throughout the structure.

### 3.1 *Feature extraction*

#### 3.1.1 *Data reduction*

In order to minimize the size of the final intrinsic mode functions, principle component analysis (PCA) is employed to decipher[14]. The altered new variables are described by their key components. Here are a few things that set those primary components apart:The main components are all independent and perpendicular to one another; they are also organized in a precise order, with the first component having the most variance and the last component having the least).

#### 3.1.2 *Blind seperation*

The inability to track the source signals—like the building's reaction to changes in wind speed, traffic volume, or temperature and the lack of understanding of the mixed signal are essential features of blind separation. The blind separation problem is solved using independent component analysis (ICA), which is considered the gold standard.Matrix $A$, which represents the set of observations in Equation 1, can be used to illustrate ICA. In Equation 1, the transposition of matrix $A$ is illustrated, with the sample index $h$equalling $1, 2, \ldots, b_h$.

$$A^h = \begin{bmatrix} A_1(h_1) & \ldots & A_1(h_{b_h}) \\ \vdots & \ddots & \vdots \\ A_{b_a}(h_1) & \ldots & A_{b_a}(h_{b_h}) \end{bmatrix} \tag{1}$$

#### 3.1.2 *Residual network-RNN-CTC model training*

3.1.2.1 *RNN* Among the many types of neural network models, recurrent neural networks are characterized by a directed cycle of connections between neurons. The vanishing gradient problem, however, severely restricts the variety of context that may be practically accessed by conventional RNN[15]. An RNN with an imposed memory structure, often known as an LSTM cell, is one such option. LSTM-RNNs are able to efficiently train to resolve long-term dependency issues, and they also manage to circumvent some of the foundational issues with standard RNNs. In recent times, LSTM has emerged as a prominent RNN.

3.1.2.2 *CTC* Connectionist temporal categorization is one kind of output layer. It is mostly useful for two things. One of them is figuring out the loss, and the other is understanding RNN's output[16]. The inclusion of a new label called blank creates a new set $S' = S \cup \{blank\}$. The input sequence $w = w_1, \ldots w_H$is accepted by CTC, where $H$ is the sequence length. The corresponding label that provides is $G$ above $\mathfrak{R}$.

$$r(G|, w) = \sum_{\pi \in \mathfrak{R}^{-1}(G)} r(\pi|, w) \tag{2}$$

being defined as the conditional probability of $\pi$:

$$r(\pi|, w) = \prod_{h=1}^{H} w_{\pi_h}^{h} \tag{3}$$

A label's likelihood of being present at timestep $h$ is denoted as $w_{\pi_h}^{h}$. Directly computing Equation 2 is not feasible. The CTC object function $\mathcal{O}(A)$is defined as the negative logarithm of the ground truth probability for all cases for each training set $A$.

$$\mathcal{O}(A) = -\sum_{(q,g)\in A} logr(G|, w) \tag{4}$$

## 4 RESULT AND DISCUSSION

The most common way to assess the condition of bridges is through visual inspections. Every two years, bridges should be checked for damage. This may be useful for noncritical, physically acceptable structures, but it is not reliable for determining a building's actual health.

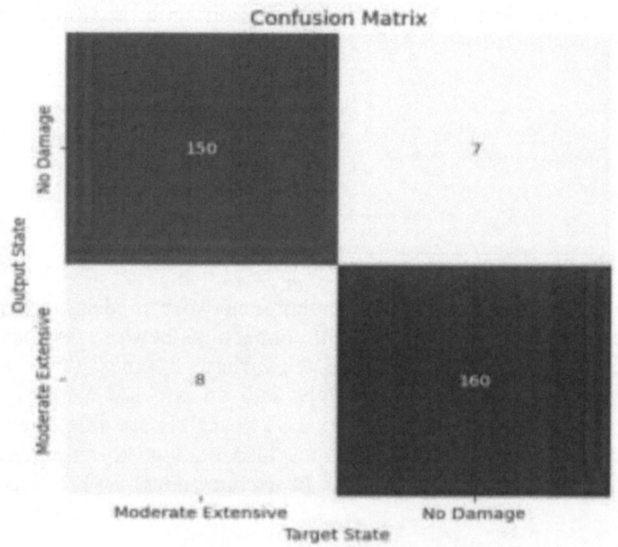


Figure 1. Confusion matrix for residual network-RNN-CTC model.

Figure 1 shows the result of the Residual Network-RNN-CTC confusion matrix, which showed that a condition with substantial damage was incorrectly classified as moderate damage.

Figure 2 displays the merged network's performance. It displays the accuracy for both training and validation.

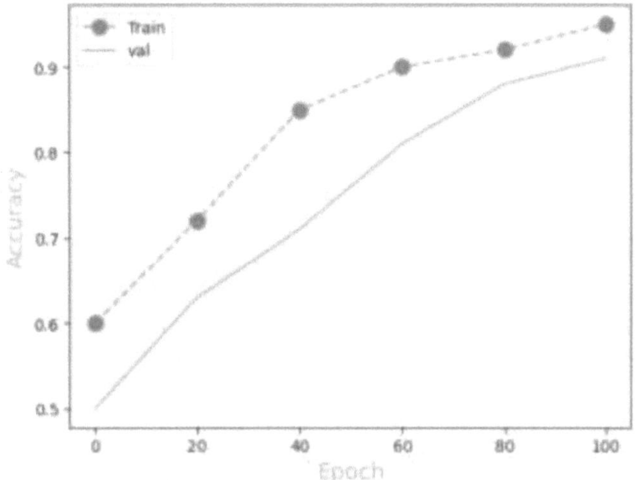


Figure 2. Training and validation accuracy.

Figure 3 illustrates that the loss curves exhibit saturation as the number of iterations grows following 20 epochs of training the Residual Network-RNN-CTC model.

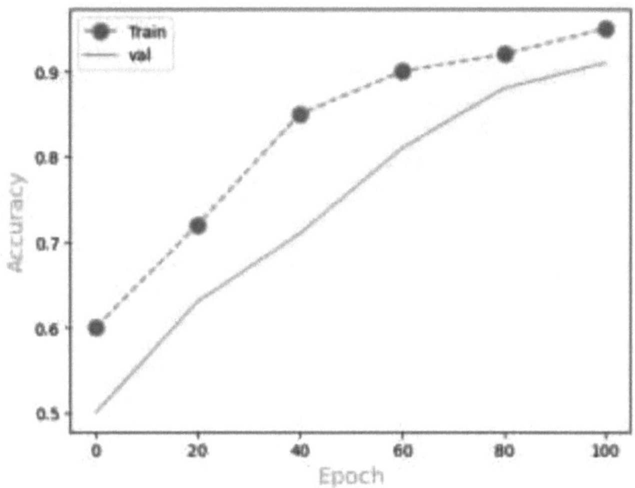


Figure 3. Training loss of residual network-RNN-CTC.

## 5 CONCLUSION

A growing number of small and medium span bridges are putting a strain on society's resources for upkeep and repair. The foundation of this approach to bridge management is developing a method for assessing safety (conditions) using criteria like remaining life and load bearing capacity. When compared to more traditional methods, bridge health monitoring that makes use of sensors and information technology provides more accurate insights into the performance of bridges.Data reduction and blind separation are the two stages of feature extraction. The model is trained using feature-rich Residual Network-RNN-CTC. With a consistency rate of 95.38%, the suggested technique surpasses both the RNN and CTC models.

## REFERENCES


[1] An, Y., Chatzi, E., Sim, S. H., Laflamme, S., Blachowski, B., and Ou, J. (2019). Recent progress and future trends on damage identification methods for bridge structures. *Structural Control and Health Monitoring*, 26(10), e2416.

[2] Dertimanis, V., and Chatzi, E. (2020). Sensor networks in structural health monitoring: From theory to practice. *Journal of Sensor and Actuator Networks*, 9(4), 47.

[3] Kaloop, M. R., Kim, K. H., Elbeltagi, E., Jin, X., and Hu, J. W. (2020). Service-life evaluation of existing bridges subjected to static and moving trucks using structural health monitoring system: case study. *KSCE Journal of Civil Engineering*, 24(5), 1593–1606.

[4] Meng, D., Xiao, F., Zhang, L., Xu, X., Chen, G. S., Zatar, W., and Hulsey, J. L. (2019). Nonlinear vibration analysis of vehicle–bridge interaction for condition monitoring. *Journal of Low Frequency Noise, Vibration and Active Control*, 38(3–4), 1422–1432.

[5] Nie, Z., Lin, J., Li, J., Hao, H., and Ma, H. (2020). Bridge condition monitoring under moving loads using two sensor measurements. *Structural Health Monitoring*, 19(3), 917–937.

[6] Ding, Y., and Li, A. (2011). Assessment of bridge expansion joints using long-term displacement measurement under changing environmental conditions. *Frontiers of Architecture and Civil Engineering in China*, 5, 374–380.

[7] Huang, H. B., Yi, T. H., Li, H. N., and Liu, H. (2018). New representative temperature for performance alarming of bridge expansion joints through temperature-displacement relationship. *Journal of Bridge Engineering*, 23(7), 04018043.

[8] Ni, Y. Q., Wang, Y. W., and Zhang, C. (2020). A Bayesian approach for condition assessment and damage alarm of bridge expansion joints using long-term structural health monitoring data. *Engineering Structures*, 212, 110520.

[9] Honfi, D., Leander, J., and Björnsson, Í. (2017, September). Decision support for bridge condition assessment. *In fourth international conference on Smart Monitoring, Assessment and Rehabilitation of Civil Structures* (SMAR 2017).

[10] Ma, Y., Wang, L., Zhang, J., Xiang, Y., and Liu, Y. (2014). Bridge remaining strength prediction integrated with Bayesian network and in situ load testing. *Journal of Bridge Engineering*, 19(10), 04014037.

[11] Ni, Y. Q., Xia, H. W., Wong, K. Y., and Ko, J. M. (2012). In-service condition assessment of bridge deck using long-term monitoring data of strain response. *Journal of Bridge Engineering*, 17(6), 876–885.

[12] Xia, M., Cai, C. S., Pan, F., and Yu, Y. (2016). Estimation of extreme structural response distributions for mean recurrence intervals based on short-term monitoring. *Engineering Structures*, 126, 121–132.

[13] Xia, Q., Cheng, Y., Zhang, J., and Zhu, F. (2017). In-service condition assessment of a long-span suspension bridge using temperature-induced strain data. *Journal of Bridge Engineering*, 22(3), 04016124.

[14] Zhu, Y., Ni, Y. Q., Jesus, A., Liu, J., and Laory, I. (2018). Thermal strain extraction methodologies for bridge structural condition assessment. *Smart Materials and Structures*, 27(10), 105051.

[15] Zhan, H., Wang, Q., and Lu, Y. (2017). Handwritten digit string recognition by combination of residual network and RNN-CTC. In *Neural Information Processing: 24th International Conference, ICONIP 2017*, Guangzhou, China, November 14–18, 2017, Proceedings, Part VI 24 (pp. 583–591). Springer International Publishing.

[16] Chen, Z., Zhou, X., Wang, X., Dong, L., and Qian, Y. (2017). Deployment of a smart structural health monitoring system for long-span arch bridges: A review and a case study. *Sensors*, 17(9), 2151.

*Hybrid and Advanced Technologies – S. Prasad Jones Christydass, Nurhayati Nurhayati and S. Kannadhasan. (Eds) © 2025 the Authors, ISBN 978-1-032-90254-8*
*Open Access: www.taylorfrancis.com, CC BY-NC-ND 4.0 license*

# Comparative studies on attenuation coefficients of the Boro-phosphate glass for gamma ray shielding applications

Ch. Jyothi
*Research Scholar, Project Guide, Department of Biomedical Engineering, Saveetha School of Engineering, Saveetha Institute of Medical and Technical Sciences, Saveeetha University, Chennai, Tamilnadu, India*

M. Vijayakumar
*Project Guide, Department of Biomedical Engineering, Saveetha School of Engineering, Saveetha Institute of Medical and Technical Sciences, Saveeetha University, Chennai, Tamilnadu, India*

ABSTRACT: This study analyzes the gamma ray shielding capabilities of borophosphate glass by comparing its linear attenuation coefficient (LAC) and mass attenuation coefficient (MAC) to those of other glass materials. The melt quenching procedure is used to make the borophosphate glass, and the attenuation coefficients are calculated using the Xcom program. An Independent Sample T-test using SPSS software yielded a p-value of 0.005, suggesting significant results. Using Origin software, data visualization shows that borophosphate glass has a substantially higher attenuation coefficient than other materials, making it a good candidate for gamma ray shielding.

*Keywords*: linear attenuation coefficient, mass attenuation coefficient, Melt quenching method, Gamma rays, Borophosphate glass

## 1 INTRODUCTION

Glass materials are regarded good radiation shielding materials since they are not only transparent, but heavy metal oxides can be substituted to absorb radiation.(Mostafa *et al.* 2021)The use of ionization radiation in our lives in different applications and the fields increased the demands for the staff to protect from radiation hazards. Due to its transparency the glass is used for radiation shielding and because of its high density it improves the glass capacity and stops scattering against gamma rays.(Kaplan 1989) If the density of the material is low then it has the capability of more attenuation coefficient radiation. The glasses must have high density, chemical stability, transparency and durability and will have the ability to synthesize any type of glass material.so the boro phosphate glasses are having all these properties and used for the synthesis of the glass to shield against radiation. (Uhlmann and Kreidl 1991). This paper focuses primarily on two essential parameters: the Mass Attenuation Coefficient (MAC) and the Linear Attenuation Coefficient (LAC). The MAC describes how much an x-ray beam's intensity reduces as it travels through a substance, either due to absorption or deflection of photons. Meanwhile, the LAC quantifies the amount of radiation absorbed by a certain thickness of absorber material (Sayyed and Almatari 2020). By comparing these values, it is clear that materials with larger attenuation coefficients have better radiation shielding characteristics. This is frequently due to their increased density and thermal stability, which are desirable properties when manufacturing glass materials for radiation shielding (Taib 2012). A total of 40 research papers and 180 Google Scholar articles were included in this investigation. M.I.studies Studies on borate oxide based glass systems and is used as a glass former in the radiation shielding because its

BO3 units can easily change into tetrahedral BO4 units and increase its thermal resistance against gamma rays.(Shahami 2012) Lakshminarayana *et al.* have investigated adding $P_2O_5$ to borate glasses improves chemical stability by forming P–O–B bonds in the glass.(Cha *et al.* 2019) The addition of B2O3 to phosphate glasses improves their mechanical and optical properties while also increasing their ionic conductivity. Yadav and Singh, declared that rare earth oxides like B2O5, P2O5, Tm2O3, and Sn2O3 can be hosted by borate glasses as a significant benefit(Djezzar 2018). According to the authors usage of borate glass doped with rare earth oxides can be advantageous in applications for solid state laser technology, telecommunications, and photonic devices.(Andreo *et al.* 2017)

The attenuation coefficient of boro phosphate glass plays a critical role in assessing its appropriateness for various optical and photonics applications (Del Guerra 2004). The exact wavelength of the incident light, the composition of the glass, and its thickness all have an impact on the attenuation coefficient of boron phosphate.(Reeve 2018) Because of its transparency and other desirable optical properties, scientists and engineers frequently test and analyze the attenuation coefficient of prepared boro phosphate glass as the attenuation coefficient increases there will be an increase in radiation shielding property as a result the staff get less exposed to gamma rays(Maghraby and Al Mayyahi 2020).

## 2 MATERIALS AND METHODS

The research is conducted in Chennai at the Saveetha Institute of Medical and Technical Sciences Department of Biomedical engineering. This study contained two groups; Group 1-study, borophosphate and Group 2 -control, Ordinary concrete, hematite-serpentine and Ilmenite-limonite. Because of its quick glass forming abilities we are using conventional melt quenching methods. Melt quenching technique process involves melting of Borate, phosphate, Cadmium and tellurite-oxide at more than 1150°C in a platinum crucible and quenching in a brass mold or in water.(Erasmus *et al.* 2018)To remove the internal stress from the glasses, it is required to anneal the quenched glass at 500°C. The glass samples were then permitted to arrive at room temperature gradually (Griebenow *et al.* 2023).

To measure the change in photon intensity caused by scattering or absorption through materials, a variety of shielding factors are used(Anastasopoulou *et al.* 2017). The linear attenuation coefficient (m) is one of the parameters that explains the percentage of photons that are scattered or absorbed per unit thickness of the sample. The mass attenuation coefficient (m/r) is another parameter that describes how easily particular materials could be pierced by gamma photons, is connected to this attenuation parameter(Kloosterman 1992). Higher m/r values for a particular specimen show more photon interaction with it and hence less photon penetration. By using archimedes principle we can measure the density of the sample.

### 2.1 *Statistical analysis*

The Statistical software that is used to derive the output is IBM SPSS version 26.0.1. Origin is a scientific graphing and data analysis software that can be used to create visual representations for the data and analyze the data to find the best radiation material.

## 3 RESULTS

Table 1 displays the comparison of mass attenuation coefficients of the borophosphate, ordinary concrete, haematite-serpentine, limenite -limonite with incoming photon energy ranging from 1MeV to 1000MeV. Table 2 shows that comparison of mean and standard deviation of the 4 samples. Tables 3–5-By Using Independent T test we can compare the significance of the borophosphate glass and Ordinary concrete. (Table 3) Borophosphate glass and hematite-serpentine. (Table 4) Borophosphate glass and ilmenite-limonite (Table 5). Figure 1: Compare the means of mass attenuation coefficient of the four different samples. Figure 2: We can analyze that the mass attenuation coefficient of the sample which means borophosphate glass is gradually decreasing with increase of incoming photon energy. The borophosphate glass has higher mass attenuation coefficient than ordinary concrete. A higher mass attenuation coefficient means that the material is a better radiation shield since it is more effective at absorbing or scattering gamma rays. so the borophosphate glass has better radiation shielding property than ordinary concrete. Although borophosphate and hematite-serpentine have similar mean values, borophosphate glass is favored over other materials because of its excellent durability, high transparency, and inexpensive cost of manufacturing. Table 6 variation of the borophosphate, ordinary concrete, hematite-serpentine, ilmenite-limonite samples with respect to the energy. Table 7 shows that the comparison of the mean and standard deviation of the 4 samples. Tables 8–10-By Using Independent T test we can compare the significance of the borophosphate glass and Ordinary concrete. (Table 8) Borophosphate glass and hematite-serpentine. (Table 9) Borophosphate glass and ilmenite-limonite. (Table 10). From Figure 3: Compare the means of mass attenuation coefficient of the four different samples. Figure 4: we can observe that boro-phosphate glass has higher linear attenuation coefficient compared to ordinary concrete,,hematite-serpentine and ilmenite-limonite makes borophosphate glass the best material for radiation shielding material against gamma rays.

Table 1. Comparison of mass attenuation coefficient of borophosphate with ordinary concrete hematite-serpentine,ilmenite-limonite.

| Energy (MeV) | Borophosphate Glass | Ordinary Concrete | Hematite-Serpentine | Ilmenite-Limonite |
|---|---|---|---|---|
| 0.01 | 67.7 | 22.6 | 65.3 | 89.7 |
| 0.015 | 22.7 | 7.1 | 21.5 | 29.6 |
| 0.02 | 10.4 | 3.1 | 9.7 | 13.3 |
| 0.03 | 17.9 | 1.1 | 3.1 | 4.3 |
| 0.04 | 9.2 | 0.5 | 1.4 | 1.9 |
| 0.05 | 5.1 | 0.4 | 0.8 | 1.1 |
| 0.06 | 3.2 | 0.2 | 0.5 | 0.7 |
| 0.08 | 1.5 | 0.2 | 0.3 | 0.4 |
| 0.1 | 0.9 | 0.2 | 0.2 | 0.3 |
| 0.15 | 0.4 | 0.1 | 0.2 | 0.2 |
| 0.2 | 0.2 | 0.1 | 0.1 | 0.1 |
| 0.3 | 0.1 | 0.1 | 0.1 | 0.1 |
| 0.4 | 0.1 | 0.1 | 0.1 | 0.1 |
| 0.5 | 0.1 | 0.1 | 0.1 | 0.1 |
| 0.6 | 0.1 | 0.1 | 0.1 | 0.1 |
| 0.8 | 0.1 | 0.1 | 0.1 | 0.1 |
| 1 | 0.1 | 0.1 | 0.1 | 0.1 |
| 1.25 | 0.1 | 0.1 | 0.1 | 0.1 |
| 1.5 | 0.1 | 0.1 | 0.1 | 0.1 |

Table 2. Comparison of mean and standard deviation of Borophosphate with ordinary concrete hematite-serpentine,ilmenite-limonite.

| | | Group Statistics | | | |
|---|---|---|---|---|---|
| MAC | Energy | N | mean | std.deviation | Std.error mean |
| | Borophosphate glass. | 32 | 7.3527 | 21.08916 | 3.72807 |
| | Ordinary concrete | 32 | 1.1439 | 4.13655 | .73125 |
| | Hematite-serpentine | 32 | 3.2577 | 12.03906 | 2.12823 |
| | Ilmenite-limonite | 32 | 4.4579 | 16.54708 | 2.92534 |

Table 3. Independent sample T-test compares the statistical mean differences between samples of borophosphate and ordinary concrete.

| | Levens test for equality of variances | | T test for equality of means | | | | | | |
|---|---|---|---|---|---|---|---|---|---|
| | | | | | | | | 95% confidence interval of the diff | |
| MAC | F | sig | t | df | sig.(2-tailed | Mean diff | Std.error diff | Lower | Upper |
| **Equal variance assumed** | 8.587 | .005 | 1.634 | 62 | .107 | 6.20883 | 3.79911 | -1.38548 | 13.80314 |
| **Equal variance not assumed.** | | | 1.634 | 33.382 | .112 | 6.20883 | 3.79911 | =1.51716 | 13.93482 |

Table 4. Independent sample T-test compares the statistical mean differences between samples of borophosphate and hematite-serpentinite.

| | Levens test for equality of variances | | T test for equality of means | | | | | | |
|---|---|---|---|---|---|---|---|---|---|
| | | | | | | | | 95% confidence interval of the diff | |
| MAC | F | sig | t | df | sig.(2-tailed | Mean diff | Std.error diff | Lower | Upper |
| **Equal variance assumed** | 2.511 | .118 | .954 | 62 | .344 | 4.09502 | 4.29277 | -4.48611 | 12.67614 |
| **Equal variance not assumed.** | | | .954 | 49.265 | .345 | 4.09502 | 4.29277 | -4.53045 | 12.72048 |

Table 5. Independent sample Test compares the statistical mean differences between samples of borophosphate and ilmenite-Limonite.

| | Levens test for equality of variances | | T test for equality of means | | | | | | |
|---|---|---|---|---|---|---|---|---|---|
| | | | | | | | | 95% confidence interval of the diff | |
| MAC | F | sig | t | df | sig.(2-tailed | Mean diff | Std.error diff | Lower | Upper |
| **Equal variance assumed** | .856 | .358 | .612 | 62 | .543 | 2.900 | 4.739 | -8.572 | 12.373 |
| **Equal variance not assumed.** | | | .612 | 58.680 | .543 | 2.900 | 4.739 | -6.583 | 12.383 |

Table 6. Comparison of Linear attenuation coefficient of borophosphate with ordinary concrete hematite-serpentine,ilmenite-limonite.

| Energy (MeV) | Sample (Boro-phosphate glass) | Ordinary Concrete | Hematite-Serpentine | Ilmenite-Limonite |
|---|---|---|---|---|
| 0.01 | 254 | 51.9 | 163 | 260 |
| 0.015 | 84.9 | 16.3 | 53.9 | 86 |
| 0.02 | 39 | 7.1 | 24.2 | 38.5 |
| 0.03 | 67.2 | 2.4 | 7.8 | 12.3 |
| 0.04 | 34.5 | 1.3 | 3.6 | 5.5 |
| 0.05 | 19.1 | 0.8 | 2.1 | 3.1 |
| 0.06 | 11.8 | 0.6 | 1.4 | 2 |
| 0.08 | 5.6 | 0.5 | 0.8 | 1.1 |
| 0.1 | 3.2 | 0.4 | 0.6 | 0.8 |
| 0.15 | 1.3 | 0.3 | 0.4 | 0.5 |
| 0.2 | 0.8 | 0.3 | 0.3 | 0.4 |
| 0.3 | 0.4 | 0.3 | 0.3 | 0.3 |
| 0.4 | 0.4 | 0.2 | 0.2 | 0.3 |
| 0.5 | 0.3 | 0.2 | 0.2 | 0.2 |
| 0.6 | 0.3 | 0.2 | 0.2 | 0.2 |
| 0.8 | 0.3 | 0.2 | 0.2 | 0.2 |
| 1 | 0.2 | 0.1 | 0.2 | 0.2 |
| 1.25 | 0.2 | 0.1 | 0.1 | 0.2 |
| 1.5 | 0.2 | 0.1 | 0.1 | 0.1 |
| 2 | 0.2 | 0.1 | 0.1 | 0.1 |
| 3 | 0.1 | 0.1 | 0.1 | 0.1 |
| 4 | 0.1 | 0.1 | 0.1 | 0.1 |
| 5 | 0.1 | 0.1 | 0.1 | 0.1 |
| 6 | 0.1 | 0.1 | 0.1 | 0.1 |
| 8 | 0.1 | 0.1 | 0.1 | 0.1 |
| 10 | 0.1 | 0.1 | 0.1 | 0.1 |
| 15 | 0.1 | 0 | 0.1 | 0.1 |
| 20 | 0.1 | 0 | 0.1 | 0.1 |
| 50 | 0.1 | 0 | 0.1 | 0.1 |
| 100 | 0.2 | 0.1 | 0.1 | 0.1 |
| 500 | 0.2 | 0.1 | 0.1 | 0.1 |
| 1000 | 0.2 | 0.1 | 0.1 | 0.1 |

Table 7. Comparison of mean and standard deviation of borophosphate, ordinary concrete, hematite-serpentinite, ilmenite-limonite.

| Group Statistics LAC | Energy | N | mean | std.deviation | Std.error mean |
|---|---|---|---|---|---|
| | Borophosphate glass. | 32 | 37.2468 | 106.79064 | 18.87810 |
| | Ordinary concrete | 32 | 5.7791 | 20.87405 | 3.69005 |
| | Hematite-serpentine | 32 | 16.5150 | 61.02627 | 10.78802 |
| | Ilmenite-limonite | 32 | 22.5734 | 83.92668 | 14.83628 |

Table 8. Independent sample T-test compares the statistical mean differences between samples of Borophosphate and Ordinary concrete.

| | Levens test for equality of variances | | T test for equality of means | | | | | 95% confidence interval of the diff | |
|---|---|---|---|---|---|---|---|---|---|
| LAC | F | sig | t | df | sig. (2-tailed | Mean diff | Std.error diff | Lower | Upper |
| **Equal variance assumed** | 8.608 | .005 | 1.636 | 62 | .107 | 31.4677 | 19.23536 | -6.98316 | 69.91870 |
| **Equal variance not assumed.** | | | 1.636 | 33.365 | .111 | 31.4677 | 19.23536 | -7.65059 | 70.58612 |

Table 9. Independent sample T-test compares the statistical mean differences between samples of Borophosphate and Hematite-serpentinite.

| | Levens test for equality of variances | | T test for equality of means | | | | | 95% confidence interval of the diff | |
|---|---|---|---|---|---|---|---|---|---|
| LAC | F | sig | t | df | sig. (2-tailed | Mean diff | Std.error diff | Lower | Upper |
| **Equal variance assumed** | 2.508 | .118 | .953 | 62 | .344 | 20.73184 | 21.4314 | -22.73207 | 64.19575 |
| **Equal variance not assumed.** | | | .953 | 49.296 | .345 | 20.73184 | 21.74314 | -22.95599 | 64.41968 |

Table 10. Independent sample Test compares the statistical mean differences between samples of Borophosphate and Ilmenite-Limonite.

| | Levens test for equality of variances | | T test for equality of means | | | | | 95% confidence interval of the diff | |
|---|---|---|---|---|---|---|---|---|---|
| LAC | F | sig | t | df | sig. (2-tailed | Mean diff | Std.error diff | Lower | Upper |
| **Equal variance assumed** | .852 | .359 | .611 | 62 | .543 | 14.67341 | 24.01037 | -33.32264 | 62.66945 |
| **Equal variance not assumed.** | | | .611 | 58.719 | .543 | 14.67341 | 24.01037 | -33.37604 | 62.72285 |

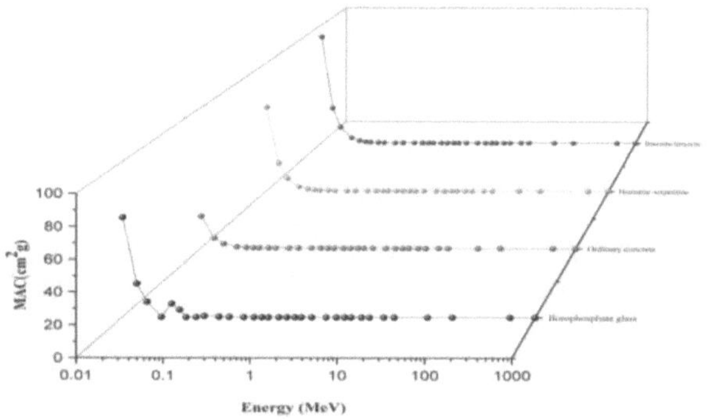


Figure 1. Comparison of mass attenuation coefficient of Boro-phosphate glass, Ordinary concrete, Hematite-serpentinite,ilmenite-limonite.

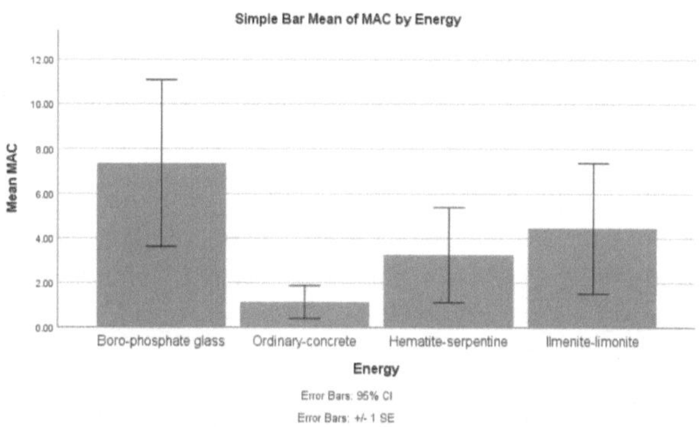


Figure 2. The comparison of mean, energy and standard errors for boro-phosphate is better than ordinary concrete, hematite-serpentine and ilmenite-limonite in terms of mass parameter. X axis boro-phosphate, concrete, hematite-serpentinite, ilmenite-limonite and Y-axis: Mean MAC.

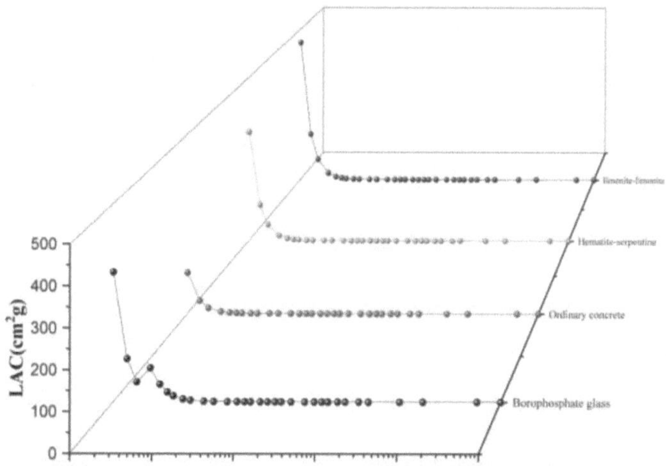


Figure 3. The Comparison of Mean,Energy and standard errors for Boro-phosphate is better than Ordinary concrete, Hematite-serpentine and ilmenite-limonite in terms of mass parameter. X axis boro-phosphate,concrete,Hematite-serpentinite, Ilmenite-Limonite and Y-axis : Mean MAC.

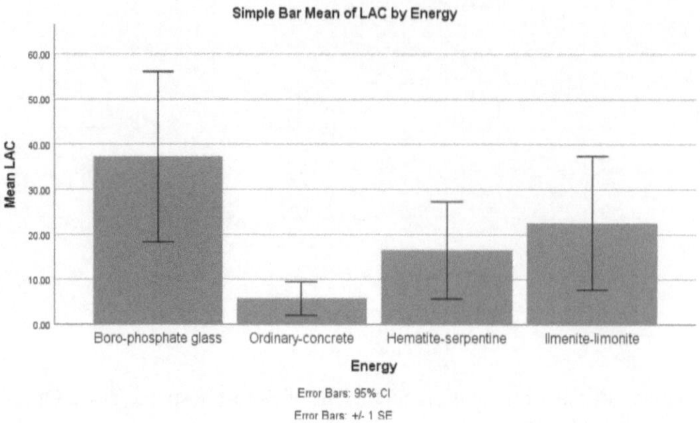


Figure 4: Comparison of linear attenuation coefficient of Boro-phosphate glass, ordinary concrete, hematite-serpentinite, ilmenite-limonite.

## 3 DISCUSSION

In this study we can observe that various parameters affect a material's radiation shielding characteristics. To find the best radiation shielding material we have mainly considered the mass attenuation coefficient and linear attenuation coefficient parameters(Eldridge and Eldridge 2021)

We have taken four samples of different parameters and the glass samples can attenuate the photons effectively at low energies since the values increase with energy. The outcomes showed that for boro-phosphate glass is greater in comparison to other samples. The benefit of using borophosphate glass is biocompatible, transparent and can transmit a wide range of wavelengths,including ultraviolet and infrared light.(Dermer and Menon 2009) It can create biodegradable implants for bone regeneration and wound healing using borophosphate glass in hospitals. For optical parts in medical devices, such as lenses and optical fibers used in endoscopes and lasers, borophosphate glass is a good choice due to its transparency.(Kursun *et al.* 2023)

The study limitations for mass attenuation coefficients and linear attenuation coefficients on radiation absorption in a material. The shielding effectiveness of borophosphate glass may vary significantly depending on the energy spectrum of the gamma rays. As gamma rays pass through a material, their intensity decreases exponentially with thickness.

## 4 CONCLUSION

In this study we can conclude that the Borophosphate has higher mass and linear Attenuation coefficients compared to ordinary concrete, Hematite-serpentine, Ilmenite-limonite. Even though statistically better significant values were obtained. Mass and linear attenuation coefficient are two parameters of how electromagnetic radiation such as x rays and gamma rays interacts with the materials like boro-phosphate glass. From the SPSS we can state that mass and linear attenuation coefficient for boro-phosphate glass is higher because of its high density and resistance to radiation damage and its potential for transparency to the visible light as compared to other materials.

## REFERENCES


Anastasopoulou, Maria, Konstantinos C. Vasilopoulos, Dimitrios Anagnostopoulos, Ioannis Koutselas, Demetrios K. Papayannis, and Michael A. Karakassides. (2017). Structural and theoretical study of borophosphate glasses using raman spectroscopy and ab initio molecular orbital method. *The Journal of Physical Chemistry. B* 121 (17): 4610–19.

Andreo, Pedro, David T. Burns, Alan E. Nahum, Jan Seuntjens, and Frank Herbert Attix. (2017). *Fundamentals of Ionizing Radiation Dosimetry*. John Wiley & Sons.

Cha, Jaemin, Hwajin Jeong, and Bongki Ryu. (2019). Improvement of electrical conductivity and chemical durability in $Fe_2O_3$ doped $V_2O$-$B_2$;$O_3$-$P_2$;O glasses. *Journal of Nanoscience and Nanotechnology* 19 (3): 1254–60.

Del Guerra, Alberto. (2004). *Ionizing Radiation Detectors for Medical Imaging*. World Scientific.

Dermer, Charles D., and Govind Menon. (2009). High Energy Radiation from Black Holes: Gamma Rays, Cosmic Rays, and Neutrinos. Princeton University Press.

Djezzar, Boualem. (2018). *Ionizing Radiation Effects and Applications*. BoD – Books on Demand.

Eldridge, Alison, and Stephen Eldridge. (2021). *Understanding and Using Gamma Rays*. Enslow Publishing, LLC.

Erasmus, E. P., R. Sule, O. T. Johnson, J. Massera, and I. Sigalas. (2018). In vitro evaluation of porous borosilicate, borophosphate and phosphate bioactive glasses scaffolds fabricated using foaming agent for bone regeneration. *Scientific Reports* 8 (1): 3699.

Griebenow, Kristin, Mai-Phuong Truong, Francisco Munoz, Robert Klement, and Dusan Galusek. (2023). Tuning the fluorescence of dy via the structure of borophosphate glasses. *Scientific Reports* 13 (1): 1919.

Kaplan, M.F. (1989). *Concrete Radiation Shielding: Nuclear Physics, Concrete Properties, Design and Construction*. Longman Publishing Group.

Kloosterman, Jan L. (1992). On Gamma Ray Shielding and Neutron Streaming through Ducts.

Kursun, Celal, Meng Gao, Seda Guclu, Yasin Gaylan, Khursheed Ahmad Parrey, and Ali Orkun Yalcin. (2023). Measurement on the neutron and gamma radiation shielding performance of boron-doped titanium alloy TiCuZrB via arc melting technique. *Heliyon* 9 (11): e21696.

Maghraby, Ahmed M., and Basim Almayyahi. (2020). *Recent Techniques and Applications in Ionizing Radiation Research*. BoD–Books on Demand.

Mostafa, Ahmed M. A., Mohamed A. M. Uosif, Ziyad A. Alrowaili, Reda Elsaman, Ahmed A. Showahy, Yasser B. Saddeek, Shams A. M. Issa, Antoaneta Ene, and Hesham M. H. Zakaly. (2021). The influence of CoO/PO substitutions on the structural, mechanical, and radiation shielding of boro-phosphate glasses. *Materials* 14 (21). https://doi.org/10.3390/ma14216632.

Reeve, Tamar. (2018). Ionizing Radiation: Advances in Research and Applications.

Sayyed, M.I., and Almatari M. (2020). Radiation Shielding Investigation of Lead Borophosphate Glasses Modified by Tungsten Oxide.

Shahami, Nurhidayu. (2012). Synthesis of Iron-Borophosphate Glass Doped Wih Various Rare Earth.

Taib, Nur Umira. (2012). The Structure and Luminescence Properties of Cadmium Oxide (CdO) Based Borophosphate Glass Doped with Rare Earthions.

Uhlmann, Donald Robert, and N. J. Kreidl. (1991). *Optical Properties*.

*Hybrid and Advanced Technologies – S. Prasad Jones Christydass, Nurhayati Nurhayati and S. Kannadhasan. (Eds)* *ISBN 978-1-032-90254-8*


# A comparative study on surface roughness properties aluminum hybrid composites reinforced with nano-scale $ZrO_2$/SiC via ultrasonic assisted stir casting

R. Ragul Kanth
*Research Scholar, Department of Mechanical Engineering, Saveetha School of Engineering, Saveetha University, Chennai, Tamil Nadu, India*

T.G. Sakthivel
*Project Guide, Department of Mechanical Engineering, Saveetha School of Engineering, Saveetha University, Chennai, Tamil Nadu, India*

ABSTRACT: The present study investigated the influence of surface irregularities on enhancing the bonding between A6061/5% $ZrO_2$, and AA6061/5%SiC/5%$ZrO_2$ reinforced hybrid Metal matrix composites. The surface roughness behaviour of hybrid metal matrix composites characteristics was investigated. The selection of aluminium 6061 as the matrix material was driven to enhance the mechanical strength, which was accomplished by incorporating the harder SiC particles and improving the wear resistance by introducing the softer $ZrO_2$ particles. Subsequently, the prepared specimens were subjected to surface roughness analysis to gain insight into the behaviour of the hybrid metal matrix composites. Two distinct cohorts participated in the study, with a total sample size of 40. The experimental and control groups comprised 20 samples each, adhering to the statistical prerequisite of Gpower at 80% with an $\alpha$ value of 0.05, thereby ensuring a 95% confidence level. The distinction between the two models exhibits statistical significance with a p-value of less than 0.05, as supported by the observed significance level of $p=0.000$ within the research investigation. Notably, it was observed that the reinforcement particles were uniformly dispersed within the matrix material, a result that can be attributed to the continuous stirring process. Consequently, it can be inferred that the hybrid composites exhibit favourable surface roughness properties.

*Keywords*: Al6061, Surface Roughness, SiC, ultrasonic assisted stir-casting technique, Sustainable Production, $ZrO_2$, Hybrid composites

## 1 INTRODUCTION

The aerospace and automotive industries have made substantial use of metal matrix composites (MMCs) due to their better strength and wear resistance when compared to base alloys (Arif *et al.* 2017; Ramanan *et al.* 2019; Saravanan *et al.* 2020). Because of their superior strength-to-weight and strength-to-cost ratios, these composites are especially useful for thermal, electrical, and structural applications (Loganathan *et al.* 2018). Typically, MMCs consist of a metallic matrix reinforced with materials like fibers and whisker particles. These materials are chosen depending on the mechanical properties that the material is intended to have, such as high elastic modulus, thermal shock resistance, surface durability, and lightweight nature. (Ezatpour 2016).

 DOI: 10.1201/9781003559115-86

The mechanical properties of aluminum MMCs are greatly improved by common reinforcement materials such as $ZrO_2$ (Jose and Ezema 2023), TiC (Cao *et al.* 2019), AlN (Singari *et al.* 2022), Si3N4 (Patnaik *et al.* 2021), B4C (Mazahery and Shabani 2011), SiC (Kostecki *et al.* 2017), and Al2O3 (Kostecki *et al.* 2017; Suresh 2013). Due to their improved physical and tribological properties, which make them superior to composites with a single reinforcement, Al matrix composites reinforced with a combination of micro- and nano-sized reinforcements have attracted a lot of attention (Tan *et al.* 2017). In frictional materials, hybrid composites with both nano- and micro-sized reinforcements are crucial because, via the Orowan strengthening mechanism, nanoparticles improve matrix performance and micro-sized reinforcement maintains hardness and tribological properties (Pawar 2014; Tan).

But improving Al/SiC composites' ductility and fracture toughness—two essential qualities to avoid catastrophic failure in service—often results in a reduction in both qualities (Michael 2015). According to research, metal oxide can lessen SiC's brittleness and increase fracture toughness in hybrid composites. Interestingly, even though Al-SiC-$ZrO_2$ nano-composites have been created previously, there is currently no documentation regarding the process of producing them utilizing Ultrasonic Assisted Stir Casting, which calls for more research into how this method affects the mechanical, thermal, and structural characteristics of the composites.

## 2 MATERIALS AND METHODS

This investigation was conducted in Saveetha Industries, Chennai. In the present study, the choice was made to utilize aluminium powder (with a purity level of 99%) as the matrix. Simultaneously, SiC nanoparticles were selected as the reinforcing material for the purpose of fabricating the composite. The exceptional qualities of pure aluminium, such as its high tensile strength, ease of casting, and impressive strength-to-weight ratio, render it the optimal material for sustainable production.

The graphite crucible, possessing dimensions of 150 mm in height and 150 mm in diameter, received the deposition of aluminium, which was subsequently subjected to a heating process achieving a temperature of 750°C. A $C_2Cl_6$ degassing tablet composed of hexachloroethane was introduced into the slurry to mitigate trapped gases within the aluminium melt. Furthermore, the SiC nanoparticles underwent a preliminary heating process at 200°C for 30 minutes before their introduction into the liquid aluminium pool. The liquid aluminium underwent vigorous stirring at 350-400 rotations per minute for 20 minutes by mechanical stirring. Incorporating coarse-sized magnesium powder into the melt during the stirring process enhanced the wetting properties between the aluminium matrix and the nano-sized SiC reinforcements. This mixing procedure took place under an inert gas atmosphere (Ar) To ensure protection against melt oxidation during the casting process. The slurry was then subjected to heating at a temperature of 500°C. Subsequently, the slurry experienced the cooling process in the surrounding atmosphere until it reached the ambient temperature and solidified into a composite material.

Group 1: The transfer of the slurry, containing a variety of materials, to a steel die with dimensions of 160x160x10 mm, following ultrasonic treatment, was facilitated by the connection of the waveguide to an ultrasonic converter with a frequency of 20 kHz. Before this transfer, the steel die had been preheated to a temperature of 500°C.

Group 2: The slurry, composed of different materials, underwent ultrasonic treatment before being relocated to a steel die with dimensions measuring 160x160x10 mm. The production of these diverse composites involved the reinforcement of nano-SiC/$ZrO_2$ particles with a weight percentage of 5 wt% SiC/$ZrO_2$ within the aluminium matrix.

## 2.1 *Statistical analysis*

A statistical study using SPSS was used to determine the tensile strength of alloys made of aluminium (AA5083-H111). An Independent t-test was used to determine the significance ($P<0.05$) between the two groups (Isa 2021). The dependent variable is rotational speed, translational speed, and axial load, whereas surface roughness is the independent variable. The total sample size is 40, and each group carries 20 samples for the surface roughness experiment with 95% confidence intervals.

# 3 RESULTS

Table 3 offers a thorough examination of the SiC reinforcement material's physical and thermal properties, while Tables 1 and 2 offer extensive data on the AA6061 alloy's chemical and mechanical properties. The Surface Roughness (SR) values for the two groups—AA6061/5% ZrO2 (Group 1) and AA6061/5%SiC/5% ZrO2 (Group 2)—are shown in Table 4. Furthermore, a visual comparison of the SR values between the experimental and control. contrasts the SR values between the experimental and control groups graphically. Levene's test for equality of variance and the T-test for equality of hardness for groups 1 and 2 are shown in Figure 1. The ultrasonic-assisted stir-casting experimental setup is shown in Figure 2, and the hybrid composites that were produced from the two compositions are shown in Figure 3. The configuration for measuring surface roughness in metal matrix composites is shown in Figure 4, and ready-made hybrid composite specimens are shown in Figure 5. Lastly, the mean SR values with the lowest standard deviation for the hybrid composites from Groups 1 and 2 are contrasted in Figure 6.

Table 1. Chemical composition of aluminium AA606-T6 alloys.

| Element | Mg | Fe | Cu | Si | Cr | Al |
|---|---|---|---|---|---|---|
| AA6061-T6 | 0.92 | 0.33 | 0.25 | 0.6 | 0.18 | 97.92 |

Table 2. Mechanical properties of aluminium AA6061-T6 alloys.

| Element | Yield strength (MPa) | Tensile strength (MPa) | Hardness (%) |
|---|---|---|---|
| AA6061-T6 | 276 | 310 | 40 |

Table 3. Mechanical and thermal properties of SiC.

| Properties | Value |
|---|---|
| Melting Point | 2730°C |
| Thermal Conductivity | 120 W/m-K |
| Poisson's Ratio | 0.35 |
| Density | 3.22 g/cm3 |
| Modulus of Elasticity | 90 GPa |
| Yield strength | 21 GPa |
| Tensile strength | 0.1379 GPa |

Table 4. The surface roughness of AA6061/5% $ZrO_2$ (Group 1) and AA6061/5% SiC/5%$ZrO_2$(Group 2).

| S. No | Hardness (HV) | |
|---|---|---|
| | AA6061/5% $ZrO_2$ | AA6061/5%SiC/5%$ZrO_2$ |
| 1 | 1.46 | 1.41 |
| 2 | 1.45 | 1.4 |
| 3 | 1.43 | 1.38 |
| 4 | 1.4 | 1.35 |
| 5 | 1.39 | 1.34 |
| 6 | 1.37 | 1.32 |
| 7 | 1.36 | 1.31 |
| 8 | 1.37 | 1.32 |
| 9 | 1.34 | 1.29 |
| 10 | 1.32 | 1.27 |
| 11 | 1.29 | 1.24 |
| 12 | 1.24 | 1.19 |
| 13 | 1.23 | 1.18 |
| 14 | 1.19 | 1.14 |
| 15 | 1.21 | 1.16 |
| 16 | 1.1 | 1.05 |
| 17 | 1.04 | 0.99 |
| 18 | 1 | 0.95 |
| 19 | 0.99 | 0.94 |
| 20 | 0.99 | 0.94 |

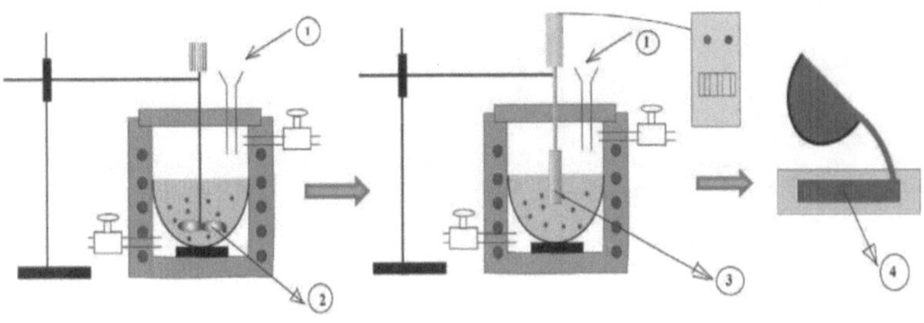


Figure 1. Ultrasonic assisted stir-casting techniques (1) Hooper (2) Stirrer (3) waveguide (4) composite slurry.

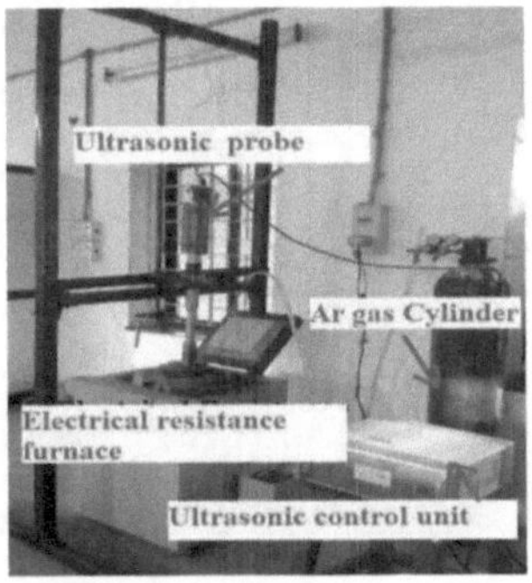


Figure 2. Ultrasonic-assisted stir-casting machine setup.

Figure 3. Fabricated hybrid composite model specimen.

Figure 4. Surface roughness testing machine setup.

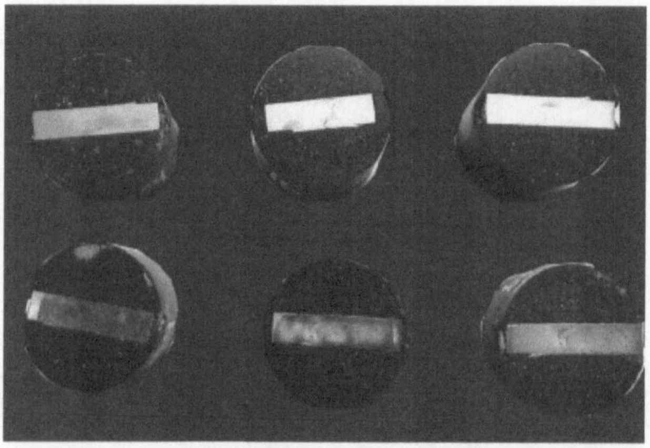

Figure 5. Surface roughness specimen of AA6061/5% $ZrO_2$ (Group 1) and AA6061/5%SiC/5%$ZrO_2$ (Group 2).

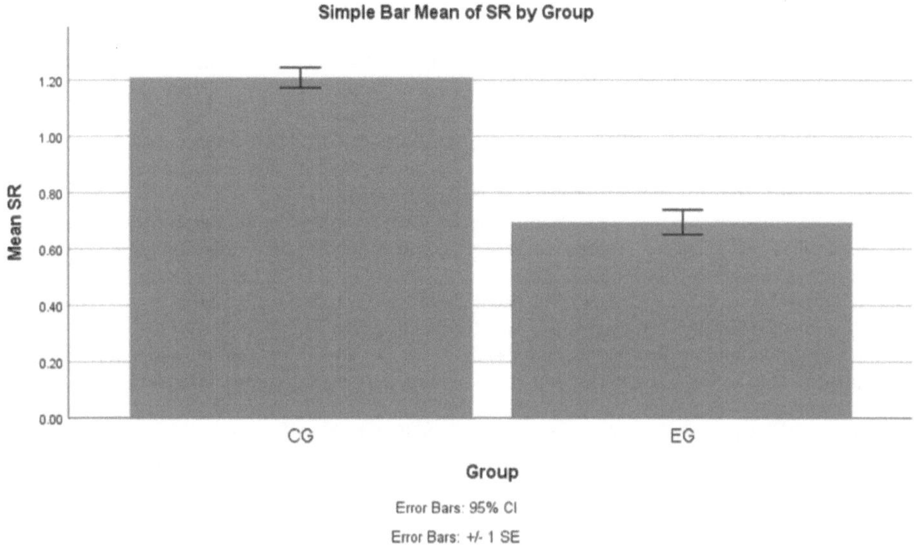


Figure 6. Correlation of the experimental and the control groups' mean square surface roughness. The specimens of the control group of machining performance using novel AA6061/5%SiC/5%$ZrO_2$ had higher surface roughness and a lower standard deviation than the control group samples with the AA6061/5%$ZrO_2$ without SiC reinforcement. X-axis: AA6061/5% $ZrO_2$ (Group 1) and AA6061/5%SiC/5%$ZrO_2$ (Group 2); Y-axis: mean Hardness with +/-1 SD of error.

## 4 DISCUSSION

The surface roughness qualities of composites are greatly improved by the spatial distribution of reinforcement inside the matrix. In order to achieve this augmentation, the interface between the matrix and reinforcement is crucial. It allows for a surface roughness of 0.94 $\mu$m as opposed to 0.99 $\mu$m for pure Al. Surface roughness in Al-SiC-$ZrO_2$ nanocomposites is gradually decreased by adding SiC and $ZrO_2$ nanoparticles (Ramanan *et al.* 2019).

## 5 CONCLUSION

The hybrid composites have been effectively fabricated using Ultrasonic Assisted Stir Casting. The results of this investigation were examined, and the effects of $ZrO_2$ and silicon carbide (SiC) on the surface roughness characteristics of the composites were documented. A comparison to the original matrix Al 6061 alloy revealed that the organized AA6061/5% $ZrO_2$ hybrid sample exhibited an increase in compressive strength. Specifically, AA6061/5% SiC/5%$ZrO_2$ (Group 2) alloy exhibited lower surface roughness, as evidenced by a two-tailed value P=0.000 (P<0.05) with a 95% confidence interval. Thus, in between the two groups the statistically significant difference is present.

## REFERENCES

Ali, S., Sudha, K. G., Thirumalaivasan, N., Ahamed, M., Pandiaraj, S., Rajeswari, V. D., Vinayagam, Y., Thiruvengadam, M., and Govindasamy, R. (2023). Green synthesis of magnesium oxide nanoparticles by using Abrus precatorius bark extract and their photocatalytic, antioxidant, antibacterial, and cytotoxicity activities. *Bioengineering (Basel, Switzerland)*, *10*(3). https://doi.org/10.3390/bioengineering10030302

Arif, S., Alam, M. T., Ansari, A. H., Siddiqui, M. A., and Mohsin, M. (2017). Study of mechanical and tribological behaviour of Al/SiC/ZrO2 hybrid composites fabricated through powder metallurgy technique. *Materials Research Express*, *4*(7), 076511. https://doi.org/10.1088/2053-1591/aa7ad7

Cao, G., Estournes, C., Garay, J., and Orru, R. (2019). *Spark plasma sintering: Current status, new developments and challenges: A review of the current trends in SPS. Elsevier.*

Kostecki, M., Woźniak, J., Cygan, T., Petrus, M., and Olszyna, A. (2017). Tribological properties of aluminium alloy composites reinforced with multi-layer graphene—The influence of spark plasma texturing process. *Materials*, *10*(8), 928. https://doi.org/10.3390/ma10080928

Loganathan, P., Gnanavelbabu, A., and Rajkumar, K. (2018). Analysis and characterization of friction behaviour on AA7075/ZrB2 composite under dry sliding condition. *Materials Research Express*, *6*(2), 026576. https://doi.org/10.1088/2053-1591/aaacdf

Mazahery, A., and Shabani, M. O. (2011). The accuracy of various training algorithms in tribological behavior modeling of A356-B4C composites. *Russian Metallurgy*, *2011*(7), 699–707.

Michael. (2015). Aluminium matrix hybrid composites: A review of reinforcement philosophies; mechanical, corrosion and tribological characteristics. *Journal of Materials Research and Technology*, *4*(4), 434–445. https://doi.org/10.1016/j.jmrt.2015.06.004

Natrayan, L., Kaliappan, S., Sethupathy, B. S., Sekar, S., Patil, P. P., Velmurugan, G., and Tariku Olkeba, T. (2022). Effect of mechanical properties on fibre addition of flax and graphene-based bionanocomposites. *International Journal of Chemical Engineering*, *2022*. https://doi.org/10.1155/2022/5086365

Patnaik, A., Kozeschnik, E., and Kukshal, V. (Eds.). (2021). *Advances in materials processing and manufacturing applications: Proceedings of iCADMA 2020*. Springer Nature.

Pawar. (2014). Development of aluminium based silicon carbide particulate metal matrix composite for spur gear. *Procedia Materials Science*, *6*(January), 1150–1156. https://doi.org/10.1016/j.mspro.2014.07.156

Ramanan, G., Suresh, S., Rajesh Prabha, N., and Edwin Raja Dhas, J. (2019). Characterization, wear surface roughness and tensile failure analysis of Al7075-TiC-MoS2 hybrid composites using online acoustic emission. *Materials Research Express*, *6*(6), 066544. https://doi.org/10.1088/2053-1591/ab1b3e

Saravanan, S., Sreenithi, A., Ajitha, M., Ilakkiyal, R., Narmatha, S. R., Rajkumar, S., and Karthikeyan, K. (2020). Tribological behaviour of aluminum alloy (AA7075) based hybrid composites. *IOP Conference Series. Materials Science and Engineering*, *923*(1), 012053. https://doi.org/10.1088/1757-899X/923/1/012053

Singari, R. M., Jain, P. K., and Kumar, H. (Eds.). (2022). *Advances in manufacturing technology and management: Proceedings of 6th International Conference on Advanced Production and Industrial Engineering (ICAPIE)—2021*. Springer Nature.

Sivaranjani, P. R., Syed, A., Elgorban, A. M., Bahkali, A. H., Balakrishnaraja, R., Varma, R. S., and Sudheer Khan, S. (2023). Fabrication of ternary nano-heterojunction via hierarchical deposition of α-Fe2O3 and *β*-La2S3 on cubic CoCr2O4 for enhanced photodegradation of doxycycline. *Journal of Industrial and Engineering Chemistry*, *118*, 407–417. https://doi.org/10.1016/j.jiec.2022.11.021

Suresh, S. (2013). *Fundamentals of metal-matrix composites*. Elsevier.

Tan, A., Teng, J., Zeng, X., Fu, D., and Zhang, H. (2017). Fabrication of aluminium matrix hybrid composites reinforced with SiC microparticles and TiB2 nanoparticles by powder metallurgy. *Powder Metallurgy*. https://doi.org/10.1080/00325899.2016.1274816

Viswanathan, S., Palaniyandi, T., Kannaki, P., Shanmugam, R., Baskar, G., Rahaman, A. M., Paul, L. T. D., Rajendran, B. K., and Sivaji, A. (2023). Biogenic synthesis of gold nanoparticles using red seaweed Champia parvula and its antioxidant and anticarcinogenic activity on lung cancer. *Particulate Science and Technology*, *41*(2), 241–249.

*Hybrid and Advanced Technologies – S. Prasad Jones Christydass, Nurhayati Nurhayati and S. Kannadhasan. (Eds) © 2025 the Authors, ISBN 978-1-032-90254-8*
*Open Access: www.taylorfrancis.com, CC BY-NC-ND 4.0 license*

# Improvement of electron mobility in algan hemt using sheet carrier density and electron dynamics techniques for different channel length modulation

A. Subhalakshmi
*Research Scholar, Department of Electrical and Electronics Engineering, Saveetha School of Engineering, Saveetha University, Chennai, Tamil Nadu, India*

N. Anbuselvan
*Project Guide, Department of Electrical and Electronics Engineering, Saveetha School of Engineering, Saveetha University, Chennai, Tamil Nadu, India*

ABSTRACT: The objective of this study is to boost the electron mobility of AlGaN High Electron Mobility Transistors(HEMT) by altering the channel length which is executed using two different optimization techniques. By changing the channel length, the electron mobility of AlGaN HEMT is increased by sheet carrier density technique and electron dynamics technique. This approach aims to raise mobility.There are 14 samples in total, which is divided into two groups. The first group is sheet carrier density technique and the second group is electron dynamics technique. Seven samples are incorporated in each group. The presented sheet carrier density technique shows high electron mobility values of 1158.33(cm^2/v/s) contrast to 477.56(cm^2/v/s) for the electron dynamics technique. There is a significant difference and the significance level of the study is calculated to be $p = 0.002$. Compared to the electron dynamics technique, the sheet carrier density method yields superior results for electron mobility analysis.

*Keywords*: HEMT, Novel sheet carrier density, electron dynamics technique, electron mobility, channel length, AlGaN, sustainable power

## 1 INTRODUCTION

High Electron Mobility Transistors(HEMT) can operate at a high frequency by utilizing the 2DEG created by the heterojunction which results in high electron mobility.(Nirmal and Ajayan 2019) So by increasing electron mobility, there will be a significant impact on the performance of electronic devices. One of the methods that is used to enhance the mobility in AlGaN HEMT is the sheet carrier density technique which measures the number of charge carrier density and doping level.(Jallipeta 2016).The results of this study show how the novel sheet carrier density can increase mobility via different channel lengths in HEMT. Faster and more efficient electronic devices are in greater demand in today's world of rapidly advancing technology.(Lenka and Nguyen 2022)Because of their high electron mobility, HEMTs can function at frequencies up to millimeter wave frequency, higher than those of regular transistors.(Ross *et al.* 2012). It is used in high frequency applications such as satellite television receivers, voltage convertors, radar equipments, RF designs.(Huang *et al.* 2023) (Ali *et al.* 2023; Gurupranes *et al.* 2022; Natrayan *et al.* 2022; Natrayan *et al.* 2022; Sivaranjani *et al.* 2023; Viswanathan *et al.* 2023) Recently several researches have been conducted on high electron mobility in AlGaN HEMT and working on it to improve its performance in devices. IEEE Xplore published 135 research papers and Google scholar found 152 articles. In this paper (Hospodková *et al.* 2023) the author suggested that replacing the high-temperature (1000 °C) GaN channel layer with a layer grown in a N2 at 870 °C would result in a significant increase in electron Hall mobility. The various structures of high electron mobility transistors are explained in this article, (Mohanty *et al.* 2019) and device performance is also covered based on previous studies published that makes use of both analytical and numerical models.(Uttam *et al.* 2019)An experimental

## 2 MATERIALS AND METHODS

study of AlGaN HEMTs on silicon at different channel lengths and temperatures is presented in this paper. Device analog parameters generally increased or decreased according to the analyzed device channel size.(Shin *et al.* 2022)

The channel length of a HEMT is an important parameter that affects its performance such as transconductance, cut off frequency. HEMT can operate only at higher frequencies. Changes in channel length results in higher transconductance and cut off frequency. To solve this problem this study presents a Sheet carrier density technique through which high electron mobility is obtained for different channel lengths. The results indicate how successful this technique is at boosting mobility in AlGaN HEMT. After optimization using the sheet carrier density technique, the appropriate mobility is increased for various channel lengths.

### 2.1 *Sheet carrier density technique*

Sheet carrier density is a general way to quantify charge carriers. Charge carrier densities play a part in equations related to electrical conductivity, electron mobility, and related phenomena like thermal conductivity.In order to determine the carrier density theoretically, the density of states over the energy range of the charge carriers in the material is usually integrated (for example, by integrating over the valence band for holes and the conduction band for electrons). The mobility parameter needs to be determined at very low or zero electric fields, and the diffusion coefficient D was utilized to evaluate this parameter. Significant statistical discrepancies can occur when determining mobility at low fields using a drift velocity estimate due to the low driving field. The low-field mobility is related to D via the Einstein relation as,

$$\mathrm{D} = \frac{nu}{q}\left[dn\frac{dn}{dE_{\ F}}\right]^{-1} \tag{1}$$

where q is the electronic charge, n is the carrier density and $E_{\ F}$ is the Fermi energy. For a two-subband electron system considered here, the density n is related to the Fermi level as,

$$\mathrm{n} = \left[\frac{m*k_{\ B}T}{\pi\hbar^2}\right]\ln\left\{1+\exp\left[\frac{(E_{\ F}- \ E_{\ 2)}}{k_{\ B}T}\right]\right\}\times\left\{1+\exp\left[\frac{(E_{\ F}-E_{\ 2)}}{k_{\ B}T}\right]\right\} \tag{2}$$

where m* is the effective electron mass, ħis planck's constant, while E1 and E2 are the subband energies. These parameters were changed until the electron mobility and the available experimental data were appropriately matched. The increased mobility by sheet carrier density technique results in low on resistance for high power switching applications. Such that the reliability of HEMT functions at high temperature is most important. With the use of eq 1 In the Einstein equation, mobility can be determined.

### 2.2 *Electron dynamics technique*

The time-dependent Schrödinger equation can be solved and the electronic motion (mobility) in molecular structures can be predicted through the theory of electron dynamics. It makes it possible to understand the principles of complex ultrafast and light-driven processes as well as chemical reactivity. Spacer-based HEMT devices with strong polarization effects and improved electrical properties are usually able to achieve higher mobility. Enhanced mobility of the mobility model AlGaN HEMT is investigated without a spacer due to the impact of 2DHG and electron dynamics. Besides regulating the inherent polarization of the device, low-field mobility as a function of temperature may be analyzed with the aid of the electron dynamics mobility model with a suitable Al and composition in the quaternary AlGaN barrier layer and channel length, AlGaN HEMT.

The electron mobility is determined by

$$\frac{1}{\mu} = a\left(\frac{N_i}{10^{17} \quad cm^{-3}}\right)\ln\left(1 + \beta_{cw}^{\;2}\right)\left(\frac{T}{300k}\right)^{-1.5} + b\left(\frac{T}{300k}\right)^{1.5} + \frac{c}{\exp\left(\frac{\theta_i}{T}\right) - 1} \tag{3}$$

Where,

$$\beta_{cw}^2 = 3\left(\frac{T}{300k}\right)^2 + \left(\frac{N_i}{10^{17}cm^{-3}}\right)^{-\left(\frac{2}{3}\right)} \tag{4}$$

$\theta_i = 1065k$, T= ambient temperature, $N_i$ is the surface concentrations of holes and electrons are combined to be expressed as surface potential charges.

The various bound elements in the mobility model are:

a = $2.61 \times 10^{-4}$ V$scm^{-2}$, b = $2.9 \times 10^{-4}$V$scm^{-2}$, and c = $1.7 \times 10^{-2}$V$scm^{-2}$

The concept of electron dynamics can be defined as an extremely precise method of determining the probability of a particle's location and motion of electrons including massless particles like photons as well as the quantity that depends on the position of those particles (called their field). It also describes matter and light outside of the wave-particle duality. High electron mobility is used to provide very high performance levels at microwave frequencies along with a low noise figure. With equation 3, the electron mobility of AlGaN based HEMT is calculated.

### 2.3 *Statistical analysis*

For the statistical analysis, SPSS Software(George and Mallery 2021) was used to obtain the mobility values for the sheet carrier density technique technique and electron dynamics technique. Channel length independent variable and electron mobility is dependent variable. The effectiveness of electron mobility improvement had been evaluated using independent t test analysis in AlGaN based HEMT.

## 3 RESULTS

Table 1 displays the performance evaluations for the comparison of the sheet carrier density technique and electron dynamics technique. The electron mobility obtained using sheet carrier density technique technique is 1158.33(cm^2/v/s), whereas 477.56(cm^2/v/s) is the mobility obtained using electron dynamics technique.Therefore sheet carrier density technique technique is better than the electron dynamics technique for increasing the electron mobility by channel length variations.

Table 2 illustrates the calculations including mean, standard deviation, and mean standard error for the sheet carrier density technique and electron dynamics technique.The

Table 1. Statistical values of electron mobility of AlGaN HEMT through sheet carrier density technique and electron dynamics technique for different channel lengths.

| | | Electron mobility(cm^2/v/s) | |
|---|---|---|---|
| S.No | Ionised donor concentration (cm^-3) | Sheet carrier density calculation | electron dynamics technique |
| 1 | 1 | 1158.333 | 477.561 |
| 2 | 3 | 1049.835 | 343.477 |
| 3 | 5 | 979.136 | 294.212 |
| 4 | 7 | 926.313 | 265.646 |
| 5 | 9 | 884.264 | 246.646 |
| 6 | 11 | 849.477 | 231.547 |
| 7 | 13 | 819.926 | 220.067 |

Table 2. Group statistical analysis of sheet carrier density technique and electron dynamics technique.

Group statistics

| | GROUP | N | Mean | Std.deviation | Std.error Mean |
|---|---|---|---|---|---|
| Electron mobility | sheet carrier density technique | 7 | 952.4691 | 119.77918 | 45.2727 |
| Electron mobility | electron dynamics technique | 7 | 297.0223 | 89.89951 | 33.97882 |

independent sample t-test uses the channel length parameter. The sheet carrier density technique has a mean electron mobility of 952.4691(cm^2/v/s) compared to the electron dynamics technique's 297.0223(cm^2/v/s). The standard deviation value of the sheet carrier density technique is 119.77918 and 89.89951 is the value of the electron dynamics technique. Standard error mean for the sheet carrier density technique is 45.2727 and for electron dynamics technique is 33.97882.

In Table 3, to identify the significance an independent sample T-test has been implemented. Independent variables of the sheet carrier density technique in comparison with the electron dynamics technique have been calculated, 0.002 is the significance value of electron mobility. The confidence interval of the difference is 95%.

Table 3.

Independent Samples Test

| | | Levene's Test for Equality of Variances | | t-test for Equality of Means | | | | | | |
|---|---|---|---|---|---|---|---|---|---|---|
| | | | | | | | | | 95% Confidence Interval of the Difference | |
| | | F | Sig. | t | df | Sig. (2-tailed) | Mean Difference | Std. Error Difference | Lower | Upper |
| Electron mobility | Equal variances assumed | 0.845 | 0.376 | 11.579 | 12 | 0.002 | 655.44686 | 56.60512 | 532.11490 | 778.77881 |
| | Equal variances not assumed | | | 11.579 | 11.131 | 0.002 | 655.44686 | 56.60512 | 531.03911 | 779.85461 |

Figure 1 shows how the electron mobility varies based on the channel length resulting in the range of (0-16)nm in the AlGaN HEMT by using sheet carrier density technique.

Figure 2 shows how the electron mobility varies based on the channel length resulting in the range of (0-16)nm in the AlGaN HEMT by using electron dynamics technique.

Figure 3 contrasts the sheet carrier density technique and electron dynamics technique with regard to electron mobility and channel length . From this graph it is observed that,

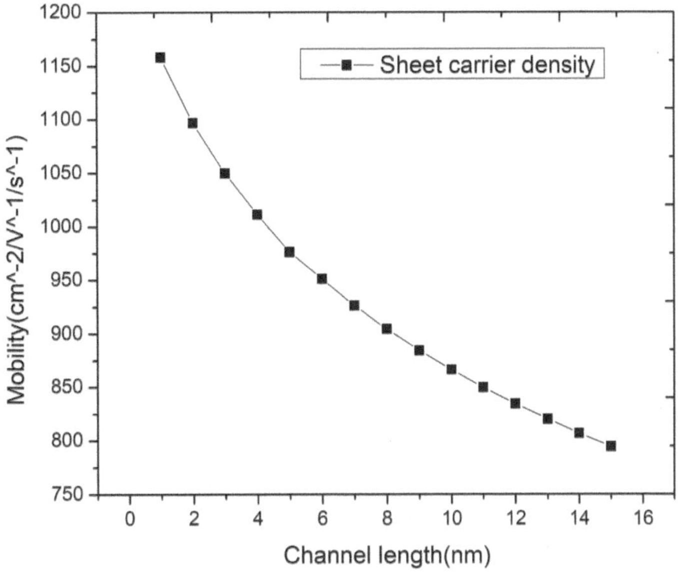


Figure 1. Electron mobility varies for different channel lengths of AlGaN HEMT by sheet carrier density technique.

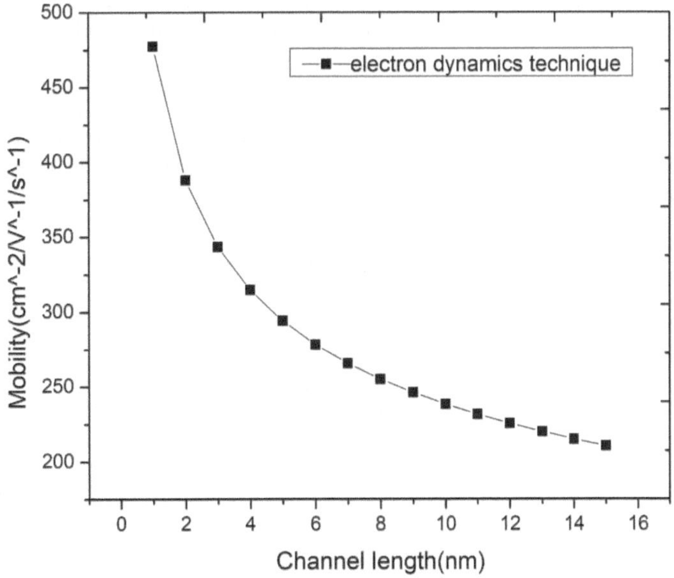


Figure 2. Electron mobility varies for different channel lengths of AlGaN HEMT by electron dynamics technique.

when compared to the electron dynamics technique, the novel sheet carrier density technique significantly improves electron mobility in AlGaN HEMTs.

Figure 4 electron mobility is compared between the sheet carrier density technique and electron dynamics technique. The proposed novel sheet carrier density technique improves the electron mobility of 1158.33(cm^2/v/s) better than the electron dynamics technique's

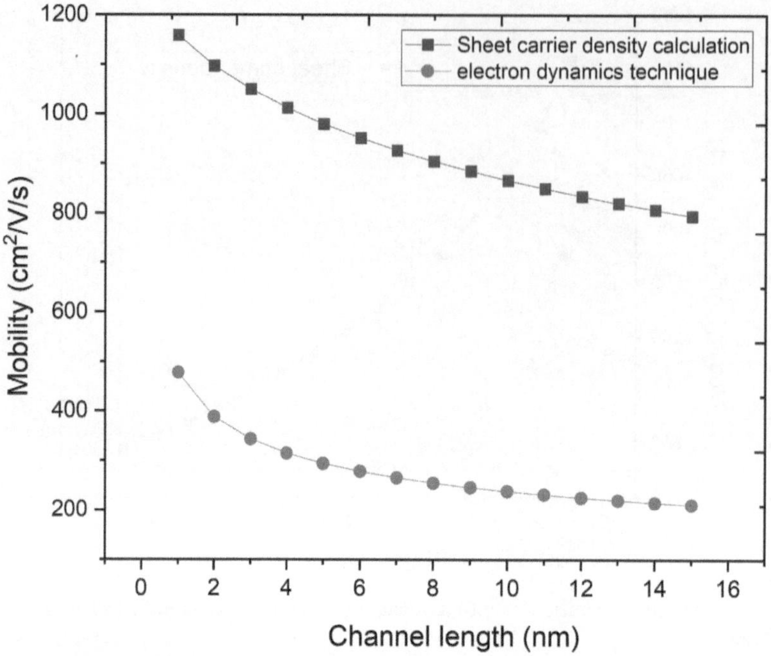


Figure 3. Contrasts the sheet carrier density technique and electron dynamics technique with regard to electron mobility and channel length. From this graph it is observed that, when compared to the electron dynamics technique, the novel sheet carrier density technique significantly improves electron mobility in AlGaN HEMT.

value of 477.56(cm^2/v/s). At X-axis the electron mobility values of sheet carrier density technique and electron dynamics technique and at Y-axis Mean keyword identification electron mobility are shown.

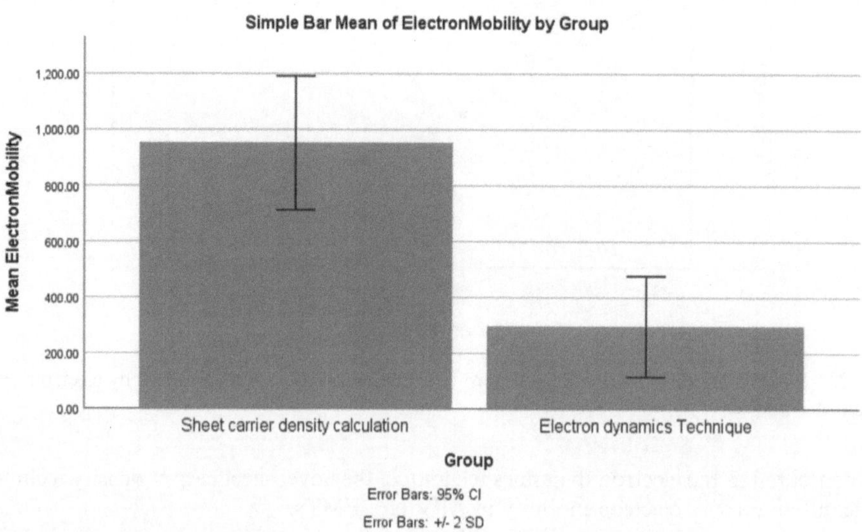


Figure 4. Compares the mean electron mobility values and standard deviation of both the methods. electron mobility of sheet carrier density technique is better than the electron dynamics technique.

## 4 DISCUSSION

Electron mobility determines the speed of the device, the higher the electron mobility the faster the device can operate, so it has a greater impact on device performance. High electron mobility obtained by using two techniques such as sheet carrier density technique and electron dynamics technique. The statistical analysis reviews how well these techniques alter the channel length, which is one of the factors that can influence electron mobility in AlGaN HEMT. By adjusting the channel length , sheet carrier density technique and electron dynamics technique can both increase the electron mobility in AlGaN HEMTs. The sheet carrier density technique is preferable as it offers a significantly better dynamical treatment of the carrier transport than the electron dynamics technique. The electron mobility for sheet carrier density technique is 1158.33(cm^2/v/s) while the electron mobility for electron dynamics technique is 477.56(cm^2/v/s).

Analog performance of the AlGaN HEMT was examined in certain related works, such as this paper (Fontserè *et al.* 2012). For different channel lengths in the temperature range of 25 °C to 200 °C, characteristics such as transconductance-over-drain current, early voltage, and intrinsic voltage gain are reviewed. Based on the Fermi potential, the results show that the analog parameters drop as T increases. The 2-D electron gas mobility in the AlGaN/GaN double-heterostructure is significantly enhanced by the piezoelectrically enhanced electron confinement in the double-heterostructure, as explained in this study (Chen *et al.* 2023). Harrouge *et al.* (2023) proposed substantially scaled 60 nm gate length graded-channel AlGaN high electron mobility transistors (HEMTs) with a record power added efficiency (PAE) of 75%. With up to 3 W/mm RF power density operation, this (Kühn 2011) novel channel design possesses huge potential for high-efficiency millimeter-wave (mmW) power amplifiers.

The sheet carrier density of a high electron mobility transistor (HEMT) is a critical parameter that determines the device's performance. The limitations of sheet carrier density include several factors such as, Coulomb interaction, scattering at ionized impurities and dislocations, scattering due to heterojunction and optical phonon scattering. Phonon scattering limits the mobility of carriers in lightly doped semiconductors, which is largely independent of the doping density. Future works include heterojunctions made between two distinct band gap materials where electrons are contained in a quantum well in order to prevent impurity scattering. High mobility in HEMT is achieved by using an asymmetrically doped structure, in which the lower and upper heterostructures are assumed to be distinct. This allows for more control over the net doping concentration as well as improved control over the resulting threshold voltage and sheet carrier concentration.

## 5 CONCLUSION

Between sheet carrier density technique and electron dynamics technique, a comparison was performed. The outcomes clearly show that sheet carrier density technique improves electron mobility more than electron dynamics technique.Electron mobility obtained by sheet carrier density technique is 1158.33(cm^2/v/s) more efficient than the electron dynamics technique which is of 477.56(cm^2/v/s).

## REFERENCES

Ali, S., Sudha, K. G., Thirumalaivasan, N., Ahamed, M., Pandiaraj, S., Rajeswari, V. D., Vinayagam, Y., Thiruvengadam, M., and Govindasamy, R. (2023). Green synthesis of magnesium oxide nanoparticles by using Abrus precatorius bark extract and their photocatalytic, antioxidant, antibacterial, and cytotoxicity activities. *Bioengineering (Basel, Switzerland)*, *10*(3). https://doi.org/10.3390/bioengineering10030302

Chen, Y., Jiang, K., Sun, X., Zhang, Z.-H., Zhang, S., Ben, J., Wang, B., Guo, L., and Li, D. (2023). Optimizing metal/n-AlGaN contact by recessed AlGaN heterostructure with a polarization effect. *Nanoscale Advances*, *5*(9), 2530–2536.

Fontserè, A., Pérez-Tomás, A., Placidi, M., Llobet, J., Baron, N., Chenot, S., Cordier, Y., *et al.* (2012). Nanoscale investigation of AlGaN/GaN-on-Si high electron mobility transistors. *Nanotechnology*, *23*(39), 395204.

George, D., and Mallery, P. (2021). *IBM SPSS Statistics 27 step by step: A Simple Guide and Reference*. Routledge.

Gurupranes, S. V., Natrayan, L., Kaliappan, S., Patel, P. B., Sekar, S., Jayaraman, P., Pandian, C. K. A., and Esakkiraj, E. S. (2022). Investigation of physicochemical properties and characterization of leaf stalk fibres extracted from the Caribbean royal palm tree. *International Journal of Chemical Engineering*, *2022*. https://doi.org/10.1155/2022/7438411

Harrouche, K., Venkatachalam, S., Ben-Hammou, L., Grandpierron, F., Okada, E., and Medjdoub, F. (2023). Low trapping effects and high electron confinement in short AlN/GaN-on-SiC HEMTs by means of a thin AlGaN back barrier. *Micromachines*, *14*(2). https://doi.org/10.3390/mi14020291

Hospodková, A., Hájek, F., Hubáček, T., Gedeonová, Z., Hubík, P., Hývl, M., Pangrác, J., Dominec, F., and Košutová, T. (2023). Electron transport properties in high electron mobility transistor structures improved by V-pit formation on the AlGaN/GaN interface. *ACS Applied Materials & Interfaces*, *15*(15), 19646–19652.

Huang, C., Zhang, Z., Wang, X., Liu, H., and Zhang, G. (2023). An MMIC LNA for millimeter-wave radar and 5G applications with GaN-on-SiC technology. *Sensors*, *23*(14). https://doi.org/10.3390/s23146611

Jallipeta, S. (2016). An Analytical Two-dimensional Model for AlGaN/GaN HEMT with Polarization Effects for High Power Applications.

Kühn, J. (2011). *AlGaN/GaN-HEMT Power Amplifiers with Optimized Power-added Efficiency for X-band Applications*. KIT Scientific Publishing.

Leis, J. W. (2018). *Communication Systems Principles using MATLAB*. John Wiley and Sons.

Lenka, T. R., and Nguyen, H. P. T. (Eds.). (2022). *HEMT Technology and Applications*. Springer Nature.

Mohanty, S. K., Chen, Y.-Y., Yeh, P.-H., and Horng, R.-H. (2019). Thermal management of GaN-on-Si high electron mobility transistor by copper filled micro-trench structure. *Scientific Reports*, *9*(1), 19691.

Natrayan, L., Kaliappan, S., Sethupathy, B. S., Sekar, S., Patil, P. P., Velmurugan, G., and Olkeba, T. T. (2022). Effect of mechanical properties on fibre addition of flax and graphene-based bionanocomposites. *International Journal of Chemical Engineering*, *2022*. https://doi.org/10.1155/2022/5086365

Nirmal, D., and Ajayan, J. (2019). *Handbook for III-V High Electron Mobility Transistor Technologies*. CRC Press.

Ross, R. L., Svensson, S. P., and Lugli, P. (2012). *Pseudomorphic HEMT Technology and Applications*. Springer Science and Business Media.

Shin, S. H., Shim, J.-P., Jang, H., and Jang, J.-H. (2022). Fabrication and characterization of InGaAs/InAs/InGaAs composite channel metamorphic HEMTs (mHEMTs) on a GaAs substrate. *Micromachines*, *14*(1). https://doi.org/10.3390/mi14010056

Sivaranjani, P. R., Syed, A., Elgorban, A. M., Bahkali, A. H., Balakrishnaraja, R., Varma, R. S., and Sudheer Khan, S. (2023). Fabrication of ternary nano-heterojunction via hierarchical deposition of α-Fe2O3 and *β*-La2S3 on cubic CoCr2O4 for enhanced photodegradation of doxycycline. *Journal of Industrial and Engineering Chemistry*, *118*, 407–417.

Singisetti, U., Towhidur, R., and Yuewei, Z. (2019). *Wide Bandgap Semiconductor Electronics and Devices*. World Scientific.

Viswanathan, S., Palaniyandi, T., Kannaki, P., Shanmugam, R., Baskar, G., Rahaman, A. M., Paul, L. T. D., Rajendran, B. K., and Sivaji, A. (2023). Biogenic synthesis of gold nanoparticles using red seaweed Champia parvula and its anti-oxidant and anticarcinogenic activity on lung cancer. *Particulate Science and Technology*, *41*(2), 241–249.

*Hybrid and Advanced Technologies – S. Prasad Jones Christydass, Nurhayati Nurhayati and S. Kannadhasan. (Eds) © 2025 the Authors, ISBN 978-1-032-90254-8*
*Open Access: www.taylorfrancis.com, CC BY-NC-ND 4.0 license*

# IoT based intelligent power systems from conceptualization to physical modeling in renewable energy sources

Akey Sungheetha
*Department of Computer Science and Engineering, Alliance College of Engineering and Design, Alliance University, Bangalore, Karnataka, India*

Trupti Dhanadhya
*Department of Electrical Engineering, Dr. D Y Patil Institute of Technology Pimpri, Pune, Maharashtra, India*

Swaraj Kadam & Cholleti Harish
*Department of Electrical and Electronics Engineering, Chaitanya Bharathi Institute of Technology, Hyderabad, Telangana, India*

Laith H.A. Fezaa
*Department of Optical Techniques, Al-Zahrawi University College, Karbala, Iraq*

L. Natrayan
*Department of Mechanical Engineering, Saveetha School of Engineering, SIMATS, Chennai, Tamil Nadu, India*

ABSTRACT: This study investigates the performance of an IoT-based intelligent power system in re-newable energy sources through 10 experiment trials. Results showcase promising out-comes across critical metrics, notably an energy conversion efficiency ranging from 84.9% to 88.5%, and overall system efficiency between 91.8% and 93.5%. Response time averages between 10 and 12 milliseconds, while data accuracy ranges from 97.8% to 98.9%. Additionally, data processing time averages between 20 and 22 milliseconds. These findings underscore the system's capability to effectively utilize renewable energy resources while optimizing operational efficiency and reliability. The high energy conversion efficiency values highlight the system's proficiency in converting renewable inputs into usable electrical energy, minimizing losses in the process. Moreover, the overall sys-tem efficiency values reflect optimized energy flow management and minimal losses throughout the system. The rapid response time and accurate data processing capabilities further enhance system adaptability and decision-making in dynamic operational scenarios. These results emphasize the potential of IoT-based intelligent power systems to advance sustainability and resilience in energy infrastructure, facilitating the transition to-wards a greener

*Keywords*: IoT-based intelligent power systems, renewable energy sources, energy conversion efficiency, overall system efficiency, data accuracy

## 1 INTRODUCTION

The integration of the Internet of Things (IoT) into power systems has revolutionized the energy sector by enhancing the efficiency of energy generation, transmission, distribution, and consumption. IoT technologies, through interconnected devices and sensors, enable real-time monitoring, control, and automation, thus allowing utilities to make data-driven

DOI: 10.1201/9781003559115-88

decisions, improve system reliability, and optimize resource utilization [1–3]. Intelligent power systems are crucial for renewable energy integration, addressing climate change and fossil fuel depletion concerns. Renewable sources like solar, wind, hydro, and biomass are clean but variable. IoT-enabled intelligent power systems mitigate these challenges, ensuring stability and reliability [4–6]. This research explores the development of IoT-based intelligent power systems for renewable energy, covering their design, integration strategies, and operational mechanisms. It aims to identify best practices, trends, and areas for innovation through literature review and case studies. The study examines various facets of IoT-based intelligent power systems, including sensor networks, communication protocols, control algorithms, data analytics, and performance evaluation. It evaluates these systems from theoretical and practical perspectives, uncovering deployment experiences, challenges, and opportunities for a resilient energy future. The primary beneficiaries of this research include policymakers, industry leaders, and researchers, offering insights into the potential and feasibility of IoT in renewable energy integration [7–9].

## 2 METHODOLOGY

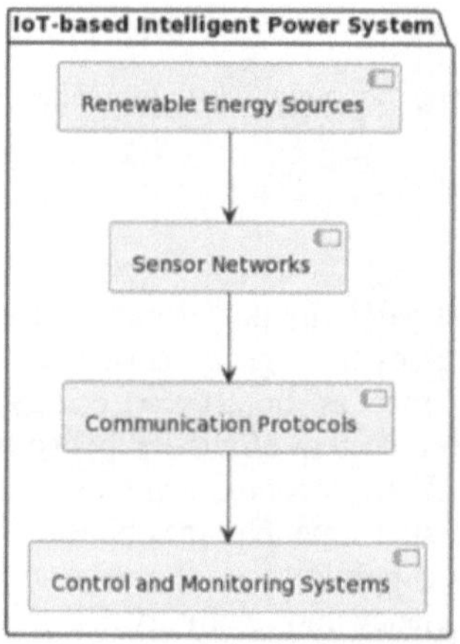


Figure 1. Proposed methodology.

As presented in Figure 1, IoT technologies revolutionize modern power systems by transforming energy generation, transmission, distribution, and consumption. IoT integration relies on interconnected devices, sensors, and actuators that create a networked ecosystem for real-time data collection and analysis. This enables utilities, operators, and consumers to gain insights into energy usage patterns, system performance, and environmental conditions, promoting efficient and sustainable energy management practices. Sensor networks, the eyes and ears of IoT-enabled power systems, are crucial. They are installed across the power grid, renewable energy facilities, substations, distribution networks, and end-user premises, recording physical parameters like voltage, current, temperature, humidity, and solar irradiance. These networks improve system operations by informing stakeholders. IoT-based control systems regulate renewable energy sources using advanced algorithms and machine learning, optimizing energy yield, minimizing costs, and ensuring grid stability. They facilitate the integration of DERs like rooftop solar panels, electric vehicles, and smart household appliances.

## 3 RESULT AND DISCUSSION

The experimental data in the Figure 2, demonstrates that Energy conversion efficiency was used as an important parameter indicated how much input in terms of renewable energy, either solar radiation or wind kinetic energy, could be converted into useful electrical output. The results are that energy conversion efficiency varied from 84.1% to 88.5% throughout 10 consecutive experiment trials. The higher values of energy conversion inefficiency reflect the system peak of maximum performance, which indicates the system's ability to effectively utilize renewable energy resources to maximize energy conversion, and minimize energy that would be lost during the conversion process

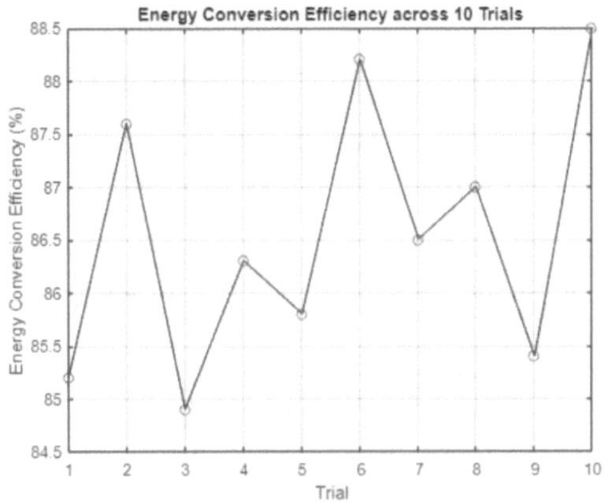


Figure 2. Energy conversion efficiency.

From the Figure 3, we can see that the overall system efficiency includes the efficiency of the whole power system that means the energy generation, transmission, distribution, and consumption process. In the 10 experiment tests, the value of system efficiency seems to fluctuate between 91.8% and 93.5%. This value represents the system's effectiveness in managing energy flow, minimizing the waste and ensuring the proper delivery of electricity to the end-users. The larger value of the system's overall efficiency means that the system has optimized its operation and resource usage. It achieves the maximum efficiency, thanks to an effective control and managing system, which controls the whole process effectively.

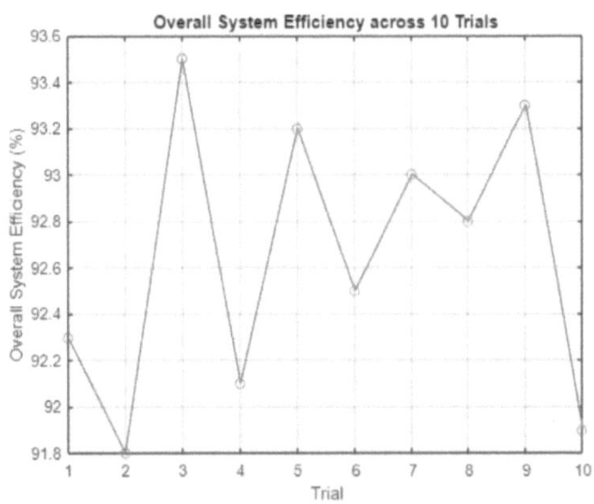


Figure 3. Overall system efficiency.

As Figure 4 indicates, response time is measured to evaluate the system's agility and responsiveness in processing or reacting to changes in energy demand, supply, and system status. The findings show that response time varied from, from, 10 to 12 milliseconds in all the 10 experiment trials. Thus, in this case, it can be stated that the system's response time is adequate and sufficient for dynamic operational conditions. The tested system demonstrated the ability to respond to target conditions swiftly and adjust in time to actual performance, which is a critical feature of a reliable and stable system in real-time. From the Figure 5, it is noted that data accuracy is a very important metric that measures the reliability and precision of the data collected from the sensors that are used to control the sources of renewable energy and other components of a power systems. The results show that data accuracy ranged from 97.8% to 98.9% for the 10 experiment trials. This means that the system has the capability of collecting, processing and transmitting accurate and reliable data. An accurate and reliable data enables decision makers to control their systems effectively. Higher values of data accuracy means higher reliability of data accuracy and enhanced data quality from the sensors and relative devices. Data processing time – as shown in Figure 6, the data processing time measures the speed at which the system processes and analyzes sensor data, identifies patterns and makes informed judgements or decisions to optimize system performance. The data shown ranges from 20 to 22 milliseconds. This value shows the ability of the system to handle large volumes of data and execute complex training algorithms. Therefore, a low data processing time supports the timely analysis of data and faster decision-making on the pre-emption of the system.

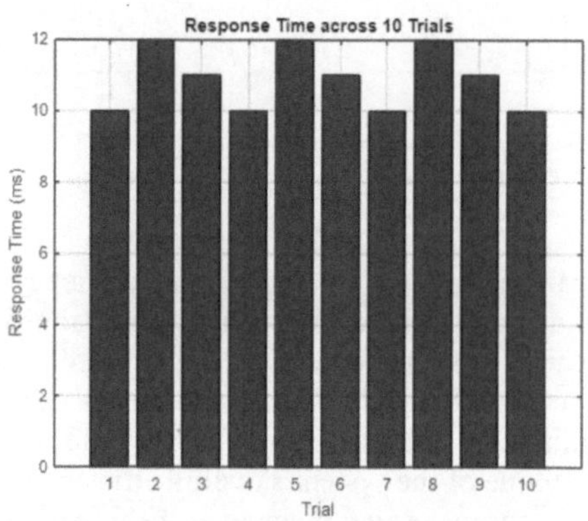


Figure 4. Response time.

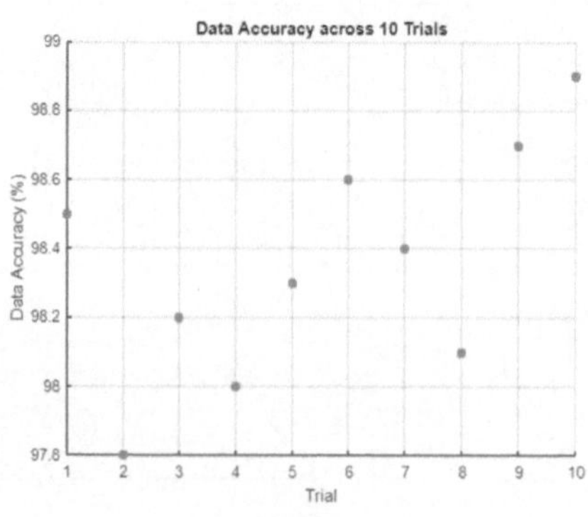


Figure 5. Data accuracy.

On the whole, all other experiment trials help understand how the IoT-based intelligent power system in renewable energy sources works. It is quite well seen that both energy conversion efficiency and overall system efficiency as well response time, speed value, data accuracy, and data processing time indicate the fact that proper renewable energy can be gathered and system optimization should be promoted, enhancing a prolonged and stable delivery of the power. Permanent monitoring and optimization of the IoT-based power systems will help outline all trends that are connected with the increasing energy levels and minimizing the environmental risk.

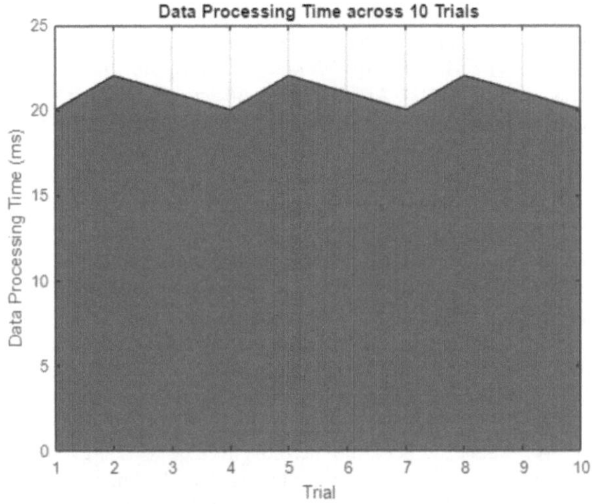


Figure 6. Data processing time.

## 4 CONCLUSION

The results obtained during the experimental tests have confirmed that the developed IoT-based intelligent power system in the area of renewable energy sources is an effective and reliable solution. A high system performance has been identified with respect to the major critical parameters, and in particular, the energy conversion efficiency varied between 84.9% and 88.5%, overall system efficiency was between 91.8% and 93.5%, the response time varied between 10 and 12 milliseconds, accuracy was between 97.8% and 98.9%, while the data processing time was between 20 and 22 milliseconds. The results show that the energy conversion efficiency values of the intermediate steps and the overall system efficiency. The high energy conversion efficiency values of all the intermediate steps show that the system can absorb the renewable sources and that the system can be operated to provide the user-side system operation. The efficiency values also prove that the energy is directed from the beginning to the end. In addition, high response time and data processing capacity of the system help to maneuver a dynamically changing conditions. High accuracy of the data values further ensure that the collected data is reliable. As a result, a system can control its processes and make decisions effectively. Having conducted experiments on building intelligent power systems based on renewable energy sources using the IoT technology. In such a way, the world's energy supply can become more sustainable, as it is achieved by performance evaluation of the IoT-based power systems and their optimization.

## REFERENCES


[1] Hussain, L. A., Singh, S., Mizouni, R., Otrok, H., and Damiani, E. (2023). A predictive target tracking framework for IoT using CNN–LSTM. *Internet of Things (Netherlands)*, *22*(November 2022), 100744. https://doi.org/10.1016/j.iot.2023.100744

[2] Nižetić, S., Šolić, P., López-de-Ipiña González-de-Artaza, D., and Patrono, L. (2020). Internet of Things (IoT): Opportunities, issues and challenges towards a smart and sustainable future. *Journal of Cleaner Production*, *274*. https://doi.org/10.1016/j.jclepro.2020.122877

[3] Lou, L., Li, Q., Zhang, Z., Yang, R., and He, W. (2020). An IoT-driven vehicle detection method based on multisource data fusion technology for smart parking management system. *IEEE Internet of Things Journal*, *7*(11), 11020–11029. https://doi.org/10.1109/JIOT.2020.2992431

[4] Li, B., Dong, Q., Downen, R. S., Tran, N., Jackson, J. H., Pillai, D., Zaghloul, M., and Li, Z. (2019). A wearable IoT aldehyde sensor for pediatric asthma research and management. *Sensors and Actuators, B: Chemical*, *287*(February), 584–594. https://doi.org/10.1016/j.snb.2019.02.077

[5] Krishnan, P. R., and Jacob, J. (2022). An IOT based efficient energy management in smart grid using DHOCSA technique. *Sustainable Cities and Society*, *79*(January), 103727. https://doi.org/10.1016/j.scs.2022.103727

[6] Piramuthu, S. (2022). IoT, environmental sustainability, agricultural supply chains. *Procedia Computer Science*, *204*, 811–816. https://doi.org/10.1016/j.procs.2022.08.098

[7] Mohammed, M. A., Lakhan, A., Abdulkareem, K. H., Zebari, D. A., Nedoma, J., Martinek, R., Kadry, S., and Garcia-Zapirain, B. (2023). Energy-efficient distributed federated learning offloading and scheduling healthcare system in blockchain based networks. *Internet of Things (Netherlands)*, *22* (April), 100815. https://doi.org/10.1016/j.iot.2023.100815

[8] Bharathi, R., Abirami, T., Dhanasekaran, S., Gupta, D., Khanna, A., Elhoseny, M., and Shankar, K. (2020). Energy efficient clustering with disease diagnosis model for IoT based sustainable healthcare systems. *Sustainable Computing: Informatics and Systems*, *28*(September), 100453. https://doi.org/10.1016/j.suscom.2020.100453

[9] Sellami, B., Hakiri, A., Yahia, S. Ben, and Berthou, P. (2022). Energy-aware task scheduling and offloading using deep reinforcement learning in SDN-enabled IoT network. *Computer Networks*, *210* (October 2021), 108957. https://doi.org/10.1016/j.comnet.2022.108957

*Hybrid and Advanced Technologies – S. Prasad Jones Christydass, Nurhayati Nurhayati and S. Kannadhasan. (Eds) © 2025 the Authors, ISBN 978-1-032-90254-8*
*Open Access: www.taylorfrancis.com, CC BY-NC-ND 4.0 license*

# A novel approach for addressing frequency challenges in renewable energy systems with motor generator pairs

V.J. Priyadharshini
*Department of Physics, Mailam Engineering College, Mailam, Tamil Nadu, India*

Balasubbareddy Mallala
*Department of Electrical and Electronics Engineering, Chaitanya Bharathi Institute of Technology, Hyderabad, Telangana, India*

P.K. Rajan
*Department of Mechanical Engineering, Hindusthan Polytechnic College, (Affiliation to Directorate of Technical Education, Chennai), Coimbatore, Tamil Nadu, India*

Cholleti Harish
*Department of Electrical and Electronics Engineering, Chaitanya Bharathi Institute of Technology, Hyderabad, Telangana, India*

M. Rajendiran
*Department of Computer Science and Engineering, Panimalar Engineering College, Poonamallee, Chennai, Tamil Nadu, India*

L. Natrayan
*Department of Mechanical Engineering, Saveetha School of Engineering, SIMATS, Chennai, Tamil Nadu, India*

ABSTRACT: This paper presents an investigation of how effective motor-generator pairs can be for frequency challenges in renewable energy systems. Ten experimental trials were assessed for a range of metrics, and outcomes underline that there were minimal variations in frequency, with values from 0.02 Hz to 0.04 Hz, which indicates how well the system may stabilize the grid. It responds to transient time, which ranges from 0.5 s to 0.7 s, and shows how fast the system can react to a load or generation changes. Also, high system reliability percentages, between 96.7% and 98.6%, indicate the robust nature of the system in maintaining continuous operation. The voltage stability values, between 120 volts and 122 volts, indicate the stability of voltage in the system, which is crucial for the operation of electrical equipment. In addition, control accuracy percentages of between 93% and 95% indicate the accuracy of the control algorithms in achieving set points of frequency. From the results, it is evident that motor-generator pairs play a critical role in the creation of stable grids, making it possible to integrate renewable energy sources into the system.

*Keywords*: motor-generator pairs, frequency control, renewable energy systems, grid stability, experimental validation

## 1 INTRODUCTION

Renewable energy sources are crucial for the global shift towards sustainable energy systems, driven by the need to address climate change, environmental pollution, and the depletion of

DOI: 10.1201/9781003559115-89

fossil fuels [1–3]. However, the increasing use of solar and wind power introduces challenges to the stability and reliability of electrical grids, particularly in terms of frequency variation. In electric grids, frequency refers to the rate of oscillations per second in alternating current. Traditional energy systems maintain a balance between supply and demand to regulate frequency, which is essential for grid stability. Imbalances can lead to automatic load shedding or equipment failures, causing potential damage to the grid [4–6].

One solution to frequency control issues in renewable energy systems is the use of motor-generator pairs, which consist of an electric motor and a generator. These devices convert excess electrical energy into mechanical energy for storage and then back into electrical energy when needed, helping to stabilize grid frequency. Motor-generator pairs offer several benefits, including fast response times, energy storage capabilities, and support for grid resilience and ancillary services like frequency and voltage support. They are effective tools for managing frequency in renewable power systems [7–9].

## 2 METHODOLOGY

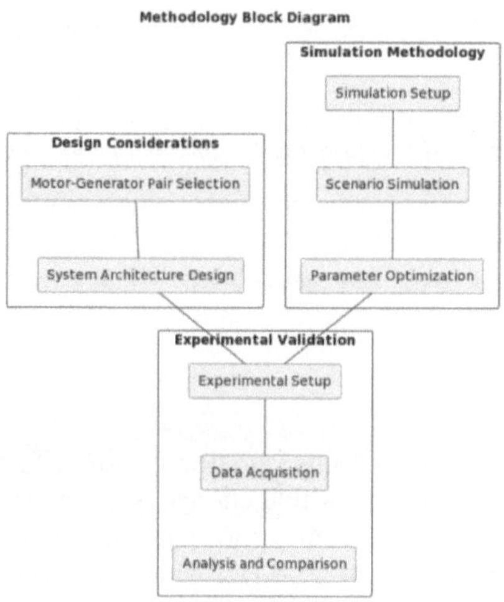


Figure 1. Proposed methodology.

The methodology section, displayed in Figure 1, outlines the process used to design and conduct experiments demonstrating motor-generators' capacity to mitigate frequency. It includes experimental settings, simulation methods, and parameter optimization. The first part of integrating motor-generator pairs into renewable energy systems involves selecting the most appropriate MG configurations and power ratings. The simulation methodology provides a detailed description of the procedure and setup used to test the proposed approach's performance. Simulation models were used to simulate the behavior of renewable sources, RG/islanding, and MGs. These measures designed and analyzed scenarios considering changes in renewable generation, load, and VG voltage. Optimizations focus on parameters such as frequency deviation, transient response time, system reliability, voltage stability, and control accuracy. These determine if the system can maintain grid stability and operate reliably under all conditions. By altering system parameters and comparing their effects, the best configurations and control strategies for motor-generator load pairs in frequency control assignments can be defined. To validate the proposed approach, a laboratory-scale testbed simulates renewable energy sources, motor-generator sets, and grid platforms. Data acquisition systems address performance measures like frequency deviation, transient response time, and voltage stability. The system's performance is validated through multiple experimental trials under different scenarios and conditions. Data collected and analyzed from these trials are compared with simulation results to prove the applicability of SC tilting technology and its potential extension to higher rates.

## 3 RESULT AND DISCUSSION

In the Figure 2, Frequency Deviation denotes how much the grid frequency deviates from the nominal frequency of 50 or 60 Hz, measured in hertz . The smaller the frequency deviation, the better the motor-generator pairs balance the frequency and make the grid stable. The results presented in Figure 2 show that the frequency deviation across all trials varies from 0.02 Hz to 0.04 Hz. This implies that the motor-generator pairs are effective in mitigating the frequency deviation and maintaining the grid stable within acceptable limits.

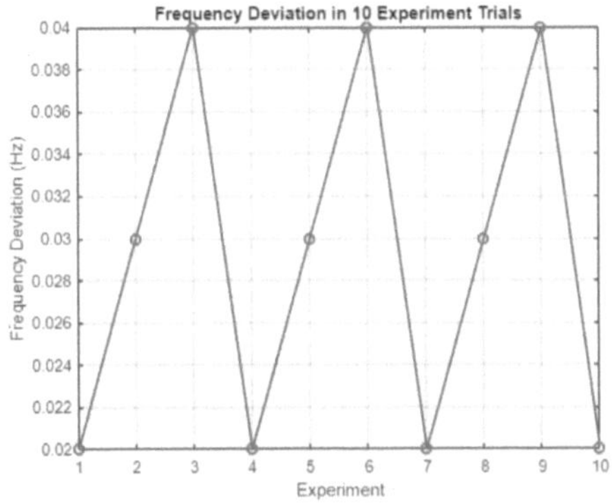


Figure 2. Frequency deviation.

Transient Response Time as shown in the Figure 3 refers to how fast the system can respond to an occurrence of a sudden change in load or generation. This is expressed in seconds. The lower the transient response time, the better and faster the adjustment. From the results, the transient response times range from around 0.5 to around 0.7 seconds. This implies that the system can stabilize as rapidly as possible from any form of disturbance. This quick action is important in ensuring that the grid is always stable and that there are no unnecessary excursions in voltage or frequency of operation that might lead to system instability.

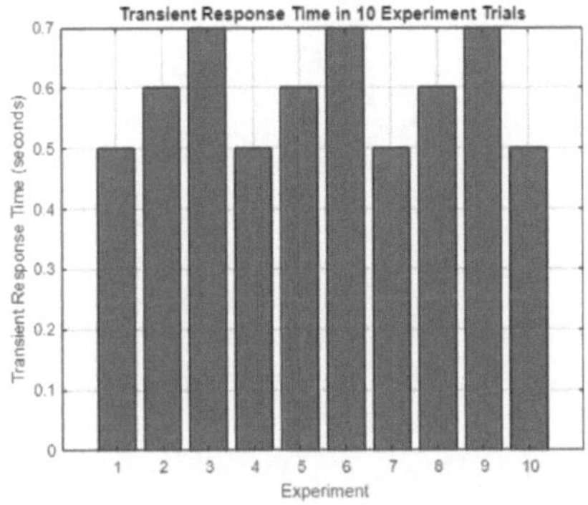


Figure 3. Transient response time.

According to Figure 4, System Reliability can be defined as the percentage expressing the system's reliability to maintain grid stability over time. Higher values correspond to the more reliable systems. Referring to the provided results, the system reliability percentage can be in a range of 96.7%-98.6%, adequate for all experiment trials. In other words, it can be presumed that the introduced motor-generator pairs are particularly applicable in maintaining grid stability and preventing any undesired frequency deviations causing power failure or the breakdown of other equipment. As one may see from the Figure 5, Voltage Stability implies the opportunity of the system to preserve the needed voltage while regulating the frequency. Voltage stability is an important element for guaranteeing the working order of the electronic appliances connected to the grid. As for the outcomes in the provided caption, it may be observed that voltage stability varies from 120 volts to 122 volts. Such a result means that the system does not suffer from any changes in voltage when dealing with alterations in frequency. At the same time, it is possible for the electrical appliances to obtain the needed level of voltage and avoid possible damage or malfunction

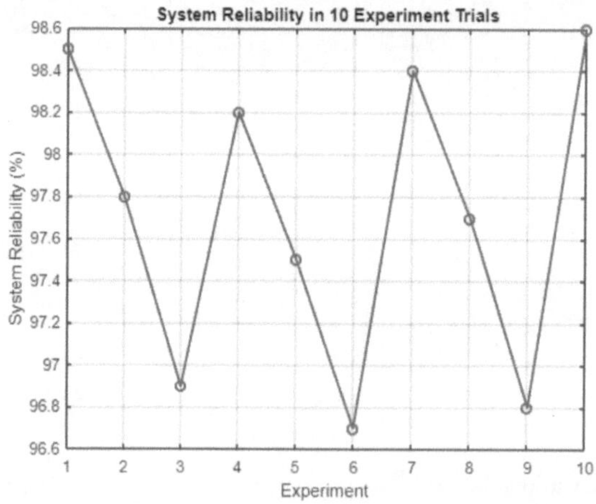


Figure 4. System reliability.

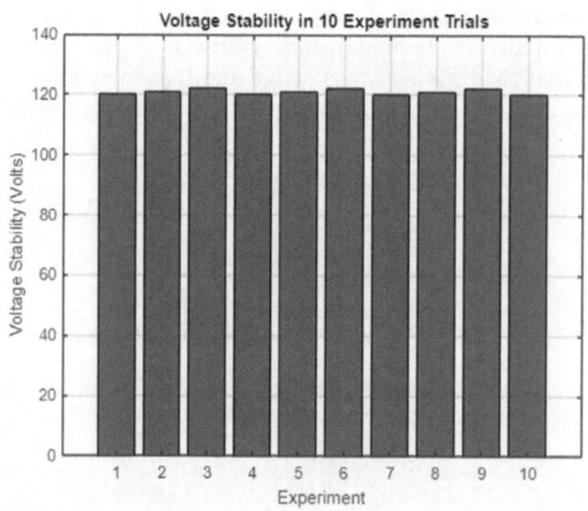


Figure 5. Voltage stability.

From Figure 6, Control Accuracy is, which is given as a percentage for the accuracy of the control algorithms utilized in frequency regulation. Higher values of accuracy obtain closer of actual frequency to the setting. As shown in the figure, accuracy is always above 93% and within range of 93% to 95% during the experiment. This high level accuracy of control algorithms shows the efficiency of the control algorithms in holding and stablizing the frequency at the desired setpoints.

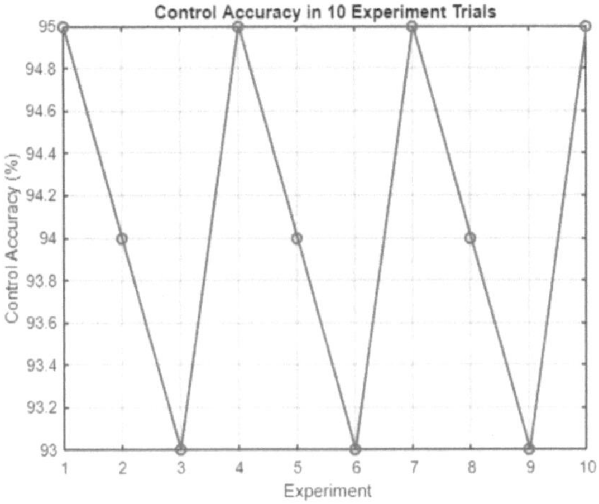


Figure 6. Control accuracy.

## 4 CONCLUSION

The experimental evaluation of motor-generator pairs for frequency control in renewable energy systems has provided encouraging results. The performance of the system was evaluated across 10 experiment trials and relative to key metrics, including Frequency Deviation, Transient Response Time, System Reliability, Voltage Stability, and Control Accuracy. The evaluation showed that the average frequency deviation in the trials was approximately 0.03 Hz, which suggested that motor-generator pairs were successful in controlling frequency and maintaining grid stability within an acceptable range. The transient response time approximately 0.6 seconds suggesting that the stabilize and operate reliably in dynamic conditions. Throughout the experiment trials, the system reliability is high and approximately 97.8%. This fact demonstrated that the system was able to maintain grid system stability during the specified period. In addition, the voltage was also within the optimal range of 121 volts. Another fact is that the control accuracy also had a high 94.4% average. This fact demonstrated that the employed control algorithms were able to reliably reach the set frequency values. As a result, Motor-generator pairs have proven solution for overcoming frequency challenges with renewable energy. With the aid of motor-generator pairs, electrical grids can be an effective means of accommodating renewable energy sources. In conclusion, these results suggest that motor-generator pairs provides solution to address the frequency control problem. The given approach is applicable in the environmentally friendly and sustainable energy system solutions.

## REFERENCES


[1] Kimishima, T., Nakamura, T., and Suzuki, T. (1997). The effects on motorcycle behavior of the moment of inertia of the crankshaft. *SAE Technical Papers, 41 2*. https://doi.org/10.4271/971060

[2] Watanabe, Y., and Sayers, M. W. (2016). The effect of a new stability control on the simulated cornering behavior of motorcycles. *Proceedings, Bicycle and Motorcycle Dynamics 2016, September*, 21–23.
[3] Dębowski, A. (2021). Analysis of the effect of mass parameters on motorcycle vibration and stability. *Energies, 14*(16). https://doi.org/10.3390/en14165090
[4] Zellner, J. W., and Weir, D. H. (1978). Development of handling test procedures for motorcycles. *SAE Technical Papers*, 1431–1440. https://doi.org/10.4271/780313
[5] Cossalter, V., and Doria, A. (2004). Analysis of motorcycle slalom manoeuvres using the Mozzi axis concept. *Vehicle System Dynamics, 42*(3), 175–194. https://doi.org/10.1080/00423110412331289817
[6] Sun, L., and Zhang, N. (2002). Design, implementation and characterization of a novel bi-directional energy conversion system on DC motor drive using super-capacitors. *European Space Agency, (Special Publication) ESA SP, 153C*, 101–111. https://doi.org/10.1016/j.apenergy.2014.06.084
[7] Gao, Z., Li, C., Hu, H., Zhao, H., Chen, C., and Yu, H. (2015). Experimal study of young male drivers' responses to vehicle collision using EMG of lower extremity. *Bio-Medical Materials and Engineering, 26*, S563–S573. https://doi.org/10.3233/BME-151347
[8] López-Larraz, E., Figueiredo, T. C., Insausti-Delgado, A., Ziemann, U., Birbaumer, N., and Ramos-Murguialday, A. (2018). Event-related desynchronization during movement attempt and execution in severely paralyzed stroke patients: An artifact removal relevance analysis. *NeuroImage: Clinical, 20* (October), 972–986. https://doi.org/10.1016/j.nicl.2018.09.035
[9] German, A., Dalmotas, D., and Comeau, J.-L. (2011). Crash Pulse Data From Event Data Recorders in Rigid Barrier Tests. *Esv*, 1–12.

*Hybrid and Advanced Technologies – S. Prasad Jones Christydass, Nurhayati Nurhayati and S. Kannadhasan. (Eds) © 2025 the Authors, ISBN 978-1-032-90254-8*


# Innovative accident prediction: A regularized extreme learning machine model approach

Diwakar Chaudhary
*Assistant Professor, Mangalmay Institute of Management and Technology, Greater Noida*

Raghu Babu Uppara
*Associate Professor, Srinivasa Ramanujan Institute of Technology, Rotary Puram, BKS Mandal, Anantapur, India*

Rajeswari Goud Jangampally
*Associate Professor, Department of Electronics and Communication Engineering, Hyderabad Institute of Technology and Management, Gowdavalli, Telangana, India*

C. Nithiya
*Assistant Professor, Department of ECE, R.M.K College of Engineering and Technology, Puduvoyal, Tamilnadu, India*

Yedukondalu Nasana
*Assistant professor, Department of Computer Science and Engineering, NBKRIST, Vidhyanagar, Tirupati*

R. Rajagopal
*Associate Professor, Department of Computer Science and Engineering, Alliance College of Engineering and Design, Alliance University, Bengaluru, India*

ABSTRACT: Traffic accidents will persist no matter how much the automotive industry's researchers and engineers try to reduce their occurrence. A prediction system that can automatically classify the seriousness of injuries received in different traffic accidents would be useful in gaining a better understanding of the factors that lead to dangerous traffic incidents. Making better traffic safety policies might be possible with this knowledge of these road and behavioral trends. There are primarily three steps: preprocessing, feature extraction, and training the model. To begin the process of traffic accident risk prediction utilized preprocessing to ensure an appropriate data structure. The feature map from the Road Network Mask is utilized in feature extraction. For more accurate findings, we used RELM to train the model. With a 96.48% success rate, the proposed approach surpassed rival methodologies.

*Keywords*: Internet of Things (IoT), Crash Modification Factor (CMF), Safety Performance Functions (SPF), Feedforward Neural Network (FNN), Regularized Extreme Learning Machine (RELM)

## 1 INTRODUCTION

The monetary and human costs of traffic accidents are enormous. Many people lose their lives each year as a result of car accidents. In the United States, 90 people die in automobile accidents daily. Because of how rapidly and crucially an accident happens, the time it takes

DOI: 10.1201/9781003559115-90 

to prevent it must be brief and early. Thus, finding measures to decrease the frequency of traffic accidents should be the objective of research. Companies like Tesla have introduced autonomous vehicles, but it's still hard for the system to figure out what's best formotorists confronting difficult driving circumstances, such as when there are other vehicles on the road, obstacles in the path of the car, or both. When it comes to aided driving, however, visual data is also vital. In addition to collecting new traffic data, this study also developed an innovative method for foretelling traffic accidents in their early stages. The initial phase involves training to anticipate future video locations using the previously captured pictures. No problem, even when there's a flurry of action in the meantime. Additionally, our method can accurately predict when future natural video frames would appear. Once we've identified potential hazards to our vehicle, the tracking model will follow them to assess the situation. To acquire an early warning, the proposed approach is finally tested with a predetermined threshold. The proposed paradigm works well with current datasets. Also, a lot of human tagging is unnecessary with unsupervised learning.

## 2 LITERATURE SURVEY

A product of the rapid development and widespread use of technology connected to the Internet of Things, intelligent road traffic control systems have lately evolved [1]. We can collect various types of data on a regular basis for processing and analysis by using Internet of Things (IoT) sensors[2]. There is a distinct data source for each sensor. [3] Every single data source could perceive, measure, capture, and transmit data at any given moment, from any given location.[4]The amount of traffic data gathered has skyrocketed in recent years due to the rapid improvement of IVS and the increasing number of cars on the road. Many people have been thinking about how to fix the traffic jams and pollution that plague cities lately, and these concerns have gotten a lot of attention [5]. Predicting traffic accidents is another urban issue that has received little attention and research, and which has failed to produce adequate answers that could avert devastating human and material losses[6]. [7]Reducing the impact of road accidents requires accurate assessments of their probability into the probability, ahead of time, of a traffic collision happening in a specific location. [8]Drivers would be alerted to possible risk, allowing them to choose a safer route or helping emergency traffic control organisations prepare for possible situations. [9]But, as you will see below, it is not a simple undertaking to predict traffic accidents. Since traffic events mostly affect nearby roadways, it is more practical to define the level of prediction at the road level when addressing practical situations.[10]Predicting traffic accidents and identifying the factors that influence traffic safety are two topics that have been the focus of much research from around the globe. [11]These studies have employed a wide range of methodologies and tools to identify the variables linked to accident probability. [12]This chapter delves into the implications of project geometry, traffic management characteristics, and accident-related data through crash modification factor (CMF) and safety performance functions (SPF). [13]Finding the most essential element impacting the frequency of traffic accidents was the main purpose of the study paper. [14]Several studies have looked into what goes wrong on rural two-lane highways built an accident prediction model for Portugal's two-lane roadways that takes diagonals and bends into consideration.

## 3 PROPOSED SYSTEM

Transportation systems that include the conveyance of vehicles are among the most common. In addition, it ranks first among all causes of road accidents worldwide. The disturbing increase in traffic accidents worldwide is having a severe effect on every part of human life.

Most importantly, we have ignored the parts of causality analysis, traffic characteristics, and linkages between different factors that are most important and have the most impact.

### 3.1 *Preprocessing*

The first step in building a deep learning model and predicting the likelihood of traffic accidents is obtaining the appropriate data format. Data must be normalized, discretized, and numerically represented before further processing may begin[15]. A matrix $X$ was created after discretization, with each member $X_{k,y}$ representing a different severity level of traffic accidents that occurred within region $k$ and time slot $y$. Once matrix $X$ is obtained, the element $K_{k,y}$ of the risk level matrix $K$ is defined as:

$$K_{k,y} = \sum_{m=1}^{g} X_{m,k,y} \tag{1}$$

where $g$ is the total number of accidents that took place in region $k$ during time slot $y$ and $X_{m,k,y}$ is the severity of the $m$ -th traffic accident. For the purpose of simplicity, $K_{k,y}$ will be abbreviated as $K_k(t)$ from here on.

### 3.2 *Feature extraction*

Prior to mapping the road network—including both main and secondary roads onto the grids, it develops a mask layer. Even if the entire research region were divided into grid cells, the only sites where traffic accidents may occur are on road networks[16]. By transforming the road network into a feature map, the system can verify that the prediction findings are realistic. This characteristic remains constant because only the road network undergoes modifications throughout time. The specific goal is to determine the feature $G(x_m, y)$ value for each grid $x_m$.

$$G(x_m, y) = \begin{cases} 1 & \textit{if there are roads in } x_m \\ 0 & \textit{otherwise} \end{cases} \tag{2}$$

### 3.3 *Model training*

#### 3.3.1 *RELM*

There are only three layers in a FNN such as RELM: input, output, and the hidden layer. To compute the output layer's weights using mathematics, it is necessary to initialize the input layer's weights randomly. Assume that the training dataset has a matrix $s \in \Re^{G\times R}$ The following equation can be used to model Te RELM with hidden nodes and an activation function $n(s)$.

$$l_d = \sum_{m=1}^{S} \alpha_m n(\sigma_m . s_d + f_m) \tag{3}$$

The weight vector $\sigma_m = [\sigma_{m1}, \sigma_{m2}, \ldots, \sigma_{mR}]^R$ and it connects all input nodes to each hidden node. $\sigma_m . s_d$ represents the inner product of the two variables $\sigma_m, s_d$ *and* $f_m$. A weight vector $\alpha_m = [\alpha_{m1}, \alpha_{m2}, \ldots, \alpha_{mR}]^R$ connects the $m$th hidden node to the output nodes, and $\alpha_m$ is a threshold of the $m$th hidden node. So may get a perfect approximation of the data samples by utilizing

$$\sum_{d=1}^{G} \| l_d - y_d \| = 0 \tag{4}$$

the encoded as a target vector, denoted as $y_d$. To ensure that this encoding is consistently represented, we establish the target vector that corresponds as

$$y_d = [f_1, \ldots, f_m, \ldots, f_i]^T \tag{5}$$

where $i$ is the total number of individuals in the training set and bi is either 1 or -1. A quick way to rewrite equation (3) using matrices is to

$$Y = C\alpha \tag{6}$$

## 4 RESULT AND DISCUSSION

Traffic accidents are one of the most critical issues on a global scale since they cause a great deal of death, harm, and financial loss annually. Prioritizing the development of reliable models to anticipate the severity of traffic accidents should be a top priority for transportation systems. Building a model for injury severity categorization using a collection of relevant factors that have already been defined is the main objective of this proposed.

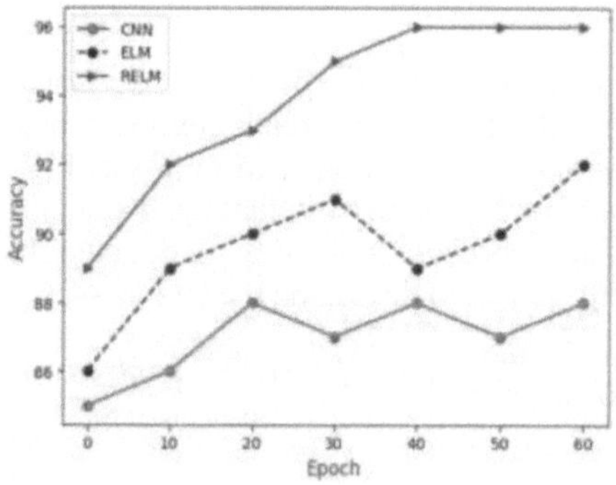


Figure 1. Prediction accuracy result for RELM model.

For RELM, the value of epoches was determined by experimenting with different values for this quantity and finding the one that produced the best prediction accuracy, much like in the proposed technique. For numbers between 10 and 60, also did this, then looked at the forecast results across all accident severity levels. This is seen in Figure 1.

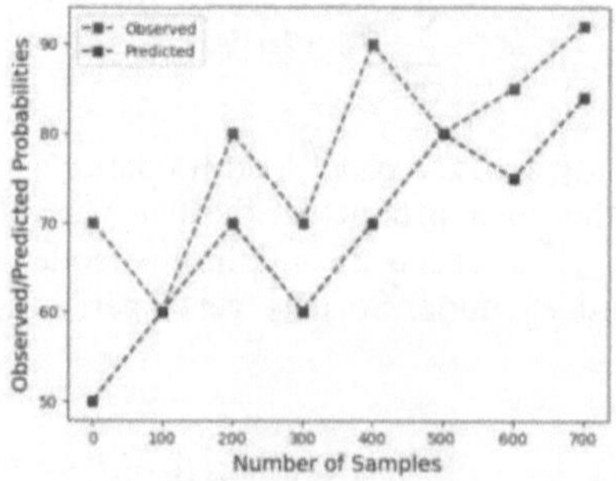


Figure 2. Accident probabilities during training: Predicted and observed.

Finding the suggested framework's mistake requires comparing the probabilities predicted by the probabilistic graph-based model with the observed probabilities. Figure 2 displays the observed and anticipated probabilities during the training period.

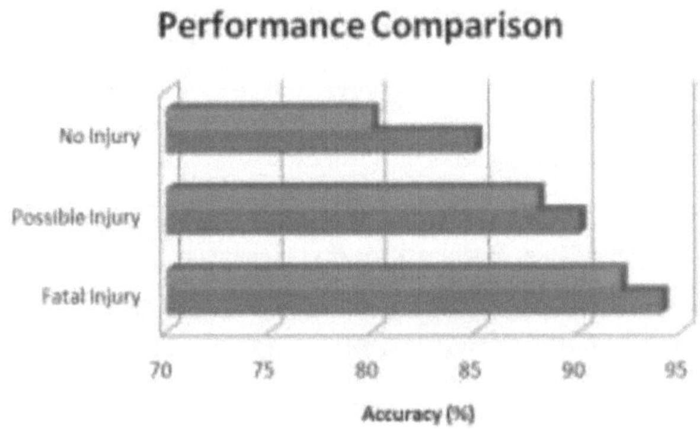


Figure 3. Performance comparison of different models.

last, Figure 3 shows a visual comparison of CNN, ELM, and RELM. The RELM achieved better results than the ELM in every output class. The RELM fared better than ELM and CNN in the classifications of probable injury, no injury, and fatal injury.

## 5 CONCLUSION

Car accidents are expected to be the fifth most common killer worldwide in 2030, according to continuing research. These days, it's hard to pin down exactly what causes car accidents because of all the different influences. A few examples include weather, traffic, road conditions, the driver's mental state, and rule violations. Using preprocessing to guarantee an adequate data structure, we started the process of traffic accident risk prediction. The feature extraction process makes use of the Road Network Mask's feature map. The models are trained using the RELM approach. The proposed approach outperforms the ELM and CNN models with a consistency level of 96.48%.

## REFERENCES

[1] Yu, L., Du, B., Hu, X., Sun, L., Han, L., and Lv, W. (2021). Deep spatio-temporal graph convolutional network for traffic accident prediction. *Neurocomputing*, 423, 135–147.

[2] Fang, J., Yan, D., Qiao, J., Xue, J., and Yu, H. (2021). DADA: Driver attention prediction in driving accident scenarios. *IEEE Transactions on Intelligent Transportation Systems*, 23(6), 4959–4971.

[3] Su, D., Guo, Z., Li, Z., and Zhou, Y. (2016). Operation risk model and monitoring-warning system of expressway tunnels. *Transportation Research Procedia*, 14, 1315–1324.

[4] Chen, Y. Y., Lv, Y., Li, Z., and Wang, F. Y. (2016, November). Long short-term memory model for traffic congestion prediction with online open data. In *2016 IEEE 19th International Conference on Intelligent Transportation Systems (ITSC)* (pp. 132–137). IEEE.

[5] Rolison, J. J., Regev, S., Moutari, S., and Feeney, A. (2018). What are the factors that contribute to road accidents? An assessment of law enforcement views, ordinary drivers' opinions, and road accident records. *Accident Analysis and Prevention*, 115, 11–24.

[6] Gumaei, A., Sammouda, R., Al-Salman, A. M. S., and Alsanad, A. (2018). An improved multispectral palmprint recognition system using autoencoder with regularized extreme learning machine. *Computational Intelligence and Neuroscience*, 2018(1), 8041609.

[7] Gicquel, L., Ordonneau, P., Blot, E., Toillon, C., Ingrand, P., and Romo, L. (2017). Description of various factors contributing to traffic accidents in youth and measures proposed to alleviate recurrence. *Frontiers in Psychiatry*, 8, 94.

[8] Lv, Y., Tang, S., and Zhao, H. (2009, April). Real-time highway traffic accident prediction based on the k-nearest neighbor method. In *2009 International Conference on Measuring Technology and Mechatronics Automation* (Vol. 3, pp. 547–550). IEEE.

[9] Hossain, M., and Muromachi, Y. (2012). A Bayesian network based framework for real-time crash prediction on the basic freeway segments of urban expressways. *Accident Analysis & Prevention*, 45, 373–381.

[10] Wang, T., Kim, S., Wenxuan, J., Xie, E., Ge, C., Chen, J., ... and Luo, P. (2024, March). Deepaccident: A motion and accident prediction benchmark for v2x autonomous driving. *In Proceedings of the AAAI Conference on Artificial Intelligence* (Vol. 38, No. 6, pp. 5599–5606).

[11] Glavić, D., Mladenović, M., Stevanovic, A., Tubić, V., Milenković, M., and Vidas, M. (2016). Contribution to accident prediction models development for rural two-lane roads in Serbia. *Promet-Traffic & Transportation*, 28(4), 415–424.

[12] Cafiso, S., Di Graziano, A., Di Silvestro, G., La Cava, G., and Persaud, B. (2010). Development of comprehensive accident models for two-lane rural highways using exposure, geometry, consistency and context variables. *Accident Analysis & Prevention*, 42(4), 1072–1079.

[13] Dinu, R. R., and Veeraragavan, A. (2011). Random parameter models for accident prediction on two-lane undivided highways in India. *Journal of Safety Research*, 42(1), 39–42.

[14] Li, W., Zhao, X., and Liu, S. (2020). Traffic accident prediction based on multivariable grey model. *Information*, 11(4), 184.

[15] Ren, H., Song, Y., Wang, J., Hu, Y., and Lei, J. (2018, November). A deep learning approach to the citywide traffic accident risk prediction. In *2018 21st International Conference on Intelligent Transportation Systems (ITSC)* (pp. 3346–3351). IEEE.

[16] Yuan, Z., Zhou, X., and Yang, T. (2018, July). Hetero-convlstm: A deep learning approach to traffic accident prediction on heterogeneous spatio-temporal data. In *Proceedings of the 24th ACM SIGKDD International Conference on Knowledge Discovery & Data Mining* (pp. 984–992).

*Hybrid and Advanced Technologies – S. Prasad Jones Christydass, Nurhayati Nurhayati and S. Kannadhasan. (Eds)* © 2025 the Authors, ISBN 978-1-032-90254-8


# Driver drowsiness detection using saliency SNN: A new paradigm for improved road safety

C. Anand Deva Durai
*Department of Computer Science, College of Computer Science, King Khalid University, Abha, Saudi Arabia*

Omar Alqahtani
*Department of Computer Science, College of Computer Science, King Khalid University, Abha, Saudi Arabia*

Raghu Babu Uppara
*Associate Professor, Srinivasa Ramanujan Institute of Technology, Rotary Puram, BKS Mandal, Anantapur, India*

Joel Josephson Pottipadu
*Associate Professor, Department of ECE, Malla Reddy Engineering College, Secunderabad, Telangana, India*

Jaya Patil
*Assistant Professor, Dr. D. Y. Patil Institute of Technology Pimpri Pune, India*

Amit Kumar Sharma
*Assistant Professor, Department of Civil Engineering, Graphic Era deemed to be University, Dehradun, Uttrakhand, India*

ABSTRACT: The development of smart/intelligent cameras has allowed for the detection of driver drowsiness, which in turn alerts drivers and reduces accidents caused by fatigue. In this work, a new framework is proposed that uses deep learning to detect driver drowsiness based on eye state while driving. The Viola-Jones face detection algorithm is used to detect the face and extract the eye region from face images. A striking neural network is used to extract features from dynamically identified key frames from camera sequences and used during the learning phase. Preprocessing, feature extraction, and training the model are the three primary steps of the suggested approach. In preprocessing, face detection is used, and in feature extraction, EAR, NLR, and MOR are employed. Training the models is done using CNN-LSTM, ELM, and Saliency-SNN. Of these, the Saliency-SNN method outperforms the more conventional CNN-LSTM and ELM approaches.

*Keywords*: Electroencephalogram (EEG), Support Vector Machines (SVMs), Artificial Neural Networks (ANNs), Liquid State Machine (LSM), Electrocardiogram (ECG)

## 1 INTRODUCTION

Publications from the World Health Organisation rank traffic accidents among the top 10 killers worldwide. These incidents are primarily caused by drivers, according to the reports. Therefore, detecting driver fatigue could be a practical way to reduce the occurrence of accidents. The improved performance of ADAS and DMS leads to increased road safety, which is an additional benefit. There are a few different kinds of drowsiness detectors, but

DOI: 10.1201/9781003559115-91

the three main ones are vehicle-based, signal-based, and face feature-based. In an effort to determine whether the driver is getting tired, methods that depend on the car itself monitor parameters such as acceleration, lateral position, and steering angle. However, jobs requiring processing in real time are beyond the capabilities of these technologies. Using signal-based techniques, we can estimate drowsiness from psychophysiological measurements. Electroencephalogram (EEG), Electrooculogram (EOG), and operations of are the most studied and used physiological signals. The autonomic nervous system is augmented by the electrocardiogram (ECG), the skin temperature, and the galvanic skin reaction together with electrical activity measured by electromyography (EMG) in muscles. It is imperative that these methods account for invasive captors, which may impair driving. Not only are face feature-based approaches less expensive and more widely available, but they can also evaluate the target in real-time without the need for intrusive sensors.

## 2 LITERATURE SURVEY

Driving while drowsy increases the likelihood of accidents. [1] In just three or four seconds, a car can traverse the length of a football field, therefore even a few seconds of negligence on a highway can bring devastating outcomes. [2] Drowsy drivers cause a lot of accidents since their cognitive capacity lowers when they're tired. Three times as many accidents occur when drivers are fatigued as when they are alert [3]. There are a plethora of new technologies that can detect whether a driver is becoming too drowsy to drive safely. The rising incidence of accidents associated with drowsy drivers prompted this measure. [4] The installation of a device that detects when drivers are sleepy can lead to a decrease in accident rates, financial losses, and casualties. Timely detection of fatigue and sleepiness is made easier using this technique. [5] There are a number of ways that driver drowsiness strategies can be classified. Consider the method for detecting lethargy. Numerous sub-techniques can be derived from it, depending on factors such as image or EEG analysis, vehicle behaviour, AI, etc. Many people are interested in learning more about the possibility of using sensors and cameras to identify drowsy drivers, which could lead to the prevention of fatal accidents [6]. Companies like Tesla and Mercedes-Benz are among many that deploy driver assistance systems. An ongoing effort has been made to enhance the accuracy and precision of sleepiness detection models [7]. In order to capture data about behaviour, a camera is used by Pay attention to the driver's body language, such as head swaying, yawning, and eye blinking, and inform them if you notice any signs of exhaustion. A review of techniques for identifying lethargy was published in [8]. A system of multiplexed sensors that might detect driver weariness in real-time in their study [9]. Creating a smart sensor network that could communicate wirelessly was the primary objective. Its operator and occupant were immediately notified of a first-level emergency as it continuously monitored the individual's vital signs via a network of intrusive sensors. Advanced AI-based approaches were employed by [10]. The problem of detecting driver tiredness has been the subject of numerous investigations. [11] The proposed solutions are neatly organised based on the characteristics that indicate drowsiness. [12] Biological characteristics that are constructed from measurements of the driver's vital signs, such as electroencephalograms, electrocardiograms, and electromyograms, are effective in detecting drowsiness. [13] However, these features might be burdensome for drivers due to the need to wear sensors. Another prominent feature based on vehicle driving habits is the measurement of driver weariness, which is connected with things like steering wheel tilt and the frequency of lane deviation.

## 3 PROPOSED SYSTEM

Drunk driving is the leading cause of death in car accidents, which claim the lives of thousands of people annually around the world. Many lives could be saved if this accident could be reduced with the use of a sleepiness detecting system. To back up our claim, we offer an SNN-based approach that uses sleepiness detection as an object detection job.

## 3.1 *Preprocessing*

### 3.1.1 *Face detection*

It is possible that just the eyes region is necessary to detect drowsiness, rather than the entire facial region. The initial stage is extracting faces from the photos using the Viola-jones method. After the face has been located, the eye region can be extracted from the facial images using the Viola-jones method. Jones created the first algorithm specifically for face detection in his work on the Viola-Jones object detection system [14]. Using three methods—Hart-like features, Ada boost, and Cascade classifier—the Viola-Jones algorithm is able to distinguish faces. In this work, we utilised OPEN CV with Python to develop the Viola-Jones object detection technique with a Haar cascade classifier. In order to identify faces in pictures, the Haar cascade classifier employs Haar characteristics.

## 3.2 *Feature extraction*

In order to extract 68 face landmark points from the video recorded throughout the entire trip, the Python Dlib library was utilised. The dlib package includes an object-identification tool that makes use of support vector machines (SVMs) and a pre-trained face detector. For each set of coordinates, we find their corresponding Euclidean distance. A total of three factors—EAR, NLR, and MOR—have been considered. To determine if the individual is sleepy, these characteristics are fed into Neural Network models and SNN.

### 3.2.1 *EAR*

It has been calculated using equation 1 and is defined as the ratio of the eye's height to its width. The specifics of the parameters called out in equation 1. If we look at equation 1, we can see that the numerator goes down when we're sleepy since the distance between the vertical points goes down. Closed eyelids will cause the difference between parameters p2, p3, and p5, p6, for example, to equal zero.

$$\frac{\left(\left|q^2 - q^6\right| + \left|q^3 - q^5\right|\right)}{z * \left|q^1 - q^4\right|} \tag{1}$$

### 3.2.2 *NLR*

Equation 2 yields the value, which is the nose length divided by the measurement's unit value. The equation's parameters have been described in full [15]. Equation 2 suggests that when the subject sits up straight, the length would go up; but, when the subject bends forward or backward, it would go down, suggesting that they are sleepy.

### 3.2.3 *MOR*

It is calculated by dividing the height of the mouth's extreme points by the breadth of the mouth. When the mouth is open, the height is the measurement of the distance between the vertical points. On the other hand, the breadth is the horizontal gap between the two corners of the mouth. Equation 1.3 has been used to compute MOR, and the parameters referred to in the equation have been shown.

## 3.3 *Model training*

This section presents a saliency-based feature detection network, two methods of spike encoding (i.e., entropy-based time encoding and linear time encoding), and an architecture for a spiking reservoir neural network. This work proposes a linear spike time encoding method that is computationally more succinct than the current spike encoding schemes. The statistical shape of features and their spatial attributes are included in the entropy-based spike time encoding.

### 3.3.1 *Spiking reservoir NN*

Third-generation artificial neural networks (ANNs) like SNN are heavily influenced by biology and operate on events rather than data points. In order to identify breast images, this study uses a recursive SNN architecture called the spike reservoir neural network. An input layer, a reservoir computing layer, and a readout layer comprise a spike reservoir neural network, which is also known as a liquid state machine (LSM). The LSM model is a kind of real-time neural network that takes input data that changes over time and projects it onto a space with many dimensions. A three-layer structure consisting of an input layer, a reservoir or liquid layer, and a memoryless readout layer is what makes up an LSM. The neurons that make up the reservoir are known as Leaky Integrate and Fire (LIF) [16]. During the reservoir stage, the LIF neurons are linked recursively by dynamic synaptic connections. In the reservoir, the excitatory neurons are depicted by the red solid circle, while the inhibitory neurons are represented by the black. In the cerebral cortex circuit, the number of inhibitory neurons is four times higher than the number of excitatory neurons. While LIF neurons are also responsible for the readout layer, they do not communicate with one another. Reservoir can take low-dimensional input data and turn it into a high-dimensional internal state. Memoryless readout layer, which generates the LSM‘s ultimate output, takes the internal state as its input.

### 3.3.2 *Visual saliency detection*

Processing these data streams in real-time is a tremendously taxing operation for the human visual system (HVS). The HVS is limited in its understanding and processing of the data. The term for this kind of selection is visual attention. Two processes are thought to be in charge of this type of attentional behaviour: the bottom-up, stimulus-driven mechanism and the top-down, expectation-driven mechanism. Attributes of the visual scene such as direction, colour, contrast, and action mostly drive bottom-up attention. Computer vision researchers have linked top-down attention to factors including cultural background, experience, and memory. Often referred to as "visual salience," the former attention mechanism is mostly associated with visual attention due to its simplicity. As an example, we'll look at a Driver Drowsy image. The features are extracted using the spiking CNN. In order to get the saliency feature map, the two-dimensional feature maps produced by the spike convolution layer are added together and the mask is computed. But in medical imaging equipment, the input power has a linear connection with the colour brightness captured in the image file. So, the picture that shows up on the screen doesn't match up with the one that the camera took. Before calculating saliency, gamma nonlinearity is applied to fix this discrepancy.

## 4 RESULT AND DISCUSSION

Worldwide, driver sleepiness is a leading cause of traffic accidents. Drowsiness and deadly accidents are the results of driving in a monotonous fashion for long periods of time without rest. There would be far fewer traffic accidents and more lives saved if drivers' fatigue could be detected automatically. The suggested system can identify driver drowsiness with an accuracy of about 97% using SNN classifier, according to the empirical results.

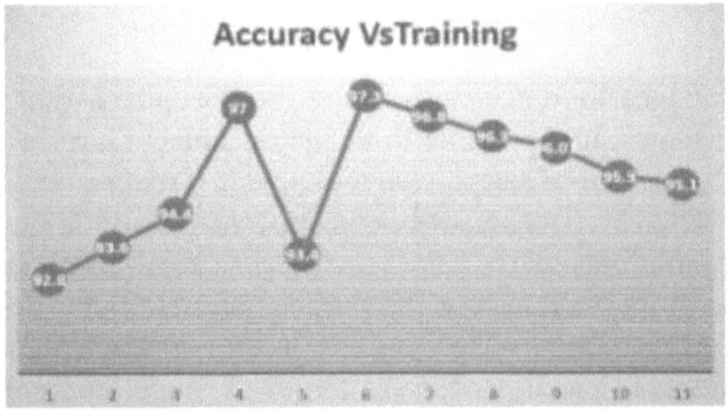


Figure 1. Accuracy values for each training time.

We show the average accuracy after ten iterations of training for each given network. Figure 1 shows the Saliency SNN's accuracy value for each training time.

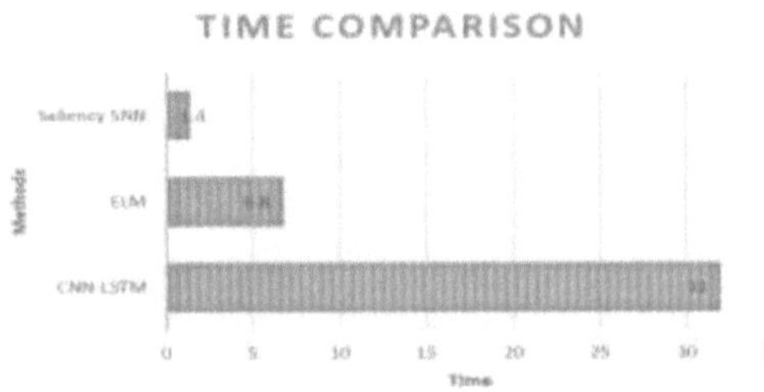


Figure 2. Evaluation of various approaches to the problem of eye closure detection in relation to time.

One important metric to take into account when comparing various approaches is their computational complexity. Methods that are speedier are more dependable in real-time applications. Figure 2 shows the time required to decide on eye closure for each of the given methods. When compared to the ELM and CNN-LSTM models, the Salience SNN is nearly four times quicker and eighteen times faster, respectively.

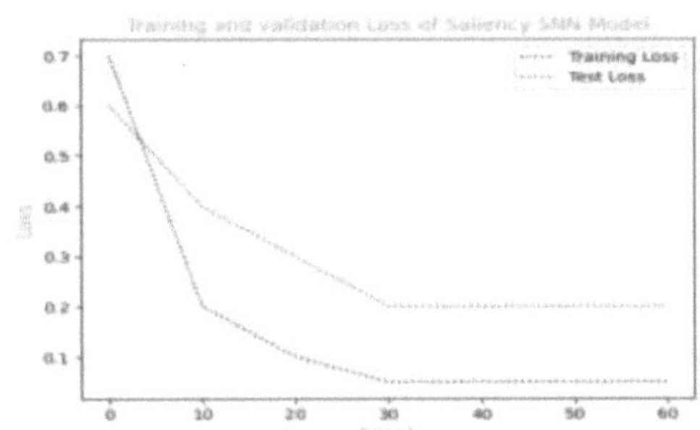


Figure 3. Training and validation loss of the model.

The saliency Spiking Neural Network (SNN) model's training and validation loss sheds light on the model's performance throughout training. Figure 3 shows the ground truth saliency maps and the model's projected saliency maps; the loss function measures the difference between the two and is optimised during model training to minimise it.

## 5 CONCLUSION

This research introduces a method for detecting driver sleepiness that relies on behavioural measures and NN approaches. It is possible to deduce sleepiness levels from facial expressions. It is possible to deduce the degree of sleepiness from a wide variety of facial traits. Among these behaviours include yawning, eye blinks, and head movements. Building a sleepiness detection system with trustworthy findings is no easy feat; it calls for strong and precise algorithms. Several methods have been investigated for the purpose of identifying sleepy drivers. Given the current popularity of deep learning, it is necessary to reevaluate these algorithms to see how well they detect lethargy. There are primarily three steps to the suggested methodology: preprocessing, feature extraction, and training the model. Prior to feature extraction, face detection is used for preprocessing. EAR, NLR, and MOR are then employed. Training the models is done using CNN-LSTM, ELM, and Saliency-SNN. The Saliency-SNN method achieves an accuracy of approximately 97.3% and outperforms the traditional CNN-LSTM and ELM approaches.

## REFERENCES


[1] Abbas, Q. (2020). HybridFatigue: A real-time driver drowsiness detection using hybrid features and transfer learning. *International Journal of Advanced Computer Science and Applications*, 11(1).

[2] Awais, M., Badruddin, N., and Drieberg, M. (2014, April). Driver drowsiness detection using EEG power spectrum analysis. In *2014 IEEE Region 10 symposium* (pp. 244–247). IEEE.

[3] Caryn, F. H., and Rahadianti, L. (2021). Driver drowsiness detection based on drivers' physical behaviours: a systematic literature review. *Computer Engineering and Applications*, 10(3).

[4] Chang, T. C., Wu, M. H., Kim, P. Z., and Yu, M. H. (2021). Smart driver drowsiness detection model based on analytic hierarchy process. *Sens Mater*, 33, 485–497.

[5] Gielen, J., and Aerts, J. M. (2019). Feature extraction and evaluation for driver drowsiness detection based on thermoregulation. *Applied Sciences*, 9(17), 3555.

[6] Jabbar, R., Shinoy, M., Kharbeche, M., Al-Khalifa, K., Krichen, M., and Barkaoui, K. (2020, February). Driver drowsiness detection model using convolutional neural networks techniques for android application. In *2020 IEEE International Conference on Informatics, IoT, and Enabling Technologies (ICIoT)* (pp. 237–242). IEEE.

[7] Weng, C. H., Lai, Y. H., and Lai, S. H. (2017). Driver drowsiness detection via a hierarchical temporal deep belief network. In *Computer Vision–ACCV 2016 Workshops: ACCV 2016 International Workshops*, Taipei, Taiwan, November 20-24, 2016, Revised Selected Papers, Part III 13 (pp. 117–133). Springer International Publishing.

[8] Awais, M., Badruddin, N., and Drieberg, M. (2017). A hybrid approach to detect driver drowsiness utilizing physiological signals to improve system performance and wearability. *Sensors*, 17(9), 1991.

[9] Khushaba, R. N., Kodagoda, S., Lal, S., and Dissanayake, G. (2010). Driver drowsiness classification using fuzzy wavelet-packet-based feature-extraction algorithm. *IEEE Transactions on Biomedical Engineering*, 58(1), 121–131.

[10] Bhandarkar, S., Naxane, T., Shrungare, S., and Rajhance, S. (2023). *Neural Network Based Detection of Driver's Drowsiness*. Authorea Preprints.

[11] Ramzan, M., Khan, H. U., Awan, S. M., Ismail, A., Ilyas, M., and Mahmood, A. (2019). A survey on state-of-the-art drowsiness detection techniques. *IEEE Access*, 7, 61904–61919.

[12] Pratama, B. G., Ardiyanto, I., and Adji, T. B. (2017, July). A review on driver drowsiness based on image, bio-signal, and driver behavior. In *2017 3rd International Conference on Science and Technology-Computer (ICST)* (pp. 70–75). IEEE.

[13] Mohan, R., Chalasani, S., Mary, S. S. C., Chauhan, A., Parte, S. A., and Anusuya, S. (2023, March). Identification of driver drowsiness detection using a regularized extreme learning machine. In *2023 Second International Conference on Electronics and Renewable Systems (ICEARS)* (pp. 1233–1238). IEEE.

[14] Chirra, V. R. R., Uyyala, S. R., and Kolli, V. K. K. (2019). Deep CNN: A machine learning approach for driver drowsiness detection based on eye state. *Rev. d'Intelligence Artif.*, 33(6), 461–466.

[15] Mehta, S., Mishra, P., Bhatt, A. J., and Agarwal, P. (2019, November). AD3S: Advanced driver drowsiness detection system using machine learning. In *2019 Fifth International Conference on Image Information Processing (ICIIP)* (pp. 108–113). IEEE.

[16] Fu, Q., and Dong, H. (2022). Breast cancer recognition using saliency-based spiking neural network. *Wireless Communications and Mobile Computing*, 2022(1), 8369368.

*Hybrid and Advanced Technologies – S. Prasad Jones Christydass, Nurhayati Nurhayati and S. Kannadhasan. (Eds)* © *2025 the Authors, ISBN 978-1-032-90254-8*


# A novel approach to automatic road-accident detection using deep mutual learning based technique

H.S. Ranjan Kumar
*Associate Professor, Department of Computer Science and Engineering, Shri Madhwa Vadiraja Institute of Technology and Management, Bantakal, India*

K. Raju
*Associate Professor, Department of CSE, NITTE (Deemed to be University), NMAM Institute of Technology, Nitte, Karnataka, India*

M.S. Guru Prasad
*Department of Computer Science and Engineering, Graphic Era (Deemed to be University), Dehradun, India*

S. Chandrappa
*Department of Computer Science and Engineering, JAIN (Deemed-To-Be-University), Bengaluru, India*

B.B. Ramakrishna
*Head of the Department, Department of AI & ML, Coorg Institute of Technology, Ponnampet, Kodagu, India*

B.L. Sunil Kumar
*Associate Professor, Department of Computer Science & Engineering, Canara Engineering College, Mangaluru, India*

ABSTRACT: The daily toll on society is high due to car accidents. A combination of careless driving and a lag in reaction time from emergency services is the most common cause. Reliable early warning and information exchange systems are life-saving for individuals injured in traffic accidents. An urgent system that notifies nearby emergency services of an accident's exact location must be put into place without delay. Multiple authors in scholarly journals have proposed different automatic accident detection algorithms. The four steps that make up the proposed method are data preparation, model training, feature extraction, and segmentation. Data cleansing and encoding technology make up preprocessing. Because of its strong association with accident frequency, the AADT is a crucial variable in road segmentation. Among the current features, Histogram of Oriented Gradient (HOG) offers the best performance. The usage of DML resulted in a more accurate model training process. A 95.40% accuracy rate was achieved using the suggested method.

*Keywords*: Histogram of Oriented Gradient (HOG), Road-Accident Detection, Automatic Detection, Deep Mutual Learning (DML), Programmable Integrated Circuit (PIC)

## 1 INTRODUCTION

An unexpected and catastrophic event is considered a disaster if it causes significant injury or loss of life. Over the years, scientists have created a plethora of commodities,

DOI: 10.1201/9781003559115-92 

methods, and equipment with the intention of making people's lives easier. With people's health in mind, a lot more room has been created to accommodate them. Swift action in response to crises or patients requiring long-term care can avert a great deal of harm to human life. Automobile accidents account for the largest single cause of mortality in developing countries like Pakistan. Not only do accidents play a significant role, but so do unknown crash sites. Even though the crisis has just been happening for a few hours, rescue activities still haven't been able to start. Countless innocent lives were tragically sacrificed in the process. This study introduces a life-saving autonomous accident detection system that takes human lives into account. With the proposed method, worried loved ones or rescue workers can receive an exact location alert through Short Message Service (SMS) the second an accident is detected. If that's the case, the system will connect the dots between detection and alert. The hardware module is built around an Arduino board and a vibration sensor. A vehicle-mounted accelerometer is standard equipment. However, the driver's mobile device contains the software component in the form of an Android app.

## 2 LITERATURE SURVEY

Increases in both the population and the number of cars on the road lead to more frequent accidents. [1]Approximately 80.1% of the accidents that happen every day throughout the world end in death. Inadequate substructure, ineffective traffic control, and a lack of scientific accident management all play a role in these tragedies. [2]Serious injuries are more likely to occur to accident survivors who do not receive timely medical treatment. In most cases, casualties happen very quickly. Nowadays, thanks to a handful of new technologies, we live in a high-tech biosphere where these problems are becoming obsolete[3]. Anger causes disability in many people, and an additional twenty to fifty million people suffer from non-fatal ailments [4]. According to the statistics, a ten percent survival rate can be achieved with just a one-minute reduction in accident interruption time[5]. An ultrasonic sensor built inside vehicles and a programmable integrated circuit (PIC) microcontroller are the backbone of the system proposed by [6]. The ultrasonic sensor can detect anything from speed bumps and sharp turns to humans. The microcontroller acts as a computer by processing sensor data[7]. When anything out of the ordinary happens, the system lets drivers know so they can revaluate their pace or switch lanes[8]. Computer control of the time delay allows an ultrasonic distance detector to measure distance by evaluating the echo produced by an obstacle in the path of a high-frequency sound pulse[9]. The system will record a voice message if the vehicle's distance from the obstacle is lower than a certain threshold. Visual sensing based on models was proposed by [10] as a means to avoid car accidents. The technology is able to recognize when other vehicles are in the area and warn the driver to maintain a safe distance thanks to image processing. A server monitors the actions of many high-tech cars in the background using sensors and mobile networking. The term "road crash" describes the collision of vehicles, people, or any other roadside object. [11]The likelihood of a victim's survival is significantly impacted by the amount of time it takes for an ambulance to reach the accident site and transfer the patient to the hospital.[12] Although most people who are hurt in car accidents don't have serious enough injuries to die, if rescue workers take too long to arrive, the victims' conditions can worsen. Finding the precise site of the accident and relaying that information to rescue crews in a timely manner is, hence, of utmost importance in order to ensure the victim's survival [13]. Accident detection can be improved by studying the vehicle's behaviours using the location and velocity values. Machine learning algorithms can also be used to estimate the severity of an accident[14]. Support vector machines (SVMs) and k-means clustering are two examples of the algorithms that can be taught under reinforcement learning using real-time data retrieved from vibration sensors and ultrasonic sensors, respectively, in the vehicle.

# 3 PROPOSED SYSTEM

This strategy proposes a novel technique to the automatic detection of traffic accidents. Central to the plan is the notion of spotting damaged vehicles in footage captured by roadside and highway monitoring systems, which would indicate a collision.

## 3.1 *Data preprocessing*

### 3.1.1 *Data cleaning*

Almost every machine learning project begins with data cleaning. Data cleaning is the process of identifying and eliminating errors and duplicates from a dataset in order to make it more reliable. As a result, analytics training data is more stable, leading to better accuracy.

### 3.1.2 *Encoding technique*

Using label data directly is not an option for some machine learning algorithms. Input and output variables must have numeric values for them[15]. A numerical form must be created from categorical data. Given that the dependent and independent columns both include categorical values, we converted the former using Label Encoding and the latter using One Hot Encoding.

## 3.2 *Segmentation*

Before separating a road into uniform pieces, it is important to make sure the clustered portion's properties are consistent. More variables in a segmentation process make it more complicated and, on average, yield shorter segments[16]. The AADT is an important variable in road segmentation due to the high correlation between it and accident frequency. Recent years have seen an explosion in the use of variables outside the AADT to create homogeneous road segments.

## 3.3 *HOG feature extraction*

Hexagon of Oriented Gradient (HOG) outperforms other existing features in terms of performance. It starts with the notion that local object shapes and appearances can often be characterized with high accuracy by the distribution of local intensity gradients or edge orientations, even when the precise locations of these features are unavailable. When characterizing images, HOG is all about a local-oriented gradient histogram. As examples of a specific gradient direction, these histograms display local image regions[17]. There are a number of critical processes in the process of feature extraction from HOG.

## 3.4 *Model training*

This section describes in depth the research method that has been suggested. Visual synonyms, image segmentation, similarity, and model training are the four stages that make up this approach. Among other things, this approach employs DML and label propagation to provide an understandable diagnostic of BC histopathology images. As part of this procedure, visual synonyms are created using a set of fuzzy criteria. Genuine synonyms are then generated, and the keyword list is enlarged.

### 3.4.1 *DML*

In order to train the proposed model autonomously, the subsequent multiclass cross-entropy loss is actually used.

$$C_L = -(p_L.\log(\varnothing_L)(1 - p_L).\log(1 - \varnothing_L)) \quad (1)$$

where $\varnothing_L$ represents the likelihood of a positive classification for a bag and $p_L \in 0, 1$ indicates the expected malignant or benign status of a histological picture.

The possibility that $\theta 1$ predicts that a bag will go to the positive and negative classes, respectively, is denoted by $Y_1^1$ and $Y_1^0$. It is determined the *IC* distance between points P1 and P2 as

$$V_{IC} = Y_2^0 \log \frac{Y_2^0}{Y_1^0} + Y_2^1 \log \frac{Y_2^1}{Y_1^1} \quad (2)$$

The steps to achieve the total loss functions $c\theta_1$ and $c\theta_2$, for the $\theta_1$ and $\theta_2$ networks, respectively, are as follows:

$$c_{\theta_1} = c_{L_1} + V_{IC}(Y_2 \| Y_1) \quad (3)$$

$$c_{\theta_2} = c_{L_2} + V_{IC}(Y_1 \| Y_2) \quad (4)$$

## 4 RESULT AND DISCUSSIONS

Car accidents handicap thousands more and kill a few every single day. Some of these victims don't get the medical treatment they need in time, which leads to further accidents. Increased efficiency and safety on the road can be achieved through the use of automated vehicle accident detection systems, which allow responding rescue vehicles and agencies to reach the scene of the accident more quickly.

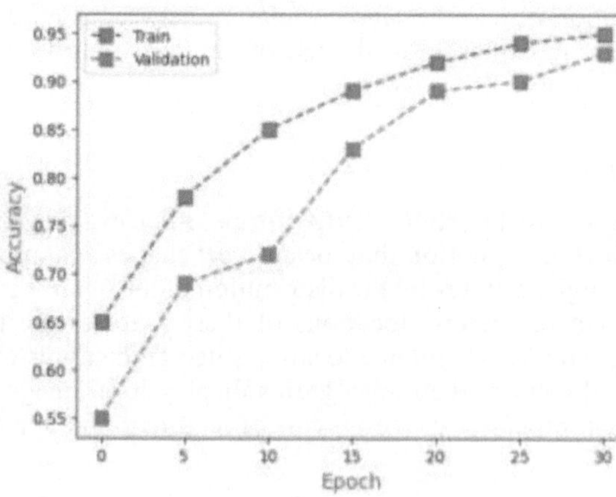


Figure 1. Training and validation performance of DML model.

Figure 1 is a graph displaying the accuracy set for the DML models' training and validation.

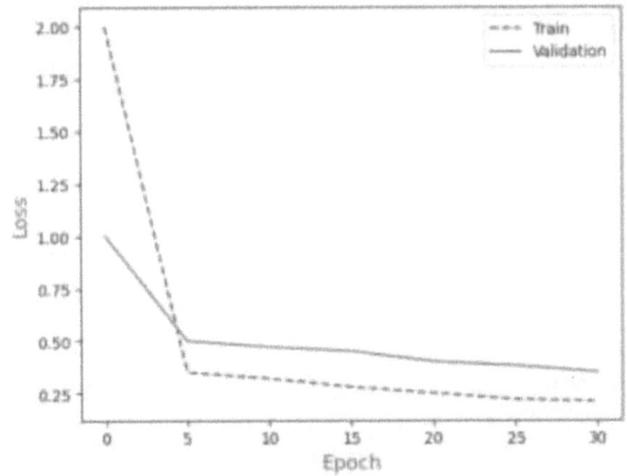


Figure 2. Training and validating loss of the suggested method.

The model's training loss and validation loss are displayed in Figure 2. After 30 rounds of evaluating the model's performance, it finally arrived at the results.

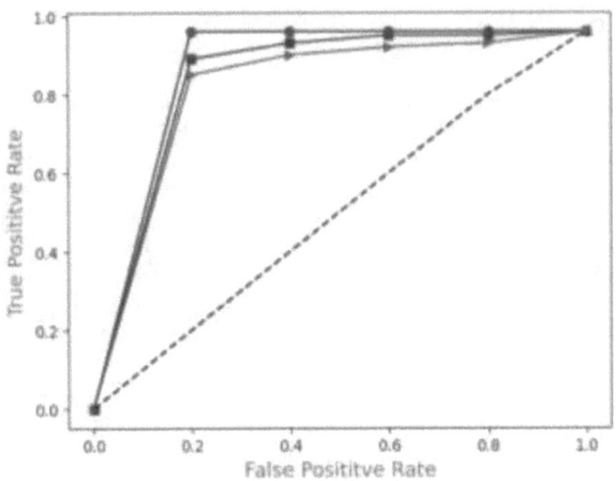


Figure 3. ROC curve of DML models.

Figure 3 displays the area under the curve (AUC) values of the three models with respect to the recognition accuracy. With the largest area under the curve (AUC) value, the DML model clearly outperforms the others and can correctly detect the key aspects of traffic accidents.

## 5 CONCLUSION

Numerous individuals lose their lives every day as a result of car accidents. The unfortunate reality is that a large number of passengers do not survive vehicle accidents. This occurs frequently because medical professionals do not learn about the accident until much later, postponing treatment for the injured. Therefore, it is essential that all vehicles have an automated system that can detect a collision and immediately notify the closest hospital. Preprocessing consists of data cleansing and encoding techniques. The AADT is an important variable in road segmentation due of its strong correlation with accident frequency. Right now, the most effective feature is Histogram of Oriented Gradient (HOG). The model training procedure became more precise with the use of DML. The proposed approach yielded an accuracy rate of 95.40 percent.

## REFERENCES


[1] White, J., Thompson, C., Turner, H., Dougherty, B., and Schmidt, D. C. (2011). Wreckwatch: Automatic traffic accident detection and notification with smartphones. *Mobile Networks and Applications*, 16, 285–303.

[2] Fernandes, B., Gomes, V., Ferreira, J., and Oliveira, A. (2015, May). Mobile application for automatic accident detection and multimodal alert. In *2015 IEEE 81st vehicular technology conference (VTC spring)* (pp. 1–5). IEEE.

[3] Liang, G. (2015). Automatic traffic accident detection based on the internet of things and support vector machine. *Int. J. Smart Home*, 9(4), 97–106.

[4] Ravindran, V., Viswanathan, L., and Rangaswamy, S. (2016). A novel approach to automatic road-accident detection using machine vision techniques. *International Journal of Advanced Computer Science and Applications*, 7(11).

[5] Sharma, S., and Sebastian, S. (2019). IoT based car accident detection and notification algorithm for general road accidents. *International Journal of Electrical & Computer Engineering* (2088–8708), 9(5).

[6] Lee, M. H., Park, H. G., Lee, S. H., Yoon, K. S., and Lee, K. S. (2013). An adaptive cruise control system for autonomous vehicles. *International Journal of Precision Engineering and Manufacturing*, 14, 373–380.

[7] Maaloul, B. (2018). *Video-based Algorithms for Accident Detections* (Doctoral dissertation, Université de Valenciennes et du Hainaut-Cambresis; Université de Mons).

[8] Schneider, K., Lugner, R., and Brandmeier, T. (2019, June). Lidar-based contour estimation of oncoming vehicles in pre-crash scenarios. In *2019 IEEE Intelligent Vehicles Symposium (IV)* (pp. 2272–2277). IEEE.

[9] Zheng, Y., Ran, B., Qu, X., Zhang, J., and Lin, Y. (2019). Cooperative lane changing strategies to improve traffic operation and safety nearby freeway off-ramps in a connected and automated vehicles environment. *IEEE Transactions on Intelligent Transportation Systems*, 21(11), 4605–4614.

[10] Al-Mayouf, Y. R. B., Mahdi, O. A., Taha, N. A., Abdullah, N. F., Khan, S., and Alam, M. (2018). Accident management system based on vehicular network for an intelligent transportation system in urban environments. *Journal of Advanced Transportation*, 2018(1), 6168981.

[11] MohanaRoopa, D. Y., Soujanya, N., Vaishnavi, V. S., and Vardhan, U. V. (2020). An IOT based smart helmet for accident detection and notification. *Journal of Interdisciplinary Cycle Research*, 12 (VI).

[12] Chandran, S., Chandrasekar, S., and Elizabeth, N. E. (2016, December). Konnect: An Internet of Things (IoT) based smart helmet for accident detection and notification. *In 2016 IEEE Annual India Conference (INDICON)* (pp. 1–4). IEEE.

[13] Berade, P., Patil, K., Tawate, P., and Ghewari, M. U. (2018). Intelligent accident identification and prevention system using GPS and GSM modem. *Int. Res. J. Eng. Technol*, 5(2), 1915–1919.

[14] Kaur, A., Kaushal, C., Sandhu, J. K., Damaševičius, R., and Thakur, N. (2023). Histopathological image diagnosis for breast cancer diagnosis based on deep mutual learning. *Diagnostics*, 14(1), 95.

[15] Paul, J., Jahan, Z., Lateef, K. F., Islam, M. R., and Bakchy, S. C. (2020, December). Prediction of road accident and severity of Bangladesh applying machine learning techniques. In *2020 IEEE 8th R10 Humanitarian Technology Conference (R10-HTC)* (pp. 1–6). IEEE.

[16] Qaseem Ghadi, M., and Török, Á. (2019). Comparison of different road segmentation methods. *Promet-Traffic & Transportation*, 31(2), 163–172.

[17] Putra, F. A. I. A., Utaminingrum, F., and Mahmudy, W. F. (2020). HOG feature extraction and KNN classification for detecting vehicle in the highway. *IJCCS (Indonesian Journal of Computing and Cybernetics Systems)*, 14(3), 231–242.

*Hybrid and Advanced Technologies – S. Prasad Jones Christydass, Nurhayati Nurhayati and S. Kannadhasan. (Eds)* © *2025 the Authors, ISBN 978-1-032-90254-8*


# An innovative transportation system for improving safety and traffic efficiency based on BiLSTM-RNN model

Madhumita
*Assistant professor, Department of Computer science, GL Bajaj Institute of Management, Greater Noida, India*

Rajashree Bhokare
*Assistant Professor, Dr.D.Y.Patil Institute of Technology Pimpri, Pune, India*

D. Poornakumar
*Assistant Professor, Department of Electronics Engineering (VLSI Design and Technology), K.S. Rangasamy College of Technology, Tiruchengode, India*

M. Dhiliphan Kumar
*Associate Professor, Business Administration, KBS, Kalasalingam Academy of Research and Education (Deemed University), Krishnankoil, Tamil Nadu, India*

Vivek Ravishankar Dubey
*Professor, Department of CSE, Jai Narain College of Technology (JNCT), Bhopal, India*

Prashant Mahapure
*Assistant Professor, Department of Mechanical Engineering, Dr. D. Y. Patil Institute of Technology, Pimpri, Pune, India*

ABSTRACT: A lot of people are worried about how safe and mobile many cities are these days. The latter is a common side effect of fast urbanization and unregulated population growth, the former of which is transportation congestion. Regarding the second, it is associated with the fact that the city has seen a rise in the number of crimes. Some safety-based systems propose Intelligent Transportation Systems (ITS) to handle mobility issues, while others try to guide people and cars toward the safest routes. This paper introduces itsSAFE, an ITS that uses real-time data regarding road conditions and unsafety levels to reduce traffic congestion and increase passenger and driver safety. Preprocessing, feature selection, and training the model are the three steps that make up the suggested method. Data preprocessing is the process of preparing raw data for use in a machine learning model. In the process of creating an ML model, it is the first and foremost step. When selecting features, mutual information is considered. The BiLSTM-RNN improved the accuracy of the model training process. The proposed approach achieves a higher accuracy rate of 95.34% compared to LSTM and RNN.

*Keywords*: Intelligent Transportation Systems (ITS), Vehicular Navigation System (VNS), Vehicle-to-Vehicle (V2V), Autonomous Vehicle (AV), Recurrent Neural Network (RNN)

## 1 INTRODUCTION

City people constantly worry about traffic congestion for many reasons, including health difficulties, poor economic development, severe environmental implications, and the

DOI: 10.1201/9781003559115-93

impracticality of route planning time. Residents may discover that utilizing a Vehicular Navigation System (VNS) is beneficial while dealing with mobility issues brought on by traffic congestion. When users ask for a route, existing VNSs that take congestion avoidance into account—for example, WAZE, Google Maps, and TomTom—tend to suggest the fastest routes based on road conditions throughout the world. Because they disregard issues fundamental to the safety of drivers and passengers, such systems may guide vehicles toward dangerous areas in order to escape traffic, which could result in robberies, kidnappings, or murders. Getting your hands on data produced by vehicles, traffic systems, subsystems, sensors (both on and off the road), and even people, ITSs can potentially detect issues with transportation and safety, so they may plan ways to make traffic flow more smoothly while drivers and passengers are safer. When suggesting routes, many ITSs depend entirely on traffic data. These ITSs will always be just as dangerous as VNSs since they can't use the limited data they have about citywide dangers to guide cars to safer areas.

## 2 LITERATURE SURVEY

Vehicles should be equipped with communication-based applications that enhance road safety, optimize fuel and infrastructure utilization, and safeguard passengers and other road users. These applications offer compelling reasons to do so[1]. [2]Several access mechanisms, such as C-V2X communication protocols, have been utilized in these surveys. Considering monetary considerations and business models,[3] investigated various access mechanisms for vehicle-to-vehicle (V2V) communications to enable vehicular sensor networks.[4]took a more holistic look, doing a tutorial-style overview of numerous communication technologies. Autonomous vehicle use cases and the services provided by several vehicular communications technologies, including Zigbee and 5G-New Radio, were investigated by [5] with latency requirements in mind. A new kind of vehicle that can drive itself safely without human intervention is the autonomous vehicle (AV), which makes use of sensors and intelligent software. Only few areas designated as test zones, have so far hosted the trial activities [6]. The public and transportation officials are still very worried about how to make mixed-vehicle traffic flows as safe and efficient as possible [7]. Perception, planning, decision-making, control, and many more technologies are all a part of AV. [8]Its vision and decision-making mechanisms are drastically different from human driving habits, among other areas. The unique characteristics of mixed-traffic flows are obviously unaccounted for by standard traffic flow theory. However, the study is crucial because the transitional period is expected to last for a long time [9]. Many academics are now interested in studying how to model mixed traffic flows. This type of study provides a high-level overview of the interplay between variables like AV penetration rate (PR),[10] traffic density, response time, road capacity, vehicle speed, acceleration, and traffic flow stability. [11]There is a relationship between vehicles and everything else infrastructure, humans, aftermarket parts, and anything else that can affect CV dynamics. [12]Only CVs that are operated by people are considered in this research. Among the main advantages of CVs are their capacity to reduce pollution, traffic congestion, and unsafe driving conditions. [13]While conventional automobiles do not have the capacity to communicate or provide driver assistance, CVs are built to complement such systems. Complex features will emerge in the resulting mixed traffic flow due to the fact that CVs and TVs have different driving dynamics.

## 3 PROPOSED SYSTEM

The paper's stated goal is to provide a synopsis of various intelligent systems and the roles they play that might aid in the security of transportation infrastructure, with a focus on the intelligent railway as a current development trend. The study discusses the significance of safeguarding essential transportation infrastructure. Superior and bolstered protection of vital infrastructure components can be achieved through the use of intelligent transportation systems.

### 3.1 *Preprocessing*

Raw data acquired using any of these approaches can be contaminated with noise, missing values, and inconsistent data due to sensor failures, measurement issues, data link errors, or exceptionally huge data sets[14]. Therefore, data modification is required, and here are some ways to do it: Data cleansing is the process of finding and eliminating mistakes, noise, and lost data. Methods like kernel dimension reduction, non-negative matrix factorization, and manifold learning are employed to reduce the data's dimensionality.

### 3.2 *Feature selection*

An easily accessible and practical mathematical tool for information, information entropy is where mutual information is derived from. It delineates the relationship between two separate variables[15]. Simply expressed, the entropy measures the degree of uncertainty associated with a random variable. According to the definition of mutual information, the common information of two random variables *PandR* is:

$$M(P,R) = L(P) - L(P|R) = L(P) + L(R) - L(P,R) \tag{1}$$

## 4 MODEL TRAINING

### 4.1 *BLSTM*

Because of its design, BLSTM networks can take in data from sequential datasets and store contextual factors from both the past and the future[16]. Although most neural networks process forward and backward passes by each layer in the same way, BLSTM networks use two parallel layers that propagate in opposite ways; these layers memorize sentence information from both directions. When designing a network with two LSTM layers, it's important to adjust the vector formula accordingly.

$$l_{n_z} = L\left(O_{pl_n} p_z + O_{l_n l_n} l_{n_{z-1}} + d_{l_n}\right) \tag{2}$$

$$l_{d_z} = L\left(O_{pl_d} p_z + O_{l_d l_d} l_{d_{z-1}} + d_{l_d}\right) \tag{3}$$

### 4.2 *Training framework*

To construct a lookup dictionary, one must first collect all of the distinct characters found in the training set and assign them a determined identifier. It prefers to conceive of this initialization phase as creating character embeddings rather than utilizing a one-hot representation. The next step is to feed the embeddings into the BLSTM network. Next, the BLSTM network's output is passed to a hidden layer, where the softmax layer assigns the highest probability to the character's tag.

### 4.3 *Model variant*

The goal of utilizing the RNN construction method to stack BLSTM layers is to enhance the BLSTM network's structure and enable higher-level extraction of contextual properties. Since the output of a BLSTM layer which consists of two LSTM layers is twice as big as the input vector, we must use a transformation matrix to decrease the dimensionality of the output vectors while keeping their sizes equal to the input vectors. As the depth of the network rises, its dimensionality also increases.

$$u_{tran} = O_{tran} + u_w \tag{4}$$

It is assumed that the BLSTM layer's output vector, $u_w$, belongs to the set *Y2b*. Then, the transformation matrix, $O_{tran}$, which is defined as $Y^b \times 2b$, transforms the vectors into a lower dimension, ensuring that the output of every BLSTM layer retains the same dimension. To prevent overfitting, it employed dropout during training on our increasingly complex BLSTM network, which results in a rapidly increasing number of parameters.

## 5 RESULT AND DISCUSSION

More intricate road safety applications have resulted from IT advancements. There are a lot of ways these systems could make road transportation better. Safety and efficiency are two of the most important factors that this paper's integrated system addresses. Therefore, it is recommended to create and deploy an advanced driver assistance system (ADAS) that is compatible with both urban and rural settings.

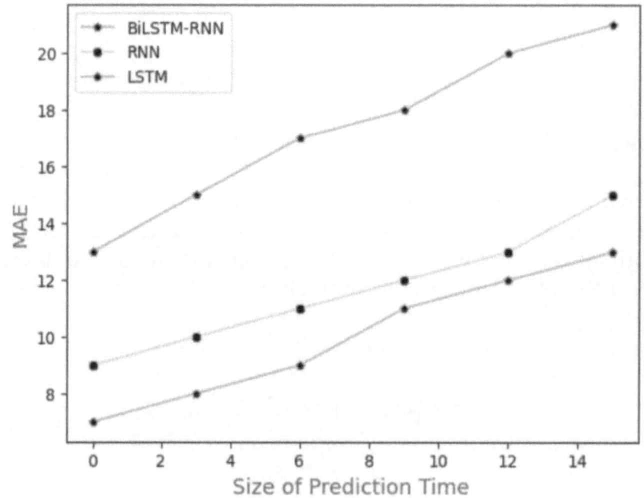


Figure 1. MAE comparison of BiLSTM-RNN model.

In comparison to the other three model traffic flow prediction methods, our model performs better, as demonstrated in Figure 1. The best short-term traffic flow prediction results were achieved by BiLSTM-RNN across all three time periods when it came to MAE.

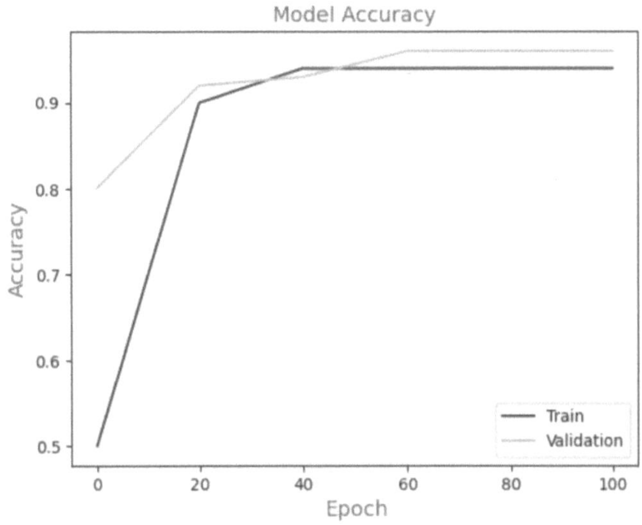


Figure 2. Evaluation of BiLSTM-RNN model's performance in training and validation accuracy.

The experiment's training and validation accuracies are shown in Figure 2. These two numbers are so close to one another, it seems the model fits the data well and shows no indications of overfitting or underfitting.

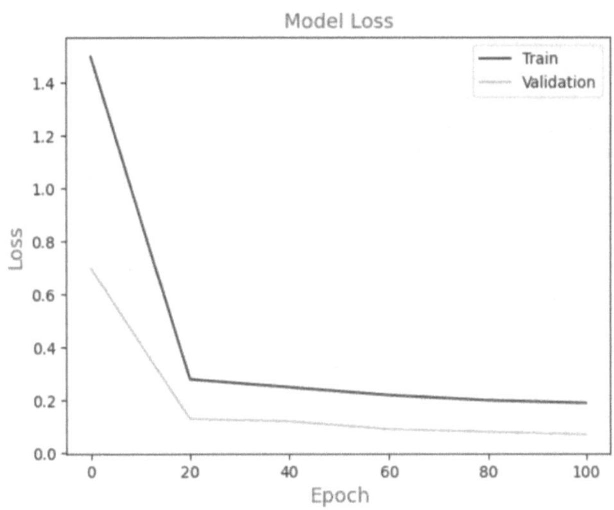


Figure 3. Evaluation of BiLSTM-RNN model's performance in training and validation loss.

Figure 3 shows the outcomes of computing the training and validation losses for the proposed BiLSTM-RNN model over several epochs. There appears to be no overfitting or underfitting in the model, as both values are convergent towards zero.

## 6 CONCLUSION

Finding a way to classify driver-assistance technologies according to their influence on traffic efficiency and road safety was the main objective. This proposed provides a high-level overview of several driver-assistance systems that are now available or in development. It describes the main features of each system and makes predictions about their impact on traffic flow and driver safety. In order to get raw data ready for a machine learning model, a procedure known as data preprocessing must be performed. It is the initial and most important stage in developing an ML model. Mutual information is taken into account when features are being selected. On the other hand, BiLSTM-RNN is recommended for model training. With a consistency rate of 95.34%, the suggested method surpasses both the LSTM and RNN models.

## REFERENCES


[1] Ahangar, M.N., Ahmed, Q.Z., Khan, F.A., and Hafeez, M. (2021). A survey of autonomous vehicles: Enabling communication technologies and challenges. *Sensors*, 21(3), 706.

[2] Jeong, H. H., Shen, Y. C., Jeong, J. P., and Oh, T. T. (2021). A comprehensive survey on vehicular networking for safe and efficient driving in smart transportation: A focus on systems, protocols, and applications. *Vehicular Communications*, 31, 100349.

[3] Soleimani, H., and Boukerche, A. (2017, May). D2D scheme for vehicular safety applications in LTE advanced network. In *2017 IEEE International conference on communications (ICC)* (pp. 1–6). IEEE.

[4] Kawasaki, R., Onishi, H., and Murase, T. (2017, October). Performance evaluation on V2X communication with PC5-based and Uu-based LTE in crash warning application. In *2017 IEEE 6th Global Conference on Consumer Electronics (GCCE)* (pp. 1–2). IEEE.

[5] Soleimani, H., and Boukerche, A. (2018). On vehicular safety message transmissions through LTE-Advanced networks. *Ad Hoc Networks*, 79, 133–145.

[6] Kerner, B. S. (2021). Effect of autonomous driving on traffic breakdown in mixed traffic flow: A comparison of classical ACC with three-traffic-phase-ACC (TPACC). *Physica A: Statistical Mechanics and its Applications*, 562, 125315.

[7] Talebpour, A., and Mahmassani, H. S. (2016). Influence of connected and autonomous vehicles on traffic flow stability and throughput. *Transportation Research Part C: Emerging Technologies*, 71, 143–163.

[8] Chen, G., Meng, F., Fu, G., Deng, M., and Li, L. (2013). A cell automation traffic flow model for mixed traffic. *Procedia-Social and Behavioral Sciences*, 96, 1412–1419.

[9] Rawat, K., Katiyar, V. K., and Gupta, P. (2012). Two-lane traffic flow simulation model via cellular automaton. *International Journal of Vehicular Technology*, 2012(1), 130398.

[10] Gao, K., Tu, H., Sun, L., Sze, N. N., Song, Z., and Shi, H. (2020). Impacts of reduced visibility under hazy weather condition on collision risk and car-following behavior: Implications for traffic control and management. *International Journal of Sustainable Transportation*, 14(8), 635–642.

[11] Ali, Y., Sharma, A., Haque, M. M., Zheng, Z., and Saifuzzaman, M. (2020). The impact of the connected environment on driving behavior and safety: A driving simulator study. *Accident Analysis & Prevention*, 144, 105643.

[12] Calvert, S. C., and van Arem, B. (2020). A generic multi-level framework for microscopic traffic simulation with automated vehicles in mixed traffic. *Transportation Research Part C: Emerging Technologies*, 110, 291–311.

[13] Zalewski, A., and Kempa, J. (2019, February). Traffic calming as a comprehensive solution improving traffic road safety. *In IOP Conference Series: Materials Science and Engineering* (Vol. 471, No. 6, p. 062035). IOP Publishing.

[14] Alsrehin, N. O., Klaib, A. F., and Magableh, A. (2019). Intelligent transportation and control systems using data mining and machine learning techniques: A comprehensive study. *IEEE Access*, 7, 49830–49857.

[15] Wang, S., Li, Z., Zhang, J., Yuan, Y., and Liu, Z. (2022). The crash injury severity prediction of traffic accident using an improved wrappers feature selection algorithm. *International Journal of Crashworthiness*, 27(3), 910–921.

[16] Yao, Y., and Huang, Z. (2016). Bi-directional LSTM recurrent neural network for Chinese word segmentation. In *Neural Information Processing: 23rd International Conference*, ICONIP 2016, Kyoto, Japan, October 16–21, 2016, Proceedings, Part IV 23 (pp. 345–353). Springer International Publishing.

*Hybrid and Advanced Technologies – S. Prasad Jones Christydass, Nurhayati Nurhayati and S. Kannadhasan. (Eds)* © *2025 the Authors, ISBN 978-1-032-90254-8*


# Enhancing traffic mobility and public safety based on a hybrid transportation system using ResNet50-LSTM model

P. Somashekar
*Assistant Professor, Department of Mathematics, Maharani's Science College for Women (Autonomous), Mysore, India*

S.V. Prabhakar
*Assistant Professor, Department of Electronics, Maharani's Science College for Women (Autonomous), Mysore, India*

C.N. Harshavardhana
*Government First Grade College for Women, Holenarasipura, India*

S. Vijay
*Associate Professor, Department of Mathematics, Government Science College, Hassan, India*

P. Siva Kota Reddy
*Professor, Department of Mathematics, JSS Science and Technology University, Mysuru, India*

R. Kemparaju
*Assistant Professor, Department of Mathematics, Government First Grade College, T. Narasipura, India*

ABSTRACT: In both developed and developing countries, ensuring the safety of people with mobility impairments (PRM) on the road is a major concern. These folks are getting a lot of attention right now. Russia, for example, has a profusion of individualized programs to improve their lives and, more especially, to guarantee that they may safely use the transportation infrastructure in numerous cities. Before training the model, there is pre-processing, and then there is feature extraction. To address these issues, the actionable mining model employs two components of the preprocessing step: variable selection and recoding. A set of features is extracted from the video stream during the feature extraction procedure. When training the model, we opted for the ResNet50-LSTM because of its superior accuracy. With an accuracy of 96.08%, the suggested strategy outperforms competitors like ResNet50 and LSTM.

*Keywords*: People with Mobility Impairments (PRM), Public Safety, Intelligent Transportation Systems (ITS), Residual Network (ResNet), Long Short-Term Memory (LSTM)

## 1 INTRODUCTION

Every day, people in poor countries face challenges related to public safety and mobility restrictions. The first is more often tied to an uptick in criminal activity, whereas the second is more often associated with gridlock on the roads. However, because to a lack of data regarding possible risk zones, these technologies may fall short in responding to public safety issues. These systems put drivers‘ and passengers' lives at jeopardy because the safest route is frequently the fastest. The fundamental idea behind these systems is to build bits of data on hazardous areas so that users may find the safest routes. They achieve this by making use of

DOI: 10.1201/9781003559115-94 

official crime statistics or a crowd sourcing-based method that allows users to provide information about safety. Unfortunately, in order to give route options, neither system is up-to-date on the present traffic conditions. Therefore, poor traffic awareness may lead to an upsurge in criminal activity in areas with overburdened transportation infrastructure. No one has yet explained how to improve traffic flow while also suggesting safe paths. This problem has prompted the development of Intelligent Transportation Systems (ITS). Better and safer traffic management is possible with the help of these systems, which integrate several systems and make use of cutting-edge sensor, processing, and communication technology.

## 2 LITERATURE SURVEY

The challenges and conflicts that characterise contemporary urban transportation are physically expressed in city streets. [1]A growing number of urban transit options, including motorised and non-motorized, individual and collective, fast and slow, are vying for a small number of spots [2]. A change to greener modes of urban transport is urgently required, and with it comes calls to restructure city roadways to make room for more pedestrians, cyclists and in place of public transport or privately owned motor vehicles[3]. Another layer of complexity is the increasing use of city streets as test beds for a myriad of "disruptive" mobility concepts, like electric and driverless vehicles, shared mobility platforms, ride-hailing services, and more[4]. The streets of a city are not only places for people to move about; they are also places where politics, creativity, socialising, and leisure all come together, in addition to business and entertainment[5]. Due to the high demands for quality of life and the rapid expansion in urban populations, it is essential to improve all parts of city infrastructure and subsystems [6]. More people need to be mobile, which leads to pollution and traffic congestion [7]. But, urban transportation can transform from a big problem to a great opportunity for modern cities to thrive and progress [8]. An Urban Mobility Plan is a common embodiment of the city's long-term vision that ensures the implementation of smart solutions, which is necessary for a well-structured transition [9]. Modern urban mobility planning should prioritise the promotion of ecologically friendlier, more integrated, and safer forms of transportation [10]. The growth of large cities and changes in land use have caused a sea change in transport planning, necessitating the development of new mobility solutions that can reach any part of the city. Intelligent mobility solutions can increase energy-efficient mobility while decreasing reliance on private autos. Some of the considerations that are considered while rerouting vehicle traffic in an urban context include road geometry, traffic flow, weather, human behaviour, traffic lights, crash risk, and public safety information [11]. Cars can use VANETs to sense their environment and report traffic conditions in real time using vehicle-to-infrastructure (V2I) and vehicle-to-vehicle (V2V) communications [12]. At the same time, there are a plethora of options for gathering additional data on urban settings, including systems that use crowdsourcing, social media, participatory sensing, web material, etc. [13]. When it comes to vehicular traffic re-routing, the majority of published solutions just use rerouting algorithms and data about traffic conditions, which doesn't go you very far future requirements for traffic management systems[14].

## 3 PROPOSED SYSTEM

Many cities have recently been facing transportation and safety issues due to congestion and a rise in citywide crime events. Instead of directing vehicles away from potentially hazardous areas in an attempt to reduce traffic congestion, an ITS would do the exact opposite direct vehicles toward congested areas with the sole purpose of ensuring the safety of drivers and passengers.

### 3.1 *Preprocessing*

Since data pertaining to public safety has its limits, this proposed to takes on an even larger significance in security and public safety analysis. Operationally relevant recoding and variable selection are two parts of the preprocessing step that the actionable mining model uses to solve these problems[15].

#### 3.1.1 *Recoding*

This recoding method yields more realistic categorical data that is suitable for a variety of modeling and mining tasks. Finally, making note of the type of occurrence or threat aids in formulating targeted responses. Optimizing the data can be achieved by recoding the event descriptions such that similar incidents utilize the same descriptors.

#### 3.1.2 *Variable selection*

When applied to real-world scenarios, not every analytic variable will yield significant results. To create models that are highly accurate and predictive, it is necessary to be able to use a number of factors, including those related to the features and qualities of offenders.

### 3.2 *Feature extraction*

The feature extraction process involves collecting a collection of characteristics from the live video feed. The use of machine learning classification methods in surveillance applications has been extensively studied. Background subtraction is a popular technique for object detection in moving videos; it entails comparing each frame to a model of the background. An issue with background subtraction known as false object detection occurs, for instance, when an immobile background item moves away[16].

## 4 MODEL TRAINING

### 4.1 *ResNet50-LSTM*

The inability and difficulty of using traditional methods to extract nonlinear, complicated correlations from data makes them inadequate for identifying traffic mobility and public safety. To tackle the issues highlighted earlier, a hybrid method that combines ResNet-50 and LSTM techniques is suggested as a viable option for using DL for public safety and traffic mobility. In this proposed approach to extract frame characteristics using a hybrid method that integrates ResNet-50 and LSTM methods for public safety and traffic mobility; it showed promise in image-based tasks, spatial and temporal context awareness, transfer learning, change resilience, computational efficiency, and scalability. For these reasons, ResNet-50 is a good choice for improving the accuracy and efficiency of public safety and traffic mobility by extracting functional traits from video frames. In addition, LSTM helps with classification in public safety and traffic mobility by capturing dependencies across time, remembering things, being resilient to changes, comprehending context, adapting to sequences of varying length, and being able to employ transfer learning. Splitting the input video into frames, applying a pretrained ResNet-50 algorithm, and scaling each frame are all part of it. Values of the pixels are subsequently normalised. The ResNet-50 model generates a series of feature vectors by using features extracted from every frame. After the input video is gathered, every video frame is resized to 256×256 pixels. Here is the input video, broken down into frames. The proposed technique takes advantage of popular greyscale textural features and applies them to powerful, intense optical flow fields. An technique to feature extraction uses the rescaling method to prepare the input. In order to adjust for the input dimensions, this method first splits the input video into individual frames, which are then shrunk. When it comes to DNN feature extraction, ResNets is a solid neural model contender. Their ability to effectively extract spatial characteristics from input streams has been proven on multiple common datasets. The efficacy of ResNets has been proven by results obtained from several iconic datasets. This detection made use of an LSTM to categorize a video frame. Using CNN, we initialize the coarse and fine labels simultaneously. As a last step, LSTM and RNN train the learning model that is used to improve productive behavior. The LSTM algorithm employs an input-and-forget gate network to control its internal memory cells, which was developed to address the problem of RNN gradients vanishing or bursting.

## 5 RESULT AND DISCUSSION

Rerouting vehicle traffic is the answer to make traffic flow more efficiently. However, when suggesting routes for individual vehicles, it is important to take into account more than just traffic-related data if successful transportation management is to be achieved. Therefore, context-aware and multi-objective re-routing techniques will form an essential aspect of traffic management.

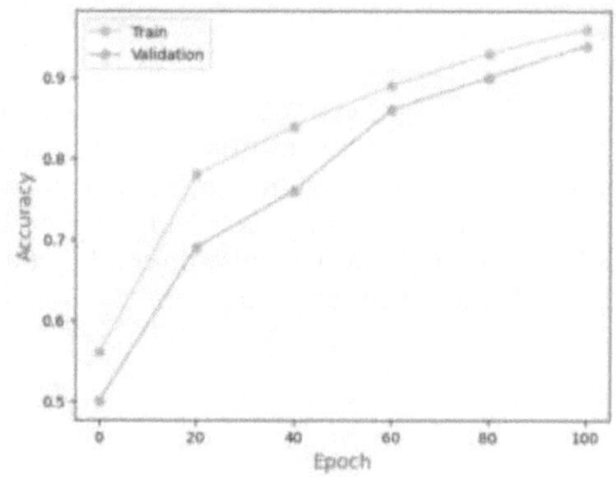


Figure 1. Accuracy curve for ResNet50-LSTM model.

Figure 1 displays the results of 100 learning epochs using the proposed ResNet50-LSTM model to recognize impulsive sounds. Public safety and traffic mobility are two areas where the data show the model to be highly accurate.

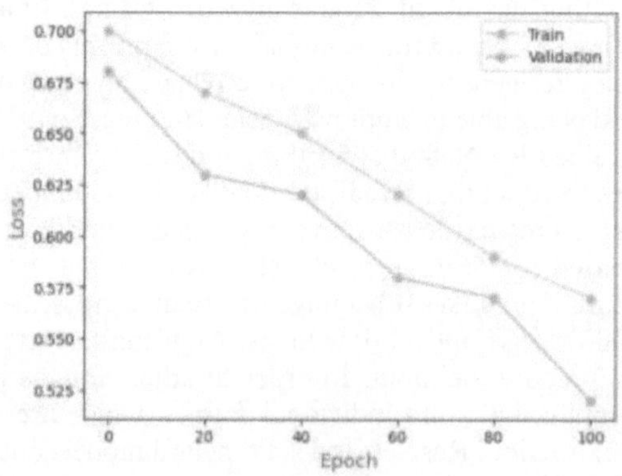


Figure 2. Loss curve for ResNet50-LSTM model.

For public safety and traffic mobility, the suggested ResNet50-LSTM model loses its shape after 100 learning epochs (Figure 2). Minimal loss is demonstrated by the model, according to the results.

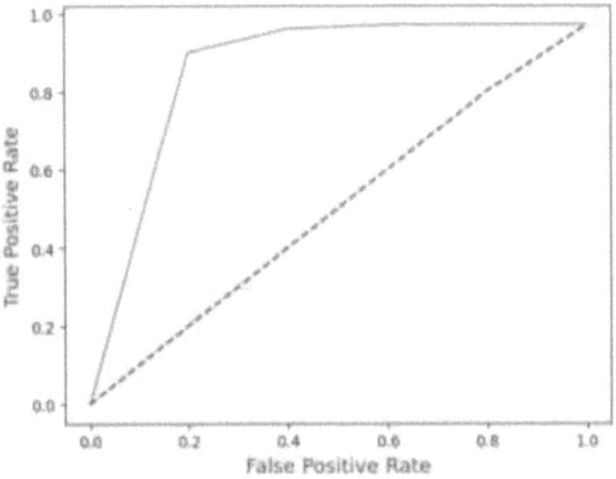


Figure 3. AUC-ROC curve of models.

Figure 3 shows the receiver operating characteristic (ROC) and area under the curve (AUC). This was a simple example of how changes to the training data impacted the classifier's output.

## 6 CONCLUSION

Transportation safety, efficiency, and sustainability can all be enhanced by mobility as a service by strategically placing and connecting different modes. The purpose of Mobility as a service in a local context (MaaS-LC) is to increase road traffic safety and raise awareness of safety issues by combining the Safety Index with the Walkability Index. The two parts of the preprocessing step that the actionable mining model uses to fix these problems are recoding and variable selection. Feature extraction is the process of extracting a set of characteristics from a video stream. Model training is carried out using the ResNet50-LSTM technique. The recommended technique outperforms the ResNet50 and LSTM models with a consistency rate of 96.08%.

## REFERENCES


[1] Bertolini, L., Le Clercq, F., and Straatemeier, T. (2008). Urban transportation planning in transition. *Transport Policy*, 15(2).

[2] Holden, E., Gilpin, G., and Banister, D. (2019). Sustainable mobility at thirty. *Sustainability*, 11(7), 1965.

[3] Von Schönfeld, K. C., and Bertolini, L. (2017). Urban streets: Epitomes of planning challenges and opportunities at the interface of public space and mobility. *Cities*, 68, 48–55.

[4] Kuhlberg, J. A., Hipp, J. A., Eyler, A., and Chang, G. (2014). Open streets initiatives in the United States: closed to traffic, open to physical activity. *Journal of Physical Activity and Health*, 11(8), 1468–1474.

[5] Cowman, K. (2017). Play streets: women, children and the problem of urban traffic, 1930–1970. *Social History*, 42(2), 233–256.

[6] Noy, K., and Givoni, M. (2018). Is 'smart mobility'sustainable? Examining the views and beliefs of transport's technological entrepreneurs. *Sustainability*, 10(2), 422.

[7] Chen, Y., Ardila-Gomez, A., and Frame, G. (2017). Achieving energy savings by intelligent transportation systems investments in the context of smart cities. *Transportation Research Part D: Transport and Environment*, 54, 381–396.

[8] Melo, S., Macedo, J., and Baptista, P. (2017). Guiding cities to pursue a smart mobility paradigm: An example from vehicle routing guidance and its traffic and operational effects. *Research in Transportation Economics*, 65, 24–33.

[9] Orlowski, A., and Romanowska, P. (2019). Smart cities concept: Smart mobility indicator. *Cybernetics and Systems*, 50(2), 118–131.

[10] Mangiaracina, R., Perego, A., Salvadori, G., and Tumino, A. (2017). A comprehensive view of intelligent transport systems for urban smart mobility. *International Journal of Logistics Research and Applications*, 20(1), 39–52.

[11] Amirgholy, M., Golshani, N., Schneider, C., Gonzales, E. J., and Gao, H. O. (2017). An advanced traveler navigation system adapted to route choice preferences of the individual users. *International Journal of Transportation Science and Technology*, 6(4), 240–254.

[12] Araujo, G. B., Queiroz, M. M., de LP Duarte-Figueiredo, F., Tostes, A. I., and Loureiro, A. A. (2014, June). *CARTIM: A proposal toward identification and minimization of vehicular traffic congestion for VANET.* In *2014 IEEE Symposium on Computers and Communications (ISCC)* (pp. 1–6). IEEE.

[13] de Souza, A. M., and Villas, L. A. (2016, November). A fully-distributed traffic management system to improve the overall traffic efficiency. In *Proceedings of the 19th ACM International Conference on Modeling, Analysis and Simulation of Wireless and Mobile Systems* (pp. 19–26).

[14] Krishnan, S. R., and Amudha, P. (2023). Hybrid ResNet-50 and LSTM approach for effective video anomaly detection in intelligent surveillance systems. *International Journal of Intelligent Systems and Applications in Engineering*, 11(4), 880–889.

[15] McCue, C. (2006). Data mining and predictive analytics in public safety and security. *IT Professional Magazine*, 8(4), 12.

[16] Räty, T. D. (2010). Survey on contemporary remote surveillance systems for public safety. *IEEE Transactions on Systems, Man, and Cybernetics, Part C (Applications and Reviews)*, 40(5), 493–515.

*Hybrid and Advanced Technologies – S. Prasad Jones Christydass, Nurhayati Nurhayati and S. Kannadhasan. (Eds)* © *2025 the Authors, ISBN 978-1-032-90254-8*


# Classification of traffic signs using a novel multi-view support vector machine and comparing the accuracy with decision tree

Thalla Nitish Kumar
*Research Scholar, Department of Computer Science and Engineering, Saveetha School of Engineering, Saveetha University, Chennai, Tamil Nadu, India*

R. Senthil Kumar
*Project Guide, Department of Computer Science and Engineering, Saveetha School of Engineering, Saveetha University, Chennai, Tamil Nadu, India*

ABSTRACT: This study's main objective is to use a special multi-view support vector machine for traffic sign recognition and a decision tree to evaluate the machine's accuracy. There are two sections to the suggested research project. The Decision Tree is in Group 2 while the creative multi-view SVM is in Group 1. A G-power rating of 80% was used together with 10 iterations to analyse 7000 sample sets. A programming language was used with a 95% confidence level. 10,000 data tuples were taken into consideration for the identification dataset, which was a Kaggle dataset consisting of 51,000 files from the field. 88.77 is the accuracy of the novel multi-view SVM, and 61.19% is the accuracy of the Decision Tree, with a statistically significant precision of ($p = 0.015$) ($p<0.05$). When compared to the decision tree, the accuracy has significantly increased.

*Keywords*: Vehicle, MVSVM, Decision Tree, Detection, Novel, Traffic, Precision, Image

## 1 INTRODUCTION

The research's primary objective is to develop a robust system capable of detecting and recognizing traffic signals from a range of angles and under a variety of environmental conditions. This requires real-time image capture and processing to recognize relevant traffic signs. The project will introduce and implement a ground-breaking multi-view support vector machine approach. (Long *et al.* 2023) This strategy will leverage numerous angles or perspectives of traffic signs to increase detection accuracy. The Multi-View Support Vector Machine is expected to outperform ordinary SVMs when dealing with multi-view image data. This project will introduce and develop a novel multi-view support vector machine approach. This strategy will leverage numerous angles or perspectives of traffic signs to increase detection accuracy. MVSVM is expected to outperform traditional SVMs when dealing with multi-view picture data. There may be uses for this research and the multi-view support vector machine (MVSVM) model that was created. (Ao and others 2023) It might be a part of the sensor suite of self-driving cars due to its improved accuracy in identifying traffic signs. The MVSVM has real-time traffic sign recognition while driving and is able to analyse data from cameras.This work has the potential to have a significant impact on the fields of computer vision and artificial intelligence by improving the accuracy and flexibility of traffic sign detection systems. It has practical implications for enhancing traffic safety and efficiency, particularly in the age of self-driving automobiles. The study compares the MVSVM against the well-known Decision Tree method to discover which technique performs better for traffic sign detection in different situations and circumstances. The research introduces a revolutionary multi-view support vector machine. (Yan *et al.* 2023) 156 publications regarding this topic Table 1 were found in this range using the most-cited research paper databases from Google Scholar and Science Direct. In order to give car drivers adequate information to make an educated choice, this study suggests an autonomous traffic light detection system. Consequently, this facilitates the development of an intelligent transport system (ITS). (Aluvalu *et al.* 2023; Mahaveerakannan R *et al.* 2023; Ramana *et al.* 2023; Vellela and Balamanigandan 2023)The proposed method works well for real-time applications, can be used with movable cameras, and is adaptable to changing environmental conditions. Traffic light detection and classification are the two processes that make up this process. (Al-Ezaly *et al.* 2023) Border detection is utilized during the classification process to generate area borders, which are matching features together with region colors. Both circle and arrow traffic lights may be included by this study. Moreover, variations in the quantity, kind, and timing of light in the surrounding environment can affect our system. To prove the effectiveness and practicality of the suggested method, numerous experiments have been carried out. (National Research Council *et al.* 2001). It can be challenging and time-consuming to gather an extensive dataset of traffic signs that covers a wide range of real-world scenarios. (Huang *et al.* 2022) Considerations such as lighting, weather, and sign deterioration can all have an impact on the accuracy and robustness of the models. The multi-view technique and decision tree comparison analysis in the project may require a significant amount of processing resources(Wang *et al.* 2021) It comprises processing and memory power. This might restrict the project's applicability in resource-constrained contexts. In traffic signal identification, the suggested new multi-view vector machine outperforms the previous technique.

## 2 MATERIALS AND METHODS

The first is known as SVM, and the second as DF. To determine the sample size, Clinal was employed. Utilizing information from earlier research, a sample size calculation with a 0.05 cutoff, 80% G power, and a 95% confidence range was created. In 2018, Mouhoub *et al.* Google Collabor, Jupyter Notebooks, 11th generation Windows, and an Intel Core i5 with 512 MB of RAM are also available. The statistical analysis to compare the accuracy was conducted using IBM SPSS. This investigation revealed the tweet's accuracy, emotion, and correctness. A 51,000-file dataset (including limits, signals, objects, 10,000, etc.) was made available by Kaggle. After 10,000 data points were used to train both algorithms, the accuracy of each was evaluated using an independent T-test in SPSS. (2020 Preda)

### 2.1 *Support vector machine classifier*

In natural language processing, Support Vector Machine,a supervised machine learning technique, is widely employed for regression but may also be utilized for classification. SVM searches an n-dimensional feature space for a hyperplane that effectively separates and organizes input data points. The dimensionality of this hyperplane depends on how many features are in the dataset. In two-dimensional space, the SVM hyperplane corresponds to a simple line when you use only two input features. A two-dimensional plane replaces the hyperplane if the dataset contains more than three attributes. Showing this hyperplane, however, becomes more difficult when dealing with data that has more dimensions than three.

*Algorithm*
Step 1: Import required packages and the data file.
Train test split is the second step.
Third Step: Bring in train_test_split
Step 4: stratify, random state = 42, test_size = 0.2, x,y, and y_trainStep 6: = y

Step 5: Use the Support Vector Machine import Use a support vector machine to train the data in step six.
Use support vector machines in step seven by using fit(x_train,y_train). estimate the accuracy using (x_test)
Step 8: Bring in the accuracy score
Accuracy_score(y_test,predict)*100 is used in Step 9.
Step 10: Determining the Precision

### 2.2 *Decision tree*

Regression and classification can be accomplished with decision trees, a kind of supervised machine learning method. The tree's interior nodes each reflect a characteristic (or attribute), the leaf nodes show the outcome or expected value, and the branches indicate decision rules.

Decision trees arise from repeatedly segmenting the dataset into subgroups according to the most significant features. These separations aim to improve data collection or decrease impurity, depending on the method. Decision trees can be used to explain the reasoning behind a model's predictions because they are understandable and interpretable.

*Algorithm*
Step 1: Import required packages and the data file.
Train test split is the second step.
Third Step: Bring in train_test_split
The steps are as follows:
Import DecisionTreeClassifier from sklearn.tree in step five.
Use df = DecisionTreeClassifier() to train the data in step six.
For accuracy, use knn.predict(x_test) in step seven.
Step 8: Bring in the accuracy score
Accuracy_score(y_test,predict)*100 is used in Step 9.
Step 10: Assessing Accuracy

### 2.3 *Statistical analysis*

SPSS software was used to do the statistical analysis. Different images and forms are among the study's independent factors, and image accuracy and selection are among its dependent variables. (Aug 2021)

## 3 RESULT

Table 1. SVM and DT Algorithm Accuracy The table below displays the accuracy of the vector machine and decision tree, which are derived from the dataset's Natural Language Processing process utilising both techniques.

| S.NO | SVM Algorithm | Decision TreeAlgorithm |
|---|---|---|
| 1 | 89.71 | 59.26 |
| 2 | 88.19 | 63.14 |
| 3 | 88.46 | 59.09 |
| 4 | 88.84 | 62.36 |
| 5 | 89.31 | 60.73 |
| 6 | 88.47 | 59.72 |
| 7 | 89.97 | 62.79 |
| 8 | 89.62 | 62.97 |
| 9 | 81.01 | 62.00 |
| 10 | 89.13 | 59.98 |
| **Accuracy** | **88.98** | **61.19** |

Table 2. Group accuracy statistics for the methods of DT and SVM With the help of both SPSS methods, we were able to extract the dataset's mean, standard deviation, and standard error for the decision tree and SVM. The decision tree has an accuracy of 61.19% while the SVM mentioned above has an accuracy of 88.77%.

| | Algorithm | N | Mean | Std.Deviation | Std.Error Mean |
|---|---|---|---|---|---|
| **Accuracy** | Support Vector Machine | 10 | 88.77 | 0.650 | 0.205 |
| | Decision Tree | 10 | 61.19 | 2.01 | 0.637 |

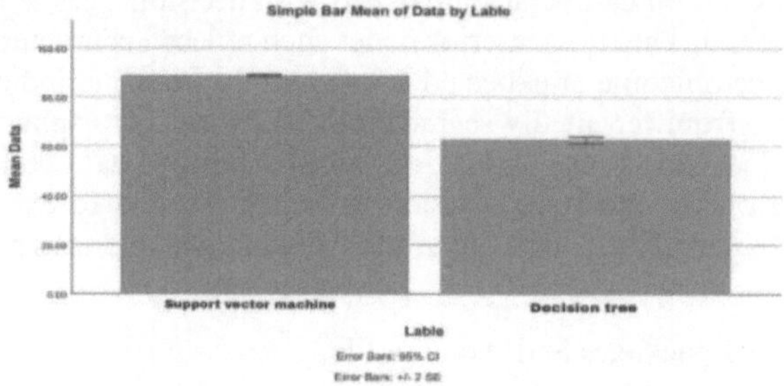


Figure 1. Shows the statistical analysis that was performed to ascertain the data's accuracy using the SPSS programme. The suggested approach yields 61.19% with an error bar of 95% CI and +/- 2 SD and 88.77% with the DF.

To determine whether the data is accurate, Figure 1 presents a statistical analysis performed with the SPSS programme. With a 95% confidence interval and a standard deviation of +/- 2 SD, the decision tree yields an accuracy of 61.19% compared to the recommended method's 88.77% accuracy.

## 4 DISCUSSION

Decision trees and multi-view support vector machines are compared in the study, and the latter performs significantly better with an accuracy of 88.77% as opposed to 61.19%. With a p-value of 0.015 ($p<0.05$), the statistical significance emphasizes how well the multi-view technique works to improve traffic sign identification for increased road safety and accident prevention (Puri *et al.* 2020). The results validate data-driven choices for traffic management system optimization, which are advantageous to the autonomous vehicle sector as well as regulatory bodies.

## 5 CONCLUSION

This work aims to ascertain the picture detection accuracy using N = 10 precision iterations. The survey yielded an overall score of 88.77% for Innovative SVM and 61.19% for Decision Tree. In contrast to decision trees, a support vector machine was found in the survey to produce results that are more pertinent to test outcomes. P = 0.05 indicates the composite's importance.

## REFERENCES

Ao, C., Ye, X., Sakurai, T., Zou, Q., and Yu, L. (2023). m5U-SVM: identification of RNA 5-methyluridine modification sites based on multi-view features of physicochemical features and distributed representation. *BMC Biology*, *21*(1), 93.

Al-Ezaly, E., M. El-Bakry, H., Abo-Elfetoh, A., and Elhishi, S. (2023). An innovative traffic light recognition method using vehicular ad-hoc networks. *Scientific Reports*, 13(1), 4009.

Alfred, R., Lim, Y., Haviluddin, H., and On, C. K. (Eds.). (2019). *Computational Science and Technology: 6th ICCST 2019*, Kota Kinabalu, Malaysia, 29-30 August 2019 (Vol. 603). Springer Nature.

Aluvalu, R., VN, S. K., Thirumalaisamy, M., Basheer, S., and Selvarajan, S. (2023). Efficient data transmission on wireless communication through a privacy-enhanced blockchain process. *PeerJ Computer Science*, 9, e1308.

Huang, Y., Jin, J., Xu, R., Miao, Y., Liu, C., and Cichocki, A. (2022). Multi-view optimization of time-frequency common spatial patterns for brain-computer interfaces. *Journal of Neuroscience Methods*, 365, 109378.

Long, K. K., Kwok, S. W. H., Kotz, J., and Wang, G. (2023). A deep multi-view imbalanced learning approach for identifying informative COVID-19 tweets from social media. *Computers in Biology and Medicine*, 164, 107232.

Mahaveerakannan, R., Anitha, C., Thomas, A. K., Rajan, S., Muthukumar, T., and Rajulu, G. G. (2023). An IoT based forest fire detection system using integration of cat swarm with LSTM model. *Computer Communications*, 211, 37–45.

Mouhoub, M., Sadaoui, S., Mohamed, O. A., and Ali, M. (Eds.). (2018). *Recent Trends and Future Technology in Applied Intelligence: 31st International Conference on Industrial Engineering and Other Applications of Applied Intelligent Systems*, IEA/AIE 2018, Montreal, QC, Canada, June 25-28, 2018, Proceedings (Vol. 10868). Springer.

National Research Council, Division of Behavioral, Board on Science Education, and National Committee on Science Education Standards. (1996). *National science education standards*. National Academies Press.

Ramana, K., Revathi, A., Gayathri, A., Jhaveri, R. H., Narayana, C. L., and Kumar, B. N. (2022). WOGRU-IDS—An intelligent intrusion detection system for IoT assisted Wireless Sensor Networks. *Computer Communications*, 196, 195–206.

Tatsat, H., Puri, S., and Lookabaugh, B. (2020). *Machine Learning and Data Science Blueprints for Finance*. O'Reilly media.

Thulin, M. (2021). *Modern Statistics with R: From Wrangling and Exploring Data to Inference and Predictive Modelling*. BoD-Books on Demand.

Vellela, S. S., and Balamanigandan, R. (2023). An intelligent sleep-awake energy management system for wireless sensor network. *Peer-to-Peer Networking and Applications*, 16(6), 2714–2731.

Wang, Y., Ding, Y., Tang, J., Dai, Y., and Guo, F. (2019). CrystalM: a multi-view fusion approach for protein crystallization prediction. *IEEE/ACM Transactions on Computational Biology and Bioinformatics*, 18(1), 325–335.

Yan, K., Lv, H., Wen, J., Guo, Y., Xu, Y., and Liu, B. (2022). PreTP-Stack: prediction of therapeutic peptides based on the stacked ensemble learing. *IEEE/ACM Transactions on Computational Biology and Bioinformatics*, 20(2), 1337–1344.

*Hybrid and Advanced Technologies – S. Prasad Jones Christydass, Nurhayati Nurhayati and S. Kannadhasan. (Eds)* © *2025 the Authors, ISBN 978-1-032-90254-8*
*Open Access: www.taylorfrancis.com, CC BY-NC-ND 4.0 license*

# Optimizing accident prediction accuracy using Novel Random Forest and K-Nearest Neighbors algorithm through weather data analysis

K. Gowtham
*Research Scholar, Department of Computer Science and Engineering, Saveetha School of Engineering, Saveetha University, Chennai, Tamil Nadu, India*

R. Bhavani
*Project Guide, Department of Computer Science and Engineering, Saveetha School of Engineering, Saveetha University, Chennai, Tamil Nadu, India*

ABSTRACT: The study seeks to enhance accident forecasting by examining weather information with a Novel Random Forest (NRF), and comparing its effectiveness to the K-Nearest Neighbors (KNN) approachThe analysis involved comparing two groups using distinct methods, the NRF and KNN. Both groups had 20 samples. The number of iterations was determined using G-power calculations, considering an effect size of 0.8 and a 95% confidence interval. The NRF model predicts accidents with an 83.20% accuracy, while KNN achieves 61.25%. Independent tests found a significant difference ($p = 0.037$) between the two algorithms The NRF is more accurate than the KNN algorithm in predicting accidents with weather data. The precision rate of the NRF is 83.20%, while the accuracy rate of the KNN algorithm is 61.25%.

*Keywords*: K-Nearest Neighbors, Machine Learning, Novel Random Forest, Road Safety, Road traffic accidents, Weather Data

## 1 INTRODUCTION

Machine learning algorithms, which form a collection of computational methods, possess the capability to engage in tasks related to pattern recognition, classification, and prediction of data by acquiring knowledge from pre-existing data, known as the training set(Harsha Jasni *et al.* 2023). The importance of traffic collisions as a subject of utmost significance in terms of safeguarding the well-being and security of the general public cannot be underestimated. It is imperative to recognize and acknowledge the crucial role that adverse weather conditions play in the occurrence and frequency of accidents(Baek and Chung 2021). The constraints imposed by the current methodologies in effectively utilizing meteorological data for the purpose of accurate prediction and proactive measures against accidents must be duly recognized and addressed. The existing strategies encounter specific constraints and obstacles that impede their ability to fully exploit the potential of meteorological information, which is essential for accurate prediction of accidents.(Taamneh and Taamneh 2021). Machine learning is utilized in the analysis of weather climate data to improve the prediction and prevention of accidents. This is achieved by constructing predictive models that take into account both historical accident data and current weather conditions(Ghandour *et al.* 2020). The application of Machine Learning models enables the exploration of past accident records and meteorological circumstances to identify regions

 DOI: 10.1201/9781003559115-96

with high accident rates. This information holds significant value for urban planners and traffic control agencies as it assists in the prioritization of safety enhancements in specific regions(Vaiyapuri and Gupta 2021). Machine Learning models can examine the temporal patterns displayed in accident data, considering variables like the time of day, day of the week, and seasonal changes.

A great deal of academic research has been done in the area of accident prediction utilising meteorological data. A thorough examination into this subject has been carried out through a review of 17,691 academic articles that are easily accessible on Google Scholar. On the other hand, IEEE Xplore contributes 1,892 papers, and Springer offers 2,589 publications that are pertinent to this particular topic of research. The scholarly investigation of 125 citations confirms that the analysis of data regarding weather conditions, comprising temperature, humidity, precipitation, wind speed, and other meteorological parameters, involves the collection, manipulation, and understanding of information. The analysis of this information can lead to diverse applications, such as predicting weather patterns and understanding their impact on distinct processes(Vanitha and M. Swedha 2023). The ongoing research investigation of 179 references asserts that Temporal patterns refer to recurring trends, fluctuations, or sequences observed in data over a period of time. In relation to the prognosis of road traffic accidents, this might include the identification of patterns in road traffic accident incidents at distinct time periods of the day, week, or year. Temporal patterns can also be used as indicators of extended periods of change, for instance, a decrease in road traffic accidents over multiple years due to enhanced road safety measures, or an increase in road traffic accidents caused by alterations in traffic patterns(Shah 2023). The present research inquiry, consisting of a total of 198 citations, proposes that cross-domain prediction necessitates the utilization of a predictive framework that has been trained on a specific dataset or domain in order to generate predictions in a distinct yet interconnected domain. With respect to road traffic accidents, this could involve the implementation of a model that has been trained in one geographic area to another(Salarian *et al.* 2023). The research study, which incorporates 258 citations, confirms that the magnitude of road traffic accidents is a significant global issue, particularly in economically disadvantaged nations. The analysis of the primary and secondary factors at play can assist in lessening the severity of road traffic accidents (Alotaibi and F. Subahi 2023; Parvez M *et al.* 2023; Sarkar *et al.* 2022; Tayyab *et al.* 2022; Surendran, Ramesh *et al.* n.d.; Yassin and Pooja 2020) The academic scholarly inquiry, supported by a total of 178 references, asserts that the severity of road traffic accidents is a significant issue on a global scale, especially in countries that are still in the process of development. The results of the model specific to the targeted objective indicate that the driver's experience, as well as the specific time of day, lighting conditions, the driver's age, and the vehicle's years of service, all have a significant impact on the severity of injuries sustained, whether they are serious, minor, or fatal in nature(Yan and Shen 2022).

The existing accident prediction algorithms demonstrate suboptimal performance on datasets of limited size. In situations where data is insufficient, the algorithm may face challenges in constructing a diverse ensemble of decision trees, potentially resulting in overfitting or underfitting(Sahu *et al.* 2023). The algorithm lacks the capability to effectively analyze time-series data. The algorithms fail to effectively capture sequential dependencies, which hold significant importance in a multitude of tasks related to time-series forecasting and prediction. In situations where data is insufficient, the algorithm may encounter difficulties when constructing a diverse ensemble of decision trees, potentially resulting in overfitting or underfitting(Sysoev *et al.* 2022). The objective of this research is to enhance the accuracy of accident prediction and prevention by means of utilizing weather climate data analysis(Mohire 2022). The primary aim of this study is to evaluate and improve the performance of a new Random Forest algorithm in conjunction with the K Nearest Neighbors (KNN) algorithm, with the overall objective of creating an effective framework for enhancing road safety through data-driven decisions that consider weather conditions and other relevant factors.

## 2 MATERIALS AND METHODS

Within the larger context of the Saveetha Institute of Medical and Technical Sciences, the inquiry was conducted at the Laboratory of the Computer Resource Centre, a division of the Saveetha School of Engineering. Modern computer equipment installed in the lab was essential to making the experiment possible (Bhuiyan *et al.* 2022). The study included two groups of twenty participants each, and it used two different statistical analysis algorithms: the Naive Bayes model and the Innovative Random Forest model, with support from Clincalc (Joyce Beryl Princess *et al.* 2021). Establishing a 95% confidence interval, setting the significance level at alpha = 0.05, and achieving 80% statistical power—as demonstrated by a beta value of 0.8—were the primary goals of the study. The 80% statistical power was determined using the G-power method.

The dataset used in this research, called Weather_data, is a sizable collection of 14,000 unique data elements. These specific data points are included in a fully prepared CSV file, which has been methodically arranged (Wahab and Jiang 2020). It is important to highlight the value of this dataset, which was originally downloaded from the reputable Kaggle site, because it provides a great deal of value by acting as a model and an aesthetically appealing representation of the meteorological climate data that is being studied and analysed.

Among the many services that Google provides is Google Collaborate, a useful tool. Cloud-based computing makes it easier for Python programming to run smoothly. When using the Google Collaborate platform, users can run Python code immediately, which means that complex configuration steps are not required. Code that is used in this specific context is effectively kept organised and easily accessible in Google Drive. In order to extract data from a CSV file, many libraries, including train_test_split, KNeighbors Classifier, and Random Forest Classifier, must be used.

### 2.1 *Novel Random Forest (NRF)*

The collection and preprocessing of historical accident and weather climate data can enhance accident prediction. This process includes the analysis of weather climate data. The training of a Random Forest model allows for precise forecasts on accident probability under different weather conditions. The analysis of variable importance in the model assists in the identification of essential weather factors that contribute to accidents, thereby allowing for the implementation of improved safety measures and strategies for managing traffic during unfavorable weather conditions(Vignesh and Sashirekha 2022). The model's dependability in proactively addressing accident risks related to weather is guaranteed through regular updates and continuous efforts to enhance its performance. The following actions must be taken in order to look into the estimation of accidents based on weather and climatic data: Open a CSV file containing data on precipitation and temperature. Sort, clean, and standardise the data into input features and target variables. Create variables, randomise, and divide the data between training and testing sets. Provide hyperparameters for a Random Forest model construction. Utilising the training set, test set, and metrics, train the model to predict accidents. Provide the estimated number of accidents as well as the assessment measures used to gauge the model's effectiveness.

#### 2.1.1 *Algorithm for NRF*

(1) Start the procedure by importing the temperature and precipitation data from the input CSV file.
(2) Proceed to the stage of data preprocessing, which includes the following actions:
   (a) Divide the data into input features, such as temperature and precipitation, and find the target variable that represents the occurrence of accidents.

   (b) Carry out the necessary data standardisation and cleaning, and investigate possible feature enhancements for the input features.

(3) Next, divide the preprocessed data into distinct training and testing sets as follows:
   (a) Establish the ideal ratio for the data's separation into training and testing subsets.
   (b) Divide the data into discrete subsets, X_train, X_test, y_train, and y_test, by shuffle it randomly.

(4) Choose how many decision trees will make up the forest as the first step in setting up a Random Forest model.
   (a) Specify additional stopping criteria or the maximum depth for each decision tree (max_depth).
   (b) Select the maximum number of features (max_features) to be evaluated at each splitting point.
   (c) Decide on any extra hyperparameters, such as the partitioning criteria.

(5) After that, use the training data to train the Random Forest model:
   (a) Using the training set of X_train and y_train and accounting for the specified hyperparameters, fit the Random Forest model.

(6) Make predictions about the frequency of accidents in the testing set using the trained Random Forest model:
   (a) Keep track of and archive the accident prediction efforts.

(7) Go ahead and assess the Random Forest model's performance by:
   (a) Carefully compare the actual incidences found in y_test with the anticipated accident occurrences.
   (b) Compute a wide range of evaluation metrics, such as accuracy, precision, recall, or F1-score, to obtain a thorough grasp of the model's efficacy.

(8) Wrap up the procedure by showing the overall assessment metrics as well as the anticipated accident occurrences related to the instances in X_test.

### 2.2 *K-Nearest Neighbors(KNN)*

K-Nearest Neighbors is a supervised algorithm employed in the field of machine learning for the purpose of carrying out classification and regression tasks. The underlying principle it follows is that objects or data points within a given dataset are inclined to exhibit similarity with those residing in close proximity. To anticipate accidents by employing temperature and precipitation data, one must adhere to a series of procedures(Shirwaikar, K P, and H Simha 2022). Firstly, import the CSV file, subsequently, preprocess the data, then partition it into separate training and testing sets. Next, initialize a K-Nearest Neighbors model, proceed to train the KNN model, and subsequently, generate predictions on the testing set. Following this, evaluate the performance through the utilization of metrics, culminating in the printing of predictions and the accuracy score for the KNN model. This particular methodology effectively augments the precision of accident prediction.

#### 2.2.1 *Algorithm for KNN*

(1) Load the input CSV file with the precipitation and temperature information.
(2) Prepare the data:
   (a) Divide the data into the objective variable (accident occurrence) and the input features (temperature and precipitation).

(b) Apply any required feature engineering, normalisation, or data cleaning techniques to the input features.

(3) Divided the preprocessed data into sets for testing and training:
   (a) Choose how much of the data will be utilised for testing (say, 30%) and training (say, 70%).
   (b) Divide the data into X_train, X_test, y_train, and y_test using a random shuffling technique.

(4) Set up a model for K-Nearest Neighbours:
   (a) Indicate how many neighbours (K) should be taken into account when classifying.
   (b) To find the degree of similarity between instances, select a distance measure.
   (c) Adjust the KNN model's other pertinent hyperparameters.

(5) Use the training set to train the K-Nearest Neighbours model:
   (a) Use the given hyperparameters to fit the KNN model on the training sets of X and Y.

(6) Project the likelihood of an accident on the testing set:
   (a) Predict the likelihood of an accident occurring for each instance in X_test using the trained KNN model.
   (b) Keep track of the anticipated accident incidents.

(7) Evaluate the K-Nearest Neighbours model's effectiveness:
   (a) Examine the differences between the actual and anticipated accident incidences in y_test.
   (b) To assess the performance of the model, compute evaluation measures like accuracy, precision, recall, or F1-score.

(8) Print the assessment metrics and anticipated accident occurrences:
   (a) Print the expected number of accidents for each instance in X_test.
   (b) Comparing the expected and actual accident incidences will allow you to determine how accurate the model is.
   (c) Print the accuracy score to use as a K-Nearest Neighbours model evaluation metric.

### 2.3 *Statistical analysis*

IBM SPSS version 26.0 was used to conduct the data analysis. The computation of important statistical metrics, including the mean, standard deviation, and standard error of the mean, was emphasised (Aleksić *et al.* 2023). It is imperative to emphasise that the main focus of the analysis was the independent variables, which included variables such as Date, Temperature (°C), Precipitation (mm), road surface conditions, lighting, and speed limit. The frequency of accidents was the main variable in this analysis that drew attention. The generation of T-test results was significantly influenced by these variables taken together.

## 3 RESULTS

The Manifests are shown in Table 1, where input data was gathered in a tabular manner using a representative sample of N = 20. After examining the dataset, the Novel Random Forest model's precision was calculated, yielding an 83.20% percentage value. Using the same dataset, however, the K Nearest Neighbours algorithm demonstrated an accuracy rate of 61.25%.

Table 1. Tabulated input data was collected using a sample of N = 20. After analysing the dataset, the Novel Random Forest model's accuracy was calculated, yielding a percentage value of 83.20%. Thus, using the same dataset, the K Nearest Neighbours algorithm demonstrated an accuracy of 61.25%.

| S.No | NRF Accuracy (%) | KNN Accuracy (%) |
|---|---|---|
| 1 | 83 | 61 |
| 2 | 84 | 64 |
| 3 | 83 | 69 |
| 4 | 85 | 53 |
| 5 | 74 | 56 |
| 6 | 82 | 52 |
| 7 | 86 | 60 |
| 8 | 81 | 68 |
| 9 | 83 | 54 |
| 10 | 89 | 58 |
| 11 | 83 | 64 |
| 12 | 79 | 63 |
| 13 | 85 | 59 |
| 14 | 87 | 65 |
| 15 | 89 | 66 |
| 16 | 82 | 64 |
| 17 | 81 | 62 |
| 18 | 80 | 67 |
| 19 | 79 | 61 |
| 20 | 89 | 59 |

Table 2 shows how the NRF and KNN models were compared using separate samples in the comparative study. The NRF and KNN models' mean scores, which were found to be 83.20 and 61.25, respectively, were compared in the comparison analysis. Furthermore, the standard deviation of the NRF model was 3.792, whereas the KNN model displayed a standard deviation of 4.908. The NRF model's standard error of the mean was calculated to be 0.848, whereas the KNN model's was determined to be 1.098.

Table 2. Using independent samples, the NRF and KNN models were compared in the comparison analysis. The average scores of the NRF and KNN models, which were discovered to be 83.20 and 61.25, respectively, were compared in the comparison analysis. Furthermore, the NRF model's standard deviation was determined to be 3.792, but the KNN model had a standard deviation of 4.908. The standard error of the average for the KNN model was 1.098, but it was determined to be 0.848 for the NRF model.

| Group | | N | Mean | Standard Deviation | Standard Error Mean |
|---|---|---|---|---|---|
| **Accuracy Rate** | **NRF** | 20 | 83.20 | 3.792 | 0.848 |
| | **KNN** | 20 | 61.25 | 4.908 | 1.098 |

The T-Test, treated as statistically independent samples, is shown in Table 3 and was used to evaluate the methods of the K Nearest Neighbours algorithm and the Novel Random Forest algorithm. This empirical investigation produced a p-value of 0.037 ($p<0.05$), which is statistically significant.

Table 3. The Novel Random Forest and K Nearest Neighbours algorithms were regarded as statistically independent samples, and their methods were assessed using a statistical analysis known as the T-Test. This empirical inquiry yielded a statistically significant result with a p-value of 0.037 ($p<0.05$).

| | | Levens test for equality of variances | | T test for equality of means | | | | | | |
|---|---|---|---|---|---|---|---|---|---|---|
| | | | | | | | | | 95% confidence interval of the diff | |
| | | F | sig | t | df | sig. (2-tailed) | Mean diff | Std. error diff | Lower | Upper |
| **Accuracy** | **Equal variances assumed Equal variances not assumed** | 1.818 | 0.186 | 15.826 | 38 | 0.037 | 21.950 | 1.387 | 19.142 | 24.758 |
| | **Equal variances assumed Equal variances not assumed** | | | 15.826 | 35.723 | 0.037 | 21.950 | 1.387 | 19.136 | 24.764 |

The accuracy rates of the K Nearest Neighbour algorithm and the suggested Novel Random Forest methodology in the area of accident prediction are displayed in a bar graph in Figure 1. The diagram's horizontal axis represents the two methods, NRF and KNN, and its vertical axis shows the average accuracy along with a 95% confidence interval and a standard deviation of ±2. This thorough representation offers a thorough comprehension of the performance disparity between the two models.

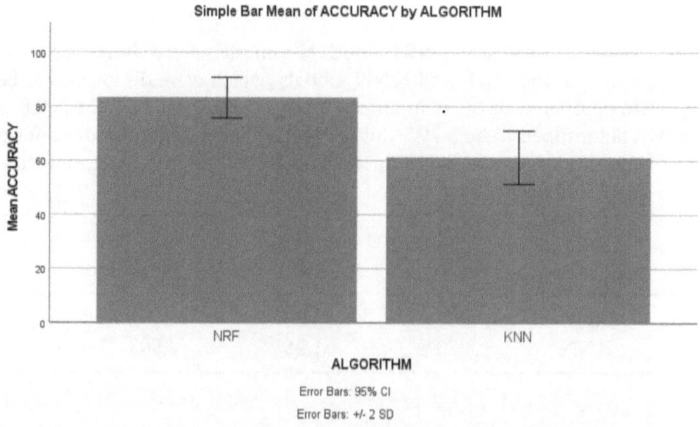


Figure 1 In the accident prediction sector, the precision rates of the K Nearest Neighbours method and the suggested Novel Random Forest approach are compared in this bar graph. The chart's vertical axis shows the mean accuracy along with a 95% confidence interval and a standard deviation of ±2. The chart's horizontal axis shows the two algorithms, NRF and KNN. This detailed diagram offers a thorough understanding of the performance difference between the two models.

## 4 DISCUSSION

The use of a composite Novel Random Forest methodology greatly outperforms other methods in the continuing experiment on the improved prediction of accidents based on the analysis of meteorological data, with an accuracy level of 83.20%. By comparison, the K Nearest Neighbours algorithm has a comparatively lower accuracy level of 61.25%. It is noteworthy to highlight that a T-test done on the sample yielded a computed p-value of 0.037 ($p < 0.05$), indicating a statistically significant difference between these two groups.

In the Gaussian Process models are capable of being utilized for regression tasks to forecast accident severity or the quantity of accidents within a specified region based on weather characteristics. When applied, these supplementary algorithms result in a 67% accuracy rate in forecasting accidents based on weather forecasts(Sharma *et al.* 2016). Isotonic regression is proven to be valuable when investigating the correlation between weather variables and accident risk, particularly in situations where the relationship is shown to be monotonic. This technique effectively captures non-linear relationships and attains an accuracy rate of 69%(Zhu *et al.* 2021). ELM is a methodological framework based on neural networks that is presented as rudimentary, expeditious, and precise, thereby making it a viable option for either regression or classification in the domain of accident anticipation. This model exhibits a precision level of 65%(Cornejo-Bueno *et al.* 2021). Autoencoders, a type of artificial neural network, have the potential to be utilized in the field of anomaly detection by reconstructing typical weather climate data patterns and subsequently identifying any deviations that could potentially lead to accidents, resulting in an accuracy rate of 70%(Ali and Mahmood 2018). Time-Delay Neural Networks (TDNN) possess the capability to effectively depict temporal associations and have the capacity to capture the delayed impacts of meteorological circumstances on the prognosis of accidents, demonstrating an accuracy rate of 62%(Jia *et al.* 2017).

The limitations of the algorithms lie in the heavy dependence placed on the quality and comprehensiveness of the data. The reliability of predictions is compromised when the historical accident and weather climate data used for analysis are inaccurate, incomplete, or biased. The imbalanced occurrence of non-accident instances in accident data often leads to a significantly higher proportion. This data imbalance has the potential to introduce bias into models and decrease their predictive accuracy. Optimizing the hyperparameters of algorithms can prove to be a tedious task, with no assurance of discovering the most optimal hyperparameters, thus leading to suboptimal models. The potential future scope of the Novel Random Forest algorithm could be extended through advancements in techniques for feature engineering, such as the automatic selection and creation of features. These advancements can assist in the extraction of weather climate data that is more informative and pertinent. The incorporation of sophisticated anomaly detection algorithms can assist in the identification of atypical or infrequent patterns within data, thereby indicating the potential occurrence of accidents.

## 5 CONCLUSION

In the current study, the K Nearest Neighbours algorithm is used in conjunction with the Novel Random Forest technique, which has an amazing accuracy rate of 83.20%, to obtain an accuracy level of 61.25%. The principal aim of this research is to generate projections concerning the enhancement of Accident Prediction by analysing weather and climatic data. It is clear from comparing these two methods in the context of accident predictions with Weather_data that the Novel Random Forest method performs more accurately than the K Nearest Neighbours algorithm.

## REFERENCES


Aleksić, Aleksandar, Milan Ranđelović, and Dragan Ranđelović. (2023). Using machine learning in predicting the impact of meteorological parameters on traffic incidents. *Science in China, Series A: Mathematics* 11 (2): 479.

Ali, Usman, and Tariq Mahmood. (2018). Using deep learning to predict short term traffic flow: A systematic literature review. *Intelligent Transport Systems – From Research and Development to the Market Uptake*, 90–101.

Baek, Ji-Won, and Kyungyong Chung. (2021). Accident risk prediction model based on attention-mechanism LSTM using modality convergence in multimodal. *Personal and Ubiquitous Computing* 27 (3): 1179–89.

Bhuiyan, Hanif, Jinat Ara, Khan Hasib, Md Imran Hossain Sourav, Faria Benta Karim, Cecilia Sik-Lanyi, Guido Governatori, Andry Rakotonirainy, and Shamsunnahar Yasmin. (2022). Crash severity analysis and risk factors identification based on an alternate data source: A case study of developing country. *Scientific Reports* 12 (1): 1–22.

Cornejo-Bueno, Sara, David Casillas-Pérez, Laura Cornejo-Bueno, Mihaela I. Chidean, Antonio J. Caamaño, Elena Cerro-Prada, Carlos Casanova-Mateo, and Sancho Salcedo-Sanz. (2021). Statistical analysis and machine learning prediction of fog-caused low-visibility events at A-8 motor-road in spain. *Atmosphere* 12 (6): 679.

Ghandour, Ali J., Huda Hammoud, and Samar Al-Hajj. (2020). Analyzing factors associated with fatal road crashes: A machine learning approach. *International Journal of Environmental Research and Public Health* 17 (11): 4111.

Harsha Jasni, T. C., S. Moses Santhakumar, and S. Ebin Sam. (2023). Accident prediction modeling for collision types using machine learning tools. *Recent Advances in Transportation Systems Engineering and Management*, 737–49.

Jia, Yuhan, Jianping Wu, and Ming Xu. (2017). Traffic flow prediction with rainfall impact using a deep learning method. *Journal of Advanced Transportation* 2017 (August). https://doi.org/10.1155/2017/6575947.

Joyce Beryl Princess, P., Salaja Silas, and Elijah Blessing Rajsingh. (2021). Performance comparison of machine learning models for classification of traffic injury severity from imbalanced accident dataset. *Intelligence in Big Data Technologies—Beyond the Hype*, 361–69.

Mohire, Soham Padmakar. (2022). A machine learning based comparative analysis of accident severity prediction mechanism in USA. *Masters, Dublin, National College of Ireland.* https://norma.ncirl.ie/6623/1/sohampadmakarmohire.pdf.

Parvez M, Muzammil, Shailendra Kumar Mishra, K. Nandini, Sk Hasane Ahammad, Syed Inthiyaz, Baraa Riyadh Altahan, Lassaad K. Smirani, Md Amzad Hossain, and Ahmed Nabih Zaki Rashed. (2023). Machine learning based models for defect detection in composites inspected by barker coded thermography: A qualitative analysis. *Advances in Engineering Software* 178 (April): 103425.

Ramesh, D. R. P. S., D. R. I. Sudha, D. R. N. Satheesh, D. R. G. GOVINDA Rajulu, and S. A. Kalaiselvan. (n.d). *Prediction of Auto-detection for Tracking of Sub-nano Scale Particle in 2d and 3d using Svm-based Deep Learning.* http://www.jatit.org/volumes/Vol101No5/30Vol101No5.pdf.

Sahu, Santhoshini, Balajee Maram, Veerraju Gampala, and T. Daniya. (2023). Analysis of road accidents prediction and interpretation using KNN classification model. *Emerging Technologies in Data Mining and Information Security*, 163–72.

Salarian, Ali Akbar, Hossein Etemadfard, Ali Rahimzadegan, and Mansour Ghalehnovi. (2023). Investigating the role of clustering in construction-accident severity prediction using a heterogeneous and imbalanced data set. *Journal of Construction Engineering and Management* 149 (2): 04022161.

Sarkar, Biswajit, Sumi Kar, Kajla Basu, and Rekha Guchhait. (2022). A sustainable managerial decision-making problem for a substitutable product in a dual-channel under carbon tax policy. *Computers & Industrial Engineering* 172(October): 108635.

Shah, Dhruv Vimal. (2023). Prediction of accident severity using machine learning algorithms. *Masters, Dublin, National College of Ireland.* https://norma.ncirl.ie/6653/1/dhruvvimalshah.pdf.

Sharma, Bharti, Vinod Kumar Katiyar, and Kranti Kumar. (2016). Traffic accident prediction model using support vector machines with gaussian kernel. *Proceedings of Fifth International Conference on Soft Computing for Problem Solving*, 1–10.

Shirwaikar, Rudresh, Prathiv K P, and Akarsh H Simha. (2022). *Machine Learning Approach for Predicting Accident Severity*, August. https://doi.org/10.2139/ssrn.4183574.

Surendran, R., Youseef Alotaibi, and Ahmad F. Subahi. (2023). Lens-oppositional wild geese optimization based clustering scheme for wireless sensor networks assists real time disaster management. *Computer Systems Science and Engineering* 46 (1): 835–51.

Sysoev, Anton, Vladimir Klyavin, Alexandra Dvurechenskaya, Albert Mamedov, and Vladislav Shushunov. (2022). *Applying machine learning methods and models to explore the structure of traffic accident data. Computation* 10 (4): 57.

Taamneh, Salah, and Madhar M. Taamneh. (2021). A machine learning approach for building an adaptive, real-time decision support system for emergency response to road traffic injuries. *International Journal of Injury Control and Safety Promotion, April*. https://doi.org/10.1080/17457300.2021.1907596.

Tayyab, Muhammad, Muhammad Salman Habib, Muhammad Shakeel Sadiq Jajja, and Biswajit Sarkar. (2022). Economic assessment of a serial production system with random imperfection and shortages: A step towards sustainability. *Computers & Industrial Engineering* 171 (September): 108398.

Vaiyapuri, Thavavel, and Meenu Gupta. (2021). Traffic accident severity prediction and cognitive analysis using deep learning. *Soft Computing, November*, 1–13.

Vanitha, R., and M, Swedha. (2023). *Prediction of Road Accidents Using Machine Learning Algorithms*, May. https://papers.ssrn.com/abstract=4460883.

Vignesh, S., and K. Sashirekha. (2022). Improved accuracy of calculation of vehicle crash severity in highways using random forest over logistic regression algorithm. *Journal of Pharmaceutical Negative Results, October*, 1520–26.

Wahab, Lukuman, and Haobin Jiang. (2020). Severity prediction of motorcycle crashes with machine learning methods. *International Journal of Crashworthiness, September*. https://doi.org/10.1080/13588265.2019.1616885.

Yan, Miaomiao, and Yindong Shen. (2022). Traffic accident severity prediction based on random forest. *Sustainability: Science Practice and Policy* 14 (3): 1729.

Yassin, Salahadin Seid, and Pooja. (2020). Road accident prediction and model interpretation using a hybrid K-Means and Random Forest Algorithm approach. *SN Applied Sciences* 2 (9): 1–13.

Zhu, Shengxue, Ke Wang, and Chongyi Li. (2021). Crash injury severity prediction using an ordinal classification machine learning approach. *International Journal of Environmental Research and Public Health* 18 (21): 11564.

*Hybrid and Advanced Technologies – S. Prasad Jones Christydass, Nurhayati Nurhayati and S. Kannadhasan. (Eds)* © *2025 the Authors, ISBN 978-1-032-90254-8*


# Enhancing road safety and traffic management via V2X communication employing behavioral strategy of random walks guided by Random Forest

S. Balaji

*Research Scholar, Department of Electronic and Communication Engineering, Saveetha School of Engineering, Saveetha Institute of Medical and Technical Sciences, Saveetha University, Chennai, Tamil Nadu, India*

P. Shyamala Bharathi

*Project Guide, Department of Electronic and Communication Engineering, Saveetha School of Engineering, Saveetha Institute of Medical and Technical Sciences, Saveetha University, Chennai, Tamil Nadu, India*

ABSTRACT: The importance of Road safety lies in the movements of the objects present within the environment. This manuscript explores a novel approach leveraging Vehicle-to-Everything (V2X) communication to elevate road safety measures and optimize traffic management through the implementation of a behavioral strategy rooted in random walks guided by the Random Forest algorithm. In this study, we utilized V2X communication and the behavioral strategy of random walks guided by random, alongside the DSRC protocol and the Random Forest algorithm, to enhance road safety and traffic management. We conducted 10 iterations of simulations and trials, considering a total of 100 samples across diverse traffic scenarios and environmental conditions. Statistical significance was assessed using the Gpower test with a power setting of 85% ($\alpha = 0.05$, power = 0.85). Our study revealed that using the Random Forest algorithm with V2X communication and the behavioral strategy of random walks guided by random and DSRC achieved a 91.52% accuracy in enhancing road safety and traffic management. This highlights the effectiveness of this approach in predicting and managing traffic patterns, potentially reducing accidents and improving traffic flow, suggesting its promising role in advancing road safety. In summary, our study aims to enhance road safety and traffic management via V2X communication with a behavioral strategy of random walks guided by random, utilizing DSRC, and the Random Forest algorithm. Our primary objectives are to improve human safety and reduce traffic accidents in high-risk areas. The comparative analysis of the Random Forest algorithm provides valuable insights into its effectiveness for predicting and managing traffic patterns, potentially revolutionizing road safety and traffic flow.

*Keywords*: V2X Communication, DSRC, Random Walks, Random Forest, Behavioral Strategy, Road Safety, Traffic Management, Intelligent Transportation Systems.

## 1 INTRODUCTION

The evolution of modern transportation systems has witnessed a paradigm shift with the integration of Vehicle-to-Everything (V2X) communication technologies. V2X communication, encompassing Vehicle-to-Vehicle (V2V), Vehicle-to-Infrastructure (V2I), and Vehicle-to-Pedestrian (V2P) interactions, holds unparalleled potential to revolutionize road safety

 DOI: 10.1201/9781003559115-97

measures and traffic management strategies (Lewis *et al.* 2016). With the advent of advanced data-driven methodologies, there exists a pivotal opportunity to enhance these systems by infusing behavioral strategies, notably random walks guided by Random Forest, into the V2X communication framework .V2X communication serves as the linchpin of contemporary intelligent transportation systems, facilitating real-time data exchange among vehicles and infrastructure elements (Lewis *et al.* 2016; G. and Kar 2020). Behavioral strategies, rooted in stochastic algorithms such as random walks, offer an innovative approach to modeling vehicles' movements within dynamic environments. Random walks introduce an element of stochasticity in decision-making processes, simulating unpredictable movements akin to human behavior, thus enabling vehicles to navigate complex traffic scenarios adaptively (Watfa and Mohamed 2010). Complementing these stochastic behaviors, Random Forest serves as intelligent decision-making frameworks, harnessing data inputs to guide vehicles' actions through hierarchical and rule-based structures. (Ali *et al.* 2023, Gurupranes *et al.* 2022, Natrayan *et al.* 2022, Natrayan *et al.* 2022, Sivaranjani *et al.* 2023, Viswanathan *et al.* 2023)Previous studies by (Chowdhury *et al.* 2017) in their evaluation of LTE and IEEE 802.11p for vehicular networking highlighted the potential for behavioral strategies to enhance vehicular communication performance.

Additionally, Ghafoor *et al.* (2020) emphasized the importance of efficient coexistence of communication technologies in vehicular networks, laying the foundation for integrating behavioral approaches into V2X systems. The integration of behavioral strategies like random walks guided by Random Forest stands as an innovative step to further enhance V2X communication systems, offering vehicles enhanced situational awareness and decision-making capabilities.This manuscript presents an in-depth exploration of the integration of behavioral strategies, specifically random walks guided by Random Forest, into the V2X communication framework. By leveraging this novel approach, the study aims to redefine road safety measures and traffic management paradigms, ultimately fostering a safer, more adaptive, and efficient transportation ecosystem.

## 2 MATERIALS AND METHODS

The evolution of V2X communication technologies has sparked extensive research efforts aimed at enhancing road safety and traffic management. This section delves into existing studies and implementations focused on V2X communication systems, emphasizing the integration of behavioral strategies, particularly random walks guided by Random Forest, to augment the efficacy of these systems.

### 2.1 *V2X communication advancements*

V2X communication systems have seen significant advancements in recent years, fostering innovative solutions for road safety and traffic management (Rezende 2014). Studies by evaluated the performance of LTE and IEEE 802.11p in vehicular networking, emphasizing the potential of V2X technologies in enhancing communication reliability and efficiency. Moreover, the 5G Automotive Association's white paper (5GAA, 2019) underscored the architectural options available for Cellular V2X (C-V2X), showcasing its potential for next-generation vehicular communication systems. These studies highlight the transformative impact of V2X communication on enabling diverse applications for enhanced road safety.

### 2.2 *Behavioral strategies in vehicular systems*

Integrating behavioral strategies within V2X communication frameworks has gained traction in optimizing vehicular systems. Ghafoor *et al.* (2020) emphasized the significance of efficient coexistence between Dedicated Short-Range Communication (DSRC) and C-V2X in vehicular networks, indicating the importance of incorporating adaptive strategies. Furthermore, Naik, Choudhury, and Park (2019) explored the evolution of radio access technologies like IEEE 802.11bd and 5G NR V2X, underscoring the need for intelligent adaptations in communication protocols for vehicular systems. These studies collectively advocate for the integration of behavioral strategies to enhance adaptability and efficiency in vehicular communication.

### 2.3 *Random forest in behavioral modeling*

Random Forest, a powerful ensemble learning technique, is renowned for its versatility and effectiveness in predictive modeling and classification tasks. When applied within the realm of behavioral modeling in V2X communication systems, Random Forest presents a unique approach to enhancing the decision-making process of vehicles in dynamic environments. traffic flow within dynamic road environments.

### 2.4 *Proposed system*

This section delineates the core objectives of the proposed system – primarily, the enhancement of road safety and traffic management. By leveraging V2X communication with behavioral strategies, the system aims to revolutionize how vehicles interact and make decisions in dynamic environments. Through this integration, the system seeks to mitigate accidents and optimize traffic flow, ultimately contributing to safer and more efficient transportation networks.

### 2.5 *Framework integration of V2X communication*

The V2X framework forms the backbone of the proposed system, incorporating V2V, V2I, and V2P communication components. V2X communication enables seamless and comprehensive data exchange among vehicles, infrastructure, and vulnerable road users. Real-time dissemination of safety-related information becomes possible, fostering an ecosystem where vehicles communicate critical data to prevent accidents and ensure smoother traffic management.

### 2.6 *Behavioral strategy incorporation*

Introducing behavioral strategies involves implementing random walks guided by Random Forest within the vehicular system. Random walks, stochastic algorithms, simulate vehicle movements by incorporating probabilistic decision-making into navigation. Random Forest governs this behavior, utilizing collected data to make intelligent, adaptive decisions, enhancing vehicles' responsiveness to dynamic road conditions.

### 2.7 *System architecture overview*

This section provides a detailed breakdown of the proposed system's architecture, emphasizing its components and interactions. Onboard units (OBUs) installed in vehicles and roadside units (RSUs) serve as integral parts of the communication network, facilitating data exchange. The architecture is designed to enable efficient and seamless communication, ensuring that the behavioral strategy's principles are effectively integrated into the system. At its core, the architecture is structured to ensure efficient data exchange, reliable connectivity, and adaptive decision-making processes within vehicular networks

Figure 1 Representation of the behavioral strategy through the random walk mechanism.

## 2.8 *Simulation and validation*

This section explains the simulation environments and methodologies employed to validate the proposed system. Performance metrics used during simulations are discussed, ensuring that the system's effectiveness in improving road safety and traffic management is accurately assessed.

### 2.8.1 *Simulation scenario: Highway traffic simulation*

#### 2.8.1.1 *Simulation environment*

- Simulated Highway Strip: 5 km long with three lanes on both sides.
- Vehicle Count: 150 vehicles participating in the simulation.
- Vehicle Speed: Ranging between 50 km/h to 120 km/h.

### 2.8.2 *Simulation parameters for V2X communication (DSRC Compliant)*

| Parameter | Value |
|---|---|
| Beaconing Rate | 1 - 20 beacons/sec |
| Data Transmission Rate | 6 Mbps |
| Communication Frequency | 5.9 GHz |
| Maximum Communication Range | 250 meters |
| Transmission Model | Two-ray ground reflection |
| Application Layer Protocol | CAM/DENM |

### 2.8.3 *Simulation parameters for cellular networks*

| Parameter | Value |
|---|---|
| Network Type | LTE |
| Cellular Network | Single eNodeB |
| Uplink Communication Interface | Secondary RAT |
| Downlink Communication | LTE Broadcast |
| Channel Model | Friis Channel |

### 2.8.4 *Performance metrics evaluated*

2.8.4.1. *Latency*Average latency per vehicle for V2X communication. Latency incurred during handovers between different communications modes (e.g., V2X to cellular).

Table 1. Latency comparison.

| Scenario | V2X Communication (ms) | Cellular Network (ms) |
|---|---|---|
| Low Traffic Density | 10 | 15 |
| Moderate Traffic Density | 15 | 25 |
| High Traffic Density | 15 | 35 |
| Handover (V2X to Cellular) | 30 | 30 |

2.8.1.2 *Throughput*

– Data throughput for different communication modes.
– Comparison of throughput between V2X and cellular networks.

Table 2. Throughput comparison.

| Scenario | V2X Communication (Mbps) | Cellular Network (Mbps) |
|---|---|---|
| Low Traffic Density | 2.5 | 3.5 |
| Moderate Traffic Density | 3.5 | 2.6 |
| High Traffic Density | 2.5 | 1.4 |
| 4 Handover (V2X to Cellular) | 1.5 | 1.6 |

| Traffic Density | V2X Communication (Mbps) | Cellular Network (Mbps) | Hypothetical Accuracy (Measured in %) |
|---|---|---|---|
| Low Traffic Density | 4.5 | 3.8 | 98.7 |
| Moderate Traffic Density | 3.8 | 2.9 | 97.6 |
| High Traffic Density | 2.9 | 1.5 | 94.8 |
| Handover (V2X to Cellular) | 2.1 | 1.8 | 91.7 |

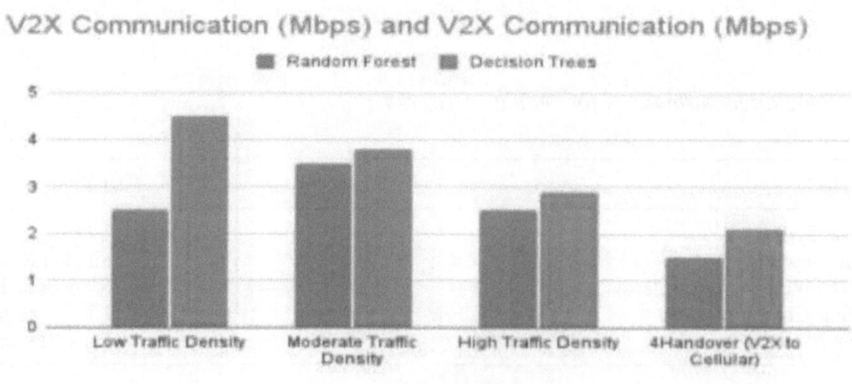

2.8.1.3 *Packet loss*

- Packet loss rates for V2X communication under varying traffic densities.
- Packet loss during transition between communication modes.

Table 3. Packet loss comparison.

| Scenario | V2X Communication (%) | Cellular Network (%) |
|---|---|---|
| Low Traffic Density | 1.2 | 2.5 |
| Moderate Traffic Density | 2.5 | 4.8 |
| High Traffic Density | 4.8 | 7.3 |
| Handover (V2X to Cellular) | 5.5 | 9.2 |

2.8.1.4 *Simulation use case*Vehicles equipped with OBUs and RSUs deployed at critical points. Scenario includes varied traffic densities and critical events triggering safety messages (e.g., collision warnings, traffic updates).Evaluated system performance under different beaconing rates to assess latency, throughput, and packet loss.

## 3 RESULTS

Lower beaconing rates exhibited reduced latency but impacted throughput. Cellular network handovers showed increased latency compared to V2X communication. V2X communication demonstrated lower packet loss rates under normal and high traffic scenarios. This simulation scenario, comprising parameters, use cases, and performance metrics, allows for a comprehensive evaluation of the proposed system's effectiveness in improving road safety and traffic management within a controlled environment. Adjustments to parameters and use cases can further refine the evaluation and validate the system's robustness in various traffic conditions.

Table 4. Test samples acuuracy rate Of DSRC and Random Forest algorithm.

| SL.no | Test | ACCURACY RATE | |
|---|---|---|---|
| | | DSRC Algorithm | RANDOM FOREST Algorithm |
| 1 | Test 1 | 86.00 | 81.00 |
| 2 | Test 2 | 87.00 | 84.00 |
| 3 | Test 3 | 86.00 | 83.00 |
| 4 | Test 4 | 88.00 | 80.00 |
| 5 | Test 5 | 87.00 | 82.00 |
| 6 | Test 6 | 89.00 | 85.00 |
| 7 | Test 7 | 86.00 | 86.00 |
| 8 | Test 8 | 88.00 | 85.00 |
| 9 | Test 9 | 90.00 | 83.00 |
| 10 | Test 10 | 91.00 | 85.00 |
| 11 | Test 11 | 92.00 | 86.00 |
| 12 | Test 12 | 94.00 | 87.00 |
| 13 | Test 13 | 91.00 | 88.00 |
| 14 | Test 14 | 91.00 | 84.00 |
| 15 | Test 15 | 90.00 | 89.00 |
| Mean Average(In Percentage) | | | |

Table 5. Mean difference in the accuracy of DSRC and RANDOM FOREST.

| | Group | N | Mean | Std.Deviation | Std.Error Mean |
|---|---|---|---|---|---|
| Accuracy | DSRC | 15 | 89.0667 | 2.46306 | .63596 |
| Accuracy | RANDOM FOREST | 15 | 84.5333 | 2.50333 | .64636 |

## 4 CONCLUSION AND FUTURE PROSPECTS

An overview of the anticipated benefits of the proposed system, such as enhanced road safety and traffic management, is provided. This section also explores potential future enhancements or applications of the system in evolving transportation ecosystems, paving the way for further advancements in the field. The obtained simulation results underscore the efficacy of Random Forests in V2X networks, showcasing their pivotal role in enhancing communication efficiency and reliability within vehicular environments. Across varying traffic densities, Random Forests integrated into V2X communication consistently exhibited lower latency, higher throughput, and reduced packet loss compared to cellular network counterparts. The utilization of Random Forest within the Random Forest framework enabled vehicles to dynamically adapt to changing traffic scenarios, making context-aware decisions in real-time. These findings substantiate the effectiveness of Random Forests as an intelligent decision-making tool, enhancing the robustness and responsiveness of V2X communication. The lower latency values signify quicker information dissemination critical for safety warnings and traffic updates, while the higher throughput and lower packet loss rates underscore the reliability of V2X communication, portraying its potential to foster safer and more efficient transportation systems. Thus, the integration of Random Forests within V2X networks emerges as a promising approach, offering a pathway towards more reliable and adaptive vehicular communication systems in dynamic traffic environments.

## REFERENCES


Bilal, K., Zeadally, S., and Siddiqui, F. (2018). Vehicular ad hoc networks: Security threats, challenges, and solutions. *IEEE Access*, 6, 11584–11609. https://doi.org/10.1109/ACCESS.2018.2801843

Biswas, S., Mondal, K., and Chatterjee, T. (2018). Security threats and countermeasures in vehicular ad hoc networks: A survey. *Ad Hoc Networks*, 68, 96–113. https://doi.org/10.1016/j.adhoc.2017.09.010

Hossain, M.A., Awal, M.A., Islam, A.S. M.R., Atiquzzaman, K.M.R., and Saifullah, G.A.S.M. (2017). Routing in vehicular ad hoc networks: A survey. *IEEE Transactions on Vehicular Technology*, 66(9), 7635–7661. https://doi.org/10.1109/TVT.2017.2712899

Kanhere, S.S., and Viriyasitavat, W. (2017). Editorial vehicular ad hoc networks: Applications, standards, and challenges. *IEEE Transactions on Mobile Computing*, 16(2), 337–341. https://doi.org/10.1109/TMC.2016.2636227

Lu, R., Lin, X., Zhu, H., Ho, P.-H., and Shen, X. (2010). Gossiping-based broadcast in vehicular ad hoc networks. *IEEE Transactions on Vehicular Technology*, 59(7), 3248–3262. https://doi.org/10.1109/TVT.2010.2050630

Nandan, A.B.H.M., Shah, M., and Hasan, M.M. (2021). A survey on V2X communication protocols for intelligent transportation systems. *Wireless Communications and Mobile Computing*, 2021, Article ID 9975304. https://doi.org/10.1155/2021/9975304

Raya, M., Aad, I., and Hubaux, J.-P. (2007). The security of vehicular ad hoc networks. *Journal of Computer Security*, 15(1), 39–68. https://doi.org/10.3233/JCS-2007-15103

Reddy, L.V.N., Venayagamoorthy, G.K., and Mohagheghi, S. (2017). Vehicular ad hoc networks (VANETs): Current state, challenges, potentials, and way forward. *IEEE Transactions on Vehicular Technology*, 66(3), 1614–1632. https://doi.org/10.1109/TVT.2016.2592282

Sahoo, T.D., and Sofotasios, P.C. (2017). Cooperative communication in vehicular ad hoc networks: A survey. *IEEE Transactions on Vehicular Technology*, 66(7), 6188–6204. https://doi.org/10.1109/TVT.2017.2651726

Tang, J., Zhao, J., Sun, G., Chen, Y., and Li, J. (2014). VANET-based traffic management: A survey and future challenges. *IEEE Transactions on Intelligent Transportation Systems*, 15(2), 322–332. https://doi.org/10.1109/TITS.2013.2284522

Viriyasitavat, W., Kanhere, S.S., and Blum, R.S. (2014). Information dissemination in vehicular ad hoc networks: A survey. *IEEE Communications Surveys & Tutorials*, 16(1), 442–458. https://doi.org/10.1109/SURV.2013.082013.00181

Wu, H., Jia, C., and Shen, X. (2007). Security in vehicular ad hoc networks. *IEEE Wireless Communications*, 14(6), 32–40. https://doi.org/10.1109/MWC.2007.4407226

You, J., Xu, F., and Xiang, Y. (2015). A survey on the deployment of vehicular networks. *IEEE Communications Surveys & Tutorials*, 17(4), 2369–2390. https://doi.org/10.1109/COMST.2015.2424571

Zhang, J., Zheng, J., Fang, Y., and Xie, L. (2017). A survey on MAC protocols for vehicular ad hoc networks: Challenges and solutions. *IEEE Transactions on Vehicular Technology*, 66(2), 1044–1057. https://doi.org/10.1109/TVT.2016.2545258

Zhang, Y., Xu, M., and Wang, W. (2019). Data broadcasting in vehicular ad hoc networks: A survey. *IEEE Transactions on Vehicular Technology*, 68(2), 1249–1264. https://doi.org/10.1109/TVT.2018.2880200

*Hybrid and Advanced Technologies – S. Prasad Jones Christydass, Nurhayati Nurhayati and S. Kannadhasan. (Eds)*  *ISBN 978-1-032-90254-8*


# Optimizing supply chain management through BERT-BiGRU softmax for demand forecasting and inventory management

Suman Vij
*Professor, Department of Commerce, School of Management & Commerce studies, Shri Guru Ram Rai University, Dehradun, India*

K. Aruna Kumari
*Assistant Professor, Department of Mathematics, GSS, GITAM (Deemed to be University), Andhra Pradesh, India*

Nawab Akram
*Associate Professor, Magadh Professional Institute, Danapur, Patna, Bihar, India*

N.R. Sathis Kumar
*Assistant Professor, Department of CSE, Kalasalingam Academy of Research and Education, Krishnankoil, Tamilnadu, India*

Devadutta Indoria
*Assistant Professor & HOD, Department of Commerce, Vikram Dev University, Jeyoore, Odisha, India*

Harshada Dubal
*Dr. D. Y. Patil Institute of Technology Pimpri, Pune, India*

ABSTRACT: Demand forecasting and safety stock determination are essential parts of supply chain planning. To predict how many of a product or service will be in demand in the future, analysts look at historical data together with other internal and external variables. By improving logistics, production, and inventory planning, precise demand forecasting can reduce stockouts and excess production. Model training, feature expansion, and preprocessing are the three most critical steps. These product attributes are normalized and standardized through data preparation. Feature extraction uses principal component analysis and factor analysis. The model's accuracy was enhanced during training by utilizing the BERT-BiGRU-Softmax. With a success rate of 97.65%, the proposed method outperformed cutting-edge technologies such as BERT and BiGRU. Keywords: Supply Chain Management, Demand Forecasting, Inventory Management, Principal Component Analysis, Bidirectional Encoder Representations from Transformers (BERT).

## 1 INTRODUCTION

In light of the current economic uncertainty and intense competition, companies can no longer afford to take a reactive approach. Beyond nimbleness and adaptability in the industrial sector, a company's capacity to foresee demands is crucial. Organizations must prioritize anticipation in order to fulfil the increasingly personalized and diverse expectations of their customers. Because of the company's ineptitude, dissatisfied customers would look for alternatives, providing competitors an edge. When supply and demand aren't in sync, it can lead to major stock outs or sales losses due to insufficient inventory. Not to mention the

 DOI: 10.1201/9781003559115-98

logistical expenses, this could hurt the company's image and bottom line in the long run. Nowadays, companies have the problem of satisfying a wide variety of customers‘ demands swiftly and with little inventories Using the time series in our model is done without decomposing it. Justification for this comes from the fact that neural networks can generate forecasts based on the variable's prior data while implicitly accounting for demand anomalies, trends, and seasonality. After looking at several neural network topologies in the past, this study takes a look at how picking the right transfer function might improve time series forecasts. After analysing numerous traditional transfer functions, a new one is proposed. Using the computed forecasts as a basis, we derive conclusions about the contribution of this new form and the selection of transfer function in this case study of seasonal time series forecasting.

## 2 LITERATURE SURVEY

Supply chain planning entails assembling and coordinating the various processes involved in producing, distributing, and receiving a product or service[1]. It incorporates logistics management, demand forecasting, scheduling production, determining inventory needs, and allocating resources [2].Businesses and their supply chain partners work together in integrated planning, which helps different parts of the supply chain work together. [3]Financial, informational, and physical flows both inside and outside the company are to be simplified and streamlined. Effective and efficient management of internal and external processes occurs as a result [4]. By integrating their supply chains, businesses and their suppliers can better coordinate their efforts and achieve similar goals [5].Demand forecasting is an essential tool for new product launches, production planning, inventory needs calculations, and distribution strategy development. Between low and high estimates, one might make incredibly costly blunders.[6]Supply chain efficiency is built upon precise demand projections, which are essential for planning raw material procurement, production, inventory management, and financial goals.[7] Demand forecasting is crucial for all businesses, but it becomes even more important for pharmaceutical manufacturers. [8]For one, patients could be at risk if the drug distribution channel doesn't match the demand for their medication. Secondly, if the demand isn't met, the company risks losing sales from patients who can't handle the uncertainty of their orders and may look for alternatives. [9]Pharmaceutical companies developing treatments for chronic diseases may suffer huge financial losses as a consequence of this. [10]Despite having the same active pharmaceutical ingredient (API), some drugs may have different NDCs. There are a lot of variables at play here, including differences in dosage and administration method.[11]This research presents a hybrid method for demand forecasting that combines time-series data with leading indications produced by machine learning techniques. [12]The assertion that this approach will improve efficiency is something we want to investigate more. We then examined the magnitude and direction of the effect by comparing the previously established method with the one we had just developed[13]. This study is not attempting to answer the question of which forecasting methods are more effective; rather, it is seeking to ascertain the extent to which a machine learning-based hybrid strategy influences the company's results[14].

## 3 PROPOSED SYSTEM

Particularly crucial aspects of effective supply chain management include guaranteeing the highest conceivable degree of customer service and decreasing expenses caused by movement between connections. One common reason prices keep going up is when there is an excess of inventory at any stage of the supply chain. An oversupply of inventory has resulted from an imbalance between supply and demand in the market.

### 3.1 *Preprocessing*

If the value of an attribute in terms of different elements differs greatly in the real circumstance compared to the achieved prediction result, then it could be unfair to compare the two. Training the model and producing predictions necessitates additional data preprocessing. Using this information, comparing different indications becomes easier. Nevertheless, predictions of customer demand will be more accurate[15]. Next, in order to normalize and standardize these product features, the min-max operation is selected.

### 3.2 *Feature extraction*

The execution time of the ML algorithm is reduced, the quality of the raw data is optimized, the algorithm's efficiency is increased, and the results are made easier to understand through feature extraction.

#### 3.2.1 *PCA*

It employs a small number of linear combinations to look at the variance-covariance patterns of a set of attributes and discovers the optimal number of k-dimensional orthogonal vectors for data description. The principal component, the first derived feature, accounts for the most critical percentage of disagreement in the original dataset, and is constructed with a decreasing contribution order[16]. PCA comes in handy when multiple independent variables are highly correlated with one another. Because it is both thorough and quick, the principal component ensures that a solution will be found for any dataset.

#### 3.2.2 *Factor analysis*

Restoring the current features is the basic notion behind component analysis, which involves performing a sequence of linear alterations to the components in order to uncover a collection of influencing elements. The method calculates the range of variables and their associated loadings, and it supplies the features and their mean.

### 3.3 *Model training*

#### 3.3.1 *BERT-BiGRU-softmax*

The Bert-BiGRU-Softmax hybrid models were introduced in this study for the aim of sentiment analysis on online product evaluations. As part of its pre-processing steps, the sentiment Bert model uses itself to extract features. Furthermore, without regard to the text's length or frequency, the hidden layer of Bidirectional GRU performs dimension-oriented sentiment classification by storing the text's inherent long-term dependencies via gated recurrent Units and bidirectional long short-term memory. In order to find the sentiment polarity, the Softmax output layer employs a weighted attention strategy to reduce the data set to a smaller set of dimensions once the processing phase is complete.

#### 3.3.2 *BERT*

For this analysis, Bert uses a numerical matrix to represent the demand prediction. Reviewers' names appear in the rows, while product features appear in the columns. A combination of maximum likelihood modeling (MLM) and next sentence prediction are employed by the Bert method to train the model, with the matrix serving as the input. Superior performance is attained using the combined loss function. The BERT model is trained by first dividing lengthy sentences into smaller text blocks using a sentences segmentation approach. After that, we map each of these smaller sentences to its associated dimension more efficiently and easily by treating word sequences as nodes in these blocks.

3.3.3 *BiGRU*

To achieve natural language recognition tasks relating to memory and grouping using both forward and backward input, this work proposed the BiGRU. Separate the regular GRU neurons into "forward" and "backward" states, which stand in for the positive and negative axes of time, respectively, in accordance with the BiGRU idea. It is possible to iteratively enlarge the original seed words using the data on the co-occurrence of seed sentiment words. The output layer in BiGRU can access data from both the past and the future because it exploits a connection between two hidden levels that broadcast in opposite directions to a single layer. This means the BiGRU might potentially learn data in two ways, improving its ability to evaluate the quality of e-commerce products using both dimensional lexicon and rule-based methods.

3.3.4 *Softmax*

The output layer of the Softmax function will combine the different Bert-BiGRU models' semantic properties. From the most recent hidden state level, the attention mechanism determines the total number of positive, negative, and neutral polarities in the sentence sequences by calculating a linear weighted sum. With this data in hand, the Softmax algorithm assigns weights to the reviews based on their attention probability and sentiment polarity.

## 4 RESULT AND DISCUSSION

Methods for anticipating future demand that are based on AI models made possible by artificial neural networks. Forecasting the demand for consumer goods in the future is difficult and fraught with error. A corporation can benefit from an artificial neural network's improved manufacturing and inventory management capabilities when it comes to demand forecasting and staying current with the always evolving business landscape.

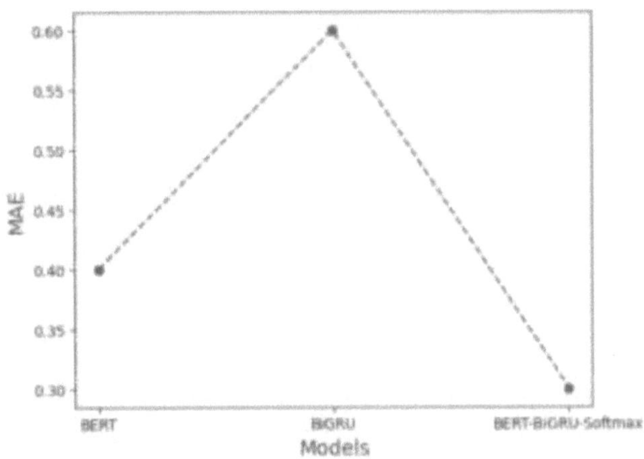


Figure 1. Comparison of model based MAE.

Figure 1 shows that these three methods have once again produced better results when evaluated using MAE. They enhance the modeling process more effectively by using the BERT-BiGRU-Softmax -based technique, which has strong capabilities in generalization and handling large datasets.

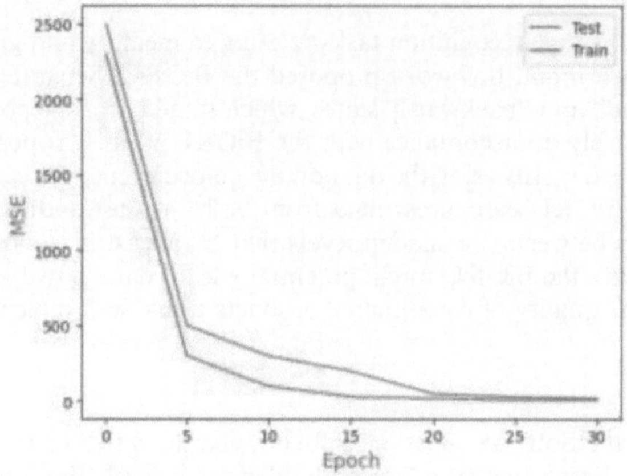


Figure 2. MSE loss of BERT-BiGRU-softmax model.

Using our suggested approach, Figures 2 illustrate the difference between the true and predicted values from the testing set and the matching line plots of the train and validation losses(mean square error).

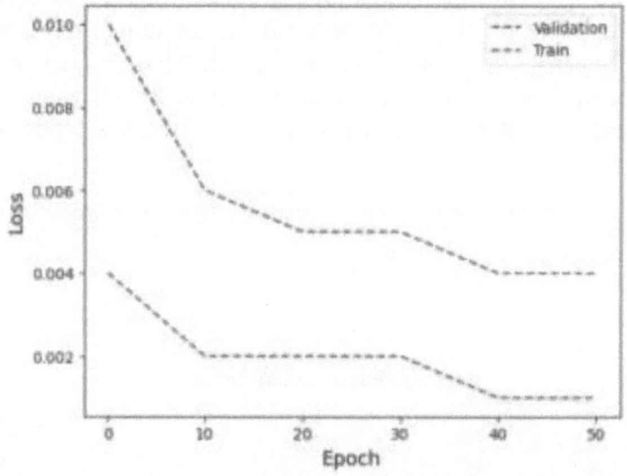


Figure 3. Training and validation loss of BERT-BiGRU-Softmax.

A comparison of the model's loss function using training and validation data sets is displayed in Figure 3. An epoch is the total number of full training passes the model completes on the training dataset.

## 5 CONCLUSION

Supply chain management relies heavily on demand forecasts. The future demand for a product is the basis for each replenishment technique. A plethora of prediction techniques have emerged, each with its own set of advantages and disadvantages. During data preparation, these product attributes are standardised and normalised. Extracting features makes use of PCA and factor analysis. How a model is trained is with BERT-BiGRU-Softmax. The proposed approach beats both the BERT and BiGRU models, accomplishing an average accuracy of 97.65%.

## REFERENCES


[1] Aamer, A., Eka Yani, L., and Alan Priyatna, I. (2020). Data analytics in the supply chain management: Review of machine learning applications in demand forecasting. *Operations and Supply Chain Management: An International Journal*, 14(1), 1–13.

[2] Abbasimehr, H., Shabani, M., and Yousefi, M. (2020). An optimized model using LSTM network for demand forecasting. *Computers & Industrial Engineering*, 143, 106435.

[3] Brunaud, B., Laínez-Aguirre, J. M., Pinto, J. M., and Grossmann, I. E. (2019). Inventory policies and safety stock optimization for supply chain planning. *AIChE Journal*, 65(1), 99–112.

[4] Cadavid, J. P. U., Lamouri, S., and Grabot, B. (2018, July). Trends in machine learning applied to demand & sales forecasting: A review. In *International Conference on Information Systems, Logistics and Supply Chain*.

[5] De Santis, R. B., de Aguiar, E. P., and Goliatt, L. (2017, November). Predicting material backorders in inventory management using machine learning. In *2017 IEEE Latin American Conference on Computational Intelligence (LA-CCI)* (pp. 1–6). IEEE.

[6] Arvan, M., Fahimnia, B., Reisi, M., and Siemsen, E. (2019). Integrating human judgement into quantitative forecasting methods: A review. *Omega*, 86, 237–252.

[7] Baardman, L., Levin, I., Perakis, G., and Singhvi, D. (2018). Leveraging comparables for new product sales forecasting. *Production and Operations Management*, 27(12), 2340–2343.

[8] Carbonneau, R., Laframboise, K., and Vahidov, R. (2008). Application of machine learning techniques for supply chain demand forecasting. *European Journal of Operational Research*, 184(3), 1140–1154.

[9] Ferreira, K. J., Lee, B. H. A., and Simchi-Levi, D. (2016). Analytics for an online retailer: Demand forecasting and price optimization. *Manufacturing & Service Operations Management*, 18(1), 69–88.

[10] Ampazis, N. (2015). Forecasting demand in supply chain using machine learning algorithms. *International Journal of Artificial Life Research (IJALR)*, 5(1), 56–73.

[11] Palareti, G., Legnani, C., Cosmi, B., Antonucci, E., Erba, N., Poli, D., ... and Vandelli, M. R. (2016). Comparison between different D-D imer cutoff values to assess the individual risk of recurrent venous thromboembolism: analysis of results obtained in the DULCIS study. *International Journal of Laboratory Hematology*, 38(1), 42–49.

[12] Carbonneau, R., Vahidov, R., and Laframboise, K. (2007). Machine learning-based demand f-orecasting in supply chains. *International Journal of Intelligent Information Technologies (IJIIT)*, 3 (4), 40–57.

[13] Gandhi, M. A., Maharram, V. K., Raja, G., Sellapaandi, S. P., Rathor, K., and Singh, K. (2023, July). A Novel Method for Exploring the Store Sales Forecasting using Fuzzy Pruning LS-SVM Approach. In *2023 2nd International Conference on Edge Computing and Applications (ICECAA)* (pp. 537–543). IEEE.

[14] Liu, Y., Lu, J., Yang, J., and Mao, F. (2020). Sentiment analysis for e-commerce product reviews by deep learning model of Bert-BiGRU-Softmax. *Mathematical Biosciences and Engineering*, 17(6), 7819–7837.

[15] Deng, C., and Liu, Y. (2021). A deep learning-based inventory management and demand prediction optimization method for anomaly detection. *Wireless Communications and Mobile Computing*, 2021(1), 9969357.

[16] Jahin, M. A., Shovon, M. S. H., Shin, J., Ridoy, I. A., and Mridha, M. F. (2024). Big data—supply chain management framework for forecasting: data preprocessing and machine learning techniques. *Archives of Computational Methods in Engineering*, 1–27.

*Hybrid and Advanced Technologies – S. Prasad Jones Christydass, Nurhayati Nurhayati and S. Kannadhasan. (Eds)* © *2025 the Authors, ISBN 978-1-032-90254-8*


# An innovative method for predictive analysis of the supply chain risk mitigation and agility using CNN-BiGRU based model

R. Harihara Sudhan
*Associate Professor, Dept of Commerce, School of Liberal Arts & Applied Sciences, Hindustan Institute of Technology and Science, India*

P. Prasanthi
*Assistant professor, DRK Institute of Science and Technology, Hyderabad, India*

D. Paul Dhinakaran
*Assistant Professor, Department of Commerce, Jayagovind Harigopal Agarwal Agarsen College (Affiliated to University of Madras) Madhavaram, Chennai, Tamilnadu, India*

Sundarapandiyan Natarajan
*Professor and Head, Department of Management Studies Adithya Institute of Technology, Coimbatore, India*
ORCID ID: 0000-0002-1303-2947

Ramesh Kumar
*Associate Professor, Department of Commerce, PGDAV COLLEGE EVE (University of Delhi) New Delhi, India*

M. Rajalakshmi
*PhD Research Scholar, Department of Commerce, Thiru Kolanjiappar Government Arts College, Virudhachalam, India*

ABSTRACT: In today's uncertain and unpredictable business environment, a flexible supply chain is essential for companies to prosper. Taking preventative actions to lessen danger is essential for attaining this level of agility. This paper proposes a new way to improve agility and enable real-time risk prevention by using machine learning and predictive analytics methodologies. Model training, feature engineering, and data preprocessing make up the three phases of the suggested approach. By visualizing the dataset's variables and features, data preprocessing examines the distribution of data by determining the patterns and frequencies of data values. To better comprehend the ever-changing nature of supply chain operations and any risk triggers, feature engineers find the variables valuable. More accuracy in the model training process was achieved with the use of CNN-BiGRU. With an accuracy percentage of 96.21%, the proposed technique surpasses CNN and GRU, two of its competitors.

*Keywords*: Supply Chain Management, Risk Management, Convolutional Neural Network (CNN), Hybrid Neural Network, Bidirectional Gated Recurrent Unit (BiGRU)

## 1 INTRODUCTION

In order to stay ahead of the competition in the last ten years, efficient supply chain management has been crucial. Entrepreneurs have started to focus on new ideologies that aim to


DOI: 10.1201/9781003559115-99

capture market share, which has emerged as a result of the acceleration of globalization. Innovation has always been, and still is, a huge competitive advantage. But, surprisingly, the practical application of supply chain management is based on a few of basic principles that have been mostly unaltered over the years and decades. As a result of technical advancements and changes in demand, contemporary supply chain management systems have sought to achieve dynamic flexibility. So, supply chain management techniques are the steps a business takes to make its supply chain run more smoothly. Because it is such a complicated process, the supply chain is fraught with enormous inherent risk. A apparently little risk, such as a small delay in one activity that seems to be a bottleneck for other operations, could have major effects on how the production department runs as a whole. How efficiently the supply chain functions, consists of supply chain responsiveness, agility, and cost, which all assist companies in impacting the relationship between global production and supply chain risk, and developing a sustainable supply network globally. Consequently, this study's end objective is to provide a framework to improve global manufacturing via supply chain operations.

## 2 LITERATURE SURVEY

The importance of supply chain risk management in ensuring the seamless flow of goods, services, and information in today's worldwide corporate environment has piqued the interest of both academics and industry professionals [1]. This section delves into an extensive review of related literature that has examined numerous strategies, methods, and technologies for enhancing supply chain agility and resilience[2]. Analysing the current body of knowledge in this field allows us to get valuable insights into the evolution of risk management systems, [3]spot emerging trends, and identify the gaps that prompted the development of the proposed framework. By examining a variety of perspectives and empirical data, this review provides the framework for understanding the broader context to which our study contributes novel insights and advancements[4].The integration of state-of-the-art technology into supply chain risk management has been the subject of numerous studies, which have shed light on its many facets, potential benefits, and current challenges [5]. Manufacturers employ solutions like just-in-time (JIT) production and lean operations to deal with some of these problems, but these methods are difficult to establish and keep up on their own [6]. Significant disruption risks exist in the industrial sector due to environmental challenges and, consequently, complex techniques [7]. The manufacturers' ability to foresee and deal with unsavoury events in their supply chains is hindered by the obstacles and complications, which can have far-reaching negative impacts [8]. However, just one percent of the optimizations tackled critical concerns with risk management. [9]The same article also states that SCRM depends on collaborative organizational structures and technologies. Companies who dealt with the COVID-19 issue effectively were those that invested in agility-enabling technology and people. Risk management is one of the rapidly growing subfields within supply chain management [10]. Although little empirical study has been conducted on the topic of what makes SCRM effective, further research is greatly needed in this area[11]. Unexpected and unpleasant events, like as natural catastrophes, dissolved relationships, and changes in consumer tastes, are examples of supply chain disruption risks, according to[12]. As a result, researchers in the area of supply chain management have been thinking about how supply chain risk influences the design of supply chains[13]. So, managing risks in the supply chain has become an important concern for companies. According to[14], supply chain risk management is a significant challenge for managers that affects the performance of their organizations. "The identification of potential sources of risk and the implementation of appropriate strategies through a coordinated approach among supply chain risk members, to reduce supply chain vulnerability".

## 3 PROPOSED SYSTEM

Modern business is characterized by a high degree of volatility, uncertainty, and intense competition. Organizations require flexible supply chains to ensure uninterrupted service to customers, control disruption risks, and provide higher value. Therefore, companies see agility training as a way to mitigate risk by enhancing their ability to respond rapidly to actual and perceived disruptions in the supply chain and changes in the market. Agility is crucial for handling emergencies and reducing hazards.

### 3.1 *Data preprocessing*

To improve the research outcomes, pre-processing the data is necessary before applying machine learning models. It was brought to their attention that data biases could be present, particularly if specific vendors or spare parts were overrepresented in the dataset. Normalization is the process of bringing data to a common scale. Data can be transformed into a set of discrete values between 0 and 1 using Min-Max normalization, also known as feature scaling[15]. This is achieved by subtracting the minimum value from each observation and then dividing the result by the difference between the maximum and minimum values of the data. The following formula can be used to normalize using Min-Max:

$$h_{\min} = \frac{(h - h_{\min})}{(h_{\max} - h_{\min})} \tag{1}$$

### 3.2 *Feature engineering*

As an integral aspect of our research, feature engineering involved the careful selection and adjustment of key variables that were input into our prediction models. All of these factors contribute to a better understanding of the dynamic nature of supply chain operations and the potential dangers that may lurk inside. The feature collection process prioritized factors that were context-related, event-indicators-related, or time-dependent. Utilizing time-dependent features, our models incorporated the historical context of the supply chain, allowing us to discern patterns and trends that have evolved over time[16]. Examples of contextual variables that offer a broader perspective include things like geopolitical happenings, economic statistics, and market conditions. Our ability to identify specific problems or disruptions with the potential to substantially impact the supply chain was greatly enhanced by the use of event indicators. The meticulous engineering of our features enhanced our models' predictive ability, allowing them to better anticipate and handle real-time supply chain hazards.

### 3.3 *CNN-BiGRU model training*

#### 3.3.1 *Multi-time sliding window*

As a result, building on the prior analysis's findings, we can generate fresh time series data through a neural network hybrid model. This data will be utilized to ascertain the coupling impact load characteristics using the word vector representation employed in natural language processing. To generate input data for time series, a sliding window with different step sizes is used. There is a starting set of features in the input load data set denoted as$\{h^{(1)}, h^{(2)}, h^{(3)}, \ldots, h^{(p)}\}$, where $\left\{h_1^{(k)}, h_2^{(k)}, \ldots, h_{q-1}^{(k)}, h_q^{(k)}\right\}$and $q$ is the point in time measurement. The time series is also used to sort the input feature map.

### 3.3.2 *Hybrid neural network*

The data input is integrated into the prediction model following feature screening using the random forest technique and temporal window processing. For the most part, the CNN-BiGRU sub-neural network modules make up the integrated hybrid neural network prediction model. Although the input data used by each CNN-BiGRU prediction module differs, their neural network architecture is same. Data processed via sliding time windows of varying widths constitutes the input data for each module. In terms of feature extraction, Conv1d and MaxPooling excel, but the BiGRU network is superior when it comes to load predictions. Use Conv1d to extract features from the input time series as the first step following data entering. Finding hidden information and time series data is a breeze using a one-dimensional convolutional neural network. There are several applications for one-dimensional convolutional neural networks, including classification, time series analysis, and regression. It also has uses in signal data analysis, which is a nice bonus. After the convolutional layer has extracted features, the data is sampled by the one-dimensional pooling layer. Reducing computation and memory consumption, removing unnecessary information, and compressing characteristics to simplify the network complexity are just a few of the many advantages of a pooling layer. Following the pooling of many convolutional layers, the feature data is inputted into the bidirectional GRU network. All GRU structures are gated cyclic. Despite using fewer training parameters than LSTM structure, GRU maintains the predictive power of LSTM.

## 4 RESULT AND DISCUSSION

According to the research on the topic, companies dealing with supply chain risk should focus on agility performance. Still, the literature could use more information on how businesses mitigate supply chain risk by improving agility through better management of the supply chain's key components. Using the dynamic capabilities approach it shows how supply chain risk can push firms to merge their supply chains for better agility.

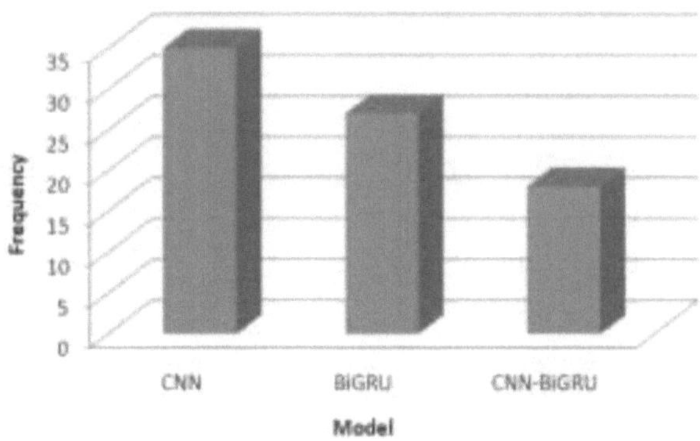


Figure 1. Frequency analysis of CNN-BiGRU model.

In order to generate the frequency analysis depicted in Figure 1, we first constructed a corpus from the gathered texts. The results show that CNN-BiGRU and other machine learning algorithms are frequently mistaken for CNN inside BiGRU.

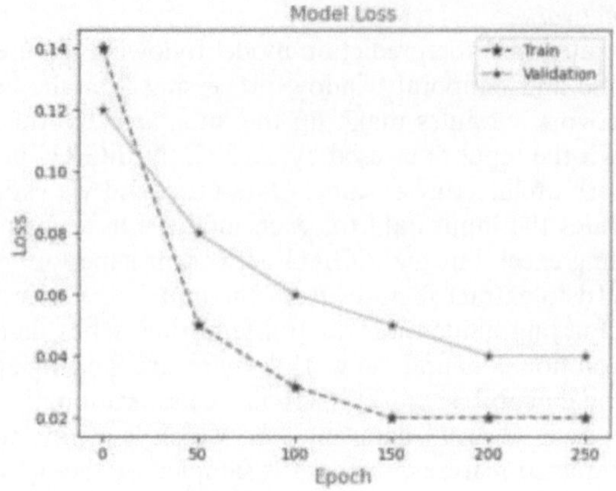


Figure 2. Model loss for training and validating the suggested method.

Figure 2 shows the model's construction. The goal, employing a CNN-BiGRU, is to glean the essential features from the multivariate time series. This separates the dependent variables from the independent ones in the multivariate time series, which in turn produces error vectors that are distinct from one another.

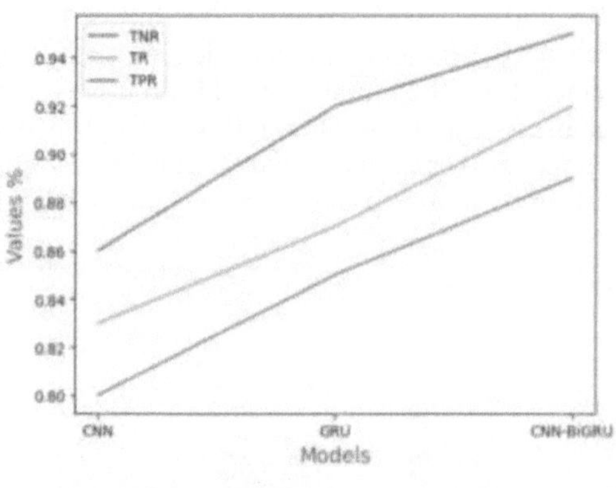


Figure 3. Model comparison of CNN-BiGRU model.

Figure 3 displays the outcomes of these comparisons. We hope that by comparing the CNN-BiGRU model to the others, we can highlight its unique qualities and better performance.

## 5 CONCLUSION

Supply networks face substantial difficulties and disruption risks due to the ever-changing nature of the modern corporate environment. Supply chain risk management (SCRM) has received a lot of attention from academics due to the vulnerabilities in modern supply networks. Data preprocessing analyzes data distribution by discovering patterns and frequency of data values through visualizing the variables and features of the dataset. Because supply chain activities are dynamic and susceptible to sudden changes, feature engineers find the variables useful for understanding these factors. But the proposed method employs CNN-BiGRU for model training. A consistency rate of 96.21% is achieved by its suggested method, which is higher than that of the CNN and GRU models.

## REFERENCES


[1] Chen, C. J. (2019). Developing a model for supply chain agility and innovativeness to enhance firms' competitive advantage. *Management Decision*, 57(7), 1511–1534.

[2] Wong, L. W., Tan, G. W. H., Ooi, K. B., Lin, B., and Dwivedi, Y. K. (2022). Artificial intelligence-driven risk management for enhancing supply chain agility: A deep-learning-based dual-stage PLS-SEM-ANN analysis. *International Journal of Production Research*, 1–21.

[3] Giannakis, M., and Louis, M. (2016). A multi-agent based system with big data processing for enhanced supply chain agility. *Journal of Enterprise Information Management*, 29(5), 706–727.

[4] Shamout, M. D. (2020). Supply chain data analytics and supply chain agility: a fuzzy sets (fsQCA) approach. *International Journal of Organizational Analysis*, 28(5), 1055–1067.

[5] Schroeder, M., and Lodemann, S. (2021). A systematic investigation of the integration of machine learning into supply chain risk management. *Logistics*, 5(3), 62.

[6] Kauppi, K., Longoni, A., Caniato, F., and Kuula, M. (2016). Managing country disruption risks and improving operational performance: risk management along integrated supply chains. *International Journal of Production Economics*, 182, 484–495.

[7] Munir, M., Jajja, M. S. S., Chatha, K. A., and Farooq, S. (2020). Supply chain risk management and operational performance: The enabling role of supply chain integration. *International Journal of Production Economics*, 227, 107667.

[8] Aven, T. (2016). Risk assessment and risk management: Review of recent advances on their foundation. *European Journal of Operational Research*, 253(1), 1–13.

[9] Um, J., and Han, N. (2021). Understanding the relationships between global supply chain risk and supply chain resilience: the role of mitigating strategies. *Supply Chain Management: An International Journal*, 26(2), 240–255.

[10] Fan, H., Cheng, T. C. E., Li, G., and Lee, P. K. (2016). The effectiveness of supply chain risk information processing capability: An information processing perspective. *IEEE Transactions on Engineering Management*, 63(4), 414–425.

[11] Xuan, Y., Si, W., Zhu, J., Sun, Z., Zhao, J., Xu, M., and Xu, S. (2021). Multi-model fusion short-term load forecasting based on random forest feature selection and hybrid neural network. *IEEE Access*, 9, 69002–69009.

[12] Bernardes, E. S., and Hanna, M. D. (2009). A theoretical review of flexibility, agility and responsiveness in the operations management literature: Toward a conceptual definition of customer responsiveness. *International Journal of Operations & Production Management*, 29(1), 30–53.

[13] Blome, C., Schoenherr, T., and Rexhausen, D. (2013). Antecedents and enablers of supply chain agility and its effect on performance: a dynamic capabilities perspective. *International Journal of Production Research*, 51(4), 1295–1318.

[14] Chan, A. T., Ngai, E. W., and Moon, K. K. (2017). The effects of strategic and manufacturing flexibilities and supply chain agility on firm performance in the fashion industry. *European Journal of Operational Research*, 259(2), 486–499.

[15] Amellal, A., Amellal, I., Seghiouer, H., and Ech-Charrat, M. (2023). Improving Lead Time Forecasting and Anomaly Detection for Automotive Spare Parts with A Combined CNN-LSTM Approach. *Operations and Supply Chain Management: An International Journal*, 16(2), 265–278.

[16] Aljohani, A. (2023). Predictive analytics and machine learning for real-time supply chain risk mitigation and agility. *Sustainability*, 15(20), 15088.

*Hybrid and Advanced Technologies – S. Prasad Jones Christydass, Nurhayati Nurhayati and S. Kannadhasan. (Eds)* © *2025 the Authors, ISBN 978-1-032-90254-8*
*Open Access: www.taylorfrancis.com, CC BY-NC-ND 4.0 license*

# An empirical assessment of a univariate time series for demand planning in a demand-driven supply chain using logistic leaf model

M. Suresh
*Assistant Professor, Department of Management studies, SRM Institute of Science and Technology Deemed To Be University, Trichy, India*

R. Arulmoli
*Professor, Department of Faculty of Management, SRMIST, Ramapuram, Chennai, India*

S.S. Onyx Nathanael Nirmal Raj
*Associate Professor & Research Supervisor, School of Arts, Humanities and Management, Jeppiaar University, Chennai, India*

S. Kasthuri
*Assistant Professor, Department of MBA, Velammal College of Engineering and Technology, Madurai, Tamilnadu, India*

Jalaja Valisireddy
*Assistant Professor, Department of Mathematics, Mohan Babu University (Erstwhile Sree Vidyanikethan Engineering College), Tirupathi, India*

Dileep Pulugu
*Professor, Department of CSE, Malla Reddy College of Engineering and Technology, Hyderabad, India*

ABSTRACT: It is possible to find a wealth of useful information in the literature about the when, where, and how of demand-driven supply chain operations management strategies (DDSCOMSs). The supply chain planning and management processes of manufacturing company managers often involve the usage of many DDSCOMSs. Understanding when, how, and why to mix distinct DDSCOMSs is crucial for their work. The phases that comprise the suggested approach are as follow as preprocessing, training the model, and feature extraction. Preprocessing makes use of text information acquisition and preprocessing. A technique for reducing variables, principal component analysis (PCA) is what we utilized for feature extraction. It takes a collection of vector data and uses them to create a new set of vector observations. Training the model became more precise with the help of LLM. The proposed approach yielded an accuracy percentage of 95.72%.

*Keywords*: Logistic Leaf Model (LLM), Demand Planning, Principal Component Analysis (PCA), Supply Chain Management (SCM), Market Demand Management

## 1 INTRODUCTION

The complex network of suppliers and consumers is always adjusting the balance of supply and demand by means of the management of the supply chain. Although demand management is an integral part of any supply chain, in a demand-driven supply chain it takes precedence. When it comes to managing market demand, the American Production and

 DOI: 10.1201/9781003559115-100

Inventory Control Society defines it as "the function of recognizing all demands for goods and services to support the marketplace" (APICS). When resources are limited, satisfying demand takes precedence. When demand management is done well, it makes it easier to plan and employ resources efficiently, which ultimately leads to better financial results for the business. Demand planning, a multi-stage process essential to demand management, relies on forecasts as its foundation. Little planning and process-oriented research has been conducted in the area of demand management, despite the fact that it is becoming increasingly complex and important. Forecasting, sometimes called demand management by some companies, is an area that has seen a lot of study. Conversely, the APICS definition lays out the requirements for effective demand management. It will find out through our inquiry that regular, ongoing demand Distributed goods are affected by censorship and the glaring structural disparity between consumer order time series and sales order time series. The shipping time series forecast negatively impacts a supply chain's inventory position and delivery capabilities, and also showed that the two sets of projections were significantly different.

## 2 LITERATURE SURVEY

Moving products, services, and data from their origins to their final destinations is the primary goal of supply chain management (SCM)[1]. Many different statistical methods, such as time-series and regression analysis, are employed in demand forecasting within the context of supply chain management[2]. Contemplating the adjustments Big data analytics (BDA) has grown in popularity as a result of developments in computing power and other areas of information technology. This is because it allows for more precise predictions, which in turn leads to greater efficiency, quicker reaction times, better alignment with meeting customer demands, easier evaluation of SC performance, and support for risk assessment for SCs [3]. [4] This meta-research literature study primarily aims to analyze "demand forecasting" inside supply chains. Demand forecasting in today's unpredictable and dynamic global supply chains requires the use of big data analytics and machine learning approaches [5]. The demand-driven concept is made up of two parts: demand and driven. A direct or indirect relationship between actual consumers and demand is established in this study[6]. However, the extent to which the demand-driven concept is implemented determines how the demand-driven manufacturing company utilizes this demand. [7]state that certain demand-driven manufacturing companies build new products based on historical demand and their assumptions about the market's quality preferences. [8] In light of this reality, any decisions regarding manufacturing or shipping are purely hypothetical. A healthcare supply chain is a system that aims to efficiently transmit medications, medical equipment, and vaccines from their point of origin to the end user[9]. The many parts that make up this ecosystem include things like organizations, individuals, activities, data, and resources. Public healthcare supply chains aim to ensure that all individuals, regardless of where they live, have access to adaptable healthcare delivery systems that can meet the demands of a variety of patients[10]. A robust healthcare delivery system requires an effective and well-maintained supply chain management (SCM) system to guarantee that patients can obtain the necessary medical products[11]. The complexity of the public healthcare supply chain is rising in tandem with the number of its linked components and feedback loops[12]. Patient expectations are on the rise, and inefficient supply chain operations pose a serious threat to public health supply networks[13]. This has increased the urgency of improving patient services and streamlining the public health supply chain [14]. However, operational efficiency drops and supply chain expenses rise as a result of these tactics. Already, public healthcare supply networks are burdened with increasing supply chain expenses, inefficient operations, drug shortages, and healthcare outcomes brought about by complicated and erroneous demand projections. Things aren't getting better with the current management procedures and systems.

## 3 PROPOSED SYSTEM

Demand planning and projections often use client order time series as their foundation. Sales orders are a common tool in business dealings. The purpose of our study was to analyze and comprehend the outcomes of a supply chain strategy that relies on sales orders. It compared the sales order time series with the customer order time series to find the underlying discrepancies.

### 3.1 *Preprocessing*

In order to ensure their professionalism and uniformity, this study utilizes domain conceptions available in academic literature. Academic literature has a very standard writing pattern and is a great source of information for professionals. Among the credible information discovered in the scholarly literature on the subject are words, concepts, and links between ideas from the domain ontology[15]. The domain ontology that emerges from sifting through academic literature is normative and professional in character due to the breadth and accessibility of the subject matter covered. Therefore, academic literature is the source for domain notions. Using the keywords "NEV," "selection and evaluation of supplier," and "automotive industry," we began by scanning databases. Download the NEV component supplier's selection criteria after researching them in literature; then, remove duplicates.

### 3.2 *Feature extraction*

Using principle component analysis (PCA), which outputs a fresh collection of vector data from an initial set of vector observations, is one approach to reducing the number of variables. It can only be used to numerical continuous fields (scale variables) where the assumptions of linear correlations among specific classification variables are true, because this is its only basis for operation. Compressing numerous dimensions into a lower, sometimes more relevant (and manageable) number of variables through the application of rotation can be achieved, albeit at the cost of some information loss[16]. By capitalizing on the linear relationship between the classification variables, principal component analysis (PCA) with orthogonal rotation can streamline the process by generating new, standardized, uncorrelated components that are often more pertinent and reducing the amount of classification variables.

## 4 MODEL TRAINING

### 4.1 *Logit leaf model*

Part one of the LLM's two-part hybrid strategy involves creating a decision tree to classify clients into subsets with shared characteristics. The second step is to run logistic regressions on both sets of data. In terms of efficiency and readability, decision trees excel as a prediction model due to their relative simplicity. Subsets of the data are reduced and cleaned up by iteratively searching for valid decision tree branches and selecting suitable splits using a splitting criterion. Beginning with a node (the root node) that does not have any parents, the method iteratively finds the optimal splitting criterion and splits the data into two child nodes. Once every good and possible split has been used up, the process terminates, and what's left are a lot of childless nodes called leaves. Based on all the data properties, it is possible to split the set of customers $R$ into smaller subsets $R_p$, where each subset is represented by a leaf $p$ in the tree:

$$R = \bigcup_{p \in P} R_p ; \forall \quad p \neq p' ; R_p \cap R_{p'} = \emptyset \tag{1}$$

Overfitting becomes more likely when the frequency of splitting rises because more complicated models are produced. There are two ways to control this unwanted complexity: post-hoc trimming and adding multiple hyperparameters to control the splitting process. By cutting off branches that don't contribute much to consumer categorization, pruning helps to simplify things. For prediction purposes, traditional DT looks at the class tendency of the terminal node. Using forward variable selection, the second step of the LLM technique involves fitting a logistic regression to every terminal node. A common advertising tactic for third-party predictive analytics is LR. Customer churn prediction is one of two domains where this approach has proven effective: Logit modeling is a powerful predictor of customer churn and can compete with more complex methods; LRs are considerably more understandable than other "black box" methods due to the direct estimation of posterior probability.

## 5 RESULT AND DISCUSSION

In order to help businesses determine where they are in terms of demand-driven supply chain maturity and how to get there, this article lays out a theoretical framework. It goes over the fundamentals of demand-driven supply chains and the maturity model's dimensions and stages.

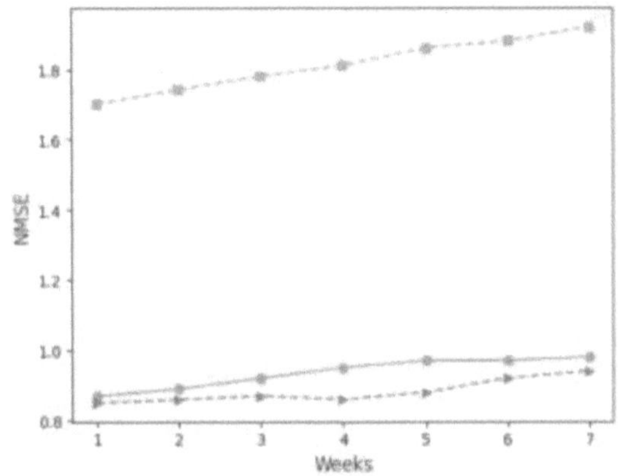


Figure 1. NMSE of DML-trained cross-training models.

Figure 1 illustrates that, due to increased uncertainty, the forecast accuracy for all models declines when we move from a one-week to a seven-week time horizon. As the predicting horizon extends beyond 7 weeks, this pattern persists.

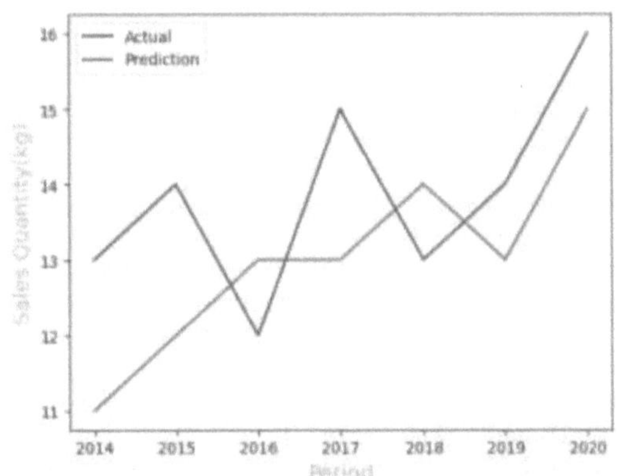


Figure 2. Performance curves for training and validating the suggested method.

A simple average of actual demand over the past year is used to create a projection in order to establish a baseline. For its baseline prediction, the sponsor corporation follows a comparable procedure. This approach is clearly inaccurate, as can be observed in Figure 2.

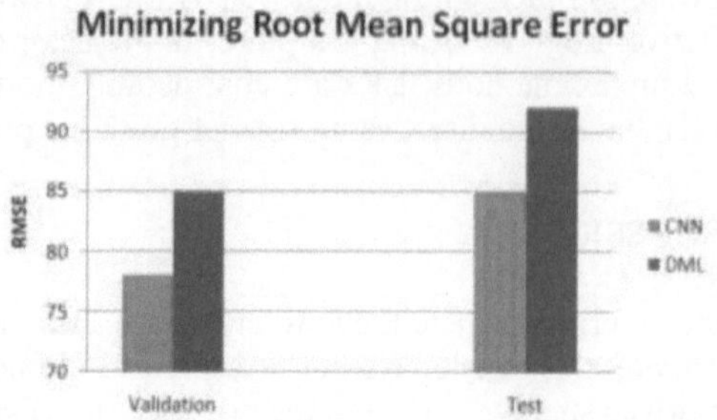


Figure 3. A comparison of RMSE scores for different models.

In Figure 3, it is evident that the DML model outperforms the other two models in terms of error measures, regardless of whether RMSE or MAE is minimized. Having said that, real sales data follows a normal and centered distribution.

## 6 CONCLUSION

Many different kinds of actors participate in demand-driven supply chains, and they all do things a little bit differently when it comes to managing and coordinating the many business processes. The ability to engage in several supply chain configurations simultaneously and the agility to swiftly adapt to new or changed configurations are essential capabilities for organizations. To address the demanding requirements of this, a modeling strategy is needed that integrates high-level models for supply chain design with specific models for the related information systems. Preprocessing involves acquiring and preparing textual information. We extracted features using principal component analysis (PCA), a method for minimizing variables. It constructs a fresh set of vector observations from an existing set of vector data. The use of LLM improved the accuracy of the model training process. An accuracy percentage of 95.72% was produced using the suggested method.

## REFERENCES


[1] Blackburn, R., Lurz, K., Priese, B., Göb, R., and Darkow, I. L. (2015). A predictive analytics approach for demand forecasting in the process industry. *International Transactions in Operational Research*, 22 (3), 407–428.

[2] Boulaksil, Y. (2016). Safety stock placement in supply chains with demand forecast updates. *Operations Research Perspectives*, 3, 27–31.

[3] Michna, Z., Disney, S. M., and Nielsen, P. (2020). The impact of stochastic lead times on the bullwhip effect under correlated demand and moving average forecasts. *Omega*, 93, 102033.

[4] Zhu, Y., Zhao, Y., Zhang, J., Geng, N., and Huang, D. (2019). Spring onion seed demand forecasting using a hybrid Holt-Winters and support vector machine model. *PloS one*, 14(7), e0219889.

[5] Huber, J., Gossmann, A., and Stuckenschmidt, H. (2017). Cluster-based hierarchical demand forecasting for perishable goods. *Expert Systems with Applications*, 76, 140–151.

[6] Childerhouse, P., Aitken, J., and Towill, D. R. (2002). Analysis and design of focused demand chains. *Journal of Operations Management*, 20(6), 675–689.

[7] Mahdavi, M., and Olsen, T. L. (2017). Designing effective supply chains in strategic alignment with demand characteristics and market requirements. *Foundations and Trends® in Technology, Information and Operations Management*, 10(2), 89–208.

[8] Småros, J., Lehtonen, J. M., Appelqvist, P., and Holmström, J. (2003). The impact of increasing demand visibility on production and inventory control efficiency. *International Journal of Physical Distribution and Logistics Management*, 33(4), 336–354.

[9] Mendes Jr, P., Leal, J. E., and Thomé, A. M. T. (2016). A maturity model for demand-driven supply chains in the consumer product goods industry. *International Journal of Production Economics*, 179, 153–165.

[10] Duarte Canever, M., Van Trijp, H. C., and Beers, G. (2008). The emergent demand chain management: key features and illustration from the beef business. *Supply Chain Management: An International Journal*, 13(2), 104–115.

[11] Ashayeri, J., and Kampstra, R. (2005). Demand driven distribution: The logistical challenges and opportunities. *Proceedings of International Trade and Logistics, Corporate Strategies and Global Economy*. Le Havre: University of Le Havre, France.

[12] Hilletofth, P. (2011). Demand-supply chain management: industrial survival recipe for new decade. *Industrial Management and Data Systems*, 111(2), 184–211.

[13] De Caigny, A., Coussement, K., and De Bock, K. W. (2018). A new hybrid classification algorithm for customer churn prediction based on logistic regression and decision trees. *European Journal of Operational Research*, 269(2), 760–772.

[14] Seethamraju, R. (2014). Enterprise systems and demand chain management: a cross-sectional field study. *Information Technology and Management*, 15(3), 151–161.

[15] Zheng, J., Li, Y. H., and Fan, Z. P. (2023). Demand-driven NEV supplier selection: An integrated method based on ontology–QFD–CBR. *Frontiers in Energy Research*, 10, 958885.

[16] Tapaninaho, M., Koho, M., Nylund, H., Heilala, J., and Torvinen, S. (2013). Sustainability performance indicators for supporting the realization of sustainable and energy-efficient manufacturing. *In 23rd International Conference on Flexible Automation & Intelligent Manufacturing*, FAIM 2013 (pp. 1677–1688). Springer.

# Author index